Principles of Modern Physics

Principles of Modern Physics

ROBERT B. LEIGHTON

Professor of Physics
California Institute of Technology

McGRAW-HILL BOOK COMPANY, INC.

NEW YORK   TORONTO   LONDON

1959

PRINCIPLES OF MODERN PHYSICS

II

Preface

For many years the undergraduate curriculum in physics at the California Institute of Technology has included a course in atomic and nuclear physics in the fourth (senior) year. In recent years this course has been taken not only by all fourth-year physics majors and many mathematics and astronomy majors but, to an increasing extent, also by first-year graduate students in physics whose preparation in one or more of the fields of modern physics has been weak and by advanced graduate students who are majoring in other fields of science or engineering. The objectives of this course have thus become threefold: to provide graduating physics majors who do not plan to pursue graduate work with a sound basis in modern physics for careers in applied physics; to give those who do go on for advanced degrees a coherent working familiarity with many fields, some of which they will study more deeply in their graduate work; to offer nonphysics graduate students a useful over-all view of modern physics which will provide a convenient point of departure for advanced study of special topics needed in their particular fields of specialization.

In this book I have attempted to meet some of the rather severe and often conflicting requirements that such a course places upon a textbook: It should cover a quite broad range of subjects, but at sufficient depth to leave the student with more than a mere "survey" view of modern physics. The various topics must not be oversimplified to the point where their physical essence is lost, but neither should the physics be hidden in an overrigorous or too general treatment. The mathematical and physical background of the student should be freely drawn upon, but material new to him should be presented in considerable detail. Unfortunately, these requirements and many others not only fight among themselves but also mean different things to different people. I can only hope that my aim has been reasonably accurate.

The general plan of treatment is to concentrate first on the broad fundamental principles which underlie most of modern physics as we know it, and then to see how these fundamental principles operate to yield the observed complex behavior of matter. The approach is largely expository and analytical rather than historical and discursive. Such historical features as are included are limited to brief introductory paragraphs to provide a general orientation for a given topic. I have tried to keep the level of the treatment uniform, but of course the material on nuclear physics is necessarily rather more qualitative than that on atomic physics, if only because no correct basic theory corresponding to the non-relativistic quantum mechanics is yet available in this field. To a certain extent it has been possible to begin a topic at a relatively elementary level and then work up gradually to its more complex aspects; this should make the book a suitable text for a fourth-, fifth-, or perhaps even a sixth-year course if certain topics are pursued in the literature beyond the level attained in the text.

Although this volume was undertaken with a one-year course for fourth-year physics majors uppermost in mind, it seemed desirable at several points to extend the range of subject matter and also the depth of treatment beyond the limits which one might ordinarily consider appropriate for the fourth year, with the result that the book is distinctly too long to cover completely in a one-year course. This was done to provide a more complete treatment of an important field, to allow some latitude of choice on the part of the instructor of what material to include, to make the book better suited to the needs of a first-year graduate course, or to make it more valuable for later reference use. Even so, it was necessary to omit whole fields of great importance, and to abridge considerably the treatment of other fields, simply for lack of space. It will hardly be necessary to point out to an expert in any field that his specialty is either entirely omitted or, if included, is not treated exhaustively, authoritatively, or, in some cases, even adequately. I have, however, striven to make the treatment correct as far as it goes and to avoid outright errors or falsehoods; I shall be grateful to receive any constructive criticism in this regard.

For the benefit of those who may contemplate using this book, it may be well to list as clearly as possible the preparatory material that is (perhaps optimistically) assumed. Briefly, this comprises classical mechanics, through Lagrange's equations, and rigid-body dynamics, including the tensor of inertia; electromagnetism through Maxwell's equations, including the radiation from an oscillating electric dipole and preferably oscillating cavities; and mathematics through ordinary differential equations (including power-series solutions), some partial differential equations (including separation of variables), orthogonal functions, vector analysis, and some tensor analysis. In addition it should ordinarily be expected that a student will have had at least a little contact with the *facts* of modern physics: the atomic theory of matter, atomic line spectra, heat radiation, kinetic theory, and other topics—perhaps at the level of a second-year physics or chemistry course. This is not to say that a student who is lacking one or more of these items cannot successfully attack the present material, since many of the essential mathematical techniques are presented in detail in the text. However, he may well find himself doing considerable outside reading.

The exercises that will be found distributed throughout the book are of two general types: those which involve filling in steps that are sometimes left out of the text or verifying some stated conclusion, and those in which the material being discussed is to be extended or applied to a physical problem. Of these, the former serve the purpose of saving precious space in deriving results in the text, while at the same time offering the reader an opportunity to test his mastery of the material; the latter broaden his experience or physical insight, and should also provide valuable experience with the numerical magnitudes of atomic and nuclear phenomena. The exercises are decidedly not of uniform difficulty, nor is it essential that they all be worked. Some of them can be done almost by inspection, while others will require considerable effort of even a good student. In any case, I am firmly of the opinion that a student's grasp of material of this kind is not adequately measured by "memory" or "discussion" questions or "plug-in" exercises, but is best attained and retained by an intensive contact with problems which require analytical thought, rather than the mere application of a formula, for their solution.

Concerning the incidental but important matter of symbols, units, and nomenclature: Rationalized MKSA units are used officially throughout the text, but in many instances where a result is more familiar in some other system of units, the more familiar form is also quoted. The exercises do not adhere to any single system of units, but rather tend to be phrased in familiar terms and be capable of solution by the use of whatever system of units is most comfortable for the student. Wherever possible, the symbols and nomenclature recommended in Documents 6 and 7 of the SUN commission of the IUPAP have been adopted,[1] although not without some possibility of inconsistencies resulting from proofreading blindness due to contrary habits of long standing.

Finally, I should like to express my sincere thanks to the many teachers, colleagues, students, and secretaries whose help and encouragement have been indispensable to the completion of a rather formidable task.

Robert B. Leighton

[1] *Physics Today,* **9**(11), 23 (1956); **10**(3), 30 (1957).

Contents

Symbols and Abbreviations
for Units of Physical Quantities

[As recommended in Document 6 of SUN commission of IUPAP, *Physics Today*, **9** (11), 23 (1956).]

MKSA units		Other quantities	
meter	m	micron	μ
second	s	angstrom	Å
kilogram	kg	liter	l
newton	N	minute	min
joule	J	hour	h
watt	W	day	d
ampere	A	atmosphere	atm
coulomb	C	calorie	cal
volt	V	kilocalorie	kcal
farad	F	steradian	sr
ohm	Ω	electron volt	eV
henry	H	10^3 " "	keV
weber	Wb	10^6 " "	MeV
		10^9 " "	GeV*
		atomic mass unit	amu
		10^{-3} amu	mmu
cgs units		electromagnetic units	emu
		electrostatic units	esu
centimeter	cm		
second	s		
gram	g	**Prefixes**	
dyne	dyn		
erg	erg	giga (10^9)	G
degree Kelvin	°K	mega (10^6)	M
degree Celsius	°C	kilo (10^3)	k
oersted	Oe	centi (10^{-2})	c
gauss	G	milli (10^{-3})	m
maxwell	Mx	micro (10^{-6})	μ

* In the American literature, BeV is used to denote 10^9 eV.

1

The Theory of Relativity

We begin our study of modern physics with a brief treatment of the theory of relativity. This is a fitting starting place for several reasons. The theory of relativity was the first branch of modern physics to become firmly established in essentially its present form and has had far-reaching effects upon our concept of the basic framework of all physical measurement, *space and time*. Many of its results can be easily deduced by using only elementary algebra and differential calculus, and most physical laws exhibit a striking and deeply significant symmetry with respect to space and time when expressed in the "language" of relativity. And finally, since we shall need many of the results of special relativity in our later work, it is well to treat the subject comprehensively at the beginning rather than piecemeal as the need arises.

The theory of relativity deals with the way in which observers who are in a state of motion relative to one another describe physical phenomena. In this chapter we shall consider the theory of relativity as it applies to the simplest case, in which two observers are in relative motion in a straight line at constant speed. This restricted case is commonly called the *special theory of relativity*. The general theory of relativity, which applies to the case of arbitrary relative motion, will be mentioned only briefly.

1-1. The Galilean Transformation; the Principle of Relativity

It has long been accepted that there is no meaning to absolute translational motion through space. This idea is supported by our everyday experience with the laws of mechanics as they apply inside a uniformly moving automobile, train, or airplane; for in such a "moving" reference

1

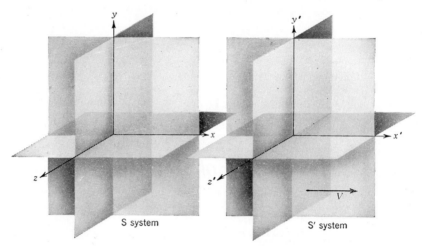

FIG. 1-1. Illustrating the uniform relative motion of two coordinate systems along their common x-axes. The system x', y', z' is called the S'-system and is said to be moving at constant speed V along the positive x-direction with respect to the x, y, z system, or S-system.

frame these dynamical laws appear the same as in a laboratory "at rest" on the earth. One can describe the situation analytically by saying that the laws of motion are *covariant*—i.e., they retain the same form—with respect to a transformation of coordinates to a uniformly moving system. If the relative motion of the two systems is directed along their common x-axes, as shown in Fig. 1-1, the transformation to be considered is the so-called *Galilean transformation:*

$$x' = x - Vt \qquad y' = y \qquad z' = z \qquad t' = t \tag{1}$$

If one makes the plausible assumption that the mechanical force is the same for all observers, this transformation leaves Newton's laws in the same form for all observers. Since $d^2x'/dt'^2 = d^2x/dt^2$, a given force produces the same *acceleration*, independently of whatever uniform translational velocity the coordinate system used as a reference frame may possess, and it therefore seems impossible to detect absolute motion of an object through space on the basis of its response to impressed forces. This property of the laws of mechanics leads naturally to the supposition that the universe may be so constituted that it is impossible *by any kind of experiment whatever* to detect absolute motion through space. This hypothesis is called the *principle of relativity.*

The principle of relativity was an accepted theory of physics for over two centuries. But when Maxwell, in 1865, formulated his dynamical theory of the electromagnetic field, it appeared that absolute motion through space might be detectable by *optical* means. For out of Maxwell's equations there emerged the new and surprising result that *electro-*

magnetic waves ought to exist in empty space.   Maxwell found the speed of propagation of these waves to be equal to the ratio of the electromagnetic to the electrostatic units of charge.   This ratio was so nearly equal to the measured speed of light that ʋe concluded that light must itself be an electromagnetic disturbance of the type described by his equations.

Now, all wave motion with which we are familiar possesses a *medium* for its propagation.   This medium must ordinarily possess both *inertia* and *elasticity* if the speed of propagation is to be other than zero or infinity. The speed of light is far greater than that of any other known wave motion, so that the inertia-like property of space must be very tiny and its "elastic" shear rigidity correspondingly very great.   The idea that all space is filled with an electrically rigid medium called the *luminiferous ether* whose ordinary mechanical density and viscosity are so small that the planets and even much smaller bodies can move through it without hindrance gained universal acceptance through Maxwell's work.

It seems quite clear that, if space is not really empty, but filled with a rigid medium, there might be some meaning to absolute motion after all. And it even appears possible that our speed through this medium might

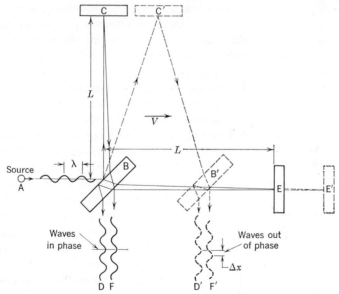

FIG. 1-2. Schematic diagram of the apparatus used in the Michelson-Morley experiment.   The optical path lengths ABCBD and ABEBF may be assumed equal when the instrument is "at rest" in the ether.   If it is then set in motion toward the right with speed V, light takes a longer time to traverse the path BE'B' than the corresponding path BC'B', and a phase difference is thereby introduced into the interference pattern of the two beams at D', F'.   In the actual experiment, the fringe pattern at D', F' was measured as a function of the *azimuthal orientation* of the entire instrument, which was mounted on a stone slab and floated in mercury.

be measured by comparing the speed of light in different directions. Such an experiment was carried out by Michelson (1881) and by Michelson and Morley (1887). In the latter famous experiment[1] the times of traversal of a light ray through equal paths parallel and perpendicular to a supposed direction of motion through the ether were compared by measuring the phase difference of a "split" monochromatic light beam in traversing an interferometer, as shown in Fig. 1-2. This experiment was sufficiently sensitive that a speed of about 10 km s^{-1} should have been detectable; yet, in spite of the fact that the earth's orbital speed around the sun alone amounts to 30 km s^{-1}, *no effect was observed*. Many attempts were made to explain the null result of the Michelson-Morley experiment without wholly giving up the idea of an ether:

1. It was suggested that bodies moving through space might drag the ether along locally, so that the ether immediately surrounding the body would be at rest with respect to it. The speed of light would then be locally isotropic. This explanation is unsatisfactory because light approaching the body from a distance would be affected differently than was found to be the case in the observed *aberration of starlight* due to the earth's orbital motion about the sun (Fig. 1-3).

2. The possibility that the velocity of light adds vectorially to that of the source was also considered. If this were the case, the Michelson-Morley experiment would be explained, but other difficulties would appear. Light arriving from distant binary star systems would exhibit certain effects because of the difference in radial velocity of the two components of the binary system. One such effect would be a spurious eccentricity in the apparent orbits of double stars, which would be detectable statistically. In addition, the basic idea of an ether is contradicted by such an assumption, since a wave propagates at a speed determined by the *medium* and not by the source.

3. The most serious proposal advanced was that bodies which move through the ether suffer a change of shape just sufficient to make the speed of light *appear* to be the same in various directions.

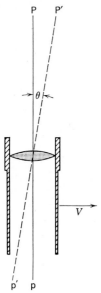

FIG. 1-3. Illustrating the phenomenon of the aberration of starlight. Because of the transverse motion of the observer, the light received from a star P appears to come from a direction P'P' at an angle $\theta \approx V/c$ away from the true direction PP. The orbital speed of the earth around the sun introduces an angle θ of 20″.5.

[1] Michelson and Morley, *Silliman J.*, **34**, 333, 427 (1887).

The change that is needed is a contraction in the direction of motion of the body, such that a body of length L_0 at rest is shrunk to a length $L_0(1 - V^2/c^2)^{\frac{1}{2}}$ when in motion at speed V parallel to its length. This effect, called the Lorentz-FitzGerald contraction, was supposed to follow from Maxwell's equations; no one was successful in proving that it actually did, however. Furthermore, later experiments[1] have shown that a simple length contraction is not alone sufficient; a *time-dilation* effect is also necessary. This would be most difficult for Maxwell's equations to provide, but follows naturally from Einstein's theory.

Einstein[2] finally proposed a radically different approach to the problem posed by the Michelson-Morley experiment. He explained its null result simply by returning to the *principle of relativity*, which directly asserts the impossibility of detecting absolute motion through space. He found, however, that this requires that our notions of space and time as independent, universal quantities must be fundamentally modified, and that time must in fact be treated on an equal basis with the three length dimensions of space, instead of as an independent universal scalar parameter. These new properties of space and time lead in turn to radically different results in kinematics and in dynamics, and several totally new and unexpected fundamental relationships which connect familiar physical quantities with one another appear. Many of these relationships have been of such a striking nature that their direct confirmation by experiment has served to establish beyond any doubt that Einstein's theory penetrates deeply indeed into the basic structure of our universe.

EXERCISE

1-1. Assume that monochromatic light of wavelength λ is used in the Michelson interferometer of Fig. 1-2, and that the apparatus is moving to the right at speed V through a fixed ether. Find the phase difference between the two beams arriving at D′, F′. *Ans.:* $2\pi c \, \Delta t/\lambda \approx 2\pi L V^2/\lambda c^2$.

1-2. The Postulates of Special Relativity

The special theory of relativity may be built up from the following two fundamental postulates:

1. A properly formulated description of a physical phenomenon can contain no reference, either explicit or implicit, to an absolute speed of translational motion of the coordinate frame in which the phenomenon is described.

[1] Kennedy and Thorndike, *Phys. Rev.*, **42**, 400 (1932).
[2] *Ann. Physik*, **17**, 891 (1905).

2. The speed of light is the same for any two observers who are in uniform rectilinear relative motion, and is independent of the motion of the source.

The first of these postulates may be regarded as requiring that a correctly formulated physical law be expressible in the same *form* in any two coordinate systems which are in uniform relative translational motion, since, if this were not possible, the difference in the form of the physical law as a function of what reference frame is used could constitute a basis for selecting a "preferred" reference frame. An expression which retains the same *appearance* in different coordinate systems is said to be *covariant* with respect to the coordinate transformation which carries one system into the other. We may thus test the correctness of a physical law by examining its transformation properties between systems in uniform relative motion. This of course assumes that the transformation law for the coordinates is itself correctly known.

If we proceed to test the covariance of the laws of physics as suggested above, we find that Newton's laws are covariant with respect to the Galilean transformation, but that Maxwell's equations are not.

The truth of the latter statement is relatively easy to demonstrate for the equation

$$\nabla \cdot D = \rho \tag{1}$$

in the absence of polarizable matter: First, the operator $\nabla\cdot$ means that certain derivatives of a vector field are to be taken at a given instant of time. Since $t' = t$, this is then equivalent to taking the divergence in the S′ system at each corresponding point at the same instant. Thus we have the equality

$$\nabla \cdot D = \nabla' \cdot D \tag{2}$$

Next, we observe that a charge must have the same magnitude for all observers, since it could be measured by each observer by measuring the force it exerts upon an equal charge. (If the charge took on a different value for different observers, this could be used as a basis for defining the absolute speed of a system.) We therefore must have

$$\rho' = \frac{\Delta q'}{\Delta x' \, \Delta y' \, \Delta z'} = \frac{\Delta q}{\Delta x \, \Delta y \, \Delta z} = \rho \tag{3}$$

since $\Delta x' \, \Delta y' \, \Delta z' = \Delta x \, \Delta y \, \Delta z$ at any fixed time t.

Finally, the force equations $f = q(E + v \times B) = q(E' + v' \times B')$ require that

$$E = E' - V \times B' \tag{4}$$

where $V = v - v'$ is the vector relative velocity of S′ with respect to S.

We thus find that

$$D = D' - \varepsilon_0 V \times B'$$

in a region where $P = P' = 0$.

Therefore, in terms of quantities measured only in S′,

$$\nabla' \cdot D' = \rho' + \varepsilon_0 \nabla' \cdot V \times B' \tag{5}$$

The extra term on the right is not in general equal to zero, and since it depends upon V, this provides a means of defining a special reference system. One could, for example, assert with considerable philosophical justification that the system of reference for which Maxwell's equations have their simplest form is a system in a state of absolute rest. This is in effect the basis of the ether hypothesis.

One might next wonder whether the above conflict with the principle of relativity could be caused by Maxwell's equations themselves being in error. Perhaps there is some other set of equations, to which Maxwell's equations are a close approximation, which *can* be written in covariant form. That this is not a possible solution to the problem within the framework of Einstein's postulates can be shown by considering a very simple physical experiment, the description of which by observers O and O′ will be shown to require the abandonment of the *Galilean transformation*, rather than of Maxwell's equations.

Let systems S and S′ be in uniform relative motion such that at some time the two origins coincide. At this instant, which the observers will define as $t = t' = 0$, let a short light pulse be emitted from the origin and let us imagine that O and O′ have each set up photoelectric cells and time recorders at various points fixed in their respective systems in order to be able to follow the progress of the wave pulse as it expands in all directions from the source. From the two postulates of special relativity, the two observers must find the same *form* for the equation of the wavefront in his system. That is, O will say that

$$x^2 + y^2 + z^2 - c^2 t^2 = 0 \tag{6}$$

is the equation of the wavefront, while O′ must find it to be

$$x'^2 + y'^2 + z'^2 - c^2 t'^2 = 0 \tag{7}$$

Now, direct substitution reveals that the Galilean transformation 1-1(1) cannot be reconciled with the above description of this physical experiment. It is easy to see why this is so; Eqs. (6) and (7) simply assert that each observer will find, at any instant, that the shape of the wavefront is a sphere of radius ct and ct', respectively, with its center fixed at his own origin of coordinates. Clearly, the Galilean transformation says that the wave pulse always has a spherical wavefront, but also that, if the center of the expanding sphere remains at a fixed point in one system, *it must move at speed V in the other.*

1-3. The Lorentz Transformation

From the preceding discussion, we conclude that we must either give up the two postulates we have set down or abandon the Galilean transformation. If we abandon the postulates, we must face the many problems posed by the Michelson-Morley experiment, the measurements of the aberration of starlight, and other puzzling experimental results. On the other hand, to abandon the Galilean transformation, which appeals so strongly to our sense of what is "right," seems even less palatable. The difficulty is that our sense of what is "right" is based upon countless centuries of experience with nature at a level where a "feeling" for rigid-body dynamics has carried with it great survival value—but all we can really expect of nature is that what she has taught us in this way is *nearly* "right." Let us therefore try to replace the Galilean transformation with a new relationship between the two reference frames which will satisfy the requirements of Eqs. 1-2(6) and (7) and which is in agreement with the Galilean transformation if the two systems are moving very slowly with respect to each other.

Inspection of Eqs. 1-2(6) and (7) and the Galilean transformation 1-1(1) reveals that y and z give no trouble, but that there are some terms involving x and t that must somehow be made to disappear. This must of course be done without disturbing the combination $x - Vt$ in the transformation, since the speed of one system with respect to the other must mean the rate at which a point, fixed in one system, appears to move past the other system. If we also require that straight lines in one system transform into straight lines in the other, we can at most modify the first equation by a constant factor. Thus we shall try to retain the x transformation in the relatively simple form

$$x' = \gamma(x - Vt) \tag{1}$$

where γ is a factor which is very nearly equal to unity under the conditions of everyday experience. Further thought reveals that *the equation $t' = t$ cannot be correct,* since no rearrangement of space coordinates alone can give wave pulses that are simultaneously concentric spheres in *both* systems. The simplest modification of the time transformation is one which contains only x and t linearly:

$$t' = At + Bx \tag{2}$$

where A should be nearly unity and B nearly zero under ordinary familiar conditions.   If we now insert these two modified forms into Eq. 1-2(7) and set this equal to Eq. 1-2(6), we have

$$0 = (\gamma^2 - B^2c^2)x^2 + y^2 + z^2 + (\gamma^2V^2 - A^2c^2)t^2 - 2(ABc^2 + \gamma^2V)xt$$
$$= x^2 + y^2 + z^2 - c^2t^2 \tag{3}$$

By inspection, we find that

$$\gamma^2 - B^2c^2 = 1 \qquad A^2c^2 - \gamma^2V^2 = c^2 \qquad \text{and} \qquad ABc^2 + \gamma^2V = 0 \tag{4}$$

These three equations suffice to determine A, B, and γ in terms of V; the result is

$$\gamma = A = \left(1 - \frac{V^2}{c^2}\right)^{-\frac{1}{2}} \qquad B = -\frac{V\gamma}{c^2}$$

If we call $\beta = V/c$, the new transformation equations we have just found may be written

$$x' = \gamma(x - \beta ct) \qquad z' = z$$
$$y' = y \qquad\qquad t' = \gamma\left(t - \frac{\beta x}{c}\right) \tag{5}$$

where $\beta = V/c$ and $\gamma = (1 - V^2/c^2)^{-\frac{1}{2}}$.

A linear transformation of coordinates and time which preserves Eq. 1-2(6) in covariant form is called a *Lorentz transformation*.   The above transformation is the Lorentz transformation which applies to the special case in which the relative motion of S and S' is in the x-direction of each system.   In Sec. 1-6 we shall consider a more general form of the Lorentz transformation, but for the present we shall restrict our attention to the special form (5).

EXERCISES

1-2. Solve Eq. (4) for A, B, and γ.

1-3. Solve Eq. (5) for x, y, z, and t.

Ans.:
$$x = \gamma(x' + \beta ct') \qquad z = z'$$
$$y = y' \qquad\qquad t = \gamma\left(t' + \frac{\beta x'}{c}\right) \tag{6}$$

1-4.   Relativistic Kinematics

The transformation 1-3(5) was derived by using postulates which assert the correctness of the principle of relativity and the constancy of the speed of light but make no direct mention of space and time.   Yet this transformation may be regarded as defining a certain "geometry" of space and time which must exist if the postulates are valid.   Let us now examine some of these "geometrical" properties of space and time implied by the Lorentz transformation.

First, it is important to understand clearly the meaning of a measurement of *spatial position* and of *time* from the standpoint of different observers, and of the comparison of these measurements. It may be assumed that two observers can compare the lengths of their meter sticks, and the rates of their clocks, at a time when they are not in relative motion. The observers can then acquire a relative velocity V by a completely symmetrical process. Each then lays out a rectangular Cartesian coordinate system with x-axis in the direction of relative motion and mounts clocks at various points on this coordinate framework. The clocks within a given system may be synchronized by sending light signals from a "standard" clock at the origin, allowing for the propagation time required for the light pulse to reach each clock within the system. The various clocks will of course appear to read different times, as seen from the origin, but they will read exactly the same when the propagation time is taken into account. The observers may then "synchronize" the two systems by defining the instant $t = t' = 0$ to be that instant at which the origins of the two systems coincide as they pass each other.

With their coordinates and clocks laid out and "synchronized," the two observers proceed to compare the positions and times of various events that occur. An *event* is any physical occurrence that is localized in space and in time. A flash of light of extremely short duration from a small source, the coincidence of the end of a meter stick with a fiducial point on a scale, and the disintegration of a radioactive atom are all examples of events. For definiteness we shall often employ the short light flash, since we may then imagine that the location and time of such an event are established by photographing the coordinates and a nearby clock by the light of a flash. Indeed, *each observer may read the coordinates and time of an event in terms of the scales and clocks of both systems*.

A. The Lorentz-FitzGerald Contraction; the Time Dilation. Suppose first that the two observers wish to compare the lengths of their meter sticks when these sticks are oriented *perpendicular* to the direction of motion. This can be done by mounting each meter stick with one end at the origin and the other end on the y-axis of its own system. As these meter sticks slide past one another, a photograph of the ends of the sticks can be taken. By symmetry, as well as by Eq. 1-3(5), the result must be that the meter sticks appear equally long to the two observers.

If the two observers next wish to compare their meter sticks when they are oriented in a direction *parallel* to their relative motion, the situation is slightly more complicated. Let the ends of the meter stick of the

S′-system be photographed simultaneously in the S-system. The two events under consideration are then the locations of the two ends of the stick at the time $t = t_0$. We have from Eq. 1-3(5)

(a) $x_1' = \gamma(x_1 - Vt_0)$

(b) $x_2' = \gamma(x_2 - Vt_0)$ (1)

as the coordinates of the ends of the meter stick at this time. But we know that, since the meter stick is stationary in the S′-system, $x_2' - x_1' = L_0 = 1$ m. Also, since the ends were photographed at the same time t_0 in the S-system, $x_1 - x_2 = L$, the length of the stick in this system. We thus have

$$L_0 = \gamma L \qquad \text{or} \qquad L = L_0 \left(1 - \frac{V^2}{c^2}\right)^{\frac{1}{2}} \qquad (2)$$

so that in the S-system the meter stick of the S′-system appears *shorter* by the factor $(1 - V^2/c^2)^{\frac{1}{2}}$. Any physical object appears shorter when it is in motion than it does when it is at rest. This contraction along the direction of motion is called the Lorentz-FitzGerald contraction; it was first introduced to explain the results of the Michelson-Morley experiment. We now find this contraction to be a basic property of space and time, rather than of electrical matter.

Suppose next that the observers wish to compare the *time interval* between two events which occur at the *same spatial location in the* S′-*system*. In this system let the time interval be $t_2' - t_1' = T_0$. In the S-system these two events occur at *different places* as well as at different times, and we find from Eq. 1-3(6) that

$$t_2 - t_1 = \gamma(t_2' - t_1') \qquad \text{or} \qquad T = T_0 \left(1 - \frac{V^2}{c^2}\right)^{-\frac{1}{2}} \qquad (3)$$

The two observers therefore disagree on the length of this time interval. An interval T_0 between two events at a *fixed point* in one system appears longer in a system with respect to which the first system is in motion. Since the two events in question might be taken to be the interval between two successive ticks of a clock in one system, the above analysis shows that *each* observer will find that a clock in the other's system is running at a *slower rate* than his own. This effect is called the *relativistic time dilation*. It has been verified experimentally in various ways, mostly in studies of rapidly moving radioactive particles and excited atoms.

EXERCISES

1-4. Show that the result (2) is also obtained if the S-observer measures *how long* it takes the S′-meter stick to pass by a fixed point and then multiplies this result by V.

1-5. Show that the time dilation (3) is also obtained if the S-observer measures the x coordinate of the two copunctal S′-events, and divides the difference of these coordinates by V.

The two results just derived may appear somewhat paradoxical at first sight; for how can *each* observer find that the other observer's meter sticks are short and his clocks running slow? The resolution of this apparent paradox lies in the realization that the experiments by which the observers make these comparisons are *symmetrical* but *not identical*. Thus the two observers cannot directly compare the lengths of their meter sticks when they are laid in the x-direction, because this requires making a *simultaneous* comparison of the locations of the ends of the sticks.   But in the Lorentz transformation, the term "simultaneous" no longer has the universal significance that was assumed in writing the Galilean transformation: Events which appear simultaneous in one system may not appear so in another.   An analogous situation exists in the case of time measurements.   If the rate of an S-clock is to be measured by O′, the time interval in S′ between two successive "ticks" of the clock fixed in S must be measured.   But during this interval the clock will have moved so that the single S-clock *must be compared with two different S′-clocks*.

Another puzzling aspect of the time dilation is the detailed mechanism by which the rate of a clock changes when it is in motion.   A clock is, after all, a physical device whose motions are governed by the laws of nature.   It seems strange indeed that our conclusions regarding the rates of clocks should be true of *all* clocks, no matter what their design.   Why should a wristwatch behave in just the same way as a revolving body or an oscillating electrical circuit?   The answer is to be found *in the form taken by the laws of nature* rather than in a detailed analysis of each possible kind of clock.   For, if the laws which govern the operation of clocks can be written in a form which is covariant with respect to the Lorentz transformation, then it will automatically be true that the rates of all clocks governed by these laws will change in the same way under the Lorentz transformation.   Since all ordinary clocks obey the laws of mechanics and electrodynamics, all we really need prove is that *these laws* are covariant with respect to the Lorentz transformation.   We shall do this shortly.   It will still be instructive, however, to examine some particular type of clock in sufficient detail that the actual mechanism of the time dilation will be apparent for that case.

Consider a clock composed of an electric-flash tube, a photoelectric cell, and a mirror arranged as shown in Fig. 1-4.   The flash tube and photocell are side by side with a light baffle between them, and a circuit is so arranged that, when the photocell receives a pulse of light, it causes the flash tube to emit another pulse of light with negligible delay.   The "period" of this clock is then equal to the time required for a light pulse to travel from the flash tube to the mirror and back to the photocell. This time is just

$$\tau' = \frac{2D}{c} \qquad (4)$$

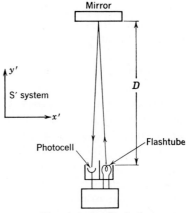

FIG. 1-4. An S'-clock.

Now let us consider the operation of this clock as viewed by an observer O moving to the left at speed V. To him, the clock is moving to the right at speed V, and he analyzes its operation as shown in Fig. 1-5. Because the clock is moving, the light must proceed in a

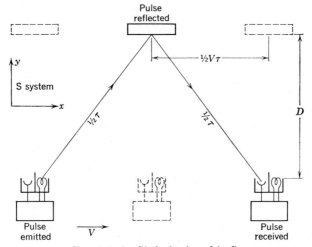

FIG. 1-5. An S'-clock viewed in S.

diagonal path from the flash tube to the mirror and thence to the photocell. By the second postulate, the light travels on this diagonal path at the same speed c as it travels directly forth and back in S'. It must therefore *take longer* for the light to return to the photocell as seen by S than as seen by S'.

EXERCISES

1-6. Complete the analysis of the operation of the above clock and show that one obtains the time dilation formula (3) as a result.

1-7. Suppose the clock were oriented with D *along* the direction of motion. Show that the same time dilation is obtained in this case. Remember to include the Lorentz-FitzGerald contraction.

1-8. Consider two events in the S-system that occur at different points (x_1, y_1, z_1) and (x_2, y_2, z_2) but at the *same time* t_0. Show that these two events are *not* simultaneous in the S'-system but are separated by a time interval $\Delta t' = -V \Delta x / c^2 (1 - V^2/c^2)^{1/2}$.

1-9. A cosmic-ray μ-meson is moving vertically through the atmosphere with a speed $v = 0.99c$. Its mean life expectancy against radioactive decay into an electron and two neutrinos is 2.22 μs, as measured in its own "rest" system. What will be its mean life expectancy as viewed by an observer on the earth? *Ans.: 15.7 μs.*

1-10. A stick of length L is at rest in one system and is oriented at an angle θ with respect to the x-axis. What are the apparent length and orientation angle of this stick as viewed by an observer moving at speed V with respect to the first system?

$$Ans.: \qquad L' = L \left(\frac{\cos^2 \theta}{\gamma^2} + \sin^2 \theta \right)^{1/2} \qquad \tan \theta' = \gamma \tan \theta$$

1-11. An airplane 10.0 m long is flying at 300 m s^{-1}. How much shorter will this airplane appear to be to an observer on the ground? How long will it take for the pilot's watch to "lose" 1 μs with respect to the clock of a ground observer? *Ans.: $\Delta L = 5 \times 10^{-6}$ μ, $T = 23$d.*

1-12. Consider a radiating atom to be a clock that "ticks" ν_0 times per second and emits a light wave with each tick. By taking into account the time dilation and the radial component of velocity of the atom with respect to an observer, derive the relativistic Doppler formula.

$$Ans.: \qquad \nu = \frac{\nu_0}{\gamma(1 + \beta \cos \theta)} \qquad\qquad (5)$$

1-13. A star is moving away from the earth at 50 km s^{-1}. By how many angstrom units will the Hα line ($\lambda = 6563$ Å) be shifted, and in which direction?

B. The Transformation of Velocity. The Lorentz transformation 1-3(5) and also Eqs. 1-4(2) and (3) imply a singularity in the relative measurements made by two observers when their relative speed approaches that of light; and if speeds *greater* than c were attainable, these equations would imply *imaginary* values for certain coordinates and times as measured by one observer or the other. We can easily show that a relative velocity greater than that of light *cannot be attained* by one physical system with respect to another by any stepwise process of compounding successive relative speeds, as long as no individual relative speed exceeds that of light.

Consider two systems in a state of motion described by Eqs. 1-3(5) and (6).   Let a particle or other physical object be projected in the x-direction at a speed u_x' with respect to the system S', so that this object is moving faster than u_x' with respect to S.   How fast will it appear to move, as viewed from S?   From Eq. 1-3(6) we have

$$dx = \gamma(dx' + V\, dt') \qquad dt = \gamma\left(dt' + \frac{V\, dx'}{c^2}\right)$$

so that
$$u_x = \frac{dx}{dt} = \frac{dx' + V\, dt'}{dt' + V\, dx'/c^2} = \frac{dx'/dt' + V}{1 + V\, dx'/c^2\, dt'} \tag{6}$$

or
$$u_x = \frac{u_x' + V}{1 + u_x'\, V/c^2} \tag{7}$$

Thus it appears that u_x is *less* than the sum of u_x' and V and is less than c if both u_x' and V are less than c.

EXERCISES

1-14. Find the law of transformation of velocities for an arbitrary direction of motion of the particle in the S'-system.

Ans.: $u_x = \dfrac{u_x' + V}{1 + u_x'\, V/c^2} \qquad u_y = \dfrac{u_y'}{\gamma(1 + u_x'\, V/c^2)}$

$$u_z = \frac{u_z'}{\gamma(1 + u_x'\, V/c^2)} \tag{8}$$

1-15. A particle appears to move with speed u at an angle θ with respect to the x-axis in a certain system.   At what speed and angle will this particle appear to move in a second system moving with speed V with respect to the first?   Why does this answer differ from that of Exercise 1-10?

1-16. An observer sees two particles traveling in opposite directions, each with a speed of $0.99c$.   What is the speed of one particle with respect to the other?   *Ans.:* $0.99995c$.

The above results illustrate another of the strange new properties possessed by space and time in Einstein's theory: *the speed of light must be regarded as an upper limit to physically attainable speeds for material bodies.*   We shall later find that it would indeed require an infinite amount of energy to accelerate even so tiny a particle as an electron to the speed of light.

1-5.   Review of Tensor Analysis

Before continuing our discussion of the results of special relativity, we shall retrace our steps somewhat and examine this subject from a new point of view which will provide us with a deeper insight into the new relationship of space and time revealed by the Lorentz transformation. According to this new viewpoint, time and the three dimensions of space are regarded as the four rectangular coordinates of a four-dimensional Euclidean space in which a four-dimensional vector and tensor analysis can be defined. In order to see that such a view is possible and useful, we must first recall briefly the principles of vector and tensor analysis in ordinary three-dimensional Euclidean space. To keep the mathematics simple, we shall consider rectangular Cartesian coordinates only, designating the three coordinates as x_1, x_2, and x_3. Since we shall presently wish to extend most of the results to the case of four dimensions, we shall often indicate in parentheses the form exhibited by tensors in n dimensions.

A. Scalars, Vectors, and Tensors. In describing the phenomena of nature it is customary to introduce a system of coordinates by means of which the relative locations and orientations of various interacting objects can be expressed, and to write the natural laws in terms of measurements made in this coordinate system. It seems quite clear, however, that the objects in nature must actually possess intrinsic properties and interactions not related to a particular coordinate system. We therefore seek to express the natural laws in such form that these intrinsic relationships stand out clearly from the extrinsic properties of a particular measurement frame. Tensor analysis furnishes one means of doing this.

Tensor analysis, in its physical applications, deals with the relationship between measurements of the intrinsic properties of natural phenomena made in different coordinate frames. In the restricted case we are considering, the different coordinate frames are rectangular Cartesian coordinate systems having various arbitrary *angular orientations*. We find from direct observation that quantities of physical interest fall into several categories according to how many independent parts or *components* are needed to specify them completely and how the magnitudes of these components as measured in two coordinate frames are related to one another:

1. The simplest quantities of physical interest have but a single component whose magnitude is independent of the orientation of the coordinate system used for measurement. Quantities of this type are called *scalars*. The mass or charge of a particle and the speed of light are examples of scalars.

2. Next in complexity are physical quantities which, in three- (or n-) dimensional space, require for their specification three (or n) components whose magnitudes depend in a certain way upon the orientation of the coordinate system that is used. Such quantities are called *vectors*, and each of the three (or n) components is associated with a

different one of the three (or n) rectangular directions. The electric-dipole moment of a molecule, the momentum of a particle, the electric field, the velocity field within a moving fluid, and the position coordinates of a point with respect to a fixed origin are all familiar examples of vectors.

Of primary importance for our later purposes are the so-called *transformation properties* of a vector. It is readily verified, using the geometrical notion of a vector as a directed line segment together with the rules for the resolution and addition of vectors, that the components of a given vector A in two coordinate systems S and S′ are related by the equations (Fig. 1-6)

$$A'_1 = A_1 \cos (x'_1,x_1) + A_2 \cos (x'_1,x_2) + A_3 \cos (x'_1,x_3)$$
$$A'_2 = A_1 \cos (x'_2,x_1) + A_2 \cos (x'_2,x_2) + A_3 \cos (x'_2,x_3) \qquad (1)$$
$$A'_3 = A_1 \cos (x'_3,x_1) + A_2 \cos (x'_3,x_2) + A_3 \cos (x'_3,x_3)$$

Equations (1) may be written in the abbreviated form

$$A'_i = \sum_{k=1}^{3} c_{ik} A_k \qquad i = 1, 2, 3 \qquad (2)$$

where the quantities c_{ik} are the direction cosines which appear above. In this notation, the subscript i refers to one of the axes of S′, and k to one of the axes of S. Of course we might equally well have written the equations giving the components of A in S in terms of those in S′:

$$A_r = \sum_{s=1}^{3} c'_{rs} A'_s \qquad r = 1, 2, 3$$

$$(3)$$

where, now, the subscript r refers to some axis of S and s to some axis of S′.

3. There also exist physical quantities having more components and more complex transformation properties than vectors. Such quantities are called *tensors*. There are various kinds of tensors, and we shall study some of their properties. For purposes of illustration, we may

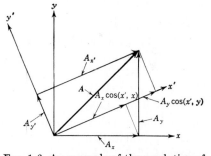

Fig. 1-6. An example of the resolution of a vector into rectangular components in two coordinate systems. For clarity, the coordinate systems are assumed to have common z-axes, and the vector is taken to lie in the xy plane. The expression $A_{x'} = A_x \cos (x',x) + A_y \cos (x',y)$ is interpreted geometrically as the resolution of the vectors iA_x and jA_y into component vectors $i'A_x \cos (x',x)$ and $i'A_y \cos (x',y)$ along x', whose sum is the component $i'A_{x'}$ of A along x'.

list the tensor of inertia, the elastic stress and strain tensors, and the elastic constants of crystalline media as examples of tensors.

The number of components needed to define completely a tensor quantity determines the *rank* of the tensor: a tensor of rank r has 3^r (n^r) components in three (n) dimensions. Scalars and vectors are themselves tensors, of rank zero and one, respectively.

The various components of a tensor quantity are identified with the three (n) space directions in a way that grows in complexity with increasing rank. The components of a second-rank tensor are associated with all possible sets of two space axes, those of a third-rank tensor with all possible sets of three space axes, and so on.

The particular set of axes associated with a given component of a tensor is denoted by *subscript indices* equal in number to the rank of the tensor. Thus T_{13} denotes that component of a second-rank tensor associated with the x_1 and x_3 axes; and H_{211}, that component of a third-rank tensor associated with the x_2, x_1 and the x_1 axes.

Just as for vectors, the components of a higher-rank tensor depend upon the orientation of the coordinate axes. If the components of a second-rank tensor are measured in the two systems S and S' whose axes possess the mutual direction cosines c_{ik} $(i,k = 1,2,3)$ with respect to each other, the components in these two systems are related by the equations

$$T'_{rs} = \sum_{i=1}^{3} \sum_{j=1}^{3} c_{ri} c_{sj} T_{ij} \qquad r, s = 1, 2, 3 \qquad (4)$$

B. The Rotation Matrix. Tensor analysis deals both with the relationship between the components of a given tensor as measured in differently oriented coordinate systems and with relationships between tensors themselves, independently of any particular coordinate system. We shall first examine the transformation properties of tensors.

We shall be dealing with the so-called *linear orthogonal transformation of coordinates*, which corresponds to a simple rigid-body rotation of axes from one orientation to another. In such a transformation the coordinates (x'_1, x'_2, x'_3) of a point in one frame S' are related to the coordinates (x_1, x_2, x_3) of this point in another frame S by the equations

$$\begin{aligned}
x'_1 &= c_{11} x_1 + c_{12} x_2 + c_{13} x_3 \\
x'_2 &= c_{21} x_1 + c_{22} x_2 + c_{23} x_3 \\
x'_3 &= c_{31} x_1 + c_{32} x_2 + c_{33} x_3
\end{aligned} \qquad (5)$$

where the quantities c_{ij} are the direction cosines $\cos (x'_i, x_j)$ of the various axes of the S'-system with respect to the axes of the S-system. For convenience in tabulating and applying these quantities, they are often written in matrix form:

$$
c_{ij} = \begin{array}{c} (x_i') \\ \downarrow \end{array} \overset{(x_j) \rightarrow}{\begin{pmatrix} c_{11} & c_{12} & c_{13} \\ c_{21} & c_{22} & c_{23} \\ c_{31} & c_{32} & c_{33} \end{pmatrix}}
\tag{6}
$$

This is called the *rotation matrix* of the transformation. These direction cosines, which collectively define the relative orientation of the two systems, are not all independent. It is easily seen, by considering the distance Δr between any two fixed points P_1 and P_2 as expressed in S and S', that

$$
(\Delta r)^2 = (\Delta x_1')^2 + (\Delta x_2')^2 + (\Delta x_3')^2 = (\Delta x_1)^2 + (\Delta x_2)^2 + (\Delta x_3)^2 \tag{7}
$$

By inserting the transformation (5) into this equation, the following relations will be seen to hold:

$$
\sum_{i=1}^{3} c_{ij}\, c_{ij} = 1 \qquad\qquad j = 1, 2, 3 \tag{8}
$$

and
$$
\sum_{i=1}^{3} c_{ij}\, c_{ik} = 0 \qquad j \neq k \qquad j, k = 1, 2, 3 \tag{9}
$$

so that only *three* of the c_{ij} are in fact independent. These relations are of fundamental importance in tensor analysis and will be used repeatedly in what follows.

EXERCISES

1-17. Define $c_{ij}' = \cos(x_i, x_j')$. Then $c_{ij}' = c_{ji}$. Multiply each of the Eqs. (5) by an appropriate c_{ij}' and add the three equations to obtain expressions for the coordinates of S in terms of those of S'.

Ans.:
$$
\begin{aligned}
x_1 &= c_{11}'\, x_1' + c_{12}'\, x_2' + c_{13}'\, x_3' \\
x_2 &= c_{21}'\, x_1' + c_{22}'\, x_2' + c_{23}'\, x_3' \\
x_3 &= c_{31}'\, x_1' + c_{32}'\, x_2' + c_{33}'\, x_3'
\end{aligned}
\tag{10}
$$

1-18. Show that

$$
\sum_{j=1}^{3} c_{ij}\, c_{ij} = 1 \qquad\qquad i = 1, 2, 3 \tag{11}
$$

and
$$
\sum_{j=1}^{3} c_{ij}\, c_{kj} = 0 \qquad i \neq k \qquad i, k = 1, 2, 3 \tag{12}
$$

1-19. Show, by considering the volume of a unit cube as measured in S and S' or otherwise, that

$$\det c_{ij} = \begin{vmatrix} c_{11} & c_{12} & c_{13} \\ c_{21} & c_{22} & c_{23} \\ c_{31} & c_{32} & c_{33} \end{vmatrix} = 1 \tag{13}$$

1-20. Solve Eq. (5) by determinants for the unprimed coordinates and thus show that each direction cosine c_{ij} is equal to its cofactor in the determinant of the preceding exercise.

C. The Summation Convention. In tensor analysis certain characteristic combinations of indices and summation signs occur so regularly that an abbreviated notation was introduced by Einstein to represent them. This abbreviated form of writing, which greatly simplifies the manipulation of tensor equations and reveals more clearly their content, is embodied in the following rules, which must be observed in writing and interpreting tensor equations:

1. If a given letter index appears only once in each term of a tensor equation, it is called a *free index* and the equation holds true for each of the three (n) possible values this index may assume.
2. Any letter index which appears exactly twice in a given term of a tensor equation is called a *summation index* and is understood to imply a summation of this term over all three (n) possible values of this index.
3. No letter index may appear more than twice in a given term.

In order to illustrate the application of these rules and to demonstrate the simplicity of tensor manipulations, we shall re-express Eq. (5) in tensor form and shall derive some of the other formulas given above. In tensor notation, Eq. (5) takes the simple form

$$x_i' = c_{ij}\, x_j \tag{14}$$

where, by rule 1, the equation stands for the three equations obtained by setting i equal to 1, 2, or 3 and, by rule 2, each of these equations consists of a sum of three terms, one term for each value of j.

The relation (7) takes the tensor form

$$\Delta x_i'\, \Delta x_i' = \Delta x_k\, \Delta x_k \tag{15}$$

If we substitute for $\Delta x_i'$ the equivalent expression [from Eq. (14)]

$$\Delta x_i' = c_{ij}\, \Delta x_j = c_{im}\, \Delta x_m$$

we have

$$c_{ij}\, \Delta x_j\, c_{im}\, \Delta x_m = \Delta x_k\, \Delta x_k \tag{16}$$

We have used different summation indices i and k on the two sides of

Eq. (15) to emphasize that each side represents a *sum of three terms* and that the equation is true for the sum but not necessarily for the individual terms. This procedure is not strictly necessary but is usually desirable for the above reason. A summation index [such as i, j, k, or m in Eqs. (15) and (16)] in tensor equations is also called a *dummy index* and may be changed at will to any other symbol, provided rule 3 is not thereby violated. In Eq. (16), on the other hand, we *must* use different indices, j and m, for the summations involved in the two expressions for $\Delta x_i'$ because these summations are independent of one another. This is one reason for adopting rule 3.

If we regroup the terms on the left side of Eq. (16), we have

$$c_{ij}\, c_{im}\, \Delta x_j\, \Delta x_m = \Delta x_k\, \Delta x_k$$

On the left, $3 \times 3 \times 3 = 27$ terms occur, and among these terms each of the 9 possible products of pairs of the Δx's is represented, each having as a coefficient a sum of three products of the c_{ij}'s. Now, because of the special situation that this equation is true for *arbitrary values of each* Δx_i, the coefficient of $\Delta x_j\, \Delta x_m$ on the left must be equal to the coefficient of the corresponding term on the right. Thus, if $j = m$, the coefficient must be unity; and if $j \neq m$, it must be zero. This is expressed by the equation

$$c_{ij}\, c_{im} = \delta_{jm} \tag{17}$$

where δ_{jm} is the *Kronecker delta*, which is defined as being unity if its indices are equal and zero otherwise. Note that the above simple equation represents the *nine* equations (8) and (9) because the two free indices j and m may each have any one of three values.

We shall now derive Eq. (10) by tensor manipulation. We start with Eq. (14), $x_i' = c_{ij}\, x_j$, and multiply each side by c_{ki}' (which automatically requires summation on i by rule 2), whence

$$c_{ki}'\, x_i' = c_{ki}'\, c_{ij}\, x_j$$
$$= c_{ik}\, c_{ij}\, x_j$$

since $c_{ki}' = c_{ik}$. But, by Eq. (17), this is just $c_{ki}'\, x_i' = \delta_{kj}\, x_j = x_k$, or

$$x_k = c_{ki}'\, x_i' \tag{18}$$

which is the tensor form of Eq. (10).

EXERCISE

1-21. Derive Eqs. (11) and (12), using tensor notation.

D. The General Law of Tensor Transformation. We shall adopt the following definition of a three-dimensional (n-dimensional) Cartesian tensor:

Definition:

A tensor of rank r is a quantity having 3^r (or n^r) components in any given rectangular Cartesian coordinate system S.   Each component is represented by an algebraic symbol having r free indices, each index being associated with some axis of the given coordinate system. The components of the tensor in one coordinate system S' are related to its components in another system S by the relations

$$T'_{ijk\ldots m} = c_{is}\, c_{jt}\, c_{ku} \cdot\,\cdot\,\cdot\ c_{mv}\, T_{stu\ldots v} \qquad (19)$$

where the quantities c_{mn} are the direction cosines, $\cos{(x'_m, x_n)}$, between the various axes of S' and S.

According to this definition, any quantity known to be a tensor *must* transform according to Eq. (19) and, just as important, any quantity having the proper number of components which transform by Eq. (19) *must* be a tensor.

EXERCISES

1-22. In the system S' define the nine quantities δ'_{ij} to be equal to unity if $i = j$, and zero otherwise.   In system S, define δ_{rs} similarly. Show that the quantities so defined are the components of a tensor of rank two.

1-23. The quantities c_{ij} are nine in number and have two free indices but are not the components of a second-rank tensor.   Why?

A tensor of rank zero has but a single component which has the same value in all coordinate systems, and it is called a scalar or a *scalar invariant*.   A tensor of rank one has three (or n) components which transform as do the rectangular components of a line segment, and it is called a vector.   Second-rank tensors are sometimes referred to as *dyadics*.

E. Tensor Addition, Multiplication, and Contraction. We shall now briefly consider some of the elementary properties of Cartesian tensors that follow from the foregoing definition.   First, consider the set of quantities that are obtained by adding together the corresponding components of two tensors of equal rank.   This set of quantities is defined as the *sum* of the two tensors and is itself a tensor of the same rank.   That is, if $A_{ij\ldots k}$ and $B_{ij\ldots k}$ are the components of two tensors, then the quantities $C_{ij\ldots k}$ which are defined by the equation

$$C_{ij\ldots k} = A_{ij\ldots k} + B_{ij\ldots k} \qquad (20)$$

are the components of a tensor of the same rank as $A_{ij\ldots k}$ and $B_{ij\ldots k}$. To prove this, we need only show that the components of $C_{ij\ldots k}$ transform as do the components of a tensor:

$$
\begin{aligned}
C'_{rs\ldots t} &= A'_{rs\ldots t} + B'_{rs\ldots t} \\
&= c_{ri}\,c_{sj}\ldots c_{tk}\,A_{ij\ldots k} + c_{ri}\,c_{sj}\ldots c_{tk}\,B_{ij\ldots k} \\
&= c_{ri}\,c_{sj}\ldots c_{tk}(A_{ij\ldots k} + B_{ij\ldots k}) \\
&= c_{ri}\,c_{sj}\ldots c_{tk}\,C_{ij\ldots k}
\end{aligned}
$$

$$\text{Q.E.D}$$

For first-rank tensors this is, of course, just the familiar law of vector addition.

Next let us consider the *multiplication* of tensors. If every component of a tensor of rank b is multiplied by every component of a tensor of rank c, the 3^{b+c} quantities so formed constitute a tensor of rank $b + c$. In symbolic form, this means that the quantities

$$
C_{ij\ldots krs\ldots t} = A_{ij\ldots k}B_{rs\ldots t} \tag{21}
$$

are the components of a tensor of rank equal to the sum of the ranks of $A_{ij\ldots k}$ and $B_{rs\ldots t}$. The proof of this consists again in showing that the quantities $C_{ij\ldots krs\ldots t}$ defined in this way transform in accordance with the prescribed law for tensors:

$$
\begin{aligned}
C'_{fg\ldots hlm\ldots n} &= A'_{fg\ldots h}B'_{lm\ldots n} \\
&= (c_{fi}\,c_{gj}\,\cdots\,c_{hk}\,A_{ij\ldots k})(c_{lr}\,c_{ms}\,\cdots\,c_{nt}\,B_{rs\ldots t}) \\
&= c_{fi}\,c_{gj}\,\cdots\,c_{hk}\,c_{lr}\,c_{ms}\,\cdots\,c_{nt}\,(A_{ij\ldots k}B_{rs\ldots t}) \\
&= c_{fi}\,c_{gj}\,\cdots\,c_{hk}\,c_{lr}\,c_{ms}\,\cdots\,c_{nt}\,C_{ij\ldots krs\ldots t}
\end{aligned}
$$

$$\text{Q.E.D.}$$

Thus, for example, the product of a scalar and a tensor is a tensor of the same rank; of a vector and a vector, a tensor of rank two; of a vector and a second-rank tensor, a third-rank tensor; and so on.

Next, we examine a process called *tensor contraction*, which consists in setting any two free indices of a tensor equal. By rule 2 this automatically implies summation on this index, and the resulting quantity is a tensor whose rank is less by two than that of the original tensor. Symbolically, let $A_{i\ldots j\ldots k\ldots l}$ be a tensor of rank r $(r \geq 2)$; then if any two indices, such as j and k, are set equal (and summed), the result is a tensor of rank $r - 2$. This is easily proved, starting from the tensor transformation law for $A_{i\ldots j\ldots k\ldots l}$:

$$
A'_{r\ldots s\ldots t\ldots u} = c_{ri}\,\cdots\,c_{sj}\,\cdots\,c_{tk}\,\cdots\,c_{ul}\,A_{i\ldots j\ldots k\ldots l}
$$

Thus, setting indices s and t equal and summing,

$$
A'_{r\ldots s\ldots s\ldots u} = c_{ri}\,\cdots\,c_{sj}\,\cdots\,c_{sk}\,\cdots\,c_{ul}\,A_{i\ldots j\ldots k\ldots l}
$$

But $c_{sj}\, c_{sk} = \delta_{jk}$, so the only terms that survive in the grand summation on the right are those for which $j = k$. Thus we have

$$A'_{r\ldots s\ldots s\ldots u} = c_{ri} \cdot \cdot \cdot c_{ul} A_{i\ldots j\ldots j\ldots l} \tag{22}$$

which has just the form of the transformation law for a tensor of rank $r - 2$.

The operation of tensor contraction is familiar in vector analysis as the so-called *scalar product* of two vectors. If A_i and B_j are two first-rank tensors, then $C_{ij} = A_i B_j$ is a second-rank tensor. When contracted by setting $i = j$ and summing, this becomes

$$C_{ii} = A_i B_i = D \tag{23}$$

where D is a tensor of zero rank, i.e., a scalar. In vector notation, this would be written $\boldsymbol{A} \cdot \boldsymbol{B} = D$.

F. Symmetry and Antisymmetry of Tensors. If the components of a tensor of rank two or greater are related to one another by the conditions

$$A_{i\ldots j\ldots k\ldots l} = A_{i\ldots k\ldots j\ldots l} \tag{24}$$

or

$$A_{i\ldots j\ldots k\ldots l} = -A_{i\ldots k\ldots j\ldots l} \tag{25}$$

the tensor is said to be symmetric or antisymmetric, respectively, in the indices j and k. Practically all tensors of physical interest possess symmetry of one type or the other in their various indices.

Not all of the components of a symmetric or antisymmetric tensor are independent—a symmetric tensor of rank two, for instance, has only six [or $\frac{1}{2}n(n + 1)$] independent components, while an antisymmetric second-rank tensor has but three [or $\frac{1}{2}n(n - 1)$]. A few of the more important properties of symmetric and antisymmetric tensors are brought out by the following exercises.

EXERCISES

1-24. Show that the symmetry character of a given tensor is the same in all coordinate systems.

1-25. Show that any second-rank tensor can be written as the sum of a symmetric and an antisymmetric second-rank tensor.

1-26. How many independent nonzero components has a three-dimensional third-rank tensor that is (*a*) symmetric in two indices, (*b*) antisymmetric in two indices, (*c*) symmetric in all pairs of indices (that is, $A_{ijk} = A_{ikj} = A_{jik} = A_{kji}$), (*d*) antisymmetric in all pairs of indices (that is, $A_{ijk} = -A_{ikj} = -A_{jik} = -A_{kji}$)? *Ans.*: (*a*) 18, (*b*) 9, (*c*) 10, and (*d*) 1.

1-27. (*a*) Show that a three-dimensional third-rank tensor which is antisymmetric with respect to *all* pairs of indices has but six nonzero components and that only one of these may be specified independently. (b) Let the 123 component of such a tensor be called Q in the system S and Q' in the system S'. Show that $Q' = Q$. Such a tensor is called a *pseudoscalar* because of this invariant property of its components. Although a pseudoscalar and a true scalar transform in the same way in rectangular Cartesian coordinates, there are several essential differences between them which appear in the more general form of tensor analysis. One of these differences is that a *pseudoscalar changes sign* if the coordinate system is changed from a right-handed to a left-handed system, while a true scalar does not.

1-28. Let T_{ij} be a three-dimensional antisymmetric tensor of rank two. Show that its components obey the transformation law

$$T'_{23} = c_{11} T_{23} + c_{12} T_{31} + c_{13} T_{12}$$
$$T'_{31} = c_{21} T_{23} + c_{22} T_{31} + c_{23} T_{12} \qquad (26)$$
$$T'_{12} = c_{31} T_{23} + c_{32} T_{31} + c_{33} T_{12}$$

Such a tensor is called a *pseudovector* because its three independent components transform like the components of a first-rank tensor. One of the differences between a pseudovector and a true vector is that the components of a pseudovector *change sign* if the coordinate system is changed from a right-handed to a left-handed system.

1-29. Show that the vector product of two vectors can be written as a pseudovector (three dimensions only).

G. Tensor Fields. A tensor whose components are functions of spatial position throughout a given region constitutes a *tensor field*. Tensor fields are needed to describe the intrinsic properties, physical state, and equations of motion of continuous media. In addition to the operations of addition, multiplication, and contraction described previously, other tensor operations, involving derivatives of the components of a tensor field with respect to spatial position, are of importance. Since some of these tensor-field operations are already familiar in vector analysis as the gradient, the divergence, and the curl, we shall indicate only in brief outline the form these operations take in tensor analysis.

Consider first the gradient. In vector analysis this operation produces, from a scalar field σ, a vector field whose components V_i are the directional derivatives of the scalar field along the three space axes. In tensor notation,

$$V_i = \frac{\partial \sigma}{\partial x_i} \qquad (27)$$

To show that such an operation is consonant with the definition of a tensor, we use a familiar formula of partial differentiation, written using the summation convention (rule 2)

$$\frac{\partial}{\partial x_i} = \frac{\partial x'_j}{\partial x_i}\frac{\partial}{\partial x'_j} \tag{28}$$

Thus we obtain (observing that $\partial x'_j / \partial x_i = c'_{ij}$)

$$V_i = c'_{ij} V'_j \tag{29}$$

which is the transformation law for a first-rank tensor.

The gradient can be applied to a tensor field of any rank, and it produces a tensor field of next higher rank:

$$T_{ij\ldots kl} = \frac{\partial R_{ij\ldots k}}{\partial x_l} \tag{30}$$

In vector analysis, the *divergence* of a vector field V_i produces a scalar field σ:

$$\sigma = \frac{\partial V_i}{\partial x_i} \tag{31}$$

To prove that σ so defined is a scalar, we insert into this equation both the transformation law for V_i and the previously used partial derivative transformation:

$$\sigma = \frac{\partial x'_k}{\partial x_i}\frac{\partial(c'_{ij} V'_j)}{\partial x'_k} = c'_{ij}\, c'_{ik}\frac{\partial V'_j}{\partial x'_k} = \frac{\partial V'_j}{\partial x'_j} = \sigma'$$

The divergence of a tensor of rank r $(r \geq 1)$ can be evaluated with respect to any of its indices, and the result is a tensor of rank $r - 1$:

$$T_{i\ldots jl\ldots m} = \frac{\partial R_{i\ldots jkl\ldots m}}{\partial x_k} \tag{32}$$

The vector-operator *curl* is in many respects similar to the vector product of two vectors, in that it produces an antisymmetric second-rank tensor which, because of its simple transformation properties, is commonly thought of as a vector. It is really a pseudovector, however.

The curl can be "applied" to any index of a tensor of rank r, and it produces an antisymmetric tensor of rank $r + 1$ which may be treated in three dimensions as a pseudotensor of rank r:

$$T_{i\ldots j\ldots kl} = \frac{\partial R_{i\ldots j\ldots k}}{\partial x_l} - \frac{\partial R_{i\ldots l\ldots k}}{\partial x_j} \tag{33}$$

The foregoing differential operations, which are already familiar from ordinary vector analysis, do not by any means exhaust the possible types of tensor-field operations. Indeed, it is easy to see that these are but the simplest operations involving differentiation and that, in fact, one can combine any number of differential operations according to the rules of tensor analysis with the result sure to be a tensor. Fortunately, we will not encounter tensor-field operations of any great complexity.

EXERCISES

1-30. Prove that the gradient of a tensor of rank r is a tensor of rank $r + 1$.

1-31. Prove that the divergence of a tensor of rank r is a tensor of rank $r - 1$.

1-32. Show that, in three dimensions, one may also write the curl of a vector as $V_i = -\epsilon_{ijk}(\partial U_j/\partial x_k)$, where ϵ_{ijk} is the constant pseudoscalar whose 123 component is equal to unity.

1-6. The 4-vector System of Space Time

We are now ready to reconsider special relativity, as we have mentioned previously, from a viewpoint in which time and the three dimensions of space are together regarded as forming a four-dimensional continuum in which the physical universe exists. The basis for such a viewpoint may be taken to be the Lorentz invariance of the quantity

$$(\Delta x)^2 + (\Delta y)^2 + (\Delta z)^2 - c^2(\Delta t)^2 = (\Delta x')^2 + (\Delta y')^2 + (\Delta z')^2 - c^2(\Delta t')^2 \tag{1}$$

for two events separated in space and time by amounts Δx, Δy, Δz, and Δt. The similarity between Eq. (1) and Eq. 1-5(7) suggests that (Eq. 1) be interpreted as representing, basically, a *geometrical relationship* between the quantities x, y, z, and t as measured in the S-system and the corresponding quantities x', y', z', and t', measured in the S'-system.

In order to make the similarity of Eqs. 1-5(7) and 1-6(1) more complete, we shall identify the quantities x, y, z, ict with the four rectangular Cartesian coordinates x_1, x_2, x_3, x_4 of a four-dimensional Euclidean space. There is then such close correspondence between the two forms that it is possible to generalize virtually every feature of three-dimensional tensor analysis to four dimensions. In particular, the linear orthogonal transformations which leave the quadratic form (μ is summed from 1 to 4)

$$\Delta x_\mu \, \Delta x_\mu = (\Delta s)^2 \tag{2}$$

invariant are to be regarded as rotations about the origin in four dimensions. These mathematical "rotations" in four dimensions correspond physically to an angular reorientation *combined with uniform transla-*

tional motion of the two coordinate systems–namely to the *Lorentz transformation*.[1]

The main benefits to be gained from this four-dimensional viewpoint are twofold. First, a physical relationship which is written as a four-dimensional tensor equation is, by the principles of tensor analysis, automatically *covariant with respect to the Lorentz transformation* and therefore in agreement with the postulates of special relativity. And second, the transformation properties of physical quantities that are expressible as four-dimensional tensors are relatively easy to remember and to use.

Most of the remaining discussion of special relativity will be directed toward expressing the laws of particle mechanics and electrodynamics in four-dimensional tensor form. Before doing this, however, a few aspects of the extension of tensor analysis to four dimensions and its application to special relativity should be made clear.

A. Distinction between Space and Time. According to our new viewpoint, space and time together form a four-dimensional Euclidean space whose four basic axes are, for any given observer, orthogonal and independent. A vector or tensor quantity in this space will be referred to as a 4-vector or a 4-tensor. The four axes are not, of course, completely *equivalent*, since one of these coordinates is in its physical nature different from the others; that is, x_4 must be measured by a clock rather than a meter stick. Yet the Lorentz transformation represents a rotation of axes in this four-dimensional space, so that a quantity which is measurable purely by a meter stick by one observer must be measured using *both meter sticks and clocks by another*. This property of the Lorentz transformation might lead one to expect that a rotation of axes through 90° could cause a purely spatial interval, say, Δx, to appear as a purely temporal interval of length $\Delta t' = \Delta x/c$ to another observer. But inspection of the Lorentz transformation reveals that such a rotation would correspond physically to a relative motion of the coordinate systems at the *speed of light*. Since this cannot be attained by material systems, there is no possibility of dispensing with either clocks or meter sticks in making physical measurements. Time and distance are in fact quite distinct in nature for any given observer.

The essential physical distinction between time and space appears as a difference in the algebraic sign with which the two types of quantity enter into the quadratic form (1)—a difference in sign that carries

---

[1] We shall consider only the so-called *proper* Lorentz transformations which involve only a "rotation" and not a reversal of sign of the coordinate axes. The improper transformations, which involve a reversal of one or more of the four axes, play a fundamental role in the theory of particles, and we shall consider them briefly in that connection in Chap. 20.

through into most of the equations in which 4-vectors are combined with one another. This distinction between space and time "components" can be represented mathematically in more than one way. In the method that is used here, the "time" components of all 4-vectors are imaginary; this has great simplicity because the geometry of space-time is then Euclidean. There are, however, certain difficulties with this representation, the main one being that it cannot be taken over directly into the general theory of relativity. The more desirable procedure from the standpoint of general relativity is to use the notation and results of *differential geometry*. In this notation, the properties of the space are determined by the so-called metric tensor g_{ij} which characterizes the line element ds^2 of the space. Unfortunately, the use of differential geometry demands a quite extensive mathematical preparation, while a familiarity with ordinary Cartesian tensors suffices for the present approach.

B. Notation. In what follows, we shall often use both three- and four-dimensional tensor analysis. We shall keep these two forms distinct by retaining Latin indices for three-dimensional tensors and introducing *Greek-letter indices for four-dimensional tensors.* Thus, in any equation in which they appear, the Latin indices will be assumed to run from 1 to 3; correspondingly, Greek indices will be assumed to run from 1 to 4.

C. The Lorentz Transformation as a Four-dimensional Linear Orthogonal Transformation. In Sec. 1-3 we arrived at a special case of the Lorentz transformation by making use of the invariance of the quadratic form $x_\mu x_\mu = x'_\nu x'_\nu$ for two observers in relative motion, and we now assert that this transformation may be interpreted as a four-dimensional rotation of orthogonal axes. If this assertion is true, we must of course find that the coefficients γ and $\gamma\beta$ of Eq. 1-3(5) actually do satisfy the various conditions that must be satisfied by the "direction cosines" $c_{\mu\nu}$ which connect the two sets of axes, and we do indeed find that the relations

$$c_{\mu\nu} = c'_{\nu\mu} \tag{3}$$

$$c_{\mu\nu} c_{\mu\rho} = \delta_{\nu\rho} \tag{4}$$

and $$\det c_{\mu\nu} = 1 \tag{5}$$

are true, where $c_{11} = c_{44} = \gamma$, $c_{14} = i\beta\gamma = -c_{41}$, $c_{22} = 1$, $c_{33} = 1$, and all other c's are zero. This special case of the Lorentz transformation may be expressed in the matrix form

$$
c_{\mu\nu} = \begin{array}{c} (x'_\mu) \\ \downarrow \end{array}
\overset{\displaystyle (x_\nu) \rightarrow}{
\begin{pmatrix}
\gamma & 0 & 0 & i\beta\gamma \\
0 & 1 & 0 & 0 \\
0 & 0 & 1 & 0 \\
-i\beta\gamma & 0 & 0 & \gamma
\end{pmatrix}} \tag{6}
$$

Using these "direction cosines," the Lorentz transformation assumes the tensor form

$$x'_\mu = c_{\mu\nu} x_\nu \tag{7}$$

which is of course also the transformation law for the components of *any* four-dimensional vector.

EXERCISES

1-33. Retrace the foregoing review of three-dimensional tensor analysis and extend the results to four dimensions insofar as you are able. Special care will be necessary in connection with pseudoscalars and pseudovectors.

1-34. Describe the physical situation represented by the Lorentz transformation whose rotation matrix is

$$c_{\mu\nu} = \begin{pmatrix} 1 & 1 & 0 & i \\ -1/\sqrt{2} & 1/\sqrt{2} & 0 & 0 \\ 0 & 0 & 1 & 0 \\ -i/\sqrt{2} & -i/\sqrt{2} & 0 & \sqrt{2} \end{pmatrix}$$

1-35. Consider a physical quantity that can be represented as a four-dimensional first-rank tensor (i.e., a 4-vector). Show that the four components of this vector will appear to a *given observer* as an ordinary 3-vector and a 3-scalar.

1-36. Carry out an analysis similar to that of the previous problem for the case of a four-dimensional antisymmetric second-rank tensor, and show that the six independent components of this tensor will appear as two ordinary 3-vectors to a given observer. One of these vectors is a true vector, and the other a pseudovector.

1-37. Work Exercise 1-26 for four dimensions. *Ans.:* (a) 40, (b) 24, (c) 20, and (d) 4.

1-7. Relativistic Mechanics

We shall now use some of the ideas just discussed to study the mechanics of mass particles in special relativity. Our problem is to extend somehow the familiar vector law of particle mechanics,

$$f_i = \frac{d}{dt}\left(m \frac{dx_i}{dt} \right) \tag{1}$$

to 4-vector form, for the above three-dimensional form does not satisfy the postulates of special relativity; that is, it is not covariant under the

Lorentz transformation. We note that there are two things clearly wrong with Eq. (1) as it stands:

1. This equation is an ordinary 3-vector equation; to put it into 4-vector form, a *fourth component of force and momentum must be introduced.*
2. Differentiation with respect to time, which is a valid tensor operation in prerelativistic mechanics where time is regarded as a *universal scalar parameter*, is *not* a valid four-dimensional tensor operation.

A. Proper Time. The second of the above difficulties can be surmounted if we can find a new parameter which *is* a true 4-scalar and which reduces to the time t for low velocities of relative motion of two coordinate systems. Such a parameter can indeed be found; it is called the *proper time* τ. The proper-time interval $\Delta\tau$ between two events whose space-time coordinates are $x_\alpha{}^{(1)}$ and $x_\alpha{}^{(2)}$ is defined by

$$(\Delta\tau)^2 = -\frac{1}{c^2} \Delta x_\alpha \, \Delta x_\alpha \tag{2}$$

where $\Delta x_\alpha = x_\alpha{}^{(2)} - x_\alpha{}^{(1)}$. In terms of the *space* and *time* differences, Eq. (2) is simply

$$(\Delta\tau)^2 = (\Delta t)^2 - \frac{(\Delta r)^2}{c^2} \tag{3}$$

where $\Delta t = t^{(2)} - t^{(1)}$ and $(\Delta r)^2 = (\Delta x)^2 + (\Delta y)^2 + (\Delta z)^2$. From Eq. (3) we see that $(\Delta\tau)^2$ can be positive, negative, or zero, so that $\Delta\tau$ itself can be real, imaginary, or singular. If $(\Delta\tau)^2$ is positive, $\Delta\tau$ is real and is said to be a *timelike* interval; if $(\Delta\tau)^2$ is negative, $\Delta\tau$ is imaginary and is said to be a *spacelike* interval. Since $(\Delta\tau)^2$ is a 4-scalar invariant, the time- or spacelike character of the proper time interval between two events is Lorentz invariant. If two events are separated by a timelike interval in a given reference frame, it is possible to find a new reference frame, moving at less than the speed of light with respect to the first, in which the two events occur at the *same point* in space but at *different times;* in this reference frame the measured time interval Δt between the events is equal to the proper-time interval $\Delta\tau$. On the other hand, if two events are separated by a spacelike interval, it is possible to find a new reference frame, again moving at a speed less than c with respect to the first, in which the two events occur at the *same time*, but at *different points* in space. In this frame the imaginary proper-time difference between the two events is equal in magnitude to the time that would be required for a light pulse to travel from one space point to the other. In the singular case $\Delta\tau = 0$, the two events are so related that a light pulse leaving the event having the smaller time coordinate would just arrive at the other; or, equivalently, the proper-time interval between the emission of a pulse of light and its reception is zero.

Events separated by timelike proper-time intervals can be "connected" by signals traveling at less than the speed of light, whereas events separated by a spacelike interval cannot be so connected. Such relativistic considerations are intimately involved with the ideas of physical causality used in modern theories of the structure of matter. Some of the above ideas are illustrated in Fig. 1-7.

B. 4-vector Velocity. To return to our problem of expressing the laws of particle dynamics in 4-vector form, let us recall that the trajectory of a particle as seen by a given observer is a one-parameter curve in space time. That is, if we specify one coordinate, say, t, the other three coordinates are completely determined. (Alternatively, we might specify x, which would then fix y, z, and t.) Clearly, we can put all four space-time coordinates on an equal footing, and at the same time describe the trajectory in terms of a universal scalar parameter, if we regard *all four* space-time coordinates as being specified in terms of the *four-dimensional arc length* of the trajectory. The proper-time interval $\Delta\tau$ is a measure of this arc length and is the parameter that we shall actually use (Fig. 1-8).

We are now in a position to define a 4-vector velocity U_μ by differentiat-

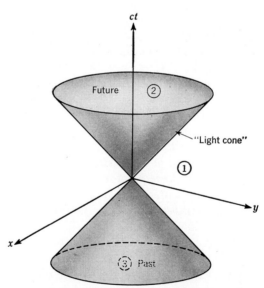

Fig. 1-7. Schematic diagram illustrating the relationships discussed in the text. Through a given event (taken here to be at the origin) there exists a hypercone called the "light cone." Events lying on the space-axis "side" of this light cone (region 1) will appear to occur simultaneously with event 1 in an appropriate Lorentz frame events lying in regions 2 and 3 lie in the future and past, respectively, for all Lorentz frames.

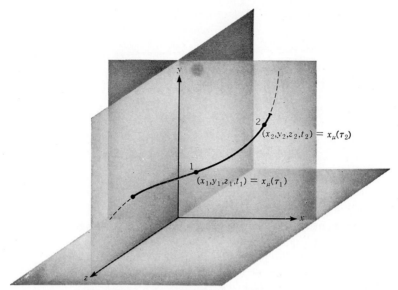

FIG. 1-8. The trajectory of a particle regarded as a parametric function of the proper time τ.

ing the space-time position 4-vector x_μ with respect to the proper time:

$$U_\mu = \frac{dx_\mu}{d\tau} \tag{4}$$

The above velocity is, by virtue of the manner in which it is defined, a valid 4-vector whose components transform according to the law $U'_\nu = c_{\nu\mu} U_\mu$, as can be verified by differentiating Eq. 1-6(7). The four components of the 4-vector velocity are

$$
\begin{aligned}
U_1 &= \frac{dx}{d\tau} = \frac{dx}{dt}\frac{dt}{d\tau} = \frac{\dot{x}}{(1 - u^2/c^2)^{1/2}} \\
U_2 &= \frac{dy}{d\tau} = \frac{dy}{dt}\frac{dt}{d\tau} = \frac{\dot{y}}{(1 - u^2/c^2)^{1/2}} \\
U_3 &= \frac{dz}{d\tau} = \frac{dz}{dt}\frac{dt}{d\tau} = \frac{\dot{z}}{(1 - u^2/c^2)^{1/2}} \\
U_4 &= \frac{ic\,dt}{d\tau} = \frac{ic}{(1 - u^2/c^2)^{1/2}}
\end{aligned}
\tag{5,}
$$

where $u^2 = \dot{x}^2 + \dot{y}^2 + \dot{z}^2$ is the square of the ordinary 3-vector velocity u of the particle.

EXERCISES

1-38. Supply the missing steps in Eq. (5).

1-39. Show that $U_\mu U_\mu = -c^2$.

C. 4-vector Momentum. A 4-vector momentum can next be defined by multiplying the above 4-vector velocity by the mass m_0 of the particle:

$$P_\mu = m_0 U_\mu \tag{6}$$

The components of the 4-vector momentum are

$$P_1 = P_x = \frac{m_0 \dot{x}}{(1 - u^2/c^2)^{1/2}} \qquad P_3 = P_z = \frac{m_0 \dot{z}}{(1 - u^2/c^2)^{1/2}}$$

$$P_2 = P_y = \frac{m_0 \dot{y}}{(1 - u^2/c^2)^{1/2}} \qquad P_4 = \frac{iW}{c} = \frac{im_0 c}{(1 - u^2/c^2)^{1/2}} \tag{7}$$

The first three components of P_μ form a 3-vector P_i called the *ordinary momentum* whose components are P_x, P_y, and P_z, and the symbol W stands for what we shall shortly call the *total energy* of the particle. The 4-vector momentum is often referred to as the *momentum-energy 4-vector.* The three "space" components of this vector reduce, for small velocities, to those of the familiar three-dimensional momentum vector, and the "time" component reduces to the "scalar" $im_0 c$, a constant except for terms of order u^2 and higher.

EXERCISE

1-40. Write the transformation equations connecting the components of ordinary momentum and the total energy in two coordinate systems in uniform relative motion with speed $V = \beta c$.

Ans.:
$$P'_x = \gamma \left(P_x - \frac{\beta W}{c} \right) \qquad P'_z = P_z$$
$$P'_y = P_y \qquad\qquad W' = \gamma(W - \beta c P_x) \tag{8}$$

D. 4-vector Force. Differentiation of the 4-vector momentum with respect to τ yields another 4-vector, sometimes called the *Minkowski force:*

$$\frac{d(m_0 \, dx_\mu/d\tau)}{d\tau} = F_\mu \tag{9}$$

Relation (9) may be taken to be a definition of *the relativistic force that acts upon a particle to change its proper momentum.* Since the "space" components of this relativistic force reduce, for low velocities, to the familiar force components of Newtonian mechanics, Eq. (9) may be taken to be an appropriate four-dimensional generalization of the laws of motion of a particle.

If Newton's *third law* is now postulated to hold for the 4-vector force, *conservation of linear momentum* (and as we shall see, *conservation of energy* also) follows immediately for particle dynamics. The dynamical laws are then in a form compatible with Einstein's postulates and with the known form of Newton's equations valid for low velocities, and a great many experiments involving the collision of particles moving at speeds comparable with the speed of light have provided overwhelming proof that the law (9), and Newton's third law, are valid for high speeds as well.

E. *Ordinary Force*. It is perhaps unfortunate that the historical development of relativity was such that the above form of the dynamical laws is not always used. Instead, one usually encounters three force components defined in the old way, as the *time* derivative of the *ordinary momentum*:

$$f_i = \frac{d}{dt} \frac{m_0 \dot{x}_i}{(1 - u^2/c^2)^{1/2}} \qquad i = 1, 2, 3 \qquad (10)$$

The above 3-vector is invariant to rotations of the space axes but is *not* Lorentz invariant, so that the dynamical laws are not in a form that clearly satisfy Einstein's postulates. It is necessary to derive special transformation laws for the components of the above force (called the *ordinary force*), and one must then remember these special transformation equations as an added piece of mental baggage. In what follows, F_μ and f_i will be used to represent the 4-vector force and the ordinary force, respectively.

EXERCISE

1-41. Write the law of transformation for the components of (a) the 4-vector force and (b) the ordinary force.

Ans.:
$$\begin{aligned}
&F_1' = \gamma(F_1 + i\beta F_4) & F_3' = F_3 \\
&F_2' = F_2 & F_4' = \gamma(F_4 - i\beta F_1)
\end{aligned} \qquad (11)$$

$$f_x' = \left(1 - \frac{u'^2}{c^2}\right)^{1/2} \frac{f_x + i\beta F_4(1 - u^2/c^2)^{1/2}}{(1 - u^2/c^2)^{1/2}(1 - \beta^2)^{1/2}}$$

$$f_y' = \frac{(1 - u'^2/c^2)^{1/2}}{(1 - u^2/c^2)^{1/2}} f_y \qquad f_z' = \frac{(1 - u'^2/c^2)^{1/2}}{(1 - u^2/c^2)^{1/2}} f_z \qquad (12)$$

F. *The Variation of "Mass" with Velocity*. In the historical development of relativity, the dynamical laws were arrived at by methods that did not utilize the concept of 4-vectors. For example, the form 1-7(7) for the "space" components of momentum was arrived at by *defining* momentum to be (1) the product of a scalar property of a particle, called its "mass," and its ordinary velocity u and (2) a quantity which is conserved in an isolated dynamical system. It was then found, by considering an ideal collision between equal elastic spheres, that the scalar "mass" had to be a function of the speed of the sphere, and indeed just the function $m_0/(1 - u^2/c^2)^{1/2}$, where m_0 is the mass of the sphere when it is at rest. Thus there appeared, as a result of this particular approach, the idea of *relativistic mass*, defined by

$$m = \frac{m_0}{(1 - u^2/c^2)^{1/2}} \tag{13}$$

m_0 is called the *rest mass* of the body, and is the same for all observers.

It is important to recognize clearly that the above concept is the result of a particular approach, or of a certain frame of mind, and does not necessarily imply any new and profound property of matter. Indeed it seems clear that the 4-vector approach, in which momentum is defined as the product of *rest mass* (an invariant scalar) and the 4-*vector velocity* (an invariant 4-vector), is much neater and philosophically more satisfying than the old historical one, since in the 4-vector approach it is clearly the new relationship between space and time, and not any property of matter alone, that accounts for the new form exhibited by the momentum. In any case it is necessary to check the final physical laws, no matter by what process they are arrived at, against the actual properties exhibited by nature. By this criterion either of the above expressions for the dynamical laws must be considered to be experimentally verified, but of course the criteria of simplicity and aesthetics speak strongly for the 4-vector form.

G. Relativistic Energy. Let us now examine more closely the "time" component of the 4-vector momentum. This component is

$$P_4 = m_0 U_4 = im_0 c \frac{dt}{d\tau} = \frac{im_0 c}{(1 - u^2/c^2)^{1/2}} \tag{14}$$

and is experimentally found to be conserved in particle collisions, when summed over all the particles. We see that, for zero speed, P_4 has the value $im_0 c$ and that it increases *quadratically with u* for small speeds. In fact, if we expand Eq. (1) by the binomial theorem, we obtain

$$P_4 = im_0 c \left[1 + \frac{1}{2} \frac{u^2}{c^2} + \frac{3}{8} \frac{u^4}{c^4} + \cdots \right] \tag{15}$$

The dominant *variable* term in this expression is precisely i/c times the quantity which we recognize as the *nonrelativistic kinetic energy*, $\frac{1}{2}m_0 u^2$. Thus the conservation of P_4 for a system of colliding elastic particles is equivalent for low velocities to the prerelativistic law of the conservation of *kinetic energy* in such collisions. Therefore, in order to preserve the nomenclature in such form that the relativistic laws can be "joined on" to the nonrelativistic ones in the low-speed regime, we shall multiply P_4 by c/i and call the result the *total energy W* of the particle.   Thus we have

$$W = \frac{m_0 c^2}{(1 - u^2/c^2)^{1/2}} = \frac{cP_4}{i} \tag{16}$$

Furthermore, we shall regard this total energy as being composed of an intrinsic, constant part $m_0 c^2$, which we call the *self-energy*, or *rest* energy, of the particle, and a variable part which the particle possesses by virtue of its speed.   This latter part we call the kinetic energy T.   Thus we may write

$$W = m_0 c^2 + T \tag{17}$$

It should be carefully noted that the kinetic energy is equal to $\frac{1}{2} m_0 u^2$ *only for speeds that are small compared with c.*

EXERCISE

1-42. Expand T in powers of u/c.   At what value of u/c is the true kinetic energy equal to (a) 1.01 times the nonrelativistic value, (b) 1.1 times, (c) 2 times, (d) 10 times?   *Ans.:* (a) $u/c = 0.115$, (b) $u/c = 0.347$, (c) $u/c = 0.786$, and (d) $u/c = 0.984$.

H. The Conservation of Momentum and of Mass Energy.   We have seen that the conservation of the time component of 4-vector momentum is equivalent to the prerelativistic law of conservation of kinetic energy in elastic collisions, i.e., collisions in which the number and rest masses of the particles are the same after as before the collision.   But it was predicted by Einstein, and has been brilliantly verified by experiment, that there can occur physical processes in which a more general form of the conservation of 4-vector momentum holds, where the number and the rest masses of the particles before and after a collision are *different*.   That is, the conservation of 4-vector momentum is valid in the form

$$\sum_{i=1}^{N} P_\mu(i) = \sum_{j=1}^{N'} P_\mu(j) \tag{18}$$

where the summation on the left is to be carried out over all the particles initially present before a collision, decay, or other interaction, and the summation on the right, over all the particles present afterward.   Furthermore, the conservation of P_4 implies that *rest mass* and *kinetic energy* need not separately be conserved but *may be converted, one into the other*.   This latter more general energy-conservation law is known as the *law of conservation of mass energy*.   It has been verified quantitatively both in the particle reactions of nuclear physics (Chap. 17) and in the decay of unstable particles into lighter fragments, and today it is one of the best-established results of special relativity.

1-43. Make use of the Lorentz invariance of the "length" of a 4-vector to establish the relation

$$W^2 = P^2 c^2 + M_0{}^2 c^4 \tag{19}$$

for a particle of rest mass M_0, ordinary momentum P_i and total energy W $(P^2 = P_i P_i)$. W, Pc, $M_0 c^2$, and T are thus related as shown in Fig. 1-9.

1-44. For the particle of the previous exercise, show that

$$Pc = \beta W \tag{20}$$

where $\beta = u/c$, u is the speed of the particle, and P is the magnitude of the ordinary momentum.

1-45. A photon may be regarded as a particle of zero rest mass and total energy $W = h\nu$. Use the transformation law for the momentum 4-vector to derive the relativistic Doppler formula 1-4(5).

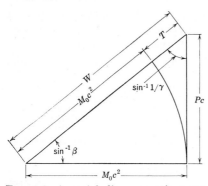

1-46. A photon is emitted at an angle θ' by a source stationary in the S'-system and received at an angle θ in the S-system. Use the above method to find the relation between θ' and θ. This effect is called the *aberration* of light. Evaluate the magnitude of the effect for distant starlight seen by an observer on the earth, if the angle of arrival is at right angles to the earth's orbital velocity.

FIG. 1-9. A useful diagrammatic representation of the mass, energy, and momentum of a particle.

$$Ans.: \quad \cos \theta = \frac{\cos \theta' + \beta}{1 + \beta \cos \theta'} \tag{21}$$

1-47. Using classical electromagnetic theory, show that the momentum carried by a small, localized "lump" of radiation (a "photon") is $P = W/c$, where W is the electromagnetic energy of the lump. For simplicity, consider the lump to consist of a small cross section A of a plane wave train of finite length L.

1-48. In many problems involving interacting particles, the analysis is considerably simplified if it is carried out in the *center-of-mass system* (abbreviated CM) of the interaction, which is defined as that system in which the *net ordinary momentum is zero*. If, in a given system, N particles of rest masses $m_0(i)$, momenta $p(i)$, and total energies $w(i)$ interact, find the speed and direction in which the CM system is moving through the given system.

$$Ans.: \qquad \beta_{CM} = \frac{\displaystyle\sum_{i=1}^{N} p(i)c}{\displaystyle\sum_{i=1}^{N} w(i)} \qquad (22)$$

1-49. Show that the CM system, defined as above, is really a center-of-energy system, in the sense that the particles are weighted according to their total energies rather than their rest masses.

1-50. A particle of rest mass M_1 traveling in the positive x-direction with kinetic energy T_1 collides with a stationary particle of rest mass M_2. After the collision, two different particles, whose rest masses are M_3 and M_4, emerge. Particle 3 emerges with momentum P_3 along the negative z-axis. In what direction and with what momentum does particle 4 emerge from the collision?

$$Ans.: \quad \tan \theta_4 = \frac{P_3}{(2T_1M_1 + T_1^2/c^2)^{\frac{1}{2}}} \qquad P_4 = (P_3^2 + 2T_1M_1 + T_1^2/c^2)^{\frac{1}{2}}$$

1-51. A particle of rest mass M moving with speed $\beta_0 c$ collides inelastically with (i.e., sticks to) a stationary particle of rest mass m. Find the speed of the composite particle. *Ans.*: $\beta' = \beta_0[\gamma_0 M/(\gamma_0 M + m)]$.

1-52. Find the rest mass of the composite particle of the previous problem in terms of the kinetic energy T of the incident particle and the rest masses M and m of the two particles.

$$Ans.: \qquad M_c = (M + m)\left[1 + \frac{2mT}{(M + m)^2 c^2}\right]^{\frac{1}{2}}$$

1-53. The Berkeley Bevatron was designed to have sufficient energy to produce negative protons by bombarding hydrogen gas with protons. One possible reaction is $p + p \rightarrow p + p + (p + \bar{p})$, where $(p + \bar{p})$ represents a proton–negative proton pair produced in the collision. With what minimum kinetic energy, in units of $M_p c^2$, must the moving proton strike the proton at rest in order to make the reaction work? *Hint:* Solve the problem in the CM system and then transfer to the laboratory system. *Ans.*: $T_{\min} = 6M_p c^2$.

1-54. An electron-positron pair can be produced by a photon (symbolized γ) striking a stationary electron, according to the reaction

$\gamma + e^- \rightarrow e^- + (e^+ + e^-)$. Find the threshold energy of the photon for this reaction. *Ans.:* $h\nu_{\min} = 4m_e c^2$.

1-55. A collision in which a photon strikes an electron and is scattered by the electron is called a *Compton collision*. If the photon is deflected through an angle θ, find the change in its wavelength. *Ans.:* $\Delta\lambda = (h/m_0 c)(1 - \cos\theta)$.

1-56. Show that the momentum P and the kinetic energy T of a particle of rest mass M are connected by the relation $P^2 c^2 = 2TMc^2 + T^2$.

1-57. A radioactive particle of rest mass M_0 is moving with momentum P_0 when it decays spontaneously into two lighter particles of rest masses M_1 and M_2. The latter particles are observed to have momenta P_1 and P_2, respectively. Find M_0.

$$\begin{aligned} Ans.:\quad M_0^2 c^4 &= (M_1 + M_2)^2 c^4 + 2W_1 W_2 - 2M_1 M_2 c^4 - 2P_1 \cdot P_2 c^2 \\ &= (M_1 + M_2)^2 c^4 + 2M_1 T_2 c^2 + 2T_1 W_2 - 2P_1 \cdot P_2 c^2 \end{aligned}$$

1-58. Show that the scalar product $F_\mu U_\mu$ is equal to zero.

1-59. Show from the result of the previous exercise that the scalar product of the ordinary force and the ordinary velocity is equal to the rate of increase of the relativistic total energy.

1-8. Relativistic Electrodynamics

Since the special theory of relativity assumes the validity of Maxwell's equations, it might appear unnecessary to examine the effect of the Lorentz transformation upon them. It is instructive, however, to express Maxwell's equations in 4-vector notation, for in this way several interesting and fundamental relationships will be seen to exist between various familiar electrical quantities.

A. Maxwell's Equations, Electromagnetic Potentials. Maxwell's equations have the following form in rationalized MKSC units:

$$(a) \qquad \nabla \times E = - \frac{\partial B}{\partial t}$$

$$(b) \qquad \nabla \cdot D = \rho$$

$$(c) \qquad \nabla \times H = J + \frac{\partial D}{\partial t} \tag{1}$$

$$(d) \qquad \nabla \cdot B = 0$$

In these equations, E and B are the electric-field intensity and magnetic induction, respectively, and are defined in a given reference frame in terms of the ordinary force f acting upon a charge q moving with ordinary velocity u:

$$f = q(E + u \times B) \tag{2}$$

The quantities D and H are defined in terms of E and B and the polarization vectors P and M:

(a)
$$D = \varepsilon_0 E + P$$

(b)
$$H = \frac{B}{\mu_0} - M \tag{3}$$

For simplicity, we shall for the most part treat cases not involving material media. The quantities ρ and J are the electric-*charge and -current densities*, respectively. These are defined by the equations

(a)
$$\rho = \lim_{\Delta V \to 0} \frac{\sum_i q_i}{\Delta V}$$

(b)
$$J = \lim_{\Delta V \to 0} \frac{\sum_i q_i u_i}{\Delta V} \tag{4}$$

where the summations are to be carried out over all charges lying within ΔV, and u_i is the ordinary velocity of q_i. The limit $\Delta V \to 0$ is intended to apply in a physical sense only—the volume element must be kept large enough that a sufficiently large number of elementary charges q_i will be contained inside it to assure that ρ and J will be, macroscopically, smooth functions of position. ρ and J are said to be the *sources* of the fields E and B.

It is usually most convenient to describe the electromagnetic field in terms of the so-called *electromagnetic potentials* A and ϕ. Equation (1*d*) permits B to be expressed as the curl of another vector:

$$B = \nabla \times A \tag{5}$$

Substitution of this expression into Eq. (1*a*) and integration then yield

$$E = -\frac{\partial A}{\partial t} - \nabla \phi \tag{6}$$

In the commonly encountered case of homogeneous linear, isotropic media, in which $D = \varepsilon E$ and $B = \mu H$, further substitution of Eqs. (5) and (6) into the remaining pair of Maxwell's equations (1*b*) and (1*c*) then leads to the following equations that must be satisfied by A and ϕ:

(a)
$$\nabla^2 \phi + \frac{\partial}{\partial t} (\nabla \cdot A) = -\frac{\rho}{\varepsilon}$$

(b)
$$-\nabla^2 A + \mu\varepsilon \frac{\partial^2 A}{\partial t^2} + \nabla(\nabla \cdot A) + \mu\varepsilon \frac{\partial}{\partial t} (\nabla \phi) = \mu J \tag{7}$$

Now Eq. (5) does not determine A uniquely for given B, nor does Eq. (6) determine ϕ uniquely for given E and A. This is the basis of the theory of *gauge transformations* of A and ϕ. One is free to impose one scalar condition upon A or ϕ, whereupon the equations (7) yield unique potentials (except for additive constants). This additional condition often is taken to be $\nabla \cdot A = 0$, but we shall find that another condition, the so-called *Lorentz condition*, is more suitable for special relativity. This condition is

$$\nabla \cdot A = -\mu\varepsilon \frac{\partial \phi}{\partial t} \tag{8}$$

which then reduces Eq. (7) to the symmetrical form

(a)
$$\nabla^2\phi - \mu\varepsilon \frac{\partial^2\phi}{\partial t^2} = -\frac{\rho}{\varepsilon}$$

(b)
$$\nabla^2 A - \mu\varepsilon \frac{\partial^2 A}{\partial t^2} = -\mu J$$

$$\tag{9}$$

The electromagnetic potentials ϕ and A are thus related by inhomogeneous wave equations to the *sources* of the fields. The solutions of these equations have the form of disturbances traveling away from the sources at a speed $v = (\mu\varepsilon)^{-\frac{1}{2}}$. In vacuo, $v = c = (\mu_0\varepsilon_0)^{-\frac{1}{2}}$.

B. The Charge-current 4-vector. As a starting point in expressing Maxwell's equations in 4-vector form, we shall show that J and ρ may be combined to form a four-dimensional current vector, J_μ. To do this, we first observe that the charge of a particle is a *four-dimensional scalar;* indeed, this is assumed in Eq. (4), and since these definitions lead to an experimentally valid description of electromagnetism, we shall retain this assumption.

Consider the measurement of the charge density ρ by two observers, O and O′, who are in relative motion. For simplicity, let the charges be at rest with respect to one of these observers, say, O. Call ρ_0 the charge density observed by O. Now O and O′ will agree on the amount of charge contained within a small element of volume $\Delta x\, \Delta y\, \Delta z$ (as defined by O). The volume element itself is modified by the Lorentz contraction, however, so that $\Delta V' = \Delta V/\gamma$. Thus the charge densities are related by

$$\rho' = \gamma\rho_0 \tag{10}$$

which is the same transformation law as for a *time interval*, as given by Eq. 1-4(3).

The transformation equations for the components of the current density may be established most simply by writing Eq. (4b) as $J = \rho u$ (we assume for simplicity that all charges in a given volume element have the same ordinary velocity). Then, using Eqs. (10) and 1-7(5), we see that the three components of the ordinary current-density vector are just equal to the three "space" components of $\rho_0 U_\mu$. Thus, if we define a four-dimensional current density J_μ by the equation

$$J_\mu = \rho_0 U_\mu \tag{11}$$

where ρ_0 is the density of charge in a system in which the charges are at rest and U_μ is the 4-vector velocity of the charges, then the three "space" components of J_μ are equal to the three components of the ordinary current density J, and J_4 is equal to $ic\rho$:

$$J_\mu = (J_x, J_y, J_z, ic\rho) \tag{12}$$

EXERCISE

1-60. Show that the equation of continuity of electric charge is expressed by the 4-vector divergence equation

$$\frac{\partial J_\mu}{\partial x_\mu} = 0 \tag{13}$$

C. The 4-vector Potential. Let us next consider the potential equations (9). We first observe that the operator $\nabla^2 - (1/c^2)\,\partial^2/\partial t^2$, which is called the *d'Alembertian operator*, has a very simple form when expressed as a four-dimensional tensor operator:

$$\nabla^2 - \frac{1}{c^2}\frac{\partial^2}{\partial t^2} = \frac{\partial^2}{\partial x_\mu\,\partial x_\mu} \tag{14}$$

(This is sometimes abbreviated $\square^2$.)

Clearly, we may now combine the potentials A and ϕ to form a 4-vector potential A_μ whose components are

$$A_\mu = \left(A_x, A_y, A_z, \frac{i\phi}{c}\right) \tag{15}$$

for then Eq. (9) takes on the simple form, in vacuo,

$$\frac{\partial^2 A_\mu}{\partial x_\nu\,\partial x_\nu} = -\mu_0 J_\mu \tag{16}$$

EXERCISES

1-61. Verify Eq. (14).

1-62. Verify that Eq. (16) correctly represents Eqs. (9).

1-63. Show that the Lorentz condition has the 4-vector form

$$\frac{\partial A_\mu}{\partial x_\mu} = 0 \tag{17}$$

D. The Electromagnetic-field 4-tensor. We have so far succeeded in expressing Eq. (9) in 4-vector form, and since these equations suffice to determine the electromagnetic potentials, the Lorentz invariance of Maxwell's equations is assured. But for many purposes it is desirable to work directly with the fields $\boldsymbol{E}$ and $\boldsymbol{B}$. We shall therefore seek to express these fundamental field vectors in four-dimensional form. A hint on how to do this comes from Eqs. (5) and (6) relating the field components to the potentials. These are seen to involve derivatives of all possible pairs of components of the 4-vector potential A_μ with respect to the corresponding x_μ's, in combinations familiar to us in the vector operation of the *curl*. In our discussion of tensor analysis we saw, however, that such an operation applied to a vector really defines an *antisymmetric second-rank tensor* [Eq. 1-5(33)].

Guided by the foregoing considerations, let us define an antisymmetric second-rank tensor whose components are

$$f_{\mu\nu} = \frac{\partial A_\nu}{\partial x_\mu} - \frac{\partial A_\mu}{\partial x_\nu} \tag{18}$$

This is called the *field tensor*. By writing down each component of $f_{\mu\nu}$ and comparing with Eqs. (5) and (6), we easily arrive at a correspondence with the ordinary electric field and magnetic induction, as indicated in the following matrix:

$$f_{\mu\nu} = \begin{pmatrix} 0 & B_z & -B_y & \dfrac{-iE_x}{c} \\[2mm] -B_z & 0 & B_x & \dfrac{-iE_y}{c} \\[2mm] B_y & -B_x & 0 & \dfrac{-iE_z}{c} \\[2mm] \dfrac{iE_x}{c} & \dfrac{iE_y}{c} & \dfrac{iE_z}{c} & 0 \end{pmatrix} \tag{19}$$

EXERCISE

1-64. Carry out the procedure outlined above to obtain Eq. (19).

E. Maxwell's Equations in 4-tensor Form. We are now ready to write Maxwell's equations in terms of the field tensor (19). Two of these equations are obtained by setting the divergence of the field tensor equal to the 4-vector current density, and the remaining two equations by setting the symmetrical third-rank tensor gradient of the field tensor equal to zero:

(a)
$$\frac{\partial f_{\mu\nu}}{\partial x_\nu} = \mu_0 J_\mu$$

(b)
$$\frac{\partial f_{\nu\sigma}}{\partial x_\alpha} + \frac{\partial f_{\sigma\alpha}}{\partial x_\nu} + \frac{\partial f_{\alpha\nu}}{\partial x_\sigma} = 0 \tag{20}$$

EXERCISES

1-65. Write out a sufficient number of the components of Eq. (20) to verify that they do in fact embody Maxwell's equations in vacuo.

1-66. Show that Eq. (20*b*) is satisfied identically by any four-dimensional antisymmetric second-rank tensor field whose components are the curl of a 4-vector field, as expressed by Eq. (18). This four-dimensional result is somewhat analogous to the three-dimensional vector identities $\nabla \cdot \nabla \times A = 0$ or $\nabla \times \nabla S = 0$.

The above 4-vector form of electrodynamics shows quite clearly that Maxwell's equations are in fact covariant under Lorentz transformations, so that Einstein's postulates are not contradicted. Since these postulates refer specifically only to the *speed of light*, some other equations for electrodynamics, which also lead to a wave equation, might have been called for. Furthermore, the association of A and ϕ, of J and ρ, and of B and E, as components of 4-vectors and 4-tensors, illustrates clearly the close association—indeed, the inseparability—of magnetic and electric phenomena. It is a familiar result of prerelativity electrodynamics that what appears to be a pure magnetic field in one system must appear to be a combination of an electric field and a magnetic field in another system. We now find, however, that this interchangeability penetrates much deeper than was previously evident.

The synthesis of ordinary scalar and vector quantities into four-dimensional tensors, and the consequent simplification of the equations of physics, also demonstrates strikingly the importance of finding the "right" representation for the physical world. Some scientists maintain that the sole criterion of the correctness of a physical law is its ability to predict accurately the outcome of physical experiments, and that equally accurate formulations of a given law must be judged equally valid and equally acceptable. Others believe that there are additional criteria by which a physical law must be judged, such as its symmetry or simplicity of expression, or the economy of postulates needed to define it. This latter viewpoint has time and again not only led to a great simplification of physical laws but, more important, has been directly responsible for the discovery of new relationships between physical quantities and of new physical effects, with a consequent improvement in our understanding of nature.

EXERCISES

1-67. Write the Lorentz transformation equations for (*a*) the electromagnetic potentials and (*b*) the current and charge densities.

Ans.:
$$A'_x = \gamma\left(A_x - \frac{\beta\phi}{c}\right) \qquad A'_z = A_z$$

$$A'_y = A_y \qquad\qquad \phi' = \gamma(\phi - \beta c A_x) \tag{21}$$

$$J'_x = \gamma(J_x - \beta c\rho) \qquad J'_z = J_z$$

$$J'_y = J_y \qquad\qquad \rho' = \gamma\left(\rho - \frac{\beta J_x}{c}\right) \tag{22}$$

1-68. Apply the transformation law for a second-rank tensor to the field tensor $f_{\mu\nu}$, and thus find the Lorentz transformation equations for the electric and magnetic fields.

Ans.:
$$B'_x = B_x \qquad\qquad E'_x = E_x$$

$$B'_y = \gamma\left(B_y + \frac{\beta E_z}{c}\right) \qquad E'_y = \gamma(E_y - \beta c B_z) \tag{23}$$

$$B'_z = \gamma\left(B_z - \frac{\beta E_y}{c}\right) \qquad E'_z = \gamma(E_z + \beta c B_y)$$

These transformation equations are sometimes most conveniently expressed in the form

$$B'_\| = B_\| \qquad\qquad E'_\| = E_\|$$

$$B'_\perp = \gamma\left(B_\perp - \frac{V \times E}{c^2}\right) \qquad E'_\perp = \gamma(E_\perp + V \times B)$$

1-69. Use the above transformation equations to find the electric and magnetic fields due to a charge moving at constant speed βc. Express these fields in spherical polar coordinates with polar axis along the x-axis, and make a sketch of the electric lines of force for β nearly equal to unity (Fig. 1-10).

Ans.: (a)
$$E = \frac{q e_r (1 - \beta^2)}{4\pi\varepsilon_0 r^2 (1 - \beta^2 \sin^2 \theta)^{3/2}} \tag{24}$$

(b)
$$B = \frac{q e_\phi \beta (1 - \beta^2) \sin\theta}{4\pi\varepsilon_0 c r^2 (1 - \beta^2 \sin^2 \theta)^{3/2}}$$

1-70. Show that the equations of motion of a charged particle of charge q and rest mass m_0, moving in an electric and magnetic field, are

$$m_0 \frac{d^2 x_\mu}{d\tau^2} = q f_{\mu\nu} U_\nu \tag{25}$$

Interpret the equation corresponding to $\mu = 4$.

1-71. A charged particle moves in crossed, uniform electric and magnetic fields. Show that it is possible to transform to a moving system in which one or the other of the fields vanishes, unless $E = cB$.

1-72. A charged particle moves in a uniform magnetic field of induction $B = kB$. Show that the particle moves in a helical path at constant

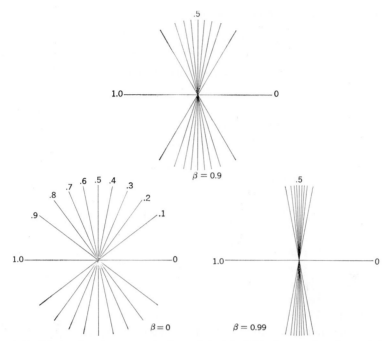

FIG. 1-10. Illustrating the concentration of electric field of a moving charge into the equatorial plane, as described by Eq. 1-8 (24a). Note that, at all speeds, the field is an inverse-square field, but is not isotropic except at very low speeds. The decimals define the proportion of the total flux lying within a circular cone whose axis is the line of motion of the charge.

speed, and find the relations between the momentum of the particle and the parameters of the helix.

Ans.:
$$P = \frac{qBR}{\cos \alpha} \tag{26}$$

where $P_z = P \sin \alpha$

R = "radius" of helix

1-73. A mass particle of charge q and rest mass m_0 moves in an electromagnetic field. By writing the Euler-Lagrangian equations for the 4-space coordinates x_μ, using τ as the independent variable, show that a suitable Lagrangian function is

$$L = \tfrac{1}{2}m_0 U_\mu U_\mu + q U_\mu A_\mu \tag{27}$$

1-74. Define a momentum p_μ conjugate to the coordinate x_μ, by the equations $p_\mu = \partial L/\partial U_\mu$ and $U_\mu = dx_\mu/d\tau$, and construct a Hamiltonian function for the particle. Show that this Hamiltonian is

$$H = \frac{(p_\mu - q A_\mu)(p_\mu - q A_\mu)}{2m_0} \tag{28}$$

and that the canonical equations are

$$\frac{dx_\mu}{d\tau} = \frac{\partial H}{\partial p_\mu} = \frac{P_\mu - qA_\mu}{m_0} \quad \text{and} \quad \frac{dp_\mu}{d\tau} = -\frac{\partial H}{\partial x_\mu} = \frac{q}{m_0}(p_\nu - qA_\nu)\frac{\partial A_\nu}{\partial x_\mu}$$

1-75. Show that the following quantities are valid four-dimensional tensors:

(a) $|A|^2 c^2 - \phi^2$ (c) $\mathbf{A} \cdot \mathbf{J} - \rho\phi$ (e) $\rho\mathbf{E} + \mathbf{J} \times \mathbf{B}$

(b) $|J|^2 - \rho^2 c^2$ (d) $\mathbf{B} \cdot \mathbf{H} - \mathbf{D} \cdot \mathbf{E}$

1-76. A particle of charge q and rest mass m_0 moves in an electromagnetic field whose 4-vector potential is independent of one of the space-time coordinates, say, x_α. Show that the motion is such that the change in the α-component of the 4-vector momentum of the particle between any two points on its space-time trajectory is

$$\Delta P_\alpha = -q\,\Delta A_\alpha \tag{29}$$

where ΔA_α is the change in A_α between these two points in space time.

1-77. A long, straight wire of radius a is coaxial with a straight, hollow circular cylinder of inner radius b. An adjustable steady current I flows along the wire and back through the cylinder. In addition, a fixed d-c potential difference ϕ exists between the wire and cylinder. If electrons are thermionically emitted at negligible speed from the wire and are attracted toward the cylinder, for what range of I will they reach the cylinder?

Ans.:
$$I < \frac{2\pi \sqrt{2m_0 q\phi + q^2\phi^2/c^2}}{\mu_0 q \ln b/a}$$

1-9. Energy Units

In dealing with the high-energy particles that come from particle accelerators, nuclear reactions, and cosmic rays, it is convenient to introduce a special system of units, called *energy units*, for the measurement of mass, energy, and momentum. In this system the quantities W, T, $M_0 c^2$, and Pc are all measured in units of 10^6 electron volts, abbreviated MeV, and the symbols M and P are used to represent $M_0 c^2$ and Pc. This change in notation simplifies the writing of many of the equations of special relativity because it eliminates most of the c's that occur in these equations. For illustration, some of the more useful equations of particle mechanics are collected together from our previous work and expressed in energy units:

1. The Lorentz transformation for energy and momentum:

$$P'_x = \gamma(P_x - \beta W) \qquad P'_z = P_z$$
$$P'_y = P_y \qquad W' = \gamma(W - \beta P_x) \tag{1}$$

2. The momentum and energy of a particle of speed $u = \beta c$:

$$P = \gamma \beta M \tag{2}$$

$$W = \gamma M \quad \text{or} \quad \gamma = \frac{W}{M} \tag{3}$$

$$P = \beta W \quad \text{or} \quad \beta = \frac{P}{W} \tag{4}$$

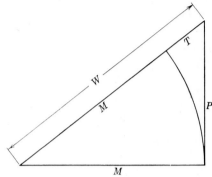

3. Relations between energy, momentum, and rest mass (Fig. 1-11):

$$W^2 = P^2 + M^2 \tag{5}$$

$$W = M + T \tag{6}$$

$$P^2 = 2MT + T^2 \tag{7}$$

4. The velocity of the center-of-mass system of a number of particles of velocities $u_i = \beta_i c$:

$$\beta_{\text{CM}} = \frac{\sum_i P_i}{\sum_i W_i} = \frac{\sum_i \beta_i W_i}{\sum_i W_i} \tag{8}$$

FIG. 1-11. Relations between energy, momentum, and mass in energy units.

The rest masses of some of the particles that are commonly encountered in particle reactions are given in Table 1-1.

TABLE 1-1. MASSES OF SOME PARTICLES

Particle	Mass in m_e	$M_0 c^2$, MeV
Electron, e............	Unity	0.510976
$\mu^{\pm}$-meson..............	206.8	105.7
π^0-meson..............	264.4	135.1
$\pi^{\pm}$-meson..............	273.2	139.6
$\tau^{\pm}$-meson..............	966.4	493.8
Proton, p..............	1836.1	938.21
Neutron, n............	1838.7	939.50
Λ^0 hyperon............	2181	1114.4
Deuteron, d...........	3671.3	1875.41
α-particle..............	7296.1	3727.07

One feature of the actual usage of energy units requires special comment; it is that, although momentum is treated as a quantity equivalent to energy, the *units* of momentum are written as MeV/c rather than MeV. This small departure from uniformity is at first somewhat confusing, but actually avoids great confusion in discussions where energy and momentum must be kept distinct. As a simple numerical illustration, an electron whose momentum is 0.511 MeV/c has a total energy of 0.723 MeV, a kinetic energy of 0.212 MeV, a β of 0.707 and a γ of 1.41.

In the following exercises, use energy units.

EXERCISES

1-78. A particle of rest mass M, moving with momentum P, collides with an electron at rest. Find the maximum momentum that the electron can have after the impact.

$$Ans.: \qquad P_e = \frac{2P[m_e + (P^2 + M^2)^{1/2}]m_e}{M^2 + m_e^2 + 2m_e(P^2 + M^2)^{1/2}}$$

1-79. Show that the relation between P in MeV/c, B in gauss, R in centimeters, and charge Z in electrons for a particle moving at right angles through a magnetic field is

$$P = 300 \times 10^{-6} ZBR \qquad (9)$$

1-80. (*a*) What is the kinetic energy of a proton whose momentum is 800 MeV/c? (*b*) If this proton collides with a proton at rest, what is the speed of the CM system? *Ans.:* (*a*) 295 MeV. (*b*) $\beta_{CM} = 0.368$.

1-81. The maximum kinetic energy that has been measured (indirectly) for a cosmic-ray primary particle is about 10^{13} MeV (i.e., about one joule!). Assume that a proton of about this energy traverses the galaxy (diameter about 10^5 light years). How long does it take, in the proton's reference frame, to pass through the galaxy? *Ans.:* About 5 min.

1-82. An electron is accelerated from rest through a potential difference of 1.02×10^6 V, and it then enters a uniform magnetic field of induction 480 gauss. If the electron moves at right angles to this field, what will be the radius of curvature of its path? *Ans.:* 10 cm.

1-83. A π°-meson decays into two γ-rays through the reaction $\pi^\circ \rightarrow \gamma + \gamma$. If a π°-meson which is moving with speed βc decays in flight into two γ-rays which travel along and opposite the path of the π°, show that the geometric mean of the energies of the two γ-rays in the laboratory system is 67.55 MeV, the energy of each γ-ray in the CM system.

1-10. The Classical Electron

For many years after the discovery of the electron, great efforts were directed toward describing its structure on the basis of classical relativistic electrodynamics. Much of this work is associated with the name of H. A. Lorentz. Some of the earliest work, which seemed partially successful at the time, attempted to identify some or all of the rest mass of the electron with the *electrostatic energy* due to its charge, through Einstein's equation $E = mc^2$.

It was believed that most, and perhaps all, of the mass of an electron is electromagnetic in nature. On the assumption that all of the mass is electromagnetic, various values were calculated for the size of the electron, using various models for the charge distribution within it. All of the values obtained for the size of the electron contained the quantity

$$r_0 = \frac{e^2}{4\pi\varepsilon_0 mc^2} \tag{1}$$

which came to be called the "classical radius" of the electron. Its numerical value is

$$r_0 = 2.81784 \times 10^{-15} \text{ m}$$

or $$r_0 = \frac{e^2}{mc^2} = 2.81784 \times 10^{-13} \text{ cm} \qquad \text{in cgs esu} \tag{2}$$

EXERCISE

1-84. Take as a model of the electron a uniformly charged spherical shell of radius R, and evaluate R as described above.   *Ans.:* $R = \frac{1}{2}r_0$.

Attempts were also made to relate the mechanical and electrical properties of the electron by treating the momentum of the electron as due, at least in part, to the *momentum in its electromagnetic field*. Since a moving electron is surrounded by an electric and a magnetic field, there must be momentum in this field. The net momentum has the same direction as the velocity of the charge and is proportional to the velocity; an equivalent *mass* can then be defined by dividing the electromagnetic momentum by the velocity. The measured mass of the electron can thus be considered to be the sum of this "electromagnetic mass" and its "intrinsic mass" (if any). The radius of the electron could, as above, be found if one knew the distribution of charge throughout it and if one knew what fraction of the measured mass was electromagnetic.

EXERCISE

1-85. Use the expressions derived in Exercise 1-69 for the electric and magnetic fields surrounding a point charge to evaluate the electromagnetic momentum of a slowly moving electron ($\beta \ll 1$). Take as a model of the electron a uniformly charged spherical shell of radius R and evaluate R as suggested above. (Recall that the momentum density carried by an electromagnetic field is $\mu\varepsilon S$, where $S = E \times H$ is Poynting's vector.)   *Ans.:* $R = \frac{2}{3}r_0$.

The ideas outlined above met with serious difficulty from several quarters. One of the most fundamental problems had to do with the failure of any simple electromagnetic model of the electron to transform properly under the Lorentz transformation. For example, the electromagnetic energy of a moving "spherical-shell" electron transforms according to

$$W = W_0\gamma(1 + \tfrac{1}{3}\beta^2) \tag{3}$$

rather than $W = \gamma W_0$, as would be required if all of the electronic mass were really electromagnetic, while the *momentum* of such a body transforms in the proper way:

$$P = \gamma\beta M_0 c \tag{4}$$

All such electromagnetic models not only disagreed on the value obtained for the mass but also gave incorrect transformation properties for the energy, the momentum, or both. This was taken to mean that not all of an electron's mass is electromagnetic, but that a certain part must be of some other type. And indeed, purely electromagnetic models of the electron cannot be valid on another ground: such a structure would be unstable under its own field and would expand without limit unless held together by some other agency. The nature and transformation properties of the "glue" that holds the electron together have never been satisfactorily formulated, and no accepted theory of the detailed structure of the electron is yet available. So far, the most satisfactory theory appears to be that due to Dirac (Chap. 20), but in this the electron is regarded as a point singularity which possesses given mass and charge and is then deduced to have also a certain spin and magnetic moment. We will find in our later work that in certain circumstances an electron seems to have a finite size and in others it appears to be a point-singularity.

EXERCISE

1-86. Carry out an exact evaluation of the field energy and momentum for an electron moving at relativistic speeds, assuming a thin spherical shell of charge for its structure. It is simplest to transform the static electric field of the shell to the moving system by Eq. 1-8(23), but to carry out the volume integration by using the coordinates of the electron's frame. (The factor γ will be needed to relate volume elements in the stationary and the moving frame.)

Ans.: $$W = \frac{e^2\gamma(1 + \beta^2/3)}{8\pi\varepsilon_0 R} \qquad P = \frac{e^2\beta\gamma}{6\pi\varepsilon_0 Rc}$$

1-11. The General Theory of Relativity

The foregoing description of special relativity has illustrated some of the profound effects relativity has had upon our theories of physical phenomena and has shown how intimately connected are many concepts which were previously thought to be quite separate. Special relativity has proved spectacularly successful, not only in resolving the conflicts that led to its formation, but also in producing *new physical concepts*, which have been beautifully verified experimentally. However, the special theory cannot be expected to contain *all* of the results that the relativity principle might yield. The investigations that have been pressed in an attempt to produce a theory of relativity whose validity is not restricted to any special coordinates or to any special state of relative motion of physical observers has led to the so-called general theory of relativity. The general theory has been in practice a *theory of gravitation;* attempts to include electromagnetism in the general theory to produce what is called a *unified field theory* have so far been unsuccessful.

It is not possible here to discuss the general theory in detail, both because of limitations of space and the mathematical preparation needed. We shall, however, state the postulatory basis for the general theory and describe briefly some of its chief results. The postulates of general relativity are as follows:

1. The principle of relativity is extended to apply literally to observers in *any* state of motion who are using *any* system of space-time coordinates. This principle is called, in general relativity, the *principle of covariance*. According to this principle, one expects to be able to write the laws of physics in a completely covariant form, much as the Lagrangian equations of ordinary classical mechanics retain the *same appearance* for any system of mechanical coordinates.

2. The second postulate of general relativity allows for the application of the principle of covariance to two observers, one of them located in a gravitational field in which free particles are all accelerated "downward" and the other situated in an "elevator," free of gravitational field but accelerated by some other means; the postulate asserts the indistinguishability of such experimental conditions. The postulate is called the *principle of equivalence*, and it can be regarded as asserting the equality of gravitational mass and inertial mass, or the possibility of removing, locally, the effects of a gravitational field by transforming to a freely falling (i.e., accelerated) coordinate system. Thus the postulate contains a specific reference to gravitational effects and states essentially that experiments performed in a gravitational field will yield the same results as will similar experiments performed in an accelerated laboratory, away from gravitational fields.

In the special theory of relativity it was found that the geometry of space time is Euclidean; or, to extend a concept familiar in two-dimensional geometry, space time is "flat." The trajectory of a free particle in such a "flat" space is a straight line with uniform speed, i.e., a *geodesic curve* in this space. It is also a familiar result of ordinary mechanics that a particle, constrained to move on a frictionless surface but otherwise free, moves in a geodesic curve on the surface. These ideas lead naturally to the supposition that, in the general theory, where all restrictions on the permissible types of relative motion are removed, the character of space time may be such as to yield a geodesic curve for the trajectory of a free particle in all cases. This would then require that space time no longer be "flat," but that it be "curved" for observers who are in accelerated laboratories or who are located in a gravitational field. This in turn leads to the notion that matter might exert its influence upon neighboring matter by producing a local "warp" or "curvature" of space time. This is the essential idea of Einstein's theory of gravitation.

3. The third postulate of Einstein's general theory of relativity puts in quantitative form the general ideas described above. It is at this point that the subject becomes too complex mathematically to pursue here; however, it can be stated that there exists from differential geometry a measure of "curvature" for any space. This measure takes the form of a fourth-rank tensor $R^{\sigma}_{\mu\nu\rho}$, called the Riemann-Christoffel tensor. This tensor forms a second-rank tensor $R_{\mu\nu}$ when it is contracted with respect to σ and ρ and a scalar R when further contracted with respect to μ and ν. Einstein's equation for gravitation is written in terms of these quantities and the metric tensor $g_{\mu\nu}$ that characterizes the line element of space time as follows:

$$R_{\mu\nu} - \tfrac{1}{2} R \, g_{\mu\nu} = -K T_{\mu\nu} \tag{1}$$

$T_{\mu\nu}$ is the energy-momentum tensor which describes the distribution of the matter and energy that produces the curvature of space, and K is the gravitational constant expressed in appropriate units. Its numerical value is

$$K = 8\pi G/c^4 = 2.073 \times 10^{-43} \text{ kg}^{-1} \text{ m}^{-1} \text{ s}^2$$

where G is the ordinary constant of gravitation.

The above formulation of general relativity has proved capable of describing all ordinary gravitational phenomena with an accuracy equal to that of the Newtonian theory, and in addition has successfully accounted for some small but definite effects that could not be explained in terms of the Newtonian theory. In particular there have been three so-called "crucial tests" of the general theory of relativity: all of them have been successfully met. They are the following:

1. The bending of light rays by gravitating bodies. A light ray follows a geodesic curve in space time corresponding to $d\tau = 0$. The bending of a light ray from a distant star in passing close to the surface of the sun affords a measure of the curvature of space in the vicinity of a massive body. The results of such tests are in good agreement with the general theory which predicts an outward displacement of $1''.75$ for star images near the limb of the sun.

2. The advance of the perihelion of the planet Mercury. It has long been known that the perihelion point of the orbit of Mercury undergoes a precession along the orbit, mostly caused by the perturbing effects of Venus and the other planets. Out of a total advance of almost $6000''$ (of arc) per century, about $40''$ per century remains unaccounted for by planetary perturbations. Einstein's theory predicts an advance of $43''$ per century, which agrees quite well with the experimental value.

3. The shift of wavelength of light emitted by an atom in a gravitational field. One of the simplest applications of the principle of equivalence is the deduction that the rate of a clock located in a region of high gravitational potential V will be decreased by an amount

$$\Delta T = \frac{T \, \Delta V}{c^2} \tag{2}$$

with respect to that of a clock situated in a region of lower gravitational potential $V - \Delta V$. This shift has been measured for "atomic clocks" located at the surface of extremely massive, small stars. For the sun, the wavelength shift of the light from such an atom amounts to $\Delta\lambda/\lambda = 2.12 \times 10^{-6}$, and for a very dense white dwarf, the companion to Sirius, it is about 30 times as great. In both cases the shift has been found to be in agreement with the theoretical value. It has also been suggested that this relation might be tested by measuring the rate of a clock carried by an artificial satellite.

It is interesting to note that only the second of the above tests was known when Einstein developed his theory. The remaining two were first predicted theoretically and then sought and found experimentally. This makes all the more impressive the excellent agreement between the general theory and the observational results. Seldom does the work of a single man meet with such spectacular success, not only in removing the inconsistencies that prompted it, but also in introducing new ideas and predicting new results which are later found experimentally. Thus, even though Einstein's form of the general theory of relativity may ultimately be replaced by a more complete and satisfactory theory, it will nevertheless always stand as a significant monument to the intellect of mankind.

REFERENCES

Bergmann, P. G.: "Introduction to the Theory of Relativity," Prentice-Hall, Inc., Englewood Cliffs, N.J., 1942.

Einstein, A.: "The Meaning of Relativity," Princeton University Press, Princeton, N.J., 1946.

Frank, P.: "Einstein: His Life and Times," Alfred A. Knopf, Inc., New York, 1947.

Gamow, G.: "Mr. Tompkins in Wonderland," The Macmillan Company, New York, 1949.

Houston, W. V.: "Principles of Mathematical Physics," 2d ed., chap. XVI, International Series in Pure and Applied Physics, McGraw-Hill Book Company, Inc., New York, 1948.

Lass, H.: "Vector and Tensor Analysis," International Series in Pure and Applied Mathematics, McGraw-Hill Book Company, Inc., New York, 1950.

Lorentz, H. A., A. Einstein, H. Minkowski, and H. Weyl: "The Principle of Relativity" (trans.), Dover Publications, New York.

Möller, C.: "The Theory of Relativity," Oxford University Press, New York, 1952.

Tolman, R. C.: "Relativity Thermodynamics and Cosmology," Oxford University Press, New York, 1934.

2

Quantum Mechanics

The nineteenth century may be regarded as marking the culmination of our understanding of the large-scale behavior of matter under the conditions of ordinary experience. That century saw not only the completion of the science of dynamics founded two centuries earlier by Galileo and Newton but also the brilliant synthesis by Maxwell of the laws of electricity, magnetism, and electromagnetic induction into a complete dynamical theory of electromagnetism. Thermodynamics became an exact science with the enunciation of the law of conservation of energy, and the foundations of the kinetic theory of matter were laid with the development of classical statistical mechanics by Maxwell, Boltzmann, Gibbs, and others. Indeed, our understanding, or potential understanding, of the laws of nature appeared so satisfactory at the close of the century that many physicists believed they foresaw an imminent decline in what is today termed "basic research."

Yet the very exactness and apparent completeness of the classical physical laws contained the seeds of their downfall when they came to be applied outside the range of direct experience. We have seen in the first chapter how Maxwell's equations led to conclusions that were contradicted by experiment, and how the resolution of these contradictions at the hands of Einstein led to entirely new and far-reaching concepts of space, time, and matter that were previously unsuspected. In like manner the laws of classical physics were unable to account for many simple, clear experimental facts concerning the behavior of atomic systems, and they have been supplanted in this domain by a new theory, called the *quantum theory*, which has been as revolutionary as was relativity in its impact on our concepts of nature.

The study of the microscopic structure and behavior of matter, which really began only with the present century, has absorbed a greater and greater proportion of the efforts of physicists and has also spread spectacularly into the fields of chemistry, engineering, and even biology. The physicist has successfully surmounted the challenge presented by the problems of the electronic structure of atoms and has partially solved some of those relating to nuclear structure. Most of the remainder of this book will be devoted to the study of the fundamental results of this half century of progress in physics.

We begin this study with an examination of some of the fundamental laws which seem to govern the behavior of the submicroscopic particles that compose matter. These quantum-mechanical laws may be regarded as being analogous to Newton's laws of motion in ordinary large-scale particle dynamics. In this chapter we shall also describe some of the more important analytical procedures by which physical phenomena are treated, using quantum mechanics. In order better to grasp the meaning of the laws of quantum mechanics, it is well to begin with a brief outline of some of the experimental situations and theoretical predicaments, failures, and inspired guesses that finally led to the quantum theory as we now know it.

2-1. Introductory Background

A. The Zeeman Effect. The discovery of the electron by J. J. Thomson in 1897 may properly be said to have provided the indispensable key to a quantitative study of the ultimate structure of matter, for almost immediately thereafter it was proved by Zeeman and Lorentz, using the *Zeeman effect*, that the electron is a direct participant in the emission of spectral radiation. Although spectroscopy was at that time a well-developed observational science, and great amounts of quite precise spectroscopic data had been amassed, the close connection between radiation and the electrical constitution of matter was only vaguely understood. Zeeman and Lorentz identified the spectral radiation emitted by an excited gas with the *electric-dipole radiation* from oscillating charged particles within individual atoms of the gas, and deduced that such radiation should be modified if the emitting atoms are situated in a magnetic field. Each spectral frequency should be "split" into three separate frequencies of magnitudes

$$\omega_1 = \omega_0 + \frac{eB}{2m} \qquad \omega_2 = \omega_0 \qquad \omega_3 = \omega_0 - \frac{eB}{2m} \qquad (1)$$

where ω_0 is the angular frequency for the undisturbed atom, B is the magnetic induction, and e/m is the ratio of charge to mass for the electrical particles whose oscillations supposedly cause the radiation. This is qualitatively just what had been seen by Zeeman in his discovery of the Zeeman effect on the D-lines of sodium in 1896. Zeeman and Lorentz also deduced that the three components of radiation should show different behavior with respect to the *angular distribution* of their intensity and the character of their *polarization*. For instance, if the radiation were viewed from a direction at right angles to that of the magnetic field, all three components should be visible, the undeviated component ω_2 being plane-polarized with the electric vector parallel to the magnetic field and the two deviated components ω_1 and ω_3 polarized at right angles to this. The intensities of ω_1 and ω_3 should furthermore each be half as great as that of ω_2. On the other hand, if the radiation were viewed in a direction *along* the magnetic field, only the shifted components ω_1 and ω_3 should be visible, and these should be circularly polarized in opposite senses. It is this latter situation which permits the *sign* of e/m to be determined, since the correlation between the sense of the circular polarization, the sign of the frequency shift, and the magnetic-field direction depends upon the sign of charge.

Zeeman analyzed the splitting of the sodium D-lines from the above point of view and found the particles responsible for the radiation to be negative and to possess a ratio of e to m equal to that of Thomson's electron, within the precision of his measurements. While this result established quite firmly that the electron plays a dominant role in the formation of line spectra, it was soon found that the Zeeman splitting of most spectral lines cannot be described, even qualitatively, by the above analysis. Some are not split at all, but most are split into more than the three components required by the classical theory. The line patterns observed in the Zeeman effect are called *normal* if there are three components whose shifts are in agreement with the classical Zeeman-Lorentz analysis, and are otherwise called *anomalous*. We shall see in Chap. 8 that the quantitative explanation of the puzzling anomalous Zeeman patterns constitutes one of the most spectacular successes of the modern quantum theory.

EXERCISES

2-1. Consider a particle of charge e and mass m which is attracted toward a fixed point with a force $\boldsymbol{F} = -m\omega_0{}^2\boldsymbol{r}$ and is also acted upon by a magnetic field $\boldsymbol{B}$ in the z-direction. Write the differential equations of motion of the particle in rectangular coordinates.

Ans.: $m\ddot{x} + m\omega_0{}^2x - eB\dot{y} = 0$ $m\ddot{y} + m\omega_0{}^2y + eB\dot{x} = 0$

$$m\ddot{z} + m\omega_0{}^2z = 0$$

2-2. Solve the differential equations of Exercise 2-1 and show that the three frequencies given by Eq. (1) are nearly correct if $eB/m \ll \omega_0$. *Hint:* The substitutions $u = x + iy$ and $w = x - iy$ will be found to simplify the solution.

2-3. Show that the three frequencies found in Exercise 2-2 may be identified with a linear oscillation parallel to the field and two circular motions in a plane perpendicular to the field. In one of these circular motions the magnetic force is directed inward; and in the other, outward. Show that the variation of intensity and polarization with the direction of emission described in the text can be accounted for in terms of these three independent motions, and find how the correlation of the polarization and the frequency shift depends upon the sign of e/m.

B. Black-body Radiation, Planck's Quantum Hypothesis. One of the first really disturbing failures of classical physics, and the first appearance of the quantum theory, resulted from attempts to describe the *thermal radiation of hot bodies* in terms of classical statistical mechanics. It is well known that at a given temperature objects made of different materials, but of similar shape and size, may radiate energy at quite different rates and that the spectral distribution of the radiated energy may be quite different for the different materials. But in spite of these differences, it is still possible to treat thermal radiation quite generally, because of a very remarkable fact: If the radiation *inside* a hollow body is viewed through a small hole in its wall, it is found that the intensity and spectral distribution of the radiation at a given temperature are quite accurately *the same for all materials* and are independent also of the size and shape of the enclosure. This radiation is called *isothermal cavity radiation*, and since it represents the *external* radiation of an ideal black body, it is also called *black-body radiation*.

The total amount of radiation and its spectral distribution were measured with sufficient accuracy by Lummer and Pringsheim (1900) to provide a critical test of several theoretical results that had been obtained at that time. Many of these results were confirmed experimentally—principally those that were derived on the basis of the laws of thermodynamics—but no classical theory was able to account for all of the experimental observations.

The general features of the spectral-energy distribution of the cavity radiation are shown in Fig. 2-1, in which the emitted energy per unit wavelength interval per unit area per unit time is plotted against wavelength for various temperatures. It is seen from this figure that the total amount of radiation (given by the area under the distribution curve) increases rapidly as the temperature increases, and that the wavelength corresponding to the maximum point of the distribution curve shifts progressively toward shorter wavelengths. The problem that faces any theory of thermal radiation is of course to describe these features quantitatively.

The most successful advances toward a quantitative description of black-body radiation, prior to 1900, were made through the application

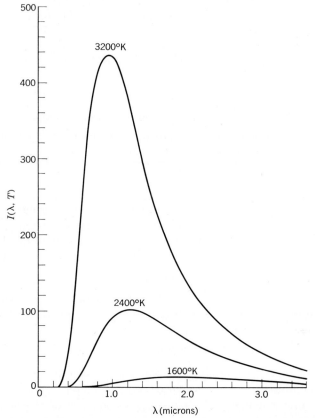

FIG. 2-1. Spectral distribution of black-body radiation at various temperatures. $I(\lambda,T)$ is the total emissive power in watts per square centimeter per micron.

of the laws of thermodynamics. The radiation inside an isothermal enclosure was treated as a thermodynamical fluid which contains energy, exerts pressure, and occupies volume, just as does a material fluid. The properties of this fluid were determined from the laws of electromagnetism. In this way several important properties of cavity radiation were established:[1]

[1] For a detailed discussion of these classical results, see F. K. Richtmyer, E. H. Kennard, and T. Lauritsen, "Introduction to Modern Physics," 5th ed., chap. 4, McGraw-Hill Book Company, Inc., New York, 1955.

1. *Kirchhoff's law* states that, for a nontransparent body of arbitrary composition exposed to isotropic radiation of wavelength λ, the fraction of the incident radiation that is absorbed, called the *monochromatic absorptivity* $A(\lambda,T)$, is equal to the *monochromatic emissivity* $E(\lambda,T)$ measured relative to that of an ideal black body. This law follows directly from the necessary equality of the absorption and emission of radiation by each surface element of the wall of an isothermal enclosure.

2. The classical expressions for the energy and momentum of plane waves provide a relation between the pressure P and energy density U for isotropic radiation:

$$P = \tfrac{1}{3}U \tag{2}$$

3. It also follows directly from classical electromagnetism that the total power e_{B} emitted by each unit area of a black body is simply related to the energy density U of cavity radiation:

$$e_{\text{B}}(T) = \tfrac{1}{4}cU(T) \tag{3}$$

where c is the speed of light.

4. The *Stefan-Boltzmann law* is established by considering a reversible Carnot engine whose working fluid is cavity radiation. This law states that the total power emitted by each unit area of black body is proportional to the fourth power of the absolute temperature:

$$e_{\text{B}}(T) = \sigma T^4 \tag{4}$$

σ is called Stefan's constant. Its value could not be deduced from classical physics, but it was known approximately from experimental measurements. It is equal to

$$\sigma = 0.56686 \times 10^{-7} \text{ W m}^{-2}\,{}^\circ\text{K}^{-4} \tag{5}$$

5. During a reversible adiabatic expansion of black-body radiation, the radiation must remain of black-body character, since otherwise a heat engine which would violate the second law of thermodynamics could be devised. By considering the Doppler effect upon radiation which strikes a perfectly reflecting moving piston, *Wien's displacement law* is then established: If cavity radiation is slowly expanded or compressed to a new volume and temperature, the radiation that was originally present at wavelength λ will, by the Doppler effect, be transformed to wavelength λ' where

$$\lambda T = \lambda'T' \tag{6}$$

and the energy density $dU = (4/c)I(\lambda,T)\,d\lambda$ will in the process be changed to a value $dU' = (4/c)I(\lambda',T')\,d\lambda'$ given by

$$\frac{dU}{dU'} = \left(\frac{T}{T'}\right)^4 \tag{7}$$

$[I(\lambda,T)$ is the monochromatic emissive power of a black body at temperature T.] Thus the problem of describing black-body radiation is reduced to finding a single function of λ and T by writing

$$I(\lambda,T) = \left(\frac{T}{T'}\right)^4 I(\lambda',T') \frac{d\lambda'}{d\lambda}$$

$$= \left(\frac{T}{T'}\right)^5 I \left(\frac{\lambda T}{T'}, T'\right)$$

$$= T^5 f(\lambda T) \tag{8}$$

The agreement of this form of the Wien displacement law with experiment is shown in Fig. 2-2, which shows $I(\lambda,T)/T^5$ plotted vs. λT for several temperatures.

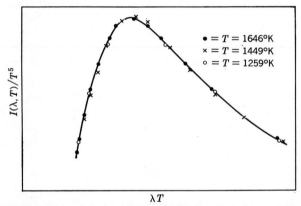

FIG. 2-2. Experimental verification of Wien's displacement law. A more extensive plot of this function can be found in Appendix B. (By permission from F. K. Richtmyer, E. H. Kennard, and T. Lauritsen, "Introduction to Modern Physics," 5th ed., McGraw-Hill Book Company, Inc., New York, 1955.)

The thermodynamical considerations described above account very satisfactorily for many of the features of black-body radiation but are unable to specify completely the spectral distribution of the radiation. The situation may be likened to that of the kinetic theory of gases, in which thermodynamical considerations provide some, but not all, of the information of interest concerning the motions of the molecules. For example, the dependence of the average energy of the molecules upon the temperature can be found by thermodynamical considerations, but the *distribution of velocities* cannot; such information can, however, be found by applying statistical mechanics to the problem. Similarly, one might expect that certain properties of the cavity radiation, such as the function $f(\lambda T)$ of Eq. (8), might be outside the reach of thermodynamics but within that of statistical mechanics. It was in the attempts to obtain a suitable function $f(\lambda T)$ from classical statistical mechanics that the inadequacy of classical mechanics for the treatment of the problem of cavity radiation became apparent.

Many of the classical statistical treatments of cavity radiation were based upon very special assumptions regarding the nature of the atomic process involved in the emission or absorption of radiation. However, one treatment—that of Rayleigh and Jeans (1900)—had a firm foundation in classical statistical mechanics. Its basic assumption was that the *equipartition law* holds for the various *electromagnetic modes of vibration* of the cavity in which the radiation is situated. The distribution of energy as a function of wavelength should thus be kT per normal mode of oscillation times the number of normal modes per unit wavelength interval.

One can see immediately that this theory *must give an impossible result;* for since the number of normal modes of vibration of a *continuous medium* is infinite (there being in this case no lower limit to the wavelength of a normal mode of vibration), both the *energy content* and the *specific heat* of any cavity should be infinite. When the theory is worked out quantitatively, this indeed turns out to be the case; but it is found that the form obtained for $I(\lambda, T)$ is asymptotically correct for *sufficiently long wavelengths*, i.e., for wavelengths much longer than $0.3/T$ cm. Numerically, the number of normal modes of vibration per unit volume and per unit wavelength interval is found to be $N(\lambda) = 8\pi/\lambda^4$, so that

$$I(\lambda, T) = \frac{c}{4} N(\lambda)(kT) = \frac{2\pi c k T}{\lambda^4} \qquad (9)$$

Planck approached the problem of cavity radiation by treating the interactions of a set of charged, one-dimensional *harmonic oscillators* with the radiation field. He reasoned that, since the composition of the wall of an isothermal enclosure does not influence the character of the cavity radiation, a simple model such as this one might as well be used. An empirical formula which accurately represents black-body-radiation measurements was known by him to be

$$I(\lambda, T) = \frac{c_1}{4} \frac{(e^{c_2/\lambda T} - 1)^{-1}}{\lambda^5} \qquad (10)$$

This empirical distribution law will be recognized as corresponding at long wavelengths to the Rayleigh-Jeans law. Planck was able to derive this law theoretically, using as a basis the following hypotheses:

1. Each oscillator *absorbs* energy from the radiation field in a continuous manner, following the laws of electrodynamics.
2. An oscillator can *radiate* energy only when its total energy is an *exact integral multiple* of a certain unit of energy for that oscillator; and when it radiates, the oscillator radiates *all* of its energy.

3. The actual radiation or nonradiation of a given oscillator as its energy passes through a critical value as described in (2) is governed by statistical chance; the ratio of the probability of nonemission to the probability of emission is proportional to the intensity of the radiation that excites the oscillator.

Planck assumed that the energies from which his oscillators could radiate were

$$E_n = nh\nu \tag{11}$$

where ν is the frequency of the oscillator, n is an integer, and h is a constant. Using this value for the energy levels and the above hypotheses, Planck determined the entropy of the oscillators by finding the *most probable* value of the energy of the oscillator in terms of the energy of the radiation. The distribution of the radiant energy was then determined by using the condition, for each wavelength, that the entropy S and the energy E are related by

$$dS = \frac{dE}{T} \quad \text{or} \quad T = \frac{dE}{dS} \tag{12}$$

where, by the second law of thermodynamics, T *must be the same for all wavelengths.* Thus he arrived at the expression

$$I(\lambda, T) = \frac{2\pi c^2 h}{\lambda^5 (e^{ch/\lambda kT} - 1)} \tag{13}$$

This equation, which contains only the single unknown constant h, represents to an extremely high degree of accuracy the actual energy distribution of cavity radiation. The constant h is called Planck's quantum of action; it has the numerical value

$$h = 6.6252 \times 10^{-34} \text{Js} \tag{14}$$

The details of Planck's arguments are presented in his book, "The Theory of Heat Radiation,"[1] and make worthwhile reading:

It is true that we shall not thereby prove that this hypothesis represents the only possible or even the most adequate expression of the elementary dynamical law of the vibration of oscillators. On the contrary I think it very probable that it may be greatly improved as regards form and contents. There is, however, no method of testing its admissibility except by the investigation of its consequences, and as long as no contradiction in itself or with experiment is discovered in it, and as long as no more adequate hypothesis can be advanced to replace it, it may justly claim a certain importance.

[1] M. Planck and M. Masius, "The Theory of Heat Radiation," p. 154, The Blakiston Division, McGraw-Hill Book Company, Inc., New York, 1914. *Quotation by permission.*

EXERCISES[1]

2-4. Show from Eq. (8) that the wavelength λ_m at which $I(\lambda,T)$ is maximum satisfies the relation

$$\lambda_m T = \text{const} \tag{15}$$

Evaluate this constant, using Eq. (13). *Ans.:* $\lambda_m T = 0.28978 \times 10^{-2}$ m °K.

2-5. A closed graphite crucible at 27°C is placed inside a furnace whose walls are maintained at a temperature of 1727°C. (*a*) Treating the crucible as a black body of surface area 40 cm², compute the initial rate at which the crucible gains heat by radiation from the furnace walls. (Assume that the crucible is everywhere convex.) (*b*) If the crucible has a mass of 100 g and a mean specific heat of 0.3 cal g⁻¹ °C⁻¹, how long will it take the crucible to reach a temperature of 100°C? Neglect the radiation of heat by the crucible over this temperature range. *Ans.:* (*a*) 3.6 kW. (*b*) 2.5 s.

2-6. A small hole of area S is made in the wall of a furnace containing cavity radiation at a temperature T. A black sphere of radius r is placed in front of the hole at a distance R from it. Neglecting the radiation from the outside walls of the furnace, show that the power absorbed by the black sphere is $P = \sigma S T^4 r^2 / R^2$.

2-7. A *gray body* is one whose absorptivity $A(\lambda,T)$ is constant (independent of λ and T). Prove that the net rate of heat transfer between two parallel gray surfaces, whose separation is small compared to their linear dimensions or their radius of curvature, is given by

$$P = \frac{\sigma(T_1{}^4 - T_2{}^4)}{(1 - 1/A_1 - 1/A_2)}$$

where A_1, T_1 and A_2, T_2 are the absorptivities and absolute temperatures of the two surfaces.

2-8. The wavelength of maximum intensity in the solar spectrum is 0.500 μ. Assuming the sun radiates as a black body, compute its surface temperature.

2-9. A gray body of absorptivity A, surface area S, and heat capacity C cools from temperature T_1 to T_2 by radiation alone in surroundings of temperature T_0, and $T_1 > T_2 > T_0$. Assuming that cooling takes place slowly so that at any time all parts of the body are at the same temperature, show that the time required for this change is

$$t = \frac{C}{2A\sigma S T_0{}^3}\left[\tan^{-1}\frac{T_1}{T_0} - \tan^{-1}\frac{T_2}{T_0} - \frac{1}{2}\ln\frac{(T_2 + T_0)(T_1 - T_0)}{(T_1 + T_0)(T_2 - T_0)}\right]$$

[1] Problems 2-5 to 2-8 are adapted by permission from N. H. Frank, "Introduction to Electricity and Optics," 2d ed., McGraw-Hill Book Company, Inc., New York, 1950.

2-10. At what temperature does the pressure of cavity radiation equal 1 atm?

2-11. Establish the relations (2) and (3), using classical electromagnetic theory.

C. The Photoelectric Effect. In his successful treatment of black-body radiation, Planck introduced the quantum hypothesis in a form such that only the oscillators, and not the radiation field, were quantized. Many scientists regarded this as merely an *ad hoc* hypothesis, and several attempts were made to arrive at Planck's radiation law without introducing the quantum hypothesis. All such attempts failed, or else required the introduction of an even less acceptable idea.

A short time after Planck's first publication of his radiation law, however, Einstein put forward the idea that the *radiation field itself might also be quantized*, and proposed his famous photoelectric equation to account for the observed facts of the photoelectric effect.[1] The quantitative verification of this equation established quite firmly the discrete nature of the radiation field and put Planck's quantum hypothesis on an even stronger foundation.

The photoelectric effect was discovered in 1887 by Hertz, in the course of his experiments in the propagation of electric waves. He observed that a spark would jump a small gap more easily if the terminals of that gap were illuminated by the light of another spark than if they were optically shielded. He established that the effect was due to ultraviolet light, that it was most pronounced if the terminals were clean and smooth, and that the negative terminal was more sensitive than the positive one.

Many experimental facts concerning the photoelectric effect were soon discovered:

1. The effect involves the emission of *negative particles* (Hallwachs, 1889).
2. The emitted particles are forcibly ejected by the light (Hallwachs, Elster and Geitel, 1889).
3. A close relationship exists between the *contact potential* of a metal and its *photosensitivity*, the more electropositive elements being sensitive to longer wavelengths (Elster and Geitel, 1889).
4. The photocurrent is *proportional to the intensity of the light* (Elster and Geitel, 1891).
5. The emitted particles are electrons (Lenard, J. J. Thomson, 1899).
6. The *kinetic energies* of the emitted electrons are *independent* of the intensity of the light, and the *number* of electrons is proportional to the intensity of the light (Lenard, 1902).
7. The emitted electrons possess a maximum kinetic energy which is greater, the shorter the wavelength of the light, and no electrons

[1] Einstein, *Ann. Physik*, **17**, 132 (1905).

whatever are emitted if the wavelength of the light exceeds a certain "threshold" value (Lenard, 1902).

8. Photoelectrons are often emitted without measurable time delay after the illumination is turned on. For very weak illumination the observed delays are consistent with those to be expected for particles ejected at random at an average rate proportional to the illumination intensity.

The above experimental facts are the most important ones relating to the photoelectric effect. Many of them have been subjected to extremely rigid tests in more recent times, and many of the relationships that were established only roughly in the early work have been established with very great precision by later workers.

The experimental results (6) and (7) are qualitatively and quantitatively extremely difficult to explain with the wave theory of light. One of the most difficult things to explain is the localization, in a single electron, of the radiant energy which, on the classical theory of radiation, must have fallen on an area covering some 10^8 atoms. Another difficulty is the existence of a sharp threshold wavelength; and yet another is the fact that the time delay in the emission of photoelectrons does not exceed 3×10^{-9} s, whereas on the wave theory of light, time delays varying with the intensity from a few microseconds up to several days would be expected.

Difficulties such as these led Einstein to propose, in 1905, that the *radiation field is quantized*, so that energy can be absorbed from it only in quanta of size $h\nu$, where h is identical with Planck's quantum of action. He thus suggested that many of the observed facts of the photoelectric effect might be described by the equation,

$$h(\nu - \nu_0) = \tfrac{1}{2}mv^2 \tag{16}$$

where ν_0 is the threshold frequency and $\tfrac{1}{2}mv^2$ is the maximum kinetic energy of the ejected electrons.

There was not sufficient quantitative evidence in 1905 to either confirm or deny the truth of Einstein's photoelectric equation. Very precise measurements were subsequently made, however, with the result that this equation was definitely and completely verified, and indeed some of this work was sufficiently precise that it furnished, for several years, the best measurement of the magnitude of Planck's constant.[1]

The success of Einstein's equation demonstrated clearly the corpuscular character of radiation and thereby raised new questions concerning the true nature of light; for it was by this time well established that radiation

---

[1] Millikan, *Phys. Rev.*, **7**, 355 (1916); "Electrons (+ and −), Protons, Photons, Neutrons, Mesotrons, and Cosmic Rays," University of Chicago Press, Chicago (revised), 1947.

is a *wave phenomenon*.   Thus there arose the puzzle of the wave-particle duality of light—a puzzle that was to become even more puzzling, with the appearance of the reverse problem of the particle-wave duality of electrons, before it became clearer.   Actually, the theoretical treatment of radiation is now considered to be in a rather satisfactory state of perfection, and we shall later examine some of the simpler features of the theory by which both the wavelike and the particle-like properties of radiation are accounted for.   This theory of radiation is called *quantum electrodynamics*.

It must be emphasized that the fact that radiation is both wavelike and particle-like does not mean that both of these aspects must be explicitly dealt with in every practical problem.   That is to say, Maxwell's equations, although not strictly valid in the most general theoretical sense, are an *exceedingly good approximation* in cases where *huge numbers of photons* are involved, as, for example, in any problem involving ordinary electric circuits.   On the other hand, Maxwell's equations are useless in cases where *individual* photons must be considered, as, for example, in the photoelectric effect.

D. *Atomic Structure, the Nuclear Atom, the Bohr Theory.*   During the latter half of the nineteenth century many lines of investigation, taken together, provided very strong experimental support for the age-old, but much disputed, atomic theory of matter.   Much of the evidence was indirect; for example, the kinetic theory of gases seemed successfully to bridge the gap between the laws of mechanics and the laws of thermodynamics, and the investigations of the chemical combining weights and of the electrochemical equivalent weights of the elements indicated a close relationship between the electrical and chemical properties of matter.   These and many other results seemed to point toward an atomic theory, but many celebrated and competent scientists remained unconvinced, even as late as 1900, of the necessity, or even the desirability, of an atomic hypothesis.   However, the discovery of the electron and its intimate connection with many rather diverse phenomena provided direct and convincing proof of the atomic constitution of matter, and the acceptance of the idea was practically complete by 1905.

Classical physics had proved so spectacularly successful in its ability to account for a tremendous variety of physical phenomena ranging from optical birefringence to the theory of planetary perturbations that it was only reasonable that the question of the structure of atoms should be approached by using the classical laws.   The available evidence seemed to lead to the following conclusions:

1. Electrons are present as constituents of all atoms and are the direct source of spectral radiation.

2. Since matter is ordinarily electrically neutral, some other constituent, carrying a positive charge, must also be found in all atoms.
3. Most of the mass of an atom must be associated with its positive charge, because of the small size of m/e for the electron compared with that of even the lightest atom. It was known that the mass of the electron was only about 1/2000 that of a hydrogen atom.
4. The absence of harmonic "overtones" of a given spectral line in the spectrum of an element indicated that the mechanical motion of an electron during the emission of radiation must be exactly simple-harmonic.
5. Classical electromagnetism requires that an accelerating charge radiate energy in proportion to the square of its acceleration. This pointed to the conclusion that *an atom which is not radiating energy must be static.*

J. J. Thomson devised an atomic model which embodied many of the known properties of atoms, and this model served a quite useful purpose as a specific basis for devising further experiments to uncover additional properties of the atom. In this theory the positive, massive part of the atom was supposed to have the form of an elastic, jellylike sphere, with the positive charge distributed uniformly throughout it. On or in this sphere the negative electrons were supposed to be fixed, and were supposed to be capable of simple harmonic vibrations about their mean equilibrium positions. The various normal modes of vibration of the electrons were supposed to account for the frequencies observed in the spectral radiation of the atom. To agree with X-ray-scattering experiments, it was assumed that the number of electrons per atom was of the same order as the atomic weight and that hydrogen, presumably the simplest atom, had but a single electron.

Thomson's atomic theory met its end as a result of the experiments of E. Rutherford (1910), in which energetic α-particles from the decay of radioactive atoms were allowed to pass through thin metal foils. The scattering of the α-particles away from their original direction of motion was measured by counting the scintillations produced by their impacts upon a zinc sulfide screen. The apparatus is shown schematically in Fig. 2-3.

Thomson's theory predicted that practically all of the α-particles should be found within a very few degrees of their original direction of motion, but whereas the actual results of the scattering experiment agreed with the expected distribution for small deflections, they were in *serious disagreement* with it for large deflections, since an impossibly large number of α-particles were observed to be scattered through *very large* angles, even greater than 90°. There is no plausible mechanism by which large angles can result from encounters with a Thomson atom. The existence of large deflections in the observed results of the scattering experiment means that, either in a single encounter or in several of the encounters experienced by an α-particle in traversing the metal foil, much stronger forces must have acted upon the α-particle than would be expected on the basis of the Thomson atomic model.

One way to obtain larger individual deflections is to make the Thomson atom smaller. To obtain sufficiently large deflections in this way, however, atoms would have to be many orders of magnitude smaller than they are known to be on the basis of crystal-lattice spacings, gaseous mean free paths, and other evidence.

Another way in which one can obtain large individual deflections is to make the *positive-charge distribution smaller without reducing the size of the negative-electron distribution by a corresponding*

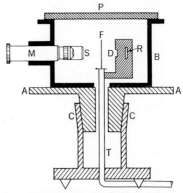

Fig. 2-3. Diagram of Rutherford's scattering apparatus. (Reproduced by permission from E. Rutherford, J. Chadwick, and C. D. Ellis, "Radiations from Radioactive Substances," Cambridge University Press, New York, 1930.)

amount.   Rutherford carried this idea to the limit in which the positive charge was reduced to a point charge at the center of a spherical shell of electrons of radius $b \approx 4 \times 10^{-10}$ m.   When an α-particle penetrates such an atom, large angular deflections will result if the α-particle passes close to the positive *nucleus* inside, and most of the deflection will take place very near this nucleus; for this reason the shielding effect of the electrons can be neglected as a first approximation.   The problem is then that of the motion of an α-particle in the inverse-square electric field of the point nucleus. This problem will be treated in some detail in connection with particle scattering (Chap. 14), but it suffices for our present purposes simply to quote the result

$$\frac{dN}{d\Omega} = \frac{N_0 n t Z_1^2 Z_2^2 e^4}{64\pi^2\varepsilon_0^2 m_1^2 v_0^4 \sin^4 \frac{1}{2}\theta} \tag{17}$$

where $dN/d\Omega$ = number of α-particles scattered into a unit solid angle at an angle θ with respect to the incident direction

N_0 = number of incident particles

n = number of scattering centers per unit volume of foil

t = thickness of foil

$Z_1 = 2$ = atomic number of incident particle

Z_2 = atomic number of foil material

e = electronic charge
ε_0 = permittivity of vacuum
m_1 = mass of incident particle
v_0 = speed of incident particle
θ = angle of deflection

The above law is called the *Rutherford law of single scattering*.[1] Various of its features were accessible to direct experimental test: The number of scattered α-particles per steradian should be:

1. Directly proportional to the *thickness t* of the foil
2. Directly proportional to the *square of the atomic number* Z_2 of the scattering nuclei in the foil
3. Inversely proportional to the *square of the kinetic energy* $\frac{1}{2}m_1v_0^2$ of the α-particles
4. Inversely proportional to the *fourth power of the sine of half the scattering angle*

The above points were very carefully tested by Geiger and Marsden (1913) by varying the thickness of the foil, its composition, the energy of the incident α-particles, and the angle of the zinc sulfide screen with respect to the incident beam. The results of this experiment were in *excellent agreement* with the theoretical predictions, and the idea of the nuclear atom was thereby firmly established.

Rutherford's scattering experiments provided very strong evidence for the validity of the nuclear model of the atom. However, this model was most strongly in conflict with the classical theory of electromagnetism. The basis of this conflict was the question of the *stability* of a nuclear atom: On the one hand, a nuclear atom could not be in *static* equilibrium under the influence of the known electrical forces between point charges, and on the other hand, *a dynamical atom would necessarily radiate energy and eventually collapse*. This conflict was resolved by Niels Bohr,[2] who in 1913 presented his successful theory of atomic hydrogen. Bohr's theory, in its final form, embodied the following ideas:

1. A hydrogen atom consists of a positive nucleus (proton) and a single electron in a state of relative circular motion under the action of their mutual electrical attraction.
2. The atom may remain for extended periods of time in a given state *without radiating electromagnetic waves, provided this state is one for which the angular momentum of the atom is an integral multiple of $\hbar$.*[‡]

---

[1] See Rutherford, *Phil. Mag.*, **21**, 669 (1911). This and other famous papers are reprinted in R. T. Beyer, "Foundations of Nuclear Physics," Dover Publications, New York, 1949.

[2] *Phil. Mag.*, **26**, 1 (1913).

[‡] Because of its regular occurrence in quantum-mechanical equations, the quantity $h/2\pi$ is abbreviated $\hbar$, called h-bar.

3. Radiation is emitted whenever the atom "jumps" from one of the "allowed" states of energy E_1 to another of lower energy E_2.

4. When radiation is emitted, its frequency is determined by the *Einstein frequency condition*

$$h\nu = E_1 - E_2 \tag{18}$$

The postulates thus recognize the results of the Rutherford scattering experiments and at the same time assert the inapplicability of classical electromagnetic theory to atomic processes. The classical theory is replaced by rules that introduce Planck's quantum h into the problem in *two separate ways:* First, it is asserted that the angular momenta, and thus the energies, of the allowed circular orbits are determined by equating the angular momentum to an integral multiple of $\hbar$. It is then further assumed that Planck's constant also determines the *frequency of the emitted radiation* through Einstein's frequency condition.

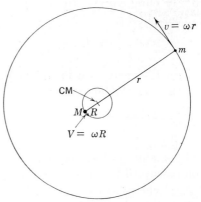

Fig. 2-4. Illustrating the relative motion of the electron and proton of a hydrogen atom about their common center of mass. Note that $mr = MR$

The determination of the energy states of the hydrogen atom proceeds as follows: The proton of mass M and the electron of mass m revolve in circular paths of radii R and r about their common center of mass with an angular velocity ω (Fig. 2-4).

1. The angular momentum is

$$(mr^2 + MR^2)\omega = mr^2\omega\left(1 + \frac{m}{M}\right) = n\hbar \qquad n = 1, 2, 3, \ldots \tag{19}$$

2. The centripetal acceleration of the electron $\omega^2 r$ is produced by the electrical attractive force

$$m\omega^2 r = \frac{e^2}{4\pi\varepsilon_0(r + R)^2} = \frac{e^2}{4\pi\varepsilon_0 r^2(1 + m/M)^2} \tag{20}$$

3. The total energy $T + V$ of the atom is

$$\begin{aligned} E_n &= \tfrac{1}{2}mr^2\omega^2 + \tfrac{1}{2}MR^2\omega^2 - \frac{e^2}{4\pi\varepsilon_0(r + R)} \\ &= \tfrac{1}{2}mr^2\omega^2\left(1 + \frac{m}{M}\right) - \frac{e^2}{4\pi\varepsilon_0 r(1 + m/M)} \\ &= -\tfrac{1}{2}mr^2\omega^2\left(1 + \frac{m}{M}\right) \qquad \text{from Eq. (20)} \end{aligned} \tag{21}$$

4. Each of the above equations may be solved for $\omega^2 r^4$:

$$
\begin{aligned}
\omega^2 r^4 &= \frac{n^2\hbar^2}{m^2(1 + m/M)^2} = A \\
&= \frac{e^2 r}{4\pi\varepsilon_0 m(1 + m/M)^2} = rB \\
&= -\frac{2E_n r^2}{m(1 + m/M)} = r^2 E_n C
\end{aligned}
\tag{22}
$$

5. By eliminating r from the right-hand sides of the above equalities, one obtains

$$
E_n = \frac{B^2}{AC} = -\frac{me^4}{32\pi^2\varepsilon_0^2 n^2\hbar^2(1 + m/M)}
\tag{23}
$$

for the energy (in joules) of the allowed orbit of angular momentum $n\hbar$. The radius r of the orbit is[1]

$$
r_n = \frac{A}{B} = \frac{4\pi\varepsilon_0 n^2\hbar^2}{me^2}
\tag{24}
$$

The smallest of these orbits, corresponding to $n = 1$, is by hypothesis the one occupied by the electron when the atom is unexcited. This state of lowest energy is called the *ground state* of the atom. For the hydrogen atom the radius of the ground-state orbit, or *first Bohr orbit*, is

$$
a_0 = \frac{4\pi\varepsilon_0\hbar^2}{me^2} = 5.29172 \times 10^{-11} \text{ m}
\tag{25}
$$

The frequency of the radiation corresponding to the transition from orbit n to n' is

$$
\nu = \frac{E_n - E_{n'}}{h} = \frac{me^4}{8\varepsilon_0^2 h^3(1 + m/M)}\left(\frac{1}{n'^2} - \frac{1}{n^2}\right)
\tag{26}
$$

and the *wave number* (number of waves per meter) is

$$
\begin{aligned}
\tilde{\nu} = \frac{\nu}{c} &= \frac{me^4}{8\varepsilon_0^2 ch^3(1 + m/M)}\left(\frac{1}{n'^2} - \frac{1}{n^2}\right) \\
&= R_{\text{H}}\left(\frac{1}{n'^2} - \frac{1}{n^2}\right)
\end{aligned}
\tag{27}
$$

The above equation is the principal measure of the success of the Bohr theory; it has exactly the form of an empirical equation deduced by Balmer (1885) from spectroscopic wavelength measurements. Balmer

---

[1] If cgs esu are used in the preceding analysis, the quantized energy values and orbit radii are given by

$$
E_n = \frac{me^4}{2n^2\hbar^2(1 + m/M)} \quad \text{and} \quad r_n = \frac{n^2\hbar^2}{me^2}
$$

found that the series of spectral lines of hydrogen that fall in the visible part of the spectrum are accurately represented by

$$\lambda = \frac{1}{\tilde{\nu}} = \frac{bn^2}{n^2 - 4} \qquad n = 3, 4, 5, \ldots \ . \qquad (28)$$

Other series of lines in the spectrum of hydrogen corresponding to other values of n' were later found. The Bohr theory provided, in the formula (27), not only an expression having the proper *form* for the various series but one that gave a *quantitatively accurate value for the coefficient* R_H, which was previously merely an empirical constant. The coefficient R is called the *Rydberg*, in recognition of Rydberg's work in sorting out many series in the measured spectra of various elements. For hydrogen, it has the value

$$R_H = \frac{me^4}{8\varepsilon_0^2 ch^3 (1 + m/M_p)} = 10{,}967{,}757.6 \ m^{-1} = 10.9677576 \ \mu^{-1} \quad (29)$$

It should also be remarked that the formulas developed for hydrogen were recognized as applying equally well, with only minor modifications, to *any* atom having but one electron circling the nucleus. Thus, if one of the two electrons that are normally associated with an atom of helium were removed, leaving the helium *singly ionized*, a similar analysis could be carried out, with the clear result

$$\begin{aligned}
\tilde{\nu} &= \frac{mZ^2 e^4}{8\varepsilon_0^2 ch^3 (1 + m/M_{He})} \left(\frac{1}{n'^2} - \frac{1}{n^2} \right) \\
&= 4R_{He} \left(\frac{1}{n'^2} - \frac{1}{n^2} \right)
\end{aligned} \qquad (30)$$

where $Z = 2$ is the atomic number of helium (that is, Ze is the nuclear charge of the atom) and M_{He} is the mass of an α-particle.

One thus concludes that the spectrum of ionized helium should be similar to that of neutral hydrogen, except that all lines are reduced in wavelength by almost exactly a factor of four. This is observed to be the case. A slight difference is of course introduced by the fact that R_{He} is not quite equal to R_H. Numerically,

$$R_{He} = 10.9722267 \ \mu^{-1} \qquad (31)$$

An analysis based upon the assumption of an *infinitely massive* nucleus of charge Ze gives

$$\tilde{\nu} = Z^2 R_\infty \left(\frac{1}{n'^2} - \frac{1}{n^2} \right) \qquad (32)$$

where $R_\infty = me^4 / 8\varepsilon_0^2 ch^3 = 10.9737309 \ \mu^{-1}$.

EXERCISE

2-12. It is of interest to compare the frequency of the emitted radiation with the frequency of revolution of the electron in its orbit. If the electron makes a unit quantum jump from n to $n - 1$, show that the frequency of the emitted radiation lies between the orbital-revolution frequencies of the upper and lower states. This is an example of *Bohr's correspondence principle*, which states that the quantum-mechanical and classical behavior of a system should be the same in the limit of large quantum numbers.

E. Spectroscopic Terms, Atomic Energy States. It was found empirically by Balmer (1885), Rydberg (1890), and by many other spectroscopists that the wave numbers (reciprocals of the wavelengths) of many of the spectral lines of an element could be classified into a number of *series.* Each series was accurately representable by an expression of the form

$$\tilde{\nu} = \tilde{\nu}_{\infty} - \frac{R}{(n + \delta)^2} \tag{33}$$

where $\tilde{\nu}_{\infty}$ represents a limiting wave number toward which the members of the series converge, n is an integer which ranges from some initial value $n_0 > 0$ to infinity, δ is a (small) correction term which can often be neglected, and R is a quantity which is *very nearly the same* for all series and elements. It was further found by Rydberg that several series of a given element could be represented in the form

$$\tilde{\nu} = R \left[\frac{1}{(n' + \delta')^2} - \frac{1}{(n + \delta)^2} \right] \tag{34}$$

where n' is an integer whose value defines the particular series, δ' is small, and $n > n'$. In this representation of the wave number as a difference between two similar quantities, the two quantities came to be called *terms;* and the numerical magnitude of each term, the *term value.*

The Bohr theory of atomic hydrogen provided a *physical basis* for the above representation and indicated that the terms correspond to certain *allowed energy states* of the atom. Using this physical picture of the meaning of the terms, it was possible to conclude that, not only those spectral lines representable in series form, but *all spectral lines arise from the transition of an atom from one allowed energy state to another.*

It is usually most convenient to represent the spectroscopic terms in graphical form on a so-called *energy-level,* or *term, diagram.* Such a diagram is shown in simplified form for atomic hydrogen in Fig. 2-5, and the spectrum, corresponding to the transitions indicated by arrows, is shown schematically in Fig. 2-6. It will be seen later that the term diagrams for most elements are quite intricate, and that even that for hydrogen is considerably more complex than is indicated in Fig. 2-5 (for example, the fine structure of the hydrogen levels is not included in this diagram). Furthermore, it is found that not all transitions are possible from a given energy state to lower energy states, but that, on the con-

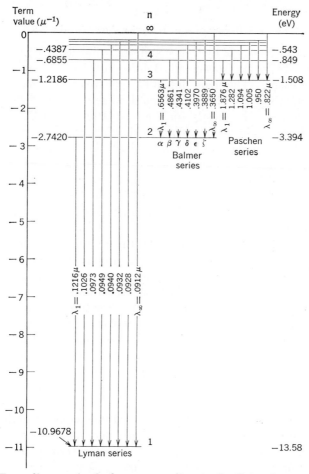

FIG. 2-5. Term diagram for hydrogen according to the Bohr theory, showing the transitions responsible for some of the spectral series of this element.

trary, there are certain *selection rules* that describe which transitions are "permitted" and which ones are "forbidden." The existence of such selection rules will later be seen to have a simple interpretation in modern quantum theory. These questions will be treated further in our later work.

F. *The Wilson-Sommerfeld Quantization Rules.* The spectacular suc-

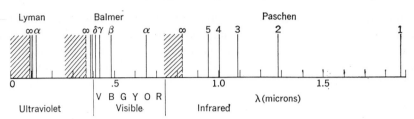

FIG. 2-6. Sketch of spectral series shown in Fig. 2-5.

cess of the Bohr theory of hydrogen and singly ionized helium encouraged many scientists to try to extend Bohr's postulates to apply to more complicated cases. The main difficulty appeared to be that of choosing what sort of orbits the electrons should be required to follow. Many suggestions were made, mostly for the case of neutral helium. Some of the models had the electrons moving in circular orbits in various orientations with respect to one another; others had the electrons vibrating back and forth around the nucleus in more or less semicircular paths. None of these models, however, could be made to yield energy levels for neutral helium that were at all comparable to the observed ones, and the one-electron atom remained the only successful treatment of atomic spectra.

Attempts were also made to apply the principles embodied in the postulates of Planck and Bohr to other problems. For example, the problem of the electrons in metals was treated by considering the quantum energy levels of a free particle inside a container having rigid walls; the harmonic oscillator was also treated. For most of these problems the quantization of the system was accomplished by using the so-called *Wilson-Sommerfeld quantization rules*. According to these rules, the phase integral of any variable over a complete cycle of its motion must be equal to an integer times h:

$$\oint p_i \, dq_i = n_i h \tag{35}$$

In this formula, p_i is the generalized momentum conjugate to the generalized coordinate q_i and $\oint$ means an integral over a complete cycle of the motion. Thus if q_i is an *azimuthal angle*, p_i is the corresponding angular momentum; or if q_i is a *position coordinate*, p_i is the corresponding component of *linear momentum*.

It is easily seen that the Bohr postulates for the hydrogen atom could be replaced by the above one, so long as circular orbits were the only ones being considered. The above rule would, however, lead to other permitted orbits of *elliptical* shape, for which

$$\oint P_\theta \, d\theta = n_\theta h \qquad \text{and} \qquad \oint P_r \, dr = n_r h \tag{36}$$

Thus, if the full three-dimensional character of the motion is considered, *three* such "phase integrals" are required to specify the orbits completely, and there are *three* "*quantum numbers*" n_r, n_θ, n_φ for the hydrogen atom according to this new approach. $n_\varphi = m$ is called the *magnetic* quantum number, $n_\theta + n_\varphi = k$ the *azimuthal* quantum number, and $n_r + n_\theta + n_\varphi = n$ the *total* quantum number.

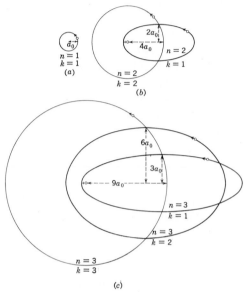

FIG. 2-7. Illustrating the character of some of the elliptical orbits which appear in Sommerfeld's extension of Bohr's theory of atomic hydrogen. (By permission from L. Pauling and E. B. Wilson, Jr., "Introduction to Quantum Mechanics," McGraw-Hill Book Company, Inc., New York, 1935.)

A theory of the hydrogen atom was worked out in considerable detail by Sommerfeld (1916), using the above quantization rules. It was found that a number of elliptical orbits of different angular momenta corresponded to each given energy and that the various *energies* were identical with those of the Bohr theory. Some of these orbits are shown in Fig. 2-7. However, when allowance was made for the relativistic "variation of mass with velocity" in the various elliptical orbits corresponding to a given energy, slight corrections which resulted in a small *separation* of the energy levels that were previously identical had to be introduced. This *fine structure* of the energy levels led to *double* lines in the theoretical spectra of hydrogen and ionized helium, the wavelength separations of the components of which were in excellent agreement with the experimentally observed fine structure of these spectra. Thus the Bohr theory, extended by the Wilson-Sommerfeld quantization rules and corrected by relativity, became a strong theory indeed. Its original inability to account for the spectra of more complicated atoms remained, however.

We shall see in later chapters that there is a close correspondence between the Wilson-Sommerfeld quantization rules and the results of the modern Schroedinger theory, but that it is only in the simplest cases, such as those mentioned above, that the Wilson-Sommerfeld rules apply. In addition, it will be seen that the successful treatment of hydrogen fine structure in this theory gave correct results only as a sheer *accident*, and that the modern explanation of the fine structure is quite different.

The foregoing brief review of early ideas of atomic structure has shown that, in the relatively short span of two decades after the discovery of the electron, a quite concrete model of the simplest atom, hydrogen, was formulated, and that the spectacular success of this model in describing the observed spectral radiation of hydrogen seemed to establish quite firmly several basic ideas that were felt to apply to all atoms:

1. Atoms consist of heavy, positively charged nuclei of very small size, surrounded by swarms of negative electrons.
2. The electrons associated with an atom are tiny material particles whose motions about the nucleus are describable in terms of ordinary classical (relativistic) mechanics.
3. But, for some reason, not every orbit that would be a possible one according to classical mechanics is actually possible.
4. The permitted orbits are defined in terms of certain *quantization conditions* whose complete form was not then quite clear.
5. The *quantum states* (4) correspond to certain *energies* of the atom.
6. Somehow the atom can remain in such quantum states for relatively long periods of time without radiating away its energy into the electromagnetic field.
7. The atom can "jump" from a given quantum state to certain of the lower states and in so doing radiate electromagnetic energy whose frequency is given by the Einstein frequency condition.

We shall soon see that, while not all the above ideas are correct, they are in many respects quite close to what is now thought to be the truth.

2-2. Primary Requisites of Quantum Mechanics

In the preceding pages we have seen how physicists were confronted with many experimental facts that demonstrated the inapplicability of classical mechanics and electromagnetism to atomic processes and how these facts were fitted into a new framework called the *quantum theory*. During this period the quantum theory was introduced in a rather negative way; that is, ordinary electromagnetic theory and Newtonian mechanics would be used in part of a problem, but at some point it would be necessary to introduce a postulate to the effect that *not all* the results of the classical theory are valid, and that instead some kind of quantum condition is in operation. This approach can be seen quite clearly in Planck's original treatment of the linear oscillator and in the Bohr theory.

It should be obvious that such a procedure cannot be regarded as a permanently satisfactory one, since no clear and complete statement of principle by which one can tell whether or not a given classical result is to be accepted has been made. Thus it must be expected that, ultimately, it will be necessary to seek a completely new and self-contained set of laws within which *both* classical mechanics and the quantum conditions are embodied and which will yield correct results for all physical problems. The new system of mechanics defined by such a set of laws is called *quantum mechanics.*

Such a reformulation of the laws of mechanics was made almost simultaneously (1926) by Schroedinger and by Heisenberg.[1] The mathematical appearance of the theories presented by these two men is quite different, but Schroedinger, and also Eckart, proved[2] that the two theories are in fact equivalent. Although certain aspects of each are embodied in current usage of the quantum theory, somewhat greater emphasis is usually placed upon Schroedinger's work because his formulation makes use of *partial differential equations*, while that of Heisenberg is built around *matrix algebra*. We shall use the Schroedinger method for all of our discussions except that of electron spin (Chap. 5). The system of quantum mechanics to be described is strictly valid only for low velocities, and thus is called *nonrelativistic quantum mechanics.*

In laying down the postulates upon which a physical theory is based, it is desirable to maintain as close a connection as possible with experimental facts; for in this way a reliable intuitive "feeling" for the theory can more easily be developed. With this idea in mind we shall now consider a number of important concepts and facts that must necessarily find a place in a correct systems of mechanics.

A. The Wave Properties of Particles. In 1924, L. de Broglie suggested[3] on purely theoretical grounds, following the analogy of the wave-particle duality of photons, that *electrons might have wave properties.* He showed that the Bohr orbits of hydrogen could be defined by the condition that the circumference of an orbit contain a *whole number of wavelengths* if the wavelength and momentum of the electron are related by the equation

$$\lambda = \frac{h}{p} \tag{1}$$

[1] Schroedinger, *Ann. Physik*, **79**, 361, 489, 784; **80**, 437; **81**, 109 (1926); Heisenberg, *Z. Physik*, **33**, 879 (1925).

[2] Schroedinger, *Ann. Physik*, **79**, 734 (1926); Eckart, *Phys. Rev.*, **28**, 711 (1926).

[3] *Phil. Mag.*, **47**, 446 (1924).

where h is Planck's constant and p is the momentum of the electron. In 1928, Davisson and Germer[1] discovered experimentally that low-energy ($\sim$200 eV) electrons striking a face of a nickel crystal actually *are* reflected in a manner that can most easily be interpreted as a *diffraction of plane waves* by the regularly spaced atoms of the crystal. In an

FIG. 2-8a. Electron diffraction of a graphite single crystal.

extensive set of experiments they established that the diffraction angle is such as to correspond to a plane wave of wavelength given by de Broglie's formula. Shortly afterward, Kikuchi (1928) succeeded in shooting a monoenergetic (68-keV) beam of electrons through a very thin sheet of mica onto a photographic plate. The diffraction pattern produced was in every way similar to an X-ray-diffraction photograph using X-rays of equal wavelength. In a similar manner G. P. Thomson obtained "powdered-crystal" diffraction photographs by shooting mono-energetic electrons (30 keV) through extremely thin gold foils. He proved that the pattern cannot be due to X-rays produced by the deceleration of the electrons in passing through the foil; he did so by showing that the pattern can be *shifted* by introducing a magnetic field into the space between the foil and the photographic plate. Figure 2-8 shows examples of electron diffraction by a single crystal and a powdered

[1] *Phys. Rev.*, **30**, 705 (1927).

crystal, and shows also the distortion of the diffraction pattern produced by a small magnet situated between the powdered crystal and the photographic plate.

Other particles than electrons also exhibit wave properties. For example, monoenergetic helium atoms (selected as to velocity by using a rotating-slit velocity selector) were successfully diffracted from a crystal of lithium fluoride, and Eq. (1) was verified to within about one per cent for these atoms.[1]

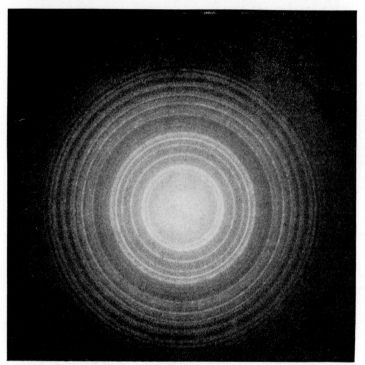

FIG. 2-8b. Electron diffraction of zinc oxide powder.

The above experiments, and many others, established beyond any possible doubt that all particles *really do* possess properties that are interpretable only in terms of waves. *This is a tremendous, undeniable fact,* which clearly must have a profound effect upon our basic ideas of the mechanics of particles. To see the sort of problems that are posed by these properties, consider the simplest possible diffraction experiment that might be used conceptually to study the wave properties of particles (Fig. 2-9). Think of a collimated beam of monoenergetic particles—say, electrons—impinging upon an opaque wall containing two narrow,

[1] Estermann, Frisch, and Stern, *Z. Physik*, **73**, 348 (1931).

parallel, closely spaced slits A and B and a detecting screen coated with a scintillating powder placed at some distance from the slits, so that particles that pass through the slits can be observed individually as they strike the screen.

Now consider two experiments that might be performed: The number of particles per unit area per unit time that strike various points on the screen might be counted either with slit A open and slit B closed or with

FIG. 2-8c. Electron diffraction of zinc oxide powder. A small magnet was introduced between the specimen and the photographic plate. The sharp, dark boundary represents the "shadow" of one pole of the magnet; the lines of force were approximately perpendicular to this boundary. The electron energy is about 40 keV. (*Photograph kindly furnished by Dr. Darwin W. Smith.*)

both slits open. The experimental fact that particles behave like waves assures us that, if the individual slits were narrower than 1 wavelength and their spacing from one another several wavelengths, the distributions of the scintillations for the two cases would appear qualitatively as shown in Fig. 2-9, where the number of scintillations per unit time and per unit area is plotted vs. the position on the screen.

These diffraction patterns are of course just what one would expect from a wave phenomenon. But now recall the process by which these

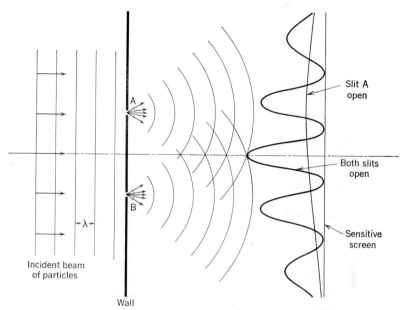

FIG. 2-9. Sketch of a conceptual experiment illustrating the diffractive interference of particles.

curves were to be obtained—namely, by *counting the impacts of individual particles upon the screen!* In the detection process each particle is localized within a *much smaller space* than that covered by the diffraction pattern, so that, at the time when each particle is detected, it must be considered to *be* a particle, localized in space. This is not the end of the puzzle; for, if a particle *is* a particle, one might think it must *always* be localized and could have arrived at the screen only by having gone through one slit or the other, but presumably *not through both.* But if the particle must go through only one slit, how can it know of the exist‑ ence of the other? To put the question in another way: If a particle can go through only one slit, then opening a second slit ought only to *increase* the chance that a particle will arrive at a given point on the screen. How, then, could the opening of the slit B cause the chance of arrival of a particle at certain points on the screen to *diminish to zero?*

Apparent paradoxes such as the above have been treated in numerous articles and books, and many of these are exceedingly interesting to study. All such paradoxes lead to the same general conclusion: that the motion of a particle must somehow be described in terms of a *wave amplitude* for its arrival at a given point in space via a given path, and that the ampli‑ tudes for various paths must combine according to the principle of super‑ position to form a *net amplitude whose square gives the probability for arrival at a given point of the screen.*

Thus we can explain experiments which exhibit the wave properties of particles only by assuming that each particle can somehow enter into constructive and destructive interference *with itself*.

EXERCISES

2-13. Show that the various circular orbits of the Bohr theory could be defined by requiring that the circumference of the orbit contain an integral number of de Broglie wavelengths. This result is in agreement with the ideas expressed above, in which a particle is visualized as being able to interfere with itself. The quantized states are thus interpretable as states in which the particle interferes constructively with itself in each successive cycle of its motion. This notion will be seen later to be quite intimately connected with present-day ideas of the nature of quantized states.

2-14. Treat the one-dimensional harmonic oscillator by using the Wilson-Sommerfeld quantization rules. Find the permitted energy levels and show that these levels can be defined for this case and, indeed, for all systems governed by the Wilson-Sommerfeld rules, by the requirement that the total phase change $\Delta\varphi$ over a cycle of the motion shall be an integral multiple of 2π: $\oint d\varphi = 2\pi n$. *Hint:* In rectangular coordinates, $d\varphi = 2\pi \, dx/\lambda$.

B. The Heisenberg Uncertainty Principle.[1] A second basic principle, which follows directly from the wave properties of matter just described, must be incorporated into a correct system of mechanics. This has to do with *indeterminacy*, or *uncertainty*, that is unavoidably introduced into the experimental measurement of physical quantities *by the measurement process itself*. One of the basic implicit assumptions of classical mechanics is that a mechanical system can, in principle, be observed with any desired degree of precision without appreciably disturbing its motion. This assumption is of course justified if one is considering the motion of objects of ordinary size, but if the motion of a very small object is to be studied, the quantum character of the object *and of the measurement process* must be allowed for.

As an illustration, consider the problem of measuring the *position* of a small particle, say, an electron. This might be done, in principle, by using a microscope. In order to obtain high resolution, it is of course necessary to use a large numerical aperture for the microscope and short-wavelength light. On the other hand, in order to disturb the electron as little as possible, one ought to make the intensity of illumination as small as possible. Therefore use only *one photon*. The experiment thus proceeds as indicated in Fig. 2-10: Very weak incident light of high

---

[1] Heisenberg, *Z. Physik*, **43**, 172 (1927).

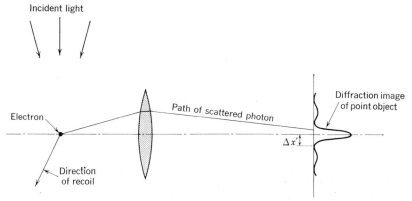

FIG. 2-10. Sketch of a "gamma-ray microscope."

photon energy $h\nu$ illuminates the electron. Presently a photon strikes the electron, is scattered, and proceeds through the microscope to some point on the diffraction pattern. The electron recoils under the impact of the photon, and moves away in some direction, as indicated by the arrow. Now, how precisely is the position of the electron determined by this experiment, and what uncertainty is introduced into its motion?

In the first place, the position of the electron is known *only approximately*, since one cannot say at *just what point* of the diffraction pattern the single photon arrived; one needs *many* photons to establish the symmetry of the pattern. One can only say that the photon most probably struck somewhere inside the first diffraction ring. This uncertainty, expressed in terms of the position of the electron, is easy to evaluate from the principles of physical optics; it is approximately equal to

$$\Delta x \approx \frac{f\lambda}{D} \qquad (2)$$

where λ is the wavelength of the light and f is the focal length and D the diameter of the microscope objective. (We omit the numerical factor that should be included in the above expression to express the fact that Δx refers to the radius of the first diffraction ring, since only the order of magnitude of Δx is now of interest.)

In the second place, the act of measuring the position of the electron has caused the electron to recoil by *an unknown amount*. This is true because *we have no way of finding out through what part of the microscope objective the photon went*. Since it *could have* gone through any part of the objective, we must admit an uncertainty in the x-component of momentum of approximately

$$\Delta p_x \approx \frac{h\nu}{c}\frac{D}{2f} \qquad (3)$$

Now, comparing the two uncertainties, we see that they contain the wavelength of the light and the dimensions of the microscope in *exactly reciprocal ways;* to make Δx small, one must accept a large Δp_x, and vice versa. The situation is stated more concisely by evaluating the *product* of the uncertainties:

$$\Delta x \, \Delta p_x \approx \frac{f\lambda}{D} \frac{h\nu}{c} \frac{D}{2f} = \frac{h}{2} \qquad (4)$$

Many other possible ways of measuring the position and the momentum of a particle can be analyzed in a similar manner, and always a relation such as the above is obtained. Also, other sets of quantities such as energy and time can be treated, with the result that similar restrictions exist for these quantities. Thus one is forced to admit that certain quantities cannot simultaneously be measured with unlimited precision, but that irreducible uncertainties are always inescapably present owing to the fact that the measuring process itself is subject to the same quantum laws as the system being measured. These minimum uncertainties are expressible in the form

$$\Delta Q \, \Delta P \approx h \qquad (5)$$

where Q and P refer to *conjugate quantities* (in the sense of Hamilton's canonical equations). Thus if Q is a rectangular coordinate, P is the corresponding component of momentum; if Q is an angular coordinate, P is the corresponding angular momentum; if Q is energy, P is the time; etc.

C. The Correspondence Principle. This principle, first stated by Bohr, asserts that the motion of a system as described by quantum mechanics and by classical mechanics must agree in the limit in which h can be neglected. That is, if the system is large enough and if our demand for accurate measurement is not too rigid, classical mechanics should furnish a good approximation to the motion of the system. This is also called the *classical limit* of quantum mechanics.

D. Quantized States and Energy Levels. As we saw in Sec. 2-1, one of the most important aspects of the Bohr theory of hydrogen was its interpretation of monochromatic spectral radiation as resulting from the transition of an atom from one *stationary state,* or *energy level,* to another. Direct experimental evidence of the existence of discrete energy levels in atoms is provided by the experiments of Franck and Hertz[1] (1914) in which atoms of mercury vapor were excited by the impact of electrons whose energy could be varied by letting them fall through an adjustable potential drop. They found that, when the energies of the electrons lay below a certain value, the impacts were purely elastic, each electron rebounding with no loss of energy. They observed a sudden onset of

[1] *Verhandl. deut. physik. Ges.,* **16,** 512 (1914).

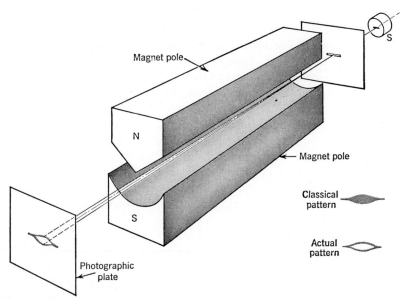

FIG. 2-11. Schematic view of Stern-Gerlach apparatus. Neutral atoms emitted from the source S are collimated into a narrow beam and directed along the sharp edge of a magnet pole piece. The inhomogeneous magnetic field exerts a transverse force upon the atoms and causes a spreading of the atomic beam.

inelastic collisions and an accompanying fluorescence of the mercury vapor, however, when the kinetic energy corresponded exactly to that required to raise a mercury atom to its first excited level, as determined by the Einstein frequency condition and the frequency of the light emitted by the atom in returning to its ground state. Many such experiments have subsequently been performed, with the result that the existence of quantized energy levels in atoms is extremely well verified.

Furthermore, since Bohr arrived at the appropriate energy levels for the hydrogen atom by postulating that *angular momentum* is quantized, it is reasonable to expect that in quantum mechanics *other quantities than energy might be subject to quantum conditions.* Direct experimental evidence for the existence of quantized angular momenta was obtained by Stern and Gerlach.[1] In their famous experiment a beam of neutral silver atoms defined by two consecutive narrow slits was allowed to pass through an inhomogeneous magnetic field, as shown in Fig. 2-11. This inhomogeneous field, acting upon the magnetic-dipole moment $\mathbf{\mu}$ of the atoms, produced a transverse force upon each atom of magnitude

$$F_z = \mu_z \frac{\partial B_z}{\partial z} \tag{6}$$

[1] *Z. Physik*, **8**, 110; **9**, 349 (1922).

where μ_z is the component of $\mathbf{\mu}$ along the field-gradient direction. Classically, the random orientation of the dipole moments would be expected to give *spreading* of the beam. Stern and Gerlach actually observed, however, a *splitting* of the beam into *two discrete components*, indicating the existence of *only two possible values for* μ_z. The existence of such quantized states must of course be included in a correct system of mechanics.

E. Selection Rules and Line Intensities. We have seen that the spectral radiation of every atom is to be interpreted to be light quanta $h\nu$ emitted as the atom "jumps" from one of its stationary states to another. By observing what spectral frequencies are present in the spectrum of an atom and searching the spectrum for various Rydberg series or other characteristic regularities, the energy levels of an atom may be deduced experimentally. Having obtained in this way an energy-level diagram, however, one finds that there exist energy states between which the system is seldom or never observed to jump. In fact, there are so many such cases, occurring with such regularity, that spectroscopists invented what are called *selection rules* to distinguish those transitions which do occur from those which do not. These selection rules can for the most part be represented graphically by arranging the various energy levels of an atom into several vertical columns, such that the observed transitions correspond to a jump from a level in one column to a level in an adjacent column. Jumps between levels in the same column or in nonadjacent columns are not ordinarily observed and are said to be *forbidden*. An energy-level diagram for sodium, arranged in this way, is shown in Fig. 2-12. (The meanings of the various symbols on this diagram will gradually become clear in the course of our later work.)

Although the idea of selection rules provides a good over-all description of permitted vs. forbidden transitions, some spectacular violations of the selection rules are sometimes observed. A correct theory of quantum mechanics must of course account physically both for the existence of selection rules and their occasional violation. *We shall find that the theory can do just this.* In addition, the intensities of the spectral lines resulting from the transitions that are permitted by the selection rules vary over a tremendous range. The validity of any quantum theory must also be measured by its ability to account quantitatively for observed intensity ratios of spectral lines.

With the foregoing list of basic requisites that must be fulfilled by quantum mechanics to guide us, we shall now consider the postulates upon which the Schroedinger theory may be based. As we do this, we shall repeatedly point out those essential features of the theory which may be identified with the requirements outlined above. This is for the specific purpose of providing a basis of reasonableness or plausibility for our steps, but is not intended as a proof of their logical necessity. Indeed, quantum mechanics can be developed from various sets of postulates and may be expressed in various quite different-appearing forms, and in any case *must be* based upon *some* set of postulates that cannot be derived directly from experimental data.   In this connection

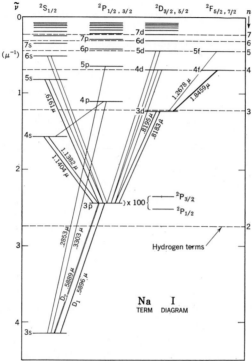

Fig. 2-12. An energy-level diagram for atomic sodium, showing some of the most intense transitions.

one must constantly guard against the tendency to let an unavoidable initial confusion about the meaning or use of a given idea become a doubt of its validity.   We shall find many examples in which quantum mechanics leads to conclusions that seem almost too fantastic to believe; yet each has withstood whatever experimental tests have been performed to check it, within the agreed limitation to nonrelativistic problems— and these tests have been stringent ones indeed.

2-3.   Postulatory Basis of Quantum Mechanics

The postulates of quantum mechanics make certain assertions concerning the probable results of measurements that might be made upon a physical system. Before considering these postulates, let us review briefly the *classical* concept of a physical system and its motion. For our present purposes, a *system* is supposed to consist of a number of particles, whose internal structures can be ignored, moving under the influence of their mutual interactions and certain other forces of "external" origin. The system is supposed to possess a sufficient number N of coordinates q_j to specify completely its configuration at any instant. These include the ordinary position coordinates of the various particles but could also involve other variables such as might be needed, for example, to specify the orientations of the magnetic moments or of the angular momenta of the particles.

It is also supposed that the system, treated on a classical basis, possesses a Lagrangian function[1] $L(q_j, \dot{q}_j, t)$. From this function a set of momenta p_j, canonically conjugate to the coordinates q_j, is defined (following the Hamiltonian treatment) by

$$p_j = \frac{\partial L}{\partial \dot{q}_j} \tag{1}$$

and the momenta are then used to eliminate the velocities $\dot{q}_j$ in the formation of the *Hamiltonian function*

$$H(q_j, p_j, t) = \sum_{j=1}^{N} p_j \dot{q}_j - L(q_j, \dot{q}_j, t) \tag{2}$$

EXERCISE

2-15. A particle of mass m is constrained to move along a straight line and is attracted toward a point on this line with a force proportional to the distance. Write the Lagrangian function for this system and carry through the above steps leading to the Hamiltonian function. *Ans.:* $H = p^2/2m + \frac{1}{2}kx^2$.

Classically, the motion of the system can be described in terms of Hamilton's canonical equations

$$\dot{p}_j = -\frac{\partial H}{\partial q_j} \quad \text{and} \quad \dot{q}_j = \frac{\partial H}{\partial p_j} \tag{3}$$

These equations define a family of trajectories in (p_j, q_j) space (called *phase space*) on one of which the system moves. If the initial values

---

[1] Although this discussion and the postulates of quantum mechanics utilize the generalized coordinate and momentum notation of Hamilton's principle, many of the applications of quantum mechanics make specific use of rectangular Cartesian coordinates and are valid only in these coordinates. For this reason it is desirable to keep rectangular Cartesian coordinates uppermost in mind in interpreting the results of the present chapter, even though most of the results are of more general validity.

of the q_j and the p_j are specified *exactly*, as is assumed to be possible in classical mechanics, then the equations of motion (3) will define an *exact trajectory* in phase space and at any later time will provide an *exact prediction* of the configuration and the momenta possessed by the system at that time.

We now begin our description of the quantum-mechanical treatment of a system by examining the basic concepts involved in describing the *state* of the system.

A. Specification of the Quantum-mechanical State of a System. In classical mechanics, the *state* of a mechanical system is specified by giving the precise numerical values of its N coordinates and N momenta. All dynamical properties possessed by the system, such as the potential energy, the kinetic energy of each particle, and the angular momentum of any group of particles, can then be found.

Further, the state at any time is uniquely determined by the state at an earlier time through the equations of motion, so that an initial specification of the state of the system serves to define all its dynamical properties for all later times. In short, *everything that can be known* about the system is contained in the specification of its state.

In quantum mechanics, the uncertainty principle limits the extent of one's knowledge concerning the coordinates and the momenta of a system, so that *complete* information, in the classical sense, cannot be obtained. Now, since the state of the system at any time must somehow depend upon an *initial state*, whose specification lacks complete precision in the classical sense, it must be true that *exactly similar systems* (systems constituted in the same way and subject to the same outside forces) started initially under *exactly similar conditions* (subject always to the limitations of the uncertainty principle) may at a later time possess *quite different values* for their various dynamical properties. Yet, if the concept of the *state* of a system is to have any meaning at all in quantum mechanics, *the state at one time ought uniquely to determine the state at a later time.*

We are thus led to the conclusion that the concept of the state of a system in quantum mechanics cannot imply *exact* knowledge of all coordinates and all momenta, but must somehow allow for the possibility that similar measurements, carried out on similar systems, might lead to different results. Our first postulate puts the above ideas into quantitative form by defining the way in which the state of a system is to be expressed:

Postulate 1:
There exist two complex *probability amplitudes* (also called *state functions* or *wave functions*) $\psi(q_j,t)$ and $\phi(p_j,t)$ which completely define

the state of a quantum-mechanical system in the following way:
If at time t the *coordinates* of the system are measured, the *probability* that these will be found to lie within the ranges q_j to $q_j + dq_j$ is

$$W_q(q_j,t) \, dq_1 \, \cdots \, dq_N = \psi^*(q_j,t)\psi(q_j,t) \, dq_1 \, \cdots \, dq_N \qquad (4)$$

while if, instead, the *momenta* of the system are measured at time t, the *probability* that they will be found to lie within the ranges p_j to $p_j + dp_j$ is

$$W_p(p_j,t) \, dp_1 \, \cdots \, dp_N = \phi^*(p_j,t)\phi(p_j,t) \, dp_1 \, \cdots \, dp_N \qquad (5)$$

Since each coordinate must be found to have *some* value as a result of a measurement, it is also usually required[1] that the total ntegrated probability be equal to unity:[2]

$$\int \, \cdots \, \int W_q(q_j,t) \, dq_1 \, \cdots \, dq_N = 1 \qquad \text{or} \qquad \int \psi^*\psi d^N q = 1\ddagger \qquad (6)$$

And similarly for the momenta, it is usually required[3] that

$$\int W_p(p_j,t)d^N p = 1 \qquad \text{or} \qquad \int \phi^*\phi d^N p = 1 \qquad (7)$$

Both ψ and ϕ must be bounded, single-valued functions, and their absolute squares must be integrable over all space.

Several features of this first postulate deserve comment. First, the prediction of the result of a physical measurement in terms of a *probability distribution* directly implies that successive measurements on systems known to be in the same state need not give the same result each time. Secondly, the requirement that ψ must be a function only of the coordinates and the time, and ϕ a function only of the momenta and the time, is necessary because exact measurement of all of the coordinates *or* all of the momenta is envisaged, and exact measurement of one of these sets of quantities precludes obtaining any information whatever regarding the other.[4]

One may regard $\psi^*\psi$ as being operationally measurable by performing an indefinitely large number of exact coordinate measurements upon

[1] An exception to this requirement will be treated in Chap. 4.

[2] In these and all other such integrals in this chapter, integration over the *entire range* of each coordinate or momentum is implied.

[3] An exception to this requirement will be treated in Chap. 4.

[4] Actually, exact measurement of any physical quantity is experimentally impossible, but we are here specifying the ideal limit toward which experimental measurements of higher and higher precision must converge. The question of how partial information concerning *both* the coordinates and the momenta, obtained by permitting some indeterminacy in each quantity, can be described in quantum mechanics is treated in more advanced texts. We shall assume that all measurements of physical quantities are made with infinite precision.

$\ddagger$ For brevity we shall henceforth use the notation $d^N q = dq_1 \, \cdots \, dq_N$ or $d^N p = dp_1 \, \cdots \, dp_N$ with a single $\int$ sign to signify integration over the complete range of *all* coordinates or momenta.

similar physical systems known to be in the same state. For example, the measurement of the diffraction-pattern intensity in an electron-diffraction experiment is in fact a measurement of $\psi^*\psi$ over the photographic plate for a huge number of individual electrons. This pattern, so obtained, may then be used to predict the relative probability of arrival at each point on the plate for a *single electron*. The goal of quantum mechanics is, for this type of experiment, to provide a *theoretical prediction* of $\psi^*\psi$ based upon a knowledge of the initial state of the electron and the nature of its interaction with the diffracting crystal.

The above postulate has defined the form in which information concerning the coordinates or the momenta of a quantum-mechanical system is to be expressed. But one wishes also to have knowledge of quantities other than coordinates and momenta. The manner in which all such information is to be obtained from ψ or ϕ will be treated in our third postulate. We may prepare the way for it by noting that one could define a probability distribution for any dynamical quantity that depends functionally only upon the coordinates or upon the momenta by using the above equations directly. The analytical expression for such a distribution might be quite complicated, however, and our needs can often be satisfied by using considerably less information than is contained in the complete distribution of a quantity. Indeed, knowledge of the *average value* and *variance* of a quantity is almost always sufficient for our purposes. For a given dynamical quantity F these are symbolized $\langle F \rangle$ and $(\Delta F)^2$ respectively. $\langle F \rangle$ is alternatively referred to as the average, the mean, or the *expectation value* of the quantity F; it is the ordinary average of an indefinitely large number of independent measurements of F carried out on identical systems known to be in the same state. If F is a function $F(q_j)$ of the coordinates only, its mean value and variance are determined from $\psi(q_j, t)$ in a straightforward manner by the equations

$$\langle F \rangle = \int F(q_j) W_q(q_j, t) \, d^N q \tag{8}$$

and
$$(\Delta F)^2 = \int [F(q_j) - \langle F \rangle]^2 W_q(q_j, t) \, d^N q$$
$$= \langle F^2 \rangle - \langle F \rangle^2 \tag{9}$$

Similarly, for a function $G(p_j)$ of the *momenta only*,

$$\langle G \rangle = \int G(p_j) W_p(p_j, t) \, d^N p \tag{10}$$

and
$$(\Delta G)^2 = \int [G(p_j) - \langle G \rangle]^2 W_p(p_j, t) \, d^N p$$
$$= \langle G^2 \rangle - \langle G \rangle^2 \tag{11}$$

EXERCISES

2-16. Supply the missing steps in the derivations of Eqs. (9) and (11).

2-17. A one-dimensional harmonic oscillator is in a state such that, at $t = 0$,

$$\psi(x) = A \exp (-x^2/2a^2) \exp (ip_0x/\hbar) \qquad (12)$$

Find: (a) $W_q(x)$, (b) A^*A, (c) $\langle x \rangle$, (d) $(\Delta x)^2$, and (e) the expectation value of the potential energy. Ans.: (a) $W_q(x) = A^*Ae^{-x^2/a^2}$, (b) $A^*A = 1/a\sqrt{\pi}$, (c) $\langle x \rangle = 0$, (d) $(\Delta x)^2 = \frac{1}{2}a^2$, and (e) $\langle V \rangle = \frac{1}{4}m\omega^2a^2$.

 B. *Relation between* $\psi(q_j,t)$ *and* $\phi(p_j,t)$. The first postulate of quantum mechanics has asserted the existence of the functions ψ and ϕ that together specify the quantum-mechanical state of a system, but it has not implied the existence of any relationship between these two quantities. However, it is clear from the discussion of the uncertainty principle that ψ and ϕ cannot be *completely* independent of one another, since in this case the uncertainty principle could be violated, e.g., by specifying that ψ and ϕ are *both* sharply peaked distributions corresponding to precisely defined coordinates *and* momenta. In addition, the wave properties of particles are such that the probability for the arrival of a particle at a given *coordinate* is described by a wave amplitude whose wavelength is related to the *momentum* of the particle by the de Broglie relation (1), which again implies some connection between ψ and ϕ. The precise form of the relationship that must exist between ψ and ϕ is specified by our second postulate of quantum mechanics:

Postulate 2:
The probability amplitudes $\psi(q_j,t)$ and $\phi(p_j,t)$ are connected by the relations

$$\phi(p_j,t) = h^{-\frac{1}{2}N} \int \psi(q_j,t) \exp \left(-i \sum_{j=1}^{N} p_jq_j/\hbar \right) d^Nq \qquad (13)$$

and $\quad \psi(q_j,t) = h^{-\frac{1}{2}N} \int \phi(p_j,t) \exp \left(+i \sum_{j=1}^{N} p_jq_j/\hbar \right) d^Np \qquad (14)$

Examination of the form of the above relations reveals that ψ and ϕ have been postulated to be *Fourier inverse* functions.[1] Thus $\phi(p_j,t)$

---

[1] Fourier inversion of a function is similar to the expansion of a function in a Fourier series, the essential difference being that the range of the variable is taken to be $-\infty$ to $+\infty$ instead of $-\pi$ to $+\pi$. One may understand the Fourier transform formulas in terms of the familiar Fourier series as follows:

Let a function $f(x)$ which is periodic in x with the period L be expanded in a Fourier series:

(a) $\qquad f(x) = \sum_{n=0}^{\infty} \left(A_n \cos \frac{2n\pi x}{L} + B_n \sin \frac{2n\pi x}{L} \right) \qquad -\frac{L}{2} < x < \frac{L}{2}$

represents a decomposition of $\psi(q_j,t)$ into (complex) sinusoidal components, with a momentum h/λ associated with each component of wavelength λ. If $\phi(p_j,t)$ is a sharply localized function, containing

---

This expansion may equally well be written in the complex exponential forms

(b)
$$\mathrm{f}(x) = \sum_{n=0}^{\infty} [A'_n e^{i(2n\pi x/L)} + B'_n e^{-i(2n\pi x/L)}]$$

or

(c)
$$\mathrm{f}(x) = \sum_{n=-\infty}^{\infty} C_n e^{i(2n\pi x/L)}$$

where $2A'_n = A_n - iB_n \qquad 2B'_n = A_n + iB_n$

and
$$\begin{aligned} C_n &= A'_n & \text{for } n > 0 \\ &= A'_0 + B'_0 & \text{for } n = 0 \\ &= B'_n & \text{for } n < 0 \end{aligned}$$

Any given coefficient, say, C_m, may be evaluated in the usual way by multiplying each side of (c) by $e^{-i(2m\pi x/L)} dx$ and integrating from $x = -\tfrac{1}{2}L$ to $x = +\tfrac{1}{2}L$:

(d)
$$\int_{-\frac{1}{2}L}^{+\frac{1}{2}L} \mathrm{f}(x) e^{-i(2m\pi x/L)} dx = LC_m$$

We wish now to let L become infinite. In this process we note that successive terms in (c) involve wavelengths which differ only infinitesimally from one another (that is, they are spaced $2\pi/L$ apart, and $2\pi/L \to 0$ as $L \to \infty$). We may first transform (c) by writing

(e)
$$\mathrm{f}(x) = \sum_{n=-\infty}^{\infty} \frac{LC_n}{2\pi} e^{i(2n\pi x/L)} \frac{2\,\Delta n\,\pi}{L}$$

where $\Delta n = 1$ for each term. If we then define a new summation variable $s = 2n\pi/L$, the summation may be written

(f)
$$\mathrm{f}(x) = \sum_{s=-\infty}^{\infty} \frac{LC_s}{2\pi} e^{isx} \Delta s \qquad \Delta s = \frac{2\pi}{L} \to 0$$

Then, noting from (d) that LC_m may remain finite as $L \to \infty$ if $\mathrm{f}(x) \to 0$ sufficiently rapidly as $x \to \pm\tfrac{1}{2}L$, we define a new function

(g)
$$LC_s \to (2\pi)^{\frac{1}{2}} \mathrm{g}(s)$$

[The factor $(2\pi)^{\frac{1}{2}}$ is introduced for symmetry reasons soon to appear.] The summation (f) may then be written in the limit as an integral:

(h)
$$\mathrm{f}(x) = (2\pi)^{-\frac{1}{2}} \int_{-\infty}^{\infty} \mathrm{g}(s) e^{isx} ds$$

and (d) then becomes

(i)
$$\mathrm{g}(s) = (2\pi)^{-\frac{1}{2}} \int_{-\infty}^{\infty} \mathrm{f}(x) e^{-isx} dx$$

$\mathrm{g}(s)$ is called the Fourier transform of $\mathrm{f}(x)$, and $\mathrm{f}(x)$ the inverse Fourier transform of $\mathrm{g}(s)$. The extension to several variables is made without difficulty.

only momenta near some value p_0, then $\psi(q_j,t)$ is a long wave train of wavelength very nearly h/p_0; conversely, if $\psi(q_j,t)$ is a sharp distribution, containing only coordinates near q_0, then the Fourier transform of this function contains a broad range of wavelengths, so that the momentum is not well defined. Such a relationship will be recognized as fitting in quite satisfactorily with both the de Broglie relation and the uncertainty principle. The following exercises illustrate these points.

EXERCISES

2-18. Find $W_q(x)$, A, $\langle x \rangle$, $(\Delta x)^2$, $\phi(p)$, $W(p)$, $\langle p \rangle$, $(\Delta p)^2$, and $\Delta x \, \Delta p$ for the one-dimensional wave function

$$\psi(x) = A \exp\left[-(x - x_0)^2/2a^2\right] \exp ip_0x/\hbar$$

Ans.: $\quad W_q(x) = A^*Ae^{-(x-x_0)^2/a^2} \qquad A^*A = \dfrac{1}{a\pi^{\frac{1}{2}}} \qquad \langle x \rangle = x_0$

$(\Delta x)^2 = \frac{1}{2}a^2$

$$\phi(p) = \frac{1}{\pi^{\frac{1}{4}}} \sqrt{\frac{a}{\hbar}} \exp\left[-(p - p_0)^2a^2/2\hbar^2\right] \exp\left[-i(p - p_0)x_0/\hbar\right]$$

$W_p(p) = \dfrac{a}{\pi^{\frac{1}{2}}\hbar} \exp\left[-(p - p_0)^2a^2/\hbar^2\right] \qquad \langle p \rangle = p_0 \qquad (\Delta p)^2 = \hbar^2/2a^2$

$\Delta x \, \Delta p = \frac{1}{2}\hbar$

2-19. Sketch the real parts of $\psi(x)$ and $\phi(p)$ of Exercise 2-18 for various values of a, and thus verify for this case the assertions in the text regarding the relation between the relative "sharpness" of the coordinate and momentum probability distributions.

2-20. Reinvert the function $\phi(p)$ of Exercise 2-18 to obtain $\psi(x)$.

2-21. Show that $\phi(p)$ of Exercise 2-18 is normalized.

2-22. Find the expectation value of the kinetic energy of a particle of mass m whose motion is instantaneously described by the wave function of Exercise 2-18. *Ans.:* $\langle T \rangle = p_0^2/2m + \hbar^2/4ma^2$.

2-23. At $t = 0$ the wave function for the electron in a hydrogen atom is $\psi(x,y,z) = A \exp\left[-(x^2 + y^2 + z^2)^{\frac{1}{2}}/a_0\right]$. Find: A, $\langle x \rangle$, $(\Delta x)^2$, and $\langle(x^2 + y^2 + z^2)^{-\frac{1}{2}}\rangle$. *Ans.:* $|A| = (\pi a_0^3)^{-\frac{1}{2}}$, $\langle x \rangle = 0$, $(\Delta x)^2 = a_0^2$, $\langle(x^2 + y^2 + z^2)^{-\frac{1}{2}}\rangle = 1/a_0$.

C. Introduction of Operators. The first two postulates of quantum mechanics assert that the state of a quantum-mechanical system is, at any instant, *completely described* by a single complex probability amplitude, $\psi(q_j,t)$, or $\phi(p_j,t)$, which means that *everything that can be known* about the system is contained in this function. The manner in which information concerning those dynamical quantities that depend functionally only upon the coordinates, or only upon the momenta, is to be obtained from

ψ and ϕ is illustrated in Eqs. (8) to (11) and in the subsequent exercises. Two questions now suggest themselves, however: First, since ϕ and ψ are uniquely determined, one by the other, is it possible to obtain information concerning functions of the *momenta* directly from $\psi(q_j,t)$ instead of from ϕ? And secondly, how does one obtain information concerning dynamical quantities that depend functionally upon *both* the coordinates and the momenta?

To investigate these questions, consider the problem of finding the expectation value of one of the momenta, say, p_s. From Eq. (10) this is equal to

$$\langle p_s \rangle = \int p_s \phi^*(p_j,t) \phi(p_j,t) \, d^N p$$

Using the second postulate, however, this may be written in the equivalent form

$$\langle p_s \rangle = h^{-N/2} \int_{p_j} \int_{q_i} p_s \phi^* \psi \, \exp \, (-i \Sigma p_j q_j / \hbar) \, d^N q \, d^N p$$

We can now eliminate p_s from the integrand by integrating by parts over the range of the coordinate q_s, calling

$$u = \psi(q_j,t) \qquad \text{and} \qquad dv = \exp \, (-i\Sigma p_j q_j/\hbar) \, dq_s$$

In this way the integral over q_s becomes

$$-\frac{\hbar}{ip_s} \psi(q_j,t) \exp \, (-i\Sigma p_j q_j/\hbar) \Big|_{-\infty}^{\infty} + \int \frac{\hbar}{ip_s} \frac{\partial \psi}{\partial q_s} \exp \, (-i\Sigma p_j q_j/\hbar) \, dq_s$$

and the uv term drops out because ψ must be zero at both $+\infty$ and $-\infty$ for cases of physical interest. Thus the expression for $\langle p_s \rangle$ becomes

$$\langle p_s \rangle = h^{-N/2} \int_{p_j} \int_{q_i} \phi^* \left(\frac{\hbar}{i}\right) \frac{\partial \psi}{\partial q_s} \exp \, (-i\Sigma p_j q_j/\hbar) \, d^N q \, d^N p$$

Again applying the second postulate, this time in the form

$$\psi^*(q_j,t) = h^{-N/2} \int \phi^* \exp \, (-i\Sigma p_j q_j/\hbar) \, d^N p$$

we finally have

$$\langle p_s \rangle = \int \psi^* \left(\frac{\hbar}{i}\right) \frac{\partial \psi}{\partial q_s} \, d^N q \tag{15}$$

A similar procedure can be followed for any integral power of p_s, with the result

$$\langle p_s^n \rangle = \int \psi^* \left(\frac{\hbar}{i}\right)^n \frac{\partial^n \psi}{\partial q_s^n} \, d^N q \tag{16}$$

The above transformation from "ϕ-language" to "ψ-language" has led to the very interesting result that the average value of a momentum component p_s can be found in ψ-language (or "coordinate language") by applying the differential operator $(\hbar/i)(\partial/\partial q_s)$ to the wave function ψ, then multiplying by ψ^*, and integrating throughout coordinate space in the same way as for finding the expectation value of a function of the coordinates. This result leads naturally to the supposition that the average value of *any* dynamical quantity $F(p_j,q_j)$ might be obtained in an analogous manner by using either ψ- or ϕ-language if an appropriate operator could be devised to represent it in an expression of the form

$$\langle F(q_j,p_j)\rangle = \int\psi^*\hat{F}_q\psi\, d^N q \tag{17}$$
$$\langle F(q_j,p_j)\rangle = \int\phi^*\hat{F}_p\phi\, d^N p \tag{18}$$

where $\hat{F}_q$ is the operator that corresponds to the quantity $F(q_j,p_j)$ in coordinate language and $\hat{F}_p$ is the corresponding operator in momentum language. We have so far found that this is possible for the momenta, for which the operators are

$$(\hat{p}_s)_p = p_s \qquad (\hat{p}_s)_q = \frac{\hbar}{i}\frac{\partial}{\partial q_s} \qquad (\hat{p}_s{}^n)_p = p_s{}^n \tag{19}$$

and
$$(\hat{p}_s{}^n)_q = \left(\frac{\hbar}{i}\right)^n\frac{\partial^n}{\partial q_s{}^n} \tag{20}$$

EXERCISES

2-24. Supply the missing steps in the derivation of Eq. (15).

2-25. Follow the procedure used to establish Eqs. (15) and (16) to show that $(\hat{q}_s)_q = q_s$, $(\hat{q}_s)_p = -(\hbar/i)(\partial/\partial p_s)$, $(\hat{q}_s{}^n)_q = q_s{}^n$, and $(\hat{q}_s{}^n)_p = (-\hbar/i)^n(\partial^n/\partial p_s{}^n)$.

2-26. Use Eq. (16) to find $\langle p_x\rangle$ and $(\Delta p_x)^2$ for the wave functions of Exercises 2-18 and 2-23. *Ans.:* Exercise 2-23: $\langle p_x\rangle = 0$, $(\Delta p_x)^2 = \hbar^2/3a_0{}^2$.

Our third postulate of quantum mechanics provides the basis for treating dynamical quantities of arbitrary form through the use of operators:

Postulate 3:

The expectation value of any dynamical quantity $F(q_j,p_j)$ may be evaluated by using either

$$\langle F\rangle = \int\psi^*\hat{F}_q\psi\, d^N q \tag{21}$$

or
$$\langle F\rangle = \int\phi^*\hat{F}_p\phi\, d^N p \tag{22}$$

where $\hat{F}_q$ is a *linear, Hermitian operator* obtained by replacing p_j by $(\hbar/i)(\partial/\partial q_j)$ in $F(q_j,p_j)$, and $\hat{F}_p$ is a *linear, Hermitian operator* obtained by replacing q_j by $(-\hbar/i)(\partial/\partial p_j)$ in $F(q_j,p_j)$.

The necessary properties that must be possessed by quantum-mechanical operators, as prescribed in the above postulate, must now be defined. Let $\hat{F}$ be an operator, and let ψ and χ be operands which satisfy the conditions of continuity, finiteness, and integrability that must hold for all wave functions in quantum mechanics. Then the requirement that an operator $\hat{F}$ be *linear* means that

(a)
$$\hat{F}(\psi + \chi) = \hat{F}\psi + \hat{F}\chi$$
(b)
$$\hat{F}(C\psi) = C\hat{F}\psi \tag{23}$$

for any real or complex numerical multiplying factor C. A linear *Hermitian* operator is a linear operator which satisfies the equation

$$\int \psi^* \hat{F}\chi \, d^N q = \int \chi (\hat{F}\psi)^* \, d^N q \tag{24}$$

for an arbitrary choice of functions ψ and χ. Some of the more useful properties of linear Hermitian operators are outlined below.

If $\hat{F}$, $\hat{G}$, and $\hat{H}$ are linear Hermitian operators and ψ and χ are wave functions,

$$
\begin{aligned}
(\hat{F} + \hat{G})\psi &= \hat{F}\psi + \hat{G}\psi && \text{definition of } \hat{F} + \hat{G} \\
(\hat{F}\hat{G})\psi &= \hat{F}(\hat{G}\psi) && \text{definition of } \hat{F}\hat{G} \\
[\hat{F}(\hat{G} + \hat{H})]\psi &= \hat{F}(\hat{G}\psi) + \hat{F}(\hat{H}\psi) && \text{distributive law}
\end{aligned}
$$

If $\hat{F}$ and $\hat{G}$ are linear Hermitian operators, then the operators

$$\hat{S} = \tfrac{1}{2}(\hat{F}\hat{G} + \hat{G}\hat{F}) \tag{25}$$

and
$$\hat{A} = \frac{1}{2i}(\hat{F}\hat{G} - \hat{G}\hat{F}) \tag{26}$$

are also linear and Hermitian.

Note that linear Hermitian operators *do not in general commute* with each other. That is $\hat{F}\hat{G} \neq \hat{G}\hat{F}$, or $\hat{F}\hat{G} - \hat{G}\hat{F} \neq 0$, except in special cases. The operator $\hat{C} = \hat{F}\hat{G} - \hat{G}\hat{F}$ is called the *commutator operator* of $\hat{F}$ and $\hat{G}$; it plays an extremely important role in quantum mechanics, as will later be seen.

One of the most important reasons for requiring that observable dynamical quantities be represented by Hermitian operators is that the Hermitian property guarantees that the *expectation values of dynamical quantities will be real*. This is seen by writing

$$\langle F \rangle = \int \psi^* \hat{F}_d \psi \, d^N q \qquad \text{and} \qquad \langle F \rangle^* = \int \psi (\hat{F}_d \psi)^* \, d^N q$$

But by Eq. (24) with $\psi = \chi$, these two expressions are equal, so that $\langle F \rangle = \langle F \rangle^*$.

The reason for requiring that quantum-mechanical operators be linear is principally to assure that the *principle of superposition* will hold for various wave functions which satisfy the equations of motion (to be described shortly) and the boundary conditions for a given system. The Hermitian property is required to assure that expectation values of dynamical quantities be real, that the total probability $\int \psi^* \psi \, d^N q$ remain normalized to unity at all times, and for other reasons that will gradually appear.

Although the postulate just described deals only with the question of how to represent given dynamical quantities for the purpose of calculating their expectation values, we shall later find that there are certain operators which are linear and Hermitian but which do not correspond to any classical dynamical quantity; yet some of the most fundamental and distinctive results of quantum mechanics follow from their use. Examples are the electron-spin operators, the parity operator, and the exchange operator. They will be dealt with individually as occasion arises.

EXERCISES

2-27. Find the commutator operator (a) of q_s and p_s, (b) of q_s and p_r $(s \neq r)$, (c) of q_s and q_r, and (d) of p_s and p_r.

2-28. Show that q_s and $(\hbar/i)(\partial/\partial q_s)$ are linear, Hermitian operators. Remember that the wave functions corresponding to situations of physical interest are integrable over the entire range of the coordinates and vanish at infinity.

2-29. Find a linear Hermitian operator to represent the dynamical quantity $q_s p_s$.

2-30. A particle of mass m moves in a conservative force field whose potential function is $V(x,y,z)$. Find the Hamiltonian function for this particle, using rectangular Cartesian coordinates, and show that the Hamiltonian is equal to the *total energy* (i.e., kinetic plus potential) of the particle. Thus show that the operator corresponding to the *total energy* of the system is

$$(\hat{H})_q = -\frac{\hbar^2}{2m} \nabla^2 + V(x,y,z) \tag{27}$$

This operator is called the *Hamiltonian operator* of the system.

2-31. Find the operator in coordinate language that corresponds to the angular momentum of the above particle about the z-axis.

Ans.:
$$(\hat{L}_z)_q = \frac{\hbar}{i}\left(x\frac{\partial}{\partial y} - y\frac{\partial}{\partial x}\right) \tag{28}$$

2-32. Show that the operator of Exercise 2-31 is also equal to $(\hbar/i)$ $(\partial/\partial\phi)$, where ϕ is the azimuthal angle of a system of polar coordinates (cylindrical or spherical) whose polar axis is the z-axis. *Hint:* It is easier to work this exercise in one direction than in the other.

2-33. Find the commutator operator of the x- and y-components of angular momentum. *Ans.:* $\hat{L}_x\hat{L}_y - \hat{L}_y\hat{L}_x = i\hbar\hat{L}_z$.

This discussion of operators, although incomplete, will suffice to enable us to proceed with the postulatory basis of quantum mechanics. Further properties of operators will be introduced from time to time as needed. Many of the ideas presented above will become clearer as operators are used in physical problems.

D. *The Quantum-mechanical Equation of Motion.* We have so far introduced postulates that describe the manner in which the state of a physical system is to be instantaneously specified, and we have just seen, in the preceding section, how actual numerical information concerning the values that will be found for observable dynamical quantities is to be obtained from the wave functions ψ or ϕ. However, there still remains the question of how the system *moves*, that is, how the wave functions ψ and ϕ that describe the system *change with time*. Our fourth postulate of quantum mechanics deals with this question:

Postulate 4:
The wave functions ψ and ϕ change with time according to the equations

$$\hat{H}_q\psi = -\frac{\hbar}{i}\frac{\partial\psi}{\partial t} \tag{29}$$

and

$$\hat{H}_p\phi = -\frac{\hbar}{i}\frac{\partial\phi}{\partial t} \tag{30}$$

Equations (29) and (30) are both called *Schroedinger's equation including the time.* Only one of these equations is really needed, of course, since the other would follow from the second postulate; they are both written explicitly to indicate clearly the complete symmetry between coordinates and momenta that exists throughout quantum mechanics, a symmetry that recalls to mind the corresponding symmetry in Hamilton's form of the motion of classical systems and that runs much deeper than is evident in our present sketchy treatment.

Another type of symmetry is evident in the above equations. We found previously that momenta are represented in coordinate language by operators of the form $p_x \rightarrow (\hbar/i)(\partial/\partial x)$. The above postulate may be regarded as extending this equivalence to the fourth component of the 4-vector momentum, $p_4 = iW/c \rightarrow (\hbar/i)(\partial/ic\,\partial t)$, where in the nonrelativistic theory we replace W by the kinetic plus potential energy of the system.

Although the Schroedinger equation constitutes the heart of quantum mechanics, we shall defer further comment upon its significance until we have seen some of its applications. Real understanding can come only with intimate working contact with it.

The above postulates provide a basis for a nonrelativistic quantum-mechanical treatment of systems of particles whose configuration can be specified in terms of coordinates and momenta in the classical sense. There are now known to exist certain observable quantities that have no classical analogue, however, and special assumptions are needed to introduce these quantities into the above framework. (Examples of such quantities are *electron spin* and the *parity* of wave functions.) Before introducing these further complications, we shall examine some of the consequences of the material already presented. Having introduced the postulates of quantum mechanics in a form in which the canonical symmetry between coordinates and momenta is explicitly exhibited, it is no longer necessary for us to deal with both the coordinate and momentum languages in each of our subsequent discussions. Inasmuch as the coordinate representation is of somewhat greater utility and is in wider use, we shall drop the momentum language with the brief comment that it finds application in many special problems that we shall not encounter and could be used equally as satisfactorily as the ψ-language in many problems where we shall use the more familiar coordinate form. In what follows we shall also drop the q subscript on operators.

2-4. Some Elementary Properties of Quantum Mechanics

Before proceeding with the application of quantum mechanics to specific physical problems, we shall first examine some of the simpler properties of this system of mechanics. It is well to begin by verifying that the above postulates actually do provide a theoretical basis for the various experimental facts outlined in Sec. 2-3. We shall now treat some of these points.

A. The Wave Properties of Particles. It has already been shown in connection with the first two postulates that the introduction of *probability amplitudes* for the coordinates and the momenta, with a Fourier-inverse relation between the two, provides a description of the wave properties of particles that satisfies the de Broglie wavelength relation as well as the uncertainty principle. We now see that *it is the wave function ψ that describes the wave properties of a particle*, and the apparent paradox presented by the diffraction experiment of Fig. 2-9 is explained qualitatively as follows: The incident particles, being of homogeneous momentum, are described by a sinusoidal wave function $\psi_i = A \exp(ip_0 z/\hbar)$. Most of the particles are stopped by the potential barrier (wall), but a few pass through the slits and are described in the space between the slits and the scintillating screen by a wave function that is the sum of two coherent cylindrical waves expanding from the two slits, just as for optical diffraction. It is the interference between these two *coherent waves* that gives rise to the observed diffraction pattern, the *intensity* of particles arriving at a given point being proportional to the probability *density* $W_q(x)$ at that point. Any attempt to detect which of the two slits a given particle actually went through requires the introduction of some new interaction into the problem, with a resultant loss of coherence of the wave functions and a corresponding change in the observed result of the experiment.

B. The Uncertainty Principle. It has already been stated several times that the postulates presented above give a satisfactory account of the principle of uncertainty, and this has been specifically illustrated in several of the exercises. It can be shown quite generally, using these postulates, that a minimum uncertainty product $\Delta F \, \Delta G$ exists for any two dynamical quantities $F(q_j, p_j)$ and $G(q_j, p_j)$ whose operators $\hat{F}$ and $\hat{G}$ do not commute with one another. In any measurement of F and G, uncertainties will exist in the measured values such that $\Delta F \, \Delta G$ will exceed, and at best can be equal to, one-half the absolute average value of the commutator of $\hat{F}$ and $\hat{G}$. That is,

$$\Delta F \, \Delta G \geq \tfrac{1}{2} |\int \psi^* (\hat{F}\hat{G} - \hat{G}\hat{F}) \psi \, d^N q| \tag{1}$$

It may be mentioned as an illustration of this result that the one-dimensional wave function

$$\psi(x) = \frac{1}{a^{1/2} \pi^{1/4}} \exp\left[-\frac{(x - x_0)^2}{2a^2} + \frac{i p_0 x}{\hbar} \right]$$

which was treated in Exercise 2-18, exhibits the *minimum possible uncertainty product* $\Delta x \, \Delta p$. This wave function may therefore be regarded as corresponding as closely as possible to the classical statement that a one-dimensional particle is located at coordinate x_0 and is moving with momentum p_0.

EXERCISE

2-34. Use the above relation to find the minimum uncertainty product of x and y, y and p_z, x and p_x, and p_x and p_y.

C. The Correspondence Principle. One of the most important facts to establish for the system of quantum mechanics described by our postulates is that classical mechanics is valid as an approximation for cases in which the finite size of h can be ignored. This can be proved quite generally,[1] but we shall consider only a simple case, in the interest of brevity. Consider a particle of mass m moving under a system of

[1] Ehrenfest, *Z. Physik*, **45**, 455 (1927).

forces described by the Hamiltonian function

$$H = \frac{p^2}{2m} + V(x,y,z)$$

where

$$p^2 = p_x{}^2 + p_y{}^2 + p_z{}^2 \tag{2}$$

Let us calculate the time rate of change of the expectation value of one of the coordinates, say, x. We have for this latter quantity

$$\langle x \rangle = \int \psi^*(x,y,z,t) x \psi(x,y,z,t) \; dx \; dy \; dz$$

so that

$$\frac{d\langle x \rangle}{dt} = \int \left[\frac{\partial \psi^*}{\partial t} x \psi + \psi^* x \left(\frac{\partial \psi}{\partial t} \right) \right] dx \; dy \; dz$$

$$= \int \left[\frac{i}{\hbar} (\hat{H}\psi)^* x \psi + \psi^* x \left(-\frac{i}{\hbar} \right) \hat{H}\psi \right] dx \; dy \; dz$$

$$= \int \psi^* \frac{i}{\hbar} (\hat{H} x - x \hat{H}) \psi \; dx \; dy \; dz \tag{3}$$

Thus the time rate of change of $\langle x \rangle$, i.e., the *x-velocity*, is equal to the expectation value of $i/\hbar$ times the commutator of the x-coordinate and the Hamiltonian operators.

We must now compute the commutator of x and $\hat{H}$:

$$\hat{H} = -\frac{\hbar^2}{2m} \left(\frac{\partial^2}{\partial x^2} + \frac{\partial^2}{\partial y^2} + \frac{\partial^2}{\partial z^2} \right) + V(x,y,z)$$

so that

$$\hat{H} x - x \hat{H} = -\frac{\hbar^2}{2m} \left(\frac{\partial^2}{\partial x^2} x - x \frac{\partial^2}{\partial x^2} \right)$$

$$= -2 \frac{\hbar^2}{2m} \frac{\partial}{\partial x}$$

$$= -\frac{\hbar^2}{m} \frac{i}{\hbar} \hat{p}_x$$

Thus

$$\frac{d\langle x \rangle}{dt} = \frac{\langle p_x \rangle}{m} \tag{4}$$

That is, the *rate of change of the expectation value of x is equal to the expectation value of the momentum divided by the mass.*

EXERCISES

2-35. Supply the missing steps in the derivation of Eqs. (3) and (4).

2-36. Show that, for any dynamical quantity $F(q_j, p_j)$,

$$\frac{d\langle F \rangle}{dt} = \int \psi^* \frac{i}{\hbar} (\hat{H}\hat{F} - \hat{F}\hat{H}) \psi \; d^N q \tag{5}$$

On the other hand, we obtain for the rate of change of the x-component of momentum,

$$\frac{d\langle p_x \rangle}{dt} = \int \psi^* \frac{i}{\hbar} (\hat{H}\hat{p}_x - \hat{p}_x\hat{H})\psi \, dx \, dy \, dz$$

$$= \int \psi^* \left(\hat{H} \frac{\partial}{\partial x} - \frac{\partial}{\partial x} \hat{H} \right) \psi \, dx \, dy \, dz$$

$$= \int \psi^* \left(-\frac{\partial V}{\partial x} \right) \psi \, dx \, dy \, dz$$

$$= \left\langle -\frac{\partial V}{\partial x} \right\rangle \tag{6}$$

That is, *the rate of change of the expectation value of the momentum is equal to the negative of the expectation value of the gradient of the potential energy.*

The above derived relationships will be recognized as being the ordinary classical equations of motion of this system in Hamiltonian form, if $\langle x \rangle$ is taken to represent the classical x-coordinate of the particle, $\langle p_x \rangle$ its classical x-momentum, and $\langle -\partial V/\partial x \rangle$ the classical x-force. That is, the above equations become just the classical equations of motion for any situation where the *uncertainty* in the values of the various quantities can be neglected, namely, where the finite size of h can be neglected.

EXERCISES

2-37. Supply the missing steps in the derivation of Eq. (6).

2-38. The smallest particle that can be seen in an ordinary microscope is about 0.1 μ in diameter. Assuming a specific gravity of 2.0, calculate the minimum uncertainty in the velocity of such a particle if the uncertainty in its position is 1 per cent of its diameter.

D. Quantized States. We now investigate the manner in which quantum mechanics describes the existence of "quantized states" in which certain quantities such as energy and angular momentum have discrete, precisely defined values. We have seen that it is a characteristic feature of quantum mechanics that dynamical quantities do not in general have precisely defined values in a system in a given state, but that there is usually a distribution, or "spread," of values that will be found as a result of measurement. This does not mean, however, that a system *cannot be in* a state in which a given quantity has a precisely defined value, but only that this is not the *most general* situation.

There are actually two questions that must be considered: (1) What kind of wave function corresponds to a situation in which some given dynamical quantity has a precisely defined value? (2) If a system is, at some initial time, known to be in a state described by one of these wave functions, what happens to the system as time proceeds? As we shall see, a special significance attaches to those states in which certain dynamical quantities have precisely defined values that *do not change with time.*

Taking now the first question, it is easy to demonstrate that a wave function $\psi(q_j)$ that satisfies the equation

$$\hat{F}\psi = F_0\psi \tag{7}$$

where F_0 is a real constant, is one for which the dynamical quantity $F(q_j, p_j)$ has the precise value F_0. To show this, we need only evaluate $\langle F \rangle$ and $\langle F^2 \rangle$:

$$
\begin{aligned}
\langle F \rangle &= \int \psi^* \hat{F}\psi \, d^N q \\
&= \int \psi^* F_0 \psi \, d^N q \\
&= F_0 \int \psi^* \psi \, d^N q \\
&= F_0
\end{aligned}
\tag{8}
$$

and

$$
\begin{aligned}
\langle F^2 \rangle &= \int \psi^* \hat{F}(\hat{F}\psi) \, d^N q \\
&= \int \psi^* \hat{F}(F_0 \psi) \, d^N q \\
&= F_0 \int \psi^* \hat{F}\psi \, d^N q \\
&= F_0^2
\end{aligned}
\tag{9}
$$

Thus

$$(\Delta F)^2 = \langle F^2 \rangle - \langle F \rangle^2 = F_0^2 - F_0^2 = 0 \tag{10}$$

so that there is *no uncertainty in the value of F*; that is, F has the precise value F_0.

Equation (7) is called an *eigenvalue equation;* the number F_0 is called an *eigenvalue* of the dynamical quantity $F(q_j, p_j)$, and the function $\psi(q_j)$ that satisfies the equation is called an *eigenfunction corresponding to the eigenvalue F_0.* A given number F_0 is a *permissible eigenvalue* for the dynamical quantity $F(q_j, p_j)$ if a wave function $\psi(q_j)$ can be found which:

1. Satisfies the eigenvalue equation (7) for this value of F_0
2. Is a single-valued function of spatial position
3. Vanishes at infinity
4. Can be normalized to unit total probability according to Eq. 2-3(6)

EXERCISE

2-39. Justify each step in the derivations of Eqs. (8) and (9).

To illustrate some of the features of eigenvalue equations, let us consider first the problem of finding the eigenvalues and the corresponding eigenfunctions for a coordinate q_s and a momentum p_s. Taking first the coordinate, the eigenvalue equation is, in coordinate language,

$$q_s \psi(q_j) = q_0 \psi(q_j) \qquad \text{or} \qquad (q_s - q_0)\psi(q_j) = 0$$

To find for what values of q_0 this equation has a solution, consider the dependence of $\psi(q_j)$ upon q_s for a given value of q_0. We see that, if

$q_s \neq q_0$, then $\psi(q_j)$ must equal 0; but if $q_s = q_0$, then $\psi(q_j)$ can have *any* value, subject only to the condition that $\int \psi^* \psi \, d^N q = 1$. Actually, at $q_s = q_0$, $\psi(q_j)$ must *approach infinity*, if the integral of its absolute square is to be finite. Since this argument can be followed for any finite value of q_0, it is clear that for this case a *continuous spectrum* of eigenvalues is obtained; that is, *the coordinate q_s can have any finite value q_0.*

Actually, the function $\psi(q_j)$ defined by the above equation is an improper function. In an actual situation, the coordinate q_s could not be measured with infinite precision, and so the wave function defined by an actual measurement would not really be singular. But since the present formulation of quantum mechanics represents a limit toward which actual measurements must converge as the precision of measurement becomes greater and greater, this singular function must be admitted as a possible eigenfunction. Its absolute square is called a *delta function* and is symbolized $\delta(q_s - q_0)$. This latter function is defined by the equations

$$\delta(q_s - q_0) = 0 \qquad q_s \neq q_0 \tag{11}$$

and
$$\int_{-\infty}^{\infty} \delta(q_s - q_0) \, dq_s = 1 \tag{12}$$

The probability density of Exercise 2-18 becomes a delta function in the limit $a \to 0$.

It is easy to see that the delta function actually does represent a situation for which the value q_0 is *sure to be obtained* for the coordinate q_s since, by the definition of the meaning of $W_q(q_j)$, there is zero probability of obtaining any value of q_s *not equal* to q_0, and an infinite probability density for obtaining a value very near q_0, with a unit total probability of obtaining *some* value of q_s.

Consider next the eigenvalue equation for one of the momenta, say, p_s. The eigenvalue equation for this quantity is, in coordinate language,

$$\frac{\hbar}{i} \frac{\partial}{\partial q_s} \psi(q_j) = p_0 \psi(q_j) \qquad \text{or} \qquad \frac{\partial \psi}{\partial q_s} = \frac{i p_0}{\hbar} \psi$$

This equation may easily be integrated to obtain

$$\psi(q_j) = F(q_{j \neq s}) \exp (i p_0 q_s / \hbar) \tag{13}$$

where $F(q_{j \neq s})$ is a function which does not depend upon q_s. We thus see that any value of p_0 is a possible eigenvalue for the momentum p_s (provided that q_s is single-valued from $-\infty$ to $+\infty$) and that the corresponding eigenfunction is a sinusoidal function of q_s with *"wavelength"* h/p_0. This will be recognized also as being a limiting case of the wave function of Exercise 2-18 ($a \to \infty$).

EXERCISES

2-40. Show that $\int_{-\infty}^{\infty} f(x)\,\delta(x - x_0)\,dx = f(x_0)$.

2-41. Treat the above two cases in momentum language; that is, solve the eigenvalue equations to obtain the eigenfunctions corresponding to the above coordinate-language eigenfunctions.

We shall now examine one of the simplest physical quantities that possesses a discrete spectrum of eigenvalues—the z-component of the angular momentum of a particle about a fixed origin. The operator for this quantity was evaluated in Exercises 2-31 and 2-32. Using the result of the latter exercise, we find the eigenvalue equation for L_z to be

$$\hat{L}_z\psi = \frac{\hbar}{i}\frac{\partial\psi}{\partial\varphi} = L_{z0}\psi \tag{14}$$

which may be integrated immediately to obtain

$$\psi(q_j) = u(q_{j\neq\varphi})\exp{(iL_{z0}\varphi/\hbar)} \tag{15}$$

Equation (15) is thus similar to Eq. (13), but we are now dealing with a coordinate φ which is *not single-valued from* $-\infty$ *to* $+\infty$. That is, a given point in space may be defined by a number of different values of φ. But we have postulated the wave function to be a single-valued function of position, which requires that $\psi(\varphi + 2\pi) = \psi(\varphi)$. This then in turn requires that $2\pi(L_{z0}/\hbar) = 2\pi m$, where m is any positive or negative integer or zero. We therefore conclude that *the z-component of angular momentum can have only the values*

$$L_{z0} = m\hbar \qquad m = 0,\ \pm 1,\ \pm 2,\ \pm 3,\ \ldots\ . \tag{16}$$

This result is seen to be in agreement with the postulates of the Bohr theory and with the Wilson-Sommerfeld quantization rules; but now this quantum condition arises in a more natural manner—indeed in the same way as do the normal modes of vibration of classical physical systems—*as the solution of a boundary-value problem*. It is this requirement that ψ satisfy *boundary conditions* that yields discrete eigenvalues in quantum mechanics. In Chap. 5 we shall consider the eigenvalues of angular momentum more extensively.

Next, we consider a situation that repeatedly arises in quantum mechanics, wherein a given eigenfunction of one physical quantity is also an eigenfunction of some other physical quantity. Let us suppose that an arbitrary wave function can be expressed as a linear combination of wave functions $\psi_s(q_j)$ all of which satisfy *both* of the eigenvalue equations

$$\hat{F}\psi_s = F_s\psi_s \qquad \text{and} \qquad \hat{G}\psi_s = G_s\psi_s$$

for two dynamical quantities $F(q_j,p_j)$ and $G(q_j,p_j)$. We must therefore also be able to write

$$\hat{F}\hat{G}\psi_s = \hat{F}(G_s\psi_s) = G_s\hat{F}\psi_s = G_sF_s\psi_s$$

and

$$\hat{G}\hat{F}\psi_s = \hat{G}(F_s\psi_s) = F_s\hat{G}\psi_s = F_sG_s\psi_s$$

But $F_sG_s = G_sF_s$, since F_s and G_s are ordinary numbers that multiply ψ_s. Therefore we must also have

$$\hat{F}\hat{G}\psi_s = \hat{G}\hat{F}\psi_s \qquad \text{or} \qquad (\hat{F}\hat{G} - \hat{G}\hat{F})\psi_s = 0 \tag{17}$$

That is, *the operators $\hat{F}$ and $\hat{G}$ must commute*. We thus arrive at the important result that two quantities can simultaneously be precisely defined *only if their operators commute*. (It can also be shown that this condition is sufficient as well as necessary.) For example, any number of the coordinates, or any number of the momenta, or any combination of coordinates and nonconjugate momenta can simultaneously be precisely known; but a coordinate and its conjugate momentum cannot, since their operators do not commute.

In the above paragraphs some of the properties of eigenfunctions have been presented and examples have been given of the solution of the eigenvalue equation for certain important dynamical quantities. We now consider the second of the two questions that were offered at the beginning of this section; namely, if a system is known to be, at a certain instant, in a state $\psi(q_j)$ such that some dynamical quantity $F(q_j,p_j)$ is then precisely defined, how does the state $\psi(q_j)$ subsequently change with time? To obtain an answer to this question for a specific case, one would have to integrate the Schroedinger equation, but certain important information can be obtained by considering how the *expectation value* of $F(q_j,p_j)$ changes with time.

Equation (5) shows that $\langle F \rangle$ will generally change with time unless the operator $\hat{F}$ commutes with the Hamiltonian operator $\hat{H}$. Similarly, $\langle F^2 \rangle$ will generally change with time unless $\hat{F}$ and $\hat{H}$ commute, and the rate of change of $\langle F^2 \rangle$ and of $\langle F \rangle$ will generally be such as to yield a nonzero value for $d(\Delta F)^2/dt$, so that the uncertainty in F will *increase* with time; that is, F *will not remain precisely defined in value*. On the other hand, if $\hat{F}$ and $\hat{H}$ commute, then $\langle F \rangle$ remains equal to F_0, and $\langle F^2 \rangle$ to $F_0{}^2$, so that F *remains precisely defined* at the value F_0. The system is then said to be in a *stationary state*.

One of the most important dynamical quantities whose eigenfunctions represent stationary states is the *energy*, whose corresponding operator is the Hamiltonian operator itself. For this quantity the actual time dependence of ψ can be found explicitly. The eigenvalue equation for the total energy is

$$\hat{H}\psi(q_j,t) = E_0\psi(q_j,t) \tag{18}$$

From the time-dependent Schroedinger equation 2-3(29) we may further write, however,

$$\hat{H}\psi = -\frac{\hbar}{i}\frac{\partial\psi}{\partial t} = E_0\psi \tag{19}$$

so that the second equality may be solved explicitly to yield

$$\psi(q_j,t) = u(q_j)\exp\left[-(iE_0/\hbar)t\right] \tag{20}$$

Using the form (20) for ψ and the form of the Hamiltonian operator that corresponds to a single particle moving in a potential $V(q_j)$, Schroedinger's equation may now be written

$$-\frac{\hbar^2}{2m}\nabla^2 u + (V - E_0)u = 0 \tag{21}$$

This is called the *time-independent Schroedinger equation* for a particle. Many examples of this form of Schroedinger's equation will be found in our later work.

EXERCISES

2-42. Show that, if $\hat{F}$ and $\hat{H}$ commute, both $d\langle F\rangle/dt$ and $d\langle F^2\rangle/dt$ are zero, so that $F(q_j,p_j)$ remains precisely defined in value.

2-43. Integrate Eq. (19) to obtain Eq. (20).

2-44. Verify that Eq. (21) is the equation that must be satisfied by the function $u(q_j)$ for a single particle.

E. Conservation of Probability, Probability Current. In the first postulate of quantum mechanics it was stated that the total integrated probability $\Pi = \int\psi^*\psi\,d^N q$ must be equal to unity. On the other hand the fourth postulate established the manner in which ψ changes with time at a given location. It remains to be demonstrated that the changes in ψ which follow in accordance with the Schroedinger equation are in fact consistent with the requirement that the *total probability must not change with time.*

Let us evaluate the time derivative of the integrated probability:

$$\frac{\partial\Pi}{\partial t} = \frac{\partial}{\partial t}\int\psi^*\psi\,d^N q$$
$$= \int\left(\frac{\partial\psi^*}{\partial t}\psi + \psi^*\frac{\partial\psi}{\partial t}\right)d^N q \tag{22}$$

The partial derivatives in the integrand may be eliminated through the use of the Schroedinger equation, with the result

$$\frac{\partial\Pi}{\partial t} = \int\frac{i}{\hbar}\left[(\hat{H}^*\psi^*)\psi - \psi^*(\hat{H}\psi)\right]d^N q$$

But because the Hamiltonian operator is required to be Hermitian, the integrals of the two terms exactly cancel one another, so that there remains

$$\frac{\partial \Pi}{\partial t} = 0 \tag{23}$$

This result illustrates once again the need for requiring quantum-mechanical operators to be Hermitian.

We have thus established that the total probability remains constant in time. But it is also possible to go somewhat deeper into the matter and establish an equation of continuity for the probability density, in analogy with the classical continuity equations for electric charge and fluid flow. As a simple illustration we may consider the probability density for a single particle whose Hamiltonian operator is

$$\hat{H} = -\frac{\hbar^2}{2m} \nabla^2 + V(x,y,z) \tag{24}$$

Following the same procedure as for $\partial \Pi / \partial t$, we may evaluate $\partial W / \partial t$:

$$\frac{\partial W}{\partial t} = \frac{i}{\hbar} [(\hat{H}^* \psi^*)\psi - \psi^*(\hat{H}\psi)] \tag{25}$$

But using the explicit form (24) for the Hamiltonian operator, this is equivalent to

$$\frac{\partial W}{\partial t} + \frac{\hbar}{2mi} (\psi^* \nabla^2 \psi - \psi \nabla^2 \psi^*) = 0$$

or, using a familiar vector identity, to

$$\frac{\partial W}{\partial t} + \nabla \cdot \frac{\hbar}{2mi} (\psi^* \nabla \psi - \psi \nabla \psi^*) = 0 \tag{26}$$

The quantity $S = \frac{1}{2}(\hbar/mi)(\psi^* \nabla \psi - \psi \nabla \psi^*)$ may be regarded as a vector *probability current* whose integral over a closed surface is equal to the rate of change of the probability that the particle will be found inside this surface. Notice that $(\hbar/mi)\nabla$ is just the operator for the *velocity* of the particle. The equation of continuity may thus be written in the familiar form

$$\frac{\partial W}{\partial t} + \nabla \cdot S = 0 \tag{27}$$

EXERCISE

2-45. Supply the missing steps in the derivation of Eq. (26).

2-5. Some Useful General Theorems

The discussion of the preceding section has served to demonstrate the ability of quantum mechanics to deal with the apparently contradictory requirements set out in Sec. 2-2, and it has also illustrated some of the simpler operations through which one goes in order to solve certain problems using quantum mechanics.   In the present brief introduction to the subject, it is of course impossible to describe the full apparatus by which one can attack the most general problems; the treatment is, on the contrary, limited mainly to a presentation of the minimum essentials that are necessary to treat the particular problems that arise in our study of modern physics.   With this idea in mind, the following general theorems are now presented.   They will be useful at one stage or another in our later work.

A. Orthogonality of Eigenfunctions.   One of the most important properties possessed by the eigenfunctions that correspond to the various eigenvalues of any given dynamical quantity is that *these eigenfunctions form a complete orthogonal set of functions.*   This means that any wave function, even if it does not correspond to a precise value for this dynamical quantity, can be expressed as a linear combination of these orthogonal functions.   The orthogonality of the various functions is demonstrated as follows: Let the functions ψ_n and ψ_m be two eigenfunctions of a dynamical quantity $F(q_j, p_j)$ corresponding to the eigenvalues F_n and F_m:

(a)
$$\hat{F}\psi_n = F_n\psi_n$$
(b)
$$\hat{F}\psi_m = F_m\psi_m \tag{1}$$

Now multiply Eq. (1a) by ψ_m^* and (Eq. (1b) by ψ_n^*, subtract the complex conjugate of the resulting (b) equation from the (a) equation, and integrate the result over all coordinates:

$$\int [\psi_m^*\hat{F}\psi_n - \psi_n(\hat{F}\psi_m)^*]\, d^N q = (F_n - F_m^*)\int \psi_m^*\psi_n\, d^N q$$

By the Hermitian property of the operator $\hat{F}$, we find that

$$\int [\psi_m^*\hat{F}\psi_n - \psi_n(\hat{F}\psi_m)^*]\, d^N q = 0$$
so that
$$(F_m^* - F_n)\int \psi_m^*\psi_n\, d^N q = 0 \tag{2}$$

Thus if $m \neq n$, either $F_m^* = F_n$ or $\int \psi_m^*\psi_n\, d^N q = 0$.   The first case, in which $F_m^* = F_n$, is called *degeneracy;* it will be discussed later.   The more general case, in which $F_m^* \neq F_n$, then requires that the functions ψ_m and ψ_n be *orthogonal* over the range of the coordinate q_j.   On the other hand, if $m = n$, then $F_m^* = F_m$, since the integrand is then everywhere positive.   This demonstrates once again that the eigenvalues of an observable dynamical quantity are *real*, which is one of the reasons for postulating that quantum-mechanical operators must be *Hermitian*.

We can now establish some important properties possessed by an expansion of any given wave function in terms of the orthogonal eigenfunctions corresponding to a dynamical quantity $F(q_j,p_j)$. Suppose that the given wave function is represented at a certain time by the expansion

$$\psi(q_j) = \sum_{s=0}^{\infty} a_s \psi_s(q_j) \tag{3}$$

where $\psi_s(q_j)$ is the (normalized) eigenfunction corresponding to the eigenvalue F_s and a_s is the coefficient of this term at the given instant. Assume that there is no degeneracy, so that the eigenvalues of F are all distinct. Then we must have for the total probability

$$\int \psi^* \psi \, d^N q = \sum_{s=0}^{\infty} \sum_{r=0}^{\infty} a_s^* a_r \int \psi_s^* \psi_r \, d^N q$$

$$= \sum_{s=0}^{\infty} a_s^* a_s = 1 \tag{4}$$

Thus *the sum of the absolute squares of the coefficients in the expansion must equal unity at all times.*

Now let us inquire into the possible results of a measurement of $F(q_j,p_j)$ upon a system that is in the state $\psi(q_j,t)$ described by Eq. (3). If we evaluate the expectation value of various powers of F, these are given by

$$\langle F^n \rangle = \int \psi^* \hat{F}^n \psi \, d^N q$$

$$= \sum_{s=0}^{\infty} \sum_{r=0}^{\infty} a_s^* a_r \int \psi_s^* \hat{F}^n \psi_r \, d^N q$$

$$= \sum_{s=0}^{\infty} F_s^n a_s^* a_s \tag{5}$$

Thus *the expectation value of F^n is expressible as a sum over the various possible eigenvalues F_s^n that can be found for F^n, each multiplied by a probability $a_s^* a_s$ that a measurement of the quantity F^n will yield that eigenvalue.*

This result carries the important and fundamental implication that a *precise measurement of any dynamical quantity $F(q_j,p_j)$ can yield only one of the possible eigenvalues of F even if it is carried out upon a system that is not in an eigenstate of F.* Successive measurements of this quantity may very well yield *different ones* of the possible eigenvalues, but *no measurement can yield a value other than an eigenvalue.*[1]

---

[1] This interpretation of Eq. (5) is based upon an important result of statistics: that the totality of moments of a distribution function uniquely determine that function.

EXERCISES

2-46. Supply the missing steps in the derivation of Eq. (2).

2-47. Supply the missing steps leading to Eq. (4).

2-48. Supply the missing steps leading to Eq. (5).

B. Degenerate Eigenstates. It often happens in problems of practical interest that more than one eigenfunction ψ_s exists for a given eigenvalue F_s. In this case the integral (2) is not necessarily zero, and the eigenfunctions corresponding to this eigenvalue are therefore not necessarily orthogonal to one another. A state for which the above situation exists is called a *degenerate eigenstate,* and the number of linearly independent eigenfunctions that exist for the given eigenvalue is called the *degree of degeneracy* of the state. For example, if a certain eigenvalue F_s possesses three different (i.e., linearly independent) eigenfunctions, the state is said to be *three-fold degenerate.*

Although the independent eigenfunctions of a degenerate eigenstate are not necessarily orthogonal to one another, it is always possible to construct an equal number of eigenfunctions which *are* orthogonal to one another, using suitable linear combinations of the nonorthogonal eigenfunctions. This construction of an orthogonal set of eigenfunctions can actually be carried out in an infinite number of ways. Particular ones of these various possible sets of orthogonal degenerate eigenfunctions will often be found to have a special significance under certain conditions. For example, a degenerate eigenstate can often be slightly modified by the application of an additional interaction to the problem (as in the Zeeman effect) which results in a slight *separation of the various eigenvalues;* that is, the degeneracy can be *removed* under certain circumstances. It is then found that the wave functions that correspond to the new, distinct set of eigenvalues are a *particular* linear combination of the original (nonorthogonal) eigenfunctions. This particular set of orthogonal eigenfunctions can thus be regarded as having a special significance, even in the absence of the perturbing interaction.

Degeneracy also plays an important role in quantum statistical mechanics; for in statistical treatments each individual eigenstate is regarded as having the same importance. Thus a degenerate eigenstate must be recognized as being actually a number of distinguishable eigenstates, so that a *"statistical weight"* equal to the *degree of degeneracy* must be assigned to it.

The above discussion can be illustrated by a simple classical analogy that is mathematically identical with the quantum-mechanical case. Consider a circular vibrating membrane of uniform surface density and isotropic tension. The shapes of the various modes of vibration of this membrane correspond to the eigenfunctions, and the frequencies of these

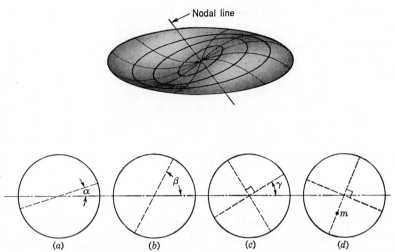

FIG. 2-13. A degenerate mode of vibration of a circular membrane. (*a*), (*b*) The nodal lines of two linearly independent but nonorthogonal modes of vibration. (*c*) The nodal lines of one of the infinitely many pairs of *orthogonal* independent modes. (Equally suitable modes can be found for any angle γ.) (*d*) A small mass *m* attached to the membrane removes the degeneracy of this family of modes and requires that the nodal lines of the orthogonal modes used to describe the perturbed system be chosen as shown.

modes to the eigenvalues, of the quantum-mechanical quantity under consideration. Some of the modes are circularly symmetric, and these modes are nondegenerate (except for a remote possibility that the roots of two Bessel functions might be related in just the right way to make the frequencies of two unrelated modes coincide). Other modes, however, do not show circular symmetry. These modes depend upon the polar angle according to the functions $\sin(m\varphi + \delta)$, where *m* is any positive integer. Each of these modes is twofold degenerate with respect to the frequency. As a simple illustration we shall often refer to the modes of lowest frequency for $m = 1$. Each of these two modes possesses a diametral nodal line, with the membrane displaced "upward" on one side of this line and "downward" on the other side, as indicated in Fig. 2-13.

The displacement of the membrane at a given radius varies as $\sin(\varphi - \alpha)$ (Fig. 2-13*a*). A linearly independent mode whose nodal line is oriented at any other angle β can then be taken to represent the second mode (Fig. 2-13*b*). No further linearly independent eigenfunction can be found for this case. However, these two modes, although linearly independent, are not *orthogonal* to one another unless $\beta - \alpha$ is an odd integer times $\pi/2$, as shown in Fig. 2-13*c*. There are clearly an infinite number of choices for the orientation of such mutually orthogonal

modes.   But if a small mass is attached to the membrane at some point
(other than the center), one is no longer free to choose the angular orientation of the two modes arbitrarily, because the mass *removes the degener-*
acy of this mode.   Clearly, any mode whose nodal line passes through the
point of attachment of the mass will be unaffected, and all other modes
will suffer a slight reduction in frequency.   This requires that the orientation shown in Fig. 2-13d for the mutually orthogonal modes be chosen
to describe the system when the perturbing effect of the mass is to be
taken into account.   We shall repeatedly meet with this situation when
we consider the effect of perturbations upon atomic systems.

EXERCISE

2-49. Show that the eigenfunction for any choice of orientation of the
nodal line can be expressed as a linear combination of $\sin (\varphi - \alpha)$ and
$\sin (\varphi - \beta)$, if $(\alpha \neq \beta)$.

C. Energy Levels of a Perturbed System.   In treating a complicated
quantum-mechanical system, it is often found that the Hamiltonian
operator can be expressed as a sum of several parts, one of which is very
much "larger" than the rest in the sense that the behavior of the system
is almost entirely governed by it.   In finding the energy levels and wave
functions for such a system, it is seldom possible to treat the entire system
exactly; but it is often possible to treat the *major part* of the system, ignoring the smaller terms, and then to evaluate the small changes in the energy
levels and in the wave functions brought about by introducing the
remaining terms of the Hamiltonian function as *small perturbations.*   To
illustrate a procedure that is commonly used in such a treatment, let us
suppose that the complete Hamiltonian operator can be expressed in the
form

$$\hat{H} = \hat{H}_0 + \hat{H}_1 \qquad (6)$$

and that the energy eigenvalues and eigenfunctions are known for the
system whose Hamiltonian operator is $\hat{H}_0$.   Let us then consider a
system that can be varied continuously from the unperturbed system
whose Hamiltonian is $\hat{H}_0$ to the actual system whose Hamiltonian is
$\hat{H}_0 + \hat{H}_1$.   Thus we shall let

$$\hat{H} = \hat{H}_0 + \alpha \hat{H}_1 \qquad (7)$$

where α is a parameter that can be varied at will from zero to unity.
Thus we may expect that the energy E_n and the wave function ψ_n of
the actual system will be expressible in the form

(a)
(b)

$$E_n = E_{n0} + \alpha E_{n1} + \alpha^2 E_{n2} + \cdots .$$
$$\psi_n = \psi_{n0} + \alpha \psi_{n1} + \alpha^2 \psi_{n2} + \cdots .$$

(8)

where E_{n0} and ψ_{n0} are the energy and wave functions that correspond to the unperturbed system (i.e., with $\alpha = 0$). E_{n1} and ψ_{n1} are called the *first-order* corrections to the energy and the wave functions; E_{n2} and ψ_{n2}, the *second-order* corrections, etc. If the various energy corrections E_{n1}, E_{n2}, and so on diminish sufficiently rapidly with increasing order, the above expressions will converge for $\alpha = 1$ and will correctly describe the perturbed system.

These corrections can be evaluated in a stepwise manner. We shall concern ourselves with only the first-order correction to the *energy*, since this quantity is of the greatest interest to us in our later work. The general procedure is to apply the Hamiltonian operator (7) to the wave function (8b), and thus to write Schroedinger's time-independent wave equation in the form

$$(\hat{H} - E_n)\psi = 0$$
$$= (\hat{H}_0 + \alpha\hat{H}_1 - E_{n0} - \alpha E_{n1} + \cdots)(\psi_{n0} + \alpha\psi_{n1} + \cdots)$$

or, collecting the coefficients of the various powers of α,

$$(\hat{H}_0 - E_{n0})\psi_{n0} + \alpha[(\hat{H}_1 - E_{n1})\psi_{n0} + (\hat{H}_0 - E_{n0})\psi_{n1}]$$
$$+ \alpha^2[\quad] + \cdots = 0$$

Now, since the above equation must be true for all values of α, the coefficient of each power of α must vanish separately. Thus we find

$$(\hat{H}_0 - E_{n0})\psi_{n0} = 0$$
$$(\hat{H}_1 - E_{n1})\psi_{n0} + (\hat{H}_0 - E_{n0})\psi_{n1} = 0 \qquad \text{etc.} \qquad (9)$$

The first equation involves only the zero-order energy and wave function, the second equation only the zero-order quantities and the first-order corrections, and so on, each equation in turn bringing in the next higher corrections. We assume that the zero-order energies and wave functions are exactly known. Then to find the first-order correction to the energy, we multiply the second equation by ψ_{n0}^* and integrate over all coordinates:

$$\int(\psi_{n0}^*\hat{H}_1\psi_{n0} - E_{n1}\psi_{n0}^*\psi_{n0} + \psi_{n0}^*\hat{H}_0\psi_{n1} - E_{n0}\psi_{n0}^*\psi_{n1})\, d^N q = 0 \qquad (10)$$

But, because of the Hermitian property of quantum-mechanical operators, we have for the third term in the above expression,

$$\int\psi_{n0}^*\hat{H}_0\psi_{n1}\, d^N q = \int\psi_{n1}(\hat{H}_0\psi_{n0})^*\, d^N q$$
$$= E_{n0}\int\psi_{n1}\psi_{n0}^*\, d^N q$$

so that this term cancels the last term. Thus, since ψ_{n0} is normalized, we obtain

$$E_{n1} = \int\psi_{n0}^*\hat{H}_1\psi_{n0}\, d^N q = \langle H_1 \rangle \qquad (11)$$

so that *the first-order correction to the energy is equal to the average value of the perturbation energy in the unperturbed state* ψ_{n0}. The first-order correction to the wave function can be found by expanding ψ_{n1} in terms of the complete orthogonal set of unperturbed wave functions ψ_{n0} and evaluating the coefficients in a straightforward way. We shall not need these corrections, however, so this procedure will not be carried out here.

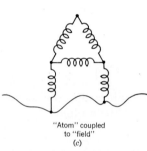

"Atom"
(a)

"Field"
(b)

"Atom" coupled
to "field"
(c)

Fig. 2-14. Schematic representation of a classical system analogous to an atom coupled to the electromagnetic field. (a) Oscillatory system having discrete modes of vibration. This is analogous to an atom. (b) An infinite string having a continuous distribution of normal modes. This is analogous to the electromagnetic field. (c) The two systems coupled by means of weak springs. This coupling is analogous to that of the charge and magnetic moments of the particle with the electromagnetic field.

Two comments must be made regarding the application of the above result to an actual problem. First, concerning the convergence of the series (8) one usually finds that the first-order correction accurately represents the effect of the perturbation if the energy shift E_{n1} is small compared with the spacing between neighboring energy levels in the vicinity of E_n. And second, when it is applied to an energy state that is degenerate, the zero-order wave functions ψ_{n0} must be chosen in a particular way, as described in Sec. 2-5B. Procedures that can be followed to find what set of degenerate eigenfunctions is appropriate for a given problem are described in textbooks on quantum mechanics.

D. Transitions between Stationary Eigenstates. As a final topic in our tabulation of important general theorems in quantum mechanics, we now consider the problem of how a system "jumps" from one energy level to another. It was the experimental observation of radiation emitted during such transitions, it will be recalled, that led to the concept of stationary states in the original Bohr theory.

We have seen that the eigenvalue equation for the total energy of a mechanical system, $\hat{H}\psi = E_0\psi = -(\hbar/i)(\partial\psi/\partial t)$, yields solutions of the form 2-4(20), $\psi(q_j,t) = u(q_j)$ exp $(-iE_0t/\hbar)$, so that, if a system is known to be in an energy eigenstate, this state will be preserved in time; that is, the system will not spontaneously jump from one state to another. This result is obtained because of the neglect of some small interaction within the idealized system which slightly modifies the various energy states in a manner similar to that treated in the previous paragraphs.

For instance, in considering the radiation emitted or absorbed by atoms, one would first treat the atom and the electromagnetic field separately and then introduce as a small perturbation the actual interaction between the two. The Hamiltonian function, thus modified, still represents the total energy of the system, and the system is still conservative when considered as a whole. However, the energy eigenstates are now modified in such a way that energy may be exchanged between the two component parts of the system through the perturbation interaction. This effect is familiar classically in the case of weakly coupled oscillatory systems—and indeed similar mathematical procedures can be used to analyze the classical and the quantum-mechanical cases. A schematic illustration of a classical system which is analogous to the important quantum-mechanical case of an atom interacting with the electromagnetic field is shown in Fig. 2-14.

A procedure that can be used in the perturbation treatment of transitions is as follows: First, the idealized, two-separate-component approximation to the actual system is treated. The Hamiltonian function for this unperturbed system is designated by H_0, and the Hamiltonian operator is $\hat{H}_0$. Now H_0 is composed of two parts, H_1 and H_2, corresponding to the two separate components of the actual system. Thus it may be

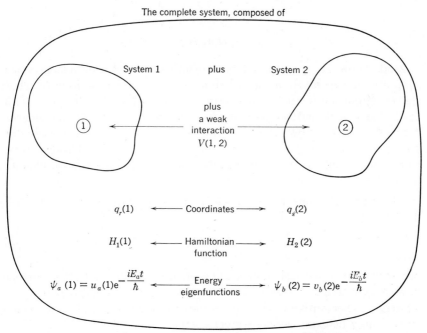

FIG. 2-15. Illustrating schematically one manner of treating time transitions.

written (see Fig. 2-15)

$$H_0 = H_1(1) + H_2(2) \tag{12}$$

and corresponding to this,

$$\hat{H}_0 = \hat{H}_1(1) + \hat{H}_2(2) \tag{13}$$

where it is understood that H_1, and therefore $\hat{H}_1$, is a function of certain coordinates [symbolized by the (1)] and similarly for H_2 and $\hat{H}_2$, *but none of the coordinates upon which H_1 depends appear in H_2, and vice versa.* This is of course what is meant by the statement that the two systems are independent. In such a case the Schroedinger equation for the energy eigenvalues of the system is separable:

$$(\hat{H}_1 + \hat{H}_2)\psi_a(1)\psi_b(2) = (E_a + E_b)\psi_a(1)\psi_b(2)$$

$$= -\frac{\hbar}{i}\frac{\partial}{\partial t}[\psi_a(1)\psi_b(2)] \tag{14}$$

where $\psi_a(1)$ and $\psi_b(2)$ satisfy the energy eigenvalue equations for the separate components:

(a)
$$\hat{H}_1\psi_a(1) = E_a\psi_a(1) = -\frac{\hbar}{i}\frac{\partial\psi_a(1)}{\partial t}$$

(b)
$$\hat{H}_2\psi_b(2) = E_b\psi_b(2) = -\frac{\hbar}{i}\frac{\partial\psi_b(2)}{\partial t}$$

$$\tag{15}$$

EXERCISE

2-50. Show (a) that the energy eigenfunctions of any system whose Hamiltonian operator can be written as a *sum* of terms, each of which depends upon a different set of coordinates, $\hat{H} = \hat{H}_1(1) + \hat{H}_2(2) + \hat{H}_3(3) + \cdots$ can be written as a *product* of eigenfunctions of the separate terms $\psi = \psi_a(1)\psi_b(2)\psi_c(3) \cdots$ and (b) that the energy corresponding to the eigenfunction ψ is $E = E_a + E_b + E_c + \cdots$. This extremely important result is used repeatedly in solving quantum-mechanical problems.

The eigenfunctions $\psi_a(1)$ and $\psi_b(2)$ may be written in the form of Eq. 2-4(20):

(a)
$$\psi_a(1) = u_a(1) \exp{(-iE_a t/\hbar)}$$
(b)
$$\psi_b(2) = v_b(2) \exp{(-iE_b t/\hbar)}$$

$$\tag{16}$$

where the functions $u_a(1)$ and $v_b(2)$ constitute complete sets of orthogonal functions in terms of which the coordinate dependence of any state of the system—either with or without an interaction between its component parts—may be expressed at a given instant.

In the absence of an interaction between the two parts of the system, any wave function of the system can of course be written in the following form, which explicitly exhibits the dependence upon the time:

$$\psi(q_j,t) = \sum_a \sum_b c_{ab}\psi_{ab} = \sum_a \sum_b c_{ab}U_{ab}(1,2) \exp(-iE_{ab}t/\hbar) \qquad (17)$$

where $U_{ab}(1,2) = u_a(1)v_b(2)$, $E_{ab} = E_a + E_b$, and the coefficients c_{ab} are constants. According to the discussion of Sec. 2-5A, $c_{ab}^{*}c_{ab}$ is the probability that a measurement of the energy of the system will yield E_a and E_b for its two component parts.

Next, one considers the complete system—including the interaction term—and writes for the new Hamiltonian function

$$H(q_j,p_j) = H_0(q_j,p_j) + V(q_j,p_j) \qquad (18)$$

where $V(q_j,p_j)$ represents the interaction energy due to the coupling between the two parts of the system. The Hamiltonian operator is then

$$\hat{H} = \hat{H}_0 + \hat{V} \qquad (19)$$

The wave function for the complete system may of course still be expressed *in terms of the orthogonal eigenfunctions of the unperturbed system,* but because of the introduction of the interaction V, the old eigenfunctions are not quite the same as the eigenfunctions of the actual system. Thus in the expansion of ψ in terms of the old eigenfunctions, the coefficients c_{ab} of Eq. (17) *will vary with time.* However, if the interaction V is weak (and we shall almost always find it to be so), then the coefficients c_{ab} will vary so *slowly* with time that it will still be meaningful to speak of the system being, at any instant, in a certain combination of eigenstates of its two component parts. Thus we write for $\psi(q_j,t)$

$$\psi(q_j,t) = \sum_a \sum_b c_{ab}(t)\psi_{ab} = \sum_a \sum_b c_{ab}(t)u_a(1)v_b(2) \exp[-i(E_a + E_b)t/\hbar]$$

$$(20)$$

where again $c_{ab}^{*}(t)c_{ab}(t)$ represents the probability that the system is in the state ψ_{ab}.

We now investigate the time behavior of the complete system by means of Schroedinger's equation in the following straightforward steps:

$$\hat{H}\psi = (\hat{H}_0 + \hat{V})\psi = -\frac{\hbar}{i}\frac{\partial \psi}{\partial t}$$

$$(\hat{H}_0 + \hat{V})\sum_a \sum_b c_{ab}(t)\psi_{ab} = -\frac{\hbar}{i}\frac{\partial}{\partial t}\sum_a \sum_b c_{ab}(t)\psi_{ab}$$

$$\sum_a \sum_b \left(c_{ab}\hat{H}_0\psi_{ab} + c_{ab}\hat{V}\psi_{ab} + \frac{\hbar}{i}\dot{c}_{ab}\psi_{ab} + c_{ab}\frac{\hbar}{i}\frac{\partial \psi_{ab}}{\partial t}\right) = 0$$

But, since $\hat{H}_0 \psi_{ab} + (\hbar/i)(\partial \psi_{ab}/\partial t) = 0$, this reduces to

$$\sum_a \sum_b \left(c_{ab} \hat{V} \psi_{ab} + \frac{\hbar}{i} \dot{c}_{ab} \psi_{ab} \right) = 0 \qquad (21)$$

To solve for the rate of change of a particular coefficient, say, $\dot{c}_{a'b'}$, we multiply the above equation by $\psi^*_{a'b'}$ and integrate over all coordinates, with the result

$$\frac{dc_{a'b'}}{dt} = -\frac{i}{\hbar} \sum_a \sum_b c_{ab} \int \psi^*_{a'b'} \hat{V} \psi_{ab} \, d^N q$$

$$= -\frac{i}{\hbar} \sum_a \sum_b c_{ab} \exp\left[-\frac{i}{\hbar}(E_{ab} - E_{a'b'})t \right] \int U^*_{a'b'} \hat{V} U_{ab} \, d^N q \qquad (22)$$

The integrals on the right are often abbreviated

$$(a'b'|V|ab) = \int U^*_{a'b'} \hat{V} U_{ab} \, d^N q \qquad (23)$$

and are called the *matrix elements* of the interaction potential V because they appear as such in the Heisenberg theory. In this way we finally obtain a set of equations for the rate of change of each coefficient $c_{a'b'}$:

$$\frac{dc_{a'b'}}{dt} = -\frac{i}{\hbar} \sum_a \sum_b c_{ab}(a'b'|V|ab) \exp \frac{i(E_{a'} + E_{b'} - E_a - E_b)t}{\hbar} \qquad (24)$$

EXERCISE

2-51. Supply the missing steps in the derivation of Eq. (24) from Eq. (21).

To see physically what is represented by Eq. (24), let us consider a simple case. Assume that at $t = 0$ the system is known to be in a state that exactly corresponds to one of the eigenstates of the unperturbed system:

$$\psi(q_j,0) = u_r(1)v_s(2) \qquad (25)$$

Thus the energy of the system will be almost exactly equal to $E_r + E_s$, the energy of the unperturbed system. For this case Eq. (24) reads, near $t = 0$,

$$\frac{dc_{a'b'}}{dt} = -\frac{i}{\hbar}(a'b'|V|rs) \exp \frac{i(E_{a'} + E_{b'} - E_r - E_s)t}{\hbar} \qquad (26)$$

where it is assumed that the coefficient c_{rs} remains nearly equal to unity and that the remaining coefficients remain negligibly small over the interval of interest. With this approximation the equations may be integrated to give, for $a'b' \neq rs$

$$c_{a'b'} = -(a'b'|V|rs) \frac{\exp\left[i(E_{a'} + E_{b'} - E_r - E_s)t/\hbar\right] - 1}{E_{a'} + E_{b'} - E_r - E_s} \tag{27}$$

Physically, this equation says that the system will not remain in the state ψ_{rs} but will gradually change to a different state. The probability that the system will be found at a later time in the state $\psi_{a'b'}$ is

$$W_{a'b'} = c^{*}_{a'b'}c_{a'b'} = 4|(a'b'|V|rs)|^2 \frac{\sin^2\left[(E_{a'} + E_{b'} - E_r - E_s)t/2\hbar\right]}{(E_{a'} + E_{b'} - E_r - E_s)^2} \tag{28}$$

This probability is thus proportional to the *square of the matrix element* $(a'b'|V|rs)$.

In Eq. (28) can be seen various features which will be of interest in a study of spectroscopic transitions. First, consider the fact that the probability of transition from a given state ψ_{rs} to another state $\psi_{a'b'}$ is proportional to the absolute square of the matrix element of the perturbation energy; this means that some of the transitions away from the state ψ_{rs} will be more probable than others, because of the virtual certainty that $(a'b'|V|ab)$ will be different for the various possible transitions. This is the principal reason for the wide range in *intensity* exhibited by the various spectral lines resulting from these transitions. Second, there is a possibility that certain matrix elements may be *exactly equal to zero* for some values of a', b', r, and s. This means that transitions between these states *will not occur*, to the degree of approximation represented by Eq. (24). This is the quantum-mechanical basis for the *selection rules* governing the transitions that may take place between various energy levels. A transition for which $(a'b'|V|rs)$ is equal to zero is said to be *first-order forbidden*.

And finally, the "resonance denominator" of the above expression will be seen to embody the condition that *energy be conserved*—within the limits of the uncertainty principle—during a transition. That is, only those transitions for which $E_{a'} + E_{b'}$ is equal, or nearly equal, to $E_r + E_s$ will be likely to occur. We shall discuss these features of atomic transitions in greater detail in connection with specific applications.

EXERCISE

2-52. Consider a number of transitions for which the matrix elements are equal, but for which the initial and final energies differ by various amounts. Let $\Delta E = E_{a'} + E_{b'} - E_r - E_s$. Sketch $W_{a'b'}$ vs. t for various values of ΔE and thus show graphically how the maximum value of $W_{a'b'}$ is related to ΔE.

2-6. Conclusion

The material contained in the brief introduction to quantum mechanics given in this chapter has been chosen with two purposes in view: first, to try to make the postulates and procedures of quantum mechanics appear as reasonable as possible from the standpoint of the requirements set forth in Sec. 2-2 and, second, to provide a sufficient basis for treating the topics to be covered in later chapters. It should be obvious that a great deal remains to be said about almost every aspect of quantum mechanics touched upon here, and that it has been necessary to trim the discussion of the many interesting philosophical questions associated with quantum mechanics to a minimum. Inevitably, such a skeletal treatment must leave the student with many incomplete ideas and many unresolved confusions in his mind. However, many of these uncertainties regarding the content and application of the material presented in this chapter will disappear as subsequent topics are discussed, for we shall constantly use this material in our later work. It should be understood that the quantum mechanics given in this chapter and the applications given in subsequent chapters are highly interdependent, in that one cannot be thoroughly understood without the other.

Finally, it should be emphasized most strongly that the adequacy and validity of quantum mechanics must not be judged solely upon the basis of its ability to deal with the phenomena for which it was specifically designed, since a sufficient number of *ad hoc* hypotheses can always be devised to deal with a limited number of facts. The real success of the Schroedinger theory is that it provides a *general foundation* for treating an almost unlimited number of physical problems, many of which involve phenomena that are qualitatively quite different from those responsible for its origin. Indeed, we shall repeatedly encounter cases where the predictions of the Schroedinger theory seem quite at variance with familiar experience, and yet agree so spectacularly with experiment that there cannot be any doubt that the theory is to be believed in preference to our so-called common sense. Our basic aim is, of course, to apply the principles of quantum mechanics to a variety of simple problems selected from a broad range of physical phenomena, to illustrate the manner in which the essential qualities of these phenomena are related to the quantum mechanics, and thus to improve our insight and "feeling" for atomic physics.

REFERENCES

Darwin, C. G.: "The New Conceptions of Matter," G. Bell and Sons, Ltd., London, 1931.

Pauling, L., and E. B. Wilson: "Introduction to Quantum Mechanics," McGraw-Hill Book Company, Inc., New York, 1935.

Rojanski, V.: "Introductory Quantum Mechanics," Prentice-Hall, Inc., Englewood Cliffs, N.J., 1938.

Schiff, L. I.: "Quantum Mechanics," 2d ed., International Series in Pure and Applied Physics, McGraw-Hill Book Company, Inc., New York, 1955.

Slater, J. C.: "Quantum Theory of Matter," International Series in Pure and Applied Physics, McGraw-Hill Book Company, Inc., New York, 1951.

Tolman, R. C.: "The Principles of Statistical Mechanics," Chaps. VII–VIII, Oxford University Press, New York, 1938.

3

The One-dimensional Harmonic Oscillator

As a first application of the principles presented in the preceding chapter we shall treat the one-dimensional harmonic oscillator. This is appropriate because, first, the harmonic oscillator is the "oldest" quantum-mechanical system, and it is interesting to see how Planck's *ad hoc* postulates concerning its quantum properties compare with the results of the Schroedinger theory. Second, it is one of the simplest problems for which discrete energy levels are obtained, and yet it finds practical application in many physical situations. For instance, the vibrations of molecules and of the atoms of crystalline solids closely approximate simple harmonic vibrations if the amplitude is sufficiently small, and the electromagnetic field is treated in many applications as a number of independent harmonic oscillators corresponding to the various classical modes of oscillation of the field. And third, many of the distinctive and characteristic features of quantum mechanics can be clearly seen in the solution of the oscillator problem.

3-1. The Hamiltonian Function; the Schroedinger Equation

In treating a physical system in quantum mechanics one almost always begins by solving the *time-independent Schroedinger equation* for the system. This is done for several reasons: The energy eigenvalues are often of direct interest in connection with spectroscopic transitions of the system; the energy eigenfunctions represent possible states in which the system may remain for extended periods of time (stationary states);
128

the energy eigenfunctions also form the zero-order approximation for the subsequent treatment of the effects of small perturbations or interactions; and finally, the energy eigenfunctions in any case constitute a special complete set of orthogonal functions in terms of which the actual state of the system may be expanded, *including its dependence upon the time.* This procedure is exactly like that used in solving a classical problem involving an oscillatory system by first finding its normal modes of vibration, since the general motion of such a system can be expressed as a linear superposition of its various normal modes, each with its appropriate time factor.

The Hamiltonian function for the harmonic oscillator was evaluated in Exercise 2-15. It is

$$H = \frac{p^2}{2m} + \frac{1}{2}\,kx^2 \tag{1}$$

The Hamiltonian operator (in coordinate language) is therefore

$$\hat{H} = -\frac{\hbar^2}{2m}\frac{\partial^2}{\partial x^2} + \frac{1}{2}\,kx^2 \tag{2}$$

and the Schroedinger equation is

$$-\frac{\hbar^2}{2m}\frac{\partial^2\psi}{\partial x^2} + \frac{1}{2}\,kx^2\psi = -\frac{\hbar}{i}\frac{\partial\psi}{\partial t} \tag{3}$$

Since we are interested in the eigenvalues of the energy of the oscillator, we write the Schroedinger equation in the time-independent form

$$-\frac{\hbar^2}{2m}\frac{\partial^2\psi}{\partial x^2} + \frac{1}{2}\,kx^2\psi = E\psi = -\frac{\hbar}{i}\frac{\partial\psi}{\partial t} \tag{4}$$

This equation will presumably have solutions for several values of E. We shall call these values E_n ($n = 0, 1, 2, \ldots$), and shall denote the corresponding eigenfunctions by ψ_n. From Eq. 2-4(20) we write

$$\psi_n(x,t) = u_n(x)\exp\left(-iE_nt/\hbar\right) \tag{5}$$

The equation satisfied by $u_n(x)$ is then the time-independent Schroedinger equation 2-4(21)

$$-\frac{\hbar^2}{2m}\frac{d^2u_n}{dx^2} + (\tfrac{1}{2}kx^2 - E_n)u_n = 0 \tag{6}$$

From the discussion of the preceding chapter we know that the functions $u_n(x)$ which satisfy Eq. (6) constitute a *complete set of orthogonal functions*, in terms of which *any* wave function for the oscillator (or for any other one-dimensional system) may be expressed. Thus the general solution of the problem of the one-dimensional simple harmonic oscillator is

$$\psi(x,t) = \sum_{n=0}^{\infty} A_n u_n(x) \exp(-iE_n t/\hbar) \tag{7}$$

EXERCISE

3-1. Suppose that $\psi(x,0)$ is known. Evaluate the coefficients A_n in the expansion (7) for this case. Assume $u_n(x)$ to be normalized.

Ans.:
$$A_s = \int_{-\infty}^{\infty} u_s^*(x)\psi(x,0)\,dx$$

3-2. Solution of the Schroedinger Equation

In order to solve the time-independent Schroedinger equation for the functions $u_n(x)$, one follows a more or less standard procedure which consists in:

1. Expressing the equation in dimensionless form
2. Determining the asymptotic behavior of $u_n(x)$ as $x \to \pm \infty$
3. Changing to a new dependent variable whose behavior at $\pm \infty$ is less violent by dividing out the asymptotic form of $u_n(x)$
4. Solving the equation for this new dependent variable by the power-series method
5. Determining for what values of the energy this power-series solution, combined with the asymptotic form, is well behaved

Taking the first step in the above-outlined procedure, we let

$$\xi = \alpha x \tag{1}$$

where α is to be determined shortly. Writing u_n' for $du_n/d\xi$, Eq. 3-1(6) becomes

$$-\frac{\hbar^2 \alpha^2}{2m} u_n'' + \left(\frac{\frac{1}{2}k\xi^2}{\alpha^2} - E_n\right) u_n = 0$$

$$\tag{2}$$

or

$$u_n'' + \left(\frac{2E_n m}{\hbar^2 \alpha^2} - \frac{km}{\hbar^2 \alpha^4} \xi^2\right) u_n = 0$$

If we now set

$$\alpha^4 = \frac{km}{\hbar^2} \quad \text{and} \quad \lambda_n = \frac{2E_n}{\hbar(k/m)^{\frac{1}{2}}} \tag{3}$$

the equation assumes the dimensionless form

$$u_n'' + (\lambda_n - \xi^2)u_n = 0 \tag{4}$$

EXERCISE

3-2. Carry out the steps leading to Eq. (4) from Eq. 3-1(6).

We now observe that, as $\xi \to \infty$, the $\lambda_n u_n$ term is negligible compared with $-\xi^2 u_n$. Thus the equation assumes the asymptotic form

$$u_n'' - \xi^2 u_n = 0 \tag{5}$$

If we were to treat ξ^2 as a *constant*, this equation would have solutions

$$u_n \to e^{\pm \xi \cdot \xi} = e^{\pm \xi^2} \tag{6}$$

We thus try a solution of the form $u_n = e^{\pm c \xi^2}$ and find that

$$4c^2 \xi^2 e^{\pm c \xi^2} \pm 2c e^{\pm c \xi^2} - \xi^2 e^{\pm c \xi^2} = 0 \tag{7}$$

so that, to the same approximation as was assumed in neglecting $\lambda_n u_n$, we see that $c = \frac{1}{2}$ gives as an asymptotic solution

$$u_n(\xi) \to e^{\pm \xi^2/2} \tag{8}$$

Because of the general requirement that an acceptable quantum-mechanical wave function must vanish at infinity, we *eliminate the positive exponent* and write $u_n(\xi)$ in the form

$$u_n(\xi) = N_n e^{-\xi^2/2} H_n(\xi) \tag{9}$$

where N_n is a normalization coefficient and $H_n(\xi)$ is a new dependent variable. Substitution of this form into Eq. (4) results in the following equation for $H_n(\xi)$:

$$H_n'' - 2\xi H_n' + (\lambda_n - 1) H_n = 0 \tag{10}$$

EXERCISE

3-3. Supply the missing steps in the derivation of Eq. (10).

We solve Eq. (10) by the power-series method, setting

$$H_n(\xi) = \sum_{r=0}^{\infty} a_r \xi^r \tag{11}$$

The equation thereby becomes

$$\sum_{r=0}^{\infty} [r(r-1)a_r \xi^{r-2} - 2r a_r \xi^r + (\lambda_n - 1)a_r \xi^r] = 0 \tag{12}$$

The coefficient of each power of ξ in this series must separately equal zero. Thus one finds the following recurrence relation:

$$(s+1)(s+2)a_{s+2} - [2s - (\lambda_n - 1)]a_s = 0 \qquad s = 0, 1, 2, \ldots . \tag{13}$$

The form of $H_n(\xi)$ is therefore completely determined for given λ_n

by two arbitrary constants, which we may take to be a_0 and a_1 in the above series solution.

EXERCISE

3-4. Carry out the power-series solution of Eq. (10) and obtain the recurrence relations (13).

If we examine the convergence of the power-series solution defined by Eqs. (11) and (13), we find that, as $s \to \infty$, $a_{s+2}/a_s \to 2/s$, so that the series converges for all finite values of ξ.   However we also observe that, in the power-series expansion of $e^{+\xi^2}$, the ratio of successive terms is $a_{s+2}/a_s \to 2/s$, so that, as $\xi \to \infty$, it appears that $H_n(\xi)$ diverges approximately as e^{ξ^2} and $u_n(\xi)$ itself approximately as $e^{\xi^2/2}$, which in turn means that $u_n(\xi)$ is not an acceptable wave function.   *The only way this situation can be avoided is to choose λ_n in such a way that the power series for $H_n(\xi)$ cuts off at some term, making $H_n(\xi)$ a polynomial.*

Inspection of Eq. (13) reveals that this can be done by choosing λ_n in such a way that the coefficient of a_s vanishes for some value of s, and at the same time setting a_0 or a_1 equal to zero.   By this procedure, a family of polynomials of alternately even and odd symmetry in ξ is produced.   These polynomials, when combined with the factor $e^{\xi^2/2}$, lead each in turn to *eigenfunctions which are alternately even and odd functions of ξ.*

The smallest value of λ for which this can be done, which we call λ_0, is $\lambda_0 = 1$.   Then setting $a_1 = 0$, as we are forced to do to make $u_0(\xi)$ vanish as $\xi \to \infty$, we have (setting $a_0 = 1$)[1] $H_0(\xi) = 1$.   The next larger value of λ we call λ_1.   From Eq. (13) it is (setting $a_1 = 2$ and $a_0 = 0$) $\lambda_1 = 3$ and $H_1(\xi) = 2\xi$.   Similarly, we have for λ_2 (setting $a_0 = -2$ and $a_1 = 0$), $\lambda_2 = 5$ and $H_2(\xi) = 4\xi^2 - 2$.   And in general, from Eq. (13) we see that, if we set

$$\lambda_n = 2n + 1 \qquad n = 0, 1, 2, \ldots \ . \tag{14}$$

$H_n(\xi)$ *will be a polynomial of degree n.*

The polynomial solutions of Eq. (10) are called the *Hermite polynomials.*   They can be defined by the series solution just obtained and in other ways.   One of the most useful expressions for $H_n(\xi)$ is by the formula

$$H_n(\xi) = (-1)^n e^{\xi^2} \frac{d^n(e^{-\xi^2})}{d\xi^n} \tag{15}$$

[1] The leading coefficients of the nonvanishing series have in each case been chosen to make our polynomial expressions agree with the customary form for the Hermite polynomials.

EXERCISES

3-5. Evaluate $H_0(\xi)$, $H_1(\xi)$, $H_2(\xi)$, $H_3(\xi)$, and $H_4(\xi)$ both from the recurrence formulas (13) and from Eq. (15), and thus verify that these formulas define the same functions, to within a multiplicative constant.

3-6. Use Eq. (15) and the differential equation (10) (with $\lambda_n = 2n + 1$) to establish the following formulas for the Hermite polynomials:

(a)
$$H'_n(\xi) = 2\xi H_n(\xi) - H_{n+1}(\xi)$$
(b)
$$H_{n+1}(\xi) - 2\xi H_n(\xi) + 2n H_{n-1}(\xi) = 0 \tag{16}$$
(c)
$$H'_n(\xi) = 2n H_{n-1}(\xi)$$

3-3. The Energy Levels and Wave Functions

We have just found that the only physically acceptable solutions of the time-independent Schroedinger equation for the harmonic oscillator have the form

$$u_n(x) = N_n e^{-\alpha^2 x^2/2} H_n(\alpha x) \tag{1}$$

where $\qquad \alpha^4 = \dfrac{km}{\hbar^2} = \dfrac{m^2\omega^2}{\hbar^2} \qquad \lambda_n = 2n + 1 = \dfrac{2E_n}{\hbar(k/m)^{\frac{1}{2}}}$

and $H_n(\alpha x)$ is the nth Hermite polynomial. The normalizing factor N_n can be shown[1] to be equal to

$$N_n = \left(\frac{\alpha}{\pi^{\frac{1}{2}} 2^n n!}\right)^{\frac{1}{2}} \tag{2}$$

These eigenfunctions correspond to certain discrete values for the energy of the linear oscillator; they are

$$E_n = (n + \tfrac{1}{2})\hbar(k/m)^{\frac{1}{2}} = (n + \tfrac{1}{2})\hbar\omega \tag{3}$$

where $\omega = (k/m)^{\frac{1}{2}}$ is the classical angular frequency of the oscillator. This equation will be recognized as corresponding to the Planck formula $E_n = nh\nu$, except for an additional $\frac{1}{2}h\nu$ energy possessed by the oscillator for each value of n. The first few normalized eigenfunctions for the linear oscillator are

(a)
$$u_0(x) = \frac{\alpha^{\frac{1}{2}}}{\pi^{\frac{1}{4}}} e^{-\alpha^2 x^2/2}$$

(b)
$$u_1(x) = \frac{\alpha^{\frac{1}{2}}}{2^{\frac{1}{2}}\pi^{\frac{1}{4}}} 2\alpha x e^{-\alpha^2 x^2/2} \tag{4}$$

(c)
$$u_2(x) = \frac{\alpha^{\frac{1}{2}}}{8^{\frac{1}{2}}\pi^{\frac{1}{4}}} (4\alpha^2 x^2 - 2)e^{-\alpha^2 x^2/2}$$

---

[1] H. Margenau and G. M. Murphy, "The Mathematics of Physics and Chemistry," 2d ed., p. 124, D. Van Nostrand Company, Inc., Princeton, N.J., 1956. (The factor $\alpha^{\frac{1}{2}}$ arises from the fact that the integral is over the coordinate x rather than the dimensionless variable ξ.)

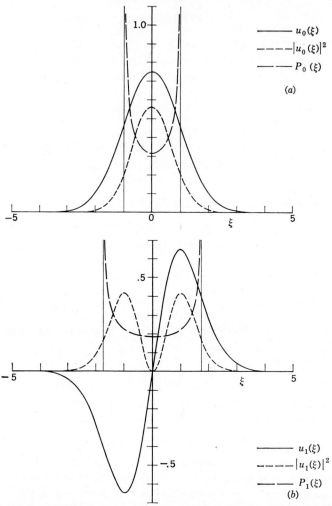

FIG. 3-1. Harmonic-oscillator wave functions. The solid curves represent the functions $\alpha^{-1/2} u_n(\alpha x)$ with $\alpha x = \xi$ for $n = 0, 1, 2, 3,$ and 10. The dotted curves represent $\alpha^{-1} u_n^* u_n$ for the same values of n. The dashed curves represent the probability distribution for a classical oscillator having the same energy as the corresponding quantum-mechanical oscillator. The vertical lines define the limits of the classical motion.

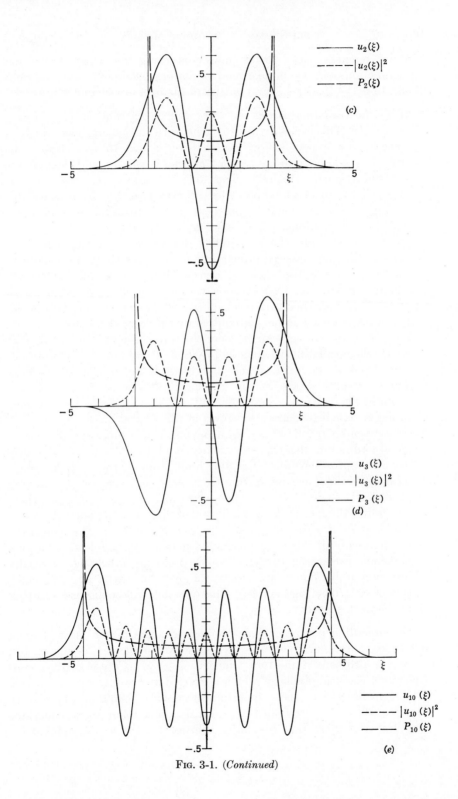

Fig. 3-1. (*Continued*)

A few remarks ought to be made concerning the energy levels and wave functions of a one-dimensional oscillator, since these wave functions illustrate many features characteristic of quantum-mechanical systems:

1. *When the oscillating particle is in its lowest energy state, it is not "standing still."* This is clearly in agreement with the uncertainty principle, since if the lowest energy state corresponded to *zero energy*, the particle would have to be *at rest at the origin*, which would then violate the uncertainty principle. The lowest energy $\frac{1}{2}\hbar\omega$ that can be attained by an oscillator is called its *zero-point energy*, because this amount of energy would be possessed by an oscillator in equilibrium with its surroundings even at *absolute-zero temperature.*

2. In its lowest state the probability density $\psi^*\psi$ for the particle is *quite unlike* the classical probability density for the same energy of oscillation, since the classical density has a minimum at $x = 0$, whereas $[u_0(x)]^2$ has its *maximum* value at $x = 0$. For large values of n the average quantum-mechanical and classical probability densities approach one another as required by the correspondence principle; however, even at very large energies there are many values of x at which the oscillating particle will *never be found* because it interferes with itself as it travels back and forth in its oscillatory motion. This is clearly shown in the graphical plots of $u_n(x)$ and $|u_n(x)|^2$ given in Fig. 3-1 for several values of n. The corresponding classical probability densities are also shown in this figure.

3. At a given energy there is a limit beyond which the oscillator cannot go according to the classical theory, and the classical probability density approaches infinity at this limit (that is, the particle spends most of its time near its maximum displacement because it is moving most slowly there). Quantum-mechanically, on the other hand, there is a finite, though perhaps small, probability that the particle will be found at *arbitrarily large distances from the origin.* This feature is characteristic of quantum mechanics and is responsible for the phenomenon called the *tunnel effect* in problems involving classically insurmountable potential barriers, e.g., α-radioactivity and field emission of electrons from metals. This question will be discussed further in the next chapter.

The preceding discussion has emphasized some of the most striking differences between a quantum-mechanical and a classical oscillator, but there are also some important similarities between these two systems. Imagine a classical oscillator which is started at $t = 0$ from $x = x_0$ with a momentum $p = p_0$ and a quantum-mechanical oscillator whose wave function at $t = 0$ is prescribed to be $\psi(x,0)$. The following relationships are typical of those that can be established for the two systems; they depend upon the fact that the energy levels are *equally spaced*, so that the time factors are all cyclic with the period $2\pi/\omega$:

1. After a full cycle ($t = 2\pi/\omega$), the classical oscillator will again be found at x_0 with momentum p_0. The quantum-mechanical oscillator will correspondingly possess the same $\psi^*\psi$ and $\phi^*\phi$ distributions at this time as at $t = 0$.
2. After a half cycle ($t = \pi/\omega$), the classical oscillator will be found at $-x_0$ with momentum $-p_0$, while the quantum-mechanical oscillator wave function will satisfy the equation $\psi(x,\pi/\omega) = -i\psi(-x,0)$, which means that $W_q(x,\pi/\omega) = W_q(-x,0)$ and $W_p(p,\pi/\omega) = W_p(-p,0)$.
3. After a quarter cycle ($t = \pi/2\omega$), the classical oscillator will be found at $x = p_0/m\omega$ with momentum $p = -m\omega x_0$, while the quantum-mechanical-oscillator probability distributions for coordinate and momentum have also changed roles: $W_q(x,\pi/2\omega) = (1/m\omega)W_p(m\omega x,0)$ and $W_p(p,\pi/2\omega) = m\omega W_q(p/m\omega,0)$.
4. The coordinate and momentum of the classical oscillator obey the equations $x = x_0 \cos \omega t + (p_0/m\omega) \sin \omega t$ and $p = p_0 \cos \omega t - m\omega x_0 \sin \omega t$. As is seen from Ehrenfest's theorem (Sec. 2-4C), the *expectation values* $\langle x \rangle$ and $\langle p \rangle$ obey similar equations for the quantum-mechanical oscillator.

EXERCISES

3-7. Show that the three wave functions given in Eq. (4) are orthogonal to one another and are normalized.

3-8. Evaluate $\langle x \rangle$, $\langle p \rangle$, $\langle x^2 \rangle$, $\langle p^2 \rangle$, and $\Delta x \, \Delta p$ for the first two wave functions of Eq. (4). *Partial ans.:* $\Delta x \, \Delta p = \frac{1}{2}\hbar, \frac{3}{2}\hbar$.

3-9. Find the classical probability density for a harmonic oscillator whose energy is $\hbar\omega/2$, and compare this with $[u_0(x)]^2$. (See Fig. 3-1.)

3-10. Evaluate the classical limit of the motion of an oscillator whose energy is $\hbar\omega/2$, and find the probability that the particle will be found outside this limit. Find also the probability that the particle will be found more than *twice this far* from the origin. *Ans.:* $P_1 = 0.158$, $P_2 = 0.0046$.

3-11. At $t = 0$, a certain harmonic oscillator is known to be described by the wave function of Eq. 2-3(12). Find the probability that the *energy* of this oscillator will be found to be $\hbar\omega/2$.

3-12. Let the wave function of a quantum-mechanical oscillator be expressed in terms of the energy eigenfunctions in the form

$$\psi(\xi,t) = e^{-i\omega t/2} \sum_{n=0}^{\infty} A_n u_n(\xi)e^{-ni\omega t}$$

Using this form, verify the statements made in (1) and (2) above.

3-13. Evaluate the expectation value of the potential energy for an energy eigenstate.   *Hint:* Use Eq. 3-2(16*b*).   *Ans.:* $\langle V \rangle = \frac{1}{2}(n + \frac{1}{2})\hbar\omega = \frac{1}{2}E_n$.

3-14. Use the result of Exercise 3-13 to evaluate $\Delta x\, \Delta p$ for an energy eigenstate.   *Ans.:* $\Delta x\, \Delta p = E_n/\omega = (n + \frac{1}{2})\hbar$.

3-15. The Hermite polynomials may also be defined by means of a *generating function*

$$e^{\xi^2-(s-\xi)^2} = \sum_{n=0}^{\infty} \frac{H_n(\xi)}{n!} s^n \tag{5}$$

so that $H_n(\xi)$ is equal to $n!$ times the coefficient of s^n in the series expansion of the exponential at the left.   Multiply this generating function by a suitable function of ξ and t and substitute $s = Ae^{-i\omega t}$ to produce a function which is a solution of the Schroedinger equation including the time.   Evaluate $\psi^*\psi$ and show that this function corresponds to a "lump" of probability of constant "shape," whose center moves like a classical oscillating particle.

Ans.: $\psi(x,t) = \dfrac{\alpha^{1/2}}{\pi^{1/4}} \exp\left(-i\omega t/2 + iA^2 \sin 2\omega t\right)$

$\exp\left[-(\xi - 2A \cos \omega t)^2/2\right] \exp\left(-2iA\xi \sin \omega t\right)$

3-16. A harmonic oscillator is initially in the state

$$\psi(x,0) = Ae^{-\alpha^2 x^2/2}\alpha x(2\alpha x + i)$$

(*a*) Find the wave function as a function of the time.   (*b*) Find the average energy of the oscillator.   *Ans.:* (*b*) $\frac{25}{14}\hbar\omega$.

3-17. Apply the first-order perturbation theory to the case of a harmonic oscillator whose force constant is changed by an amount Δk, and show that the first-order shift in the energy levels agrees with that obtained directly from Eq. (3) for $\Delta k \ll k$.

3-18. A one-dimensional harmonic oscillator is initially in the state $\psi(x,0) = A \exp\left[-(x - x_0)^2/2a^2\right] \exp(ip_0 x/\hbar)$.   What is the expectation value of its total energy at a later time $t = 6\pi/\omega$?

Ans.: $\langle E \rangle = \dfrac{p_0^2}{2m} + \dfrac{m\omega^2 x_0^2}{2} + \dfrac{1}{4} m\omega^2 a^2 + \dfrac{\hbar^2}{4ma^2}$

3-4.   Parity

The energy eigenfunctions 3-3(1) exhibit a most important property that plays a fundamental role in many physical phenomena: *these wave functions possess alternately even and odd symmetry with respect to the reflection of the coordinate through the center of attraction.* That is, for even values of n the wave functions are even functions of x, which means that $\psi(x) = \psi(-x)$; and for odd values of n are odd functions, which means that $\psi(x) = -\psi(-x)$. This symmetry property is called the *parity* and is said to be even, or $+$, for even functions, and odd, or $-$, for odd functions. Such symmetry is exhibited by all nondegenerate energy eigenfunctions for any system whose potential energy is a symmetric function of some coordinate, and it can be treated through the use of an operator in the same manner as are other quantities of physical interest. Since we shall encounter the concept of parity at several points in our later work, we shall consider it briefly at this point.

To establish the symmetry properties of energy eigenfunctions mentioned above, let us consider a case in which the potential energy, and therefore the Hamiltonian operator, is symmetric with respect to a change in sign of some coordinate, say, x:

$$H(x,y,z) = H(-x,y,z) \tag{1}$$

We then define an operator, which we call the "reflection" operator $\hat{R}(x)$, whose action upon a wave function is to replace x by $-x$ in that function. We see that $\hat{R}(x)$ has the following properties:

1. $\hat{R}[u(x)] = u(-x)$, definition of $\hat{R}(x)$.
2. $\hat{R}(x)$ corresponds to a reflection of $u(x)$ through the plane $x = 0$.
3. $\hat{R}(x)$ is a linear, Hermitian operator:

$$\hat{R}(u + v) = \hat{R}u + \hat{R}v \qquad \text{linear}$$

and

$$\int_{-\infty}^{\infty} u^*(x)\hat{R}v(x)\,dx = \int_{-\infty}^{\infty} u^*(x)v(-x)\,dx = \int_{+\infty}^{-\infty} u^*(-x)v(x)\,d(-x)$$

$$= -\int_{\infty}^{-\infty} v(x)u^*(-x)\,dx = \int_{-\infty}^{\infty} v(x)[\hat{R}u(x)]^*\,dx \qquad \text{Hermitian}$$

Therefore $\hat{R}(x)$ is, at least formally, an operator that can be used in quantum mechanics.

4. If we now inquire into what *observable quantity* is represented by the operator $\hat{R}(x)$, we see that this quantity *has no classical analogue*, since the reflection of a wave function is not a classical concept. We nevertheless find experimentally that *the physical consequences of using this operator in the same way other operators are used are actually valid*. We call the observable quantity represented by $\hat{R}(x)$ the *parity with respect to x*.

5. *The only eigenvalues of the parity are* ± 1. If we write the eigenvalue equation for the parity $\hat{R}u(x) = R_0 u(x)$ and operate a second time with $\hat{R}(x)$, we see that $\hat{R}[\hat{R}u(x)] = \hat{R}R_0 u(x) = R_0\hat{R}u(x) = R_0^2 u(x)$. But since $\hat{R}[\hat{R}u(x)] = \hat{R}[u(-x)] = u(x)$, this requires that

$$R_0^2 = +1 \qquad \text{or} \qquad R_0 = \pm 1 \tag{2}$$

6. The eigenfunctions of the parity are therefore *any even function of x or any odd function of x.*

7. If $\hat{H}(x) = \hat{H}(-x)$, then $\hat{R}$ commutes with $\hat{H}$:

$$\hat{R}(x)[\hat{H}(x)u(x)] = \hat{H}(-x)u(-x)$$
$$= \hat{H}(x)u(-x)$$
$$= \hat{H}(x)\hat{R}(x)u(x)$$

That is, $\hat{R}\hat{H} = \hat{H}\hat{R}$ or $\hat{R}\hat{H} - \hat{H}\hat{R} = 0$

Thus *the parity of a wave function does not change with time if* $\hat{H}(x) = \hat{H}(-x)$. (See Exercise 2-42.)

8. If $\hat{R}$ and $\hat{H}$ commute, it is possible to find wave functions which are simultaneously eigenfunctions of $\hat{R}$ and $\hat{H}$. This then means that a *nondegenerate* energy eigenfunction *must be* either an even function or an odd function of x and that a *degenerate* eigenfunction can be expressed as a linear combination of even and odd eigenfunctions.

Further examples and applications of these properties of eigenfunctions will be encountered frequently in our later work. We shall find that the space-symmetry character of eigenfunctions plays an essential role in such phenomena as molecular binding and excited states and atomic and nuclear radiative transitions. Indeed, the concept of parity occupies a fundamental position in our concept of the symmetry properties of the basic interactions between the various kinds of particles which occur in nature.

REFERENCES

Rojanski, V.: "Introductory Quantum Mechanics," Prentice-Hall, Inc., Englewood Cliffs, N.J., 1938.

Schiff, L. I.: "Quantum Mechanics," 2d ed., International Series in Pure and Applied Physics, McGraw-Hill Book Company, Inc., New York, 1955.

4

The Free Particle

As a second application of the quantum-mechanical principles set forth in Chap. 2, we shall consider some of the quantum properties of a *free particle* in one dimension and in three dimensions. Many of the properties of this simple case find application in approximate studies of more complex systems, and many of the characteristic features of quantum mechanics are well illustrated by it.

4-1. The Schroedinger Equation—One Dimension

A *free particle* is a particle that is subject to no forces, i.e., one that moves in a region of *constant potential energy*. The Hamiltonian function for such a particle in one dimension is thus simply

$$H = \frac{p^2}{2m} + V \tag{1}$$

where V is a constant.

The time-independent Schroedinger equation is therefore

$$-\frac{\hbar^2}{2m}\frac{d^2u}{dx^2} + Vu = Eu \tag{2}$$

This equation can easily be solved. The solution is

$$u(x) = A \, \exp\left[\pm i(2m)^{\frac{1}{2}}(E - V)^{\frac{1}{2}}x/\hbar\right] \tag{3}$$

EXERCISES

4-1. Solve the time-independent Schroedinger equation for a one-dimensional free particle. Show that this solution can be written in the form

$$u(x) = A^{\pm i\alpha x} \tag{4}$$

where $\alpha\hbar$ is the classical momentum that would be possessed by a particle of energy E.

4-2. Show that the above eigenfunctions of a one-dimensional free particle are also eigenfunctions of the momentum and that the positive exponent in Eq. (4) corresponds to a particle traveling with momentum $\alpha\hbar$ in the positive x direction.

4-2. Boundary Conditions and Normalization

The solution of the Schroedinger equation, as given above, is not complete until it has been made to satisfy the boundary condition of *vanishing at infinity* and has been normalized. Here we meet a situation somewhat different from that which was anticipated in Chap. 2, since the normalization of this wave function requires that $A = 0$. That is, the particle has a uniform, but vanishingly small, probability of being found in any unit interval from $-\infty$ to $+\infty$. Such a wave function does, however, vanish at infinity as it should.

In spite of the above difficulty, one can still obtain a useful representation for a free particle by *letting A remain finite*. In order to visualize what this means physically, we must remember that ψ is essentially a statistical quantity in the sense that it provides a prediction of average results that will be obtained if a given experiment is performed an *indefinitely large number of times*. The normalization of the total probability to unity, as introduced in the first postulate, is thus a convenience rather than a necessity, since we may ascribe meaning to the *relative* values of $\psi^*\psi$ at a number of locations even in the absence of a finite value for the integrated probability, if we imagine an experiment which is performed over and over indefinitely. Such experiments involving free particles are commonly performed by preparing a homogeneous *beam* of particles whose separation from one another in the beam is sufficiently great that they do not affect each other in the experiment. The free-particle wave function 4-1(3) with finite amplitude may thus be thought of in a loose way as corresponding to a beam of particles which contains A^*A particles per unit length along its axis. Strictly speaking, of course, the wave function for such a beam would have to contain the individual coordinates of all particles in it, so that the statistical distribution of the particles along the beam could be specified completely. Nevertheless, for convenience we shall speak of a wave function of the form

$$u(x) = A \exp{(ipx/\hbar)} \tag{1}$$

as representing a uniform beam of noninteracting particles, all moving in the positive x direction with momentum p. This representation of a beam of monoenergetic particles is extremely useful in the study of the collision and scattering of particles in atomic and nuclear physics.

EXERCISE

4-3. Find the value of $|A|$ which represents, in the above restricted sense, a beam of particles of *unit intensity*, i.e., a beam of such strength that one particle per second passes a given point on the x-axis. *Ans.:* $|A| = (m/p)^{1/2}$.

4-3. Superposition of Wave Trains

Extending the above idea, one may interpret a wave function consisting of two or more terms of the form

$$u(x) = A \exp (ip_1 x/\hbar) + B \exp (ip_2 x/\hbar) + \cdots \cdots \tag{1}$$

as representing physically a *mixture* of beams of particles—one of momentum p_1 and intensity $A^*A p_1/m$, another of momentum p_2 and intensity $B^*B p_2/m$, and so on. Indeed, the postulated relation 2-3(14) connecting $\psi(x)$ and $\phi(p)$ represents the limit of this idea, wherein a wave function of arbitrary character is written as a linear superposition of an infinite number of exponential terms of the above form. Thus, except for the number of particles involved (i.e., the type of normalization used), this form may be considered to represent an analysis of $\psi(x)$ into its *monoenergetic components*.

Further, we shall extend the above ideas to situations in which the wave function may be expressible, within a certain interval along the x-axis, as a superposition of a number of pure exponentials, but may possess some other form outside this region. In such a case it is useful still to interpret the wave function as representing, within this region, a mixture of a number of monoenergetic beams of particles, just as above. It must be remembered, however, that it is not strictly possible to define the momentum (or kinetic energy) of a particle with infinite precision and at the same time to *restrict the region* within which the particle is to have this momentum (or energy). One may also recall that the momentum of the particle can have a precisely defined value only if the eigenvalue equation $\hat{p}\psi = p_0\psi$ is satisfied *over the entire range of the coordinate*. Thus this interpretation of a wave function as representing a homogeneous beam of particles over a *partial range* of the coordinate must be considered again to be a loose, even though very useful, application of the principles of Chap. 2. We shall shortly see how this last idea may be applied.

4-4. Wave Packets; Motion of a Free Particle

A wave function that is made up of a great many waves of nearly equal momenta, superimposed in such a way as to yield a function which differs significantly from zero only in a restricted region, is called a *wave packet*. A wave packet may be taken to represent a particle whose position and momentum are both known approximately. If the uncertainty product of x and p for a wave function has its smallest possible value, this function provides a description of a particle whose coordinate and momentum are both known as accurately as is mutually possible.

An essential feature of a true wave packet is that it must contain an infinite number of waves whose wavelengths differ by infinitesimal amounts, in order that the various waves can cancel one another out at all places save in the neighborhood of some single value of x. Combinations of waves involving a finite number of components, or components whose wavelengths differ, from one to the next, by finite amounts, will correspond to an infinite, regularly repeated pattern which may have locally a concentrated form, but will have similar concentrations at each of its repetitions, ad infinitum. We shall encounter both of these types of wave packet in our later work.

A particular example of considerable interest is the wave function

$$\psi(x) = A \exp \left[-(x - x_0)^2/2a^2 \right] \exp \left(ip_0 x/\hbar \right) \tag{1}$$

which was introduced in Chap. 2 to illustrate a one-dimensional particle wave function. This function was expressed in Exercise 2-18 as a superposition of an infinite number of waves of the form (1), each having appropriate amplitude and phase relative to one another. The greatest contribution to the function (1) is made by waves corresponding to momentum near p_0, but waves of all momenta contribute to an extent that depends upon the magnitude of the parameter a.

It was also found in Exercise 2-18 that this wave function possesses the uncertainty product $\Delta x\, \Delta p = \hbar/2$, so that it represents the greatest possible simultaneous knowledge of position and momentum for a particle. This wave function is unique in this respect[1] and is therefore of some theoretical importance.

It is instructive to examine the nature of the motion of a particle in one dimension as it is described by quantum mechanics. In order that the motion shall correspond as closely as possible to the classical case of a particle whose momentum and position are both known exactly as a function of the time, we shall take the wave packet (1) as the wave function at $t = 0$ and examine the behavior as time proceeds. This

---

[1] See L. I. Schiff, "Quantum Mechanics," 2d ed., sec. 12, McGraw-Hill Book Company, Inc., New York, 1955.

could be done by solving the Schroedinger equation including the time, but such a procedure would involve considerable mathematical complication. We shall instead use the idea of a wave packet as a superposition of infinite sinusoidal wave trains to deduce the complete time dependence of the motion of the particle.

We know that a valid solution of the Schroedinger equation including the time for a free particle is

$$u(x,t) = A \exp (ipx/\hbar) \exp (-ip^2t/2m\hbar) \tag{2}$$

where $p^2/2m$ is the energy of the particle. (We have taken V to be zero for simplicity.) Therefore the most general wave function, including the time dependence, is a sum of terms of the above form with arbitrary amplitudes and phases. Since the eigenvalues are continuously distributed, this summation can be written as an integral over the possible energies, or equivalently, as an integral over all momenta:

$$\psi(x,t) = h^{-\frac{1}{2}} \int_{-\infty}^{\infty} A(p) \exp (ipx/\hbar) \exp (-ip^2t/2m\hbar) \, dp \tag{3}$$

We recognize this expression to be just that for $\psi(x,t)$ in terms of $\phi(p,t)$ as described in the second postulate of Chap. 2. Therefore, in accordance with the result of Exercise 2-18, we choose $A(p)$ to have the form

$$A(p) = \frac{(2\pi a)^{\frac{1}{2}}}{h^{\frac{1}{2}}\pi^{\frac{1}{4}}} \exp [-a^2(p - p_0)^2/2\hbar^2] \exp [-i(p - p_0)x_0/\hbar] \tag{4}$$

so that, at $t = 0$, $\psi(x,0)$ will be just the expression (1) with $|A| = (a\pi^{\frac{1}{2}})^{-\frac{1}{2}}$. Thus $\psi(x,t)$ is equal to

$$\psi(x,t) = \frac{(2\pi a)^{\frac{1}{2}}}{h^{\frac{1}{2}}\pi^{\frac{1}{4}}} \exp (ip_0x_0/\hbar) \int_{-\infty}^{\infty} \exp [-a^2(p - p_0)^2/2\hbar^2]$$
$$\exp [ip(x - x_0)/\hbar] \exp (-ip^2t/2m\hbar) \, dp \tag{5}$$

When this integral is evaluated and the result rearranged, it assumes the form

$$\psi(x,t) = \pi^{-\frac{1}{4}} \left(a + \frac{i\hbar t}{ma} \right)^{-\frac{1}{2}} \exp \left[i \left(\frac{p_0 x}{\hbar} - \frac{p_0^2 t}{2m\hbar} \right) \right]$$
$$\exp - \frac{(x - x_0 - p_0t/m)^2(1 - i\hbar t/ma^2)}{2(a^2 + \hbar^2t^2/m^2a^2)} \tag{6}$$

This will be recognized, after some scrutiny, as being basically of nearly the same form as Eq. (1). The principal features of Eq. (6) are the following:

1. The probability distribution $W_q(x,t)$ is equal to

$$W_q(x,t) = \pi^{-\frac{1}{2}} \left(a^2 + \frac{\hbar^2t^2}{m^2a^2} \right)^{-\frac{1}{2}} \exp - \frac{(x - x_0 - p_0t/m)^2}{(a^2 + \hbar^2t^2/m^2a^2)} \tag{7}$$

Thus the probability distribution is at all times Gaussian with unit total area. However, the position of its maximum, and its width, are both functions of the time.

2. The position of the maximum of $W_q(x,t)$ moves with uniform speed p_0/m, that is, with the speed corresponding to the *average value of the momentum*.

3. The width of the distribution $W_q(x,t)$ increases with time in a manner characteristic of a quantity subject to two *independent sources* of random uncertainties which are Gaussian. One of these is the coordinate uncertainty and the other is the momentum uncertainty associated with the initial wave function: The quantity $(\hbar/ma)t$ is equal to the distance traversed in time t at a speed equal to $2^{1/2}(\Delta p/m)$, where $\Delta p = \hbar/2^{1/2}a = \hbar/2\,\Delta x$ is the initial uncertainty in p. The greater the uncertainty in p, the greater is the rate of spread of the coordinate distribution $W_q(x,t)$.

4. Two complex phase factors occur in the wave function (6). The first of these, $\exp[i(p_0x/\hbar - p_0^2t/2m\hbar)]$, is the space and time phase factor that corresponds to a particle whose momentum is precisely p_0. The actual uncertainty in the momentum causes this factor to be only approximately correct for the actual wave function.

5. The rather complicated phase factor appearing in the second exponential involves the quantity $\hbar t/ma^2$, which is the ratio of the distance traversed at a speed $\hbar/ma$ during time t to the initial width a of the distribution $W_q(x,0)$. This quantity multiplies a second quantity which is equal to the square of the distance from the center $x_0 + p_0t/m$ of the distribution, measured in units of its "width" $2^{1/2}(a^2 + \hbar^2t^2/m^2a^2)^{1/2}$. This phase factor therefore is such that the wave function $\psi(x,t)$ oscillates more rapidly with x at large distances from the center of the distribution. This phase factor, whose exponent is symmetric in x, when combined with $\exp(ip_0x/\hbar)$, whose exponent is antisymmetric, yields a total rate of change of phase of ψ with x (at given time t) which is greater on one side of the distribution than on the other. This corresponds physically to the fact that, in the initial distribution $\psi(x,0)$, there was some uncertainty in p, and that, as time proceeds, the various parts of the wave function are composed of a superposition of waves which have moved to each given point at various speeds from the appropriate parts of the initial wave function. Since waves of short wavelength correspond to higher momentum, these waves will "arrive" at a given point beyond the center of the Gaussian $W_q(x,t)$ sooner than will the slower-moving long waves. Thus, the phase of the wave function changes more rapidly with x on the "forward" side of the packet than on the "backward" side. This effect is shown in Fig. 4-1.

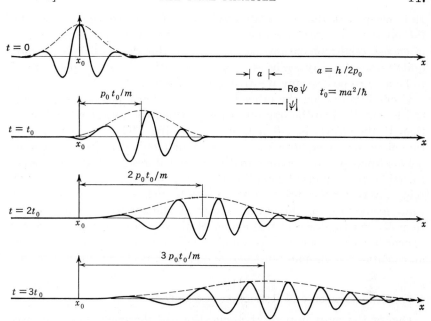

FIG. 4-1. Schematic plot of the real part of ψ, illustrating the propagation of the wave packet whose form is given by Eq. 4-4(6). The "center" of the packet moves with constant speed p_0/m, and the width increases with time. Note also that the wavelength of the oscillations becomes shorter and shorter on the "forward" side and longer and longer on the "backward" side of the packet.

EXERCISES

4-4. Show by direct substitution that the wave function (6) is a solution of Schroedinger's equation for a free particle in one dimension.

4-5. Evaluate the phase velocity of the wave function of a particle of momentum p.   *Ans.:* $v_{ph} = \frac{1}{2}(p/m)$.

4-6. Show that the group velocity of a wave packet is equal to the classical speed of the particle.   *Ans.:* $v_g = p/m$.

By the principles illustrated above, the motion of a wave packet having arbitrary initial form can be interpreted as being due to relative changes in phase of the various monoenergetic components of the packet brought about by the dispersion in the phase velocity with momentum.

4-5.   Eigenvalues of Energy and Momentum for a Free Particle

In writing the solution of the time-independent Schroedinger equation, the form 4-1(3) was chosen in order to exhibit explicitly the complex exponential character of the wave function for the case where $E > V$.

It is clear that the boundary conditions of the problem are satisfied for *any* value of E greater than V, so that there is a *continuous spectrum* of energy (and momentum) eigenvalues for this case. This is of course quite reasonable physically, since a free particle is known experimentally to be able to possess any positive kinetic energy.

If one inquires, on the other hand, whether any physical significance can be attached to the opposite case in which the total energy E is *less* than V, the answer is, offhand, in the negative. That is, a free particle cannot possess a *negative* kinetic energy or (equivalently) an *imaginary* momentum. Corresponding to this expected result, it is not possible to satisfy the boundary condition at $x = \pm \infty$ if the wave function contains any real exponential terms, so that there are no energy or momentum eigenvalues for this case. We shall find, nevertheless, that solutions having the form of real exponentials within a restricted region *do* appear in some problems involving "free" particles, and that these play an essential role in many distinctive quantum-mechanical phenomena.

4-6. Transmission and Reflection at a Barrier

One of the most important applications of the one-dimensional free-particle wave functions is to the approximate description of the behavior of a physical system whose *exact* behavior cannot be easily represented in simple form. The procedure used in such a case is to treat a model which possesses the proper qualitative features of the physical system but in which the actual *smoothly* varying potential function is replaced by a *stepwise* varying potential. The wave function of the model can then be written, within each region of constant potential, as a linear combination of exponential terms with real or imaginary exponents accordingly as $E < V$ or $E > V$, respectively. The complete wave function for the entire range of the coordinate is then obtained by matching these several solutions smoothly at the points of discontinuity of V.

EXERCISE

4-7. Find the conditions that must be satisfied by a solution of Schroedinger's equation near a point at which the potential energy undergoes an abrupt but finite change. *Ans.:* $u(x)$ and $u'(x)$ must both be continuous across the discontinuity.

As a simple example of this procedure, consider a beam of particles which impinges upon a potential "wall" of "height" V, as shown in Fig. 4-2. Such a potential function might be used to represent the physical situation of an electron inside a semi-infinite block of metal (Chap. 10). Let x be measured from the point of discontinuity.

Within each semi-infinite region the particle is free, but it suffers a sharp impulse toward the left whenever it impinges upon the discontinuity at $x = 0$. Clearly, the solution of the Schroedinger equation in the two regions may be written

(a) $$u_l(x) = Ae^{i\alpha x} + Be^{-i\alpha x} \qquad x < 0$$
(b) $$u_r(x) = Ce^{i\beta x} + De^{-i\beta x} \qquad x > 0$$
(1)

where (a) $$\alpha = \frac{(2mE)^{1/2}}{\hbar}$$

and (b) $$\beta = \frac{[2m(E - V)]^{1/2}}{\hbar}$$
(2)

Three qualitatively different cases may now be distinguished:

1. $E < 0$: In this case α and β are both imaginary, so that the exponents are real. Thus we may write

   (a) $$u_l(x) = Ae^{\alpha' x} + Be^{-\alpha' x} \qquad x < 0$$
   (b) $$u_r(x) = Ce^{\beta' x} + De^{-\beta' x} \qquad x > 0$$
   (3)

   where (a) $$\alpha' = \frac{(-2mE)^{1/2}}{\hbar}$$

   and (b) $$\beta' = \frac{[2m(V - E)]^{1/2}}{\hbar}$$
   (4)

   (The positive roots are to be used.) In order to satisfy the boundary condition at $\pm \infty$, we must clearly have $B = C = 0$. In this case,

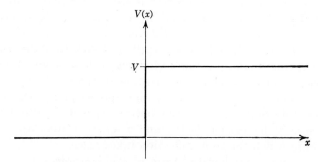

FIG. 4-2. A semi-infinite potential barrier.

however, it is not possible to satisfy the continuity condition at the point where the solutions are to be joined, since one cannot simultaneously make $u(x)$ and $u'(x)$ continuous across the junction. Therefore, *no wave function exists for this case.*

EXERCISE

4-8. Show that, if B and C are zero, it is not possible to satisfy the condition of continuity for $u(x)$ and $u'(x)$ at $x = 0$, except by putting $A = D = 0$.

2. $0 < E < V$: In this case the solutions may be written

(a) $$u_l(x) = A e^{i\alpha x} + B e^{-i\alpha x} \qquad x < 0$$
(b) $$u_r(x) = C e^{\beta' x} + D e^{-\beta' x} \qquad x > 0$$ (5)

and we must set $C = 0$ in order that u_r shall remain finite at $+\infty$. We then have two conditions at $x = 0$ with which to determine two of the remaining coefficients in terms of the third:

$$u_l(0) = u_r(0) \qquad \text{and} \qquad u_l'(0) = u_r'(0)$$

or, in terms of the coefficients A, B, and D,

$$A + B = D \qquad \text{and} \qquad i\alpha(A - B) = -\beta' D$$

Thus we may solve for B/A and D/A:

$$\frac{B}{A} = \frac{i\alpha + \beta'}{i\alpha - \beta'} \qquad \text{and} \qquad \frac{D}{A} = \frac{2i\alpha}{i\alpha - \beta'} \tag{6}$$

EXERCISE

4-9. Supply the missing steps in the derivation of Eq. (6).

These results may be interpreted physically as follows: If particles of homogeneous kinetic energy E are shot toward an abrupt potential barrier whose height V is greater than E, the beam will be *totally reflected* by this barrier (that is, $|B/A| = 1$). In this process, however, the particles *will penetrate the region which is classically forbidden* (that is, $|D/A| \neq 0$). This penetration is such that the chance of a particle being found in the forbidden region is an exponentially decreasing function of the penetration distance. This situation is closely similar to that found in the optical phenomenon of total internal reflection, since in the optical case it is also found that exponentially decreasing electric and magnetic fields exist outside the medium in which the total reflection takes place.

The solution (5a) has the form of *standing waves* in the region of lower potential. These standing waves are the result of the interference of the two wave trains. This effect is to be interpreted in the same way as were the nodes in the energy eigenfunctions for a harmonic oscillator, namely, as an interference of each particle in the beam with itself as it advances toward the potential barrier and is reflected.

3. $E > V$: In this case the wave function takes the form (1), and all four coefficients may have nonzero values. In order to obtain a physically simple situation, however, let $D = 0$. This means that no particles are approaching the barrier from the right. In this case

one may write, at $x = 0$,

$$A + B = C \quad \text{and} \quad i\alpha(A - B) = i\beta C \tag{7}$$

from which one readily obtains

$$\frac{B}{A} = \frac{\alpha - \beta}{\alpha + \beta} \quad \text{and} \quad \frac{C}{A} = \frac{2\alpha}{\alpha + \beta} \tag{8}$$

The physical situation represented by this solution is as follows: If particles of kinetic energy E are projected toward an abrupt potential barrier whose height V is less than E, a certain fraction, *but not all*, of the particles will surmount the barrier and will move beyond it with reduced kinetic energy $V - E$. On the other hand, *some of the particles will be reflected by the barrier*. This latter result is, of course, contrary to that expected classically for particles, but it is again very similar to the transmission and reflection of a classical *wave* at a refracting boundary.

EXERCISES

4-10. Supply the missing steps in the derivation of Eq. (8).

4-11. Show that the values of B/A and C/A given above satisfy the equation of continuity, that is, that the number of particles per unit time approaching the barrier is equal to the number leaving the barrier, according to the interpretation of superimposed wave trains presented in Sec. 4-3.

4-12. Treat the case in which particles are incident upon the barrier from the right. Find the ratios B/D and C/D, and thus show that *not all of the particles are transmitted, even though they gain energy in traversing the barrier*. This result is again contrary to that expected classically for the motion of particles, but is quite similar to the classical refraction and reflection of waves.

4-13. Show that it is possible to construct a perfectly transmitting barrier for particles of any specified energy greater than V by providing a transition layer between the two regions $V = 0$ and $V = V$, whose thickness and potential are suitably chosen. This is the wave-mechanical analogue of a low-reflectance coating in optics or a "matching section" in transmission-line theory.

The case just treated has brought out clearly some of the characteristic features of the quantum mechanics of particle motion; the term *wave mechanics* is often used to describe these properties of particles. The above discussion should also help the student to develop his insight regarding the behavior of particles in quantum mechanics. Clearly, the experience gained in the study of classical electrical and mechanical wave motion is quite useful in this regard.

4-7.   The Rectangular Potential Well

As a second illustration of the qualitative treatment of a physical system by the use of a stepwise varying potential, we now consider a simple one-dimensional situation which gives rise to discrete energy

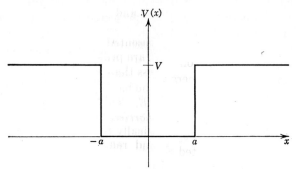

FIG. 4-3. A rectangular potential well.

levels and also to a continuum of levels.   Let the potential be equal to zero within a distance a on either side of the origin, and equal to $+V$ elsewhere, as shown in Fig. 4-3.   Such a potential energy function is called a *rectangular potential well*, or alternatively a "*square well*," and might be taken to represent a one-dimensional "atom," for example. As before, a particle moving under such a potential is subject to no forces except sharp, inward-directed impulses at $x = \pm a$.   We again consider the three cases that were discussed in the previous section:

1. $E < 0$: It is not possible to find wave functions which satisfy the finiteness conditions at $\pm \infty$ and also the continuity conditions at $x = \pm a$ for this case.
2. $0 < E < V$: It is in this energy range that the discrete energy levels lie.   The wave function may be written [see Eqs. 4-6(2) and (4)]

$$\begin{aligned} u_l(x) &= Ae^{\beta'x} + Be^{-\beta'x} & x &< -a \\ u_c(x) &= Ce^{i\alpha x} + De^{-i\alpha x} & -a &< x < +a \\ u_r(x) &= Fe^{\beta'x} + Ge^{-\beta'x} & x &> +a \end{aligned} \tag{1}$$

In order to satisfy the boundary conditions at $\pm \infty$, we must set $B = F = 0$.   We then have four continuity conditions at $x = \pm a$:

$$\begin{aligned} (a) && Ae^{-\beta'a} &= Ce^{-i\alpha a} + De^{i\alpha a} \\ (b) && \beta'Ae^{-\beta'a} &= i\alpha Ce^{-i\alpha a} - i\alpha De^{i\alpha a} \\ (c) && Ge^{-\beta'a} &= Ce^{i\alpha a} + De^{-i\alpha a} \\ (d) && -\beta'Ge^{-\beta'a} &= i\alpha Ce^{i\alpha a} - i\alpha De^{-i\alpha a} \end{aligned} \tag{2}$$

By eliminating A from the first pair of equations and G from the second pair, we easily obtain $C = \pm D$. From this we next find that, if $C = D$, then $A = G$, and if $C = -D$, then $A = -G$. *The wave functions thus exhibit the even or odd symmetry, or parity, with respect to x that is always associated with a Hamiltonian function that is symmetric in x.* It was shown in Chap. 3 that this is a fundamental property of nondegenerate energy eigenfunctions in cases such as the present one.

To proceed further, it is now necessary to observe that the above set of equations will be compatible with one another only for certain values of α and β', that is, only for certain *energies*. The symmetry which exists in the present case permits this condition to be expressed as follows: we see from the first pair of equations that, if $C = D$, we must have

$$e^{i\alpha a} + e^{-i\alpha a} = \frac{\alpha}{i\beta'} (e^{i\alpha a} - e^{-i\alpha a}) \quad \text{or} \quad \cos \alpha a = \frac{\alpha}{\beta'} \sin \alpha a \quad (3)$$

Similarly, if $C = -D$, we must have

$$\sin \alpha a = - \frac{\alpha}{\beta'} \cos \alpha a \tag{4}$$

These relations may be put into the simple form

$$\cot k = \frac{k}{(b^2 - k^2)^{1/2}} \tag{5}$$

and

$$- \tan k = \frac{k}{(b^2 - k^2)^{1/2}} \tag{6}$$

where

$$k = \frac{a}{\hbar} (2mE)^{1/2} \quad \text{and} \quad b = \frac{a}{\hbar} (2mV)^{1/2}$$

A solution of the above equations can be found graphically for given numerical values of a and V. Such a solution is shown in Fig. 4-4 for a well whose "depth" is $V = 49h^2/128ma^2$. It can be seen from this figure that, as $b \to \infty$, the roots of the transcendental equations (5) and (6) become uniformly spaced at a spacing of $\pi/2$. The energy eigenvalues for this case of an *infinitely deep* well are therefore given by

$$E_n = \frac{n^2 h^2}{32ma^2} \quad \begin{array}{c} (V = \infty) \\ n = 1, 2, 3, \ldots \end{array} \tag{7}$$

However, a well of finite depth possesses only a *finite number* of discrete energy eigenvalues, since there is no root greater than $k = b$.

The bound-state wave functions are shown in Fig. 4-5. Note the qualitative similarity of these wave functions to those for the harmonic oscillator.

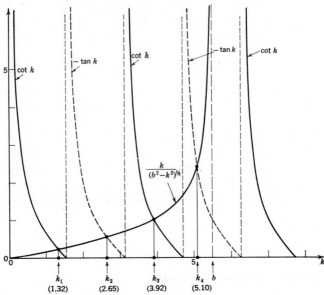

FIG. 4-4. Illustrating the solution of the transcendental equations $\cot k = k/(b^2 - k^2)^{1/2}$ and $-\tan k = k/(b^2 - k^2)^{1/2}$.

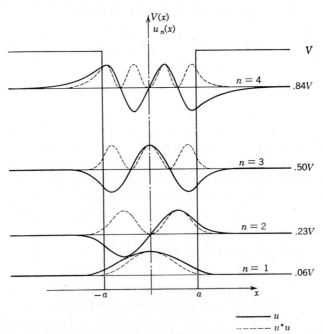

FIG. 4-5. The bound-state eigenfunctions for a rectangular potential well of depth $V = 49h^2/128\ ma^2$. Note that the penetration into the classically forbidden region $|x| > a$ is greater for the higher levels.

EXERCISES

4-14. Supply the missing steps leading to the result that $C = \pm D$ and $A = \pm G$, as outlined above.

4-15. Show that Eqs. (2c, d) are compatible if the transcendental equations (5) and (6) are satisfied.

4-16. Obtain a general expression for the number of discrete energy levels as a function of V and a. *Ans.:* N = least integer $\geq 4a(2mV)^{1/2}/h$.

4-17. Find an approximate solution of Eqs. (5) and (6) for the energy levels of a free particle moving in a rectangular potential well, and compare these with those shown in Fig. 4-4. *Ans.:* $k_n \approx \frac{1}{2}n\pi/(1 + 1/b)$.

3. $E > V$: In this case, all six coefficients may have nonzero values in the expressions

$$u_1(x) = A e^{i\beta x} + B e^{-i\beta x}$$
$$u_c(x) = C e^{i\alpha x} + D e^{-i\alpha x} \tag{8}$$

and
$$u_r(x) = F e^{i\beta x} + G e^{-i\beta x}$$

and the four continuity conditions at $x = \pm a$ may be used to reduce the number of *independent* coefficients to two. These may be identified with the amplitudes of wave trains (i.e., beams of particles) incident upon the "atom" from the left and from the right if the *scattering properties* of the "atom" are of interest, or with the amplitudes of outgoing or incoming waves if the growth or decay of short-lived unbound energy states are under study.

EXERCISE

4-18. Let a beam of particles strike an "atom" from the left (that is, let $G = 0$). Show that this beam will be *completely transmitted* (that is, $B = 0$) if the linear extent of the "atom" is an exact integral number of half wavelengths (that is, $2\alpha a = m\pi$). This is the one-dimensional analogue of the Ramsauer-Townsend effect, in which electrons of certain energies are observed to have abnormally long mean free paths in a noble gas.

4-8. Transmission through a Barrier

As a final example of useful one-dimensional free-particle problems, we shall briefly consider the reflection and transmission of a beam of particles by a *potential barrier* of the form shown in Fig. 4-6.

The procedure used in the previous examples can be applied to this problem also. One thus finds that no solution exists with $E < 0$ and that solutions exist for *all energies* with $E > 0$. If we wish to consider what happens to a beam of particles incident upon this barrier from the left, we assume solutions of the form [see Eqs. 4-6(2) and (4)]

$$E < V: \qquad u_1(x) = Ae^{i\alpha x} + Be^{-i\alpha x} \qquad x < -a$$
$$u_c(x) = Ce^{\beta' x} + De^{-\beta' x} \qquad -a < x < a \qquad (1)$$
$$u_r(x) = Fe^{i\alpha x} \qquad x > a$$

$$E > V: \qquad u_1(x) = Ae^{i\alpha x} + Be^{-i\alpha x} \qquad x < -a$$
$$u_c(x) = Ce^{i\beta x} + De^{-i\beta x} \qquad -a < x < a \qquad (2)$$
$$u_r(x) = Fe^{i\alpha x} \qquad x > a$$

The four continuity conditions at $x = \pm a$ then permit B, C, D, and F to be evaluated in terms of A. Since the expressions obtained are somewhat complicated, we shall discuss the problem qualitatively only.

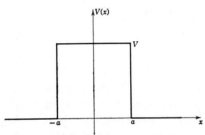

FIG. 4-6. A rectangular potential barrier.

If $E < V$, the wave function in the classically forbidden region is a combination of exponentials having real positive and negative exponents, the negative-exponent term being dominant. The beam is mostly reflected, but a certain fraction of the particles is transmitted past the barrier and proceeds with undiminished energy beyond it. This transmitted fraction increases steadily with increasing energy in this energy range. Thus we find, as a result of the impact of a particle with a barrier whose height is too great, classically, for the particle to surmount,

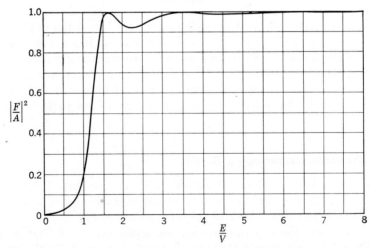

FIG. 4-7. The transmission probability for a rectangular barrier as a function of energy. The barrier height is equal to $h^2/8ma^2$. (By permission from L. I. Schiff, "Quantum Mechanics," 2d ed., McGraw-Hill Book Company, Inc., New York, 1955.)

that *there is a finite probability that the particle will actually penetrate beyond the barrier and be found on the other side.*   This result is an illustration of the very important quantum-mechanical phenomenon of *barrier penetration* which is responsible for such physical phenomena as field emission of electrons from metals and α-particle radioactivity.   This effect is also called the "*tunnel effect*," since one may say that a particle with insufficient energy to pass "over" such a barrier passes "*through*" it.

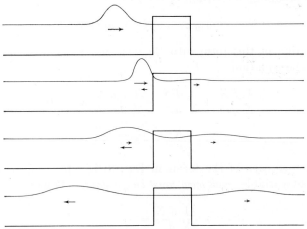

FIG. 4-8. Illustrating the penetration of a particle through a barrier (schematic).

As the energy of the bombarding particles is increased above V, the transmission probability continues to increase until it becomes unity. Further increase in the energy of the incident particles then results in a *smaller* transmission until a minimum is reached, followed by a second maximum, etc.   The transmission is complete for any energy which gives an integral number of half wavelengths inside the barrier.   This effect is shown in Fig. 4-7, which refers to a barrier whose width and height are connected by $8mVa^2 = h^2$.   The behavior of a wave packet as it strikes such a barrier is shown schematically in Fig. 4-8.

EXERCISE

4-19. Solve for F in terms of A, and thus verify that the above statements are correct.

Ans.:
$$\left| \frac{F}{A} \right| = \left[1 + \frac{V^2 \sin^2 [8m(E - V)]^{1/2}a/\hbar}{4E(E - V)} \right]^{-1/2}$$

4-9.   The Free Particle in Three Dimensions

We shall now consider briefly some of the simpler quantum-mechanical properties of a particle which is free to move in *three dimensions*.   The time-independent Schroedinger equation for the energy eigenfunctions for this case is

$$\frac{-\hbar^2}{2m} \nabla^2 u - Eu = 0 \tag{1}$$

where the potential energy has been set equal to zero and E is a constant. This equation can be solved for a perfectly free particle ($V = 0$ everywhere) by assuming a solution of the form

$$u(x,y,z) = X(x)\,Y(y)\,Z(z) \tag{2}$$

Substitution of this form into Eq. (1) and division by $u(x,y,z)$ then results in the equation

$$\frac{1}{X}\frac{d^2X}{dx^2} + \frac{1}{Y}\frac{d^2Y}{dy^2} + \frac{1}{Z}\frac{d^2Z}{dz^2} + K^2 = 0 \tag{3}$$

where

$$K^2 = \frac{2mE}{\hbar^2} \tag{4}$$

Inasmuch as each of the terms in Eq. (3) is a function of a different independent variable, *each term must be constant.* Thus let

(a)

$$\frac{1}{X}\frac{d^2X}{dx^2} = -\alpha^2$$

(b)

$$\frac{1}{Y}\frac{d^2Y}{dy^2} = -\beta^2 \tag{5}$$

(c)

$$\frac{1}{Z}\frac{d^2Z}{dz^2} = -\gamma^2$$

where

$$\alpha^2 + \beta^2 + \gamma^2 = K^2 \tag{6}$$

The solution of Eq. (5) may be written in complex exponential form:

(a) $X(x) = Ae^{i\alpha x} + A'e^{-i\alpha x}$

(b) $Y(y) = Be^{i\beta y} + B'e^{-i\beta y}$ (7)

(c) $Z(z) = Ce^{i\gamma z} + C'e^{-i\gamma z}$

or, alternatively, in the form

$$u(x,y,z) = A\,\exp\,[i(\alpha x + \beta y + \gamma z)]$$
$$= A\,\exp\,(i\mathbf{K}\cdot\mathbf{r}) \tag{8}$$

where

$$\mathbf{K} = \alpha\mathbf{i} + \beta\mathbf{j} + \gamma\mathbf{k} \tag{9}$$

Thus the solutions of Eq. (1) may be written in the form of *plane waves* of arbitrary wavelength whose wavefronts are perpendicular to a vector $\mathbf{K}$, called the *propagation vector* of the plane wave. Just as in the one-dimensional case, the energy eigenfunctions (8) are also eigenfunctions of the momentum.

EXERCISE

4-20. Show that Eq. (8) is an eigenfunction of p_x, p_y, and p_z, and therefore of p. Find the momentum p in terms of the propagation vector K.

$$Ans.: \qquad\qquad p = \hbar K \qquad\qquad (10)$$

For a perfectly free particle any value of p is permissible and the various momentum and energy eigenfunctions can be written

$$\psi_p(r,t) = A \exp (ip \cdot r/\hbar) \exp (-ip^2t/2m\hbar) \qquad (11)$$

The most general expression for $\psi(r,t)$ is then a superposition of plane waves of the form

$$\psi(r,t) = \int_{-\infty}^{\infty} \int_{-\infty}^{\infty} \int_{-\infty}^{\infty} A(p) \exp (ip \cdot r/\hbar)$$
$$\exp (-ip^2t/2m\hbar) \, dp_x \, dp_y \, dp_z \qquad (12)$$

The treatment of wave packets given in Sec. 4-4 for one dimension can be extended directly to three dimensions, with the result that the three-dimensional wave function can be expressed as a product of three independent one-dimensional wave packets each of which moves in accordance with the one-dimensional Schroedinger equation.

4-10. The Particle in a Box; Periodic Boundary Condition

We shall now treat a problem of prime importance whose solution finds application in many diverse contexts: that of a particle which is confined within a rectangular box but which is otherwise free. We shall for the present consider only the case in which the walls of the box are completely impenetrable to the particle: The potential energy is *infinite* outside the box; for simplicity we shall assume it to be zero inside the box. The Schroedinger equation and its solutions for this case are the same as for a free particle, except that there are now certain *boundary conditions* that must be satisfied by $u(r)$.

EXERCISE

4-21. By considering the behavior of the wave function for a one-dimensional rectangular potential well as the potential energy outside the well becomes infinite, or by some other means, find the boundary conditions that $u(r)$ must satisfy at the walls of the box.

$$Ans.: \qquad\qquad u(r) = 0 \qquad \text{at each wall} \qquad\qquad (1)$$

The boundary condition (1) on $u(r)$ at the box walls clearly restricts the possible choices of values of α, β, and γ (or, equivalently, of K) to those certain ones for which each of the sinusoidal factors 4-9(7) vanishes on each of the two walls appropriate to that factor. Clearly, the wave function must be of the form

$$u(\boldsymbol{r}) = A \sin \frac{l\pi x}{a} \sin \frac{m\pi y}{b} \sin \frac{n\pi z}{c} \tag{2}$$

where the box walls are defined by the six planes $x = 0$, $x = a$, $y = 0$, $y = b$, $z = 0$, and $z = c$, and l, m, and n are positive integers.

EXERCISES

4-22. Find the energy eigenvalues corresponding to the eigenfunctions (2).

Ans.:
$$E_{lmn} = \frac{\hbar^2 \pi^2}{2m} \left(\frac{l^2}{a^2} + \frac{m^2}{b^2} + \frac{n^2}{c^2} \right) \tag{3}$$

4-23. Find the absolute value of the coefficient A. *Ans.:* $|A| = (8/abc)^{1/2}$.

4-24. Rewrite Eq. (2), using a coordinate system whose origin is at the center of the box, and thus show that the eigenfunctions (2) have even or odd parity with respect to each of these x', y', and z' coordinates. The *total parity* describes the behavior of the wave function under the simultaneous change in sign of x', y', and z' together. Find how the total parity depends upon l, m, and n. *Ans.:* Total parity is even if $l + m + n - 1$ is even, odd if $l + m + n - 1$ is odd.

For certain applications it is necessary to know the relative number of energy or momentum eigenstates per unit energy or momentum interval for a free particle. For this purpose the particle is regarded as being confined inside a box whose size is so large and walls so distant that their presence does not materially influence the particle. In such a treatment it is also usually desirable to express the eigenfunctions as *traveling waves* rather than standing waves. This can be done without changing the number of eigenvalues per unit energy interval by replacing the boundary condition (1) by a less stringent one, called the *periodic boundary condition*, in which $u(\boldsymbol{r})$ is not required to vanish at each wall but merely to be periodic in each direction with periods equal to the three edges of the box. Thus in place of the eigenfunctions (2) one has

$$u(\boldsymbol{r}) = A \exp 2\pi i \left(\frac{lx}{a} + \frac{my}{b} + \frac{nz}{c} \right) = A \exp (i\boldsymbol{K} \cdot \boldsymbol{r}) \tag{4}$$

where $\boldsymbol{K} = 2\pi[(l/a)\boldsymbol{i} + (m/b)\boldsymbol{j} + (n/c)\boldsymbol{k}]$ and l, m, and n are *positive or negative integers*.

EXERCISE

4-25. (*a*) Evaluate the number of normal modes per unit wavelength interval for a vibrating uniform string of length L with fixed ends, in the limit of small wavelength. (*b*) Replace the boundary condition in part (*a*) with the condition that $y(x)$ be periodic with period L, and evaluate as above the number of traveling waves per unit wavelength interval as $\lambda \to 0$. *Ans.:* $\Delta n/\Delta\lambda \to 2L/\lambda^2$ for each case.

Exercise 4-25 shows that the periodic boundary condition in effect replaces standing waves, which are waves of equal amplitude traveling in opposite directions, with waves of *arbitrary* amplitude traveling in each direction but having wavelengths equal to integral submultiples of L rather than of $2L$. That these two representations are interchangeable and equally suitable for a given purpose depends, of course, upon the fact that the region within which a physical phenomenon is localized is small compared with the size of the region whose boundaries define the periodicity of the eigenfunctions of the free particles in question. If this condition is satisfied, either set of eigenfunctions can equally well be used as a complete set of orthogonal functions with which to construct wave packets representing particles localized within a small part of the over-all region. We shall have occasion to use each of these representations of free-particle energy states from time to time.

EXERCISES

4-26. Use Eq. (3) to derive the number of energy eigenvalues per unit energy interval for a particle in a box. This is most easily done by defining a system of rectangular Cartesian coordinates $\xi = l/a$, $\eta = m/b$, $\zeta = n/c$. Two given energies, E and $E + \Delta E$, then define two *spherical surfaces* in ξ, η, ζ space, and all points corresponding to positive integral l, m, and n values lying between these surfaces yield energies in the range ΔE. The number ΔN of such points is very nearly equal to the volume between these surfaces divided by the volume per point.

Ans.:
$$\Delta N = \frac{4\pi V}{h^3}(2m^3)^{1/2}E^{1/2}\,\Delta E \tag{5}$$

4-27. Treat the case of traveling waves described by the periodic boundary condition by the same analysis as above, and show that the same formula is obtained for ΔN.

4-28. Find the number of energy eigenstates per unit volume of *phase space* for a free particle.

Ans.:
$$\frac{\Delta N}{V\,\Delta p_x\,\Delta p_y\,\Delta p_z} = \frac{1}{h^3} \tag{6}$$

The results of the Exercises 4-26 to 4-28 are of fundamental importance in the development of quantum statistics.   We shall use them in Chap. 10 and also in connection with β-decay in Chap. 15.

This brief discussion of the free particle has served to introduce many of the important characteristic features of quantum mechanics in relatively elementary form wherein the essential physical concepts stand out most clearly.   We shall meet these same features in many places in our forthcoming study of atomic and nuclear structure.

REFERENCES

Rojanski, V.: "Introductory Quantum Mechanics," Prentice-Hall, Inc., Englewood Cliffs, N.J., 1938.
Schiff, L. I.: "Quantum Mechanics," 2d ed., International Series in Pure and Applied Physics, McGraw-Hill Book Company, Inc., New York, 1955.

5

The One-electron Atom

The work of the preceding chapters has provided us with most of the tools we shall need for our study of atomic structure using the quantum theory. In this chapter we shall investigate in some detail the structure of the simplest atoms in nature, namely, those which consist of a *nucleus and a single electron*. Examples of such atoms are neutral hydrogen (including its isotopes deuterium and tritium), singly ionized helium, doubly ionized lithium, and so on. We shall find that even these simplest atoms actually are tremendously complex, because electrons and nuclei themselves have properties other than mass and charge which must be taken into account in a complete description of atomic structure. The great advantage in introducing these properties of electrons and nuclei in connection with the one-electron atom is that many of their effects can be *analyzed quantitatively* for this simple case, whereas nearly every feature of a complicated atom defies accurate quantitative analysis beginning from first principles. We shall therefore devote considerable attention even to effects which are quite tiny in one-electron atoms, because many of these appear in magnified form in the structure of more complicated atoms and would be more difficult to analyze and to understand if introduced later and because the quantitative magnitudes of these effects in one-electron atoms may provide rigorous tests of theories of particle structure. Finally, and most important, the one-electron-atom quantum states provide an essential basis for the description of the quantum states of more complex atoms. Much of the nomenclature used in this description is adapted directly from that introduced in the study of the one-electron atom.

Our treatment of the one-electron atom will proceed along the following lines: we shall first examine the eigenstates of a simple system consisting of a nucleus and an electron, neglecting the various properties of these particles other than their charge and mass. The effects of these additional properties are quite small and will be introduced one by one as perturbations. As these small effects are introduced, we shall find it advantageous to digress now and then from our treatment of the specific case of a one-electron atom to discuss in greater detail some of the more general properties of quantum mechanics which are in operation in the one-electron atom and which play an essential role in various physical phenomena.

5-1. Hamiltonian Function and Operator for the Simple Model

Following the procedure outlined above, we take as our simple model of the one-electron atom the same as was used in the Bohr theory, namely, a single electron of mass m and charge $-e$, moving about a nucleus of mass M and charge $+Ze$. The Lagrangian and Hamiltonian functions for the classical system are first evaluated in the usual manner.

EXERCISES

5-1. Using as coordinates the rectangular coordinates ξ, η, ζ of the center of mass and the relative rectangular coordinates x, y, z of the electron with respect to the nucleus, show that the Lagrangian function for the model described above is

$$L = \tfrac{1}{2}(m + M)(\dot{\xi}^2 + \dot{\eta}^2 + \dot{\zeta}^2) + \tfrac{1}{2}m_r(\dot{x}^2 + \dot{y}^2 + \dot{z}^2)$$
$$+ \frac{Ze^2}{4\pi\varepsilon_0(x^2 + y^2 + z^2)^{\frac{1}{2}}} \quad (1)$$

where m_r is the so-called *reduced mass* $mM/(m + M)$.

5-2. Find the Hamiltonian function that corresponds to the above Lagrangian function.

$$Ans.: \quad H = \frac{p_\xi^2 + p_\eta^2 + p_\zeta^2}{2(m + M)} + \frac{p_x^2 + p_y^2 + p_z^2}{2m_r} - \frac{Ze^2}{4\pi\varepsilon_0(x^2 + y^2 + z^2)^{\frac{1}{2}}}$$
$$(2)$$

From the above Hamiltonian function, the Hamiltonian operator is obtained as prescribed by the third postulate of Chap. 2. It is

$$\hat{H} = \frac{-\hbar^2}{2(m + M)} \nabla^2_{\text{CM}} - \frac{\hbar^2}{2m_r} \nabla^2 - \frac{Ze^2}{4\pi\varepsilon_0 r} \quad (3)$$

where
$$\nabla^2_{CM} = \partial^2/\partial\xi^2 + \partial^2/\partial\eta^2 + \partial^2/\partial\zeta^2$$
$$\nabla^2 = \partial^2/\partial x^2 + \partial^2/\partial y^2 + \partial^2/\partial z^2$$
$$r^2 = x^2 + y^2 + z^2$$

This Hamiltonian operator is a sum of two parts,

$$\hat{H} = \hat{H}_{CM} + \hat{H}_{rel} \tag{4}$$

where $\hat{H}_{CM}$ depends only upon the *coordinates of the center of mass*, and $\hat{H}_{rel}$ depends only upon the *relative coordinates* of the two particles. Such a separation of the Hamiltonian operator into *independent parts* always leads to eigenfunctions which are a product of separate eigenfunctions, as was described in Exercise 2-50. Thus for the present case,

$$\psi = \psi_{CM}(\xi,\eta,\zeta)\psi_{rel}(x,y,z) \tag{5}$$

where ψ_{CM} and ψ_{rel} are wave functions corresponding to systems whose Hamiltonian operators are $\hat{H}_{CM}$ and $\hat{H}_{rel}$, respectively. That is, the complete problem can be separated, just as in the classical case, into two independent parts, one corresponding to the motion of the CM regarded as a particle of mass $m + M$ and the other to the *relative* motion of the two particles. The motion of the CM does not concern us at the present time, so we shall treat only that part of the problem relating to the relative motion.

Because of the spherical symmetry of the system it is best to write $\hat{H}_{rel}$ in *spherical polar coordinates* with polar axis parallel to the z-axis:

$$\hat{H}_{rel} = -\frac{\hbar^2}{2m_r r^2}\left[\frac{\partial}{\partial r}\left(r^2\frac{\partial}{\partial r}\right) + \frac{1}{\sin\theta}\frac{\partial}{\partial\theta}\left(\sin\theta\frac{\partial}{\partial\theta}\right) + \frac{1}{\sin^2\theta}\frac{\partial^2}{\partial\varphi^2}\right] - \frac{Ze^2}{4\pi\varepsilon_0 r} \tag{6}$$

EXERCISE

5-3. Carry out the transformation of coordinates that leads to Eq. (6).

5-2. The Schroedinger Equation

Having expressed the Hamiltonian operator in a system of coordinates that is appropriate to the problem, the Schroedinger equation may be written. Because we are primarily interested in the stationary states in which the energy of the system is precisely defined, we use the separated form of this equation. Accordingly we write

$$\psi_n(r,\theta,\varphi,t) = u_n(r,\theta,\varphi)\exp(-iE_n t/\hbar) \tag{1}$$
where
$$\hat{H}_{rel}u_n(r,\theta,\varphi) = E_n u_n(r,\theta,\varphi) \tag{2}$$

The problem thus is to find the eigenvalues E_n and the eigenfunctions $u_n(r,\theta,\varphi)$ that satisfy Eq. (2). The most general wave function for the system can then be expressed as a linear combination of ψ_n's of the form (1). Upon substitution of the Hamiltonian operator 5-1(6) into Eq. (2), the time-independent Schroedinger equation (2) becomes

$$-\frac{\hbar^2}{2m_r r^2}\left[\frac{\partial}{\partial r}\left(r^2\frac{\partial u_n}{\partial r}\right)+\frac{1}{\sin\theta}\frac{\partial}{\partial\theta}\left(\sin\theta\frac{\partial u_n}{\partial\theta}\right)+\frac{1}{\sin^2\theta}\frac{\partial^2 u_n}{\partial\varphi^2}\right]-\frac{Ze^2 u_n}{4\pi\varepsilon_0 r}$$
$$-E_n u_n = 0 \quad (3)$$

5-3.   Solution of Time-independent Schroedinger Equation

The partial differential equation 5-2(3) may be solved by the customary procedure of separating the function $u_n(r,\theta,\varphi)$ into a product of functions of a single variable. Thus we let

$$u_n(r,\theta,\varphi) = A_n R(r)\Theta(\theta)\Phi(\varphi) \quad (1)$$

where A_n is an amplitude factor independent of r, θ, and φ. Substitution into Eq. 5-2(3) and division by $AR\Theta\Phi$ then yield the equation

$$-\frac{\hbar^2}{2m_r r^2}\left[\frac{1}{R}\frac{d}{dr}\left(r^2\frac{dR}{dr}\right)+\frac{1}{\Theta\sin\theta}\frac{d}{d\theta}\left(\sin\theta\frac{d\Theta}{d\theta}\right)+\frac{1}{\Phi\sin^2\theta}\frac{d^2\Phi}{d\varphi^2}\right]-\frac{Ze^2}{4\pi\varepsilon_0 r}$$
$$-E_n = 0 \quad (2)$$

To separate the functions R, Θ, and Φ, we first multiply the entire equation by $2m_r r^2\sin^2\theta/\hbar^2$. We can then put on the right-hand side of the equation a function of φ only, and on the left-hand side a function of r and θ only:

$$-\frac{\sin^2\theta}{R}\frac{d}{dr}\left(r^2\frac{dR}{dr}\right)-\frac{\sin\theta}{\Theta}\frac{d}{d\theta}\left(\sin\theta\frac{d\Theta}{d\theta}\right)-\frac{2Ze^2 m_r r\sin^2\theta}{4\pi\varepsilon_0\hbar^2}$$
$$-\frac{2E_n m_r r^2\sin^2\theta}{\hbar^2}=\frac{1}{\Phi}\frac{d^2\Phi}{d\varphi^2} \quad (3)$$

Each side of the equation must therefore be equal to a constant. The resulting equation for R and Θ may be further separated by dividing through by $\sin^2\theta$. Using a notation for the separation constants which will later be found to be most appropriate, the equations for R, Θ, and Φ then assume the form

(a) $\qquad \dfrac{1}{r^2}\dfrac{d}{dr}\left(r^2\dfrac{dR}{dr}\right)+\dfrac{2m_r}{\hbar^2}\left[E_n+\dfrac{Ze^2}{4\pi\varepsilon_0 r}-\dfrac{\hbar^2 l(l+1)}{2m_r r^2}\right]R=0$

(b) $\qquad \dfrac{1}{\sin\theta}\dfrac{d}{d\theta}\left(\sin\theta\dfrac{d\Theta}{d\theta}\right)+\left[l(l+1)-\dfrac{m^2}{\sin^2\theta}\right]\Theta=0 \quad (4)$

(c) $\qquad \dfrac{d^2\Phi}{d\varphi^2}+m^2\Phi=0$

EXERCISE

5-4. Carry out the separation procedure that leads to Eq. (4).

To solve these equations we start with the equation for Φ and obtain at once

$$\Phi_m(\varphi) = A e^{\pm im\varphi} \tag{5}$$

We now observe that, since $u_n(r,\theta,\varphi)$ must be a single-valued function of position, the separation constant m must be a real integer:

$$m = 0, 1, 2, \ldots \tag{6}$$

EXERCISE

5-5. Show that the condition that the wave function be single-valued requires that m in Eq. (5) be a real integer which can be taken to be positive or zero.[1]

The equation for Θ may now be treated. It can be put in standard form by changing the independent variable to $\mu = \cos\theta$. Thus one obtains

$$\frac{d}{d\mu}\left[(1 - \mu^2)\frac{d\Theta}{d\mu}\right] + \left[l(l+1) - \frac{m^2}{1 - \mu^2}\right]\Theta = 0 \tag{7}$$

This is the differential equation for the *associated Legendre functions* $P_l^m(\mu)$ These functions can be defined in terms of the ordinary Legendre functions $P_l(\cos\theta)$ through the equation

$$P_l^m(\mu) = (1 - \mu^2)^{m/2}\frac{d^m P_l(\mu)}{d\mu^m} \tag{8}$$

To show that the functions defined by Eq. (8) also satisfy Eq. (7), we start with the differential equation for $P_l(\cos\theta)$:

$$(1 - \mu^2)\frac{d^2 P_l}{d\mu^2} - 2\mu\frac{dP_l}{d\mu} + l(l+1)P_l = 0 \tag{9}$$

---

[1] One may alternatively write $\Phi_m(\varphi) = A e^{im\varphi}$ and regard m as a positive or negative integer. In this case one should use $|m|$ in place of m in the associated Legendre functions introduced below. In order to avoid the clumsy appearance of $|m|$ in these functions, we shall write them as $P_l^m(\mu)$, but in our later discussion we shall often speak of m as being positive or negative.

Differentiation of this equation m times results in

$$(1 - \mu^2) \frac{d^{m+2}}{d\mu^{m+2}} P_l - 2(m + 1)\mu \frac{d^{m+1}}{d\mu^{m+1}} P_l$$

$$+ [l(l + 1) - m(m + 1)] \frac{d^m P_l}{d\mu^m} = 0$$

On the other hand, substitution of $\Theta = W \sin^m \theta$ into Eq. (7) results in

$$(1 - \mu^2) W'' - 2(m + 1)\mu W' + [l(l + 1) - m(m + 1)]W = 0$$

So that by inspection, $W = (d^m/d\mu^m) P_l$.

EXERCISE

5-6. Carry out the operations just described, and thus verify that $P_l^m (\cos \theta)$, as defined by Eq. (8), is a solution of Eq. (7) if $P_l(\cos \theta)$ is a solution of Eq. (9).

We must now solve the differential equation (9). This can be done by the power-series method,[1] by substituting

$$P_l(\mu) = \sum_{s=0}^{\infty} a_s \mu^s \tag{10}$$

into Eq. (9). We thus obtain

$$\sum_{s=0}^{\infty} [s(s - 1)a_s\mu^{s-2} - s(s - 1)a_s\mu^s - 2sa_s\mu^s + l(l + 1)a_s\mu^s] = 0 \tag{11}$$

Collection of coefficients of each power of μ then gives

$$\sum_{r=0}^{\infty} \{(r + 2)(r + 1)a_{r+2} - [r(r + 1) - l(l + 1)]a_r\}\mu^r = 0 \tag{12}$$

Since this expansion is equal to zero for all values of μ between -1 and $+1$, the coefficient of each power of μ must vanish separately. In this way the recurrence relation

$$a_{r+2} = \frac{r(r + 1) - l(l + 1)}{(r + 2)(r + 1)} a_r \tag{13}$$

is obtained. The two linearly independent solutions of Eq. (9) for given l are therefore an even and an odd function of μ, respectively, and these functions are completely specified if a_0 and a_1, the first two coefficients in Eq. (10), are given.

---

[1] Equation (7) can also be solved by power series, but the procedure is somewhat complicated by the fact that $\mu = \pm 1$ are singular points of the differential equation.

Now examination of Eq. (13) reveals that, as $r \to \infty$, $a_{r+2} \to a_r$ for any finite value of l. Thus, both of the solutions of Eq. (9) evidently exhibit logarithmic singularities as $\mu \to \pm 1$, so that *neither* appears to be an acceptable solution of Eq. (9) for application to physical problems in which μ can possess the values ± 1. As has been emphasized in numerous previous cases, it is this situation that leads to discrete eigenvalues for physical quantities, for physically acceptable wave functions can be found *only for certain values of some parameter which is related to a physical property of the system.* In the present case the parameter is the separation constant l, which we shall shortly find to be related to the *angular momentum* of the system. We see that Eq. (13) defines an acceptable function *only if l is equal to an integer.* (Only zero and positive integers need be considered, since negative integers give the same functions as do positive ones.) For each of these integral values, one of the functions defined by Eq. (13) is a *polynomial of degree l* and the other has a logarithmic singularity. This latter function, which is called $Q_l(\mu)$, is eliminated by setting its coefficient equal to zero. For successive values of l the polynomial solutions $P_l(\mu)$ are alternately even and odd functions of μ. A few of these Legendre polynomials and associated Legendre functions obtained from them by Eq. (8) are given below. Note that $m \leq l$ for these functions.

$$P_0(\mu) = 1 \quad P_1(\mu) = \mu \qquad P_2(\mu) = \tfrac{1}{2}(3\mu^2 - 1) \quad P_3(\mu) = \tfrac{1}{2}(5\mu^3 - 3\mu)$$
$$P_1^1(\mu) = (1 - \mu^2)^{1/2} \quad P_2^1(\mu) = 3\mu(1 - \mu^2)^{1/2} \quad P_3^1(\mu) = \tfrac{3}{2}(5\mu^2 - 1)(1 - \mu^2)^{1/2}$$
$$P_2^2(\mu) = 3(1 - \mu^2) \quad P_3^2(\mu) = 15\mu(1 - \mu^2)$$
$$P_3^3(\mu) = 15(1 - \mu^2)^{3/2}$$

$$\tag{14}$$

The normalizing factor for these functions is

$$N_{lm} = \left[\frac{(2l + 1)(l - m)!}{2(l + m)!} \right]^{1/2} \tag{15}$$

and with this factor the associated functions satisfy the orthogonality relations

$$\int_0^\pi N_{lm} N_{l'm} P_l^m(\cos\theta) P_{l'}^m(\cos\theta) \sin\theta \, d\theta = \delta_{ll'} \tag{16}$$

The normalized associated Legendre functions are designated

$$\Theta_{lm}(\theta) = N_{lm} P_l^m(\cos\theta)$$

Some of these functions are given analytically on page 171 and are shown graphically in Figs. 5-1 and 5-2.

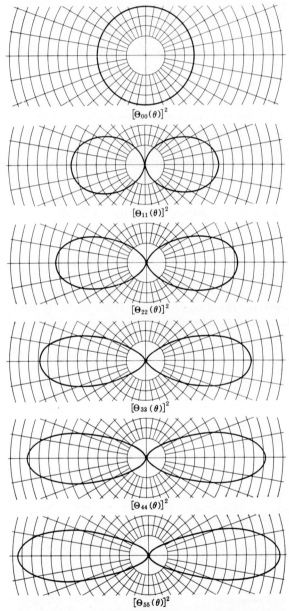

FIG. 5-1. Polar graphs of the function $[\Theta_{lm}(\theta)]^2$ for $m = \pm l$ and $l = 0, 1, 2, 3, 4,$ and 5, showing the concentration of the function about the xy plane with increasing l.   (By permission from L. Pauling and E. B. Wilson, Jr., "Introduction to Quantum Mechanics," McGraw-Hill Book Company, Inc., New York, 1935.)

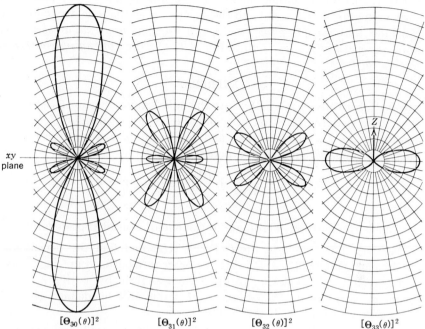

$$[\Theta_{30}(\theta)]^2 \qquad [\Theta_{31}(\theta)]^2 \qquad [\Theta_{32}(\theta)]^2 \qquad [\Theta_{33}(\theta)]^2$$

FIG. 5-2. Polar graphs of the function $[\Theta_{lm}(\theta)]^2$ for $l = 3$ and $m = 0, \pm1, \pm2,$ and ±3. (By permission from L. Pauling and E. B. Wilson, Jr., "Introduction to Quantum Mechanics," McGraw-Hill Book Company, Inc., New York, 1935.)

$$\Theta_{00} = 2^{-\frac{1}{2}}$$

$$\Theta_{10} = \tfrac{1}{2}\sqrt{6}\cos\theta \qquad \Theta_{20} = \tfrac{1}{4}\sqrt{10}\,(3\cos^2\theta - 1)$$
$$\Theta_{11} = \tfrac{1}{2}\sqrt{3}\sin\theta \qquad \Theta_{21} = \tfrac{1}{2}\sqrt{15}\sin\theta\cos\theta$$
$$\Theta_{22} = \tfrac{1}{4}\sqrt{15}\sin^2\theta$$

$$\Theta_{30} = \tfrac{3}{4}\sqrt{14}\,(\tfrac{5}{3}\cos^3\theta - \cos\theta)$$
$$\Theta_{31} = \tfrac{1}{8}\sqrt{42}\sin\theta(5\cos^2\theta - 1)$$
$$\Theta_{32} = \tfrac{1}{4}\sqrt{105}\sin^2\theta\cos\theta \tag{17}$$
$$\Theta_{33} = \tfrac{1}{8}\sqrt{70}\sin^3\theta$$

It will shortly be shown that these angular functions, and the integers l and m, are connected with various *angular-momentum states* of the system.

EXERCISES

5-7. Show that the polynomials defined by Eqs. (10) and (13) for $l = 0, 1, 2,$ and 3 agree within a multiplying factor with those given in Eq. (14).

5-8. Substitute the above polynomials into Eq. (8) and thus verify that the associated Legendre functions given in Eq. (14) satisfy Eq. (8).

Up to this point neither the energy eigenvalue E_n nor the potential-energy function $V(r)$ has entered the discussion, so that one may correctly conclude that *the foregoing results will apply to any similar system in which the potential energy is a function only of r* (i.e., *to any system involving central forces*).

EXERCISE

5-9. Substitute $\chi(r) = rR(r)$ into the radial-wave equation 5-3(4a) and thus show that the function $\chi(r)$ and the energy eigenvalue E_n can be considered as an eigenfunction and eigenvalue for a *one-dimensional* particle of mass m_r moving in a potential

$$V(r) = -\frac{Ze^2}{4\pi\varepsilon_0 r} + \frac{l(l+1)\hbar^2}{2m_r r^2} \qquad r > 0$$

and $\qquad\qquad V(r) = \infty \qquad\qquad\qquad\qquad r < 0$

where r is the position coordinate in the one-dimensional problem.

We must now evaluate the function $R(r)$ and find the permissible energy eigenvalues E_n. Following the procedure outlined in Chap. 3, we first transform the equation to dimensionless form by setting $\rho = \alpha_n r$, where α_n is a suitable constant whose value will be defined shortly. Thus $d/dr = \alpha_n d/d\rho$, so that Eq. (4a) becomes

$$\frac{1}{\rho^2}\frac{d}{d\rho}\left(\rho^2\frac{dR}{d\rho}\right) + \frac{2m_r}{\hbar^2}\left[\frac{E_n}{\alpha_n^2} + \frac{Ze^2}{4\pi\varepsilon_0\alpha_n\rho} - \frac{l(l+1)\hbar^2}{2m_r\rho^2}\right]R = 0 \quad (18)$$

The negative eigenvalues of E, corresponding to bound states, are obtained if we set $2m_r E_n/\hbar^2\alpha_n^2 = -\frac{1}{4}$ and call $2m_r Ze^2/4\pi\varepsilon_0\alpha_n\hbar^2 = \lambda_n$. That is, we let

$$\alpha_n^2 = -\frac{8m_r E_n}{\hbar^2} \qquad \text{and} \qquad \lambda_n = \frac{Ze^2}{4\pi\varepsilon_0\hbar}\left(-\frac{m_r}{2E_n}\right)^{1/2} \quad (19)$$

then the equation becomes

$$\frac{d^2R}{d\rho^2} + \frac{2}{\rho}\frac{dR}{d\rho} + \left[\frac{\lambda_n}{\rho} - \frac{1}{4} - \frac{l(l+1)}{\rho^2}\right]R = 0 \quad (20)$$

To solve this equation, we first investigate its behavior as $\rho \to \infty$. In this region the equation approaches

$$\frac{d^2R}{d\rho^2} - \frac{R}{4} = 0 \quad (21)$$

which has as its solution

$$R(\rho) = Ae^{\rho/2} + Be^{-\rho/2} \quad (22)$$

Physically, the positive exponential is not allowed, so that we assume, for all values of ρ, a solution of the form

$$R(\rho) = e^{-\rho/2}F(\rho) \quad (23)$$

Substitution of this form into Eq. (20) and cancellation of $e^{-\rho/2}$ results in the equation for $F(\rho)$

$$\frac{d^2F}{d\rho^2} + \left(\frac{2}{\rho} - 1\right)\frac{dF}{d\rho} + \left[\frac{\lambda_n - 1}{\rho} - \frac{l(l+1)}{\rho^2}\right]F = 0 \qquad (24)$$

EXERCISE

5-10. Carry out the substitutions that lead from Eq. (4a) to Eq. (24).

The above equation for $F(\rho)$ may be solved by the power-series method. In view of the regular singularity at $\rho = 0$, we assume a solution of the form

$$F(\rho) = \rho^s \sum_{k=0}^{\infty} a_k \rho^k \qquad a_0 \neq 0 \qquad (25)$$

Substituting this into Eq. (24), we find that

$$\sum_{k=0}^{\infty} [(s+k)(s+k-1)a_k\rho^{s+k-2} + 2(s+k)a_k\rho^{s+k-2} - (s+k)a_k\rho^{s+k-1}$$
$$- a_k\rho^{s+k-1} + \lambda_n a_k\rho^{s+k-1} - l(l+1)a_k\rho^{s+k-2}] = 0 \quad (26)$$

If we group together the coefficients of the various powers of ρ, we obtain

$$[s(s+1) - l(l+1)]a_0\rho^{s-2} + \sum_{t=0}^{\infty} \{[(s+t+1)(s+t+2)$$
$$- l(l+1)]a_{t+1} - (s+t+1-\lambda_n)a_t\}\rho^{s+t-1} = 0 \quad (27)$$

The coefficient of each power of ρ must vanish separately. This requires that

(a) $\qquad\qquad s(s+1) - l(l+1) = 0 \qquad$ since $a_0 \neq 0 \qquad (28)$

(b) $\quad a_{t+1} = \dfrac{(s+t+1-\lambda_n)}{(s+t+1)(s+t+2) - l(l+1)} a_t \qquad t = 0, 1, 2, \ldots$

The first of these equations is called the *indicial equation;* it determines the value of s. The remaining equations are called *recurrence relations,* by which the successive coefficients a_k are defined in terms of a_0. By inspection we see that the indicial equation requires that $s = l$ or $s = -(l+1)$. The latter value is excluded because this would give a singularity in $R(r)$ at $r = 0$. Thus $s = l$ is the only permissible value of s, so that the recurrence relations become

$$a_{t+1} = \frac{l+t+1-\lambda_n}{(l+t+1)(l+t+2) - l(l+1)} a_t \qquad t = 0, 1, 2, \ldots \quad (29)$$

We are now in a position to determine the eigenvalues E_n of the energy, by applying again the criterion that *an acceptable wave function must remain finite at infinity.* We do this by finding for what values λ_n the above recurrence formulas define a well-behaved function. Thus we test the behavior of $F(\rho)$ by evaluating the ratio a_{t+1}/a_t as $t \to \infty$

$$\frac{a_{t+1}}{a_t} \to \frac{1}{t} \tag{30}$$

Thus the function defined by Eq. (29) converges for all values of ρ for any value of λ_n. We observe, however, that the series expansion of e^ρ gives this same ratio of successive coefficients. Therefore, $e^{-\rho/2}F(\rho)$ must *increase exponentially* as $\rho \to \infty$ unless we can somehow avoid using an infinite series for $F(\rho)$. From Eq. (29) it appears that this can be done by choosing λ_n in such a way that the series terminates, that is, *by setting λ_n equal to a suitable integer*. Thus if

$$\lambda_n = n \qquad \text{and} \qquad n \geq l + 1 \tag{31}$$

the series will terminate after $n - l$ terms, and $R(\rho)$ will vanish at infinity.

EXERCISE

5-11. Carry out in detail the steps leading to Eqs. (29) and (31).

The polynomials defined by Eq. (29) with $\lambda_n = n$ and $n > l$ are closely related to the *associated Laguerre polynomials*, which are symbolized as $L_j^k(\rho)$. The associated Laguerre polynomials are solutions of the differential equation

$$\rho \frac{d^2 L_j^k}{d\rho^2} + (k + 1 - \rho) \frac{dL_j^k}{d\rho} + (j - k) L_j^k = 0 \tag{32}$$

It is easily shown that the associated Laguerre polynomials are equal to

$$L_j^k(\rho) = \frac{d^k}{d\rho^k} L_j(\rho) \tag{33}$$

where $L_j(\rho)$ is a solution of

$$\rho \frac{d^2 L_j}{d\rho^2} + (1 - \rho) \frac{dL_j}{d\rho} + jL_j = 0 \tag{34}$$

Direct substitution then reveals that a solution of Eq. (34) is

$$L_j(\rho) = e^\rho \frac{d^j}{d\rho^j} (\rho^j e^{-\rho}) \tag{35}$$

In terms of associated Laguerre polynomials, the radial function $R_{nl}(\rho)$ is equal to

$$R_{nl}(\rho) = A_{nl} e^{-\rho/2} \rho^l L_{n+l}^{2l+1}(\rho) \tag{36}$$

where $\rho = \alpha_n r$ and A_{nl} is a normalization factor equal to[1]

$$A_{nl} = \left[\left(\frac{2Z}{na_0'} \right)^3 \frac{(n - l - 1)!}{2n(n + l)!^3} \right]^{1/2} \tag{37}$$

---

[1] Schiff, L. I., "Quantum Mechanics," 2d ed., McGraw-Hill Book Company, Inc., New York, 1955.

From Eqs. (31) and (19) we have for the energy eigenvalues

$$E_n = \frac{-m_r Z^2 e^4}{32\pi^2 \varepsilon_0^2 \hbar^2 n^2} \quad \text{and} \quad \alpha_n = \frac{2m_r Z e^2}{4\pi\varepsilon_0 \hbar^2 n} = \frac{2Z}{na_0(1 + m/M)} = \frac{2Z}{na_0'} \tag{38}$$

Thus $\rho = \alpha_n r = 2Zr/na_0'$. The quantity

$$a_0 = \frac{4\pi\varepsilon_0 \hbar^2}{me^2} = 5.29172 \times 10^{-11} \text{ m} \tag{39}$$

is, as before, the radius of the first Bohr orbit [Eq. 2-1(25)].[1]

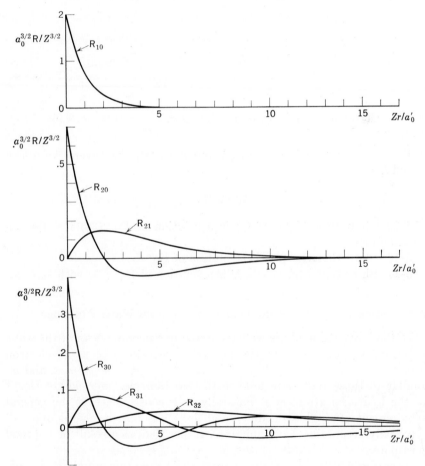

Fig. 5-3. Graphs of the radial wave functions $R_{nl}(r)$ for $n = 1, 2,$ and 3 and $l = 0, 1, 2.$

[1] Note that in the discussion of the Bohr theory the distance of the electron from the *center of mass* was designated by r. In the present discussion, however, r refers to the distance of the electron from the *nucleus*.

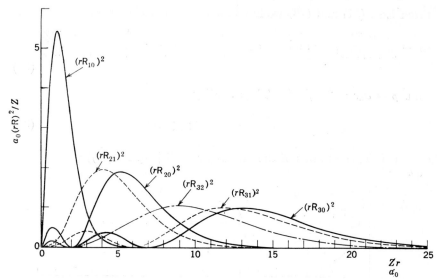

FIG. 5-4. Graphs of the radial probability density $r^2[R_{nl}(r)]^2$ for the wave functions of Fig. 5-3.

Figures 5-3 and 5-4 show $R_{nl}(r)$ and $r^2[R_{nl}(r)]^2$ for several values of n and l.

EXERCISES

5-12. Show that $(d^k/d\rho^k)L_j(\rho)$ is a solution of Eq. (32) if $L_j(\rho)$ is a solution of Eq. (34).

5-13. Show that Eq. (35) is a solution of Eq. (34).

5-14. Show that $F(\rho) = \rho^l L_{n+l}^{2l+1}$ is a solution of Eq. (24) with $\lambda_n = n$.

5-4. Some Properties of the One-electron-atom Wave Functions

In the foregoing analysis we have found one possible form of the wave functions corresponding to the bound energy states of a one-electron atom. In this process we have seen that *quantum numbers n, l,* and *m,* similar to those that were used in the old Bohr-Sommerfeld treatment of the hydrogen atom, arise naturally as a consequence of the general requirement of single-valuedness and finiteness that is placed upon all physically acceptable wave functions, rather than as a result of a special postulate.

Equation 5-3(38) shows that one of these quantum numbers, n, corresponds to the "total quantum number" of the old theory, in that it defines the same set of energy levels. We can also show that the other two quantum numbers have a significance similar to that of the azimuthal and magnetic quantum numbers of the old theory.   In the first place, we recognize that the form $e^{\pm im\varphi}$ for $\Phi(\varphi)$, with m an integer, is exactly the same as was found in Chap. 2 as an eigenfunction of the *z-component of angular momentum* of a system.   Thus we find that the *energy* eigenfunctions, as represented by the form

$$u_{nlm} = A_{nlm}R_{nl}(r)P_l^m(\cos\theta)e^{\pm im\varphi} \tag{1}$$

are also eigenfunctions of the *z-component of angular momentum*, so that a system whose wave function is one of these eigenfunctions possesses not only a total energy exactly equal to E_n *but also an angular momentum whose z-component is exactly equal to $\pm m\hbar$.*   Furthermore, if one expresses the operators corresponding to the rectangular components of angular momentum in spherical polar coordinates, and forms from them the operator for the *square of the total angular momentum*, it is found that this operator is

$$\boldsymbol{L} \cdot \boldsymbol{L} = -\hbar^2\left[\frac{1}{\sin\theta}\frac{\partial}{\partial\theta}\left(\sin\theta\frac{\partial}{\partial\theta}\right) + \frac{1}{\sin^2\theta}\frac{\partial^2}{\partial\varphi^2}\right] \tag{2}$$

Comparison of this operator with Eqs. 5-3(2) and 5-3(4b,c) shows that the functions

$$\Theta\Phi = P_l^m(\cos\theta)e^{\pm im\varphi} \tag{3}$$

are just those that would be obtained as a solution of the eigenvalue equation

$$\boldsymbol{L} \cdot \boldsymbol{L}\Theta\Phi = L_0^2\Theta\Phi \tag{4}$$

so that the *wave function u_{nlm} also corresponds to the precise value*

$$L_0^2 = l(l+1)\hbar^2$$

for the square of the total angular momentum $\boldsymbol{L}\cdot\boldsymbol{L}$.   The various energy levels of a one-electron atom are thus highly degenerate since there are, for a given value of n, n possible values of l, and for each value of l, $2l+1$ possible values of $\pm m$.[1]

EXERCISES

5-15. Write the operators corresponding to the various rectangular components of angular momentum in spherical polar coordinates and

---

[1] See footnote to Exercise 5-5.

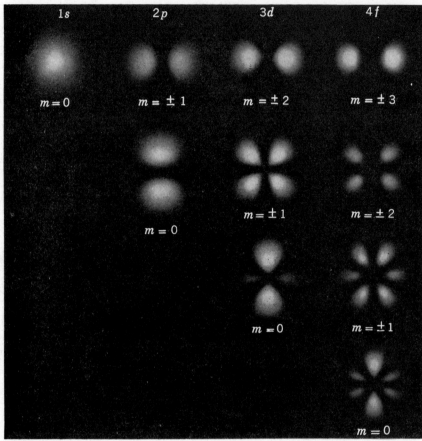

FIG. 5-5. Photographic representation of the electron probability density distribution distributions in a plane containing the polar axis, which is vertical and in the plane

combine these operators to form the operator for the square of the angular momentum.

Ans.: (a)
$$\hat{L}_x = \frac{\hbar}{i}\left(-\sin\varphi\,\frac{\partial}{\partial\theta} - \cot\theta\cos\varphi\,\frac{\partial}{\partial\varphi}\right)$$

(b)
$$\hat{L}_y = \frac{\hbar}{i}\left(\cos\varphi\,\frac{\partial}{\partial\theta} - \cot\theta\sin\varphi\,\frac{\partial}{\partial\varphi}\right) \tag{5}$$

(c)
$$\hat{L}_z = \frac{\hbar}{i}\frac{\partial}{\partial\varphi}$$

$$L \cdot L = -\hbar^2\left[\frac{1}{\sin\theta}\frac{\partial}{\partial\theta}\left(\sin\theta\,\frac{\partial}{\partial\theta}\right) + \frac{1}{\sin^2\theta}\frac{\partial^2}{\partial\varphi^2}\right] \tag{6}$$

5-16. Show that the degree of degeneracy of the nth energy level is n^2.

5-17. Write the eigenvalue equation for the square of the total angular

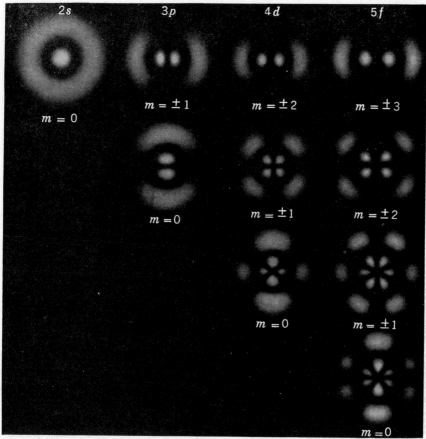

u^*u for several energy eigenstates. These may be regarded as sectional views of the
of the paper. The scale varies from figure to figure.

momentum and show that the eigenvalues and eigenfunctions are
$l(l + 1)\hbar^2$ and $P_l^m(\cos \theta)e^{\pm im\varphi}$, respectively.

We have found that the quantum-mechanical description of the energy
states of a one-electron atom, using the Schroedinger theory, is similar
in many respects to that of the old Bohr-Sommerfeld theory. However, there are also important differences between these theories, mainly
having to do with the question of the exact state of motion of the electron
within the atom. In the old theory, it will be recalled, the electron was
pictured as executing in the atom certain elliptical orbits whose size,
eccentricity, and orientation were determined by the quantization rules,
but which were otherwise described by ordinary classical mechanics.
In the new theory, on the other hand, the electron *cannot be pictured as*

being in a definite orbit at all but can only be described as being at a given location *with a certain probability.*

The three-dimensional character of the probability distribution u^*u is shown in Fig. 5-5 for several values of n, l, and m. Each of these quantum states is axially symmetric about the polar axis but shows complicated dependence upon θ and r, there being various conical and spherical nodal surfaces at which the electron is *never found.* These nodal surfaces are similar in nature to the nodal points found in the one-dimensional examples treated in the preceding chapters. They are the nodal surfaces of the "standing-wave pattern" taken on by the wave function of the atom in that energy eigenstate.

It is interesting to see how the prediction of the Schroedinger theory regarding the position of the electron compares with that of the old quantum theory. This can best be done by studying certain of the u_{nlm} wave functions which represent special states of the atom. First let us consider the *circular* orbits of the old Bohr theory. From Eq. 2-1(24) we see that the radii of these orbits were

$$r_n = \frac{n^2 a_0'}{Z} \tag{7}$$

In order to describe the corresponding state in the Schroedinger theory, we must use, for a given value of n, the largest permissible value of l— that is, we must make the *total angular momentum as large as possible*— and in order to orient the "orbit" in the equatorial plane of the spherical-coordinate system, we must let this angular momentum have its *maximum possible z-component.* Thus we set $l = m = n - 1$, so that

$$u_{n,n-1,n-1} = A_{n,n-1,n-1}\rho^{n-1}e^{-\rho/2}P_{n-1}^{n-1}(\cos\theta)e^{i(n-1)\varphi} \tag{8}$$

where $A_{n,n-1,n-1}$ is a constant needed for normalizing the wave function, $L_{2n-1}^{2n-1}(\rho) = \text{const}$, and $P_{n-1}^{n-1}(\cos\theta) = \sin^{n-1}\theta$. Also, $\rho = \alpha_n r$, where $\alpha_n = 2Z/na_0'$, so that the probability *density* is

$$u^*u = A^*A\left(\frac{2Zr}{na_0'}\right)^{2(n-1)} e^{-2Zr/na_0'}\sin^{2(n-1)}\theta \tag{9}$$

Thus in the Schroedinger theory the electron is not localized *at $r = r_n$* and *at $\theta = \pi/2$* for this quantum state, but is *most likely to be found* near $r = r_n$ and near $\theta = \pi/2$.

EXERCISE

5-18. Find the *most probable* value of r that will be found for an electron described by the wave function (8). *Hint:* Notice that the probability distribution for r is $W_r = 2\pi r^2 \int_0^\pi u^*u \sin\theta\, d\theta$. *Ans.:* $r_{\max} = n^2 a_0'/Z$.

Now let us examine the nature of the states that correspond to zero angular momentum. In this case the u_{nlm} functions are

$$u_{n00} = A_{n00}e^{-\rho/2}L_n^1(\rho) \tag{10}$$

so that, as might have been expected, all such wave functions have *spherical symmetry*. These wave functions correspond to the Bohr-Sommerfeld orbits of zero angular momentum, in which the electron travels back and forth radially in a highly eccentric ellipse. The spherical symmetry of the Schroedinger probability function may be thought of roughly as an average of these Bohr orbits over all possible angular orientations of the major axis of the ellipse. Qualitatively the Schroedinger theory agrees with the older theory also in predicting (for large n) a smaller probability of finding the electron near the nucleus than far away, since, like a comet moving around the sun in a nearly parabolic orbit, the electron should spend greater time far from the nucleus, where it is moving slowly, than near the nucleus, where it is moving fastest. But here again the two theories cannot be completely reconciled, because the Schroedinger probability density shows spherical nodal surfaces at radii where the electron is *never found*.

Much of the nomenclature used in the description of atomic structure is based upon the Bohr theory, because of the concrete picture of atomic states it portrays and because of its qualitative agreement with the Schroedinger theory. But it should be constantly borne in mind that the Bohr theory is not really capable of accounting for any but the simplest qualitative properties of atoms, whereas the Schroedinger theory is able to account quantitatively for all known features of the electronic structure of atoms.

Between the two extremes considered above, for which the probability densities correspond roughly to what was already familiar from the Bohr theory, the wave functions have quite complicated forms, as is indicated in Fig. 5-5. For example, the case $n = 2$, $l = 1$, $m = 0$ gives for the normalized wave function

$$u_{210} = \frac{(Z/a_0')^{5/2}}{4(2\pi)^{1/2}} re^{-Zr/2a_0'} \cos\theta \tag{11}$$

which has two "lobes" extending in either direction along the polar axis.

Other states show even more complicated "lobe" structures. For application to the problem of chemical binding, use is made of eigenstates which have lobes extending in directions other than along the z-axis. Such states can be described in terms of linear combinations of the u_{nlm} functions used in the foregoing analysis. For example, the function

$$u'_{211} = \frac{1}{\sqrt{2}} (u_{211} + u_{21-1})$$
$$= A_{211}re^{-Zr/2a_0'} \sin \theta \cos \varphi \qquad (12)$$

represents an energy eigenstate in which the energy is equal to E_2 and the square of the angular momentum is equal to $2\hbar^2$. The average z-component of this angular momentum is equal to zero, however. That is, if L_z is measured, it will be found to be $+\hbar$ half the time and $-\hbar$ half the time. The u^*u distribution of such a state is now *dependent upon φ as well as upon θ and r* and possesses lobes which extend along the positive and negative x-axis, just as the lobes of Eq. (11) extend along the z-axis. In fact, Eq. (12) is just Eq. (11) rotated through 90° about the y-axis. We shall later consider some of these simple lobe structures exhibited by the energy eigenfunctions in connection with the directed chemical bond. It is found that there is a strong correlation between the "lobe" structure of the wave function of an atom in its ground state and its chemical bonding properties.

One characteristic feature of the Schroedinger theory which is completely different from the older theories is that *there is a finite probability that the electron will be found in a region where it is energetically impossible, classically, for it to be.* For example, the normalized wave function for the lowest energy state is

$$u_{100} = \pi^{-\frac{1}{2}} \left(\frac{Z}{a_0'}\right)^{\frac{3}{2}} e^{-Zr/a_0'} \qquad (13)$$

According to the Bohr theory the electron cannot conceivably go farther than $2a_0'/Z$ from the nucleus, since at this distance its entire energy is potential energy. Nevertheless, computation reveals that there is actually a 0.25 chance that the electron will be found *outside* this limit.

EXERCISE

5-19. Evaluate the probability that the electron of a one-electron atom in its ground state will be found at a distance greater than $2a_0'/Z$ from the nucleus. *Ans.: $P = 13e^{-4} \approx 0.25$.*

Finally, it should be said that analysis similar to that carried out at the beginning of this section shows that *any positive value of E* yields acceptable solutions of the eigenvalue equation 5-3(4a). The totality of energy levels of a one-electron atom therefore consists of the *discrete set* of negative levels described above and a *continuum* of levels of positive energy. The negative energy levels correspond to *bound states* and the continuum to *ionized states* of the atom. We shall not have need for the analytic form of the wave functions for the ionized states, so that we shall not carry out the solution of the eigenvalue equation explicitly for this case. In certain types of problem, however, we may have to allow qualitatively for the possibility that transitions can occur between the continuum levels and the bound levels, or between continuum levels (free-bound and free-free transitions).

We conclude this section by tabulating the first few normalized wave functions

$$
\text{S:} \qquad u_{100} = \pi^{-\frac{1}{2}} \left(\frac{Z}{a_0'}\right)^{\frac{3}{2}} e^{-Zr/a_0'}
$$

$$
\text{S:} \qquad u_{200} = (32\pi)^{-\frac{1}{2}} \left(\frac{Z}{a_0'}\right)^{\frac{3}{2}} \left(2 - \frac{Zr}{a_0'}\right) e^{-Zr/2a_0'}
$$

$$
\text{P:} \qquad u_{210} = (32\pi)^{-\frac{1}{2}} \left(\frac{Z}{a_0'}\right)^{\frac{3}{2}} \frac{Zr}{a_0'} e^{-Zr/2a_0'} \cos\theta
$$

$$
\text{P:} \qquad u_{21\pm1} = (64\pi)^{-\frac{1}{2}} \left(\frac{Z}{a_0'}\right)^{\frac{3}{2}} \frac{Zr}{a_0'} e^{-Zr/2a_0'} \sin\theta\, e^{\pm i\varphi}
$$

$$
\text{S:} \qquad u_{300} = \frac{(Z/a_0')^{\frac{3}{2}}}{81(3\pi)^{\frac{1}{2}}} \left(27 - 18\frac{Zr}{a_0'} + 2\frac{Z^2r^2}{a_0'^2}\right) e^{-Zr/3a_0'}
$$

$$
\text{P:} \qquad u_{310} = \frac{2^{\frac{1}{2}}(Z/a_0')^{\frac{3}{2}}}{81\pi^{\frac{1}{2}}} \left(6 - \frac{Zr}{a_0'}\right) \frac{Zr}{a_0'} e^{-Zr/3a_0'} \cos\theta
$$

$$
\text{P:} \qquad u_{31\pm1} = \frac{(Z/a_0')^{\frac{3}{2}}}{81\pi^{\frac{1}{2}}} \left(6 - \frac{Zr}{a_0'}\right) \frac{Zr}{a_0'} e^{-Zr/3a_0'} \sin\theta\, e^{\pm i\varphi}
$$

$$
\text{D:} \qquad u_{320} = \frac{(Z/a_0')^{\frac{3}{2}}}{81(6\pi)^{\frac{1}{2}}} \frac{Z^2r^2}{a_0'^2} e^{-Zr/3a_0'}(3\cos^2\theta - 1)
$$

$$
\text{D:} \qquad u_{32\pm1} = \frac{(Z/a_0')^{\frac{3}{2}}}{81\pi^{\frac{1}{2}}} \frac{Z^2r^2}{a_0'^2} e^{-Zr/3a_0'} \sin\theta\cos\theta\, e^{\pm i\varphi}
$$

$$
\text{D:} \qquad u_{32\pm2} = \frac{(Z/a_0')^{\frac{3}{2}}}{162\pi^{\frac{1}{2}}} \frac{Z^2r^2}{a_0'^2} e^{-Zr/3a_0'} \sin^2\theta\, e^{\pm 2i\varphi}
$$

(14)

It should be recalled that the complete wave function for the quantum state (n,l,m) including the time dependence, is

$$
\psi_{nlm} = u_{nlm} \exp(-iE_n t/\hbar) \tag{15}
$$

and that *any* wave function for the one-electron atom may be represented in the form

$$
\psi = \sum_{n=0}^{\infty} \sum_{l=0}^{n-1} \sum_{m=-l}^{l} a_{nlm} u_{nlm} \exp(-iE_n t/\hbar) \tag{16}
$$

EXERCISES

5-20. Find the average value of the potential energy of a hydrogen atom in the state u_{210}. *Ans.:* $\langle V \rangle = 2E_2$.

5-21. A μ-meson is a particle of mass $206m_e$ and charge equal to that of an electron. Experiments have been performed in which a negative μ-meson is captured by an atom of some heavy element such as lead, and the radiation given off as the μ-meson cascades down through the various energy levels has been measured. Calculate the expected energy of the photon which is emitted when the meson falls from the $n = 2$ level to the $n = 1$ level in ^{206}Pb ($Z = 82$). Neglect fine-structure effects. *Ans.:* $\Delta E = 14.05$ MeV.

5-22. The wave functions for the one-electron atom have been derived on the assumption that the finite size of the nucleus can be neglected. At least one spectroscopic experiment in which the finite nuclear size must be taken into account has been performed. Take as a new model of the nucleus a uniformly charged spherical shell of radius R_0 and calculate, by first-order perturbation theory, the approximate expected shift in energy of the u_{200} state in hydrogen. Assume $R_0 \ll a_0$. *Ans.:* $\Delta E = \frac{2}{3}|E_2|(R_0^2/a_0^2)$.

5-23. The answer obtained in Exercise 5-21 is about three times larger than the experimentally measured value because of the effects of the finite size of the lead nucleus. Estimate the magnitude of this effect for the transition from u_{210} to u_{100} if the radius of the lead nucleus is 7×10^{-15} m.

5-24. Find combinations of the wave functions 5-4(1) which are energy eigenfunctions and which also possess definite parity with respect to x, y, and z. Show in general that the total parity, which describes the symmetry of the wave function under a change in sign of x, y, and z jointly, is governed by the quantum number l and is even if l is even.

5-5. Electron Spin

Our discussion of the quantum states of the one-electron atom has been restricted so far to the simplest possible model, and we have been able to solve this problem exactly in terms of, so to speak, "well-known" functions. The energy eigenvalues that were obtained in the course of this treatment are the same as those obtained in the original Bohr theory.

It was mentioned in Chap. 2 that Sommerfeld calculated certain corrections to the Bohr energy levels due to the effects of relativity upon the more eccentric orbits, and thus obtained a numerically accurate description of the so-called *fine structure* of the energy levels of a one-electron atom. This explanation of fine structure stood unquestioned for several years. In the early 1920's, however, considerable order had been introduced into the spectra of more complicated atoms by the use of the concept of energy levels first introduced by Bohr, and certain cases in which the spectrum of an atom could be roughly approximated by that of a one-electron atom—such as the alkali metals, which have a single valence electron—appeared to be appropriate cases to which to apply the Sommerfeld relativistic corrections. When this was done, it was found that Sommerfeld's *formula* gave precisely the correct shifts, even though it was obvious on physical grounds that relativistic effects should be *quite negligible* for these atoms, since the orbits involved were nearly circular and quite far from the nucleus. Thus a serious question arose regarding the true physical mechanism involved in the shift of the energy levels. In addition, spectroscopists found that some energy states appeared to require *half-integral quantum numbers* for their description.

In 1925, just before Schroedinger's theory appeared, Uhlenbeck and Goudsmit showed that the observed energy shifts could be obtained by assuming that electrons possess *angular momentum* and a *magnetic moment* which would interact with the field of the nucleus, and that this mechanism would correctly apply to both the one-electron atom and the alkali-metal atoms. Their hypothesis proved to be correct, and an intrinsic angular momentum is now regarded as one of the fundamental properties of a particle.

According to this hypothesis an electron is supposed to possess an intrinsic angular momentum or "spin," which we shall designate by the vector σ, and an intrinsic magnetic moment, which we shall designate by the vector μ. The magnitude of the spin angular momentum is assumed to be such as to correspond to *half-unit* quantum numbers for its square and its z-component: Thus we have for the *square* of σ a *single possible eigenvalue*

$$\sigma \cdot \sigma = s(s + 1)\hbar^2 \tag{1}$$

where $s = \frac{1}{2}$, that is, $\sigma \cdot \sigma = 3\hbar^2/4$, and for the component of σ along an axis, say, the z-axis, *two* possible eigenvalues

$$\sigma_z = m_s\hbar \tag{2}$$

where $m_s = +\frac{1}{2}$ or $-\frac{1}{2}$. The spin magnetic moment is assumed to be proportional to σ with the proportionality constant $-e/m$. Thus

$$\mu = -\frac{e}{m}\sigma \tag{3}$$

so that the component of μ along any axis, say, the z-axis, can have only the two eigenvalues

$$\mu_z = \pm \frac{e\hbar}{2m} \tag{4}$$

The latter quantity is called a *Bohr magneton*, and is usually designated μ_B or β.

EXERCISES

5-25. The magnetic moment of a plane loop is defined as the product of the current flowing in the loop times the area of the loop. Show that the dimensions of μ_B agree with such a definition.

5-26. An electron of charge e and mass m, circulating in a Bohr orbit around a nucleus, constitutes a "current loop" whose magnetic moment may be calculated classically. Assume the charge and mass of the electron to be distributed uniformly around a circular orbit and let the angular momentum of the electron be l. Show that the magnetic moment of the electron is then

$$M = -\frac{1}{2}\frac{e}{m}l \tag{5}$$

The properties with which the electron has been endowed by the above assumptions are rather remarkable for several reasons:

1. The spin angular momentum has been assumed to correspond to *half-integral* quantum numbers s and m_s in contrast with the results we have obtained for the orbital angular momentum of the one-electron atom where *integral* values of the quantum numbers l and m_l were found.[1]

2. A rectangular component of the spin angular momentum is supposed to have *only two eigenvalues* $+\frac{1}{2}\hbar$ and $-\frac{1}{2}\hbar$. Such a situation is unusual in quantum mechanics, for a given dynamical quantity ordinarily possesses an *infinite number* of eigenvalues which may be discrete or continuous, or partly of one type and partly of the other.

3. The electron spin is a quantum-mechanical quantity having *no classical analogue*, since in the correspondence-principle limit in which the finite size of h can be neglected, the spin angular momentum itself becomes negligible.

4. The proportionality factor between the *spin* magnetic moment and angular momentum [Eq. (3)] *is twice as large* as that connecting the *orbital* magnetic moment and angular momentum [Eq. (5)].

The only justification that can be *or need be* given for introducing these new ideas into physical theory is, of course, that they provide a means by which certain physical phenomena can be treated theoretically, and without which these phenomena cannot be treated successfully at all.

---

[1] With the introduction of electron spin, it becomes necessary to distinguish clearly its z-component $m_s\hbar$ from that of the *orbital* angular momentum. We shall therefore henceforth use the symbol m_l to designate the quantum number of L_z. m_l thus stands for the m introduced in the preceding section. We shall regard m_l as taking on the $2l + 1$ values from $-l$ to $+l$.

It also happens, however, that the particular case of electron spin is at present on an especially firm foundation because of another circumstance. In the development of a relativistically correct system of quantum mechanics, Dirac sought to devise a linear, Hermitian operator to represent the total energy

$$W = (P^2c^2 + m^2c^4)^{1/2} + V(x,y,z,t) \tag{6}$$

of a particle in space time. His solution to this problem resulted in the so-called *Dirac equation* for a particle, in which certain *matrix operators* appear. In solving this equation it was found that the properties of these matrices correspond exactly to an *intrinsic angular momentum* of $\frac{1}{2}\sqrt{3}\,\hbar$ for the particle under consideration and to a magnetic moment equal to $e\hbar/2m$. Further, it was found that such a particle ought to exist also in so-called *negative energy states* which appear experimentally as particles of *positive charge*. The success of the Dirac equation in thus accounting both for the spin and the magnetic moment of electrons, and also for the existence of positrons (or *antielectrons*) has made the electron, including the above-described spin and magnetic-moment properties, the most completely understood particle in physics to date. We shall consider these matters further in Chap. 20.

We now must consider how the above assumptions can be fitted into quantum mechanics, and more particularly into the present treatment of the one-electron atom. First, the existence of a spin angular momentum suggests the introduction of three new "variables" $\sigma_x, \sigma_y, \sigma_z$ to define the projection of this spin along the coordinate axes and three corresponding operators $\acute{\sigma}_x, \acute{\sigma}_y, \acute{\sigma}_z$. The spin variables may take on only the discrete values $\pm\hbar/2$, and the spin operators must have the property that an eigenvalue equation such as

$$\acute{\sigma}_z S = \sigma_0 S \tag{7}$$

has but *two* eigenvalues, $\sigma_0 = \pm\frac{1}{2}\hbar$, and *two* corresponding eigenfunctions of the discrete variable σ_z, which may be designated $S_{+\frac{1}{2}}$ and $S_{-\frac{1}{2}}$. These two eigenfunctions must be a complete orthogonal set of spin eigenfunctions, so that the spin dependence of any wave function may be expressed as a linear combination

$$S(\sigma_z) = aS_{+\frac{1}{2}}(\sigma_z) + bS_{-\frac{1}{2}}(\sigma_z) \tag{8}$$

These two eigenfunctions must be orthogonal to each other and for convenience are assumed to be normalized. For the present case in which the spin "variable" can have only discrete values $\pm\frac{1}{2}\hbar$, these conditions take the form of a *summation* over the two possible values of σ_z

$$\sum_{\sigma_z=-\frac{1}{2}\hbar}^{+\frac{1}{2}\hbar} S^*_{ms}(\sigma_z)S_{ms'}(\sigma_z) = \delta_{m_s m_s'} \tag{9}$$

With this substitution of a summation over a finite number of discrete values of the spin orientation σ_z for an integral over an infinite number of values of the space coordinates, the properties and applications of these spin eigenfunctions are in every way analogous to those of ordinary space-coordinate wave functions, as illustrated by the following exercises.

EXERCISES

5-27. Find the average value of σ_z for the wave function (8).

Ans.: $\qquad \langle \sigma_z \rangle = \sum_{\sigma_z=-\frac{1}{2}\hbar}^{+\frac{1}{2}\hbar} S^*(\sigma_z)\acute{\sigma}_z S(\sigma_z) = \frac{1}{2}\hbar(a^*a - b^*b)$

5-28. Suppose $S(\sigma_z)$ is given.   Find a and b.

Ans.: $\qquad a = \sum_{\sigma_z} S^*_{+\frac{1}{2}}(\sigma_z)S(\sigma_z) \qquad b = \sum_{\sigma_z} S^*_{-\frac{1}{2}}(\sigma_z)S(\sigma_z) \tag{10}$

5-29. A system is in the spin state (8).   Find the probability that a measurement of σ_z will yield the value $+\frac{1}{2}\hbar$.   *Ans.: $P_+ = a^*a$.*

Although the above-described properties of the spin operators and eigenfunctions are sufficient for most of our later work, it will be of interest to examine briefly one of the several possible mathematical forms of the theory of electron spin. This theory, due to Pauli, is commonly used and is very similar in appearance to the Heisenberg and Dirac theories. In the Pauli theory, the coefficients a and b of Eq. (8) are represented as a two-component *column matrix.*[1]

$$S = \begin{pmatrix} a \\ b \end{pmatrix} \tag{11}$$

and the spin operators $\acute{\sigma}_x$, $\acute{\sigma}_y$, and $\acute{\sigma}_z$ are represented as 2×2 square matrices:

$$\acute{\sigma}_x = \frac{1}{2}\hbar \begin{pmatrix} 0 & 1 \\ 1 & 0 \end{pmatrix} \qquad \acute{\sigma}_y = \frac{1}{2}\hbar \begin{pmatrix} 0 & -i \\ i & 0 \end{pmatrix} \qquad \acute{\sigma}_z = \frac{1}{2}\hbar \begin{pmatrix} 1 & 0 \\ 0 & -1 \end{pmatrix} \tag{12}$$

[1] For a concise review of matrix algebra, see H. Margenau and G. M. Murphy, "The Mathematics of Physics and Chemistry," 2d ed., chap. 10, D. Van Nostrand Company, Inc., Princeton, N.J., 1956.

These operators operate upon wave functions by *matrix multiplication*.[1] For example,

$$\hat{\sigma}_z S = \tfrac{1}{2}\hbar \begin{pmatrix} 1 & 0 \\ 0 & -1 \end{pmatrix} \begin{pmatrix} a \\ b \end{pmatrix} = \tfrac{1}{2}\hbar \begin{pmatrix} a \\ -b \end{pmatrix} \tag{13}$$

The eigenfunctions of σ_z are clearly equal to

$$S_{+\frac{1}{2}} = \begin{pmatrix} 1 \\ 0 \end{pmatrix} \quad \text{and} \quad S_{-\frac{1}{2}} = \begin{pmatrix} 0 \\ 1 \end{pmatrix} \tag{14}$$

EXERCISE

5-30. Show that the above representations of $S_{+\frac{1}{2}}$ and $S_{-\frac{1}{2}}$ satisfy the eigenvalue equation (7) with the eigenvalues $+\frac{1}{2}\hbar$ and $-\frac{1}{2}\hbar$.

In the Pauli theory (as in the Heisenberg theory) the *Hermitian adjoint* matrix plays the role of the complex conjugate function ψ^* of the Schroedinger theory. The Hermitian adjoint matrix is obtained by interchanging rows and columns and then replacing each element of the resulting matrix by its complex conjugate. Thus if $S = \begin{pmatrix} a \\ b \end{pmatrix}$, then $S\dagger = (a^*\ b^*)$ is the Hermitian adjoint of S.

The product of a column matrix by its Hermitian adjoint is a real matrix of but one row and column, which corresponds to $\int \psi^* \psi\, dq$:

$$S\dagger S = (a^*\ b^*) \begin{pmatrix} a \\ b \end{pmatrix} = a^*a + b^*b = 1 \tag{15}$$

---

[1] The product of two matrices a_{ij} and b_{kl} is a matrix whose elements c_{mn} are

$$c_{mn} = \sum_s a_{ms} b_{sn}$$

The number of *columns* in a_{ij} must be equal to the number of *rows* in b_{kl}, and the resultant matrix c_{mn} will have the same number of *rows* as does a_{ij} and the same number of *columns* as does b_{kl}. This is indicated schematically by the following specific example:

$$\begin{pmatrix} a_{11} & a_{12} \\ a_{21} & a_{22} \\ a_{31} & a_{32} \end{pmatrix} \begin{pmatrix} b_{11} & b_{12} & b_{13} & b_{14} \\ b_{21} & b_{22} & b_{23} & b_{24} \end{pmatrix} = \begin{pmatrix} c_{11} & c_{12} & c_{13} & c_{14} \\ c_{21} & c_{22} & c_{23} & c_{24} \\ c_{31} & c_{32} & c_{33} & c_{34} \end{pmatrix}$$

$$c_{13} = \sum_s a_{1s} b_{s3} = a_{11}b_{13} + a_{12}b_{23}$$

(Elements of the first *row* of a_{ij} are multiplied by corresponding elements of the third *column* of b_{kl} and summed to yield the element c_{13}.)

The average value of one of the spin variables is designated by

$$\langle \sigma_i \rangle = S \dagger \dot{\sigma}_i S \tag{16}$$

For example,

$$\langle \sigma_z \rangle = (a^* \ b^*) \tfrac{1}{2}\hbar \begin{pmatrix} 1 & 0 \\ 0 & -1 \end{pmatrix} \begin{pmatrix} a \\ b \end{pmatrix}$$

$$= (a^* \ b^*) \tfrac{1}{2}\hbar \begin{pmatrix} a \\ -b \end{pmatrix} = \tfrac{1}{2}\hbar(a^*a - b^*b) \tag{17}$$

Some of the properties of the Pauli spin operators and eigenfunctions are illustrated in the following exercises.

EXERCISES

5-31. Find the average values of σ_x and σ_y for the spin function $S = \begin{pmatrix} a \\ b \end{pmatrix}$.   Ans.: $\langle \sigma_x \rangle = \tfrac{1}{2}\hbar(a^*b + b^*a)$ and $\langle \sigma_y \rangle = -\tfrac{1}{2}i\hbar(a^*b - b^*a)$.

5-32. Write and solve the eigenvalue equations for σ_x and σ_y, and express the eigenfunctions as linear combinations of $S_{+\frac{1}{2}}$ and $S_{-\frac{1}{2}}$.

Ans.:

$$S^{(x)}_{+\frac{1}{2}} = \frac{A}{2^{\frac{1}{2}}} \begin{pmatrix} 1 \\ 1 \end{pmatrix} = \frac{A}{2^{\frac{1}{2}}} (S_{+\frac{1}{2}} + S_{-\frac{1}{2}})$$

$$S^{(x)}_{-\frac{1}{2}} = \frac{B}{2^{\frac{1}{2}}} \begin{pmatrix} 1 \\ -1 \end{pmatrix} = \frac{B}{2^{\frac{1}{2}}} (S_{+\frac{1}{2}} - S_{-\frac{1}{2}})$$

$$S^{(y)}_{+\frac{1}{2}} = \frac{C}{2^{\frac{1}{2}}} \begin{pmatrix} -i \\ 1 \end{pmatrix} = \frac{C}{2^{\frac{1}{2}}} (-iS_{+\frac{1}{2}} + S_{-\frac{1}{2}})$$

$$S^{(y)}_{-\frac{1}{2}} = \frac{D}{2^{\frac{1}{2}}} \begin{pmatrix} i \\ 1 \end{pmatrix} = \frac{D}{2^{\frac{1}{2}}} (iS_{+\frac{1}{2}} + S_{-\frac{1}{2}})$$

where A, B, C, and D are arbitrary complex coefficients of unit absolute value.

5-33. Find the Pauli operator for the square of the spin angular momentum, and thus show that any normalized spin function is an eigenfunction of $\dot{\sigma} \cdot \dot{\sigma}$ with the eigenvalue $\tfrac{3}{4}\hbar^2$.

Ans.: $\dot{\sigma} \cdot \dot{\sigma} = \dot{\sigma}_x{}^2 + \dot{\sigma}_y{}^2 + \dot{\sigma}_z{}^2 = \tfrac{3}{4}\hbar^2 \begin{pmatrix} 1 & 0 \\ 0 & 1 \end{pmatrix}$

The foregoing brief outline of the Pauli theory of spin, although quite incomplete, serves to indicate one manner in which the assumed spin properties of electrons can be represented mathematically. Actually, we shall have little occasion to use the spin operators or wave functions explicitly, but their general properties will often be used in our later work. In some of these applications it will be possible to neglect outright the interactions of the electron magnetic moment with the rest of the physical system, and in others the effects of these interactions will be treated as small perturbations. In either of these cases it is necessary to include in the wave function a term which represents the spin states of the electrons of the system. In the absence of an inter-action between the spin and the other coordinates this can be done by multiplying the wave function obtained before the introduction of spin by some linear combination of the spin eigenfunctions $S_{+\frac{1}{2}}$ and $S_{-\frac{1}{2}}$. For example, the energy eigenstates of the one-electron atom, including spin but neglecting the interaction between this spin and the other variables, require *five* quantum numbers to describe them: n, l, m_l, s, and m_s. Such an eigenfunction would be written

$$u_{nlsm_lm_s}(r,\theta,\varphi,\sigma_z) = A_{nlm_lm_s}R_{nl}(r)P_l^{m_l}(\cos\theta)e^{\pm im_l\varphi}S_{m_s}(\sigma_z) \qquad (18)$$

where the quantum number s is equal to $\frac{1}{2}$, m_s can take on either of the two values $+\frac{1}{2}$ or $-\frac{1}{2}$, and $A_{nlm_lm_s}$ is a normalizing factor for the wave function.

5-6. Spin-Orbit Forces

We must now consider what kinds of interaction exist between the electron spin and the other variables of a system. The assumed existence of a magnetic moment for the electron implies an interaction of this moment with any magnetic fields that may be present. These might arise from external sources, as in the Zeeman effect; from the presence of other particles having magnetic moments; from the bodily circulatory motion of other particles in "orbits"; or from the motion of the electron itself through an *electric* field, which appears partly as a *magnetic* field in its own rest frame. We thus expect to find terms having to do with these various possible interactions in the complete Hamiltonian function.

The last of the spin interactions mentioned above is the one which is responsible for the energy-level shifts in the alkali metals that led origi-nally to the spin hypothesis itself; and since this interaction is also partially responsible for the energy-level shifts in a one-electron atom, we shall now consider it in this connection. The magnetic field seen by the electron because of its own orbital motion through the nuclear electric field is, from Eq. 1-8(23),

$$\boldsymbol{B'} = -\frac{1}{c^2}\boldsymbol{v}\times\boldsymbol{E} = \frac{\boldsymbol{l}}{mec^2r}\frac{dV}{dr} \qquad (1)$$

where $V = -Ze^2/4\pi\varepsilon_0 r$ is the potential *energy* of the electron and $\boldsymbol{l}$ is the orbital angular momentum. The energy of orientation of a magnetic moment $\boldsymbol{\mu}$ in such a field is, classically,

$$W = -\boldsymbol{\mu}\cdot\boldsymbol{B} = \frac{\boldsymbol{\mathfrak{d}}\cdot\boldsymbol{l}}{m^2c^2r}\frac{dV}{dr} \qquad (2)$$

EXERCISE

5-34. Supply the missing steps in the derivation of Eq. (2).

This interaction in which the spin is, in effect, coupled with the orbital angular momentum is called a *spin-orbit interaction,* and the classical forces that result from differentiating this energy with respect to the position coordinates are called *spin-orbit forces.*

Before introducing the above spin-orbit term into the Hamiltonian function, one must inquire whether there are any other spin-orbit effects of magnitude comparable to the above one. There is one such effect, due to relativistic kinematics, whose magnitude is half as great as that of Eq. (2) but of opposite sign. It is called the *Thomas precession,*[1] and it exists whether or not the electron is assumed to possess a magnetic moment. Basically, the Thomas precession can be attributed to the time dilation between the rest frames of the electron and the proton, which causes observers in these two frames to disagree on the time required for one particle to make a complete revolution about the other. If an observer on the electron calls this time interval T, then an observer on the proton will find it to be $T' = \gamma T$, where $\gamma = (1 - v^2/c^2)^{-\frac{1}{2}}$ and v is the speed of the electron around the proton. (We assume circular motion for simplicity.) The orbital angular velocities measured by these observers will thus be $2\pi/T$ and $2\pi/T'$, respectively. Now, in the rest frame of the electron, the spin-angular-momentum vector maintains its direction in space, so that to the observer on the proton this spin vector appears to precess at a rate equal to the difference between the two angular velocities $2\pi/T$ and $2\pi/T'$. This precessional angular frequency is equal to

$$\Omega_k = \frac{2\pi}{T'}\left[\left(1 - \frac{v^2}{c^2}\right)^{-\frac{1}{2}} - 1\right] \approx \frac{2\pi}{T'}\frac{v^2}{2c^2}$$

But

$$\frac{2\pi}{T'} = \omega = \frac{|l|}{mr^2} \quad \text{and} \quad \frac{mv^2}{r} = -\frac{dV}{dr}$$

so that

$$\Omega_k = -\frac{1}{2}\frac{|l|}{m^2c^2r}\frac{dV}{dr} \tag{3}$$

On the other hand, the classical *Larmor precession* of the electron in the magnetic field B' of Eq. (1), obtained by setting the torque $\mathbf{\mu} \times B'$ equal to the rate of change of the spin angular momentum, is

$$\mathbf{\mu} \times B' = \frac{d\mathbf{\sigma}}{dt} = \Omega_l \times \mathbf{\sigma}$$

so that

$$\Omega_l = \frac{eB'}{m} = \frac{|l|}{m^2c^2r}\frac{dV}{dr} \tag{4}$$

[1] Thomas, *Nature,* **117,** 514 (1926).

This classical frequency is thus twice as large as the kinematic precession Ω_k and should therefore correspond to twice as large a term in the Hamiltonian function. We therefore introduce into the Hamiltonian function the net spin-orbit energy, i.e., one-half of Eq. (2), so that we have

$$\hat{H} = -\frac{\hbar^2}{2m}\nabla^2 - \frac{Ze^2}{4\pi\varepsilon_0 r} + \frac{1}{2}\frac{Ze^2}{4\pi\varepsilon_0 m^2 c^2 r^3}\,\mathbf{\acute{o}}\, {\overset{\cdot}{\cdot}}\, \mathbf{l} \qquad (5)$$

where $\mathbf{\acute{o}}\, {\overset{\cdot}{\cdot}}\, \mathbf{l}$ is the appropriate operator corresponding to $\mathbf{\acute{o}} \cdot \mathbf{l}$. Because of the small size of the additional term, the perturbation treatment described in Sec. 2.5C can be used to evaluate the shift in the energy levels.

EXERCISE

5-35. Show that the ratio of the spin-orbit energy to the electrostatic potential energy is of order E_n/mc^2. (Assume circular Bohr orbits as a rough approximation.)

In order to evaluate the first-order perturbation shift in the energy levels, we observe that the spin-orbit term is a product of a function of the angular momenta $\mathbf{\acute{o}}$ and $\mathbf{l}$ times a function of r. Since the unperturbed eigenfunctions are also expressed as such a product, we find that the perturbation-energy integral 2-5(11) is likewise a *product* of the average of $1/r^3$ taken over R_{nl} times the average of $\mathbf{\acute{o}} \cdot \mathbf{l}$ taken over the *angular* part of an unperturbed wave function:

$$\begin{aligned}
\Delta E_s &= \frac{1}{2}\frac{Ze^2}{4\pi\varepsilon_0 m^2 c^2}\int_0^\infty\int_0^\pi\int_0^{2\pi} R^*\Theta^*\Phi^*S^*\frac{\mathbf{\acute{o}}\,{\overset{\cdot}{\cdot}}\,\mathbf{l}}{r^3}R\Theta\Phi S r^2 \sin\theta\, dr\, d\theta\, d\varphi \\
&= \frac{1}{2}\frac{Ze^2}{4\pi\varepsilon_0 m^2 c^2}\left[\int_0^\infty R^2(r)\frac{1}{r^3}r^2\, dr\right] \\
&\qquad\qquad\qquad \left(\int_0^\pi\int_0^{2\pi}\Theta^*\Phi^*S^*\mathbf{\acute{o}}\,{\overset{\cdot}{\cdot}}\,\mathbf{l}\Theta\Phi S \sin\theta\, d\theta\, d\varphi\right) \\
&= \frac{1}{2}\frac{Ze^2}{4\pi\varepsilon_0 m^2 c^2}\left\langle\frac{1}{r^3}\right\rangle\langle\mathbf{\acute{o}}\cdot\mathbf{l}\rangle \qquad (6)
\end{aligned}$$

EXERCISE

5-36. Verify the derivation of Eq. (6).

The first of these averages can be carried out in a straightforward way through the use of a generating function representation[1] for the

---

[1] For details of this procedure, see L. Pauling and E. B. Wilson, "Introduction to Quantum Mechanics," appendix VII, McGraw-Hill Book Company, Inc., New York, 1935.

functions $R_{nl}(r)$.   The result is

$$\left\langle \frac{1}{r^3} \right\rangle = \frac{Z^3}{a_0{}^3 n^3 l(l + \frac{1}{2})(l + 1)} \tag{7}$$

provided that $l > 0$.

In order to evaluate the second of these averages, we must first examine the effects of spin somewhat more closely.   It was shown in Chap. 2 that any dynamical quantity whose operator commutes with the Hamiltonian operator is one whose expectation value and uncertainty *do not change with the time;* that is, such quantities are *constants of the motion.* The linear momentum of the center of mass, which appears in Eq. 5-1(2), is such a quantity; also, the three rectangular components and the square of the orbital angular momentum were found in Sec. 5-4 to be such quantities.   It is a general result of quantum mechanics that *every dynamical variable which is classically a constant of the motion is also a quantum-mechanical constant of the motion.*

With the introduction of a spin angular momentum, we would thus expect that, in the absence of an interaction of this spin with another agency, **σ** should be a constant of the motion.   Because of the spin-orbit interactions, however, this is not quite true; furthermore, the *orbital* angular momentum is no longer constant either, because the mutual interaction torque **μ** ✕ **B′** changes the directions of both the spin angular momentum and the orbital angular momentum.   On the other hand it would be expected classically that the square of the *total* angular momentum *j*,

$$\boldsymbol{j} \cdot \boldsymbol{j} = (\boldsymbol{\sigma} + \boldsymbol{l}) \cdot (\boldsymbol{\sigma} + \boldsymbol{l}) \tag{8}$$

as well as the three rectangular components j_x, j_y, and j_z should all be constants of the motion.

Now, in using the first-order perturbation theory to evaluate $\langle \boldsymbol{\sigma} \cdot \boldsymbol{l} \rangle$, we must recognize that we are dealing with degenerate energy levels and that we may therefore be required to use a particular set of degenerate zero-order wave functions as a starting point for the perturbation procedure.   While there exist quite general methods for finding suitable zero-order functions,[1] we shall use a more physical approach which makes use of the general result that any quantity other than energy which is a constant of the motion both in the perturbed and unperturbed systems does not change as the perturbation is applied.   We have just observed that the total angular momentum *j* is such a quantity with respect to spin-orbit effects, so that we must use, as zero-order wave functions, those combinations of the $u_{nlsm_lm_s}$ functions which correspond to the same values

---

[1] See H. Margenau and G. M. Murphy, "The Mathematics of Physics and Chemistry," 2d ed., pp. 388 ff., D. Van Nostrand Company, Inc., Princeton, N.J., 1956.

of j before and after the perturbation is included.   Before we can proceed further with our investigation of spin-orbit effects, we must therefore inquire into what values of j can be formed by combining two quantized angular momenta, $\mathbf{d}$ and $\mathbf{l}$.

5-7. The Quantization of Angular Momentum

In the preceding discussion we have found that, whereas the orbital angular momentum l and the spin angular momentum $\mathbf{d}$ are not separately constants of the motion in a one-electron atom, their sum j *is* constant. This means that eigenstates can be found in which the energy, the square of j, and any one of its rectangular components are all precisely defined. We wish now to investigate what possible precise values $j \cdot j$ and, say, j_z can possess.   We shall first treat this problem on a general level and then consider our immediate application as a special case, for in this way some important and fundamental properties of angular momentum will be exhibited.

Let us therefore assume that there exists a vector dynamical quantity J whose rectangular components are represented by the operators $\hat{J}_x$, $\hat{J}_y$, and $\hat{J}_z$.   Also, let $J \cdot J = \hat{J}_x{}^2 + \hat{J}_y{}^2 + \hat{J}_z{}^2$.   Now we shall agree to interpret as an angular momentum any such quantity whose operators satisfy the commutation relations

(a)
(b)
(c)

$$\hat{J}_x\hat{J}_y - \hat{J}_y\hat{J}_x = i\hbar\hat{J}_z$$
$$\hat{J}_y\hat{J}_z - \hat{J}_z\hat{J}_y = i\hbar\hat{J}_x$$
$$\hat{J}_z\hat{J}_x - \hat{J}_x\hat{J}_z = i\hbar\hat{J}_y$$

$$(1)$$

EXERCISES

5-37. Show that the operators for the three rectangular components of the orbital angular momentum satisfy the relations (1).

5-38. Show that the Pauli spin operators 5-5(12) satisfy the relations (1).

5-39. Show that $J \cdot J$ commutes with $\hat{J}_x$, $\hat{J}_y$, and $\hat{J}_z$.

Now suppose there is a wave function $\psi_{\alpha\beta}$ which corresponds to precise values α and β^2 for J_z and $J \cdot J$.   That is, $\psi_{\alpha\beta}$ satisfies the two eigenvalue equations

(a)
(b)

$$\hat{J}_z\psi_{\alpha\beta} = \alpha\psi_{\alpha\beta}$$
$$J \cdot J\psi_{\alpha\beta} = \beta^2\psi_{\alpha\beta}$$

$$(2)$$

Our object is to find what possible values α and β^2 can have.   To this end we construct the operators $\hat{J}_x + i\hat{J}_y$ and $\hat{J}_x - i\hat{J}_y$ and consider the functions $\psi_{\alpha\beta}^{+}$ and $\psi_{\alpha\beta}^{-}$ which are obtained if we operate upon $\psi_{\alpha\beta}$ with these operators.

These operators both commute with $\boldsymbol{J} \cdot \boldsymbol{J}$ and have the further commutation properties that

$$\hat{J}_{z}(\hat{J}_{x} + i\hat{J}_{y}) = (\hat{J}_{x} + i\hat{J}_{y})(\hat{J}_{z} + \hbar)$$

and
$$\hat{J}_{z}(\hat{J}_{x} - i\hat{J}_{y}) = (\hat{J}_{x} - i\hat{J}_{y})(\hat{J}_{z} - \hbar) \tag{3}$$

EXERCISE

5-40. Verify the commutation relations (3).

Consider first the result of applying the operator $\hat{J}_{x} + i\hat{J}_{y}$ to $\psi_{\alpha\beta}$:

$$\psi_{\alpha\beta}^{+} = (\hat{J}_{x} + i\hat{J}_{y})\psi_{\alpha\beta} \tag{4}$$

If this function is not identically equal to zero, it must be another eigenfunction of J_{z} and $\boldsymbol{J} \cdot \boldsymbol{J}$, with the eigenvalues $\alpha + \hbar$ and β^{2}. We show this as follows:

$$\begin{aligned}
\hat{J}_{z}\psi_{\alpha\beta}^{+} &= \hat{J}_{z}(\hat{J}_{x} + i\hat{J}_{y})\psi_{\alpha\beta} \\
&= (\hat{J}_{x} + i\hat{J}_{y})(\hat{J}_{z} + \hbar)\psi_{\alpha\beta} \\
&= (\hat{J}_{x} + i\hat{J}_{y})(\alpha + \hbar)\psi_{\alpha\beta} \\
&= (\alpha + \hbar)(\hat{J}_{x} + i\hat{J}_{y})\psi_{\alpha\beta} \\
&= (\alpha + \hbar)\psi_{\alpha\beta}^{+}
\end{aligned} \tag{5}$$

EXERCISE

5-41. Show that $\boldsymbol{J} \cdot \boldsymbol{J}\psi_{\alpha\beta}^{+} = \beta^{2}\psi_{\alpha\beta}^{+}$.

We have just found that, if α is an eigenvalue of J_{z}, then $\alpha + \hbar$ *is also an eigenvalue*, unless $\psi_{\alpha\beta}^{+}$ is identically zero. A similar application of $\hat{J}_{x} - i\hat{J}_{y}$ leads to the conclusion that $\alpha - \hbar$ is also an eigenvalue of J_{z}, corresponding to the same value β^{2} for $\boldsymbol{J} \cdot \boldsymbol{J}$, unless $\psi_{\alpha\beta}^{-}$ is equal to zero.

EXERCISE

5-42. Let $\psi_{\alpha\beta}^{-} = (\hat{J}_{x} - i\hat{J}_{y})\psi_{\alpha\beta}$. Show that $\hat{J}_{z}\psi_{\alpha\beta}^{-} = (\alpha - \hbar)\psi_{\alpha\beta}^{-}$ and $\boldsymbol{J} \cdot \boldsymbol{J}\psi_{\alpha\beta}^{-} = \beta^{2}\psi_{\alpha\beta}^{-}$.

By extending the above procedure, we find that $\ldots \ldots, \alpha - 2\hbar, \alpha - \hbar$, $\alpha, \alpha + \hbar, \alpha + 2\hbar, \ldots \ldots$ are all eigenvalues of J_{z}, that this series must extend in each direction until a further application of $\hat{J}_{x} + i\hat{J}_{y}$ or $\hat{J}_{x} - i\hat{J}_{y}$ gives zero, and that all of these eigenvalues correspond to the same value β^{2} for $\boldsymbol{J} \cdot \boldsymbol{J}$. It is clear that this series *must* terminate at some positive eigenvalue A such that $A^{2} < \beta^{2}$, since $\langle J_{x}^{2} \rangle$, $\langle J_{y}^{2} \rangle$, and $\langle J_{z}^{2} \rangle$

must all be positive and $\langle \mathbf{J} \cdot \mathbf{J} \rangle = \langle J_x{}^2 \rangle + \langle J_y{}^2 \rangle + \langle J_z{}^2 \rangle$. Let the eigenfunction corresponding to this greatest positive value be designated $\psi_{A\beta}$. Then we must have

(a)
$$\hat{J}_z \psi_{A\beta} = A \psi_{A\beta}$$
(b)
$$\mathbf{J} \colon \mathbf{J} \psi_{A\beta} = \beta^2 \psi_{A\beta} \tag{6}$$
(c)
$$(\hat{J}_x + i\hat{J}_y) \psi_{A\beta} = 0$$

If we apply the operator $(\hat{J}_x - i\hat{J}_y)$ to the last of the above equations, we obtain

$$(\hat{J}_x - i\hat{J}_y)(\hat{J}_x + i\hat{J}_y)\psi_{A\beta} \equiv 0 \equiv (\hat{J}_x{}^2 + \hat{J}_y{}^2 - \hbar \hat{J}_z)\psi_{A\beta} \tag{7}$$

EXERCISE

5-43. Verify Eq. (7).

Now, if we write Eq. (6b) in the form

$$\mathbf{J} \colon \mathbf{J} \psi_{A\beta} = \beta^2 \psi_{A\beta} = (\hat{J}_x{}^2 + \hat{J}_y{}^2 + \hat{J}_z{}^2)\psi_{A\beta}$$

we may use Eq. (7) to eliminate $\hat{J}_x{}^2 + \hat{J}_y{}^2$ and thus obtain

$$\beta^2 = A^2 + A\hbar \tag{8}$$

A similar treatment of the lowermost value A' of J_z leads to the corresponding equation

$$\beta^2 = A'^2 - A'\hbar \tag{9}$$

EXERCISE

5-44. Carry through the necessary steps in the derivation of Eqs. (8) and (9).

The two equations, (8) and (9), now contain all the information needed to find the possible values of J_z and $\mathbf{J} \cdot \mathbf{J}$. First, it is clear by inspection that $A' = -A$. This means that the possible values of J_z extend from $+A$ to $-A$ in steps of magnitude $\hbar$. *This is possible only if A is itself an integer or half-integer times $\hbar$.* Now let us put $A = j\hbar$, where j is integral or half-integral, into Eq. (8). Then we find

$$\beta^2 = j(j + 1)\hbar^2 \tag{10}$$

We have thus established that *angular momentum can possess only the quantized values $j(j + 1)\hbar^2$ for its square and $m_j\hbar$ for its z-component, where m_j can have any of the integrally spaced values from $-j$ to $+j$ and j is either integral or half-integral.*
We must next consider the *addition* of angular momenta. Let $\mathbf{J}_1$

and J_2 be two angular momenta which are independent of each other (i.e., whose operators commute) and of other variables in a system. The above results can then clearly be applied to these angular momenta separately, and furthermore *to their vector sum*, since the operators $\hat{J}_x = \hat{J}_{1x} + \hat{J}_{2x}$, etc., will surely also satisfy the commutation requirements (1). Therefore, if $J_1 \cdot J_1 = j_1(j_1 + 1)\hbar^2$ and $J_2 \cdot J_2 = j_2(j_2 + 1)\hbar^2$, where j_1 and j_2 are given integral or half-integral quantum numbers, then we must also have for the square of J

$$J \cdot J = j(j + 1)\hbar^2$$

and for its z-component

$$J_z = m_j \hbar \qquad -j \leq m_j \leq +j$$

where j is an integer or half-integer. But, if m_{j1} and m_{j2} are the quantum numbers for the z-components of J_1 and J_2, we must clearly have

$$m_j = m_{j1} + m_{j2} \tag{11}$$

We may use this relation to define the integral or half-integral character of j, for the integral or half-integral character of j_1, j_2, and j must clearly be the same as of m_{j1}, m_{j2}, and m_j, respectively. Thus, if j_1 and j_2 are both integral or half-integral, m_j and therefore j must be integral; and if j_1 is integral and j_2 half-integral, or vice versa, then j must be half-integral. Then, using the triangle inequalities

$$||J_1| - |J_2|| \leq |J| \leq |J_1| + |J_2|$$

we find that j must lie within the range

$$|j_1 - j_2| \leq j \leq j_1 + j_2 \tag{12}$$

These results can be summarized by the rule that *the sum of two angular momenta whose quantum numbers are j_1 and j_2 must possess a quantum number j equal to one of the integrally spaced values in the range $|j_1 - j_2| \leq j \leq j_1 + j_2$, and for a given value of j the z-component of the sum can have any of the values $m_j \hbar$, where $-j \leq m_j \leq +j$.*

Thus, for the present case in which l is an integer and the spin quantum number is equal to $\frac{1}{2}$, the total angular-momentum quantum number can have only the values $j = l \pm \frac{1}{2}$ and m_j the values

(a) $m_j = -(l + \frac{1}{2}), -(l - \frac{1}{2}), \ldots, (l - \frac{1}{2}), (l + \frac{1}{2})$
if $j = l + \frac{1}{2}$, and   (13)
(b) $m_j = -(l - \frac{1}{2}), -(l - \frac{3}{2}), \ldots, (l - \frac{3}{2}), (l - \frac{1}{2})$
if $j = l - \frac{1}{2}$.

5-8.   Spin-Orbit Fine Structure

We are now in a position to return to the problem of the spin-orbit interaction and to evaluate the average value of $\boldsymbol{\sigma} \cdot \boldsymbol{l}$ over the unperturbed wave functions. To find the proper form for the operator $\boldsymbol{\sigma} \cdot \boldsymbol{l}$ we proceed as follows: the total angular momentum $\boldsymbol{j}$ is equal, classically, to $\boldsymbol{j} = \boldsymbol{l} + \boldsymbol{\sigma}$, so that

$$\boldsymbol{j} \cdot \boldsymbol{j} = (\boldsymbol{l} + \boldsymbol{\sigma}) \cdot (\boldsymbol{l} + \boldsymbol{\sigma}) = \boldsymbol{l} \cdot \boldsymbol{l} + \boldsymbol{\sigma} \cdot \boldsymbol{\sigma} + 2\,\boldsymbol{\sigma} \cdot \boldsymbol{l} \qquad (1)$$

or $\qquad\qquad \boldsymbol{\sigma} \cdot \boldsymbol{l} = \tfrac{1}{2}(\boldsymbol{j} \cdot \boldsymbol{j} - \boldsymbol{l} \cdot \boldsymbol{l} - \boldsymbol{\sigma} \cdot \boldsymbol{\sigma})$

Thus we deduce that the *operator* for $\boldsymbol{\sigma} \cdot \boldsymbol{l}$ is equal to the *operator* for $\tfrac{1}{2}(\boldsymbol{j} \cdot \boldsymbol{j} - \boldsymbol{l} \cdot \boldsymbol{l} - \boldsymbol{\sigma} \cdot \boldsymbol{\sigma})$ and therefore that the average value of $\boldsymbol{\sigma} \cdot \boldsymbol{l}$ over an unperturbed wave function is equal to the average value of $\tfrac{1}{2}(\boldsymbol{j} \cdot \boldsymbol{j} - \boldsymbol{l} \cdot \boldsymbol{l} - \boldsymbol{\sigma} \cdot \boldsymbol{\sigma})$ for this function. The latter quantity is easily evaluated for an unperturbed wave function described by the quantum numbers n, l, s, j, and m_j, since for such a wave function these three quantities have the precise values

(a) $$\qquad\qquad \boldsymbol{j} \cdot \boldsymbol{j} = j(j + 1)\hbar^2$$
(b) $$\qquad\qquad \boldsymbol{l} \cdot \boldsymbol{l} = l(l + 1)\hbar^2 \qquad (2)$$
(c) $$\qquad\qquad \boldsymbol{\sigma} \cdot \boldsymbol{\sigma} = s(s + 1)\hbar^2$$

Thus we obtain

$$\langle \boldsymbol{\sigma} \cdot \boldsymbol{l} \rangle = \tfrac{1}{2}[j(j + 1) - l(l + 1) - \tfrac{3}{4}]\hbar^2 \qquad (3)$$

EXERCISE

5-45. Show that, for a one-electron atom in an eigenstate described by n, l, s, j, and m_j,

$$\langle \boldsymbol{\sigma} \cdot \boldsymbol{l} \rangle = \tfrac{1}{2}l\hbar^2 \qquad \text{or} \qquad -\tfrac{1}{2}(l + 1)\hbar^2$$

depending upon which of the two values of j is involved.

Combining the above results with Eq. 5-6(7), we then have for the spin-orbit energy shift Eq. 5-6(6) the values

(a) $$\quad \Delta E_s = \frac{Z^4 e^2 l \hbar^2}{16\pi\varepsilon_0 m^2 c^2 a_0^3 n^3 l(l + \tfrac{1}{2})(l + 1)} \qquad j = l + \tfrac{1}{2}$$

(b) $$\quad \Delta E_s' = \frac{-Z^4 e^2(l + 1)\hbar^2}{16\pi\varepsilon_0 m^2 c^2 a_0^3 n^3 l(l + \tfrac{1}{2})(l + 1)} \qquad j = l - \tfrac{1}{2}$$

$$(4)$$

for the two possible j values for given n and l.

These first-order energy shifts can be put in a simpler-appearing form by inserting the value of a_0, the first Bohr radius, and defining

$\alpha = e^2/4\pi\varepsilon_0\hbar c$.   Thus one obtains

(a) $\Delta E_s = \dfrac{Z^2|E_n|\alpha^2}{n(2l+1)(l+1)}$ $j = l + \frac{1}{2}$

(5)

(b) $\Delta E'_s = \dfrac{-Z^2|E_n|\alpha^2}{nl(2l+1)}$ $j = l - \frac{1}{2}, l \neq 0$

These equations show that the effect of the spin-orbit interaction is to remove part of the degeneracy that was present in the unperturbed system, splitting the original degenerate level for each value of n into several degenerate levels, slightly displaced from one another. The displacements of the levels from their unperturbed positions are quite small compared with the separations between adjacent unperturbed levels, so that the spectral lines resulting from transitions between the various levels have frequencies only very slightly different from those described by the Bohr theory. Thus each of the lines of the Balmer series, for example, ought to be *multiple* instead of single. A spectrograph of high resolution reveals this to be the case; each spectral line is said to possess *fine structure*. (Actually, the observed fine structure of hydrogen is not as would be expected on the basis of the above analysis. This is due to the fact that other effects, which have so far been neglected, must be included in the problem. We shall treat these further effects shortly.)

The fact that the fine-structure displacement of the energy levels is, in fact, small compared with the separations between adjacent unperturbed levels is due largely to the small size of the quantity α which appears in the above equations. α is called the *fine-structure constant* because it appears in all first-order perturbation formulas for atomic energy levels. It is a *pure number* and has the value

$$\alpha = \frac{e^2}{4\pi\varepsilon_0\hbar c} = \frac{1}{137.0377} \approx \frac{1}{137}$$

(6)

or $\alpha = \dfrac{e^2}{\hbar c} = \dfrac{1}{137.0377}$ cgs esu

Thus $\alpha^2 \approx 1/20{,}000$, so that the fine structure is indeed a small effect. Yet spectroscopy affords so sensitive a means of comparing energy levels that even so small an effect is quite easily observable, and in fact effects of even much smaller size have been measured with considerable precision. It is partly for this reason that spectroscopic data have provided the most rigorous quantitative tests of atomic theories.

5-9.   Relativistic Corrections

In order to compare the fine structure of hydrogen as deduced from the Schroedinger theory with the observed fine structure, it is necessary to include in the Hamiltonian function all effects which can appreciably influence the energy levels of the atom. The principal effect which has so far been omitted is that due to the nonrelativistic character of the quantum theory that has been applied to the problem. It is possible to obtain the corrections due to relativistic effects without actually using a relativistically correct quantum theory, as follows: The Hamiltonian function of a classical system is equal to the total energy (kinetic plus potential) if the system is conservative. The relativistic Hamiltonian function of a particle is (using time as the independent variable)

$$H = (p^2c^2 + m^2c^4)^{1/2} - mc^2 + V \tag{1}$$

If $p \ll mc$, this may be expanded to give

$$H = mc^2 \left(1 + \frac{p^2}{m^2c^2}\right)^{1/2} - mc^2 + V$$
$$= \frac{p^2}{2m} - \frac{p^4}{8m^3c^2} + \cdots + V \tag{2}$$

Thus the largest correction term appearing in the expression is $p^4/8m^3c^2$. This term may be regarded as a perturbation upon the idealized, purely classical problem, and the first-order energy shifts can be found as usual by evaluating the average value of $p^4/8m^3c^2$ over the unperturbed wave functions. Thus

$$\Delta E_r = -\int \psi_n^* \left(\frac{\hbar^4}{8m^3c^2}\right) \nabla^4 \psi_n \, d^N q \tag{3}$$

This integral can be evaluated,[1] using again the wave functions described by the quantum numbers n, l, s, j, and m_j. The result is

$$\Delta E_r = \frac{Z^2|E_n|\alpha^2}{4n^2} \left(3 - \frac{4n}{l + \frac{1}{2}}\right) \tag{4}$$

5-10. Hydrogen Fine Structure

Now, since first-order perturbation corrections due to different effects combine linearly, the net shift of the energy levels can be found by combining the shifts just obtained with those given by Eq. 5-8(5) with the result, for both values of j,

$$\Delta E = \Delta E_s + \Delta E_r = \frac{Z^2|E_n|\alpha^2}{4n^2} \left(3 - \frac{4n}{j + \frac{1}{2}}\right) \tag{1}$$

---

[1] Note that $p^4/8m^3c^2 = T^2/2mc^2 = (E_n - V)^2/2mc^2$. Thus
$$\Delta E_r = -(1/2mc^2)[E_n^2 - 2E_n \langle V \rangle + \langle V^2 \rangle].$$

EXERCISE

5-46. Combine Eqs. 5-8(5) and 5-9(4) for each of the two possible values of j to obtain Eq. (1).

The final formula (1) for the first-order perturbation shifts of the energy levels of a one-electron atom is identical, within terms of order α^4, with the formula obtained by Sommerfeld on the basis of the old quantum theory including the relativity correction. That is, the *numerical values* of the energy shifts are the same, and the *same total number of different energy levels* is obtained. It is important to note, however, that the *total number of states, including degeneracy,* is now almost *twice as large* as was obtained from the Sommerfeld theory. This is true, of course, because electron spin and its quantum numbers have now been introduced, so that the total number of combinations of quantum numbers that can be used to specify a state is now almost twice as great as before. It is possible to distinguish experimentally between these two theories, because the effect of a magnetic field or an electric field upon the energy levels of an atom would be *qualitatively different* for the two. The Schroedinger theory agrees with such experiments, but the old quantum theory does not.

The fine-structure formula (1) gives a complete quantitative description of the various energy levels of a one-electron atom to the approximations so far included. It is customary and desirable to represent these levels graphically on a *term diagram* similar to that used in Chap. 2 in connection with the Bohr theory. Such a diagram is shown in Fig. 5-6. Also included are the levels with their quantum-number designations given by the old quantum theory.

This term-value diagram exhibits many of the features of the fine structure of the hydrogen energy levels quite clearly; three of these are listed below.

1. The fact that the scale representing the shift of the energy levels from their positions given by the Bohr theory is 10,000 times *larger* on this diagram than is the scale by which the Bohr levels themselves are located shows strikingly that the fine structure is, indeed, *fine* and that the approximations that were used in obtaining these shifts are thus quite justified.

2. The rapid decrease in the size of the shifts with increasing n, and with increasing j for given n, shows that it is for the lowest levels of hydrogen that the fine structure will have its greatest effect upon the spectrum. Heavier one-electron atoms will, of course, show effects Z^4 times as large as for hydrogen, so that the spectra of such elements

may have considerably more noticeable fine structure. This latter effect will appear later in connection with X-ray spectra.

3. The rapid increase in the *multiplicity* of the energy levels with increasing n is shown clearly on the diagram, as is the fact that there are almost twice as many levels in the Schroedinger theory as in the Bohr

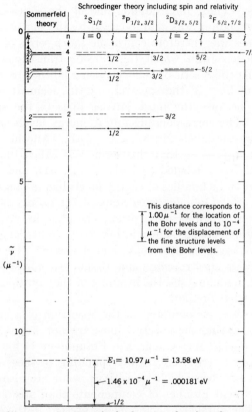

FIG. 5-6. Term diagrams for atomic hydrogen, showing fine structure. On this diagram the dashed lines represent the unperturbed energy levels as given by the Bohr formula 5-3(38). The fine structure of these levels is shown by the short solid lines, the scale representing the energy shifts being 10,000 times magnified with respect to that of the Bohr levels. The quantum numbers defining the various levels are given, both for the Schroedinger theory and for the old quantum theory, and the conventional spectroscopic designations of the terms are indicated along the top.

theory. This does not necessarily indicate, however, a correspondingly rapid increase in the multiplicity of the *spectral lines* with increasing n and n', since we shall shortly find that the *selection rules* which govern the probability of transition between states usually permit only those transitions for which l changes by *one unit*.

5-11.   Spectroscopic Term Classification, Selection Rules

Also shown on the term diagram of Fig. 5-6 are the conventional spectroscopic designations for the various angular-momentum states. This classification system arose historically as an empirical correlation of the various series of spectral lines of an atom with corresponding families of energy levels.   Thus, for example, one of the series of lines in the spectrum of, say, sodium was characterized by the *sharpness* of the lines. This series of lines was therefore called the *sharp* series of lines, and the set of energy levels from which the various transitions of the series started was called the S series, the levels being labeled 1S, 2S, . . . , nS. Another series of lines in the spectrum of this element was called the *principal* series because the most intense lines in the spectrum were members of it.   The corresponding initial energy levels were called the P series and bore the labels 2P, 3P, . . . , nP.   Another series of lines appeared rather *diffuse*, so that this series was called the D series and the energy levels were labeled 3D, 4D, . . . , nD, and so on.   Later it was found that these families of energy levels had in fact been classified in terms of the *orbital angular momentum* of the atom, the S series corresponding to zero angular momentum, the P series to one unit, the D series to two units, etc., as shown in Fig. 2-12.   This same classification system is still in use in spectroscopy, and it has even been extended to apply to molecular spectroscopy and to nuclear physics.   Since it is quite important to understand the meaning of the notation, we shall now consider it somewhat further.

Many of the series encountered in the spectrum of a given atom are *multiplet series;* that is, the spectral lines are not single, but multiple. Thus there are *doublet* series (e.g., the Fraunhofer D lines of sodium), *triplet* series, *quartet* series, etc.   Corresponding to these spectral lines, the *energy levels* between which transitions occur are also multiple, but often with a different number of components than are observed in the spectral line.   The number of components present in the energy level is called the *multiplicity* of the level; it is usually indicated by a *preceding superscript* to the capital letter which characterizes the orbital angular momentum.   Thus ^{2}P refers to a *doublet level* with $l = 1$.   The *multiplicity* of a level is equal, in terms of the quantum numbers already discussed, to the number of different j values that can be formed from given l and s values.   For the one-electron atom, for which s is equal to $\frac{1}{2}$, this number is *two*, unless $l = 0$, in which case it is *one*.

In Chap. 8 we shall find that the energy eigenstates of almost all atoms can be described in terms of a quantum number L which defines the net *orbital* angular momentum $\boldsymbol{L}$ of all of the electrons, a quantum number S which defines the net *spin* angular momentum $\boldsymbol{S}$ of all of the electrons, and a quantum number J which defines the *total* angular momentum $J = L + S$. *L always is an integer*, and S is *integral or half-integral* depending upon whether the total number of electrons in the atom is *even* or *odd*. In this general case the multiplicity of the level is equal to $2S + 1$ or $2L + 1$, whichever is smaller.

EXERCISE

5-47. Show, from the discussion of angular momentum in Sec. 5-7, that the multiplicity of a level, defined as the number of different J-values that can be formed from given L and S values, is $2L + 1$ or $2S + 1$, whichever is smaller.

Since it is most often the case that S is smaller than L, so that the multiplicity is equal to $2S + 1$, the preceding superscript has come to be *always associated with the value of S*. Thus *this index is always equal to* $2S + 1$, and is equal to the true multiplicity of the level only if S is less than, or equal to, L.

Finally, the value of the total angular-momentum quantum number J for a given level is indicated by a *following subscript*. The complete notation can thus be represented as

$$^{2S+1}(L)_J \tag{1}$$

where the numerical values of L are correlated with the letter notation according to the table (note that J is omitted):

L......	0	1	2	3	4	5	6	7	8	9	10 . . . .
Letter...	S	P	D	F	G	H	I	K	L	M	N . . . .

EXERCISES

5-48. What are the values of L, S, and J and the multiplicities of the levels having the following term designations: $^1S_0$, $^3D_2$, $^4P_{5/2}$, $^1D_2$, $^2F_{7/2}$, $^6I_{13/2}$, $^2S_{1/2}$, $^3G_3$, $^5P_1$?

5-49. What types of terms can result from the following values of L and S? (a) $L = 1$, $S = \frac{1}{2}$; (b) $L = 3$, $S = 1$; (c) $L = 2$, $S = \frac{7}{2}$; (d) $L = 5$, $S = \frac{3}{2}$; and (e) $L = 3$, $S = 3$. *Partial ans.:* (a) $^2P_{1/2}$, $^2P_{3/2}$.

Certain minor qualifications concerning the general applicability of the above notation will be added later. It will be found that this classification scheme is applicable to most atoms but that, for heavy atoms, a different type of fine-structure splitting of the unperturbed levels occurs, and for it a different classification scheme is desirable.

The energy levels, including the fine-structure splitting, described in the foregoing analysis are in excellent agreement with what is observed, but it is found experimentally that some of the transitions that might be expected to take place between these levels do not, in fact, occur. This situation has been mentioned before, in connection with both the Sommerfeld treatment of hydrogen and the treatment of temporal changes under the Schroedinger theory, and it is generally described by saying that there are *selection rules* that govern whether a given transition will actually occur or not. In the next chapter we shall examine the basis for these selection rules in the Schroedinger theory of the one-electron atom, and shall at the same time consider some of the more general aspects of radiative transitions. It will be helpful in that discussion, however, to be already familiar with the general nature of the result that is supposed to emerge. It is found by direct observation that the transitions which account for virtually all of the observed spectral lines of a one-electron atom are the so-called *electric-dipole* transitions, whose selection rules are

$$\Delta l = \pm 1 \qquad\qquad\qquad\qquad \Delta m_j = \pm 1 \quad \text{or} \quad 0$$
$$\Delta j = \pm 1 \quad \text{or} \quad 0 \quad \text{but } j = 0 \nleftrightarrow j = 0 \qquad \Delta n \quad \text{unrestricted} \quad (2)$$

The transitions governed by these selection rules will thus be seen to occur *between adjacent columns* in Fig. 5-6, which is, of course, one of the reasons for constructing energy-level diagrams in this form. Other types of transition, not satisfying the above selection rules, may also occur, but these do not ordinarily contribute significantly to the spectrum of a one-electron atom. In some atoms, however, some of the strongest lines in the spectrum may for special reasons be due to these less common types of transition.

EXERCISE

5-50. Use the above selection rules to find what transitions are permitted in the Balmer series of hydrogen: (*a*) Indicate diagrammatically the transitions that are permitted by these rules. (*b*) How many components of different frequency should be present in the Hα line, and what are their designations in the notation $n'^2 P_{3/2} - n^2 S_{1/2}$?

5-12. Nuclear Spin; Hyperfine Structure

As a final topic in this discussion of the Schroedinger theory of the one-electron atom, we shall describe briefly the phenomenon of *nuclear spin* and its effects upon the spectrum of an atom. It was suggested by Pauli (1924) that certain features of the so-called *hyperfine structure* (abbreviated hfs) of the spectral lines of some elements might be explained in terms of an *intrinsic angular momentum and magnetic moment in the nucleus of the atom.* It has since been shown that atomic nuclei actually do possess spin angular momentum analogous to that of electrons, and that they also possess magnetic moments. The magnitude of the nuclear spin of the proton in hydrogen is the same as for electrons, namely $I \cdot I = \frac{1}{2}(\frac{1}{2} + 1)\hbar^2 = \frac{3}{4}\hbar^2$ and $I_z = \pm\frac{1}{2}\hbar$, where I denotes the nuclear-spin angular-momentum vector. The magnitude of the nuclear magnetic moment, on the other hand, is *much smaller* than that of the electron. One might indeed expect, in analogy with the relation 5-5(3) connecting the magnetic moment with the spin of the electron, that the magnetic moment of a proton would be

$$\mathbf{u}' = \frac{e}{M_\mathrm{p}} I \tag{1}$$

and that its component along any given axis would be

$$\mathbf{u}'_z = \frac{e}{M_\mathrm{p}} I_z = \frac{e\hbar}{2M_\mathrm{p}} \tag{2}$$

and would thus be 1836 times smaller than the electron magnetic moment. Actually, the proton magnetic moment is of this order of magnitude but is approximately 2.8 times larger than (1). The quantity defined by (2) is called a *nuclear magneton* and is denoted by the symbol μ_N. In terms of this unit, the proton magnetic moment is found experimentally to be

$$\mu_\mathrm{p} = 2.79275\mu_\mathrm{N} \tag{3}$$

where $\mu_\mathrm{N} = e\hbar/2M_\mathrm{p} = 0.505038 \times 10^{-26}$ A m².

We shall regard the spins and magnetic moments of nuclei as being given, and reserve for a later chapter any discussion of the rules by which the nuclear spins and magnetic moments are compounded from the intrinsic spins and moments of the particles of which a nucleus is composed. It suffices to say that the nuclear-spin quantum number I can take on *integral* or *half-integral* values and that it is a *constant* for a given nucleus (because it corresponds to the nuclear ground state, and the nucleus does not become excited at the energies involved in ordinary atomic transitions).

The existence of nuclear spins and magnetic moments may be expected to lead to the following modifications in the Schroedinger theory of the one-electron atom:

1. *Nuclear-spin variables, operators, and quantum numbers must be introduced into the theory.* This can be done in the same way as was done for electron spin, with only minor changes resulting from the fact that the nuclear magnetic moments are not related to the nuclear spins in

as simple a way as are these quantities for the electron. The nuclear spin is called I, its components I_x, I_y, I_z, and the corresponding quantum numbers, I and m_I. For a proton, $I = \frac{1}{2}$ and m_I can equal $+\frac{1}{2}$ or $-\frac{1}{2}$. In the absence of appreciable interaction between the nuclear spin and the remainder of the atom, *additional degeneracy of the unperturbed levels will thus be introduced.*

2. *Interactions between the nuclear magnetic moment and various other quantities must be considered.* Thus one expects to have terms in the Hamiltonian corresponding to the interaction of the nuclear magnetic moment with (a) the magnetic field due to the motion of the electron (this corresponds to the spin-orbit interaction of the electron), Eq. 5-6(2); (b) the electron magnetic moment; and (c) an outside magnetic field if one is present.

3. *A further splitting of the levels due to the nuclear-spin interaction terms will occur.* Because the nuclear magnetic moments are approximately 1000 times smaller than the Bohr magneton, this splitting will, correspondingly, be about 1000 times smaller than the ordinary fine-structure splitting. (Hence the name "hyperfine structure.")

4. *The new total angular momentum vector F will now be the sum of three component angular momenta,* thus:

$$F = L + S + I = J + I \tag{4}$$

where J is the orbital angular momentum plus electron-spin angular momentum. F can take on such values that

$$F \cdot F = F(F + 1)\hbar^2 \tag{5}$$

where the quantum number F takes on integrally spaced values from the smallest positive integer or half-integer that can be formed from L, S, and I, up to $L + S + I$. Thus if $L = 1$, $S = \frac{1}{2}$, and $I = 3$, F can take on values from $\frac{3}{2}$ to $\frac{9}{2}$, inclusive. As a practical matter, the degenerate levels defined by the various possible values of J will have already been split by the ordinary fine-structure effects, so that the further hyperfine splitting of each J level will be governed by the condition that the total angular-momentum quantum number F can take on the integrally spaced values between $|J - I|$ and $(J + I)$. Thus, in the above example the level having $J = \frac{1}{2}$ would have a hyperfine structure consisting of two levels having F quantum numbers of $\frac{5}{2}$ and $\frac{7}{2}$, and the level corresponding to $J = \frac{3}{2}$ would be split into four levels, having F quantum numbers of $\frac{3}{2}$, $\frac{5}{2}$, $\frac{7}{2}$, and $\frac{9}{2}$.

5. *Selection rules will exist, governing the possible changes in the total angular momentum F.* These selection rules turn out to be

(a) $\qquad\qquad\qquad\qquad \Delta F = 0, \pm 1$

(b) $\qquad\qquad\qquad\qquad \Delta M_F = 0, \pm 1 \qquad\qquad (6)$

Hyperfine structure is exceedingly difficult to observe in hydrogen, and has not been observed spectroscopically at all. Modern microwave techniques used in connection with radioastronomy have, however, recently led to the detection of the radiation emitted by atomic hydrogen in interstellar space as it undergoes the transition between the two hyperfine levels in its ground state. This radiation has a frequency of 1420 megacycles per second and is being used to study the distribution and motions of hydrogen in interstellar and intergalactic space.

We shall not go further into the numerical details of hyperfine structure, both because it is not a subject of primary importance for our later considerations and because the treatment of hyperfine structure is closely similar to that of ordinary fine structure. The most important consequences of nuclear spin for atomic physics are the *additional degeneracy* this quantity introduces into the atomic energy levels, which is important in some statistical considerations and in certain effects related to the exclusion principle, which will be described later.

EXERCISES

5-51. Show that the classical interaction energy of two magnetic dipoles $\mathbf{\mu}_1$ and $\mathbf{\mu}_2$ is equal to

$$W = \frac{\mu_0}{4\pi} \left[\frac{\mathbf{\mu}_1 \cdot \mathbf{\mu}_2}{r^3} - 3 \frac{(\mathbf{\mu}_1 \cdot \mathbf{r})(\mathbf{\mu}_2 \cdot \mathbf{r})}{r^5} \right] \tag{7}$$

5-52. (*a*) Evaluate classically the maximum interaction energy of a nuclear magneton and a Bohr magneton if they are separated by a distance equal to the first Bohr radius. (*b*) Evaluate classically the orientation energy of a nuclear magneton in the magnetic field due to an electron revolving in the lowest circular Bohr orbit of hydrogen, and compare this with (*a*).

5-53. A certain fine-structure multiplet of levels corresponds to $L = 2$, $S = 1$, and $I = \frac{3}{2}$. Find the number of components and give their spectroscopic designations for the fine-structure multiplet, and find also into how many components each of these levels will be further split by hyperfine-structure splitting. Allowing for the degeneracy of each hyperfine level due to the quantum number M_F, what is the total degeneracy of each of the fine-structure levels? *Ans.:* One of the fine-structure levels is a $^3D_2$ level. It is split into four components by hfs splitting. The degeneracies of these hfs levels are 2, 4, 6, and 8, making a total degeneracy of 20 for the $^3D_2$ level.

5-54. Two levels in an atom whose nuclear spin is $I = 3$ have the designations $^2D_{3/2}$ and $^2P_{1/2}$. Find the expected number of components in the hyperfine structure of the corresponding spectral line.

The treatment of the one-electron atom given here has illustrated the quantum-mechanical basis for many of the observed features of atomic excited states. We shall later apply many of the results obtained analytically in the present chapter to more complicated atoms for which only an approximate analytical treatment can be made. In particular, the qualitative features of the spin-orbit effects will be carried over to the many-electron atom without appreciable modification.

REFERENCES

Pauling, L., and E. B. Wilson: "Introduction to Quantum Mechanics," McGraw-Hill Book Company, Inc., New York, 1935.

Rojanski, V.: "Introductory Quantum Mechanics," Prentice-Hall, Inc., Englewood Cliffs, N.J., 1938.

Schiff, L. I.: "Quantum Mechanics," 2d ed., International Series in Pure and Applied Physics, McGraw-Hill Book Company, Inc., New York, 1955.

White, H. E.: "Introduction to Atomic Spectra," International Series in Pure and Applied Physics, McGraw-Hill Book Company, Inc., New York, 1934.

6

Radiation and Radiative Transitions

No phenomenon plays a more distinctive role in shaping the familiar properties of the universe than does *electromagnetism*. Indeed, our understanding of nature depends quite fundamentally upon our first understanding the electromagnetic field and its interaction with material systems. While the large-scale actions of radiation are described with quite high precision by Maxwell's equations, we know that individual atomic processes in which radiation plays a role must be treated by using quantum mechanics and that in such a treatment the quantum properties of the radiation field, as well as of matter, must be taken into account.

A precise description of electromagnetic radiation and its interactions, including both the macroscopic limit of continuous waves and the microscopic limit of quanta, is provided by the theory of *quantum electrodynamics*. This theory is at present believed to be capable, in principle, of describing to any desired degree of precision *all observable properties of matter except those which depend upon gravitation or nuclear forces.* That is, we believe that all of the data appearing in handbooks of physics and chemistry, with the exception of the fundamental natural constants and the isotopic masses, spins, abundances, and other properties specific to nuclei, *could be deduced from first principles* if sufficiently powerful mathematical techniques were available.[1]

In the present chapter we shall examine some of the more elementary aspects of quantum electrodynamics, using an approximation which describes accurately the most common types of interaction of the electromagnetic field with material systems in the nonrelativistic limit. Rather than try to obtain this approximation from the exact theory, however, we shall approach it as an extension of Schroedinger quantum mechanics

---

[1] Quantum electrodynamics might well be abbreviated Q.E.D.!

to the description of continuous systems. In this way the analogy with the classical theory of normal coordinates will be most strongly apparent and will contribute most effectively to our understanding of the quantum properties of radiation. In Chap. 20 we shall consider some of the features of relativistic quantum electrodynamics.

6-1. Normal Coordinates and Hamiltonian Representation of a Continuous Elastic System

In the quantum-mechanical treatment of vibratory systems and of radiation, each of the normal modes of oscillation of the system is regarded as an independent quantized harmonic oscillator. The basis of such a representation lies in the classical theory of *normal coordinates* of vibrating systems. According to this theory, the coordinates that one ordinarily uses to describe a vibrating system, e.g., the lateral displacement y of a string as a function of the distance x along the string, are replaced by a new set of independent coordinates which suffice equally well to specify the instantaneous configuration of the system but which vary with time in an especially simple way, namely, *sinusoidally*. We shall now recall briefly the nature of normal coordinates for continuous systems, using for clarity the familiar example of a stretched string of uniform density ρ with fixed ends, vibrating in a plane.

If the lateral displacement y is expressed as a Fourier series

$$y(x,t) = \sum_{s=1}^{\infty} \phi_s(t) \sin \frac{s\pi x}{L} \tag{1}$$

the *normal coordinates* are simply the time-dependent coefficients $\phi_s(t)$. This equation may be regarded as an infinite set of linear equations relating the coordinates $y(x)$ to new coordinates ϕ_s, there being one equation for each different value of x. (Note that x is to be regarded not as a coordinate in the sense of defining the displacement of the string, but as a *parameter* which describes what infinitesimal segment of the string has the displacement y.) Alternatively, the normal coordinates may be expressed in terms of the coordinates $y(x)$. One does this by solving Eq. (1) by Fourier inversion: Multiply both sides of the equation by $\sin (r\pi x/L)$ and integrate from 0 to L. The orthogonality of the sine functions over this range causes all terms on the right except that for $s = r$ to vanish, with the result

$$\phi_r(t) = \frac{2}{L} \int_0^L y(x,t) \sin \frac{r\pi x}{L} \, dx \tag{2}$$

Now let us verify that the quantities ϕ_r possess the properties required of normal coordinates:

1. *Normal coordinates are coordinates which suffice to define the instantaneous configuration of the system.* This is true because a specification of the instantaneous value of each ϕ_r uniquely defines the displacement of every segment of the string at that instant.

2. *The potential and kinetic energies of the system are expressible as sums of squares of the normal coordinates and of their first time derivatives, respectively, with constant coefficients.* The kinetic energy is equal to

$$T = \int_0^L \frac{1}{2} \rho \left(\frac{\partial y}{\partial t} \right)^2 dx \tag{3}$$

which, by substitution of $\partial y / \partial t$ from Eq. (1) in the form

$$\frac{\partial y}{\partial t} = \sum_{s=1}^{\infty} \dot{\phi}_s \sin \frac{s\pi x}{L} \tag{4}$$

gives

$$T = \frac{1}{2}\rho \int_0^L \sum_{s=1}^{\infty} \sum_{r=1}^{\infty} \dot{\phi}_s \dot{\phi}_r \sin \frac{s\pi x}{L} \sin \frac{r\pi x}{L} dx$$

The orthogonality of the sine functions then leads directly to

$$T = \sum_{s=1}^{\infty} \frac{1}{4}\rho L \dot{\phi}_s^2 \tag{5}$$

Similar treatment of the potential energy

$$V = \int_0^L \frac{1}{2} \tau \left(\frac{\partial y}{\partial x} \right)^2 dx \tag{6}$$

leads to the expression

$$V = \sum_{s=1}^{\infty} \frac{1}{4} \frac{\tau}{L} s^2 \pi^2 \phi_s^2 \tag{7}$$

where τ is the tension of the string.

3. *The equation of motion of a normal coordinate is that of a harmonic oscillator.* If we form the Lagrangian function for the normal coordinates, we obtain

$$L = T - V = \sum_{s=1}^{\infty} \left(\frac{1}{4}\rho L \dot{\phi}_s^2 - \frac{1}{4} \frac{\tau}{L} s^2 \pi^2 \phi_s^2 \right) \tag{8}$$

From this, the equations of motion are easily found to be

$$\ddot{\phi}_s + \frac{s^2 \pi^2}{L^2} \frac{\tau}{\rho} \phi_s = 0 \tag{9}$$

EXERCISES

6-1. Show that Eq. (6) correctly represents the potential energy of a stretched string if the slope of the string is small.

6-2. Carry out the steps which lead from Eq. (6) to Eq. (7).

6-3. Derive the equations of motion (9) from the Lagrangian function (8).

6-4. Solve the equations of motion (9) and show that Eq. (1) then represents the familiar equation of motion of a vibrating string.

The above treatment has shown how a continuous vibratory system can be represented in Lagrangian form by using normal coordinates. Such a system can also be put into Hamiltonian form by defining a set of momenta Π_s corresponding to the various normal coordinates. Thus in the usual manner we may write, for our example of the vibrating string,

$$\Pi_s = \frac{\partial L}{\partial \dot{\phi}_s} = \tfrac{1}{2}\rho L \dot{\phi}_s \tag{10}$$

and we may then construct the Hamiltonian function

$$
\begin{aligned}
H &= \sum_{s=1}^{\infty} \Pi_s \dot{\phi}_s - L \\
&= \sum_{s=1}^{\infty} \left(\frac{1}{\rho L} \Pi_s{}^2 + \frac{1}{4} \frac{\tau}{L} s^2 \pi^2 \phi_s{}^2 \right)
\end{aligned} \tag{11}
$$

6-2.   Energy Eigenfunctions for Normal Coordinates

By the foregoing procedure a continuous system can be described in a form to which the Schroedinger theory is immediately applicable. To find the Hamiltonian operator, we replace Π_s by $(\hbar/i)(\partial/\partial\phi_s)$ according to the rule set forth in the third postulate of Chap. 2, with the result, for the vibrating string,

$$\hat{H} = \sum_{s=1}^{\infty} \left(-\frac{\hbar^2}{\rho L} \frac{\partial^2}{\partial \phi_s{}^2} + \frac{1}{4} \frac{\tau}{L} s^2 \pi^2 \phi_s{}^2 \right) \tag{1}$$

This form for the Hamiltonian operator—a summation of terms, each depending only upon a single coordinate—means that the wave function for the string can be written as a *product* of wave functions

$$\Psi(\phi_j, t) = \prod_{s=1}^{\infty} \psi_{n_s}(\phi_s, t) \tag{2}$$

as was described in Exercise 2-50. These eigenfunctions are solutions
of the equations

$$\left(-\frac{\hbar^2}{\rho L}\frac{\partial^2}{\partial\phi_s^2} + \frac{1}{4}\frac{\tau}{L}s^2\pi^2\phi_s^2\right)\psi_{n_s}(\phi_s,t) = -\frac{\hbar}{i}\frac{\partial\psi_{n_s}(\phi_s,t)}{\partial t} \tag{3}$$

and each ψ_{n_s} may be in turn expressed in the separated form given by
Eq. 2-4(20)

$$\psi_{n_s}(\phi_s,t) = u_{n_s}(\phi_s)\exp(-iE_{n_s}t/\hbar) \tag{4}$$

where now the function $u_{n_s}(\phi_s)$ is a solution of the time-independent
equation

$$\frac{-\hbar^2}{\rho L}\frac{d^2u_{n_s}}{d\phi_s^2} + \frac{1}{4}\frac{\tau}{L}s^2\pi^2\phi_s^2u_{n_s} - E_{n_s}u_{n_s} = 0 \tag{5}$$

This is the same as the time-independent Schroedinger equation for a
"particle" of "mass" $m = \frac{1}{2}\rho L$ situated in a harmonic potential well
whose "force constant" k is equal to $\frac{1}{2}(\tau/L)s^2\pi^2$ and whose classical
oscillation angular frequency is $\omega_s = (s\pi/L)(\tau/\rho)^{1/2}$. The functions
$u_{n_s}(\phi_s)$ are thus the Hermite functions

$$u_{n_s}(\phi_s) = N_{n_s}H_{n_s}(\alpha_s\phi_s)\exp(-\alpha_s^2\phi_s^2/2) \tag{6}$$

where $$\alpha_s^2 = \frac{m\omega_s}{\hbar} = \frac{1}{2}\frac{s\pi}{\hbar}\sqrt{\tau\rho} \tag{7}$$

and the energy levels are given by

$$E_{n_s} = (n_s + \frac{1}{2})\hbar\omega_s = (n_s + \frac{1}{2})\frac{s\pi}{L}\hbar\sqrt{\frac{\tau}{\rho}} \tag{8}$$

where $n_s = 0, 1, 2, \ldots$ corresponds to the quantum number n of Eq.
3-3(3).

The physical situation represented by Eqs. (2), (4), (6), and (8) is as
follows: *A continuous vibratory system may exist in any of an infinite
number of energy eigenstates, in which its classical normal modes of vibration
independently possess any of the quantized energies defined by Eq. (8), the
total energy of the system being the sum of the energies contained in its
normal modes of vibration.* The most general state of the system may be
expressed as a linear combination of such eigenstates, each with its
appropriate time factor. Thus the independence of the normal modes of
vibration, which in the classical treatment may each be excited independently of one another, extends also to the quantum-mechanical case.

One unfortunately finds certain unpleasant difficulties in this quantum-
mechanical picture; one of them is that the zero-point energy of the
system—the sum of the $\frac{1}{2}\hbar\omega_s$ energies of the various normal modes—is
infinite because the number of normal coordinates of a continuous system
is infinite. This conclusion can be avoided in treating certain physical

systems, such as a "continuous" string or a vibrating crystal, which are really aggregates of discrete particles so uniformly and closely spaced that they can be regarded macroscopically as continua, by recognizing that the number of normal coordinates of such a system is actually only two or three times the (finite) number of particles comprising the system. But it might appear that no such escape exists in the treatment of the electromagnetic field, where there seems really to be no lower limit to the possible wavelength of an electromagnetic wave train. However, even here it is possible to avoid ascribing an infinite energy density to empty space by choosing the Hamiltonian representation of the field in the proper way.[1] In order to retain the analogy between the modes of vibration of mechanical systems and of the classical electromagnetic field, we shall retain the zero-point energy, but ignore it, since it is at any rate a constant property of space and does not adversely affect those results that are of interest to us. Indeed, we shall think of an excited atom situated in the dark, i.e., surrounded only by the electromagnetic field in its lowest energy state, as emitting radiation by virtue of its interaction with these zero-point field vibrations.

6-3. Quantization of the Radiation Field

The procedure by which a suitable Hamiltonian representation of the electromagnetic field is obtained is rather similar to that outlined above for a one-dimensional system but is considerably more involved. We shall therefore restrict our treatment to the essential results:[2] The magnetic vector potential for a traveling wave disturbance is expanded in a three-dimensional Fourier series throughout a cubical region of dimension L along each axis:

$$A(r,t) = \sum_{K} (e_{K1}q_{K1} + e_{K2}q_{K2}) \exp (iK \cdot r)$$
$$+ (e_{K1}q_{K1}^* + e_{K2}q_{K2}^*) \exp (-iK \cdot r) \quad (1)$$

and the scalar potential ϕ is set equal to zero.

In this expression e_{K1} and e_{K2} are two mutually perpendicular unit vectors which define two directions of polarization of a plane wave whose wavefronts are perpendicular to K. The quantities q_{K1} and q_{K2} are independent complex normal coordinates which define the amplitudes of the complex plane wave trains for each direction of polarization. They can be regarded as being the coordinates associated with an infinite number of linear harmonic oscillators, called *field oscillators*. K is called

---

[1] W. Heitler, "The Quantum Theory of Radiation," 3d ed., p. 57, Oxford University Press, New York, 1954.
[2] *Ibid.*, chap. 2.

the *propagation vector* of the plane wave. Because of the requirement
that A must be periodic along all three axes, K is restricted to the values

$$K = \frac{2\pi}{L}(li + mj + nk) \tag{2}$$

where l, m, and n are positive integers and i, j, and k are the unit vectors
along the x, y, and z axes. The wavelength corresponding to a given
term of (1) is

$$\lambda = \frac{2\pi}{|K|} \tag{3}$$

EXERCISES

6-5. Find the electric field corresponding to one term, $A_{K1} = e_{K1}q_{K1}\exp(iK \cdot r)$, of the vector potential (1). *Ans.:* $E_{K1} = -e_{K1}\dot{q}_{K1}\exp(iK \cdot r)$.

6-6. Find the magnetic induction corresponding to the above vector
potential. *Ans.:* $B_{K1} = i(K \times e_{K1})q_{K1}\exp(iK \cdot r)$.

6-7. In a plane electromagnetic wave, E_K, B_K, and K must be mutually
perpendicular. Show that this requires that $e_{K1} \cdot K = 0$ and $e_{K2} \cdot K = 0$.

6-8. Show that the divergence of A is equal to zero. This description
of transverse electromagnetic waves in terms of a divergenceless vector
potential and zero scalar potential is called the *Coulomb gauge* and is
one of the infinitely many ways in which given E and B fields can be
described in terms of a vector and a scalar potential.

6-9. Show that the form (1) is such that A, E, and B are all real, even
though the individual terms of the expansion are complex.

Classically, each of the normal coordinates satisfies an equation of the
form

$$\ddot{q}_K + \omega_K{}^2 q_K = 0 \tag{4}$$

where $\omega_K{}^2 = c^2 K^2$, and thus the complete expression for A is that of a
superposition of plane waves of all polarizations and phases traveling with
speed c parallel to each K. Each such plane wave may be symbolized

$$A_{Ki} = e_{Ki}A_{Ki}\exp[i(K \cdot r \pm \omega_K t)] \tag{5}$$

where A_{Ki} is a complex constant amplitude of A_{Ki}.

Quantum-mechanically, each normal coordinate may be described in
terms of a wave function which satisfies a Schroedinger equation similar
to Eq. (3), Sec. 6-2, with a solution of the form characteristic of a linear
harmonic oscillator:

$$\begin{aligned}
\psi_{n_K}(q_K,t) &= u_{n_K}(q_K)\exp(-iE_{n_K}t/\hbar) \\
&= N_{n_K}H_{n_K}(\alpha_K q_K)\exp[-(\alpha_K{}^2 q_K{}^2/2) - i(n_K + \tfrac{1}{2})\omega_K t] \tag{6}
\end{aligned}$$

where $\alpha_K{}^2 = \varepsilon_0 L^3 \omega_K / \hbar$. The complete radiation field is again a super-position of plane waves,[1] but now with *quantized energies*, the most general eigenstate of the radiation field being a product of individual eigen-functions of the above type:

$$\psi(q_K, t) = \prod_K u_{n_K}(q_K) \exp\left[-i(n_K + \tfrac{1}{2})\omega_K t\right]$$

$$= U_{n_{K_1} \cdots n_{K_s} \cdots} \exp\left[-i \sum_K (n_K + \tfrac{1}{2})\omega_K t\right] \qquad (7)$$

6-4. Selection Rules and Transition Probabilities for the Harmonic Oscillator with a Linear Perturbation

Without doubt the most characteristic and important phenomenon of atomic physics is the emission and absorption of electromagnetic radiation. In the most common procedure for treating this process, which was outlined at the end of Chap. 2, the material system and the radiation field are first treated as being separate and the actual interaction between them is later introduced as a small perturbation. This inter-action takes place through the charges, magnetic-dipole moments, and higher electric and magnetic moments possessed by the particles com-prising the material system. The transitions induced by the interaction depend upon its matrix elements, and a complete treatment, of course, requires knowledge of the wave functions of both the material system and the radiation field. However, the form of the interaction depends upon the normal coordinates of the radiation field in a way that is independent of the detailed nature of the material system, so that certain important general properties of radiative transitions can be deduced from a knowledge only of the general form of the interaction and the wave functions of the radiation field. We shall now consider some of these general properties for those radiative transitions that arise from the interaction of the *charges* of the particles of a material system with the radiation field. We shall find it possible to deduce the type of *selection rules* which govern the first-order transitions of the radiation field and to obtain certain information concerning the dependence of the probability of a transition upon the intensity of the radiation.

We must first establish the nature of the interaction between charged particles and the radiation field. To do this, we consider the classical

---

[1] We found in Chap. 4 that a plane wave $\exp(i\boldsymbol{K}\cdot\boldsymbol{r})$ has the same space-coordinate dependence as does the wave function of a *particle* whose momentum is $\boldsymbol{K}\hbar$. This fact and the quantization of the energies of the normal modes of vibration of the radiation field together provide the quantum-mechanical basis for the *wave-particle duality* of light.

nonrelativistic equations of motion of a charged particle in Lagrangian form. Using tensor notation, a suitable Lagrangian function is

$$L = \tfrac{1}{2}m\dot{x}_j\dot{x}_j + e\dot{x}_jA_j - e\phi \tag{1}$$

where e and m are the charge and mass of the particle, A_j is the magnetic vector potential and ϕ is the scalar potential, both of the latter being evaluated at the location of the particle.

EXERCISE

6-10. Show that the above Lagrangian function leads to the correct equations of motion for a charged particle moving in an electromagnetic field.

A Hamiltonian function for the particle may now be constructed in the usual way: The conjugate momenta are

$$p_j = \frac{\partial L}{\partial \dot{x}_j} = m\dot{x}_j + eA_j \tag{2}$$

and the Hamiltonian function is

$$H = \frac{1}{2m}(p_j - eA_j)(p_j - eA_j) + e\phi \tag{3}$$

In these expressions the vector and scalar potentials are those which exist at the position of each particle and are due to all other particles of the system as well as to external radiation. For our present purposes, we can regard ϕ as representing the mutual electrostatic interaction of the particles of an atomic system and A as being associated with external radiation.

EXERCISE

6-11. Carry out in detail the formation of the Hamiltonian function (3).

If this Hamiltonian function is written in the expanded form

$$H = \frac{1}{2m}p_jp_j + e\phi - \frac{e}{m}p_jA_j + \frac{e^2}{2m}A_jA_j \tag{4}$$

we see that the first two terms are just those that would have appeared in the treatment of the material system alone, where the scalar potential ϕ is that due to the mutual electrostatic interactions between the various particles of this system. The last two terms, on the other hand, represent the interaction of the material system with the external radiation field.   Of these, the term $(e^2/2m)A_jA_j$ is generally of negligible importance[1] so that as a close approximation

$$V_{\text{int}} = -\frac{e}{m}\,p_jA_j = -\frac{e}{m}\,\boldsymbol{p}\cdot\boldsymbol{A} \tag{5}$$

is the interaction "potential" of a charged particle in an externally applied radiation field.   If a number of particles are present in the material system, the net interaction is a *summation* of terms of the form (5) over the various particles.   For simplicity we shall assume that the material system has in it only a single charged particle, the extension to several particles being straightforward.   The matrix element of V_{int} will have the form (Sec. 2-5D)

$$(a'n'|V_{\text{int}}|an) = -\frac{e}{m}\int_{q_i}\cdots\int u_{a'}^*\hat{p}_j\times$$

$$\left(\int_{q_K}\cdots\int\cdots U_{n'}^*\hat{A}_jU_nd^\infty q_K\right)u_a\,d^Nq_i \tag{6}$$

where the functions $u_a(q_i)$ are the eigenfunctions of the material system and $U_n(q_K)$ of the radiation field.   The subscript a stands for the set of quantum numbers which defines the state of the material system, and n does the same for the radiation field.

The physical significance of the matrix element $(a'n'|V_{\text{int}}|an)$ is that its absolute square is proportional to the probability that the interaction induces a transition in the material system, carrying it from an initial quantum state u_a to a final quantum state $u_{a'}$, and also causes the radiation field to undergo a transition from $U_{n_{K_1}\cdots n_{K_s}\cdots}$ to $U_{n'_{K_1}\cdots n'_{K_s}\cdots}$. The functions u_a are characteristic of the particular material system under consideration, and the functions $U_{n_{K_1}\cdots n_{K_s}\cdots}$ are those of Eq. 6-3(7). The latter correspond physically to eigenstates of the radiation field in which its various normal modes are excited to the degrees described by the quantum numbers n_{K_1}, n_{K_2}, . . . , n_{K_s}, ... , respectively.

We shall now restrict our attention to the bracketed expression within the integral, calling it $(n'|A_j|n)$:

$$(n'|A_j|n) = \int\cdots\int U_{n'}^*\hat{A}_jU_n\,dq_{K_1}\cdots dq_{K_s}\cdots \tag{7}$$

This part of the total matrix element can be evaluated readily if we imagine the vector potential components A_j to be expressed as plane waves of the form 6-3(1).   The matrix element (7) will then be a sum of a number of terms of the following type, one for each plane electromagnetic wave of specified propagation direction and frequency $\boldsymbol{K}_s$ and

---

[1] This matter will be discussed further in Sec. 6-7.

specified polarization direction e_{K_s}

$$(n'|A_{K_s j}|n) = e_{K_s j} \exp (i\mathbf{K} \cdot \mathbf{r}) \int \cdots \int U_{n'}^* q_{K_s} U_n d^\infty q_K \tag{8}$$

where $e_{K_s j}$ is the j-component of the polarization vector e_{K_s}.

Now because the function U_n is a product of eigenfunctions of the various independent normal coordinates q_K [Eq. 6-3(7)] and since the eigenfunctions of each q_K are an orthogonal set of functions of that coordinate, the above integral may be further broken up into a product of integrals over the individual normal coordinates:

$$\int \cdots \int U_{n'}^* q_{K_s} U_n \, dq_{K_1} \cdots dq_{K_s} \cdots = \int u_{n'_{K_1}}^* (q_{K_1}) u_{n_{K_1}} (q_{K_1}) \, dq_{K_1}$$
$$\times \int u_{n'_{K_2}}^* (q_{K_2}) u_{n_{K_2}} (q_{K_2}) \, dq_{K_2}$$
$$.$$
$$.$$
$$.$$
$$\times \int u_{n'_{K_s}}^* (q_{K_s}) q_{K_s} u_{n_{K_s}} (q_{K_s}) \, dq_{K_s}$$
$$.$$
$$.$$
$$.$$
$$\times \int u_{n'_{K_t}}^* (q_{K_t}) u_{n_{K_t}} (q_{K_t}) \, dq_{K_t}$$
$$.$$
$$.$$
$$. \tag{9}$$

Next, we observe that the orthogonality of the various eigenfunctions of each normal coordinate causes each of the above integrals to equal unity if $n'_{K_1} = n_{K_1}$, $n'_{K_2} = n_{K_2}$, $\ldots$, $n'_{K_t} = n_{K_t} \cdots$, and zero otherwise, *except for the single integral over* q_{K_s}. Thus we have

$$\int \cdots \int U_{n'}^* q_{K_s} U_n \, dq_{K_1} \cdots dq_{K_s} \cdots$$
$$= \delta_{n'_{K_1} n_{K_1}} \delta_{n'_{K_2} n_{K_2}} \cdots \left(\int u_{n'_{K_s}}^* q_{K_s} u_{n_{K_s}} \, dq_{K_s} \right) \cdots \delta_{n'_{K_t} n_{K_t}} \cdots \tag{10}$$

Finally, to evaluate this last integral, we explicitly introduce the Hermite function form of $u_{n'_{K_s}}^*$ and $u_{n_{K_s}}$:

$$\int_{-\infty}^{\infty} u_{n'_{K_s}} q_{K_s} u_{n_{K_s}} \, dq_{K_s} = \frac{N_{n'_{K_s}} N_{n_{K_s}}}{\alpha_{K_s}} \int_{-\infty}^{\infty} H_{n'_{K_s}} (\xi_{K_s}) \xi_{K_s} H_{n_{K_s}} (\xi_{K_s})$$
$$\exp (-\xi_{K_s}^2) \, dq_{K_s} \tag{11}$$

where $\xi_{K_s} = \alpha_{K_s} q_{K_s}$

The recurrence formula 3-2(16b) may now be used to write

$$\int_{-\infty}^{\infty} u_{n'_{K_s}} q_{K_s} u_{n_{K_s}} \, dq_{K_s} = \frac{N_{n'_{K_s}} N_{n_{K_s}}}{2\alpha_{K_s}} \int_{-\infty}^{\infty} H_{n'_{K_s}} (H_{n_{K_s}+1} + 2n_{K_s} H_{n_{K_s}-1})$$
$$\exp (-\xi_{K_s}^2) \, dq_{K_s} \tag{12}$$

$$= \frac{N_{n'_{K_s}} N_{n_{K_s}}}{2\alpha_{K_s} N^2_{n'_{K_s}}} (\delta_{n'_{K_s}, n_{K_s}+1} + 2n_{K_s} \delta_{n'_{K_s}, n_{K_s}-1}) \tag{13}$$

$$
\begin{cases}
(a) \quad \dfrac{N_{n_{K_s}}}{2\alpha_{K_s} N_{n_{K_s}+1}} = \dfrac{(n_{K_s}+1)^{1/2}}{2^{1/2}\alpha_{K_s}} \quad (n'_{K_s} = n_{K_s}+1) \\[4mm]
(b) \quad \dfrac{n_{K_s} N_{n_{K_s}}}{\alpha_{K_s} N_{n_{K_s}-1}} = \dfrac{n_{K_s}^{1/2}}{2^{1/2}\alpha_{K_s}} \quad (n'_{K_s} = n_{K_s}-1)
\end{cases} \tag{14}
$$

EXERCISE

6-12. Carry through in detail the steps leading from Eq. (9) to Eq. (14).

The physical interpretation of the above result is as follows: If the electromagnetic field is subjected to an interaction that is proportional to the amplitude q_{K_s} of one of its vector potential components, the first-order transitions induced by this interaction are ones in which *all normal modes except the one associated with q_{K_s} remain in the same quantum state, and the normal mode associated with q_{K_s} either gains or loses a single quantum of excitation.* Furthermore, since the total vector potential is a summation of terms, each of which is linear in some q_{K_s}, the above conclusion holds also for transitions induced by the interaction (5); namely, the first-order transitions induced by the interaction (5) are ones in which *the radiation field either gains or loses a single quantum of excitation of one of its normal modes.* A relationship of this type which restricts the possible transitions to a narrow class is called a *selection rule.* The existence of selection rules is closely related to the *orthogonality of the eigenstates* of a system, as is clearly shown in the above analysis. We shall later find that a similar analysis leads to the familiar selection rules that govern the optical transitions of excited atoms.

Certain other features of radiative transitions, beyond the existence of selection rules, can be seen in the above results. The square of the total matrix element (6), which describes the probability of transition, will contain the square of one or the other of the factors (14), depending upon whether the material system is radiating or absorbing energy from the field. This factor indicates that the probability of absorption of energy by the material system is proportional to n_K, the degree of excitation of the mode which is supplying the energy to the material system. This is of course a reasonable result, since the probability of absorption ought to be greater if more radiation is applied to the system and ought to be zero if there is no available energy in the given oscillation mode of the field. But this same factor for the contrary case of *emission* of energy by the material system is proportional to $n_K + 1$, which *is also greater if there is already some energy contained in the normal mode that is to receive the energy from the material system.* This phenomenon is called *induced emission* and is analogous to the classical process of energy exchange with vibrating systems, wherein the rate of transfer of energy is proportional to the square of the amplitude of the vibration, independently of whether the vibrating system is absorbing or emitting energy. It is customary to separate the transition probability containing the factor $n_K + 1$ into two parts, one of which contains the constant (unity) and the other of which is proportional to n_K. The first of these is called the *coefficient of spontaneous emission* because an excited material system would emit radiation to this extent even if it were "in the dark." The second is called the *coefficient of induced emission* because the emission of radiation is increased by this factor by the presence of the radiation field. A third coefficient, the *coefficient of absorption*, is introduced to describe the proportionality of the *absorbed* energy to n_K. Einstein first used these coefficients, before the advent of the quantum theory, in his treatment of black-body radiation. He deduced from thermodynamical considerations that the coefficient of spontaneous emission, which he called A, plus the coefficient of absorption, which he called B, must be equal to the total coefficient of emission, which he called C. Induced emission must always be taken into account whenever excited systems are irradiated by light of the same frequency that they can emit spontaneously.[1]

The above selection rules and transition probabilities of the radiation field are similar to those for any system whose physical behavior is that of a harmonic oscillator which is subject to a perturbation proportional to its coordinate. In particular, the exchange of energy between electrons and the *elastic vibrations of a crystal lattice* follows these rules, as does the emission of radiation by a *vibrating diatomic molecule.* We shall meet with these cases in our later work.

EXERCISE

6-13. Analyze qualitatively the nature of the selection rules that would govern transitions due to the term $(e^2/2m)A_jA_j$ in Eq. (4). Show that this interaction would induce the emission or absorption of *two quanta at a time* or the absorption of one quantum and the emission of another. Such a process is intrinsically much less probable than one involving only a single quantum, but even so there are physical processes such as the scattering of light and the Raman effect which depend upon this term.

---

[1] Induced emission finds practical application in several types of atomic-beam devices in which excited atoms or molecules are sent through an oscillating cavity. In traversing the cavity they are induced to jump from their excited state to a lower state, giving up energy to the cavity.

6-5.   Some General Properties of Radiative Transitions

In the preceding section the interaction between a charged particle and the electromagnetic field was evaluated and the selection rules which govern the most common transitions of the radiation field were obtained. We shall now extend that treatment by considering how the *properties of the material system* enter into the radiation process.   We shall for the present consider this question rather generally, but we shall later treat the one-electron atom as an illustration.   We have found the matrix element governing a transition between states u_a and $u_{a'}$ of a one-particle material system and between states U_n and $U_{n'}$ of the radiation field to be

$$(a'n'|V_{\text{int}}|an) = -\frac{e}{m} \int_{q_i} \cdots \int u_{a'}^* \hat{p}_j \times$$
$$\left(\int_{q_K} \cdots \int \cdots U_{n'}^* \hat{A}_j U_n \, d^x q_K \right) u_a \, d^N q_i \quad (1)$$

The result of the treatment of the preceding section was that the above matrix element is equal to zero unless one of the normal modes of the radiation field gains or loses a single quantum of excitation.   In the former case where $n'_K = n_K + 1$, the matrix element is

$$(a',n_K + 1|V_{\text{int}}|a,n_K) = -\frac{e}{m} \frac{(n_K + 1)^{1/2}}{2^{1/2}\alpha_K} \times$$
$$\int_{q_i} \cdots \int u_{a'}^*[\hat{p}_j e_{Kj} \exp(i\boldsymbol{K} \cdot \boldsymbol{r})] u_a \, d^N q_i \quad (2)$$

and in the latter case where $n'_K = n_K - 1$, it is

$$(a',n_K - 1|V_{\text{int}}|a,n_K) = -\frac{e}{m} \frac{n_K^{1/2}}{2^{1/2}\alpha_K} \times$$
$$\int_{q_i} \cdots \int u_{a'}^*[\hat{p}_j e_{Kj} \exp(i\boldsymbol{K} \cdot \boldsymbol{r})] u_a \, d^N q_i \quad (3)$$

where $\alpha_K^2 = \varepsilon_0 L^3 \omega_K / \hbar$ and $\boldsymbol{e}_K$ is the polarization vector (electric-field direction) of the emitted or absorbed photon.   The quantity $\hat{p}_j e_{Kj} \exp(i\boldsymbol{K} \cdot \boldsymbol{r})$ may be interpreted classically as (proportional to) the interaction of a charged particle with a plane electromagnetic wave.   If the exponential term is expanded, the successive terms of the resulting expression correspond to the interactions of the various *electric- and magnetic-multipole moments* with the electric and magnetic fields and their derivatives.   The first term, $\hat{p}_j e_{Kj}$, thus represents the interaction of the electric field of the plane wave with the electric-dipole moment of the material system, as we shall shortly see; the second term, $i\hat{p}_j e_{Kj}(\boldsymbol{K} \cdot \boldsymbol{r})$, similarly represents the electric-quadripole and magnetic-dipole interaction; and so on.

Now, inasmuch as the physical extent of an atomic system is very much less than a wavelength of ordinary visible light, *the quantity $K \cdot r$ is generally quite small:*

$$K \cdot r \approx \frac{2\pi a}{\lambda} \approx \frac{e^2}{4\pi\varepsilon_0 \hbar c} \approx \frac{1}{137} \tag{4}$$

EXERCISE

6-14. Verify Eq. (4).

The smallness of $K \cdot r$ for ordinary optical wavelengths justifies the approximation that *only the electric-dipole interaction is of practical importance in atomic transitions.* This approximation is experimentally found to be valid for the vast majority of observed transitions, and it is in fact only for cases in which the electric-dipole interaction leads to a vanishing matrix element that the further terms are of any importance. Thus, as a very good approximation we may write, for atomic transitions,

$$\hat{p}_j e_{Kj} \exp\left(i K \cdot r\right) \approx \hat{p}_j e_{Kj} \tag{5}$$

and the matrix elements (2) and (3) then assume the form

$$(a', n_K \pm 1 | V_{\text{int}} | a, n_K) = -\frac{e\left(\begin{matrix} n_K + 1 \\ 0 \end{matrix}\right)^{\frac{1}{2}}}{m 2^{\frac{1}{2}} \alpha_K} \int_{q_i} \cdots \int u_{a'}^* \hat{p}_j e_{Kj} u_a \, d^N q_i \tag{6}$$

Now the operator $\hat{p}_j e_{Kj}$ is equal to

$$\hat{p}_j e_{Kj} = \hat{p}_x e_{Kx} + \hat{p}_y e_{Ky} + \hat{p}_z e_{Kz} \tag{7}$$

where e_{Kx}, e_{Ky}, and e_{Kz} are the x, y, and z components of the polarization unit vector e_K, and $\hat{p}_x = (\hbar/i)(\partial/\partial x)$, $\hat{p}_y = (\hbar/i)(\partial/\partial y)$, and $\hat{p}_z = (\hbar/i)(\partial/\partial z)$.

The above matrix element can be put in somewhat simpler form if we recall the analysis leading to Eq. 2-4(4). It was found there that

$$\hat{p}_x = i\frac{m}{\hbar}(\hat{H}x - x\hat{H}) \tag{8}$$

If we use this result, an integral of the form

$$I_x = \iiint u_{a'}^* \hat{p}_x u_a \, dx \, dy \, dz$$

can be transformed as follows:

$$I_x = \frac{im}{\hbar} \iiint u_{a'}^* (\hat{H}x - x\hat{H}) u_a \, dx \, dy \, dz$$

$$= \frac{im}{\hbar} \iiint [u_{a'}^* \hat{H}(x u_a) - u_{a'}^* x (\hat{H} u_a)] \, dx \, dy \, dz$$

$$= \frac{im}{\hbar}(E_{a'} - E_a) \iiint u_{a'}^* x u_a \, dx \, dy \, dz \tag{9}$$

Thus, the matrix element (6) is equivalent to

$$(a',n_K \pm 1|V_{\text{int}}|a,n_K) = - \frac{i\left(n_K + \begin{smallmatrix} 1 \\ 0 \end{smallmatrix}\right)^{\frac{1}{2}}}{\hbar 2^{\frac{1}{2}}\alpha_K} (E_{a'} - E_a)e_{Kj}(a'|er_j|a) \quad (10)$$

where

$$(a'|er_j|a) = \int\int\int u_{a'}^* er_j u_a \, dx \, dy \, dz \quad (11)$$

and er is the *electric-dipole moment* of the charged particle with respect to the origin.

EXERCISE

6-15. Supply the missing steps in the derivation of Eq. (9).

The above result is easily extended to a system containing a number of charged particles simply by replacing the quantity er by the total dipole moment $\boldsymbol{P} = \sum_s e_s \boldsymbol{r}_s$. We have then the result for the matrix element (6)

$$(a',n_K \pm 1|V_{\text{int}}|a,n_K) = \frac{-i\left(n_K + \begin{smallmatrix} 1 \\ 0 \end{smallmatrix}\right)^{\frac{1}{2}}}{\hbar 2^{\frac{1}{2}}\alpha_K} (E_{a'} - E_a)(a'|P_s|a) \quad (12)$$

where P_s is the component of the electric-dipole moment of the material system along the direction of the *electric field* of the emitted or absorbed photon.

The above matrix element connects the material system with a *single* normal mode of the radiation field. In an actual physical case, the atomic system is of course coupled with *all* of the normal modes. To predict the probability of emission or absorption of radiation, we must therefore allow for all normal modes, which we do as follows: In the general treatment of time transitions given in Sec. 2-5 we obtain the probability that, having started at $t = 0$ in the quantum states u_a and U_n, the two interacting systems will at time t have changed into states $u_{a'}$ and $U_{n'}$. Applied to the present problem, this probability is

$$W_{a'n'an} = 4|(a',n'|V_{\text{int}}|a,n)|^2 \frac{\sin^2\left[-(E_{a'} + E_{n'} - E_a - E_n)t/\hbar\right]}{(E_{a'} + E_{n'} - E_a - E_n)^2} \quad (13)$$

$$= 4|(a',n_K \pm 1|V_{\text{int}}|a,n_K)|^2 \frac{\sin^2\left[(E_{a'} - E_a \pm \hbar\omega_K)t/\hbar\right]}{(E_{a'} - E_a \pm \hbar\omega_K)^2} \quad (14)$$

where we have introduced the selection rule $\Delta n = \pm 1$ in obtaining Eq. (14) from Eq. (13). Now, we do not really care which normal mode or combination of normal modes is actually involved in the emission or absorption process; what we really want to know is the transition probability for the *material system*, integrated over all the possible normal modes of the radiation field that might be involved. By inspection of the denominator of Eq. (14) we see that the greatest contribution to the transition will be made by those states for which

$$\Delta E = E_{a'} - E_a \pm \hbar\omega_K \approx 0$$

that is, *by states for which the conservation of energy holds.* Indeed, as time increases, the "resonance" represented by the $\sin^2 (\Delta E t/\hbar)/(\Delta E)^2$ term becomes sharper and sharper in terms of ΔE, so that after a long time only those states will be excited for which ΔE is zero.[1]

We shall now find the probability of transition of the material system by integrating Eq. (14) over all frequencies ω. We must, of course, include a factor which describes the *density of states* of the radiation field per unit frequency range near the resonance frequency. Calling this factor $\rho(\omega)$, we then have for the total probability of transition of the material system into state $u_{a'}$ from state u_a,

$$W_{a'a} = 4 \int_{-\infty}^{\infty} |(a',n \pm 1|V_{\text{int}}|a,n)|^2 \frac{\sin^2 (\Delta E t/\hbar)}{(\Delta E)^2} \rho(\omega) \, d\omega \qquad (15)$$

We can now transform the variable of integration to $\Delta E t/\hbar = z$, so that $dz = t \, d\omega$. If t is sufficiently great, we can also set $\omega = \omega_0$, where $\omega_0 = (E_{a'} - E_a)/\hbar$, in $\rho(\omega)$ and in the matrix element, and take these outside the integral. Thus for large t

$$W_{a'a} = 4|(a',n \pm 1|V_{\text{int}}|a,n)|^2 \rho(\omega_0) \frac{t}{\hbar^2} \int_{-\infty}^{\infty} \frac{\sin^2 z}{z^2} \, dz$$

$$= 4|(a',n \pm 1|V_{\text{int}}|a,n)|^2 \rho(\omega_0) \frac{t}{\hbar^2} \pi \qquad (16)$$

From electromagnetic theory, the density of normal frequencies of a

---

[1] Although this argument appears to permit only transitions in which energy is strictly conserved, this conclusion applies only to the case of first-order transitions which take place directly from one state to another. In higher approximation, one sometimes finds that transitions which appear to violate energy conservation are possible, and that these transitions are in fact observed experimentally. These transitions occur when no *direct* transition is possible between a certain excited state A and a state B of lower energy, but when a state C of higher energy exists *to* which the system could jump if energy were available, and *from* which the lower energy state B could be reached by a *direct* transition if the system were in the higher energy state C. The probability of such a transition is found by evaluating the matrix element for the double transition A → C and C → B. In this way one finds again that over long times energy must be strictly conserved (that is, the emitted photon has just the energy $E_A - E_B$), but since the transition takes place *through the state* C, *energy could not have been conserved at all times.* Processes of this kind which seem to require a temporary violation of energy conservation are called *virtual processes.*

cubical box of side L is equal to

$$dN = \rho(\omega)\, d\omega = \frac{L^3}{\pi^2 c^3} \omega^2\, d\omega \qquad (17)$$

Inserting this value for $\rho(\omega_0)$, together with the expression (12) for the matrix element, we finally obtain

$$W_{a'a} = \frac{\left(\bar{n} + \dfrac{1}{0}\right)}{\hbar^2 \alpha_0^2} |(a'|P_s|a)|^2 \frac{\hbar^2 \omega_0^2 2\omega_0^2 L^3}{\pi^2 c^3} \frac{t}{\hbar^2} \pi$$

$$= 2 \frac{\left(\bar{n} + \dfrac{1}{0}\right)}{\pi \varepsilon_0 \hbar c^3} \omega_0^3 |(a'|P_s|a)|^2 t \qquad (18)$$

where $\alpha_0^2 = \varepsilon_0 L^3 \omega_0 / \hbar$ and $\bar{n}$ is the average excitation of the normal modes in the neighborhood of ω_0. The expression just obtained indicates that the probability that a material system has, after a time t, jumped from state u_a to state $u_{a'}$ under the stimulus of plane-polarized radiation which has an average excitation $\bar{n}$ per normal mode in the neighborhood of the resonance frequency $\omega_0 = (E_{a'} - E_a)/\hbar$, is:

1. Proportional to t
2. Proportional to $\bar{n}$ for absorption and to $\bar{n} + 1$ for emission
3. Proportional to ω_0^3
4. Proportional to the square of the matrix element of the electric-dipole moment of the material system along the direction of polarization for the two states $u_{a'}$ and u_a

The above equation can be compared with the analogous result from classical electromagnetism if we evaluate the *average power* radiated by the system. This is equal to the photon energy times the rate of transition from one state to the other. We shall evaluate this for the spontaneous emission, $\bar{n} = 0$:

$$\Pi_{\text{qm}} = \hbar\omega_0 \frac{dW_{a'a}}{dt}$$

$$= \frac{2\omega_0^4 |(a'|P_s|a)|^2}{\pi \varepsilon_0 c^3}$$

$$= \frac{2}{3} \frac{\omega_0^4}{\pi \varepsilon_0 c^3} \sum_{s=1}^{3} |(a'|P_s|a)|^2 \qquad (19)$$

(The last step takes account of the average of $|(a'|P_s|a)|^2$ over all directions of the polarization vector of the emitted radiation.) Expression

(19) is to be compared with the classical equation for the power emitted by an oscillating electric dipole $P = P_0 \cos \omega_0 t$:

$$\Pi_{cl} = \frac{1}{3} \frac{\omega_0^4}{\pi \varepsilon_0 c^3} P_0^2 \tag{20}$$

We thus see that *a quantum-mechanical system radiates energy at the same rate as a classical oscillating electric dipole whose strength is*

$$P_0^2 = 2 \sum_{s=1}^{3} |(a'|P_s|a)|^2 \tag{21}$$

$$= 2e^2(|(a'|x|a)|^2 + |(a'|y|a)|^2 + |(a'|z|a)|^2)$$

P_0 is called the *oscillator strength* of the transition $a \to a'$.

In the next section we shall evaluate the matrix elements appearing in Eq. (21) for a one-electron atom. It will then be found that these matrix elements lead to the *selection rules* which govern electric-dipole transitions of atoms.

6-6. Electric-dipole Selection Rules for a One-electron Atom

We shall now use the result of the preceding section to find the selection rules for a one-electron atom. Because we are considering only the electric-dipole transitions, we ignore the presence of electron spin and use the wave functions in the form 5-4(1)

$$u_{nlm}(r,\theta,\varphi) = A_{nlm} R_{nl}(r) P_l^m(\cos \theta) e^{im\varphi} \tag{1}$$

If we wish to find the probability of transition of the atom from the state whose quantum numbers are n, l, m to the state n', l', m', we must evaluate the three matrix elements 6-5(21). These are

$$X = (n'l'm'|x|nlm)$$

(a) $$= \int_0^\infty \int_0^\pi \int_0^{2\pi} u^*_{n'l'm'} \, r \sin \theta \cos \varphi u_{nlm} r^2 \sin \theta \, dr \, d\theta \, d\varphi$$

$$Y = (n'l'm'|y|nlm)$$

(b) $$= \int_0^\infty \int_0^\pi \int_0^{2\pi} u^*_{n'l'm'} \, r \sin \theta \sin \varphi u_{nlm} r^2 \sin \theta \, dr \, d\theta \, d\varphi \tag{2}$$

and

$$Z = (n'l'm'|z|nlm)$$

(c) $$= \int_0^\infty \int_0^\pi \int_0^{2\pi} u^*_{n'l'm'} \, r \cos \theta u_{nlm} r^2 \sin \theta \, dr \, d\theta \, d\varphi$$

Taking first the X matrix element, we have

$$X = A_{n'l'm'} A_{nlm} \left(\int_0^\infty R_{n'l'} r R_{nl} r^2 \, dr \right) \left[\int_0^\pi P_{l'}^{m'}(\cos \theta) \sin \theta P_l^m \sin \theta \, d\theta \right]$$
$$\left(\int_0^{2\pi} e^{-im'\varphi} \cos \varphi e^{im\varphi} \, d\varphi \right) \tag{3}$$

If we substitute $\cos \varphi = \frac{1}{2}(e^{i\varphi} + e^{-i\varphi})$, the φ integral becomes

$$\frac{1}{2} \int_0^{2\pi} [e^{i(m-m'+1)\varphi} + e^{i(m-m'-1)\varphi}] \, d\varphi = \begin{cases} \pi & \text{if } m' = m \pm 1 \\ 0 & \text{otherwise} \end{cases} \tag{4}$$

We then insert this result into the θ integral, and make use of the recurrence relation[1]

$$\sin \theta P_{l'}^{m-1} = \frac{P_{l'+1}^m - P_{l'-1}^m}{2l'+1} \tag{5}$$

to obtain an integral of a product of P_l^m functions having the same upper index.

The orthogonality of the $P_l^m(\cos \theta)$ functions with respect to l for equal m then leads to the result

$$\int_0^\pi P_{l'}^{m\pm1} \sin \theta P_l^m \sin \theta \, d\theta = 0 \qquad \text{unless } l' = l \pm 1 \tag{6}$$

Finally, the r-integral, with $l' = l \pm 1$, leads to no restriction upon n'.

EXERCISE

6-16. Using the above procedure, show that

and
$$\begin{array}{ll} Y = 0 & \text{unless } m' = m \pm 1 \text{ and } l' = l \pm 1 \\ Z = 0 & \text{unless } m' = m \qquad \text{and } l' = l \pm 1 \end{array} \tag{7}$$

Note:
$$\cos \theta P_l^m = \frac{[(l-m+1)P_{l+1}^m + (l+m)P_{l-1}^m]}{2l+1} \tag{8}$$

The above analysis provides the basis for the so-called *electric-dipole selection rules* for atomic transitions, which state that *only those transitions can occur for which $\Delta l = \pm 1$ and $\Delta m = \pm 1$ or 0*. One could, without great difficulty, complete the analysis outlined above and thus evaluate the actual transition probabilities for given values of n and n'. In this way the *relative intensities* of the various spectral lines of hydrogen and other one-electron atoms could be predicted for given physical conditions. Although this is a most important matter in practical spectroscopy, it is not sufficiently fundamental to merit our further consideration.

The above selection rules were obtained by ignoring the presence of the angular momentum due to electron spin. When this is taken into account, the selection rules upon l, s, j, and m_j become

$$\begin{array}{ll} \Delta l = \pm 1 \\ \Delta s = 0 \\ \Delta j = \pm 1 \quad \text{or} \quad 0 & \text{but } j = 0 \nleftrightarrow j = 0 \\ \Delta m_j = \pm 1 \quad \text{or} \quad 0 & \text{but } m_j = 0 \nleftrightarrow m_j = 0 \quad \text{if } \Delta j = 0 \end{array} \tag{9}$$

[1] This and other recurrence relations can be found in W. Magnus and F. Oberhettinger, "Formulas and Theorems for the Special Functions of Mathematical Physics," Chelsea Publishing Company, New York, 1949.

6-7. Summary of the Properties of Radiative Transitions

Since many of the important results obtained in this chapter may tend to be obscured by the somewhat ponderous mathematical procedures that were used to obtain them, we shall now briefly summarize the main properties of radiative transitions:

1. The interaction between a charged particle and the radiation field is described by the "potential"

$$V_{\text{int}} = -\frac{e}{m} \boldsymbol{p} \cdot \boldsymbol{A} + \frac{e^2}{2m} \boldsymbol{A} \cdot \boldsymbol{A} \tag{1}$$

in the sense that, if H_{mat} is the Hamiltonian function of the charged particle moving in fixed electric and magnetic fields and H_{rad} is that of the radiation field alone, the complete Hamiltonian function of the particle and radiation field together is

$$H = H_{\text{mat}} + H_{\text{rad}} + V_{\text{int}} \tag{2}$$

2. The term $(-e/m)\boldsymbol{p} \cdot \boldsymbol{A}$ leads in first order to physical processes in which the radiation field *gains or loses one quantum at a time.* These processes are described by the broad terms *emission* and *absorption.*

3. The term $(e^2/2m)A^2$ leads to physical processes in which *two quanta* participate. The most important such processes are described by the term "scattering"; these involve an incident quantum which is in some way modified by the material system—in frequency, direction of motion, or both.

4. The probability that a material system will *absorb* a quantum whose frequency is ω_0 is proportional to the average excitation level $\bar{n}_0$ of the field oscillators whose frequencies lie near ω_0. The corresponding probability that the material system will emit a quantum of this frequency is proportional to $\bar{n}_0 + 1$.

5. In atomic spectroscopy, the most common radiative transitions are of the *electric-dipole* type. The selection rules and transition probabilities for these transitions are governed by the matrix elements of the rectangular components of the electric-dipole moment $e\boldsymbol{r}$ of the atomic system:

$$X = e(a'|x|a) \qquad Y = e(a'|y|a) \qquad Z = e(a'|z|a) \tag{3}$$

6. The rate of spontaneous radiation of energy by a material system is equal to that of a classical oscillating electric dipole $P_0 \cos \omega_0 t$, whose strength P_0 is

$$P_0^2 = 2(|X|^2 + |Y|^2 + |Z|^2) \tag{4}$$

and whose angular frequency ω_0 is that given by $\hbar\omega_0 = E_{a'} - E_a$.

7. The selection rules which govern the electric-dipole transitions of a one-electron atom, ignoring spin angular momentum, are:

$$\Delta l = \pm 1 \qquad \Delta m = \pm 1, 0 \qquad \Delta n \quad \text{unrestricted} \tag{5}$$

If spin is included, the selection rules are

$$
\begin{aligned}
\Delta l &= \pm 1 \\
\Delta s &= 0 \\
\Delta j &= \pm 1 \quad \text{or} \quad 0 \qquad \text{but } j = 0 \nleftrightarrow j = 0 \\
\Delta m_j &= \pm 1 \quad \text{or} \quad 0 \qquad \text{but } m_j = 0 \nleftrightarrow m_j = 0 \quad \text{if } \Delta j = 0
\end{aligned}
\tag{6}
$$

8. Radiative transitions whose emitted radiations are analogous to those of classical oscillating magnetic dipoles, electric or magnetic quadripoles, or higher multipoles, are possible and sometimes may even account for some of the most intense lines in an atomic spectrum. In radiative transitions of atomic *nuclei*, these higher-order multipoles account for a far greater proportion of the observed radiation than they do in the case of atomic transitions.

REFERENCE

Heitler, W.: "The Quantum Theory of Radiation," 3d ed., Oxford University Press, New York, 1954.

7

The Pauli Principle, Atomic Shell Structure, and the Periodic System

The results of Chap. 5 have provided us with considerable information about and insight into the nature of the motion of a single electron around an atomic nucleus. Some of these properties of one-electron atoms, when combined with the *Pauli exclusion principle*, make it possible for us to treat one of the most interesting and illuminating phenomena of atomic physics, that of *atomic shell structure and the periodic system of the elements*. The successful qualitative interpretation of the periodic table and of the chemical similarities of the various groups of elements, not to mention the quite detailed interpretation of the specific physical and chemical properties of certain groups, stands as one of the most satisfying triumphs of the quantum theory. In this chapter we shall treat some of the more elementary aspects of these subjects.

7-1. Introductory Sketch

The periodic system of classification of the elements dates from Mendeleev's enunciation (1869) that a *periodic relation* connects the chemical properties of an element with its atomic weight. This empirical law was of great practical value, for it provided a systematic basis for the search for new elements and for the classification of elements as they were found. Mendeleev himself predicted the existence of several elements, notably gallium and germanium, which were later found and whose chemical and physical properties agreed remarkably well with his predictions.

As knowledge of the chemical properties of the elements increased and as new elements were discovered, it became necessary to modify and extend the periodic table.   For example, the discovery of the noble gases (1894–1898, 1908) by Ramsay required the addition of an entire new column to the table; and a certain group of elements, called the rare earths, seemed all to belong at a single place in the table.   In its modern form the periodic table is built around the noble gases in such a way that successive periods are separated by a noble gas, as shown in Fig. 7-1.

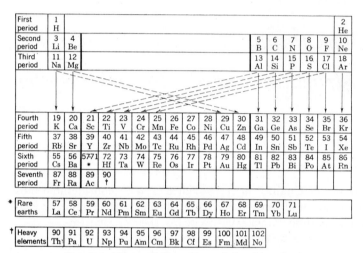

FIG. 7-1. The periodic system of the elements.

The various periods do not then contain equal numbers of elements, but the groups of chemically similar elements are, on the other hand, better defined than in some of the older forms of the table.   Thus the familiar groups of elements such as the noble gases, alkali metals, alkaline earths, and halogens are located in continuous vertical columns extending over the entire table, while the smaller but definite subgroups such as Cu-Ag-Au and Zn-Cd-Hg, etc., are located in vertical columns extending over only part of the table.   Subgroups which are related in some way (e.g., through having similar chemical valence, etc.) to one of the main groups are so indicated by arrows connecting the groups.

   With the advent of the nuclear theory of atomic structure and the Bohr theory of hydrogen, the question that naturally arose was whether the various quantized orbits of the Bohr theory, somehow extended to apply to atoms with several electrons, could provide an explanation of the form of the periodic table.   Considerable progress in this direction was made in the period up to the introduction of the modern quantum theory (1926), and indeed it is still convenient to use many of the results of this early approach, even though the Schroedinger theory is needed for quantitative considerations.

Rydberg pointed out that the atomic numbers of the noble gases were expressible in the form

$$Z = 2(1^2 + 2^2 + 2^2 + 3^2 + 3^2 + 4^2)$$

$$\downarrow \quad \downarrow \quad \downarrow \quad \downarrow \quad \downarrow \quad \downarrow$$

$$(2) \quad (10) \quad (18) \quad (36) \quad (54) \quad (86) \tag{1}$$

$$\text{He} \quad \text{Ne} \quad \text{Ar} \quad \text{Kr} \quad \text{Xe} \quad \text{Rn}$$

The members of this sequence were identified by Bohr and by Stoner (1924) with the families of orbits described by the total quantum number n of the modified Bohr theory. Since the degeneracy (including that due to the electron spin) of a given level is equal to $2n^2$, the numbers given by this expression were associated with so-called *completed shells* of orbits, where a shell is defined as a family of orbits having the same value of n. Within each shell the orbits were classified into subshells, by Stoner, according to the value of the azimuthal quantum number k of the Bohr-Sommerfeld theory. Stoner's system is shown in Table 7-1. In this table the so-called *shell notation* (K-shell, L-shell, etc.) is given, together with the values of the total quantum number n corresponding to each shell. The shells are further classified into *subshells* according to the quantum number l of the Schroedinger theory (the old azimuthal quantum number k used by Stoner is equal to $l + 1$); and the *spectroscopic* designation of the corresponding Bohr orbit is also given. (Thus the subshell 3p corresponds to $n = 3$, $l = 1$, etc.) The Rydberg expression (1) is obtained from the numbers of electrons in these subshells as shown in Table 7-2.

TABLE 7-1. QUANTUM NUMBERS FOR VARIOUS SHELLS AND SUBSHELLS

Shell	K	L		M			N				O					P		
n	1	2		3			4				5					6		
Subshell	1s	2s	2p	3s	3p	3d	4s	4p	4d	4f	5s	5p	5d	5f	5g	6s	6p	6d
l	0	0	1	0	1	2	0	1	2	3	0	1	2	3	4	0	1	2
No. of electrons	2	2	6	2	6	10	2	6	10	14	2	6	10	14	18	2	6	10

TABLE 7-2. CLOSED SUBSHELLS CORRESPONDING TO THE NOBLE GASES

Subshells	No. of electrons	Total	Element
1s	2	2	He
2s + 2p	8	10	Ne
3s + 3p	8	18	Ar
4s + 3d + 4p	18	36	Kr
5s + 4d + 5p	18	54	Xe
6s + 4f + 5d + 6p	32	86	Em

According to this scheme, then, each energy state of an atom was supposed to correspond to its electrons being in certain Bohr orbits. The electrons occupied these Bohr orbits in such a way that, in the ground state, the total energy was a minimum, *subject to the condition that only two electrons could be in an orbit corresponding to a given n, l, and m.* Pauli first enunciated this latter rule in the form of his *exclusion principle* (1925), which states that no two electrons in an atom can have the same set of quantum numbers, n, l, m_l, and m_s. The noble gases then appear to correspond to a "filled" p subshell of a given shell, and the various orbits whose "quantum-number configurations" lie between those of two noble gases are to be identified with elements lying between these gases in the periodic table.

The *electron configuration* of an element is usually described by a notation which gives the number of electrons occupying each subshell. The number of electrons in a given subshell is indicated by a superscript attached to the spectroscopic symbol for that subshell. Thus, for example, $1s^2$ means that two electrons are in the 1s subshell, $4p^3$ means that three electrons are in the 4p subshell, etc. Using this notation, the electron configurations of the first few elements are as follows:

H	1s	C	$1s^2\ 2s^2\ 2p^2$
He	$1s^2$	N	$1s^2\ 2s^2\ 2p^3$
Li	$1s^2\ 2s$	O	$1s^2\ 2s^2\ 2p^4$
Be	$1s^2\ 2s^2$	F	$1s^2\ 2s^2\ 2p^5$
B	$1s^2\ 2s^2\ 2p$	Ne	$1s^2\ 2s^2\ 2p^6$

7-2. Exchange Symmetry of Wave Functions

In order to interpret the periodic table according to the modern Schroedinger theory, it is necessary to add to the postulatory basis of quantum mechanics a rule which corresponds to the assumption, mentioned above, that only two electrons may occupy a given Bohr orbit. Before stating the postulate itself, however, a few words of preparation are needed.

In the present problem as well as in many of the phenomena we shall treat later, we are concerned with the motion of two or more *identical particles*. This situation would of course present no fundamental difficulty in classical mechanics, where one is assumed to be able, at least in principle, to follow in detail the motions of all particles and thus to keep track of which particle is which. But in quantum mechanics, the *uncertainty principle* limits our ability to follow the motions of the particles without disturbing the system, so that we can never be certain *which one* of a number of identical particles we have actually found at a given point, because the particles are in fact *physically indistinguishable*. This places certain restrictions upon the *mathematical form* that the wave function for several identical particles may have. These restrictions will now be considered.

In treating a system which contains a number of identical particles, the locations of the particles are usually defined by a number of coordinates which, for convenience, are labeled with numerical subscripts to correspond to the various particles involved, just as if these particles were distinguishable from one another, and this set of coordinates is then used in the usual manner to form the Hamiltonian operator and the Schroedinger equation. The wave function which describes the system is then a function of these various labeled coordinates. Thus if (x_1,y_1,z_1), (x_2,y_2,z_2), $\ldots$, (x_N,y_N,z_N) are the coordinates of N identical particles that comprise the given system (we assume for simplicity that there are no other particles in the system), the Hamiltonian function will be of the form

$$H = \frac{1}{2m} (p_1{}^2 + p_2{}^2 + \cdots + p_N{}^2) + V(x_1,y_1,z_1,x_2,y_2,z_2, \ldots ,x_N,y_N,z_N)$$

(1)

wherein $p_1{}^2 = p_{1x}{}^2 + p_{1y}{}^2 + p_{1z}{}^2$, etc. The Hamiltonian operator will then be

$$\hat{H} = -\frac{\hbar^2}{2m} (\nabla_1{}^2 + \nabla_2{}^2 + \cdots + \nabla_N{}^2)$$
$$+ V(x_1,y_1,z_1,x_2,y_2,z_2, \ldots ,x_N,y_N,z_N) \quad (2)$$

and the wave function will in general depend upon all coordinates

$$\psi(q_i,t) = \psi(x_1,y_1,z_1,x_2,y_2,z_2, \ldots ,x_N,y_N,z_N,t) \quad (3)$$

The restrictions that must be placed upon the form of ψ are based upon the invariance of the Hamiltonian operator (2) with respect to all possible interchanges of the identical particles of the system. For example, if particles 1 and 2 are interchanged, the Hamiltonian operator will be

$$\hat{H}' = -\frac{\hbar^2}{2m} (\nabla_2{}^2 + \nabla_1{}^2 + \cdots + \nabla_N{}^2)$$
$$+ V(x_2,y_2,z_2,x_1,y_1,z_1, \ldots ,x_N,y_N,z_N) \quad (4)$$

and since the potential energy must be the same as before (the particles behave identically in every way), it must be true that

$$\hat{H}' \equiv \hat{H} \quad (5)$$

In our treatment of parity we found that an invariance of the Hamiltonian operator with respect to a reflection of the coordinates about a center of symmetry leads to wave functions which themselves show either symmetry or antisymmetry about this same center. We shall find that the new type of invariance of $\hat{H}$ just discussed, called *exchange invariance*, leads to an analogous symmetry or antisymmetry of ψ with respect to the corresponding *exchange of particles*. Let us define an operator, which we shall call the *exchange operator*, whose action is to interchange the coordinates[1] of any pair of particles, say r and s, in any function of these coordinates upon which it operates:

$$\hat{P}_{rs}\psi(1,2, \ldots ,r, \ldots ,s, \ldots ,N)$$
$$= \psi(1,2, \ldots ,s, \ldots ,r, \ldots ,N) \quad (6)$$

(In this expression we have introduced a simplified schematic notation wherein 1 stands for all coordinates of particle 1, and similarly for 2, . . . , r, . . . , s, . . . , and N.) The exchange operator has many properties analogous to the reflection operator discussed in Chap. 3, the essential ones of which are treated in the following exercises:

EXERCISES

7-1. Show that $\hat{P}_{rs}$ is a linear, Hermitian operator.

7-2. Show that $\hat{P}_{rs}$ commutes with $\hat{H}$. This assures that eigenfunctions of $\hat{P}_{rs}$ will be constants of the motion.

7-3. Show that the only eigenvalues of $\hat{P}_{rs}$ are ± 1. *Hint:* A double application of $\hat{P}_{rs}$ must leave the eigenfunction unchanged.

The above exercises show that the eigenfunctions of $\hat{P}_{rs}$ have the property either of remaining unchanged or of changing sign when two particles of the system are interchanged:

$$\psi(1,2, \ldots ,r, \ldots ,s, \ldots ,N)$$
$$= \pm\psi(1,2, \ldots ,s, \ldots ,r, \ldots ,N) \quad (7)$$

Inasmuch as $\hat{H}$ and $\hat{P}_{rs}$ commute, we expect that the energy eigenfunctions of the system also will be eigenfunctions of $\hat{P}_{rs}$ and that the exchange symmetry expressed by Eq. (6) will be preserved in time.

So far, nothing really new has been introduced to allow for the experimental indistinguishability of identical particles, and it even seems that there need be no real restrictions upon the form of ψ, since *any* function can be written as a sum of functions which are respectively symmetric and antisymmetric under the action of $\hat{P}_{rs}$. However, *it is found experimentally that all of the eigenfunctions for a given type of particle show the*

---

[1] At the present stage of our discussion, the "coordinates" refer to the ordinary position coordinates. We shall shortly extend the meaning of the exchange of particles to include their spin coordinates, however, so that what is really meant by the particle-exchange operation is a complete exchange of *all* of the coordinates of the two particles.

same exchange symmetry, and that this exchange symmetry is related to the intrinsic angular momentum of the particle according to the following rules, which we now present as a postulate.

Exchange-symmetry Rules:

1. Identical particles having an integral quantum number for their intrinsic spins can be described only by wave functions which are *symmetric* with respect to an interchange of the space and spin coordinates of any two such identical particles:

$$\psi(1,2, \ldots ,r, \ldots ,s, \ldots ,N)$$
$$= +\psi(1,2, \ldots ,s, \ldots ,r, \ldots ,N) \quad (8)$$

2. Identical particles having a half-integral quantum number for their intrinsic spins can be described only by wave functions which are *antisymmetric* with respect to an interchange of the space and spin coordinates of any two such identical particles:

$$\psi(1,2, \ldots ,r, \ldots ,s, \ldots ,N)$$
$$= -\psi(1,2, \ldots ,s, \ldots ,r, \ldots ,N) \quad (9)$$

The particles of the first category are called *Bose particles* because they obey the Einstein-Bose statistics, which will be described in Chap. 10. Examples of Bose particles are photons (spin 1), neutral helium atoms in their ground states (spin 0), and α-particles (spin 0). The particles in the second category are called *Fermi particles* because they obey the Fermi-Dirac statistics, which will also be described in Chap. 10. Examples of Fermi particles are electrons, protons, neutrons, and μ-mesons (all spin $\frac{1}{2}$).

These rules have the effect of limiting our choice of wave functions for a system containing several identical particles to those which exhibit the proper type of exchange symmetry. This may appear to be an interesting but somewhat inconsequential limitation, but we shall find that it actually introduces *fundamental new physical effects* which are of tremendous importance in many phenomena. Some of these effects, which will be treated in greater detail in the course of our later work, are outlined below:

1. As compared with the behavior of hypothetical equal but distinguishable particles, Bose particles exhibit an additional attraction for one another and tend to be found near one another in space; Fermi particles, on the contrary, repel one another and tend not to be found near one another in space.
2. At a given temperature, the energy content and pressure of a system of Bose particles are less than, and of a system of Fermi particles

greater than, the energy content and pressure of a system of equal but distinguishable particles.

3. If the individual particles of a system are acted upon by some outside force, but *do not interact with one another*, the quantum states of the system correspond to the various particles occupying certain of the eigenstates available to a single particle. If the particles are Bose particles, they tend to occupy the same quantum states, while Fermi particles *cannot* occupy the same quantum states.

The above properties are responsible for such diverse phenomena as the saturation of chemical binding forces, the condensation of liquid helium into a superfluid state, ferromagnetism, ortho- and para-hydrogen, and the subject of the present chapter—the periodic system of the elements.

The exchange-symmetry requirements are thus responsible directly for many phenomena and indirectly for many others, but few physical situations are sufficiently transparent and mathematically tractable that they can be analyzed in exact detail. Therefore, to treat a given physical situation, it is desirable to proceed by adopting the simplest possible model which exhibits the necessary properties and which can be treated with sufficient accuracy. Furthermore, the use of a simple model is oftentimes far more effective in enhancing one's insight and intuition than is a rigorous but lengthy mathematical analysis. We shall therefore proceed with our examination of the physical consequences of the exchange-symmetry rules by considering the special case mentioned in (3), namely, a system composed of a number of identical particles which are all acted upon by the same outside force, but which, in first approximation, do not interact with one another. In fact, we can gain a great deal of information if we treat just *two* identical particles. In this case the Hamiltonian function for the two particles has the form

$$H = \frac{p_1{}^2}{2m} + V(1) + \frac{p_2{}^2}{2m} + V(2) \tag{10}$$

where $p_1{}^2/2m + V(1)$ is the Hamiltonian function of particle 1 and $p_2{}^2/2m + V(2)$ is the Hamiltonian function of particle 2, when treated separately. The Hamiltonian *operator* for the system is, by the usual rules,

$$\hat{H} = -\frac{\hbar^2}{2m} \nabla_1{}^2 + V(1) + \frac{-\hbar^2}{2m} \nabla_2{}^2 + V(2) \tag{11}$$

where

$$\nabla_i{}^2 = \frac{\partial^2}{\partial x_i{}^2} + \frac{\partial^2}{\partial y_i{}^2} + \frac{\partial^2}{\partial z_i{}^2} \quad \text{and} \quad V(i) = V(x_i, y_i, z_i, \sigma_{zi}) \quad i = 1, 2$$

and σ_{zi} is the z-component of spin angular momentum of particle i. That is, the Hamiltonian operator is a *sum of two parts:*

$$\hat{H} = \hat{H}_1 + \hat{H}_2 \qquad (12)$$

where $\hat{H}_1$ and $\hat{H}_2$ are the Hamiltonian operators for the two separate particles. Because of this separability, the energy eigenfunctions for the entire system can be expressed as a product of eigenfunctions for the individual particles (Exercise 2-50):

$$\psi(1,2) = \psi(1)\psi(2) \qquad (13)$$

Now, since the particles are identical, the forces of the system must act similarly on each; that is, the potential functions $V(1)$ and $V(2)$ must have the *same analytic form* for each particle. Thus the wave functions for the individual particles are members of *a single family of wave functions.* Let $\psi_n(i)$ denote one of these normalized energy eigenfunctions for the ith particle alone, where n stands for the totality of quantum numbers (including that for spin) that are needed to describe a given state. Using this notation, a normalized wave function for the entire system may be written in the more specific form

$$\psi(1,2) = \psi_a(1)\psi_b(2) \qquad (14)$$

This wave function describes a state in which particle 1 is in the state a and particle 2 is in the state b. This wave function, however, *is not necessarily of the postulated form* for identical particles, since we cannot be sure that $\psi(1,2) = \pm\psi(2,1)$, as required by the rules of Eq. (8) or Eq. (9). The difficulty may be remedied by writing instead

$$\psi(1,2) = A[\psi_a(1)\psi_b(2) \pm \psi_a(2)\psi_b(1)] \qquad (15)$$

which clearly *does* satisfy the rules if we take the plus sign for Bose particles and the minus sign for Fermi particles. This wave function differs from that of Eq. (14) in that it describes a state of the system in which one particle is in state a and one particle is in state b, but such that *either of the two particles is equally likely to be found in either state.*

EXERCISES

7-4. Find the real value of A which normalizes $\psi(1,2)$ in Eq. (15). *Ans.:* $A^2 = \frac{1}{2}$ if $a \neq b$, $A^2 = \frac{1}{4}$ if $a = b$ for Bose particles.

7-5. Find the expectation value of the energy of particle 1 if the state of the system is described by Eq. (15). *Ans.:* $\langle E(1) \rangle = \frac{1}{2}(E_a + E_b)$.

7-6. Find the total energy of the system, and show that this energy is precisely defined. *Ans.:* $E = E_a + E_b$.

The wave function (15), with $A = 2^{-1/2}$, is thus a normalized eigenfunction which corresponds to an energy $E_a + E_b$ for the system and which satisfies the requirements of the exchange-symmetry rules. We now note that, if both particles are assumed to be in the same state, say ψ_n, the expression (15) gives $\psi(1,2) \equiv 0$ for Fermi particles. Therefore we conclude that *two noninteracting Fermi particles cannot be in the same energy eigenstate.* Equivalently, we may say that *two noninteracting Fermi particles cannot both be in states described by the same set of quantum numbers.* Thus Pauli's original postulate is a special case of the more general exchange-symmetry rules (8) and (9).

EXERCISES

7-7. Extend the above results to apply to *three* noninteracting particles. Find an appropriate form for a normalized wave function for the system, such that one particle is in the state a, one in the state b, and one in the state c. (Assume as above that the individual functions ψ_n are normalized.) Show that $\psi(1,2,3) \equiv 0$ if any two Fermi particles are in the same state.

7-8. Show that an energy eigenfunction for a system composed of N noninteracting Fermi particles which is in the form of the determinant

$$\psi(1,2, \ldots ,N) = A \begin{vmatrix} \psi_a(1) & \psi_a(2) & \cdots & \psi_a(N) \\ \psi_b(1) & \psi_b(2) & \cdots & \psi_b(N) \\ \cdot & \cdot & & \cdot \\ \cdot & \cdot & & \cdot \\ \cdot & \cdot & & \cdot \\ \psi_n(1) & \psi_n(2) & \cdots & \psi_n(N) \end{vmatrix} \tag{16}$$

is an eigenfunction of the total energy of the system, satisfies the antisymmetry postulate, and is identically equal to zero if any two or more particles are in the same state.

7-9. Find the real value of A in the previous exercise which normalizes ψ. *Ans.:* $|A| = (N!)^{-1/2}$.

Although the results so far obtained would suffice for our treatment of the periodic table, we shall repeatedly encounter in our later work other situations in which the exchange-symmetry rules play a role. For this reason it is desirable that we now consider somewhat further the consequences of these rules. Most of the results that we shall need can be obtained from a study of the properties of a physical system composed of two identical particles which do not interact with each other and whose spins are so weakly coupled with outside influences that the coupling can be neglected. The Hamiltonian operator of such a system is then

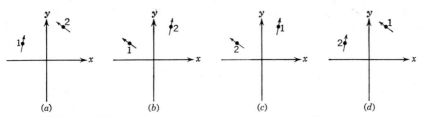

FIG. 7-2. Schematic illustration of the exchange of space and spin coordinates for two particles: (a) original configuration of the particles, (b) spin coordinates exchanged, (c) space coordinates exchanged, (d) space and spin coordinates exchanged.

invariant not only with respect to the exchange of all of the coordinates—space and spin—of the two particles but also with respect to the *separate* exchange of their space coordinates alone and their spin coordinates alone (Fig. 7-2).

Because of the assumed isolation of the spin from outside interactions, the wave function for the system will be a product of functions of the space coordinates and the spin coordinates:

$$\psi = U(q_i)\Sigma(\sigma_i) \tag{17}$$

and because of the invariance of $\hat{H}$ with respect to the separate exchange of space and spin coordinates, the functions $U(q_i)$ and $\Sigma(\sigma_i)$ *must themselves be symmetric or antisymmetric under these respective operations:*

$$U(q_i) = 2^{-\frac{1}{2}}[u_a(1)u_b(2) \pm u_a(2)u_b(1)] \tag{18}$$
and $\qquad \Sigma(\sigma_i) = 2^{-\frac{1}{2}}[s_c(1)s_d(2) \pm s_c(2)s_d(1)] \tag{19}$

where $u_m(i)$ and $s_n(i)$ are the normalized space and spin wave functions of the ith particle alone.

Finally, because of the exchange-symmetry rules, *the signs in Eqs.* (18) *and* (19) *must be correlated: for Bose particles the signs must be the same, and for Fermi particles they must be opposite.* We shall later find that this relationship is responsible for some rather important features of atomic and molecular structure.

EXERCISES

7-10. Justify the form (17) for the wave function ψ.

7-11. Establish the validity of Eqs. (18) and (19), starting from the invariance of $\hat{H}$ under the separate exchange of space and spin coordinates.

As an illustration of the physical consequences of the symmetry character of wave functions, let us evaluate the average value of the square of the distance between the two particles whose space-dependent wave function is given by Eq. (18):

$$\langle (\mathbf{r}_2 - \mathbf{r}_1)^2 \rangle = \tfrac{1}{2} \int [u_a^*(1)u_b^*(2) \pm u_a^*(2)u_b^*(1)](r_2^2 + r_1^2 - 2\mathbf{r}_1 \cdot \mathbf{r}_2)$$
$$[u_a(1)u_b(2) \pm u_a(2)u_b(1)] \, dx_1 \, dy_1 \, dz_1 \, dx_2 \, dy_2 \, dz_2 \quad (20)$$

By expanding this expression and utilizing the orthogonality of the single-particle eigenfunctions, one finds that

$$\langle (\mathbf{r}_2 - \mathbf{r}_1)^2 \rangle = \langle r^2 \rangle_a + \langle r^2 \rangle_b - 2\langle \mathbf{r} \rangle_a \cdot \langle \mathbf{r} \rangle_b \mp 2|(a|\mathbf{r}|b)|^2 \quad (21)$$

in which
$$\langle r^2 \rangle_n = \int u_n^* r^2 u_n \, dx \, dy \, dz$$
$$\langle \mathbf{r} \rangle_n = \int u_n^* \mathbf{r} u_n \, dx \, dy \, dz$$
and
$$(a|\mathbf{r}|b) = \int u_a^* \mathbf{r} u_b \, dx \, dy \, dz$$

In this expression, the first three terms are just what would be obtained if two equal but distinguishable particles were in the respective states u_a and u_b, with the wave function $U(1,2) = u_a(1)u_b(2)$. *The fourth term arises specifically from the requirement that $U(1,2)$ have the symmetric or antisymmetric form (18), and is of opposite sign for the two cases.* We thus find in this example that two identical particles whose space-dependent wave function is *symmetric* "attract" one another, in the sense that the mean square distance between them must be less than, or at most equal to, that for equal but distinguishable particles in the separate eigenstates u_a and u_b. In a corresponding sense two identical particles whose space wave function is antisymmetric "repel" one another. This result is representative of the properties of symmetric and antisymmetric wave functions and gives some insight into the physical consequences of the exclusion principle. Although these attractions and repulsions, which are called *exchange forces*, are physically just as real as if they were the result of a classical force appearing in the Hamiltonian function, they are of course nonclassical effects which arise from a deep, incompletely understood property of spin angular momentum.

EXERCISES

7-12. Derive Eq. (21) from Eq. (20).

7-13. Consider a certain physical system which consists of two identical particles with a spin of $\tfrac{1}{2}$, which as a first approximation do not interact with one another. Introduce as a small perturbation a weak mutual repulsive potential proportional to the square of the distance between them, and evaluate the first-order effect of this perturbation upon the unperturbed energy levels. Show that, if a given level is split by the perturbation, the triplet state has a lower energy than the singlet state. This result is closely analogous to the situation encountered in Russell-Saunders coupling in atomic energy states, in which a repulsion *inversely* proportional to their separation distance acts between pairs of electrons. Here also, the triplet state has lower energy than the singlet state.

In treating physical systems which contain two or more indistinguishable particles, one often wishes to know how the particles are distributed in space. Although the coordinates of the particles appear distinct from one another in the Hamiltonian function, the particles themselves of course move in the same three-dimensional space, and one cannot distinguish which one of the particles has been found in a given volume element $dx\,dy\,dz$. What we wish to know is, therefore, the probability that one of the particles is in a given volume element $dx\,dy\,dz$ *no matter where the other particles may be.* This may be called the *physical probability density,* and we shall denote it by $W(x,y,z)$. It is instructive to calculate this probability for the simple case of two noninteracting particles which was discussed above. The joint probability density distribution for the two particles is

$$U^*U = \tfrac{1}{2}[u_a^*(1)u_b^*(2) \pm u_a^*(2)u_b^*(1)][u_a(1)u_b(2) \pm u_a(2)u_b(1)] \quad (22)$$

The desired probability is then equal to

$$dP_1 = (\textstyle\iiint U^*U\,dx_1\,dy_1\,dz_1)\,dx\,dy\,dz + (\iiint U^*U\,dx_2\,dy_2\,dz_2)\,dx\,dy\,dz$$
$$= 2(\textstyle\iiint U^*U\,dx_1\,dy_1\,dz_1)\,dx\,dy\,dz = W(x,y,z)\,dx\,dy\,dz \quad (23)$$

since the two integrals are equal by symmetry. Carrying out the indicated integration over the coordinates of one particle, we find the physical probability density to be

$$W = u_a^*u_a + u_b^*u_b \quad (24)$$

The physically measurable probability distribution is therefore just the sum of the single-particle distributions for the states u_a and u_b. This result is independent of the symmetry character of the space-dependent wave function and is the same as for distinguishable particles.

EXERCISES

7-14. Verify that Eq. (23) represents the physically observable probability that one of the particles will be found in $dx\,dy\,dz$.

7-15. Carry out the integration leading to Eq. (24).

7-16. Find the probability that *both* particles will be found in the volume element $dx\,dy\,dz$.

Ans.: $dP_2 = 2u_a^*u_a u_b^*u_b\,(dx\,dy\,dz)^2$ symmetric U
               $= 0$ antisymmetric U

7-3.   The Many-electron Atom in the Schroedinger Theory

We are now ready to consider the problem of the ground state of a many-electron atom, using the Schroedinger theory. Inasmuch as we are not at present interested in obtaining a numerically exact value of the ground-state energy, we proceed as usual by neglecting the smaller terms in the Hamiltonian function and include only the major ones which materially affect the nature of the solution.

As a matter of fact, if we start by neglecting all interactions except those *between the nucleus and the individual electrons*, the results of the preceding section for noninteracting particles can be applied directly. In this case the wave functions involved are just the one-electron wave functions for an atom of atomic number Z, and the lowest energy state of the atom will be that for which the lowest possible of these one-electron energy states are occupied, subject to the condition that no two electrons are in states having the same set of quantum numbers. The order of occupation will thus be that of increasing total quantum number n, so that the Bohr-Stoner shell classification system appears to be the correct one for this situation. In this approximation, however, *there is no indication of the order in which the subshell states should be occupied by electrons.*

A better approximation, and the one that is used as a starting point in most treatments of complicated atoms, includes in the Hamiltonian function a certain part of the mutual-interaction energy of the electrons, namely, that part that can be represented by a *spherically symmetrical distribution of charge.* In this approximation, then, each electron in the atom is supposed to move in a *central field* $V_i(r)$ which consists of the field due to the nucleus and a spherically symmetric field due to the *average spatial distribution* of the remaining electrons. To compute $V_i(r)$, the charge density due to the electrons is taken to be equal to the electron charge e times the physical electron density, $W(x,y,z)$, averaged over a sphere at each given r:

$$\rho_i(r) = \frac{e \int_0^{2\pi} \int_0^\pi W_i(r,\theta,\varphi)\, \sin\theta\, d\theta\, d\varphi}{4\pi} \tag{1}$$

where

$$W_i(r,\theta,\varphi) = \sum_{j \neq i} u_j^* u_j \tag{2}$$

and u_j is the eigenfunction for the jth electron.

In one procedure[1] a set of "zero-order" wave functions $\psi_i{}^0$ is assumed for the various electrons in the atom and the total wave function is expressed as a product of these wave functions:

$$\psi^0 = \psi_1{}^0 \psi_2{}^0 \psi_3{}^0 \cdots \psi_z{}^0 \tag{3}$$

[1] Hartree, *Proc. Cambridge Phil. Soc.*, **24**, 89, 111 (1928). See also Hartree, "The Calculation of Atomic Structures," John Wiley & Sons, Inc., New York, 1957.

The wave function for each electron is then modified by evaluating the (spherically symmetric) potential $V_i^0(r)$ in which it moves, and solving numerically the radial wave equation using this function.   This yields a new radial wave function ψ_i^1 for each electron.   This process is repeated again and again, until the "output" wave functions are the same as the "input" wave functions.   That form of $V_i(r)$ which leads to wave functions which in turn lead to this same $V_i(r)$ through the above expression for the charge density is called a *self-consistent field*.   Our discussion of the results of this approximation will be restricted to certain qualitative features which can be seen rather easily and which lead to a satisfactory picture of atomic structure.   Certain features of $V_i(r)$ can be seen at once:

1. For the innermost electrons, $V_i(r)$ is approximately the potential due to the *nucleus alone* plus a constant term due to being inside the out-lying electronic-charge distribution.
2. For the outermost electrons, $V_i(r)$ is approximately the potential energy due to singly charged nucleus, i.e., a singly ionized atom.

One therefore expects, as a consequence of (1), that the wave function corresponding to the lowest state will resemble that for a *one-electron atom* of charge Ze, and as a consequence of (2), that the wave functions for the highest states will resemble *the corresponding wave functions for hydrogen*.   The states intermediate between the lowest and highest ought likewise to resemble roughly the corresponding one-electron states of an atom whose nuclear charge is $(Z - n)e$, where n is approximately the number of electrons in states of lower energy than the ith electron. In particular, the assumed spherical symmetry leads to shells and sub-shells of states, since the angular dependence of the various wave func-tions is exactly the same as for the corresponding one-electron states. Only the radial part of a wave function will differ from the corresponding one-electron wave function.

The approximation of spherical symmetry is actually quite good because of a fundamental property possessed by spherical harmonic functions: *If all possible states of given n and l are occupied by electrons, the corresponding electron density distribution is precisely spherically sym-metric.*   This means that the only lack of spherical symmetry in an atom is due to incompletely filled subshells, which usually involve only the outermost electrons.

EXERCISE

7-17. Evaluate the probability density $W(r,\theta,\varphi)$ for the following completely filled subshells, using the one-electron wave functions 5-4(14):

$n = 2, l = 1; n = 3, l = 1; n = 3, l = 2$; and thus verify the statement of the preceding paragraph for these cases.

From considerations such as those above we may now form the following approximate picture of the ground-state electronic structure of an atom: The population of electrons inside an atom can be resolved into a series of shells analogous to those envisaged in the Bohr-Stoner scheme. Each shell contains a number of electronic states, and the states are classifiable into *subshells* on the basis of their angular momenta and spin

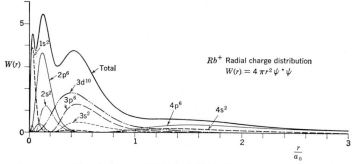

Fig. 7-3. The probability function $W(r)$ for the Rb^+ ion calculated by using the self-consistent field approximation. The radial coordinate is plotted in units of the first Bohr radius of H. Note that the 2s and 2p states, and the 3s, 3p, and 3d states, have maxima at about the same value of r, which leads to quite pronounced maxima in $W(r)$.

orientations. In the outer parts of the atom the mutual interactions of the electrons become more and more comparable in magnitude with the interaction of an electron with the screened nucleus, so that here the electron states are less well resolvable into shells and subshells.

Figure 7-3 shows the electron density $W(r)$ which results from a self-consistent field treatment of singly ionized rubidium. Note that W has maxima and minima as a function of r. These correspond to the various shells of electrons and provide a convincing substantiation of the physical reality of the shell structure of atoms.

To interpret the periodic table, we must classify the various shells and subshell states in the order of increasing energy. This we now do, with the help of the qualitative results of the first-order perturbation theory.

K-*shell*: As mentioned above, the lowest state in an atom is very nearly the 1s state $(n = 1)$ of a one-electron atom of atomic number Z. The departure from the ideal 1s state is due largely to the presence of another 1s electron of opposite spin, and less to the slight penetration of some of the outer electron states into this inner region of the atom. The elements for which the K-shell orbits are the highest populated ones are

(1) H   1s
(2) He   1s²

L-shell: The next states in order of increasing energy are rather similar to the 2s and 2p states for a one-electron atom of nuclear charge $(Z - 2)e$, since the K-shell electrons are localized near the nucleus and are spherically symmetric.[1]   There appears in this shell the first example of a general result that applies to all shells higher than the K-shell: *subshell states of smaller l lie lower in energy within a given shell.*   This is due to the fact that the wave function of an electron in a state of lower angular momentum is relatively large closer to the nucleus where the screening due to the inner-shell states is less effective.   If the one-electron L-shell states for a nuclear charge $(Z - 2)e$ are taken as a zero-order approximation and the effect of the smeared-out K-shell electrons is then introduced as a perturbation, the relative shifts of the 2s and 2p levels may be evaluated as shown in Fig. 7-4.   The result of such a treatment, which appears quite prominently in the spectra of the alkali metals, is that *the* 2s *subshell lies lower than the* 2p *subshell,* so that the configurations of the elements for which the L-shell is the highest populated shell are

(3)	Li	$1s^2\ 2s$	(7)	N	$1s^2\ 2s^2\ 2p^3$
(4)	Be	$1s^2\ 2s^2$	(8)	O	$1s^2\ 2s^2\ 2p^4$
(5)	B	$1s^2\ 2s^2\ 2p$	(9)	F	$1s^2\ 2s^2\ 2p^5$
(6)	C	$1s^2\ 2s^2\ 2p^2$	(10)	Ne	$1s^2\ 2s^2\ 2p^6$

M-shell and Higher Shells: The M-shell $(n = 3)$ contains the subshells 3s, 3p, and 3d.   As indicated above, the greater penetration toward the nucleus of the wave functions of low angular momentum leads to the expected order of occupation 3s, 3p, 3d.   In this shell, however, a new phenomenon appears: Because with increasing Z, the effective charge of the nucleus is being progressively diminished by the screening due to the inner shells of electrons and because of the $1/n^2$ dependence of the energy upon the total quantum number, the energy difference between successive outer shells is becoming *smaller and smaller.*   On the other hand, the effects of greater penetration toward the nucleus of the lower angular-momentum states are becoming *larger and larger.*   There is a point then, at which the *low angular-momentum states of the next higher shell may have a lower energy than the higher angular-momentum states of the given shell.*   This "crossover" first occurs for the 3d subshell and the 4s subshell at $Z \approx 6$.   Elements occurring after the 3p subshell is completed thus vary in their tendency to fill a 4s, as against a 3d, subshell state.   This effect has not been worked out quantitatively, but the order of filling of the various subshells has been worked out for the remainder of the table by using spectroscopic evidence.   The complete sequence in

---

[1] In all except the lightest atoms, the energies of the 2s and 2p states correspond to a one-electron atom of nuclear charge $(Z - \epsilon)e$, where ϵ is approximately 3 or 4, because of the penetration of some of the L and higher-shell electrons inside the L-shell "orbits."

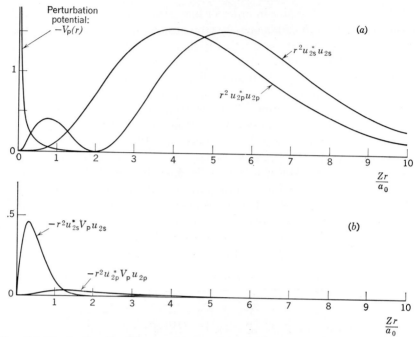

FIG. 7-4. Illustrating the difference in perturbation energy for 2s and 2p subshells. In (a) the curves for $u_{2s}^* u_{2s}$, $u_{2p}^* u_{2p}$, and V_p are shown, where u_{2s} and u_{2p} are the one-electron approximations to the "true" states and V_p is the perturbation potential, being the difference between a point charge $(Z-2)e$ and the potential due to the nucleus Ze and the K-shell electrons. In (b) the perturbation-energy integrands $u_{2s}^* V u_{2s} r^2$ and $u_{2p}^* V u_{2p} r^2$ are shown. It is clear from these curves that the 2s state will be reduced in energy more than the 2p state. Similar conclusions hold for the higher shells.

which the various subshells are filled is shown diagrammatically in the periodic table of Fig. 7-5, in which the *electron configurations* are indicated.

Some of the more prominent features of the periodic table may now be interpreted in terms of the systematics of the electron configurations:

1. The strong chemical similarity of certain groups of elements is seen to be associated with a similarity in the population of the outermost subshell of the atom.

2. The noble gases are obtained whenever a p subshell is completely occupied (exception: He $= 1s^2$). That these gases are monatomic and chemically inert may be ascribed to the facts that a completed subshell is *spherically symmetric* and that the next higher level above a p-subshell level is separated by a sizable energy gap from the p-level, thus discouraging any transformations of these atoms to a nonsymmetrical form which could enter into valence relationships with other atoms.

Periodic table (rotated figure)

	s^1	s^2
1s		1 H
2s	3 Li	4 Be
3s	11 Na	12 Mg
4s	19 K	20 Ca
5s	37 Rb	38 Sr
6s	55 Cs	56 Ba
7s	87 Fr	88 Ra

	d^1	d^2	d^3	d^4	d^5	d^6	d^7	d^8	d^9	d^{10}
3d	21 Sc	22 Ti	23 V	24 Cr $4s3d^5$	25 Mn	26 Fe	27 Co	28 Ni	29 Cu $4s^13d^{10}$	30 Zn
4d	39 Y	40 Zr	41 Nb $5s^14d^4$	42 Mo $5s^14d^5$	43 Tc	44 Ru $5s^14d^7$	45 Rh $5s^14d^8$	46 Pd $5s^04d^{10}$	47 Ag $5s^14d^{10}$	48 Cd
5d	57 La R.E.*	72 Hf	73 Ta	74 W	75 Re	76 Os	77 Ir	78 Pt $6s^15d^9$	79 Au $6s^15d^{10}$	80 Hg
6d	89 Ac	90 Th H.E.†								

					p^6: 2 He $1s^2$
	p^1	p^2	p^3	p^4	p^5	p^6
2p	5 B	6 C	7 N	8 O	9 F	10 Ne
3p	13 Al	14 Si	15 P	16 S	17 Cl	18 Ar
4p	31 Ga	32 Ge	33 As	34 Se	35 Br	36 Kr
5p	49 In	50 Sn	51 Sb	52 Te	53 I	54 Xe
6p	81 Tl	82 Pb	83 Bi	84 Po	85 At	86 Rn

Rare earths 4f

	f^1	f^2	f^3	f^4	f^5	f^6	f^7	f^8	f^9	f^{10}	f^{11}	f^{12}	f^{13}	f^{14}
	58 Ce $5d^1$	59 Pr $5d^0$	60 Nd $5d^0$	61 Pm $5d^0$?	62 Sm $5d^0$	63 Eu $5d^0$	64 Gd $5d^14f^7$	65 Tb $5d^14f^8$	66 Dy $5d^0$?	67 Ho $5d^0$?	68 Er $5d^0$?	69 Tm $5d^0$	70 Yb $5d^0$	71 Lu $5d^14f^{14}$

†Heaviest elements 5f

	f^1	f^2	f^3	f^4	f^5	f^6	f^7	f^8	f^9	f^{10}	f^{11}	f^{12}	f^{13}	f^{14}
	91 Pa $6d^1$?	92 U $6d^1$	93 Np $6d^1$?	94 Pu $6d^1$?	95 Am $6d^1$?	96 Cm $6d^1$?	97 Bk $6d^1$?	98 Cf $6d^1$?	99 Es	100 Fm	101 Md	102 No		

Examples:
22 Ti: $1s^2$ $2s^2$ $2p^6$ $3s^2$ $3p^6$ $3d^2$ $4s^2$
42 Mo: " " " " " $3d^{10}$ $4p^6$ $5s$ $4d^5$
64 Gd: " " " " " " " $4f^7$ $5d$ $6s^2$
74 W: " " " " " " " $4f^{14}$ $5d^4$ $5p^6$ $6s^2$ $5f^5$ $5d$
94 Pu: " " " " " " " $4d^{10}$ $5d^{10}$ $6p^6$ $7s^2$ $5f^5$ $6d$

FIG. 7-5. The electron configurations of the elements. Deviations from a regular order of filling of the shells are indicated in the boxes of the elements concerned.

251

3. Those elements having the strongest chemical activity occur *just before* a noble gas (halogens) and *just after* a noble gas (alkali metals). This activity is related in the first case to the fact that a completed subshell has much lower energy than one lacking one or two electrons and in the second case to the fact that the first electron outside a completed p subshell is generally relatively weakly bound. The tendency to complete a subshell can be seen also near the ends of the d subshells, where one or more than one of the electrons is removed from the s subshell for this purpose.

4. *Chemical valence* is closely related to the *number of filled states in a subshell*, particularly in p subshells. Here also occur the long series of chemically similar elements. The valence properties of the elements lying in the middle region of the d subshells are not very well defined, nor do these elements show striking chemical similarities to their counterparts in adjacent d subshells. The valence properties are most strongly defined for those elements whose incompleted subshells involve *low angular momenta*, and whose wave functions therefore extend well inward toward the nucleus and *well outward beyond the lower completed shells*, i.e., *for s and p subshells*.

5. The irregularity introduced between $_{57}$La and $_{72}$Hf by the rare earth elements is now seen to involve the filling of a 4f-*subshell*. The chemical similarity of these elements results from the fact that the 4f wave functions lie well inside the already present 6s electrons and therefore can have little effect upon the valence properties of these elements. (The *energies* of the 6s, 4f, and 5d states are nearly equal, but the spatial *extents* of the wave functions are quite different for these functions because of their widely different angular momenta.)

6. Another group of chemically similar elements, this one only partially complete because of the lack of occurrence in nature of the necessary *nuclei*, appears to be involved at $_{90}$Th.

7-4. Conclusion

The description of the periodic table that is provided by the Schroedinger theory is most satisfactory. All of the major chemical properties of the elements have at least a qualitative interpretation in terms of modern quantum theory, and many of these properties can be treated in detail in a semiquantitative way. The many regularities and some of the major irregularities in the table also receive quite reasonable interpretations in terms of the *shell structure of the atom*. We shall find in later chapters, particularly in connection with X-ray absorption spectra, that the various subshell states actually have a quantitative significance in many physical phenomena, so that the shell structure of the atom is much more than a convenient fiction having but limited application; it is a fact.

Before leaving the present discussion, two important observations should be made. First, the limitation in the extent of the periodic system of the elements is placed by the lack of *nuclei* of higher atomic number in nature. We may be sure that the theory—with the necessary relativity corrections—could provide, in principle, an appropriate electron configuration for elements having considerably larger atomic number than occur in nature.[1] And second, one feature of the subshell energy levels, as qualitatively described above, is of special importance for our later considerations of *nuclear* shell structure. This feature is that, unlike the one-electron wave functions without spin-orbit corrections, the *various energy levels are not degenerate with respect to l*. This is an illustration of the general result that, except for a few special cases such as that of an inverse-square force, the various energy levels of a central field system are *nondegenerate with respect to l*. In the present case the *l*-degeneracy of the one-electron states disappears because of the screening effect of the electron cloud, which causes the effective potential in which a given electron moves to depart from the pure Coulomb form. In the analogous case of nuclear forces, the nature of the potential function (if it exists) is not known, except that it is certainly not a Coulomb potential. Thus we expect from the outset that the energy levels of nuclei will not show *l*-degeneracy.

Finally, it is worth emphasizing again the role of the *exchange-symmetry properties* exhibited by the wave functions for several identical particles in defining the character of the periodic table and the nature of chemical interactions. If these symmetry properties did not apply, or if electrons happened to be Bose particles, the world would be a different place indeed. All atoms would be spherical and very tiny, since all the electrons would fall into the 1s state, so that chemistry as we know it would not exist. We shall see in later chapters other clear examples of phenomena whose fundamental features are defined largely by these exchange-symmetry properties.

REFERENCES

Richtmyer, F. K., E. H. Kennard, and T. Lauritsen: "Introduction to Modern Physics," 5th ed., International Series in Pure and Applied Physics, McGraw-Hill Book Company, Inc., New York, 1955.

Schiff, L. I.: "Quantum Mechanics," 2d ed., International Series in Pure and Applied Physics, McGraw-Hill Book Company, Inc., New York, 1955.

[1] The uncertainty concerning the electron configurations of some of the elements shown in Fig. 7-5 is caused by insufficient experimental study of the spectra, and not by any inability of the theory to predict, in principle, what these should be.

8

Atomic Spectroscopy

Of all the tools that have been applied to the study of the detailed structure of matter, it can fairly be said that *spectroscopy* has been applied in more ways to more problems, and has produced more fundamental information, than any other. It has been used, and is still being used, not only to extend our knowledge of the ultimate properties of matter, but also to provide accurate measurements of many quantities of quite practical interest to physicists, chemists, and engineers. It is also an indispensable tool in the study of the chemical composition and the physical condition in which matter exists in certain situations which are inaccessible to more direct observation—such as in astrophysics and atmospheric physics—and in situations of a highly transient character, as in fast chemical reactions and electrical discharges.

In modern physics the term "spectroscopy" is used in connection with a great many quite different investigative techniques, e.g., optical spectroscopy (including infrared and ultraviolet atomic and molecular spectroscopy), microwave spectroscopy, β-ray spectroscopy, γ-ray spectroscopy. In each of these techniques the quantity that is directly measured is the energy, momentum, wavelength, or frequency of a photon or electron emitted or absorbed during a transition between two states of an atomic, molecular, or nuclear system. From this information the characteristics of the *energy levels* of the system are inferred, and from these characteristics (spacing, angular momentum, multiplicity, etc.) certain basic parameters of the system can often be determined with considerable precision. In this chapter we shall examine some of the more important aspects of the original form of spectroscopy—atomic spectroscopy.

8-1. General Considerations

From what has already been said in the preceding chapters, the student should be aware of some of the more important historical facts concerning atomic spectroscopy and the part it has played in the development of the quantum mechanics. Having thus arrived at the correct mechanical laws with the help of our knowledge of the spectroscopic properties of certain simple atoms, we may now properly seek to *use these laws to improve our understanding of the spectra and the detailed structure of the more complicated atoms.*

In the preceding chapter we have seen that quantum mechanics provides a beautiful interpretation of the periodic system of the elements and of the ground state of an atom, wherein the electrons of the atom are visualized as occupying—subject to the exclusion principle—the lowest-lying of the various energy states that would exist for an electron moving in the field of the nucleus and the spherically symmetric average field due to the remaining electrons. The approximation used for that discussion was purposely kept as simple as possible consistent with the requirement that the *main features* of atomic structure should emerge from the treatment. We now wish to utilize quantum mechanics to add some finer details to the above atomic model and to illustrate further the extent of the success of the Schroedinger theory in accounting for the known features of ordinary atomic processes.

Before actually embarking upon this procedure, it is desirable to delineate the present problem more clearly by first considering qualitatively the various types of excited states in which an atom might exist, using for the present the same approximate model that was used previously. Thus if, as indicated above, the ground state of an atom is visualized as corresponding to the electrons occupying the lowest-lying of a number of possible energy states, it is clear that, if any one or more of the electrons were to occupy a higher level, the atom would be excited. The *most general* state of excitation (which is of completely negligible importance) would then correspond to the various electrons being in *any* of the energy states—including the continuum which lies above the range of the discrete states—subject only to the limitations of the exclusion principle. By far the more common situation, on the other hand, is that in which most of the electrons are in the energy states they occupy when the atom is in its ground state, but in which one or more electrons may be *completely removed* from the atom (that is, the atom is perhaps *ionized*) and one or two of the electrons are in *bound* excited states. Even with this restriction the possible states of excitation of the atom may fall into several broad classes according to the manner in which the excitation was produced. For example, we shall see later that the excitation of an atom by X-rays or by the impact of an energetic charged particle often results in the removal of electrons from an *inner shell*. This leaves the atom ionized and highly excited, and many interesting phenomena may occur in such a situation. On the other hand the atom may be excited by rather gentler means—as in an electric arc, spark, or glow discharge, or in a flame—in which case only the outermost of the electrons will be affected. Although *all* such states of excitation are in a sense a part of the same general problem, it is desirable to separate the discussion into several different categories upon the basis of the techniques used in the study of the excited states. Thus we shall study the states characteristic of *X-ray* and *microwave excitation* in later chapters. We are concerned at present only with the states of excitation in which the radiation involved falls in the visible or near-infrared or -ultraviolet regions of the spectrum; these are states of excitation in which only the outermost, *valence, electrons* are involved.

8-2. The Hamiltonian Function of a Complicated Atom

In order to deduce some of the finer details of the energy levels of a complicated atom, we proceed in a manner similar to that used in the case of the one-electron atom: namely, we consider first a problem which includes the major terms of the Hamiltonian function and whose energy levels are therefore very closely the same as for the complete problem. We then study the fine-structure splitting of these highly degenerate levels that results when the next most important terms are included as perturbations. For the extremely complicated problem now being considered, our treatment must remain rather qualitative throughout, even though a certain amount of quantitative information can be obtained by the use of advanced techniques. We shall thus be obliged to place heavy emphasis upon the similarities between the present problem and some of the features of the one-electron atom discussed in Chap. 5.

We first consider the various terms that should appear in the complete Hamiltonian function of an atom which has a nuclear charge Ze and is surrounded by N electrons. We assume for the present that $N \leq Z$; that is, the atom is not *negatively* ionized. The Hamiltonian function should thus be the sum of the following terms:

1. The kinetic energy of the electrons:

$$T = \sum_{i=1}^{N} \frac{p_i^2}{2m}$$

2. The electrostatic interaction energy of the electrons with the nucleus:

$$V_{en} = -\sum_{i=1}^{N} \frac{Ze^2}{4\pi\varepsilon_0 r_i}$$

3. The mutual electrostatic energy of the electrons:

$$V_{ee} = \sum_{i=1}^{N}\sum_{j=1}^{i-1} \frac{e^2}{4\pi\varepsilon_0 r_{ij}}$$

4. The spin-orbit energy of the electrons:

$$V_{so} = -\sum_{i=1}^{N} \frac{\mathbf{\sigma}_i \cdot \mathbf{l}_i}{m^2 r_i c^2}\frac{dV}{dr_i}$$

5. The spin-spin interactions of the electrons:

$$V_{ss} = \frac{\mu_0}{4\pi}\sum_{i=1}^{N}\sum_{j=1}^{i-1}\frac{e^2}{m^2}\left[\frac{\mathbf{\sigma}_i \cdot \mathbf{\sigma}_j}{r_{ij}^3} - 3\frac{(\mathbf{\sigma}_i \cdot \mathbf{r}_{ij})(\mathbf{\sigma}_j \cdot \mathbf{r}_{ij})}{r_{ij}^5}\right]$$

6. Interactions between the orbital magnetic moments of the electrons:

$$V_{oo} = \sum_{i=1}^{N}\sum_{j=1}^{i-1} C_{ij}\mathbf{l}_i \cdot \mathbf{l}_j$$

7. Electron spin–nuclear spin interactions

$$V_{esns} = \frac{\mu_0}{4\pi}\sum_{i=1}^{N}\frac{e}{m}\left[\frac{\mathbf{\mu}_{nuc} \cdot \mathbf{\sigma}_i}{r_i^3} - 3\frac{(\mathbf{\mu}_{nuc} \cdot \mathbf{r}_i)(\mathbf{\sigma}_i \cdot \mathbf{r}_i)}{r_i^5}\right]$$

8. Nuclear spin–electron orbital moment interactions:

$$V_{nseo} = \frac{\mu_0}{4\pi}\sum_{i=1}^{N}\frac{e}{m}\left[\frac{\mathbf{\mu}_{nuc} \cdot \mathbf{l}_i}{2\pi r_i^3}\right]$$

9. Other terms due to nuclear electric-quadripole moments, modification of the electric field near the nucleus owing to the finite size of the nucleus, interaction between the spin magnetic moments and the orbital motion of other electrons, etc.

In addition to the above terms which correspond to "true" interactions, the Hamiltonian function should also include terms to represent appreciable effects which do not correspond to a classical type of interaction, such as

10. Terms to correct for relativistic effects, e.g..

$$R = \sum_{i=1}^{N}\frac{p_i^4}{8m^3c^2}$$

The Hamiltonian function, when converted into an operator, will lead to certain wave functions for the energy states of the atom. According to the exchange-symmetry postulate of the preceding chapter, only those wave functions that are *antisymmetric with respect to the exchange of all pairs of electronic space and spin coordinates* are valid wave functions for the atom. This restriction upon the choice of acceptable wave functions has certain physically observable consequences, some of which have previously been described. For the atomic systems now being considered, the most important of these effects are, as was shown in the preceding chapter:

11. A tendency for electrons with parallel spins to avoid (or "repel") each other, and as a result of this and the electrostatic repulsion of the electrons, a strong "exchange correlation" between the spins *which tends to align electronic spins parallel to one another.*

The discussion of the preceding chapter was based upon a model which included terms (1) and (2) and a spherically symmetric "average" of term (3). We must now consider what effects the inclusion of the remaining terms will have upon the degenerate energy levels of that model. We are interested in two aspects of these effects—the *qualitative character* of each term in splitting a degenerate level into a number of distinct levels of smaller degeneracy, each characterized by a different quantized value of some "constant of the motion," and the *magnitude* of each effect.

On the basis of the perturbation theory, one can deduce the approximate magnitude of the various effects included in the above list. The analysis is too involved to be included here, but is quite straightforward for most of the effects. The final conclusion, which is in agreement with observation, is that there are two main categories into which most atoms should fall with regard to the relative magnitudes of the remaining effects:[1]

1. For most atoms, the nonspherically symmetric residual electrostatic effects of term (3) and the spin correlations of (11) are the largest of the remaining terms; the spin-orbit term (4) is considerably smaller, and the remaining terms are quite negligible.
2. For some atoms, mainly the heavier ones, the spin-orbit term (4) predominates.

We shall now consider these two cases separately.

---

[1] Some of the effects, such as the relativistic correction terms, have a very large influence upon the lowest energy states, and much less upon the higher states. These corrections will thus appreciably affect the *total ground-state energy* of the atom, but not the energies of the states available to the valence electrons measured relative to the ground-state energy.

8-3. Russell-Saunders, or LS, Coupling

The first of the above situations, in which the residual electrostatic effects of term (3) and the spin-spin correlation (11) dominate the remaining terms, was first studied by Russell and Saunders[1] and is called Russell-Saunders coupling, or LS coupling. Let us now consider what effect these terms will have upon the degenerate energy levels of the simple central-field approximation.

We recall first that the various energy levels of the central-field model were degenerate with respect to the orbital angular momenta and spins of the various electrons, since each electron was assumed to move in a *central field* and no spin interactions were considered except as a book-keeping device to prevent two electrons from being in the same state; that is, the *individual orbital and spin* angular momenta of the electrons, as well as the *net orbital* angular momentum, the *net spin* angular momentum, and the *total* angular momentum of the electrons, *were constants of the motion*. Thus the various orthogonal states of given energy were degenerate with respect to all of these quantities and might be described equally well in terms of various possible sets of quantum numbers for the v valence electrons, such as

(a) $(n_1,l_1,s_1,m_{l1},m_{s1})$, $(n_2,l_2,s_2,m_{l2},m_{s2})$, . . . , $(n_v,l_v,s_v,m_{lv},m_{sv})$

(b) (n_1,l_1,s_1,j_1,m_{j1}), (n_2,l_2,s_2,j_2,m_{j2}), . . . , (n_v,l_v,s_v,j_v,m_{jv})

(c) (n_1,l_1,s_1,j_1), (n_2,l_2,s_2,j_2), . . . , (n_v,l_v,s_v,j_v), J, M_J

(d) (n_1,l_1,s_1), (n_2,l_2,s_2), . . . , (n_v,l_v,s_v), L, S, J, M_J

$$(1)$$

The first two of these expressions will be recognized as being just the alternative quantum-number designations for the various individual one-electron energy states as described in Chap. 5, and the last two expressions the natural extension of the law of addition of independent quantized angular momenta given in that discussion to the case of several electrons. In the last two expressions, L, S, J, and M_J are the quantum numbers that define the square of the net *orbital* angular momentum of the electrons, the square of the net *spin* angular momentum, the square of the *total* angular momentum, and the *z-component* of the total angular momentum, respectively. (Note that, although fewer quantum numbers are used in the last two expressions than in the first two, the various possible *combinations* of quantum numbers in each expression necessarily define the same total number of substates of the degenerate state.)

If we now consider the effects of the residual electrostatic repulsions we see that, since these involve forces that are not directed toward the nucleus, the orbital angular momenta of the *individual* valence electrons *will no longer be constants of the motion*. Since the mutual repulsion

[1] *Astrophys. J.*, **61**, 38 (1925).

between pairs of electrons is directed along the line joining them, however, the *total orbital angular momentum will remain constant.* Similarly, the spin-spin correlations given by (11) may cause the *individual* spin angular-momentum vectors to change, but the *total* spin angular momentum will be constant.

In the case of the spin-spin correlations, the individual spin angular momenta cannot change their *magnitudes*, since the spin is an intrinsic, unchangeable property of electrons. The *directions* of the spins may change, however. On the other hand, in the case of the electrostatic repulsions, the *individual orbital* angular momenta *could* change in magnitude, *but this is found seldom to occur.*

We thus find that the introduction of these two terms into the problem leads to a situation in which only the fourth of the alternative sets of quantum numbers given above is suitable for describing the substates of the perturbed system.

As to the quantitative effects of these two terms, the effect of the spin-spin correlation is usually greater than that of the electrostatic repulsion and, for the reason given in Exercise 7-13 of the previous chapter, gives the lowest energy for the state of *largest possible S.* Further, we can see that the electrostatic energy will be a minimum if the valence electrons remain as far as possible from one another. We thus expect that the state of lowest energy would be one in which the valence-electron orbits are arranged symmetrically around the periphery of the atom, and in which this symmetrical configuration "rotates" more or less as a "rigid body." Since the individual electrons would all then be revolving in the "same direction" about the nucleus, we might expect the state of lowest energy to be the one whose quantum number L is *as large as can be formed* from the individual orbital angular-momentum quantum numbers. This is observed to be the case. The states are thus arranged in energy in the order of *decreasing L,* the largest L-value having the lowest-lying energy.

Finally, we must introduce the small spin-orbit term (4) as an additional perturbation upon the above separated but still individually degenerate levels. It is found that the total spin-orbit interaction may be represented with sufficient accuracy by an expression of the simple form

$$V_{so} = \mathrm{F}(r_i)\boldsymbol{L} \cdot \boldsymbol{S} \tag{2}$$

Using this expression for the spin-orbit interaction, we can deduce an important semiquantitative rule governing the fine-structure splitting of a given level. We follow a procedure exactly similar to the case of the one-electron atom except that the r-dependence is now so complicated that it cannot be evaluated analytically. We represent it for each given unperturbed level by a constant C which can in principle be calculated but in practice must be evaluated empirically for all but the simplest cases. The *angular* part of the spin-orbit energy can, however, be evaluated just as for the one-electron atom by writing

$$\boldsymbol{J} = \boldsymbol{L} + \boldsymbol{S} \tag{3}$$
and
$$|\boldsymbol{J}|^2 = |\boldsymbol{L}|^2 + |\mathcal{S}|^2 + 2\boldsymbol{L} \cdot \boldsymbol{S} \tag{4}$$
so that
$$\boldsymbol{L} \cdot \boldsymbol{S} = \tfrac{1}{2}(|\boldsymbol{J}|^2 - |\boldsymbol{L}|^2 - |\boldsymbol{S}|^2) \tag{5}$$

Thus the average of V_{so} over the unperturbed wave function described by the quantum numbers (1d) is equal to

$$\langle \boldsymbol{L} \cdot \boldsymbol{S} \rangle = \tfrac{1}{2}C'[J(J+1) - L(L+1) - S(S+1)] \tag{6}$$

since $|\boldsymbol{J}|^2$, $|\boldsymbol{L}|^2$, and $|\boldsymbol{S}|^2$ are all quantized constants of the motion in the unperturbed system ($C' = C\hbar^2$).

We thus see that, just as in the one-electron atom, states of different J, L, and S have different energy. The shift in energy is, for given L and S, proportional to $J(J+1)$, and the *spacing* of consecutive levels within a multiplet is

$$E_{J+1} - E_J = \tfrac{1}{2}C'[(J+1)(J+2) - J(J+1)] = C'(J+1) \tag{7}$$

Thus *the spacing between consecutive levels of a fine-structure multiplet is proportional to the larger of the two J-values involved.* This is called the *Landé interval rule.* It can be of great help in determining the J-values of observed energy levels.

To illustrate the results of the discussion up to this point, consider the specific case of an atom having two valence electrons and suppose a

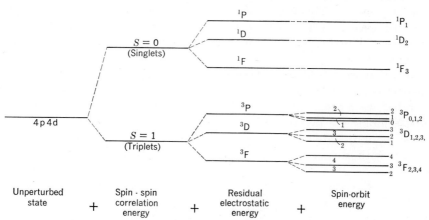

FIG. 8-1. Schematic diagram illustrating the fine-structure splitting of a level corresponding to a 4p and a 4d valence electron. The Landé-interval rule is illustrated in the spacings of the triplet levels.

certain unperturbed excited energy state of the atom corresponds to one of these electrons being in a 4p level and the other in a 4d level. The effect of the three perturbations is shown schematically in Fig. 8-1. At the left is the unperturbed level. The spin-spin correlation splits this level into two levels, the upper one having $S = 0$ (singlet state) and the lower level having $S = 1$ (triplet state). The electrostatic-energy term then splits each of these states into three states having L-values ranging from $|2 - 1|$ to $(2 + 1)$, that is, $L = 1$ (P-state), $L = 2$ (D-state), and $L = 3$ (F-state). Each of the *triplet* levels is then further split by the spin-orbit energy into three levels having J-values ranging in each case from $|L - 1|$ to $(L + 1)$. The final spectroscopic designation of the various states is also indicated in the figure.

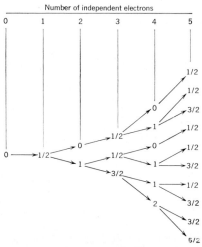

FIG. 8-2. A diagrammatic illustration of the resultant spin quantum numbers for 0, 1, 2, 3, 4, and 5 independent electrons. If a given subshell is more than half occupied, the exclusion principle will limit the number of spin combinations.

To illustrate the procedure by which the expected fine-structure character of a level having more than two electrons in unfilled shells can be deduced, consider the case of three electrons, and for definiteness let their configuration be 2p 3p 4d. We first find *what spin combinations* will be present. This is most easily done by first combining the spins of two of the electrons and then combining the third spin with each of them. Figure 8-2 illustrates the possible spin quantum numbers which can be obtained by combining several independent electron spins. We thus find that there will be two sets of doublet states ($S = \frac{1}{2}$) and one set of quartet states ($S = \frac{3}{2}$).

Next, we must combine the three orbital angular momenta. If we first combine the two p states, we see that these will lead to an S-state ($L = 0$), a P-state ($L = 1$), and a D-state ($L = 2$). Combining the d state with each of these in succession then leads to

$$\text{S} + \text{d} \rightarrow \text{D}$$
$$\text{P} + \text{d} \rightarrow \text{P, D, F}$$
$$\text{D} + \text{d} \rightarrow \text{S, P, D, F, G}$$

Finally, the introduction of the $\boldsymbol{L} \cdot \boldsymbol{S}$ coupling splits each of these levels into an appropriate number of separate levels having different J-values.

The two sets of doublet levels will thus each have as members

$$^2D_{3/2, 5/2}$$

$$^2P_{1/2, 3/2} \quad ^2D_{3/2, 5/2} \quad ^2F_{5/2, 7/2}$$

$$^2S_{1/2} \quad ^2P_{1/2, 3/2} \quad ^2D_{3/2, 5/2} \quad ^2F_{5/2, 7/2} \quad ^2G_{7/2, 9/2}$$

and the set of quartet levels the members

$$^4D_{1/2, 3/2, 5/2, 7/2}$$

$$^4P_{1/2, 3/2, 5/2} \quad ^4D_{1/2, 3/2, 5/2, 7/2} \quad ^4F_{3/2, 5/2, 7/2, 9/2}$$

$$^4S_{3/2} \quad ^4P_{1/2, 3/2, 5/2} \quad ^4D_{1/2, 3/2, 5/2, 7/2} \quad ^4F_{3/2, 5/2, 7/2, 9/2} \quad ^4G_{5/2, 7/2, 9/2, 11/2}$$

to make a total of 65 *distinct levels* in the multiplet. (Little wonder that a spectrum can be complex!) Note that, although all levels bear the name *doublet* or *quartet*, some of them have fewer than two or four components, respectively, because L is less than S for these levels.

The fine-structure splitting of a level in the case of *ideal LS coupling* may be generalized as follows for the case of v valence electrons whose configuration is $n_1l_1, n_2l_2, \ldots, n_vl_v$:

1. The unperturbed level is split by the spin-spin correlation effect into a number of well-separated levels, equal to the number of different values of the spin quantum numbers S that can be formed from the v electron spins, *allowing for exclusion-principle limitations if any given subshell is more than half filled.* If a given subshell is more than half filled, the possible values of S are the same as would be obtained for a number of electrons equal to the number of vacancies in that subshell. Of these levels, the one of highest multiplicity (largest S) lies lowest.

2. Each of the above levels is further split by the residual electrostatic effects into a number of less well separated levels, equal to the number of different values of the orbital angular-momentum quantum number L that can be formed from the individual orbital angular momenta of the v electrons, *again allowing for exclusion-principle limitations if any two or more electrons are in the same subshell.* For each value of S, the level having the largest L lies lowest.

3. Each of these resulting levels is again split by the spin-orbit effect into $2S + 1$ or $2L + 1$ levels, whichever is smaller. The Landé-interval formula governs the relative spacing of all levels within each of these multiplets, using approximately *the same value of C for all multiplets resulting from a given unperturbed level.* Within a given multiplet, the term having the smallest value of J lies lowest unless the configuration involves a more-than-half-filled subshell, in which case the largest J lies lowest. The latter are called *inverted multiplets.*

The above description shows that the fine-structure splitting of an energy level in LS coupling can be quite complicated if very many

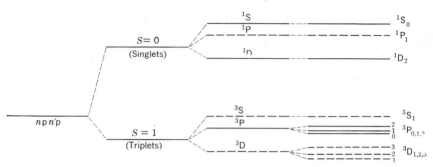

FIG. 8-3. Schematic illustration of the *LS* fine-structure splitting of a configuration involving two p electrons. The dashed levels are missing from the multiplet if the electrons are equivalent ($n = n'$).

valence electrons are involved. This complexity is somewhat reduced, of course, if some of these are in the same subshell. Electrons which are in the same subshell are called *equivalent electrons;* the exclusion principle must always be taken into account in analyzing the fine structure of a configuration involving equivalent electrons.

In case a subshell is more than half filled with electrons, the application of the exclusion principle leads to the important result that *the combinations of L and S available to these electrons are exactly the same as if only a number of electrons equal to the number of unoccupied substates in the subshell were actually present.* (Thus if five electrons were in a certain p subshell, the number of *L* and *S* combinations would be exactly the same as for *one* electron in this subshell, namely, $S = \frac{1}{2}$ and $L = 1$.) The most complicated fine structure therefore is associated with those elements having approximately *half-filled* subshells.

As an illustration of the effect of the exclusion principle upon the fine structure of a level, the fine-structure splitting is compared schematically in Fig. 8-3 for two equivalent and for two nonequivalent p electrons. The levels missing from the fine structure for the equivalent electrons are shown dotted.

This figure indicates that the ³D, ³S, and ¹P terms that are found in the fine structure of a configuration of two nonequivalent p electrons are missing from the corresponding multiplet for equivalent electrons. These terms are eliminated because of the exclusion-principle requirement that no two electrons may have the same set of quantum numbers n, l, m_l, and m_s and because *only physically distinguishable states can be counted.*

The general case of any number of equivalent electrons is best treated by the use of group theory, but unfortunately this method is beyond the scope of the present work. We can, however, deal with the particular case illustrated in Fig. 8-3 by a more elementary method. Let us tabulate the various possible values of m_{l1}, m_{s1}, m_{l2}, and m_{s2} for the two electrons, first treating them as nonequivalent (Table 8-1). We then take account of the exclusion principle and the indistinguishability of the electrons by striking out those states for which $m_{l1} = m_{l2}$ and $m_{s1} = m_{s2}$, and by eliminating duplication of indistinguishable states. Thus states 1, 4, 17, 20, 33, and 36 are removed by the first criterion, and such

TABLE 8-1. m_l AND m_s VALUES FOR TWO EQUIVALENT p ELECTRONS

m_{l1}	m_{l2}	m_{s1}	m_{s2}	Label	m_{l1}	m_{l2}	m_{s1}	m_{s2}	Label
+1	+1	+½	+½	OUT	0	−1	+	+	11
		+	−	1			+	−	12
		−	+	1			−	+	13
		−	−	OUT			−	−	14
+1	0	+	+	2	−1	+1	+	+	6
		+	−	3			+	−	8
		−	+	4			−	+	7
		−	−	5			−	−	9
+1	−1	+	+	6	−1	0	+	+	11
		+	−	7			+	−	13
		−	+	8			−	+	12
		−	−	9			−	−	14
0	+1	+	+	2	−1	−1	+	+	OUT
		+	−	4			+	−	15
		−	+	3			−	+	15
		−	−	5			−	−	OUT
0	0	+	+	OUT					
		+	−	10					
		−	+	10					
		−	−	OUT					

pairs of states as 2 and 3, 5 and 13, 6 and 15, etc., are labeled with new numbers 1, 2, 3, etc. For two equivalent p electrons, only the fifteen combinations of $M_L = m_{l1} + m_{l2}$, $M_s = m_{s1} + m_{s2}$, and $M_J = M_L + M_S$ shown in Table 8-2 are possible. We must now compare these states with the states expected in the LS multiplet: $^1S_0$, $^1P_1$, $^1D_2$, $^3S_1$, $^3P_{0,1,2}$, $^3D_{1,2,3}$. Of these, we can immediately see that the $^3D_3$ state cannot be present, since this would require M_J values of $+3$ and -3. This in turn rules out the presence of $^3D_2$ and $^3D_1$, since if one possible combination of a given L and S is present, all must be. On the other hand, state 2 requires the presence of an LS term with $L \geq 1$ and $S \geq 1$. Of the available combinations, this requires that $^3P_{0,1,2}$ be present. Furthermore, state 1 can come only from an LS term having $L \geq 2$ and $S \geq 0$; from this we conclude that the $^1D_2$ term must be present.

TABLE 8-2. M_L, M_S, AND M_J VALUES FOR TWO EQUIVALENT p ELECTRONS

Label	M_L	M_S	M_J	Label	M_L	M_S	M_J
1	+2	0	+2	9	0	−1	−1
2	+1	+1	+2	10	0	0	0
3	+1	0	+1	11	−1	+1	0
4	+1	0	+1	12	−1	0	−1
5	+1	−1	0	13	−1	0	−1
6	0	+1	+1	14	−1	−1	−2
7	0	0	0	15	−2	0	−2
8	0	0	0				

If we now count how many independent LS terms are known to be present, we have, from $^3P_2$, $^3P_1$, and $^3P_0$, $5 + 3 + 1 = 9$ states; from $^1D_2$, 5 states, or a total of 14. Furthermore, those states require M_J values of $+2, +1, 0, -1, -2; +1, 0, -1; 0$; and $+2, +1, 0, -1, -2$, respectively. These account for all of the M_J values in the table except for one, whose M_J value is zero. This then both permits and requires $^1S_0$ to be present.

By this procedure we have found that the fifteen independent states of two equivalent p electrons are to be identified with the fifteen LS states contained in $^1S_0$, $^1D_2$, and $^3P_{0,1,2}$. By a similar procedure one can deduce the fine-structure terms for other configurations involving equivalent electrons, with the results shown in Table 8-3.

If a given configuration contains both equivalent and nonequivalent electrons, one can analyze the LS coupling fine structure by starting with the equivalent electron terms and combining the other electrons, one by one, with them. Thus for the configuration $4p^2\, 5s$, the LS terms would be

$$^1S + s \rightarrow {}^2S_{1/2}$$
$$^1D + s \rightarrow {}^2D_{3/2,5/2}$$
$$^3P + s \rightarrow {}^2P_{1/2,3/2}, {}^4P_{1/2,3/2,5/2}$$

EXERCISES

8-1. Add up the degeneracies of the various LS terms appearing in Fig. 8-1 and thus verify that this is equal to the degeneracy of the 4p 4d level. *Ans.:* $g = 2^2 \times 3 \times 5 = 60$.

8-2. Add up the degeneracies of the various spin states appearing in Fig. 8-2 and thus verify that this is equal to 2^N, where N is the number of electrons whose spins are to be combined. (Each electron can independently have either of two orientations.) This expression is of course valid only if the exclusion principle does not limit the possible spin orientations.

TABLE 8-3. *LS* TERMS ARISING FROM EQUIVALENT ELECTRONS‡

In this table the numbers in the first column are the total number of independent states corresponding to the configurations given in the second column. In the third column, the preceding superscripts are the multiplicities. The exponents that appear inside the parentheses are the number of distinct terms having the given multiplicity and L value. For example, f^5, f^9: there are four ^{2}P states, five ^{2}D states, etc.

Indep states	Configuration	LS terms
(1)	s^2	^{1}S
(6)	p^1, p^5	^{2}P
(15)	p^2, p^4	^{1}(SD) ^{3}P
(20)	p^3	^{2}(PD) ^{4}S
(10)	d^1, d^9	^{2}D
(45)	d^2, d^8	^{1}(SDG) ^{3}(PF)
(120)	d^3, d^7	^{2}D ^{2}(PDFGH) ^{4}(PF)
(210)	d^4, d^6	^{1}(SDG) ^{3}(PF) ^{1}(SDFGI) ^{3}(PDFGH) ^{5}D
(252)	d^5	^{2}D ^{2}(PDFGH) ^{4}(PF) ^{2}(SDFGI) ^{4}(DG) ^{6}S
(14)	f^1, f^{13}	^{2}F
(91)	f^2, f^{12}	^{1}(SDGI) ^{3}(PFH)
(364)	f^3, f^{11}	^{2}(PD^2 F^2 G^2 H^2 IKL) ^{4}(SDFGI)
(1001)	f^4, f^{10}	^{1}(S^2 D^4 FG^4 H^2 I^3 KL^2 N) ^{3}(P^3 D^2 F^4 G^3 H^4 I^2 K^2 LM) ^{5}(SDFGI)
(2002)	f^5, f^9	^{2}(P^4 D^5 F^7 G^6 H^7 I^5 K^5 L^3 M^2 NO) ^{4}(SP^2 D^3 F^4 G^4 H^3 I^3 K^2 LM) ^{6}(PFH)
(3003)	f^6, f^8	^{1}(S^4 PD^6 F^4 G^8 H^4 I^7 K^3 L^4 M^2 N^2) ^{3}(P^6 D^5 F^9 G^7 H^9 I^6 K^6 L^3 M^3 NO) ^{5}(SPD^3 F^2 G^3 H^2 I^2 KL) ^{7}F
(3432)	f^7	^{2}(S^2 P^5 D^7 F^{10} G^{10} H^9 I^9 K^7 L^5 M^4 N^2 OQ) ^{4}(S^2 P^2 D^6 F^5 G^7 H^5 I^5 K^3 L^3 MN) ^{6}(PDFGHI) ^{8}S

‡ Gibbs, Wilber, and White, *Phys. Rev.*, **29**, 790 (1927).

8-3. By considering the combining properties of the spin angular momenta of a number of electrons, establish the *law of alternation of multiplicities* for LS coupling: "The spectral terms of successive elements in the periodic table alternate between even and odd multiplicities."

8-4. What spectral terms result from an electron configuration 3d 4f, assuming LS coupling? Indicate on a sketch the expected spacing of the various components of the multiplets.

8-5. What spectral terms result from an electron configuration 2p 3p 4p, assuming LS coupling? What degree of degeneracy is possessed by each of these levels?

8-6. What spectral terms should be present in the configuration $2p^2$ 3p?

The foregoing description of LS coupling provides a basis for predicting the *spectroscopic character of the ground state of a given atom*. The ground state ought to possess the *highest* values of L and S—and if the valence subshell is less than half full, the *smallest* value of J—that are possible under the limitations of the exclusion principle. (If the valence subshell is more than half filled, the *largest* value of J will lie lowest.)

When the exclusion principle limits the possible combinations of L and S, the ground state is found to be a state of the *highest multiplicity* (largest value of S) that can be formed under the exclusion principle. If more than one value of L is possible for this multiplicity, the *largest* of these values determines the ground state. Then, depending upon whether the valence subshell is less than (or more than) half filled, the smallest (or largest) value of J for this combination of S and L completes the identification of the ground-state term. These rules, which almost without exception are observed to hold, show clearly that the spin-spin correlation effect which tends to make the spins line up parallel to one another is the largest of the perturbation terms, followed by the residual electrostatic effects which tend to line up the orbital angular momenta, and then by the spin-orbit term.

EXERCISE

8-7. Show that the following ground-state terms satisfy the above rules: $B(^2P_{1/2})$, $Sc(^2D_{3/2})$, $Se(^3P_2)$, $Zr(^3F_2)$, $Nb(^6D_{1/2})$, $Pr(^4I_{9/2})$, $Ta(^4F_{3/2})$.

8-4. *j-j* Coupling

Although the LS coupling scheme described above provides a suitable description of the fine-structure splitting of most of the energy levels of many atoms, there is another type of coupling which occurs for some atoms. In its ideal form, this second type represents an *opposite extreme* to ideal LS coupling in the sense that the fine-structure splitting of any given level can be described in terms of a coupling which is a "mixture" of the two extreme types. The second type is called j-j coupling. It occurs in its ideal form if the spin-orbit energy [term (4)] greatly exceeds the residual electrostatic energy and the spin-spin correlation energy. In this case the largest part of the fine-structure splitting is exactly analogous, for each electron, to the spin-orbit splitting described in

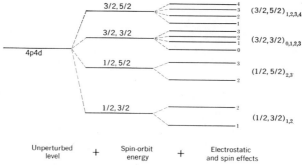

FIG. 8-4. Schematic illustration of the fine structure of a pd configuration in j-j coupling.

Chap. 5 for the one-electron atom. The effects of the various perturbations in splitting the unperturbed energy level can be described as follows:

1. The spin-orbit effects split the original level into a number of well-spaced levels, each of which corresponds to a different combination of the possible j-values for the individual valence electrons. From Eq. 5-8(5) it follows that the lowest-lying of these levels will correspond to all of the electrons having their smaller j-value ($j_i = l_i - \frac{1}{2}$). The number of levels depends upon the number of valence electrons, their orbital angular momenta (only *one* j value if $l = 0$), and the exclusion principle, and it cannot exceed 2^v.
2. The electrostatic energy and spin-spin correlation energy then further split each of the above levels into a number of levels characterized by different values of the total angular momentum J. Here again the exclusion principle may act to limit the number of possible values of J.

We thus see that a given final level is characterized by the j-values of the individual valence electrons and by the total value of J. The various levels are commonly described by the notation $(j_1, j_2, \ldots, j_v)_J$.

As an illustration of the j-j coupling scheme, let us consider the same case as was described in Fig. 8-1 for LS coupling. Figure 8-4 shows schematically how the expected fine structure for ideal j-j coupling is produced by the perturbation terms. The 4p electron can have the j-values $\frac{1}{2}$ or $\frac{3}{2}$, and the 4d electron the j-values $\frac{3}{2}$ or $\frac{5}{2}$. Thus the spin-orbit effect splits the unperturbed level into *four* levels, of which

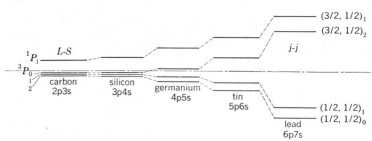

Fig. 8-5. Illustrating the gradual trend from ideal *LS* coupling in the light elements toward *j-j* coupling in the heavy elements. (Adapted by permission from H. E. White, "Introduction to Atomic Spectra," McGraw-Hill Book Company, Inc., New York, 1934.)

($\frac{1}{2}$, $\frac{3}{2}$) lies lowest and ($\frac{3}{2}$, $\frac{5}{2}$) highest. These four levels are then further split by the electrostatic and spin-spin correlation energies into the new levels shown. In each case the number of levels is equal to the number of integrally spaced values of *J* that can be formed out of the two *j*-values.

Note that the *total number* of final levels is the same as for the *LS* coupling scheme and that the *J*-value designations are also in one-to-one correspondence for the two schemes. Because of the highly schematic nature of the above example, it is not possible to define which of the above levels corresponds to each of the levels in the *LS* coupling scheme. The gradual shift in the coupling for certain levels from *LS* coupling for the light elements toward *j-j* coupling for the heavy elements has been studied for several groups of elements. One of the best illustrations of this is provided by the carbon group of elements. Figure 8-5 shows schematically this change for the lowest-lying *ps* excited level in each of the members of this group.

In actual fact, one seldom finds that the fine structure of a given configuration is exactly describable in terms of ideal *LS* or *j-j* coupling, because the relative magnitudes of the spin correlation, electrostatic, and spin-orbit energies do not satisfy the requirements of the idealized theory. Fortunately, by far the majority of cases are sufficiently close to ideal *LS* coupling that the *LS* notation is appropriate both qualitatively and quantitatively. Several cases falling between the ideal limits have been successfully treated, so that the observed departures from ideal *LS* or *j-j* coupling are well understood theoretically. Accounts of such treatments can be found in advanced textbooks on spectroscopy.

8-5. Selection Rules

The two coupling schemes described above produce a quite satisfactory semiquantitative basis for interpreting the energy levels of multielectron atoms. In order to understand the *spectral lines* that result from transitions between these levels, however, we must also know what *selection rules* govern the most common types of transition. The selection rules given below are those for *electric-dipole* transitions. Most of them can be derived by using only the general properties of the angular-momentum eigenfunctions and do not require a knowledge of the complete wave functions. A few depend upon the assumption that the valence electrons have only a weak interaction with each other. Note that they are very closely similar to the selection rules derived in Chap. 6 for the one-electron atom.

A. Selection Rules for LS Coupling

1. Transitions occur only between configurations in which *one electron* changes its state. (Only one electron "jumps" at a time.)
2. The l-value of the jumping electron must change by one unit

$$\Delta l = \pm 1$$

This is a special case of the general requirement that *the parity of the wave function must change in an electric-dipole transition*. Terms arising from configurations of odd parity are so designated by a superscript o, that is, $^3P_2^o$.

3. For the atom as a whole, the quantum numbers L, S, J, and M_J must change as follows:

$$\Delta S = 0 \qquad \Delta L = 0, \pm 1$$
$$\Delta J = 0, \pm 1 \qquad \text{but } J = 0 \to J = 0 \text{ forbidden}$$
$$\Delta M_J = 0, \pm 1 \qquad \text{but } M_J = 0 \to M_J = 0 \text{ forbidden if } \Delta J = 0$$

B. Selection Rules for j-j Coupling

1. Transitions occur only between configurations in which *one electron* changes its state. (Only one electron "jumps" at a time.)
2. The l-value of the jumping electron must change by one unit; or, more generally, *the parity must change*.
3. $\Delta j = 0, \pm 1$ for the jumping electron, and $\Delta j = 0$ for all the other electrons.
4. For the atom as a whole,

$$\Delta J = 0, \pm 1 \qquad \text{but } J = 0 \to J = 0 \text{ forbidden}$$
$$\Delta M_J = 0, \pm 1 \qquad \text{but } M_J = 0 \to M_J = 0 \text{ forbidden if } \Delta J = 0$$

The above selection rules are observed to be in operation in the vast majority of transitions. The commonest violations of these selection rules involve transitions in which *more than one electron changes its state*, or (in *LS* coupling) in which *the spin changes by one unit*. In case more than one electron jumps, the selection rules still require that the *sum* of the individual orbital angular-momentum quantum numbers change by *one unit* for an electric-dipole transition, since the parity of a state is just that of the sum of the l quantum numbers of the electrons.

8-6.   Energy-level Diagrams for Complex Atoms

One of the most satisfactory ways of describing the character of the spectrum of a given atom is by means of an *energy-level diagram* similar to that used to illustrate the fine structure of hydrogen (Fig. 5-6).   In

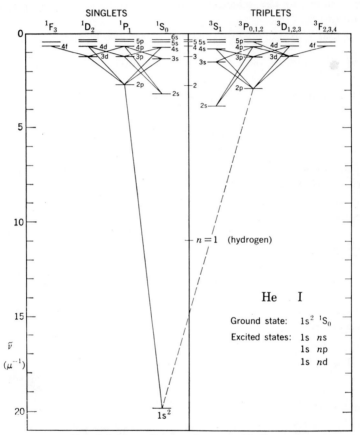

FIG. 8-6. Term diagram for neutral helium.

the following pages, diagrams are given for a number of atoms.   A partial energy-level diagram for a relatively simple "complex" atom is shown in Fig. 8-6 for helium.   This element has, in its ground state, two equivalent electrons in the 1s subshell.   Thus from the foregoing considerations we expect the ground state to be a $^1S_0$ state.   The lowest excited states involve the elevation of one of these electrons into a higher shell, so that the exclusion principle does not affect these states. For two electrons only *singlet* and *triplet* states can occur; and because of the selection rules $\Delta S = 0$, it is convenient to represent these states in *two separate groups*, with permitted transitions occurring only within each group. A transition which violates this selection rule then goes from one

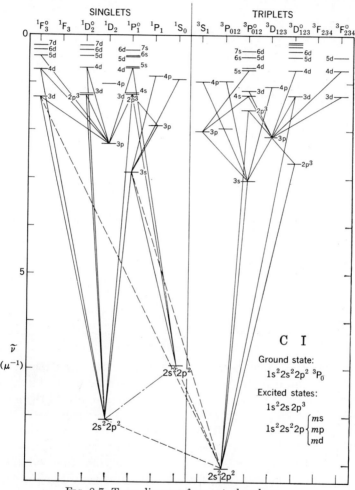

FIG. 8-7. Term diagram for neutral carbon.

group to the other and is called an *intersystem transition*. One of these is shown in Fig. 8-6. It often happens that some of the strongest lines in the spectrum of an atom are intersystem lines, because the lowest level of one group may lie above the ground-state level and the only means of getting to the ground state from this state may be through a magnetic-dipole or electric-quadripole transition. Thus even though the probabil-

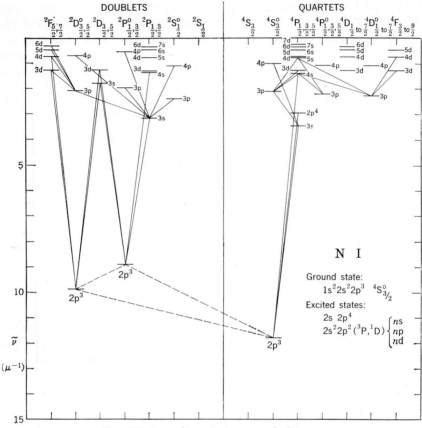

FIG. 8-8. Term diagram for neutral nitrogen.

ity of such a transition is many times smaller than that for an electric-dipole transition, it is still possible to have a strong spectral line, because atoms will continue to fall into this state from above until the number of atoms in it is so great that many transitions will occur per second in spite of the low transition probability per atom.

In Fig. 8-6, it should be noted that the higher excited levels approach those for neutral hydrogen quite closely. The transitions between these levels would thus approximate the corresponding transitions of hydrogen, except that the helium lines would show different fine structure. Energy-level diagrams for other complex atoms are shown in Figs. 8-7 and 8-8.

8-7. Regularities in Complex Spectra

From the earliest investigations of atomic spectra great effort was spent to discover quantitative or qualitative spectral regularities and to relate the spectra of neighboring elements in the periodic table. Many spectral regularities were found empirically and were used as an aid in the problem of determining the correct spectroscopic designations of the various energy levels in an atom. Modern quantum theory has provided an exceedingly satisfactory theoretical basis for these relationships and has itself aided greatly in the task of mapping the spectra of the elements. We have already seen how some of these regularities can be derived from the application of the Schroedinger theory to complex atoms. There are several regularities that we have not explicitly examined, however, and it is thus worthwhile to tabulate in a single place the principal regularities that can be observed in atomic spectra.

A. *The Rydberg-series Relationships.* The earliest general quantitative relationships between certain lines in the spectrum of many elements were discovered by Rydberg. Rydberg's series, which were described in Chap. 2, expressed empirically a fact that can also be seen to follow from the Schroedinger theory, namely, that many atomic spectral series have a form similar to the Balmer series of hydrogen. This is reasonable from the standpoint of the Schroedinger theory because, as has already been pointed out, the electric field felt by a single one of the electrons of a neutral atom at large distances from the nucleus approaches an inverse-square field due to a *singly charged* nucleus. Thus the higher excited states of any neutral atom in which only a single electron is excited ought to approach those for hydrogen. This is illustrated in Fig. 8-6 for helium, and it constitutes one of the most striking features of the spectra of the alkali metals.

Of course the lower excited states of an atom do not approach the hydrogen levels very closely, especially in the cases of the lower orbits of small total angular momentum. This regular progression of the more nearly "circular" orbits toward the hydrogen levels can be explained on the basis of the greater penetration of the states of low angular momentum toward the nucleus, where the shielding effect of the inner electrons is smaller. This effect, too, is quite prominent in the alkali spectra.

B. *The Hartley Law of Constant Doublet Separation.* Even before Rydberg's work in expressing series in analytic form, Hartley (1883) discovered that the separation between corresponding members of a multiplet line in the various members of a spectral series was *constant* when measured in terms of *reciprocal wavelength;* that is, the components of a doublet or triplet series are separated by the same frequency difference. This law is an obvious consequence of the idea of energy levels with the Einstein condition relating the frequency of emitted radiation to the energy difference of two levels, since if one energy level is, say, a doublet, then all transitions that begin or end on this level will yield two sets of lines, separated in frequency by just the energy difference of the level divided by h. Clearly, this law will also apply to widely spaced energy levels.

C. Similarities between Members of the Same Chemical Group. It was recognized very early in the study of spectra that close relationships exist between the spectra of chemically similar elements. As we have seen, the various groups of chemically similar elements are characterized by similar valence-electron configurations, and we can now appreciate that these elements will be characterized by similar electron configurations, and also by similar fine-structure splitting, for the various *excited levels*, with a possible gradual trend in the character of the coupling from one member of the group to the next.

D. The Alternation Law of Multiplicities. The alternation law of multiplicities states that spectral terms corresponding to successive (neutral) atoms in the periodic table alternate between even and odd multiplicities. This law was recognized empirically before the discovery of electron spin. As we have seen (Exercise 8-3), it is a consequence of the combining properties of the spins of the various valence electrons.

E. The Displacement Law. The displacement law was first stated by Kossel and Sommerfeld (1919): "The spectrum and energy levels of a neutral atom closely resemble the spectrum and energy levels of a *singly ionized* atom of one unit higher atomic number." This law clearly follows from the fact that the electron configurations of two such atoms would be the same. The main difference between the two cases results from the fact that one atom has one more unit of charge than the other, so that the field felt by an outer electron will be twice as great for the ionized atom as for the neutral one. We have already seen the consequences of this for the case of ionized He vs. neutral H: corresponding lines are systematically shifted toward the violet in the spectrum of the ion.

F. Isoelectronic Sequences. The displacement law can also be applied to a *series* of adjacent elements, each ionized to a successively greater degree. Such a series of atoms, whose electron configurations are identical, is called an *isoelectronic sequence*. The spectra of the various members of an isoelectronic sequence of atoms show the same type of fine-structure splitting, similar relative intensities for the various corresponding transitions, etc.

8-8. The Zeeman Effect

The analysis of the Zeeman effect given in Chap. 2 on the basis of the classical electron theory appeared to provide a valid theoretical basis for the "normal" Zeeman effect but was unable to explain the "anomalous" behavior of many spectral lines when the radiating atom is located in a magnetic field.   One of the greatest triumphs of the quantum theory lies in the beautiful quantitative interpretation it provides for the entire Zeeman effect.   We shall now briefly analyze the Zeeman effect on the basis of the Schroedinger theory.

The "Zeeman splitting," as it is called, of the energy levels of an atom results from the interaction energy of the net *magnetic moment* of the atom with an external magnetic field.   If we denote this magnetic moment by the vector $\mathbf{\mu}$, the orientation energy of $\mathbf{\mu}$ in an external field $\mathbf{B}$ is, from classical electromagnetic theory,

$$W = -\mathbf{\mu} \cdot \mathbf{B} \tag{1}$$

and since this energy is very small (that is, Zeeman splitting is ordinarily much smaller than ordinary fine-structure splitting) we may introduce this term into the Hamiltonian function of the atom as a *small perturbation*.

In order to apply the perturbation theory, we must now determine a suitable form for the *operator* corresponding to W.   This is most easily done if we assume that the levels whose splittings are to be analyzed are described by LS coupling.   We may express the net magnetic moment of the atom as the vector sum of the *orbital* magnetic moments and *spin* magnetic moments of the individual electrons [see Eqs. 5-5(3) and (5)]:

$$
\begin{aligned}
\mathbf{\mu} &= -\sum_{i=1}^{v} \left[\frac{1}{2}\left(\frac{e}{m}\right) l_i + \left(\frac{e}{m}\right) \mathbf{\delta}_i \right] \\
&= -\frac{1}{2}\left(\frac{e}{m}\right)\sum_i l_i - \left(\frac{e}{m}\right)\sum_i \mathbf{\delta}_i \\
&= -\frac{1}{2}\left(\frac{e}{m}\right)(\mathbf{L} + 2\mathbf{S}) = -\frac{1}{2}\left(\frac{e}{m}\right)(\mathbf{J} + \mathbf{S})
\end{aligned}
\tag{2}
$$

Thus we see that the net magnetic moment *is not necessarily parallel* to the total angular momentum $\mathbf{J}$ of the atom.   This is true because of the *different proportionality factors* connecting the magnetic moment with the orbital and spin angular momenta.   To keep the analysis in the simplest mathematical form, we now make the following physical argument: In the absence of an external magnetic field, the total angular momentum $\mathbf{J}$ is a constant, and $\mathbf{L}$, $\mathbf{S}$, and $\mathbf{\mu}$ "precess" around $\mathbf{J}$ at a rate depending upon the magnitude of the LS coupling.   The frequency of this precession is, in a given case, just the energy shift due to $\mathbf{L} \cdot \mathbf{S}$, divided by h.   Now, when an external field is applied, the total angular-momentum vector will no longer be constant but will precess about the applied-field direction.   For a weak applied field this precession will be *much slower* than the $\mathbf{L} \cdot \mathbf{S}$ precession, so that the *time-average* component

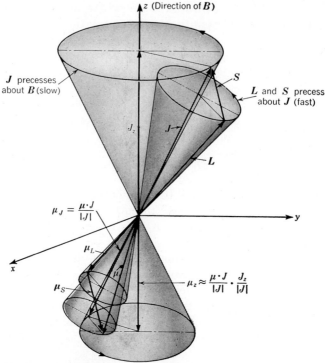

FIG. 8-9. Illustrating the approximations used in treating the Zeeman effect in relatively weak magnetic fields ($B < \sim 10^4$ gauss).

of $\mathbf{u}$ along the field is very nearly equal to the component of $\mathbf{u}$ along J multiplied by the component of J along the field. This is shown in Fig. 8-9. Thus we may write

$$
\begin{aligned}
W &= \frac{1}{2}\left(\frac{e}{m}\right)\frac{[(J + S)\cdot J](J\cdot B)}{|J|^2} \\
&= \frac{1}{2}\left(\frac{eB}{m}\right)\frac{(|J|^2 + J\cdot S)J_z}{|J|^2}
\end{aligned}
\tag{3}
$$

By the same procedure as was used to arrive at Eq. 8-3(5) we now may write

$$
J\cdot S = \tfrac{1}{2}(|J|^2 + |S|^2 - |L|^2)
\tag{4}
$$

so that we have for W,

$$
W = \frac{1}{2}\left(\frac{eB}{m}\right)\frac{[|J|^2 + \tfrac{1}{2}(|J|^2 + |S|^2 - |L|^2)]J_z}{|J|^2}
$$

The *operator* for W is thus just the operator for the quantity on the right. Since $|J|^2$, $|L|^2$, $|S|^2$, and J_z all commute with one another, there is no problem regarding the order of appearance of the various quantities, and we may write down immediately the first-order shift in the state defined by the quantum numbers L, S, J, M_J:

$$\Delta E = W = +\frac{1}{2}\left(\frac{eB}{m}\right)$$
$$\frac{\{J(J+1) + \frac{1}{2}[J(J+1) + S(S+1) - L(L+1)]\}M_J\hbar}{J(J+1)}$$
$$= +\frac{1}{2}\left(\frac{e\hbar B}{m}\right)\left[1 + \frac{J(J+1) + S(S+1) - L(L+1)}{2J(J+1)}\right]M_J$$
$$= +\frac{1}{2}\left(\frac{e\hbar B}{m}\right)gM_J \tag{5}$$

This equation indicates that a level described by quantum numbers L, S, and J will be split by the external magnetic field into a number of equally spaced levels corresponding to the various possible values of the quantum number M_J. The number of such values is $2J + 1$, since M_J takes on the integrally spaced values from $-J$ to $+J$. Thus the Zeeman effect removes the degeneracy of the levels with respect to M_J.

The quantity

$$g = 1 + \frac{J(J+1) + S(S+1) - L(L+1)}{2J(J+1)} \tag{6}$$

is called the *Landé g-factor*. If the proportionality factors connecting the magnetic moment with the orbital and spin angular momenta were equal, the g-factor would be a *constant*, independent of L, S, and J.

The splitting of the various LS levels as described by Eq. (5) into $2J + 1$ equally spaced components might still lead to a *spectral line* splitting into only *three* components as in the classical theory, if the g-factors of both of the levels involved in a transition were the same, because of the selection rule $\Delta M_J = 0, \pm 1$. This can be seen by writing the energy shift for two levels, 1 and 2, in the form

$$\Delta E_1 = \frac{1}{2}\left(\frac{e\hbar B}{m}\right)g_1 M_{J_1} \quad \text{and} \quad \Delta E_2 = \frac{1}{2}\left(\frac{e\hbar B}{m}\right)g_2 M_{J_2}$$

so that the frequency shift of a spectral transition from 1 to 2 would be

$$\Delta\omega_{12} = \frac{\Delta E_1 - \Delta E_2}{\hbar}$$
$$= \frac{1}{2}\left(\frac{eB}{m}\right)(g_1 M_{J_1} - g_2 M_{J_2}) \tag{7}$$

Thus if $g_1 = g_2$, then $\Delta\omega_{12} = \frac{1}{2}(eB/m)g_1 \Delta M_J$, giving just the three components $+\frac{1}{2}(eB/m)g_1, 0, -\frac{1}{2}(eB/m)g_1$ to the split line, in qualitative agreement with the classical theory.

It is therefore the fact that the *Landé g-factor varies* from one level to another, and not necessarily the fact that each level is split into $2J + 1$ components, that leads to an "anomalous" Zeeman pattern.

EXERCISES

8-8. Find the Zeeman structure of a spectral line which results from the transition $^4F_{3/2} - {}^4D_{5/2}$.

8-9. (*a*) By what factor will the total spread of the Zeeman pattern of the transition $^{10}H_{3/2} - {}^{10}G_{1/2}$ exceed the classical value? (*b*) How many lines will appear in the Zeeman pattern of this transition? *Ans.:* (*a*) $21\frac{1}{15}$. (*b*) 6.

8-10. A certain spectral line is known to result from a transition from a ^3D level to another level whose LS designation is unknown. The Zeeman pattern of the line is shown (to scale) in Fig. 8-10. Find (*a*) the J-value of the upper level and (*b*) the LS designation of the lower level.

FIG. 8-10.

The Schroedinger theory of the Zeeman effect correctly predicts both qualitatively and quantitatively the pattern into which any given spectral line should be split by a magnetic field. Inasmuch as each different transition possesses a characteristic pattern that may be quite complex, this fact strongly confirms the correctness of quantum mechanics, including electron spin. Indeed, much of what we now know of the spectroscopic designations of the various energy levels of atoms was either established by or checked against the observed Zeeman patterns of the spectral lines. As an illustrative example, Fig. 8-11 depicts the Zeeman splitting of the line $^3P_1 - {}^3D_2$.

8-9. The Excitation of Atoms

In our discussion of the spectroscopic transitions between the various excited levels of atoms we have almost completely ignored the question of the processes by which an atom is raised to an excited state. It is of course a familiar fact that emission spectra are observed whenever the atoms involved are in a flame or in an electric arc or spark, but we have not examined the *fundamental mechanism* by which excitation takes place in these cases. In order to give some idea of the mechanisms by which such excitation can take place, some of the commoner of these processes are listed below:

A. Excitation by the Absorption of Light. The analysis given in Chap. 6 concerning the effect of an electromagnetic field upon an atom has shown that an atom may be excited by absorbing a quantum from the electromagnetic field. In such a process the atom may reach with high probability any of the excited levels related to its initial level by an electric-dipole transition, provided that light of the appropriate frequency is available to induce this absorptive transition. A familiar case in which such an effect occurs is at the cool boundary of a flame or other source of excitation. The atoms in such a region are generally in their ground states, and radiation from the excited atoms in the center of the flame, emitted as they return to their ground states, passes through this outer region and thus provides an intense source of radiation of *just the right frequencies* to induce absorptive transitions in the outer atoms. The "cool" atoms, thus excited, later reradiate the energy they have absorbed, but this reradiation takes place in *all directions*, so that there may be a net *loss* of intensity of radiation proceeding in the original direction. This is the origin of the so-called *reversal spectrum* of a cool vapor.

B. Excitation by Collision. In a flame or other high-temperature gaseous region, the atoms may have sufficient kinetic energies to raise themselves or other atoms into excited states when they collide. Such a collision is an *inelastic* collision in the mechanical sense, since kinetic

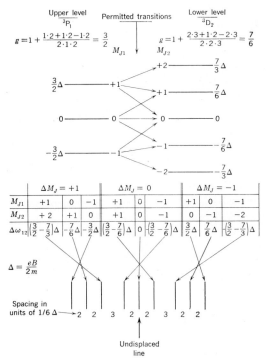

Fig. 8-11. Illustrating the Zeeman splitting of a spectral line corresponding to the transition $^3P_1 - {}^3D_2$.

energy of motion of the atom as a whole is transferred into "nonmechanical" energy of excitation of the atom. The transitions which can be induced by such means are not subject to the ordinary selection rules, of course, so that an atom may be left in a state of excitation from which it would be impossible to return to the ground state by electric-dipole transitions alone. Conversely, if an atom finds itself in a state of excitation whose mean lifetime is sufficiently long (i.e., one having a relatively small transition probability to leave this excited state) the atom may be removed from this state *by collision* before it can undergo a radiative transition. Certain transitions which have been observed spectroscopically (in gaseous nebulae) are so highly forbidden that their mean lives are *many seconds*. Such transitions cannot be observed in the laboratory because the atom would be removed from the excited state by collision long before it could radiate.

C. *Excitation by Electron Bombardment.* Somewhat related to the previous mechanism is that of excitation by electron bombardment. In this case, however, the temperature need not be very high, and the atoms need not be in equipartition equilibrium with the electrons. The electrons may themselves be produced independently of the atoms to be excited, e.g., by thermionic emission from a hot body, or may be produced as a by-product of the excitation process itself. In any case the electrons are usually accelerated by electric fields to high enough energy that they become capable of exciting an atom by colliding with it. This is one of the most important mechanisms of excitation in operation in an electric arc, spark, or low-pressure electric discharge.

Some of the early quantitative work on the excitation energies of atoms was done by bombarding gaseous atoms at low pressure with electrons of carefully controlled energy, and observing at what energies *inelastic scattering* of the electrons took place, as evidenced by a decrease in current to a collection electrode (Franck and Hertz, 1914).

D. *Ionization.* As a special case of some of the above mechanisms, an atom may be *ionized* by photoelectric absorption of a high-energy quantum or by an especially violent collision with another atom or particle. If a supply of free electrons is then available, one of these may come close enough to an atom to be captured into one of the higher bound states, and from there proceed downward toward the ground state in one or more "jumps."

8-10. The Breadth of Spectral Lines

In the discussion of the excited states of atoms so far, it has been assumed that each energy level is infinitely "sharp," that is, that each energy level corresponds to a state in which the energy of the atom is *precisely defined.* Thus one would expect that a transition between these levels would yield a photon of *precisely defined wavelength.* It is actually observed, however, that some transitions lead to very "broad" or "diffuse" lines, that is, to photons whose wavelengths are not precisely defined, but vary over a range of wavelengths. A spectroscope may be regarded as a device which resolves a photon into its various wavelength components, so that a "line profile" in which the "brightness" of the line is plotted vs. the wavelength is actually an experimental measurement of $\phi^*\phi$ for the incoming photons. We shall now consider briefly some of the major causes of this so-called "line broadening."

A. Unresolved Fine Structure. One reason why some spectral lines appear to be broad rather than narrow and sharp is that the spectroscope may not possess sufficient power to resolve these lines into their several fine-structure components. This effect, although of great practical importance, cannot be regarded as an important cause of line broadening in any fundamental sense, and we shall not consider it further.

B. Finite Lifetimes of the Excited States. Probably the most important source of line broadening from the standpoint of its theoretical implications is that due to the finite time that an atom is likely to remain in an excited state. The analysis of the interaction of an atom with an electromagnetic field given in Chap. 6 can be extended for the case of spontaneous emission to apply to finite time intervals, for which the decay of the "occupation probability" $a_n^* a_n$, of the original state must be included. It is then found that this probability *decreases exponentially* with time, with a characteristic time constant which depends upon the probabilities of transition into all of the lower-lying states. In order to measure the energy of the system with *high precision*, the indeterminacy principle demands that we use some means of measurement which does not limit the *time interval* over which the measurement is made. If the system is so perturbed, however, that it is unlikely to remain in a given state for a very long time, we must make the measurement of its energy in a *short time*, and thus accept a certain *indeterminacy* in our knowledge of its energy. Thus the shorter the time available for measuring the energy, the greater is the uncertainty in the energy.

The above schematic analysis can be carried out analytically, using the perturbation theory, and it is found that the profile of a spectral line due to a given finite "decay lifetime" of a given state is a so-called "resonance" curve with a "width" inversely proportional to the decay lifetime of the state. That is, the intensity of light having quantum energies between E and $E + dE$ is

$$dI = I(E)\, dE = \frac{I_0 \Gamma \hbar^{-1}\, dE}{\Gamma^2/4 + (E - E_0)^2/\hbar^2} \tag{1}$$

where Γ is the reciprocal of the mean life of the state, E_0 is the average photon energy, and I_0 is the total rate of emission of energy in the transition.

EXERCISES

8-11. A similar formula can be derived classically for a "damped wave train" whose electric vector is, at a fixed point,

$$E = E_0 e^{-\Gamma t/2} \sin \omega_0 t \qquad t > 0$$

and $\qquad\qquad E = 0 \qquad\qquad\qquad\qquad t < 0$

Analyze this wave train into its Fourier components, and thus show that such a wave train has a distribution of energy as a function of frequency similar to that of Eq. (1). *Hint:* Make use of the fact that Γ is ordinarily $\ll \omega_0$.

8-12. If the "width" $\Delta \bar{\nu}$ of a certain level is observed spectroscopically to be $10^{-4} \mu^{-1}$, what is the approximate lifetime of the state, in seconds?

C. Doppler Broadening. If the atoms of a gaseous source are at high temperature, a *spread* in the frequency of a given transition is produced by their randomly oriented velocities with respect to the spectrograph, as a result of the Doppler effect. This cause of line broadening is found to be sufficiently important that, for work requiring very high resolution, a low-temperature light source is used in order to keep the velocities as small as possible.

EXERCISES

8-13. The probability that the speed of a molecule of an ideal gas lies between v and $v + dv$ is given by the Maxwellian distribution

$$dP = 4\pi \left(\frac{\mu}{2\pi R T} \right)^{3/2} e^{-\mu v^2 / 2RT} v^2 \, dv \qquad (2)$$

where μ is the molecular weight, R is the universal gas constant, and T is the absolute temperature (see Exercise 10-20). Find an expression for the intensity profile of a spectral line of central frequency ν_0 which originates in such a gas.

Ans.: $\qquad I(\nu) = \frac{I_0}{\nu_0} \left(\frac{\mu c^2}{2\pi R T} \right)^{1/2} \exp \frac{-\mu c^2 (\nu - \nu_0)^2}{2 R T \nu_0^2} \qquad (3)$

8-14. Find the approximate magnitude of the Doppler broadening for an argon glow tube whose temperature is 300°K. Assume a wavelength of 0.5 μ for the radiation.

D. Collision Broadening. If the lifetime of a state is long compared to the time between successive collisions of the radiating atom with other atoms of a gaseous source, the atom may be disturbed sufficiently by the collisions that the successive "pieces" of the emitted photon do not form a single, coherent wave train. In this case the observed line breadth will correspond more nearly to the time between collisions than to the "natural" lifetime of the state, since the entire photon will effectively be a series of shorter, independent wave trains.

EXERCISE

8-15. At what pressure in centimeters of mercury should collision broadening and Doppler broadening become comparable in magnitude for the argon source of the previous problem?

E. Stark Broadening. In electric-discharge tubes, particularly those operating at high pressure and high voltage, an atom finds itself in the rather strong electric field due to a neighboring ionized atom. This electric field causes a perturbation of the energy levels of the radiating atom by the Stark effect (the electric analogue of the Zeeman effect) and thus produces a shift in the frequency of a given transition. The random nature of the situation leads to an average broadening of the spectral line. This effect can be so strong—as in high-pressure mercury lamps—that the emitted spectrum appears almost continuous.

8-11. Conclusion

The material discussed in this chapter may be regarded as the culmination of our understanding of the electronic structure of individual atoms. Most of the phenomena that have been treated here are actually in a sufficiently advanced state of development that they no longer merit great experimental or theoretical effort on the part of the physicist, but since our present objective is to examine what *is known* about atomic structure, rather than what is *not known*, these subjects are actually the most important ones for our purposes. It can be said that *every known feature of the electronic structure of atoms finds an interpretation in terms of the Schroedinger theory whose quantitative validity is limited only by purely mathematical difficulties or by the admitted limitation of the theory to nonrelativistic situations.* In view of the tremendous range of phenomena which involve individual atoms, this statement is a strong one indeed. Furthermore, we shall next find that the applicability of the theory is by no means limited to individual atoms but that it also provides an extraordinarily complete interpretation of chemical phenomena as well.

REFERENCES

Herzberg, G.: "Atomic Spectra and Atomic Structure," Prentice-Hall, Inc., Engle-wood Cliffs, N.J., 1937.

Pauling, L., and E. B. Wilson, Jr.: "Introduction to Quantum Mechanics," McGraw-Hill Book Company, Inc., New York, 1935.

Rojanski, V.: "Introductory Quantum Mechanics," Prentice-Hall, Inc., Englewood Cliffs, N.J., 1938.

Schiff, L. I.: "Quantum Mechanics," 2d ed., International Series in Pure and Applied Physics, McGraw-Hill Book Company, Inc., New York, 1955.

White, H. E.: "Introduction to Atomic Spectra," International Series in Pure and Applied Physics, McGraw-Hill Book Company, Inc., New York, 1934.

The following tables are very useful in the quantitative analysis of spectra:

Moore, C. E.: "Atomic Energy Levels," National Bureau of Standards Circular No. 467, Government Printing Office, Washington 25, D.C.

Moore, C. E.: "A Multiplet Table of Astrophysical Interest," rev. ed., Princeton University Press, Princeton N.J., 1945.

9

Molecular Binding and Molecular Spectra

Having examined the most important features of the electronic structure of atoms, we turn to a problem that may be considered to be the next logical step in our study of the quantum-mechanical properties of more and more complicated aggregates of particles. The next most complicated type of system, and one which exhibits some new and interesting properties, is that of two (or at most a few) atoms which interact with one another to form a *molecule*.

The study and systematization of the chemical properties of the elements has been, since the middle of the last century, one of the most active branches of experimental science, and many fundamental properties of matter were brought to light in the course of such investigation. In spite of the great advances that were made in the practical knowledge of the chemical properties of the elements, however, the true nature of the elementary interactions between atoms which cause them to enter into chemical relationships remained almost completely unknown until the introduction of the modern form of the quantum theory. It is true that a certain degree of success appeared to have been attained in the understanding of the forces acting between electrochemically dissimilar elements (e.g., NaCl) and that certain empirically successful expressions for the interatomic forces in various compounds had been worked out by 1910, but the *fundamental relationships* between the interatomic forces and the forces acting within the atom itself were not well understood until about 1928.

The chemical binding that exists between two elements in a compound varies with the identity of the two elements between two extreme forms: One, which occurs between highly dissimilar elements (e.g., NaCl, CaF$_2$), is called *ionic*, or *heteropolar*, binding. The other, which exists between identical or quite similar elements, is called *homopolar*, or *covalent*, binding (e.g., SiC, O$_2$, H$_2$). This latter type of binding is for us the more interesting of the two because it is active in a great many phenomena of physical interest. Therefore we shall concentrate our attention almost exclusively upon it. Since calculations of the precise numerical values of most of the properties of even the simplest molecular systems are quite complicated, our discussion must remain somewhat qualitative.

9-1. The Hydrogen Molecule Ion

The simplest example of chemical binding that occurs in nature is that of the hydrogen molecule ion, in which a single electron is "shared" between two protons. The exact solution of this problem has not been found, except for the case in which the motion of the protons is neglected, and even in this special case the mathematical equations are best treated through the use of approximation methods. Fortunately, many of the most important properties of this molecular system can be established by using only certain qualitative characteristics of the system, and our discussion will be limited mainly to these aspects of the problem.

A. The Hamiltonian Operator, the Adiabatic Approximation. We first write down the Hamiltonian function for the complete system, including the motion of the protons but neglecting all except the electrostatic interactions:

$$H = \frac{P_1{}^2}{2M} + \frac{P_2{}^2}{2M} + \frac{p^2}{2m} + \left(\frac{e^2}{4\pi\varepsilon_0}\right)\left(\frac{1}{r} - \frac{1}{r_{e1}} - \frac{1}{r_{e2}}\right) \qquad (1)$$

where $\boldsymbol{P}_1$, $\boldsymbol{P}_2$, and $\boldsymbol{p}$ are the momenta of the two protons and of the electron and r, r_{e1}, and r_{e2} are the distances between the two protons and between each proton and the electron (Fig. 9-1). The motion of the center of mass can be separated from that of the protons and of the electrons relative to the CM, just as in the one-electron atom; we ignore the CM motion. We then make use of the fact that the protons are *much more massive* than the electron to conclude that the protons will move *much more slowly* than will the electron (since forces of about the same magnitude act on all of the particles). This means two things: First, the *energy* of the system will consist almost entirely of potential energy and of the kinetic energy of the electron; that is, the $P_1{}^2/2M$ and $P_2{}^2/2M$ terms in the Hamiltonian function can be neglected[1] as a first

[1] Born and Oppenheimer, *Ann. Physik,* **84,** 457 (1927).

approximation. Second, the wave function for the motion of the electron will be, at each instant, almost the same as if the protons were motionless; that is, the rapid motion of the electron allows it to "adjust" its state continuously to correspond to the instantaneous positions of the much slower-moving protons. This is called an *adiabatic approximation*,

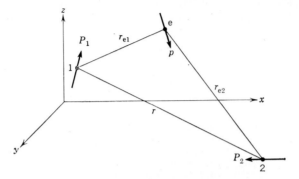

FIG. 9-1. The hydrogen molecule ion.

in which a system is subjected to such a slowly varying outside effect that the system remains instantaneously in stationary equilibrium with it.

The above considerations lead one to seek a solution of the complete problem in two parts. First, the energy levels (and the wave functions, if they are needed) corresponding to the electron moving under the influence of the two *stationary* protons are found. Then the motions of the protons are treated by using as a potential function their electrostatic energy $e^2/4\pi\varepsilon_0 r$ *plus the instantaneous total energy of the electronic motion*. The first of these problems is the one that is involved in the phenomenon of *chemical binding*, while the second is that of the *rotational and vibrational energy states of the molecule*.

Thus we consider first the problem of finding the energy levels of an electron which moves under the influence of two stationary force centers a distance r apart. The Hamiltonian function for this problem is (Fig. 9-2)

$$H_e = \frac{p^2}{2m} - \frac{e^2}{4\pi\varepsilon_0}\{[(x - \tfrac{1}{2}r)^2 + y^2 + z^2]^{-\frac{1}{2}} + [(x + \tfrac{1}{2}r)^2 + y^2 + z^2]^{-\frac{1}{2}}\}$$

(2)

Thus the time-independent Schroedinger equation for the electronic energy states is

$$\frac{-\hbar^2}{2m}\nabla^2 u_n - \frac{e^2}{4\pi\varepsilon_0}\{[(x - \tfrac{1}{2}r)^2 + y^2 + z^2]^{-\frac{1}{2}}$$
$$+ [(x + \tfrac{1}{2}r)^2 + y^2 + z^2]^{-\frac{1}{2}}\}u_n = E_n u_n \quad (3)$$

We note first that, if r is very large, the *hydrogen wave functions* 5-4(1) are solutions of this eigenvalue equation; that is, if the protons are far apart, the electron can remain very near one proton or the other, which corresponds to an *ionized* hydrogen atom and a *neutral* hydrogen atom at so great a distance from one another that they do not appreciably interact. (For example, if the electron is near proton 1, r_{e1} is much smaller than r_{e2}. This means that $e^2/4\pi\varepsilon_0 r_{e2}$ can be neglected, or can at

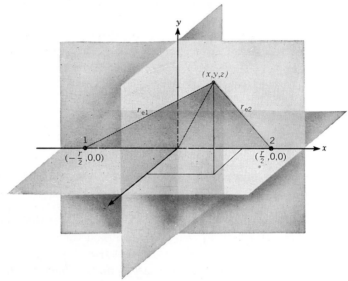

FIG. 9-2. The coordinate system used in the treatment of the hydrogen molecule ion. The two protons are situated on the x-axis at the points $(-\frac{1}{2}r, 0,0)$ and $(+\frac{1}{2}r, 0,0)$, and the electron at the point (x,y,z).

least be regarded as a small perturbation, in comparison with $e^2/4\pi\varepsilon_0 r_{e1}$. The wave functions will then be very nearly the hydrogen wave functions for proton 1.)

B. *Parity.* If r is not so large as to make the hydrogen wave functions a suitable solution of the problem, other considerations must be introduced. One such, of fundamental importance in many physical problems, has to do with that property of wave functions called the *parity*, which was discussed in Sec. 3-4. Since the Hamiltonian operator (3) is an *even function* of the x, y, and z coordinates, the energy eigenfunctions of the system must all be expressible as *even functions or as odd functions* of the coordinates. That is, since

$$H(x,y,z) = H(\pm x, \pm y, \pm z) \tag{4}$$

every nondegenerate energy eigenfunction must have the property that

$$\psi_n(x,y,z) = \pm \psi_n(\pm x, \pm y, \pm z) \tag{5}$$

and degenerate energy eigenfunctions must be expressible as *linear combinations* of even and odd functions of the coordinates. Thus, although the hydrogenic wave functions in which the electron is localized near one proton or the other are suitable solutions of Eq. (3) for large proton separations, the wave functions are degenerate and are not in the necessary symmetric or antisymmetric form with respect to x. But if r is decreased to a value where the hydrogenic wave functions are no longer even approximately correct, we must expect that the degeneracy which is present for large r may be removed and that the wave functions must then necessarily show the required symmetry or antisymmetry with respect to x. It is not difficult to construct eigenfunctions which have the necessary symmetry out of the hydrogenic functions. If $u_n(x',y,z)$ is one of the normalized hydrogenic eigenfunctions having its origin at proton 1 and $u_n(x'',y,z)$ is the same eigenfunction, except for having its origin at proton 2, then

(a) $$u_n^s(x,y,z) = 2^{-1/2}[u_n(x',y,z) + u_n(x'',y,z)]$$

(b) $$u_n^a(x,y,z) = 2^{-1/2}[u_n(x',y,z) - u_n(x'',y,z)]$$

(6)

where $x' = x + \frac{1}{2}r$ and $x'' = x - \frac{1}{2}r$, are normalized wave functions having, respectively, the same parity and opposite parity with respect to x as does $u_n(x',y,z)$ with respect to x'.

EXERCISE

9-1. Verify the above assertion.

The form (6) is an especially appropriate one for our present considerations both because it expresses in a symmetrical way our actual ignorance of *which* of the two atoms the electron may be in and because the possible removal of the degeneracy for smaller values of r will in any case *require* that the resulting *nondegenerate* eigenfunctions have even or odd parity with respect to x.

C. Qualitative Form of the Wave Functions, the Electronic Energy. We are now in a position to consider qualitatively the nature of the electronic energy eigenfunctions for any value of r. Since we shall be primarily interested in the properties of the lowest electronic states, however, we shall now restrict the discussion to the ground state and the first excited state of the molecule. In order to visualize most clearly how the wave functions and energies change with r, let us imagine that the two protons are first at a great distance from one another and that the system is in its ground state, represented by the wave functions (6) with $u_n = u_{100}$. We then imagine that r *decreases slowly to zero*, and we examine the qualitative behavior of u_0^s, u_0^a, E_0^s, and E_0^a as a function of r.

Taking first the wave functions, we can plot the value of the wave function at various points along the x-axis for several values of r, as indicated schematically in Fig. 9-3. This figure illustrates how the combination u_0^s changes from a form characteristic of the ground state of atomic hydrogen for each nucleus (the $2^{-\frac{1}{2}}$ normalization factor expresses the fact that each proton is surrounded by only "half" an electron) to a form, at $r = 0$, corresponding to the ground state u_{100} of *ionized helium*. For intermediate values of r the electron is *quite likely to be found in the region between the two protons*. On the other hand, the combination u_0^a changes from a form characteristic of the ground state of atomic hydrogen for each nucleus to a form, at $r = 0$, corresponding to the excited state u_{210} of *ionized helium*. [The x-axis of the present problem corresponds to the polar axis of the eigenfunctions 5-4(14).] In this case, the electron is *quite unlikely to be found in the region between the two protons* and in particular *will never be found in the plane midway between the protons*.

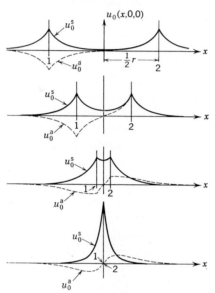

FIG. 9-3. Schematic illustration of the forms of the symmetric and antisymmetric functions u_0^s and u_0^a for various values of the interproton separation r. The ordinate is the value of the wave function at points along the internuclear axis (x-axis).

The variation of the energies of these two states as a function of r is shown schematically in Fig. 9-4.[1] These curves indicate that, whereas the electronic energy of the *odd* state does not change appreciably with r —since it has the same value at $r = 0$ as at $r = \infty$—that of the *even* state *decreases monotonically by a factor of four* as r ranges from very large values toward zero. Since the energy of the electronic state decreases as r decreases, this indicates the existence of an *attractive force* tending to draw the two protons together for the even state. *It is this force that is responsible for the binding of the hydrogen molecule ion.*

D. *Equilibrium Separation and Binding Energy.* To complete our qualitative description of the H_2^+ ion, we must now return to the discussion following Eq. 9-1(1) and evaluate the *effective potential-energy function* under which the protons move. As was mentioned in that discus-

[1] Teller, *Z. Physik*, **61**, 458 (1930).

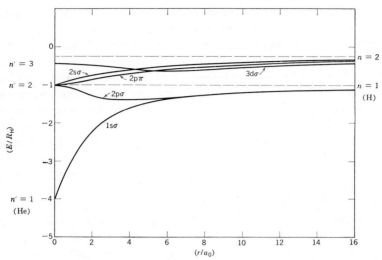

FIG. 9-4. The electronic energy (in Rydbergs) of five states of the hydrogen molecule ion H_2^+ as a function of the interproton distance r (in Bohr radii).

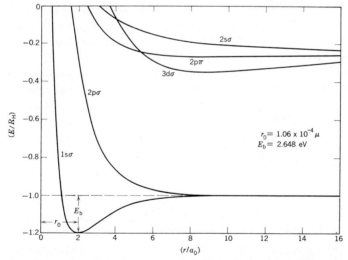

FIG. 9-5. Total energies of the lowest states of H_2^+, obtained by adding the $e^2/4\pi\epsilon_0 r$ repulsive potential energy of the protons to the electronic energy shown in Fig. 9-4.

sion, this is just $e^2/4\pi\epsilon_0 r$ plus the energy of whatever electronic eigenstate is being considered in the problem. We are at present interested in the lowest electronic eigenstates because they are the only ones for which the molecule ion can reasonably be expected to be a stable structure, so that we combine the electrostatic repulsion of the protons with the attractive energy due to the electronic states to obtain a total effective potential energy as shown in Fig. 9-5.

The physical situation represented by Fig. 9-5 is as follows: For large separations of the two protons the energy is essentially constant, but as the interproton distance is decreased, the electronic energy decreases faster than the electrostatic repulsive energy of the protons increases, leading to a net *attraction* between the two protons.   On the other hand, as the interproton separation becomes very small, the electrostatic repulsive energy overtakes and then exceeds the decrease in the electronic energy.   Thus there is a *minimum point* in the net proton potential-energy function at a certain distance of separation r_0.   This minimum represents a *point of stable equilibrium* for the protons, and the system tends to attain this equilibrium separation.   The value of the energy at this equilibrium point minus the energy at infinite separation is then the *binding energy* of the molecule.   (A small allowance must of course be made for the *zero-point* energy of vibration of the protons about this stable equilibrium position.   This motion will be described later.)   Numerical calculations of the equilibrium separation and binding energy agree within a fraction of a per cent with the experimental values.

We thus see that the H_2^+ molecule attains stability in its ground state by virtue of the "sharing" of the single electron between the two protons. It may be said that, because the electron spends most of its time *between* them, the two protons are farther from one another than from the electron, and the attraction of each of the two for the electron exceeds their mutual repulsion.   This semiclassical picture of the one-electron bond can be used as a way of making the attraction between the protons appear reasonable, but it must be emphasized that the full apparatus of quantum mechanics is needed to provide a quantitative description including the existence of a *repulsion* of the protons at close separations.

9-2.   The Neutral Hydrogen Molecule

We shall now briefly consider the case of the *neutral* hydrogen molecule, in which *two* electrons are present.   There are several features of this problem that make it similar to the previous one:

1. The Hamiltonian operator for the system can be divided into three parts, one relating to the motion of the CM, one to the electronic motion for a given proton separation, and the third to the motion of the protons under an effective potential energy consisting of their Coulomb energy and the electronic energy.

2. The Hamiltonian operator for the electronic motion is symmetrical in x (the direction of the line joining the two protons) so that the electronic eigenfunctions will here also have even or odd parity with respect to x.

3. One can express the energy eigenfunctions for large separations as

symmetric or antisymmetric combinations of the eigenfunctions for two neutral hydrogen atoms, and the eigenfunctions corresponding to a negative and positive hydrogen *ion*.

4. The electronic energy of the ground state shows a dependence upon r similar to that of Fig. 9-4, where the energy for large r is now equal to approximately *twice the ground-state energy of atomic hydrogen*, and that for $r = 0$ to the ground-state energy of *neutral helium*.

5. The effective potential energy of the two protons shows a qualitative variation with r similar to that of Fig. 9-5, with a minimum point at some separation r_0' of the protons. The observed value of r_0' is 0.7416Å, and the binding energy is 4.476 eV.

The case of the neutral molecule is thus similar in several respects to that of the ionized molecule. However, there is also an important difference between the two problems: The presence of two *identical particles* in the neutral molecule requires the application of the *exclusion principle*. The electronic eigenfunctions must now not only exhibit either *even or odd parity* with respect to each space coordinate but *must also be antisymmetric with respect to an interchange of the space and spin coordinates of the two electrons*.

These exclusion-principle requirements lead to physical effects similar to those encountered in the previous two chapters, namely, to an effective *repulsion* of the electrons if their spins are parallel and thereby to a *correlation* between the electron spins which tends to align them parallel to one another. These effects are called *exchange correlations*. They have a small effect upon the energy of the system.

We cannot go into the analytical details of the H_2 molecule, but certain features of the *ground state* of the molecule can be described qualitatively. An analysis similar to that used in the study of the H_2^+ molecule leads us to expect the ground-state electronic wave function for the neutral H_2 molecule also to be of *even parity* with respect to the space coordinates, and furthermore to have *zero spin* because it must become the lowest state of the neutral helium atom as the two protons approach one another. Both of these conclusions are observed to hold, so that, just as for H_2^+, the electrons spend most of their time between the protons.

This feature of the H_2 molecule is characteristic of homopolar binding. In this type of bond two electrons are "exchanged" between two similar atoms in a manner described by a wave function which is symmetric in the space coordinates of the two electrons but antisymmetric in the spins.

9-3. Further Properties of Chemical Binding

A. Ionic, or Heteropolar, Binding. In the preceding sections we have seen in a qualitative way the physical basis of the phenomenon of covalent binding between similar or identical atoms. The dominant feature of this type of binding is the symmetrical character of the ground-state electronic wave function with respect to the two atoms, which leads to the concept of equal sharing, or *exchange*, of the two electrons between the atoms.

In contrast to the symmetry of covalent binding, the other extreme type of binding observed to act between dissimilar elements involves a quite *unsymmetrical* distribution of the electron between the two atoms. To study ionic binding, we imagine a molecule such as NaCl and consider the electronic wave function for the system as a function of the separation distance of the two atoms. For large distances of separation, one suitable electronic wave function for the system consists of the ground-state wave functions for the two neutral atoms, and another, of the ground-state wave functions for Na^+ and Cl^-. These two states will of course generally have different energies and will thus be nondegenerate. Also, because of the dissimilarity of the two atoms, the eigenfunctions of the system will not necessarily exhibit strong symmetry or antisymmetry with respect to them. Because of the weak binding of the single s electron in Na and the strong affinity of the Cl atom for an electron to complete its p shell, the energy of the state $Na^+ + Cl^-$ is only slightly greater than that of Na and Cl at large separations. But because of the Coulomb attraction between the ions, the energy of the ionic state decreases as r^{-1}, whereas that of the neutral atoms decreases exponentially. As the separation is reduced, the Coulomb energy decreases appreciably before the exponential variation of the lower state has begun to be felt, and *the ionic energy state falls below that of the neutral atoms.* At separations comparable with the equilibrium value, the ionic energy state lies far below the neutral-atom state.

The degree of ionicity or homopolarity of a given molecule can be defined empirically in terms of whether or not the energy curve of the ionized atoms falls below that of the neutral atoms at a separation significantly greater than the equilibrium value. Given r_0 as the equilibrium separation, the molecule may be said to be predominantly ionic if the crossover point of its energy curves occurs at $r_c > \sim 2r_0$ and predominantly homopolar if $r_c < \sim 1.5r_0$. In NaCl, the crossover occurs at $r_c \approx 5r_0$, so that this molecule is purely ionic; for H_2, no crossover occurs. These features are illustrated in Fig. 9-6.

The binding between the Na and Cl atoms is thus largely due to the *electrostatic attraction* between the Na^+ and Cl^- *ions* (hence the name *ionic* binding). On the other hand, if the separation of the atoms becomes so small that the inner shells of the two atoms "overlap" one another, a strong repulsion between the atoms sets in because of the repulsion between the less well-shielded *nuclei*. Thus the effective potential-energy curve between the two atoms somewhat resembles that for similar atoms, exhibiting a minimum point for a certain separation of the atoms and a binding energy of about the same magnitude as for homopolar bonds.

B. *The "Saturation" of Chemical Bonds.*   One of the most important qualitative features of chemical forces is that of *saturation*, which means that a given atom interacts strongly with only a *limited number* of other atoms.   This saturation property of chemical forces is described empir-

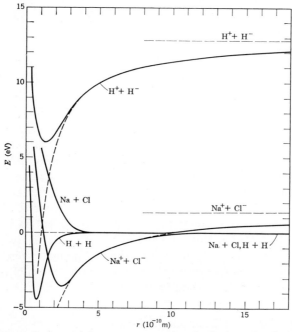

FIG. 9-6. Interatomic potential-energy curves for the ionic molecule NaCl and the homopolar molecule H_2.   The zero of energy corresponds to the neutral atoms situated at infinity.

ically by the chemical term "valence."   In the light of the preceding discussion of chemical binding, it is clear that the physical basis for this saturation property is that *an outer-shell electron is needed for each valence bond a given atom enters into in forming a molecule.*   The valence of an atom is thus limited to values equal to the number of electrons or vacancies in the outermost occupied subshell of the atom.

Another way of describing the saturation character of chemical bonds is in terms of the rapidity of the variation of the effective potential energy between pairs of atoms.   Since the energy varies for large separations as a very high inverse power of the separation distance between two atoms (as $1/r^6$ for van der Waals' forces and *exponentially* for homopolar binding) and since the atoms *repel* one another at very close distances, two atoms can be said approximately not to interact *at all* unless they are actually in "contact" with one another. This clearly restricts a given atom to interact only with its "nearest neighbors" in a compound.

It is the saturation property of chemical forces that makes it meaningful to speak of *molar* properties of a chemical compound, for it is only for strongly saturated forces that the total interaction energy of a large lump of material is *independent of its shape* and *proportional to its mass*. Gravitation is a familiar example of an *unsaturated* force.

C. The Directed-valence Bond. It was mentioned in connection with the one-electron wave functions that the "lobe structures" of the various angular-momentum wave functions are closely related to the chemical bonding properties of an atom. We can now appreciate that, if the ground state of an atom is characterized by a certain geometrical distribution of its valence electrons over the outside of the atom as described in the discussion of the residual-electrostatic-energy term in LS coupling (Sec. 8-3), the atom will tend to maintain this distribution of its valence electrons when it enters into chemical-bonding relationships with other atoms. The most spectacular example of this effect is probably that of carbon, whose four outer electrons are distributed symmetrically about the directions corresponding to the vertices of a regular tetrahedron (such a configuration can be formed by suitable linear combinations of one s and three p wave functions). When carbon enters into chemical bonds with other atoms, this tetrahedral structure has a strong influence upon the geometrical configuration of the compound. The tetrahedral structure of the diamond crystal is another manifestation of the directed-valence structure of the carbon atom.

9-4. Energy States of Diatomic Molecules

We have so far discussed qualitatively some of the quantum-mechanical features of the electronic structure of molecules, and we have seen that the formation of a stable molecule results from a rapid decrease in the electronic energy with decreasing interatomic separation, opposed by the increasing Coulomb repulsion of the atomic nuclei. We shall now endeavor to make use of these qualitative features of molecular structure to discuss in a semiquantitative way the other part of the complete problem, namely, the motions of the *nuclei* under the influence of their Coulomb repulsion and the electronic energy. The Hamiltonian function for this motion has the form, for a diatomic molecule whose nuclei have masses M_1 and M_2 [see Eq. 9-1(1)],

$$H = \frac{P_1{}^2}{2M_1} + \frac{P_2{}^2}{2M_2} + V_n(r) \tag{1}$$

where P_1 and P_2 are the momenta of the two nuclei and $V_n(r)$ is the effective potential energy between the nuclei, having a form qualitatively similar to that shown in Fig. 9-5 for a given electronic energy state n.

This Hamiltonian function can be written in a form in which the CM motion is separated from the motion about the CM, just as was done for the one-electron atom. The Hamiltonian operator for the *relative* motion of the atoms will clearly have the form of Eq. 5-1(6), except that $V_n(r)$ replaces $-Ze^2/4\pi\varepsilon_0 r$:

$$\hat{H}_{\text{rel}} = -\frac{\hbar^2}{2m_r r^2}\left[\frac{\partial}{\partial r}\left(r^2\frac{\partial}{\partial r}\right) + \frac{1}{\sin\theta}\frac{\partial}{\partial\theta}\left(\sin\theta\frac{\partial}{\partial\theta}\right) + \frac{1}{\sin^2\theta}\frac{\partial^2}{\partial\varphi^2}\right] + V_n(r)$$

(2)

where $m_r = (M_1 M_2)/(M_1 + M_2)$.

If the time-independent Schroedinger equation is now written, it is clear that exactly the same separation procedure can be used in the present case as was used for the one-electron atom, and that this procedure will lead to Eq. 5-3(4) with $V_n(r)$ replacing $-Ze^2/4\pi\varepsilon_0 r$.

Thus the present system also is characterized by *quantized orbital angular momentum and quantized z-component of angular momentum.* The radial equation is the only one that differs between the one-electron atom and the present problem. It now reads

$$\frac{1}{r^2}\frac{d}{dr}\left(r^2\frac{dR}{dr}\right) + \frac{2m_r}{\hbar^2}\left[E_s - V_n(r) - \frac{\hbar^2 K(K+1)}{2m_r r^2}\right]R = 0 \qquad (3)$$

where the *rotational quantum number K* corresponds to the quantum number l of the one-electron atom.[1] We next change the dependent variable in this equation to $\chi(r) = rR(r)$, as was done in Exercise 5-9, with the result

$$-\frac{\hbar^2}{2m_r}\frac{d^2\chi}{dr^2} + \left[V_n(r) + \frac{\hbar^2 K(K+1)}{2m_r r^2} - E_s\right]\chi = 0 \qquad (4)$$

EXERCISE

9-2. Carry out the substitution leading from Eq. (3) to Eq. (4).

We thus find that the eigenvalue equation (4) is exactly the same as would be obtained for a *one-dimensional system* consisting of a particle

---

[1] In writing Eq. (3) we assume that the only angular momentum in the problem is that due to the nuclear motion. Actually, of course, there are other possible angular momenta which can play a significant role in the complete problem. The most important of these are the electronic orbital angular momentum L and spin angular momentum S. For simplicity we shall ignore these angular momenta until we have gained some familiarity with the properties of the nuclear motion.

of mass m_r moving along a line under the influence of a potential $U_n(r)$ where

$$U_n(r) = V_n(r) + \frac{\hbar^2 K(K+1)}{2m_r r^2} \tag{5}$$

The second term on the right is sometimes called the *centrifugal potential*.

The typical form of $U_n(r)$ is shown for various angular momenta in Fig. 9-7. From this figure we can see that, for a given value of K, *the function $U_n(r)$ is approximately a simple-harmonic potential* for short

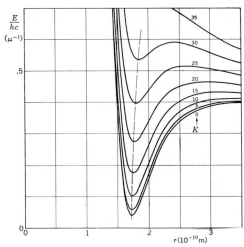

FIG. 9-7. The effective-potential curves for the diatomic molecule HgH. [After Villars and Condon, *Phys. Rev.*, **35**, 1028 (1930).]

distances on either side of its minimum point. We therefore expect the functions $\chi_s(r)$ to be similar to the eigenfunctions for a *harmonic oscillator* and the corresponding energy eigenvalues to be approximately equal to the energy at the minimum point of $U_n(r)$ plus the harmonic-oscillator values. The figure also indicates that the minimum point moves upward in energy and outward in r as K increases, and finally ceases to exist at all. That is, the molecule "stretches" as it rotates, and for sufficiently rapid rotation may even fly apart.

When actual numerical values are inserted for the various quantities appearing in Eq. (5), it is found that the effect of the rotation upon the shape of the potential curve, and therefore upon the frequency of the "harmonic" vibrations and upon the energy levels, is generally rather small for small values of K, so that the major feature of the nuclear motion is the "harmonic" vibration, with the rotational effects appearing as a fine structure upon each vibrational energy level.

The problem can be treated quantitatively in several different ways.

In one method, the potential function $V(r)$ is expanded in a Taylor series about its minimum point at $r = r_0$:

$$V_n(r) = V_{n0} + \tfrac{1}{2}k_{n0}(r - r_0)^2 + A_n(r - r_0)^3 + \cdots \tag{6}$$

where V_{n0}, k_{n0}, A_n, etc., are the constant coefficients of the Taylor-series expansion. Physically, V_{n0} is an additive constant energy which will drop out of expressions involving energy differences for given electronic excitation, and k_{n0} corresponds to the force constant of a harmonic oscillator whose potential function matches the potential $V_n(r)$ at points sufficiently near r_0.

The centrifugal-potential term $\hbar^2 K(K + 1)/2m_r r^2$ is next added to the above potential to yield $U_n(r)$, and one then expands $U_n(r)$ about *its* minimum point. Denoting this minimum point by r_K, we find that r_K increases with K according to the expression

$$\begin{aligned} r_K - r_0 &= \frac{\hbar^2 K(K + 1)}{k_{n0}m_r r_K^3} - \frac{3A_n\hbar^4 K^2(K + 1)^2}{k_{n0}^2 m_r^2 r_K^6} + \cdots \\ &\approx \frac{\hbar^2 K(K + 1)}{k_{n0}m_r r_0^3} \end{aligned} \tag{7}$$

This equation describes the "stretching" of the molecule due to its rotation. The expansion for $U_n(r)$ can be obtained by substituting

$$(r - r_0) \equiv (r - r_K) + (r_K - r_0) \tag{8}$$

into Eq. (6) and expanding $\hbar^2 K(K + 1)/2m_r r^2$ in a Taylor series about r_K. The resulting expression for $U_n(r)$ is

$$\begin{aligned} U_n(r) &= \left[V_{n0} + \frac{\hbar^2 K(K + 1)}{2m_r r_K^2} \right] + \left[\tfrac{1}{2}k_{n0} + 3\left(1 + 2\frac{A_n r_K}{k_{n0}}\right) \right. \\ &\qquad\qquad \left. \frac{\hbar^2 K(K + 1)}{2m_r r_K^4} \right] (r - r_K)^2 + \cdots \\ &= E_{nK} + \tfrac{1}{2}k_{nK}(r - r_K)^2 + \cdots \end{aligned} \tag{9}$$

As was indicated above, we expect the energy eigenvalues of the system to be approximately of the form

$$E_{nKv} = E_{nK} + E_{nv} \tag{10}$$

in which

$$E_{nK} = V_{n0} + \frac{\hbar^2 K(K + 1)}{2m_r r_K^2} \tag{11}$$

and

$$E_{nv} = \hbar \omega_{nK}(v + \tfrac{1}{2}) \tag{12}$$

where

$$\omega_{nK} = \left(\frac{k_{nK}}{m_r}\right)^{1/2} \approx \left(\frac{k_{n0}}{m_r}\right)^{1/2}$$

and v is the quantum number which defines the level of excitation of the "harmonic" vibration.

Several important and interesting features of molecular energy states are exhibited by the above equations:

1. The energy levels of a diatomic molecule are grossly characterized by three quantum numbers, n, K, and v, representing *electronic, rotational*, and *vibrational excitation*, respectively. These levels are degenerate with respect to the z-component of the rotational angular momentum, whose quantum number is M_K.

2. If $v = 0$, the energy levels for fixed n depend upon the *rotational* state of the molecule according to the equation

$$E_{nK0} = E_{nK} + \tfrac{1}{2}\hbar\omega_{nK} \approx V_{n0} + \tfrac{1}{2}\hbar\omega_{n0} + \frac{\hbar^2 K(K+1)}{2m_r r_K^2} \qquad (13)$$

The spacing of the successive rotational levels should follow the Landé-interval rule (Sec. 8-3) because of the $K(K+1)$ term. Thus the levels should become progressively farther apart, the spacing increasingly linearly with K. Aside from the constant term, these levels are the same as those of a *rigid body* of moment of inertia $m_r r_K^2$, and hence they are called the energy levels of a *rigid rotator*. Since in the present case r_K increases with increasing K, the spacing of these levels should become progressively smaller than for an ideal rigid rotator as K increases. Transitions between these levels give rise to the so-called *rotation spectrum* of a diatomic molecule. The radiation falls in the extreme infrared and is thus of such low energy that the electronic energy levels are rarely excited when the rotation spectrum is observed. Thus for the rotation spectrum, $V_n(r)$ is almost always just the electronic ground-state function $V_0(r)$.

3. Transitions which involve both the vibrational quantum number v and the rotational quantum number K give rise to the so-called *vibration-rotation spectrum* of the molecule. This radiation falls in the near infrared part of the spectrum and, like the pure rotation spectrum, usually involves only the electronic ground-state potential function $V_0(r)$. Some features of the vibration-rotation spectrum will be described shortly.

4. The frequency of the vibrational levels slowly changes with increasing K and also with v because of the dependence of the shape of the potential curve $U_n(r)$ upon K and its departure from a true harmonic potential (Fig. 9-7). Both of these effects cause the frequency to *decrease with increasing K and v*.

5. The most general type of transition is one in which the *electronic energy* changes along with the rotational and vibrational quantum numbers for the molecule. This involves a relatively large change in V_{n0} owing to the change in the electronic energy state, so that the radiation may fall in the visible or even in the ultraviolet part of the spectrum.

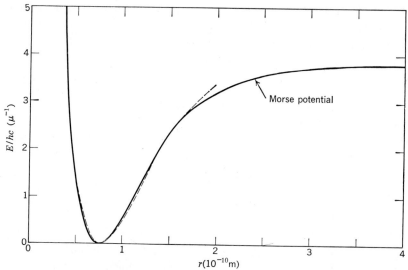

FIG. 9-8. The Morse potential 9-4(14) with B_n, β_n, and r_n adjusted to fit the observed potential curve (dashed) for the ground state of H_2. (Adapted by permission from Herzberg, "Spectra of Diatomic Molecules," copyright D. Van Nostrand Company, Inc., Princeton, N.J., 1950.)

Superimposed upon this large energy change as "fine structure" are the vibrational and rotational effects characteristic of the nuclear motion.

Before proceeding to a discussion of the spectra of diatomic molecules, we shall briefly examine a second method by which the energy levels of such systems have been studied. In this method the potential energy $U_n(r)$ is approximated by an analytic expression which on the one hand is a close approximation to the actual function, and on the other hand is sufficiently simple that eigenfunctions and eigenvalues can be obtained in closed form. Morse[1] proposed an expression of the form

$$U_{n0}(r - r_0) = B_n\{1 - \exp[-\beta_n(r - r_{n0})]\}^2 \qquad (14)$$

which possesses these desired properties. Figure 9-8 shows how closely this function approximates an actual potential curve, when the three parameters r_0, B, and β are properly chosen. The energy eigenvalues corresponding to this function are

$$E_{nv} = \hbar\omega_{n0}(v + \tfrac{1}{2}) - \frac{\hbar^2\omega_{n0}^2}{4B_n}(v + \tfrac{1}{2})^2 \qquad (15)$$

where

$$\omega_{n0} = \beta_n\left(\frac{2B_n}{m_r}\right)^{1/2} \qquad (16)$$

[1] *Phys. Rev.*, **34**, 57 (1929).

is the classical frequency of small oscillations of a mass particle m_r under the influence of this potential and the vibrational quantum number v takes on the *finite number of integral values* lying in the range

$$0 \leq 2v \leq \frac{2(2m_r B_n)^{1/2}}{\beta_n \hbar} - 1 \tag{17}$$

The Morse potential (14) has considerable advantage over the series form (6), since many terms of the latter are required to obtain a reasonable approximation to the actual potential function and the coefficients

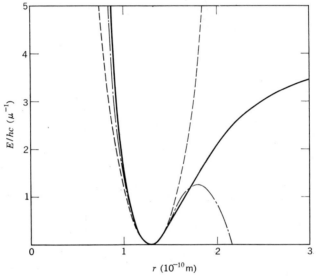

FIG. 9-9. Comparison of the potential expression 9-4(6) with the true potential energy for HCl; the two dashed curves represent the quadratic and the quadratic-plus-cubic approximations. (Adapted by permission from Herzberg, "Spectra of Diatomic Molecules," copyright D. Van Nostrand Company, Inc., Princeton, N.J., 1950.)

of the higher powers of $r - r_0$ do not often converge rapidly enough toward zero to permit the use of ordinary first-order perturbation theory. Figure 9-9 illustrates how poorly an actual potential function is matched by a series of the form (6) which includes only the constant, quadratic, and cubic terms.

On the other hand, while the Morse potential does not fit the actual potential-energy function exactly, the discrepancy is sufficiently small that first-order perturbation corrections to the energy levels are quite accurate. The energy levels exhibited by the Morse potential illustrate quite clearly two additional important properties of the vibrational states of diatomic molecules:

6. There are generally a *finite number* of discrete vibrational levels

below the binding energy B_n and a continuum of levels above this value.

7. The vibrational levels are progressively more closely spaced with increasing values of v, because of the quadratic term in $v + \frac{1}{2}$ in Eq. (15). Experimentally, the levels of a given molecule can usually be described quite accurately by a formula of the form

$$E_v = \hbar\omega_e(v + \tfrac{1}{2}) - \hbar\omega_e x_e(v + \tfrac{1}{2})^2 + \hbar\omega_e y_e(v + \tfrac{1}{2})^3 + \cdots \quad (18)$$

where ω_e is an angular frequency and $x_e, y_e, \ldots$ are coefficients which are to be appropriately chosen. Indeed, one often finds that the first and second terms alone account quite well for most of the levels of a given molecule. We shall use the expression (15) rather than (12) to represent E_{nv} for most of our forthcoming discussion, because it describes more accurately the observed physical properties of diatomic molecules.

9-5. Spectroscopy of Diatomic Molecules

Having briefly investigated the nature of the energy eigenstates of a diatomic molecular system, we are now in a position to consider the kinds of electromagnetic transitions such a system can undergo. In the interests of clarity, it is desirable that we do this in the three stages outlined in (2), (3), and (5) of the previous section, which also happen to correspond to rather widely different wavelength regions and therefore to somewhat different techniques of observation: the pure-rotation spectrum, the vibration-rotation spectrum, and the electronic-vibration-rotation spectrum.

A. The Pure-rotation Spectrum. The simplest kind of transition that a diatomic molecule can undergo is one in which *only the rotational motion changes.* Transitions of this type are of importance mainly under physical conditions in which the molecule is in its electronic ground state, and perhaps in its vibrational ground state as well. Such conditions hold, for instance, if the temperature of the surroundings is sufficiently low. These transitions give rise to the so-called *pure-rotation spectrum* of the molecule.

The radiation which is emitted when a molecule jumps from one state of rotation to a lower state—or which is absorbed when the reverse occurs—may be thought of classically as being principally due to the electric-dipole moment of the rotating molecule. Quantum-mechanically, we know that the probabilities of the commonest transitions depend upon the *matrix elements* of the electric-dipole moment for the various states concerned (Sec. 6-5). If the molecule is one which possesses an intrinsic dipole moment we expect—and indeed it is observed to be true—that electric-dipole transitions can occur easily and will account for nearly all of the observed spectral lines of the molecule; while on the other hand, if a molecule does not possess an intrinsic electric-dipole moment, we expect that electric-dipole radiation cannot be emitted or absorbed, and that such molecules will be relatively inert spectroscopically, at least insofar as the pure-rotation spectrum is concerned.   Molecules of the first type, which possess strong intrinsic electric-dipole moments, tend to be strongly polar molecules in which ionic binding is dominant, for example, KCl, NaBr, MgO, ZnS, and HF. While it is possible that an asymmetric molecule might accidentally possess a vanishing dipole moment, it is principally the so-called *homo-nuclear molecules*, like H_2, O_2, F_2, etc., which by symmetry can have no electric-dipole moment in their electronic ground states, that represent the latter class.

The selection rules for electric-dipole rotational transitions are quite analogous to those for atomic transitions.   For the case now being considered, these selection rules are simply

(*a*) $$\Delta K = +1 \quad \text{(absorption)}$$
$$= -1 \quad \text{(emission)} \tag{1}$$
(*b*) $$\Delta M_K = \pm 1 \quad \text{or} \quad 0$$

and it is, of course, essential in addition that the molecule possess an intrinsic dipole moment.

As to the type of spectrum which should be observed, we insert the above selection rules into Eq. 9-4(13) and obtain

$$\omega_K = \frac{E_{K+1} - E_K}{\hbar} = \frac{(K + 1)\hbar}{m_r r_K{}^2} \tag{2}$$

The possible rotational transitions of a diatomic molecule are thus seen to consist of a number of almost equally spaced lines,[1] each being associated with a different value of the rotational quantum number $K + 1$ of the upper state.[2]

Figure 9-10 shows the rotational energy levels of a diatomic molecule together with the transitions which correspond to Eq. (2).

EXERCISES

9-3. Derive Eq. (2).

9-4. Evaluate the classical rotational angular velocity of a diatomic

---

[1] The increase in r_K with increasing K increases the moment in inertia of the molecule and thereby progressively depresses the higher energy states and reduces the frequencies below the values given by Eq. (2).   (See Exercise 9-5.)

[2] Which of these lines are present in a given physical situation, and their relative intensities, of course depends upon how the molecules are distributed in rotational energy, i.e., upon their *temperature*.   This subject is treated in the following chapter.

molecule whose energy is $\hbar^2 K(K+1)/2m_r r_K{}^2$, and thus show that the values given by Eq. (2) lie between the classical angular velocities of the upper and lower states.

9-5. Use the expression 9-4(7) to evaluate the approximate effect of the stretching of the molecule upon the emitted frequency. Show that ω_K is very nearly equal to

$$\omega_K = \frac{\hbar(K+1)}{m_r r_0{}^2}\left[1 - \frac{2\hbar^2 K(K+1)}{k_{00} m_r r_0{}^4}\right] \tag{3}$$

9-6. LiH crystallizes in a cubic lattice of NaCl type and has a specific gravity of 0.83. Assume that the equilibrium separation r_0 of the two atoms of a diatomic molecule is about equal to the spacing between the atoms in a crystal of the same substance, and thereby evaluate the approximate wavelength in microns of the transition $K = 1 \rightarrow K = 0$ for LiH. *Ans.:* $\lambda \approx 1000\ \mu$.

B. The Vibration-rotation Spectrum. Next in complexity are transitions which involve a change in both the vibrational and rotational states of the molecule. Like the pure-rotation transitions, the vibration-rotation transitions occur far more easily if the molecule possesses an intrinsic electric-dipole moment, so that strongly polar molecules tend to have

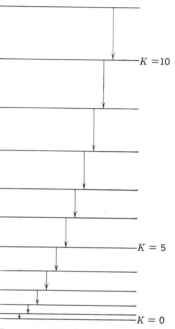

FIG. 9-10. Illustrating the transitions which lead to the pure-rotation spectrum of a molecule (schematic).

intense vibration-rotation spectra; homonuclear molecules have almost none.

The selection rules for electric-dipole vibration-rotation transitions are

(a)[1] $\qquad\qquad \Delta K = \pm 1$

(b) $\qquad\qquad \Delta M_K = \pm 1 \quad \text{or} \quad 0$

(c) $\qquad\qquad \Delta v = +1 \quad \text{(absorption)}$

$\qquad\qquad\qquad = -1 \quad \text{(emission)}$ $\qquad\qquad$ (4)

---

[1] This selection rule holds only if the orbital angular momentum of the electrons has no component along the line joining the two nuclei. The ground-state-potential functions of all but a few molecules happen to possess zero orbital electronic angular momentum, so that the given selection rules apply in the vast majority of cases. If the electronic angular-momentum component along the internuclear axis is not zero, the selection rule $\Delta K = 0$ must be added.

We again consider only the most common case, in which the molecule is in its lowest electronic state, $n = 0$, and obtain from Eqs. 9-4(10), (11), and (15) the following approximate expressions for the spectral angular frequencies for *absorption* of radiation:

$$(a) \quad \omega_{0Kv} = \frac{E_{0,K+1,v+1} - E_{0,K,v}}{\hbar} \approx \omega_{00} + \frac{(K+1)\hbar}{m_r r_K{}^2} - \frac{(v+1)^v \omega_{00}{}^2}{2B_0}$$

$$\Delta K = +1$$
$$\Delta v = +1$$
$$(\text{R branch})$$
$$(5)$$

$$(b) \quad \omega_{0Kv} = \frac{E_{0,K-1,v+1} - E_{0,K,v}}{\hbar} \approx \omega_{00} - \frac{K\hbar}{m_r r_K{}^2} - \frac{(v+1)^v \omega_{00}{}^2}{2B_0}$$

$$\Delta K = -1$$
$$\Delta v = +1$$
$$K \neq 0$$
$$(\text{P branch})$$

The two sets of spectral lines which are described by these equations for the two values of ΔK are referred to as the R-branch ($\omega_{0Kv} > \omega_{00v}$) and the P-branch ($\omega_{0Kv} < \omega_{00v}$).[1]

These equations illustrate the following principal properties of vibration-rotation spectra:

1. The relative magnitudes of the first two terms are approximately $\omega_{00} : K\omega_r$, where ω_{00} is the vibrational frequency and ω_r is the lowest rotational frequency of the molecule. Since, typically, $\omega_{00}/\omega_r \approx 100$ and K can be of the order of 10, the effects are about in the ratio $10:1$ for such a case.

2. Thus one finds a series of lines spaced approximately equally upward and downward from ω_{00v}. (For large values of K and v, however, the spacing will gradually change, becoming greater on one side of ω_{00v} and less on the other side, because of higher terms in K and v which are omitted from the above expression.)

3. The frequency ω_{00v} itself *does not appear in this series of lines*. The second of the above equations is not valid for $K = 0$, since K is the rotational quantum number of the *initial* state and $\Delta K = -1$ for this equation. (That is, $K = 0 \to K = -1$ cannot occur, since K cannot be negative.) *There is therefore a gap in the series of otherwise uniformly spaced lines.* This serves as a useful landmark to identify the "center" of a vibration-rotation multiplet.

4. The change in the spacing of the levels with increasing K and v is by no means completely described by these equations. For a quantitative description of observed results, it is necessary to introduce quadratic, cubic, and even higher dependences of E_{nKv} upon K and v.

[1] If transitions with $\Delta K = 0$ are also permitted, there are three branches to the vibration-rotation spectrum. The third branch, which is only rarely present, is called the Q-branch.

These terms arise principally from the contribution of the centrifugal potential to $U_n(r)$.

Figure 9-11 shows schematically the vibration-rotation energy levels of a diatomic molecule, and Fig. 9-12 shows the electric-dipole transitions which can occur between two vibrational states.   Figure 9-13 shows the observed absorption spectrum of HCl in the near infrared.

EXERCISES

9-7. Derive Eq. (5), and verify that the spectral lines described by these equations for given v should be equally spaced above and below ω_{00v}.   Show that ω_{00v} does not appear in this sequence of lines.

9-8. A short sequence of the lines found in the infrared absorption spectrum of HCl vapor (see Fig. 9-13) is given in the table below.   From these find (a) the force constant k_{00} of Eq. 9-4(6); (b) the equilibrium separation of the H and Cl atoms in HCl.   Ans.: (a) $k_{00} = 480$ N m^{-1}. (b) $r_0 = 1.29 \times 10^{-4}\ \mu$.

$\bar{\nu}(\mu^{-1})$	$\bar{\nu}(\mu^{-1})$
0.3016	0.2907
0.3000	0.2866
0.2982	0.2844
0.2965	0.2821
0.2946	0.2799
0.2928	0.2776

9-9. Evaluate the vibrational and fundamental rotational frequencies, in wave-number units, for DCl.   Ans.: $\bar{\nu}_{00} = 0.207\ \mu^{-1}$, $\bar{\nu}_{\text{rot}} = 0.00106\ \mu^{-1}$.

9-10. Evaluate the vibrational energy quantum $\hbar\omega_{00}$ for HCl and compare this with kT at room temperature.   Ans.: $\hbar\omega_{00} \approx \frac{1}{3}$ eV, $kT \approx \frac{1}{40}$ eV.

C. Electronic Transitions.   The most general transition of a molecule is one in which the electronic, vibrational, and rotational states all change.   In discussing this case it is desirable to introduce one additional feature to our molecular model, which will render it of somewhat broader applicability.   This feature is *the angular momentum possessed by the electron cloud.*   We have heretofore considered that the only angular momentum present in the system is that due to the nuclear motion; but the electrons may themselves possess both orbital and spin angular momentum, and this must be taken into account in treating the energy states of the molecule.   Even so, we shall still ignore some small but qualitatively important features of molecular structure, namely, those having to do with the fine-structure coupling between the various angular momenta within the system.[1]

---

[1] These fine-structure effects are rather analogous to the LS and j-j coupling effects encountered in atomic spectra.   To include them would greatly complicate the discussion without introducing any qualitatively new results.   For a more complete description of these fine-structure effects and the greek-letter notation ($^2\Sigma$, $^1\Pi$, $^3\Delta$, etc.) used to describe them, see G. Herzberg, "Spectra of Diatomic Molecules," 2d ed., chap. 5, D. Van Nostrand Company, Inc., Princeton, N.J., 1950.

We may consider the electron cloud to possess a certain angular momentum whose component along the internuclear axis may have any of the quantized values

$$|M_x| = \Omega\hbar \tag{6}$$

where Ω is a positive integer or half-integer, depending upon whether the number of electrons in the molecule is even or odd. The component of

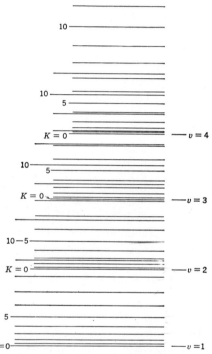

FIG. 9-11. Schematic energy-level diagram of the vibration-rotation states of a diatomic molecule.

the electronic angular momentum perpendicular to the internuclear axis is not a constant of the motion and may be ignored for the present treatment.[1] The total angular momentum of the system is thus composed of Ω along the internuclear axis and the angular momentum N due to the nuclear motion perpendicular to this, as shown in Fig. 9-14. This total angular momentum J is of course quantized with the values

$$|J|^2 = J(J + 1)\hbar^2 \tag{7}$$

where
$$J = \Omega, \ \Omega + 1, \ \Omega + 2, \ \ldots \ .$$

[1] The physical basis of this procedure is as follows: Since the Hamiltonian function

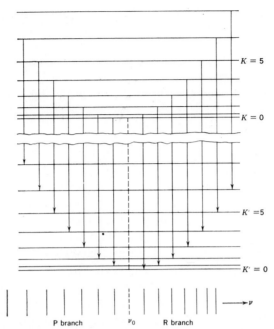

FIG. 9-12. Electric-dipole vibration-rotation transitions between two vibrational states of a diatomic molecule. (Adapted by permission from Herzberg, "Spectra of Diatomic Molecules," copyright D. Van Nostrand Company, Inc., Princeton, N.J., 1950.)

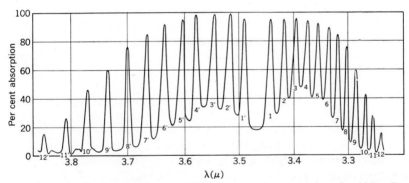

FIG. 9-13. The absorption spectrum of HCl vapor in the near infrared. [After Imes, *Astrophys. J.*, **50**, 251 (1919).]

H_e for the electronic motion is symmetrical about the internuclear axis, the axial component of electronic angular momentum is constant (except for small coupling effects which are being ignored); on the other hand, H_e does not possess symmetry about any axis perpendicular to this, so that the corresponding perpendicular component of electronic angular momentum is not necessarily constant. Ω may be treated as being rigidly fixed in the molecule and rotating with it.

We must now consider how this additional angular momentum affects the energy states of the molecules. Since the existence of an angular-momentum component along the symmetry axis endows the molecule with the properties of an *axially symmetric spinning top*, we must expect the molecule to undergo a quantum-mechanical motion analogous to the precession and nutation of a classical symmetric top. The quantum-mechanical energy levels of such a body, assumed to be perfectly rigid, have been evaluated[1] and are equal to

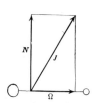

$$E_J = \frac{J(J+1)\hbar^2}{2I_N} + \frac{\Omega^2\hbar^2}{2}\left(\frac{1}{I_e} - \frac{1}{I_N}\right) \quad (8)$$

where $I_N = m_r r_K{}^2$ and I_e is the effective moment of inertia of the electrons. Because the electrons are so much less massive than the nuclei, $I_e \ll I_N$.

It is thus found that the energy levels of a "symmetric top" are equal to those for a simple rigid rotator except for two features:

FIG. 9-14. Illustrating the contribution of the electronic angular momentum Ω and the nuclear angular momentum N to the total angular momentum J.

1. An additional term proportional to Ω^2, which is constant for a given electronic state
2. The absence of values of J less than Ω

These features do not affect our previous discussion of rotation- and rotation-vibration spectra—because of the observed fact that the *ground-state* electronic states of virtually all molecules correspond to $\Omega = 0$.

We now proceed with our consideration of the electronic transitions. Allowing for the results just obtained, we may write the energy of the molecule in the abbreviated approximate form

$$E_{nJv} = V_{n0} + \gamma_n\hbar^2\Omega_n{}^2 + \alpha_n\hbar J(J+1) + (v + \tfrac{1}{2})\hbar\omega_n \quad (9)$$

In this equation the quantities V_{n0}, γ_n, Ω_n, α_n, and ω_n are all constant for a given electronic state, but each varies from one state to another. The selection rules which govern electric-dipole transitions of such a diatomic molecule are:

(a) $\Delta\Omega = \pm 1$ or 0

(b) $\Delta J = \pm 1$ or 0 but $J = 0 \nleftrightarrow J = 0$ and

                       $\Delta J = 0$ forbidden if $\Omega = 0 \rightarrow \Omega = 0$ (10)

(c) $\Delta v =$ unrestricted

[1] Reiche and Rademacher, *Z. Physik*, **39**, 444 (1926); **41**, 453 (1927); and also Kronig and Rabi, *Phys. Rev.*, **29**, 262 (1927).

There is now no strict limitation on the possible change of the vibrational quantum number v because of the participation of the electronic state in the transition. Using the above selection rules, one finds that the electronic-vibrational-rotational transitions have the angular frequencies (for *emission*)

$$\omega_{nJv} = \omega_{nn'} + (v\omega_n - v'\omega_{n'}) + (J+1)[J\alpha_n - (J+2)\alpha_{n'}] \qquad \Delta J = +1$$
$$\text{(P branch)}$$
$$= \omega_{nn'} + (v\omega_n - v'\omega_{n'}) + J(J+1)(\alpha_n - \alpha_{n'}) \qquad \Delta J = 0$$
$$J \neq 0$$
$$\text{(Q branch)}$$
$$= \omega_{nn'} + (v\omega_n - v'\omega_{n'}) + J[(J+1)\alpha_n - (J-1)\alpha_{n'}] \qquad \Delta J = -1$$
$$\text{(R branch)}$$
$$\tag{11}$$

In these equations, $\omega_{nn'}$ includes all of the effects of the change in n that are independent of J and v.

To interpret these equations, it is best to consider n and n' as being fixed and v, v', and J as being independently variable. One is then considering transitions which connect the two potential-energy curves defined by n and n'. The term $\omega_{nn'}$ is then the same for all values of v, v', and J, and furthermore is generally of such magnitude as to dominate the remaining terms, since several electron volts may be involved in the electronic transition whereas vibrational quanta amount to but a fraction of an electron volt, and rotational quanta to even less. We may thus regard the vibrational terms in Eq. (11) as a fine-structure effect upon the electronic transition, and the rotational terms in turn as a fine-structure effect upon the vibrational structure.

Taking account of the gross relative magnitudes of the terms and of the fact that ω_n and $\omega_{n'}$ differ from one another, we see that the vibrational fine structure superimposed upon an electronic transition yields a number of frequencies which fall into various sequences, or *progressions*, somewhat reminiscent of those observed in the Zeeman splitting of a spectral line. Each progression is characterized by a nearly[1] uniform spacing from one frequency to the next, and corresponding terms in the various progressions are also nearly uniformly spaced, but with a spacing different from that of the terms within a progression. These progressions may be represented conveniently in the alternative form

$$\omega_{nv} = \omega_{nn'} + \Delta v \omega_{n'} + v(\omega_n - \omega_{n'}) \tag{12}$$

where $\Delta v = v - v'$ is the change of the vibrational quantum number in the transition. The further inclusion of the rotational effects breaks up

---

[1] Quadratic terms in ω_n and $\omega_{n'}$, which have been omitted from Eq. (9) for simplicity, cause the spacing of each progression to vary with v and v'.

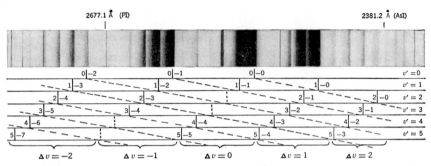

FIG. 9-15. Band spectrum of the PN molecule in emission. The various v-progressions occurring in this spectrum are identified by the diagram at the bottom of the figure. (By permission from Herzberg, "Spectra of Diatomic Molecules," copyright D. Van Nostrand Company, Inc., Princeton, N.J., 1950.)

each of the above terms into a relatively broad structure called a *band*. The totality of lines resulting from a given electronic transition then consists of a number of such bands and is called a *band system*.

The structure of each band within a band system is the same, to a fairly good approximation, and is somewhat similar to that of a vibration-rotation band. There are some rather important differences, however, between the rotational fine structures described by Eqs. (5) and (11):

1. In the former case, where the electronic state did not change, only two branches were present—the P-branch and the R-branch. However, $\Delta J = 0$ is now permitted (unless $\Omega = 0$ for both the upper and the lower electronic states), which gives rise to a third branch, called the Q-branch. In practice it is found that the Q-branch is oftentimes missing because in fact Ω frequently *is* zero for both states.

2. Whereas the spacing between lines of a vibration-rotation multiplet as described by Eq. (5) is nearly uniform, Eq. (11) contains *quadratic terms* in J with sizable coefficients. This leads to a marked nonuniformity in the line spacing, and it commonly happens that in one or

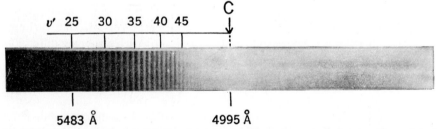

FIG. 9-16. Band spectrum of I_2 in absorption. The arrow at C denotes the point at which an absorption continuum joins the vibration progressions. (By permission from Herzberg, "Spectra of Diatomic Molecules," copyright D. Van Nostrand Company, Inc., Princeton, N.J., 1950.)

two of the branches the spacing diminishes to zero and *changes sign* with increasing J. This effect endows a band with a characteristic appearance of sharpness at one edge, where the reversal of spacing occurs, and of diffuseness in the opposite direction, where the lines gradually fade out with increasing J. The boundary of the band where this reversal occurs is called the *band head*. Depending upon the magnitudes of the various quantities involved, the diffuse shading of the band can occur toward the red or violet end of the spectrum.

3. In the present case, just as in vibration-rotation bands, there is a *missing line* in the sequence at the frequency $\omega_{nn'} + (v\omega_n - v'\omega_{n'})$. This missing line is called the *zero*, or *null*, *line;* it serves to identify

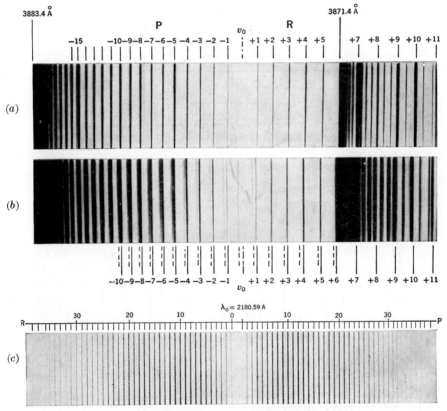

FIG. 9-17. (a) Fine structure of the CN band 0.3883μ at low temperature and (b) at intermediate temperature. The dashed lines refer to the returning limb of the P-branch. (By permission from Herzberg, "Spectra of Diatomic Molecules," copyright D. Van Nostrand Company, Inc., Princeton, N.J., 1950.) (c) Structure of the 0-0 band of CN^+ at 0.2181μ. [After Douglas and Routly, *Astrophys. J.*, **119**, 303 (1954).]

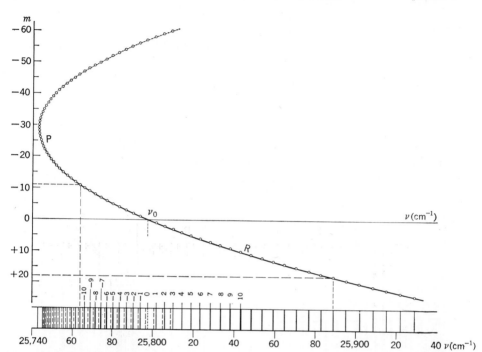

FIG. 9-18. Fortrat parabola of the CN band of Fig. 9-17 (see Exercise 9-12). In this figure, the ordinate m is equal to J for the R-branch, and to $-J$ for the P-branch. (By permission from Herzberg, "Spectra of Diatomic Molecules," copyright D. Van Nostrand Company, Inc., Princeton, N.J., 1950.)

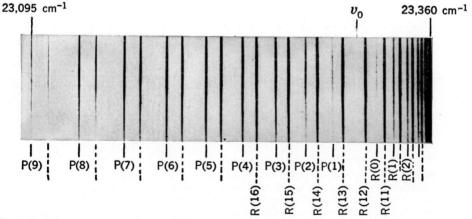

FIG. 9-19. Fine structure of the Cu H band 0.4280μ. (By permission from Herzberg, "Spectra of Diatomic Molecules," copyright D. Van Nostrand Company, Inc., Princeton, N.J., 1950.)

the origin of J within the band. All three branches converge upon this missing line. However if Ω, the quantum number for the axial component of electronic angular momentum, is not zero for both electronic states, the values of J that can occur are correspondingly limited. (Recall that $J \geq \Omega$.) This in turn restricts the possible values of J that can be used in Eq. (11) and may cause *more than one* line to be missing from the rotational structure.

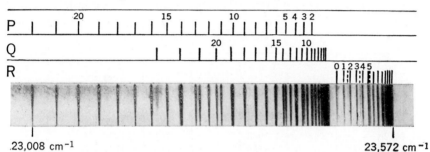

FIG. 9-20. Fine structure of the Al H band 0.4241μ. In this case the Q-branch is present, indicating that $\Omega \neq 0$ for the upper and/or lower states. (By permission from Herzberg, "Spectra of Diatomic Molecules," copyright D. Van Nostrand Company, Inc., Princeton, N.J., 1950.)

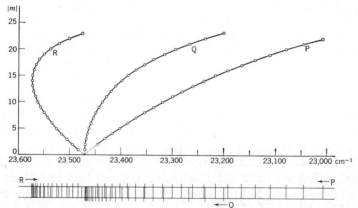

FIG. 9-21. Fortrat diagram of the Al H band of Fig. 9-20. (The direction of ν is reversed compared to Fig. 9-20.) $|m| = J$. (By permission from Herzberg, "Spectra of Diatomic Molecules," copyright D. Van Nostrand Company, Inc., Princeton, N.J., 1950.)

Some of the features of the structure of a band system as described above are illustrated for some actual cases in Figs. 9-15 to 9-21.

EXERCISES

9-11. Derive (11) from Eq. (9), using the selection rules (10).

9-12. Show that the P- and R-branches can be described by a single formula, if negative values of J are used. The curve so obtained is called a *Fortrat parabola* (see Figs. 9-18 and 9-21).

9-13. Show that the band head of the Q-branch is always at $J = -\frac{1}{2}$, while that of the P- or R-branch may be anywhere (see Fig. 9-21).

9-14. The wave numbers of some of the band heads observed in the spectrum of a diatomic molecule are given in Table 9-1 in μ^{-1}. From these data, find v and v' for each band and evaluate ω_n and ω_n' approximately. *Partial ans.:* 3.3413: v = 6, v' = 10; 4.2082: v = 6, v' = 3.

TABLE 9-1. WAVE NUMBERS

$\bar{\nu}$	v	v'	$\bar{\nu}$	v	v'	$\bar{\nu}$	v	v'
3.3413			3.6861			4.0536		
3.4411			3.7069			4.0786	1	0
3.4607			3.7713			4.1066		
3.4802			3.7933			4.1239		
3.4998			3.8155			4.1331		
3.5194			3.8377	0	1	4.1523		
3.5387			3.8519			4.1597		
3.6028			3.8756			4.1798		
3.6237			3.9467			4.1859		
3.6446			3.9699	0	0	4.2082		
3.6652			4.0288					

The foregoing treatment of the spectra of diatomic molecules, while quite brief, illustrates on the one hand how complex the energy states of even a simple structure can be, and on the other hand how powerful are the methods of quantum mechanics in reducing an almost hopeless complexity of spectral lines to a very few basic relationships. Once again we find that the Schroedinger theory is equal to the challenge offered by a phenomenon that at first may appear incapable of elementary description, and we find too that all known features of molecular spectra, including many that we have not treated, have a satisfactory interpretation on the basis of the quantum theory. Some of these further features will be considered in our later work.

9-6. Effects of Nuclear Spin, Ortho- and Para-Hydrogen

We have so far ignored the effects of nuclear spin upon the energy states and spectral transitions of a molecule. From the discussion of hyperfine structure in Chap. 5 it is clear that the effects of nuclear spin and magnetic moment upon the energy levels themselves will almost surely be so small as to be lost in the complexities of the vibration-rotation structure of a molecular transition. There is, however, a most important feature of the molecular problem that we have so far ignored: that, if the two nuclei of a diatomic molecule are *identical*, the effects of the *exclusion principle* must be taken into account in restricting the possible states the molecule can actually occupy and the transitions that it can undergo. We shall now consider this case of *homonuclear molecules* and these exclusion-principle limitations.

If the two nuclei are identical, the exchange-symmetry postulate of Chap. 7 requires that the complete wave function for the molecule be either symmetric or antisymmetric with respect to an exchange of space and spin coordinates of the two nuclei for nuclear spins that are integral or half-integral, respectively. Since to a very good approximation the spins of the nuclei can be considered to have no interaction with one another or with the electrons, the results of Chap. 7 for noninteracting particles can be applied directly to the spin functions of the nuclei. Thus if the spins of the identical nuclei have the quantum number I, there will exist $2I + 1$ linearly independent spin eigenfunctions—one for each of the values M_I of the z-component of spin. Let us call these spin functions $S_{M_I}(i)$, where $-I \leq M_I \leq +I$ and $i = 1$ or 2, to designate which nucleus is involved. The most general nuclear spin state of the system then corresponds to each of the two nuclei being in any of these $2I + 1$ spin states, the total spin wave function being either a symmetric or an antisymmetric combination of the $S_{M_I}(i)$ functions. For example, if $I = 3$, there will be seven such spin functions, and combinations such as $2^{-\frac{1}{2}}[S_3(1)S_{-2}(2) - S_3(2)S_{-2}(1)]$ are valid *antisymmetric* spin functions, while combinations such as $2^{-\frac{1}{2}}[S_3(1)S_{-2}(2) + S_3(2)S_{-2}(1)]$ or $S_2(1)S_2(2)$ are valid *symmetric* spin functions. Note that the two nuclei in the same spin state can have a symmetric spin function but cannot have an antisymmetric one, while if the nuclei are in different spin states, they can have either symmetric *or* antisymmetric spin functions.

Now, in order to satisfy the requirements of exchange symmetry or antisymmetry for a complete exchange of the two nuclei, we must have the following situation with respect to the space-coordinate wave functions:

1. If the nuclei have *integral spin* ($I = integer$), the total wave function must be *symmetric* with respect to a complete interchange of the two nuclei. This means that, if the nuclear *spin* function described above is *symmetric* with respect to an interchange of the spins, the *space*-coordinate wave function *must also be symmetric* with respect to an interchange of the space coordinates of the two nuclei. (Recall that the total wave function for the case of no interaction between the nuclear spin and other variables is a product of a space wave function and a spin wave function.) Similarly, if the spin wave func-

tion is antisymmetric, then the space wave function must also be antisymmetric.

2. If the nuclei have *half-integral spin* (I = *half-integer*), the total wave function must be *antisymmetric* with respect to a complete interchange of the two nuclei. Thus if the spin wave function is *symmetric*, the space wave function must be *antisymmetric*, and vice versa.

As was the case for the one-electron atom, the quantity which determines the space symmetry of the nuclear part of a molecular wave function is the *orbital angular momentum;* that is, wave functions corresponding to even values of the rotational quantum number K have *even* parity (even space symmetry), and those for odd values of K, *odd* parity. Thus if I is integral, symmetric spin functions are combined with even values of K (these are called *even rotational states*) and antisymmetric spin functions with odd values of K. On the other hand, if I is half-integral, the association is reversed.

An interesting case which often occurs is that for which $I = 0$. In this case the spin wave function *can only be symmetric;* thus the space wave function *must also be symmetric*. This means that half of the rotational and vibrational levels that can occur for the general case, i.e., those having odd K, are *absolutely forbidden* by the exclusion principle for the case $I = 0$.

Before applying the above considerations to the observed transitions, we must note that *a homonuclear molecule cannot possess an electric-dipole moment* because all energy eigenfunctions possess either even or odd space symmetry about the geometric center of the molecule, either of which leads to a symmetrical $\psi^*\psi$ distribution. Thus the ordinary electric-dipole transitions cannot occur for these molecules. The most probable type of radiation in such a case is either magnetic dipole or electric quadripole. In either of these cases the selection rule upon K is

$$\Delta K = 0, \pm 1, \pm 2 \tag{1}$$

Such radiation will in general be many times less intense than electric-dipole radiation (a factor of 10^{-8} for the relative intensity is not uncommon), but in the present case the rigorous impossibility of electric-dipole radiation gives the molecule no choice but to use these less probable avenues in its radiative transitions. The point is that, even though the possibility of electric-dipole radiation is absent for a homonuclear molecule, other types of transition do exist and permit *some kinds of radiation* to occur.

We are now in a position to examine the physically observable results of the preceding considerations. They may be tabulated as follows:

1. The smallness of the interaction of the nuclear spins with one another,

with the electrons of the molecule, or with other outside agencies means that the *nuclear spin wave function will be almost exactly a constant of the motion,* and in particular *will remain unchanged during a radiative transition of the molecule.*

2. Since for a given type of molecule the space symmetry of the molecular wave function is uniquely related to the *spin* symmetry, the fact that the spin wave function remains unchanged requires that the space-symmetry character also must not change during a transition of the molecule.

3. Since the space-symmetry character of the molecule is directly connected with the evenness or oddness of K, as mentioned previously, the preceding result means that states of even K can combine only with other states of even K, and similarly, odd-K states can combine only with other odd-K states.   This means that radiative transitions must involve an *even* change in K, that is, $\Delta K = 0, \pm 2, \pm 4, \ldots$, etc.   Thus if electric-dipole radiation were not already prevented by the absence of an electric-dipole moment for a homonuclear molecule, it would be prevented by this exclusion-principle requirement.

4. The above result that K must remain even or odd during a transition means that, once a molecule finds itself in a state of even K, it will tend always to remain in such a state, and similarly for odd K.

5. Because of the smallness of the nuclear spin interaction with outside agencies, the above conclusions hold also for transitions which occur as a result of collisions, etc.   The *time* required for a molecule to change its nuclear spin state depends upon the magnitude of the nuclear spin interaction, but this time is often quite long—sometimes weeks, months, or even years.

6. Thus it is meaningful to speak of the separate physical existence of *two distinct modifications* of a given diatomic molecule, one characterized by a symmetric nuclear spin function and the other by an antisymmetric spin function.   The modification corresponding to a symmetric spin state is called the *ortho*-modification, and that corresponding to an *antisymmetric* spin state, the *para*-modification.

7. Formulas analogous to 9-5(5) can be written for the selection rules $\Delta K = 0, \pm 2$, and it is thereby found that the spectral lines for the even $\rightarrow$ even transitions fall between those for the odd $\rightarrow$ odd transitions; that is, a given vibration-rotation band contains lines that alternately arise from the ortho- and the para-modifications of the molecule, respectively.

8. If $I = 0$, *only symmetric spin states can occur,* so that *only even rotational states can occur.*   Therefore, *alternate lines will be missing* from the vibration-rotation spectrum of such a molecule (for example, O_2).

9. For a given molecule, the relative intensities of the even → even and odd → odd transitions will be proportional to the relative amounts of the two modifications of the molecule present in the sample. If the sample has existed for a long time at ordinary temperatures where many rotational states are excited, the relative amounts of each modification will be proportional to the relative *statistical weights* of the two modifications. For given I, these statistical weights will be proportional to the number of *symmetric* and *antisymmetric* spin functions that can be formed out of the $2I + 1$ linearly independent spin wave functions. Since the *antisymmetric* functions are of the form

$$S_a = 2^{-\frac{1}{2}}[S_{M_{I1}}(1)S_{M_{I2}}(2) - S_{M_{I1}}(2)S_{M_{I2}}(1)]$$

where $M_{I1} \neq M_{I2}$, there will be just $N_a = \frac{1}{2}(2I + 1)(2I)$ such functions. On the other hand, there will be a *symmetric* function for each of the above antisymmetric functions, and in addition there will be $2I + 1$ symmetric functions of the type $S_s = S_{M_I}(1)S_{M_I}(2)$. Therefore, the relative statistical weights for the symmetric and antisymmetric spin states are

$$\frac{N_s}{N_a} = \frac{(I + 1)}{I} \tag{2}$$

and the ortho-modification of a molecule should be $(I + 1)/I$ times as abundant as the para-modification. The intensities of spectral transitions for these two forms should also be in this ratio. This effect is called the *alternation of intensities* in band spectra of homonuclear molecules.

10. The above relationship has been used to *measure the magnitudes of the nuclear spins* of various atomic nuclei which form homonuclear molecules. This is relatively easy if I is not too large, since if $I = 0, \frac{1}{2}, 1,$ or $\frac{3}{2}$, the relative intensities of the two series of lines should be ∞, 3, 2, or $\frac{5}{3}$, respectively, and it is then possible to decide unambiguously among these various possible values.

As an illustration of some of the above considerations, the interesting case of H_2 may be mentioned. Since the spin of a proton is $\frac{1}{2}$, there are four linearly independent spin functions. From these one can form *three symmetric* spin states and *one antisymmetric* spin state for the molecule; the former correspond to *parallel* spins, and the latter to *opposed* spins. The two forms of hydrogen are called ortho- and para-hydrogen. At ordinary temperatures, where many rotational states of the molecule are excited, the ortho form is three times as abundant as the para form; that is, ordinary H_2 gas is three-fourths ortho-hydrogen gas and one-fourth para-hydrogen. On the other hand, the lowest energy state is the

$K = 0$ rotational state, which occurs only for the para form, so that at sufficiently low temperatures hydrogen gas should eventually all become *para*-hydrogen. Experimentally this is found to occur, and with the help of a catalyzing effect of charcoal (involving a dissociation and recombination of the H_2 into the more stable form), the transformation to para-hydrogen is essentially complete in a matter of hours. When the para-hydrogen is reheated to room temperature, it retains its identity for several weeks. Various physical phenomena other than spectral transitions are different for ortho- and para-hydrogen (e.g., the heat capacity and heat conductivity), which illustrates further the reality of physical effects governed by the exclusion principle.

From the above discussion we see that, in spite of a vanishingly small physical interaction of the nuclear spin with anything else in a molecule, the very *existence* of a spin calls into play the exclusion principle, with eventual results that are quite observable. To emphasize that it is really the exclusion principle that is responsible for these effects and not the spin itself, the case of an *almost* homonuclear molecule can be cited. If the nuclei of a "homonuclear" molecule are not precisely identical but are *different isotopes of the same element*, all of the analysis leading to the wave functions and energy levels is the same, apart from a small mass difference which can be neglected for the more massive elements, but now the nuclei are *no longer indistinguishable* and the exclusion principle no longer applies. Thus all of the above requirements of the maintenance of the space symmetry of the wave functions are removed by this very tiny change of the physical system, which then leads to *quite different physical properties* for such a system.

EXERCISES

9-15. Write the equations corresponding to Eq. 9-5(5) for the selection rules $\Delta K = 0, \pm 2$. Show that the frequencies of the odd $\rightarrow$ odd transitions fall between the even $\rightarrow$ even ones.

9-16. Supply the missing steps in the derivation of Eq. (2).

9-17. ^{14}N nuclei have unit spin $I = 1$. Tabulate all of the distinguishable symmetric and antisymmetric spin wave functions, and thus verify Eq. (2) for the case of nitrogen gas.

9-18. Discuss the expected behavior of *deuterium gas* at room temperature and at low temperatures with respect to the existence and relative abundances of ortho- and para-modifications ($I = 1$).

9-7. Van der Waals' Forces

As a final example of molecular forces, we shall briefly describe a most important type of interaction which is of such a general character that it is only very slightly dependent upon the specific structure of a given molecule and which is mainly responsible for the fundamental physical phenomenon of the *condensation of a gas into a liquid or solid*. This interaction is called *van der Waals' attraction*, and it arises through the following physical mechanism: Suppose that a given molecule possesses, at some instant, an electric-dipole moment $\boldsymbol{p}$. This molecule will then be surrounded by an electric-dipole field

$$4\pi\varepsilon_0 \boldsymbol{E} = -\nabla \left(\frac{\boldsymbol{p}\cdot\boldsymbol{r}}{r^3}\right) = -\frac{\boldsymbol{p}}{r^3} + 3\left(\frac{\boldsymbol{p}\cdot\boldsymbol{r}}{r^5}\right)\boldsymbol{r} \tag{1}$$

This electric field, acting upon a second molecule in the vicinity of the first, will *induce* in the second molecule an electric-dipole moment of strength

$$\boldsymbol{p}' = 4\pi\varepsilon_0 k\boldsymbol{E} \tag{2}$$

where k is the *polarizability* of the second molecule (we assume k to be a scalar quantity, for simplicity). This induced dipole moment will in turn interact with the original dipole, the mutual energy being

$$V = -\boldsymbol{p}'\cdot\boldsymbol{E} = \frac{-k}{4\pi\varepsilon_0}(1 + 3\cos^2\theta)\frac{p^2}{r^6} \tag{3}$$

where $\boldsymbol{p}\cdot\boldsymbol{r} = pr\cos\theta$

Thus, whatever the orientation of the dipole moment of the first molecule, the interaction energy is always negative and is inversely proportional to the sixth power of the distance between the molecules. This indicates the existence of an *attractive force between the two molecules, varying as* $1/r^7$.

In arriving at the above result we assumed that one of the molecules initially possessed an electric-dipole moment. If we now ask whether this is a justifiable assumption, we might offhand think not, inasmuch as many, if not most, molecules are known to possess no intrinsic moments. However, we observe that the interaction energy is instantaneously proportional, not to the dipole moment itself, but to its *square*. Now, even a perfectly symmetrical molecule (i.e., one whose electronic distribution is perfectly symmetrical) will yield a finite value for $\langle p^2\rangle$ even though $\langle\boldsymbol{p}\rangle$ is zero. That is, every possible instantaneous position that the electrons of the molecule can occupy will lead to a dipole moment of some size and orientation for the molecule as a whole, and although these rapidly fluctuating instantaneous moments average to zero, their *mean square* must have a finite, positive value.

We thus conclude that any two molecules should exhibit a characteristic $1/r^7$ attractive force. The strength of this force in a given case depends both upon the mean-square fluctuation of the electric-dipole moment and upon the polarizability of the molecules, and will therefore show some dependence upon the nature of the molecules involved. Since the range of these characteristic parameters is not extremely wide, however, van der Waals' interaction will be rather insensitive to the type of molecule involved. This accounts for the relatively uniform properties of liquids with regard to heats of vaporization, density, etc.

EXERCISES

9-19. Establish Eq. (3).

9-20. Assume that the ground state of atomic He corresponds to two electrons moving independently under the influence of a point nucleus whose charge is $Z = \alpha$. (Each electron partially screens the He nucleus from the other electron.) (a) Evaluate α so as to give the observed ground-state energy of He. (b) Calculate the mean-square electric-dipole moment of such an atom. *Ans.:* (a) $\alpha = 1.71$. (b) $\langle p^2 \rangle = 2.06 (ea_0')^2$.

9-8. Conclusion

In this chapter we have seen in brief outline how the quantum theory accounts for some of the observed features of molecular structure and molecular spectra. The discussion has been necessarily abbreviated, and only the most obvious and important features have been mentioned.

The main features of interest to us in the present introductory treatment are of course the concepts of the establishment of a chemical bond through the sharing or *exchange* of electrons and of the essential independence of the electronic motion and the nuclear motion of a molecule, with the electronic energy acting as a potential function for the nuclear motion. This latter idea, that an effective potential energy between two particles can arise from the exchange of a smaller particle between them, plays an important role in quantum electrodynamics, where the ordinary Coulomb interaction between two charges can be described quantitatively in terms of an *exchange of photons* between the charges, and in nuclear physics, where the forces between two nucleons are thought to be due to the *exchange of mesons* between the nucleons. (The photons and mesons in these latter cases are said to be in *virtual states* because the exchange of a photon or meson in a "positive energy state" is not energetically permitted. It is a characteristic feature of quantum mechanics, however, that such processes can give rise to real physical results.)

One final remark ought to be made concerning the role of molecular spectroscopy in present-day science. Whereas we have here approached molecular spectroscopy from the standpoint of trying to deduce what properties the spectrum of a molecule ought to have, given the properties of the elementary particles of which the molecule is composed, it is clear that molecular spectroscopic measurements will themselves provide not only a confirmation of the essential correctness of the physical theory used in interpreting the spectrum *but also an accurate measurement of certain important parameters of a molecular system*, whose precise calculation might be impossibly difficult. Thus it has been found, for example, that molecular spectroscopy provides a means of measuring the binding energy, the internuclear spacing, and the local shape of the electronic energy curves for various electronic states of a molecule. In addition to this use as a means of evaluating the numerical parameters of molecular systems, molecular spectroscopy of course is a valuable tool in many applications similar to those of atomic spectroscopy, such as in obtaining information concerning physical and chemical conditions in inaccessible regions. All in all, quantum mechanics, in conjunction with the techniques of molecular spectroscopy, has provided us with an extremely satisfactory, if not always simple, interpretation of many chemical and physical phenomena.

REFERENCES

Herzberg, G.: "Spectra of Diatomic Molecules," D. Van Nostrand Company, Inc., Princeton, N.J., 1950.

Pauling, L.: "The Nature of the Chemical Bond," Cornell University Press, Ithaca, N.Y., 1944.

Schiff, L. I.: "Quantum Mechanics," 2d ed., International Series in Pure and Applied Physics, McGraw-Hill Book Company, Inc., New York, 1955.

10

Quantum Statistics

In the preceding chapters we have examined some of the more important applications of quantum mechanics to atomic and molecular systems, and we have found in general that the quantum-mechanical description of a system becomes more and more complex as the complexity of the system increases. It would appear on this basis that the quantum-mechanical treatment of a system of as great complexity as a *solid body*, or even a *gas*, would be impossibly difficult, and that the results would be hopelessly complicated. It is often the case, however, that when a system becomes sufficiently large it may, on the one hand, be so complicated that one could not even record, let alone use, information of the same type as can be obtained for simpler systems; but on the other hand, the very complexity of the system may then lead to *quite simple properties that correspond to a statistically probable behavior* of the system. Thermodynamics and the classical statistical mechanics are familiar examples. In this chapter we shall examine some of the simpler aspects of the statistical mechanics of quantum-mechanical systems, and we shall find that many of the observed properties of such systems can be described in terms of statistical distribution laws similar to those used to describe classical systems.

The field of quantum statistical mechanics is much too broad and complex for us to treat in any but a superficial manner, so that the results obtained in this chapter are necessarily restricted with respect to the kinds of situations to which they apply. In spite of this, the results to be obtained do provide an excellent approximate description of a broad range of physical phenomena, and thus furnish a reliable qualitative insight into the behavior of many actual physical systems.

10-1. Derivation of the Three Quantum Distribution Laws

The distribution laws now to be derived apply to a system which consists of a large number of *identical quantum-mechanical elements* which interact only *very weakly* with one another. This system is assumed to be slowly transferring energy between its various component elements in such a way that an *equilibrium condition* exists, in the sense that the number of elements having a given energy is not changing systematically with time. We shall treat *three kinds of systems* which together provide a description of many diverse phenomena in physics: systems composed of (*a*) *identical but distinguishable particles* or other elements, (*b*) *identical, indistinguishable particles of half-integral spin*, and (*c*) *identical, indistinguishable particles of integral spin*. In the interests of uniformity we shall think in terms of particles even for case (*a*), but the results are valid for *any kind of distinguishable elements* and are actually most often applied to the *degrees of freedom* available to a system. These degrees of freedom might be identical in their physical behavior and yet be distinguishable by means of their spatial location or orientation, e.g., the normal modes of vibration of a system of elastically coupled particles.

Consider a system composed of a large number N of identical particles which do not interact with one another. These particles are assumed to be moving under the influence of the same potential function $V(x,y,z)$ and when treated individually can exist in any of a number of states ψ_i of energy ϵ_i, where $i = 0, 1, 2, \ldots$ For example, the particles might all be acted upon by the nucleus of the same atom or might all be confined as a gas inside the same container, in which case the wave functions and energy levels would be those of a one-electron atom or of a particle inside a rigid container, respectively. We shall assume for simplicity that these one-particle levels are nondegenerate, so that they can be enumerated in the order of increasing energy and so that there is a unique correspondence between energy levels and wave functions. Degenerate levels may then be included by assigning to such levels a *statistical weight* equal to the degree of degeneracy or by regarding them as being very closely spaced, but not identical, in energy. The various energy eigenstates of the complete system will be expressible as a product [for case (*a*)], or as antisymmetric or symmetric combinations of products [for cases (*b*) and (*c*)], of the one-particle eigenfunctions:

Case (*a*) $\Psi_{i_1 i_2 \ldots i_N} = \psi_{i_1}(1)\psi_{i_2}(2) \cdots \psi_{i_N}(N)$

Case (*b*)

$$\Psi_{i_1 i_2 \ldots i_N} = (N!)^{-1/2} \begin{vmatrix} \psi_{i_1}(1) & \psi_{i_2}(1) & \cdots & \psi_{i_N}(1) \\ \psi_{i_1}(2) & \psi_{i_2}(2) & \cdots & \psi_{i_N}(2) \\ \cdot & \cdot & & \cdot \\ \cdot & \cdot & & \cdot \\ \cdot & \cdot & & \cdot \\ \psi_{i_1}(N) & \psi_{i_2}(N) & \cdots & \psi_{i_N}(N) \end{vmatrix} \quad (1)$$

Case (c)

$$\Psi_{i_1 i_2 \cdots i_N} = A[\psi_{i_1}(1)\psi_{i_2}(2) \cdots \psi_{i_N}(N) + \psi_{i_1}(2)\psi_{i_2}(1) \cdots \psi_{i_N}(N)$$
$$+ \text{ all permutations of } N \text{ particles}]$$

The *energy levels* of the complete system will of course be

$$E_{i_1 i_2 \ldots i_N} = \epsilon_{i_1} + \epsilon_{i_2} + \cdots + \epsilon_{i_N} \tag{2}$$

where ϵ_{i_r} is the energy corresponding to the wave function $\psi_{i_r}(r)$ of the rth particle.

The preceding description is of course no different, except for the assumed large number N of particles involved, than we have already used for specifying the state of a many-electron atom (Sec. 5-4). In the present case, however, we have no way of experimentally specifying, i.e., measuring, which one of its many possible states the system is in, and furthermore even a small interaction of the component elements with one another or with outside influences would in any case induce transitions between the various states so rapidly that it would be impracticable to follow these transitions in detail. We therefore make use of our previous experience with classical systems of great complexity and agree to restrict our curiosity regarding the state of the system to fit the practical confines of our limited ability to measure this state experimentally. Thus we seek to obtain expressions, analogous to the Maxwell distribution law of classical statistical mechanics, which describe the statistically probable distribution of the N particles among the various states ψ_i of the one-particle system for the three cases being considered.

Let us assume that we are able to measure the total energy of the system of N particles to a certain degree of accuracy and have found it to be E, with a rather small but finite uncertainty δE. We assume that the system is in *statistical equilibrium* at this total energy E. We then wish both to study the various ways in which the particles might be distributed among the various energy levels ϵ_i available to them and to find what is statistically the *most probable* such distribution. To do so, we proceed as follows: We first divide the entire energy range into contiguous "cells" of sizes $\Delta\epsilon_1$, $\Delta\epsilon_2$, . . . , $\Delta\epsilon_s$, ... such that each cell is very narrow in comparison to the error we are likely to make in measuring the total energy E but still of great enough extent that a large number g_s of energy states are contained within each cell. In the state of statistical equilibrium, let there be n_s particles whose energies lie in the range ϵ_s to $\epsilon_s + \Delta\epsilon_s$ of the sth cell. Clearly these various numbers n_s must satisfy the two conditions

$$\sum_{s=1}^{\infty} n_s = N \tag{3}$$

$$\sum_{s=1}^{\infty} n_s \epsilon_s = E \tag{4}$$

and in addition must satisfy the exclusion-principle restriction that not more than one particle can be in each of the g_s states of a given energy, if the particles are of half-integral spin.

Now there are a great many ways in which the N particles might be distributed among the cells, and a great many ways in which the n_s particles in a given cell might be distributed among the various energy levels of that cell. We shall call a given distribution of *cell population* n_s a *macroscopic* ("coarse-grained") distribution, and a given detailed distribution of the particles among the various *energy levels* of the cells a *microscopic* ("fine-grained") distribution. We must somehow find an appropriate index of the likelihood of occurrence of a given macroscopic distribution of particles and find for what distribution $n_s(\epsilon_s)$ this likelihood is a maximum. This distribution will then be our best estimate of the actual condition of the system.[1] We now introduce the following postulate concerning the a priori likelihood of a given microscopic distribution of the particles.

Postulate:

Every physically distinct microscopic distribution of the N particles among the various energy levels ϵ_i which satisfies both the condition that the total energy be $E + \delta E$ and the requirements of the exclusion principle, if it applies, *is equally likely to occur.*

By a physically distinct distribution we mean one whose wave function is different from that of any other distribution. In terms of the properties of the particles involved, we may alternatively say that such a distribution must in principle be *distinguishable* from any other.

Thus we may conclude that the relative likelihood of occurrence of any given macroscopic distribution $n_s(\epsilon_s)$ of the N particles among the various cells is *proportional to the number of distinguishable ways in which such a distribution can be constructed.* Calling this number $P(n_s)$, we must now compute it for each of the types of particle to be considered, and maximize P with respect to each n_s, subject to the auxiliary conditions (3) and (4). We shall now evaluate P separately for each type of particle.

A. Identical, Distinguishable Particles. In this case it is clear that P is equal to the product of the number of different ways of selecting the quotas of particles to be put into each cell out of the total collection of N particles and the number of ways in which the particles can be arranged

---

[1] Note that we are *not* inquiring into the *microscopic distribution* of the n_s particles in a given cell among the g_s energy levels in that cell, because *any* such microscopic distribution that is consistent with the basic physical properties of the system would equally well yield the correct experimental value of E. We are only inquiring into *how many* particles are in each cell, irrespective of their distribution within the cell.

within the cells. The first of these numbers can be found as follows:
Starting with the first quota n_1 to be put in the first cell, there are N ways
of choosing the first particle in this quota, and for each of these choices
there are $N - 1$ ways of choosing the second particle, etc., so that the
number of ways of choosing the first n_1 particles is just

$$P_1' = N(N - 1)(N - 2) \cdots (N - n_1 + 1) = \frac{N!}{(N - n_1)!}$$

In this expression we have counted as a different arrangement each
separate *sequence* in which the first n_1 particles could have been selected.
However, we need only to know *which* particles are in the quota n_1, but
not in *what sequence* they appear. We must therefore divide the above
number by the number of different sequences in which n_1 objects can be
arranged. This latter number is $n_1!$, so that we have for the true number
of different ways of choosing the first n_1 particles out of the total sample
of N particles,

$$P_1 = \frac{P_1'}{n_1!} = \frac{N!}{n_1!(N - n_1)!} \tag{5}$$

If we now form the second quota n_2 by the same procedure, we find

$$P_2 = \frac{(N - n_1)!}{n_2!(N - n_1 - n_2)!} \tag{6}$$

since only $N - n_1$ particles remain free to be chosen for this quota.
Similarly,

$$P_3 = \frac{(N - n_1 - n_2)!}{n_3!(N - n_1 - n_2 - n_3)!} \tag{7}$$

and so on. We thus have for the number of ways of distributing the N
particles among the various cells (remembering that all of the above
numbers are *independent*),

$$
\begin{aligned}
P(n_1, & n_2, \ldots, n_s, \ldots) \\
&= P_1 P_2 \cdots P_s \cdots \\
&= \frac{N!(N - n_1)! \cdots (N - n_1 \cdots - n_{s-1})! \cdots}{n_1!(N - n_1)!n_2!(N - n_1 - n_2)! \cdots n_s!(N - n_1 \cdots - n_s)! \cdots} \\
&= \frac{N!}{n_1!n_2! \cdots n_s! \cdots} = N! \prod_{s=1}^{\infty} \frac{1}{n_s!}
\end{aligned} \tag{8}
$$

The above formula is of course just the equation from probability
theory for the number of ways in which N distinguishable objects can be
put into an ordered array of boxes, with prescribed numbers $n_1, n_2, \ldots,$
$n_s, \ldots$ in each box.

We must now calculate the number of ways in which the n_s particles
in each cell can be arranged in the g_s energy levels included in that cell.

Clearly, since the exclusion principle does not act to limit the number of particles in each energy state, each of the n_s particles is equally likely to be in any one of the g_s states. Thus there are g_s ways in which the first particle can be put into the sth cell, and for each of these there are also g_s ways for the second particle, and so on. The total number of distinct distributions of the n_s particles among the g_s levels of the sth cell is therefore just $g_s{}^{n_s}$. We therefore have for the number of distinct microscopic distributions

Case (a)
$$P(n_s) = N! \prod_{s=1}^{\infty} \frac{g_s{}^{n_s}}{n_s!}$$
(9)

B. Identical, Indistinguishable Particles of Half-integral Spin. In this case, the indistinguishability of the particles prevents us from knowing *which ones* of the N particles have been placed in each cell, so that the only distinguishing feature of a given microscopic distribution is *which of the g_s levels* of a given cell are occupied by the n_s particles, and the exclusion principle now requires that not more than one particle occupy a given energy level.

The number of distinguishable arrangements of the n_s particles among the g_s energy levels of the sth cell may be found as follows: If we first *imagine the particles to be distinguishable*, we easily see that the "first" particle can be put in any one of the g_s levels, and for each one of these choices the "second" particle can be put in any one of the g_s − 1 remaining levels, and so on. Thus the number of sequences in which the n_s particles can be put into the g_s levels is (recall that the exclusion principle requires that $g_s \geq n_s$)

$$P'_s = g_s(g_s - 1) \cdots (g_s - n_s + 1) = \frac{g_s!}{(g_s - n_s)!}$$
(10)

However, the actual indistinguishability of the particles now requires that we shall not count as distinct distributions the various possible permutations of the particles among themselves, so that we must divide the above number by the number of permutations of n_s particles (that is, by $n_s!$). Thus we have

$$P_s = \frac{P'_s}{n_s!} = \frac{g_s!}{n_s!(g_s - n_s)!}$$
(11)

Finally, we have for the total number of microscopic distributions that can lead to the given macroscopic distribution, recalling that the various P_s are *independent*,

Case (b)
$$P(n_s) = \prod_{s=1}^{\infty} P_s = \prod_{s=1}^{\infty} \frac{g_s!}{n_s!(g_s - n_s)!}$$
(12)

C. Identical, Indistinguishable Particles of Integral Spin. In this case the indistinguishability of the particles again prevents us from knowing which of the N particles have been placed in each cell, but now the exclusion principle does not act to limit the population of a given energy level. The number of distinct arrangements of the n_s particles among the g_s energy levels may be found for this case by the following rather simple pictorial device: Consider the sth cell to consist of a linear array of $n_s + g_s - 1$ holes into which either white pegs (particles) or black pegs

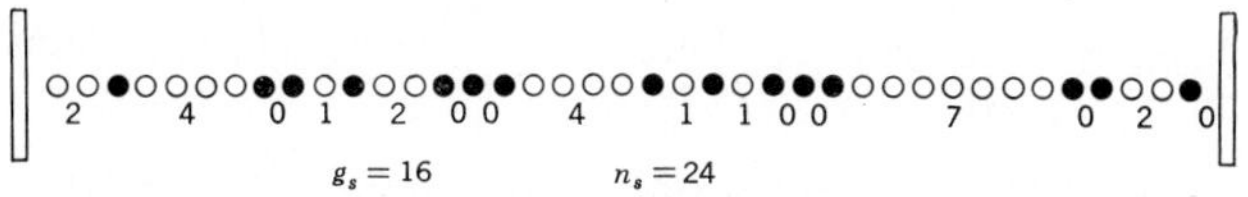

FIG. 10-1. Illustrating one method of finding the number of ways of placing n_s indistinguishable particles into g_s energy levels, with any number of particles being allowed in each level.

(partitions) can be inserted, and let us consider the various distinguishable permutations of n_s white pegs and $g_s - 1$ black pegs among the various holes. Note that $g_s - 1$ partitions are just sufficient to separate the entire cell into g_s intervals (energy levels), and that the n_s remaining holes, which are separated into groups by these partitions, then represent the distribution of the particles among the various energy levels. This is illustrated diagrammatically in Fig. 10-1. This figure shows a cell which consists of $g_s = 16$ energy levels populated by $n_s = 24$ particles. The populations of the various energy levels in this example are: 2 particles in the first energy level, 4 in the second, 0 in the third, and so on, 0 particles being in the 16th level.

Thus we see that the number of distinct permutations of the black and white pegs among the holes is just the same as the number of distinct arrangements of our n_s indistinguishable particles among the g_s energy levels. Now, the number of permutations of $n_s + g_s - 1$ *distinguishable* objects is just $(n_s + g_s - 1)!$. This, combined with the fact that in these various sequences a permutation of the particles among themselves, *or of the partitions among themselves*, does not lead to a distinguishably different arrangement, gives for our desired number of *distinct* arrangements,

$$P_s = \frac{(n_s + g_s - 1)!}{n_s!(g_s - 1)!} \tag{13}$$

Thus we have finally for this case,

Case (c)
$$P(n_s) = \prod_{s=1}^{\infty} \frac{(n_s + g_s - 1)!}{n_s!(g_s - 1)!} \tag{14}$$

D. The Maxwell-Boltzmann, Fermi-Dirac, and Einstein-Bose Distribution Laws. With the three expressions (9), (12), and (14) for the number of distinct microscopic states leading to a given macroscopic distribution, we are now prepared to determine for what distribution the number of microscopic states is a maximum for each of the three cases. For each case our problem is to determine $n_s(\epsilon_s)$ so that

$$P(n_s) = \text{maximum}$$

subject to the auxiliary conditions (3) and (4),

$$\sum_{s=1}^{\infty} n_s = N = \text{const} \quad \text{and} \quad \sum_{s=1}^{\infty} \epsilon_s n_s = E = \text{const}$$

We are dealing with such large numbers of particles ($N \approx 10^{23}$) and even with such large numbers of energy levels in a single cell ($g_s \approx 10^8$) that we may regard the various expressions as being *continuous functions* of *continuous variables* n_s and thus treat the problem by the methods of ordinary calculus. For convenience, we now change the problem to the equivalent one of maximizing the *logarithm* of $P(n_s)$, where for our three cases

Case (a)

$$\ln P(n_s) = \ln N! + \sum_{s=1}^{\infty} (n_s \ln g_s - \ln n_s!)$$

Case (b)

$$\ln P(n_s) = \sum_{s=1}^{\infty} [\ln g_s! - \ln n_s! - \ln (g_s - n_s)!] \tag{15}$$

Case (c)

$$\ln P(n_s) = \sum_{s=1}^{\infty} [\ln (n_s + g_s - 1)! - \ln n_s! - \ln (g_s - 1)!]$$

We must now maximize the value of $\ln P$ with respect to all small variations of the various cell populations which can occur while still satisfying the auxiliary conditions. That is, if each n_s changes by a small amount δn_s, we must have for the change in $\ln P$,

(a) $\delta(\ln P) = 0$

for any (small) values of δn_s that satisfy

(b) $\delta N = \sum_{s=1}^{\infty} \delta n_s = 0 \tag{16}$

(c) $\delta E = \sum_{s=1}^{\infty} \epsilon_s \, \delta n_s = 0$

This problem is easily solved through the use of Lagrange's *method of undetermined multipliers,* as follows: We first introduce two fixed parameters, α and β, called *undetermined multipliers,* and form the equation

$$\delta(\ln P) - \alpha \, \delta N - \beta \, \delta E = 0 \tag{17}$$

This equation will certainly be true for any values of α and β, if the three conditions (16) are satisfied. Treating case (*a*) first, we may write for $\delta(\ln P)$

$$\delta(\ln P) = \sum_{s=1}^{\infty} (\ln g_s - \ln n_s) \, \delta n_s \tag{18}$$

since[1] $\delta(\ln n!) \approx (\ln n) \, \delta n \qquad$ if $\delta n \ll n$

Equation (17) then takes the form, for this case,

$$\sum_{s=1}^{\infty} (\ln g_s - \ln n_s - \alpha - \beta \epsilon_s) \, \delta n_s = 0 \tag{19}$$

Now, the two auxiliary conditions may be considered to allow *all except two* of the δn_s's to be chosen arbitrarily. Suppose, to be specific, that δn_1 and δn_2 are the two *dependent* δn_s's and that all the others are independent. We then see that, if we assign such values to α and β that the coefficients of δn_1 and δn_2 vanish, *the coefficient of every δn_s must vanish,* because of the independence of the remaining δn_s's. Thus the maximum value of $\ln P$ is the one defined by the distribution $n_s(\epsilon_s)$, α, and β which satisfies the relations

Case (*a*) $\ln g_s - \ln n_s - \alpha - \beta \epsilon_s = 0$

for all values of s. Similar treatment of cases (*b*) and (*c*) leads to the two corresponding equations

Case (*b*) $- \ln n_s + \ln (g_s - n_s) - \alpha - \beta \epsilon_s = 0$
Case (*c*) $\ln (n_s + g_s) - \ln n_s - \alpha - \beta \epsilon_s = 0$ (20)

[In case (*c*) we neglect unity compared with g_s.]   Solving these equations for $n_s(\epsilon_s)$ then yields the three distributions

Case (*a*) $n_s = \dfrac{g_s}{e^{\alpha + \beta \epsilon_s}}$ Maxwell-Boltzmann

Case (*b*) $n_s = \dfrac{g_s}{e^{\alpha + \beta \epsilon_s} + 1}$ Fermi-Dirac   (21)

Case (*c*) $n_s = \dfrac{g_s}{e^{\alpha + \beta \epsilon_s} - 1}$ Einstein-Bose

[1] This result can also be obtained by using the Stirling approximation for $n!$.

The Maxwell-Boltzmann distribution is, of course, so named because it is the same distribution as was obtained in classical statistical mechanics; we now see that its occurrence in quantum mechanics is associated with the distinguishability of the particles (or other physical quantities) whose energy distribution it describes. The discreteness or continuity of the available energy levels is not involved in any essential way in this distribution law. The second and third laws are so named in recognition of the workers who first studied the effects of *indistinguishability* upon the expected distributions of particles which are or are not also affected by the exclusion principle. We shall study all of these laws in greater detail shortly.

EXERCISES

10-1. Carry out in detail the derivation of the three distribution laws (21) from Eqs. (15), (3), and (4).

10-2. Test the percentage accuracy of the relation $\delta \ln n! = \ln n \, \delta n$ for $n = 5$, 20, and 400 and $\delta n = \pm 1$, ± 2, and ± 5.

E. Evaluation of the Constant Multipliers. In order to apply the distribution laws derived above to actual physical problems, the constant multipliers α and β must be evaluated. This can be done in a formal way by inserting each of the distribution laws into the auxiliary condition equations (3) and (4). Thus we must have

$$\sum_{s=1}^{\infty} \frac{g_s}{e^{\alpha+\beta\epsilon_s} \pm 1, 0} = N \qquad \sum_{s=1}^{\infty} \frac{g_s\epsilon_s}{e^{\alpha+\beta\epsilon_s} \pm 1, 0} = E \qquad (22)$$

for the three distributions. Unfortunately these equations can seldom be solved for α and β in closed form for a given system, so that other methods must be used. It is best to regard α as being connected, by the first equation of each of the above sets, with N and with β for a given system, and thus to regard β as of more general importance. Indeed we can show that β has a *universal character* closely related to the thermodynamical quantity *temperature*, independent of the particular system being considered. In order to show this, let us briefly consider a system which consists of a mixture of two different kinds of identical particles of any of the three types being considered. In this case we divide the energy range into cells $\Delta\epsilon_s$ for the first kind of particle and into cells $\Delta\epsilon_l$ for the second kind of particle, and we then maximize the joint relative probability

$$P(n_s, n_l) = P(n_s)P(n_l) \qquad (23)$$

subject to the auxiliary conditions

(a)
$$\sum_{s=1}^{\infty} n_s = N_1$$

(b)
$$\sum_{l=?}^{\infty} n_l = N_2 \qquad (24)$$

(c)
$$\sum_{s=1}^{\infty} n_s \epsilon_s + \sum_{l=1}^{\infty} n_l \epsilon_l = E$$

where $P(n_s)$ and $P(n_l)$ are expressions for the number of distinct microscopic distributions giving rise to the macroscopic distributions $n_s(\epsilon_s)$ and $n_l(\epsilon_l)$ for two kinds of particle, N_1 and N_2 are the numbers of the two kinds of particle present in the mixture, and E is the total energy of the system. It is clear that an application of the method of undetermined multipliers to this new problem will require *three* undetermined multipliers, which we may call α_1 and α_2 for the conditions (a) and (b), and β for condition (c). This procedure will clearly lead to distributions for the two kinds of particle that are similar in form to those obtained previously, where each distribution involves *a different α but precisely the same parameter β*.

EXERCISE

10-3. Carry through the analysis outlined above and verify the above conclusion.

The above analysis not only shows that β has a universal character independent of the nature of the particles or the system but also provides us with a convenient means of evaluating this parameter. Thus let us take as one constituent of our mixture of particles a relatively small number of massive, distinguishable particles[1] inside a large rectangular container, and let the remaining particles be of any type, possibly even acted upon by different forces inside this same container. We know from the above analysis that the two kinds of particle will be described by distribution laws that are precisely the same as would describe the state of each kind of particle *in the absence of the other*, such that the total energy E is composed of two parts, E_1 and E_2, for the two constituents. (This is the statistical mechanical analogue of Dalton's *law of partial pressures* in the thermodynamics of nonreacting gas mixtures.) Having

---

[1] It is not strictly necessary that the particles be distinguishable, but only that they be sufficiently sparsely distributed that the exclusion principle does not affect the occupancy of the energy states available to the particles.

chosen the one component as we have done, however, we see that this part of the system must obey the *Maxwell-Boltzmann* statistics. Thus we must have for this part

$$n_s = \frac{g_s}{e^{\alpha_1 + \beta \epsilon_s}} \tag{25}$$

In order to evaluate α_1 and β for this constituent of the system, we must now insert the proper expression for the number of one-particle energy eigenstates g_s within the sth cell $\Delta\epsilon_s$. From Eq. 4-10(5) we find g_s to be

$$g_s = \frac{4\pi V (2m^3)^{1/2}}{h^3} \epsilon_s^{1/2} \Delta\epsilon_s$$

where $V = abc$ is the volume of the rectangular container. Inserting this value into Eq. (25), we obtain

$$n_s = \frac{4\pi V (2m^3)^{1/2}}{h^3} \epsilon_s^{1/2} \frac{\Delta\epsilon_s}{e^{\alpha_1 + \beta \epsilon_s}} \tag{26}$$

which is now in a form for which α_1 and β can be evaluated. If we treat $\Delta\epsilon_s$ as a differential quantity, we may write the two auxiliary conditions (22) in the form

$$N = \frac{4\pi V (2m^3)^{1/2}}{h^3} e^{-\alpha_1} \int_0^\infty \epsilon^{1/2} e^{-\beta\epsilon} \, d\epsilon$$

$$= \frac{4\pi V (2m^3)^{1/2}}{h^3} e^{-\alpha_1} \beta^{-3/2} \, \Gamma\left(\frac{3}{2}\right) \tag{27}$$

and

$$E_1 = \frac{4\pi V (2m^3)^{1/2}}{h^3} e^{-\alpha_1} \int_0^\infty \epsilon^{3/2} e^{-\beta\epsilon} \, d\epsilon$$

$$= \frac{4\pi V (2m^3)^{1/2}}{h^3} e^{-\alpha_1} \beta^{-5/2} \, \Gamma\left(\frac{5}{2}\right) \tag{28}$$

EXERCISES

10-4. Supply the missing steps in the derivation of Eq. (27) and (28).

10-5. Evaluate α_1 and β from the above expressions.

Ans.: $\qquad \beta = \dfrac{3N}{2E_1} \qquad \alpha_1 = \ln\dfrac{1}{2}\dfrac{4\pi V (2m^3\pi)^{1/2}}{h^3 N}\left(\dfrac{2E_1}{3N}\right)^{3/2}$

We can now complete the derivation of the quantum distribution laws by observing, on the one hand, that the above expressions yield the result

$$\frac{1}{\beta} = \frac{2E_1}{3N} = \frac{2\epsilon_{av}}{3} \tag{29}$$

for the parameter β for our partial system composed of a relatively few identical, distinguishable particles and, on the other hand, that the assumed form of the partial system is just that of an *ideal gas thermometer*, whose thermodynamical properties are well known. Thus we may make use of the fact that, for an ideal gas (i.e., one which is sufficiently *dilute* that molecular interactions do not affect its properties) the average molecular kinetic energy is just

$$\epsilon_{av} = \frac{3kT}{2} \tag{30}$$

where T is the absolute temperature and k is Boltzmann's constant. Comparing these two expressions we see that the universal quantity β is just

$$\beta = \frac{1}{kT} \tag{31}$$

Since β is the same for all types of particles in any given mixture, we must now have as the final form for the three distribution laws

Case (a) $n_s = g_s e^{-\alpha - \epsilon_s/kT}$ Maxwell-Boltzmann

Case (b) $n_s = \dfrac{g_s}{e^{\alpha + \epsilon_s/kT} + 1}$ Fermi-Dirac   (32)

Case (c) $n_s = \dfrac{g_s}{e^{\alpha + \epsilon_s/kT} - 1}$ Einstein-Bose

where the quantity g_s will depend upon the particular one-particle system under consideration, and α upon the number N of particles included in it, and the temperature.

10-2. Applications of the Quantum Distribution Laws: General Considerations

Before proceeding to direct applications of the three quantum distribution laws just obtained, it will be instructive to examine some of the more general properties exhibited by these laws, and to compare in a schematic way the behavior of systems governed by each of them. It is first desirable to emphasize again the significance of the constant β for a system. As was indicated in the previous section, β is a universal quantity, independent of the nature of the particular system being considered. Our evaluation of this parameter for a simple classical system whose properties are familiar has shown that β is essentially a measure of the *temperature* of the system. Thus the fact that β is the same for various systems "in contact" with one another (i.e., systems which are so situated that energy can be exchanged between them, as for example a mixture of particles of different types, or the physical contact, at a boundary, of two otherwise pure systems) is now seen to be equivalent to the thermodynamical result that any arrangement of physical systems in contact with one another will attain the *same temperature* when equilibrium has been attained.   That this result should emerge independently in the quantum statistics is quite gratifying.

Many of the properties of the three distribution laws can be expressed in a form that is independent of the detailed properties of a system by writing the laws in terms of the quotient

$$\frac{n_s}{g_s} = n(\epsilon) \tag{1}$$

which we call the *occupation index* of a level of energy ϵ.   This quantity is just the average number of particles per energy level at the energy ϵ, *and it is independent of the detailed distribution of the energy levels as a function of ϵ.*   The similarities and differences of the three distribution laws can now be illustrated graphically by plotting the occupation index vs. energy for various temperatures and various values of α.   Such graphs are shown schematically in Fig. 10-2.   As is indicated in Fig. 10-2a, the *Maxwell-Boltzmann distribution* is a pure exponential function at all temperatures and for all values of α.   The parameter α of course varies with the temperature, the distribution of the possible energy levels, and the number of particles in the system, but *the constant β, which determines the relative occupation index of any two levels, depends solely upon the temperature.*

Figure 10-2b shows the dependence upon energy of the occupation index for a system described by *Fermi-Dirac statistics*.   Here we find several features of interest:

1. The occupation index of a Fermi-Dirac system *never exceeds unity.*
   This can of course be seen directly from the form of this distribution law, since the denominator must always exceed unity for any positive or negative value of α.   This result, which is of course just an expression of the exclusion principle that was used in deriving the distribution, is in contrast with the results for either of the other distributions, where occupation indices exceeding unity are quite possible.

2. For most Fermi-Dirac systems, α becomes negative and rather large at very low temperatures.   This means that the exponential term will be less than unity for energies less than $-\alpha kT$.   Thus for low energies ($\epsilon \ll -\alpha kT$) the occupation index is essentially equal to unity.   As ϵ takes on values that exceed $-\alpha kT$, the occupation index rapidly diminishes toward zero over a range of ϵ of the order of a few times kT. The decrease of $n(\epsilon)$ for $\epsilon > -\alpha kT$ approaches a pure exponential decay with the same decay constant as for Maxwell-Boltzmann statistics, but with the energy scale shifted by approximately $-\alpha kT$.   The

energy $-\alpha kT$ is called the *Fermi energy* or *Fermi level* and is designated ϵ_m.

3. At very high temperatures, α becomes positive and large. In this limit the exponential factor is very much larger than unity for all energies, so that the distribution approaches a Maxwell-Boltzmann distribution at sufficiently high temperatures.

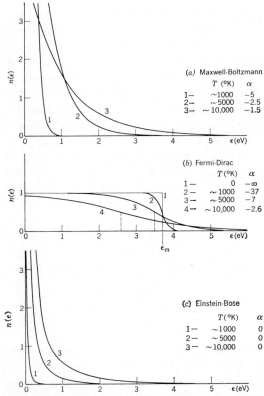

FIG. 10-2. Schematic illustration of the nature of the three distribution laws. (a) The Maxwell-Boltzmann distribution is a pure exponential curve at all temperatures and for all values of ϵ. This distribution falls off by a factor e for each kT increase in ϵ. (b) The Fermi-Dirac distribution is characterized by unit occupation index at low temperatures and low energies. At very low temperatures, $n(\epsilon)$ is equal to unity for energies less than a certain energy ϵ_m and equal to zero for energies greater than ϵ_m. At higher temperatures, $n(\epsilon)$ drops from unity toward zero less and less sharply, and at sufficiently high temperatures and low densities closely resembles the Maxwell-Boltzmann distribution. At all temperatures $n(\epsilon)$ varies approximately exponentially with $\epsilon - \epsilon_m$ for $\epsilon > \epsilon_m$, and $1 - n(\epsilon)$ varies approximately exponentially with $\epsilon_m - \epsilon$ if $\epsilon < \epsilon_m$. (c) The Einstein-Bose distribution differs significantly from the Maxwell-Boltzmann distribution only if $n(\epsilon) \gg 1$, since it is only if several particles are in the same state that the symmetric character of the Einstein-Bose wave function plays a significant role.

The physical interpretation of the limiting forms of the Fermi-Dirac distribution just described is quite clear. In the low-temperature range, the system tends to be in its lowest possible energy state. This state is one in which the lowest N levels are occupied and all higher ones are empty. At high temperatures, on the other hand, the occupation index of all levels becomes quite small, so that the exclusion-principle limitation upon the occupancy of a level has little effect upon the distribution of the particles among the various energy levels.

Figure 10-2c shows the Einstein-Bose distribution for $n(\epsilon)$. In this case the occupation index for levels of low energy, rather than being restricted to *smaller* values than for Maxwell-Boltzmann systems, actually is *greater* than for these systems. A case of fundamental interest is that of a *photon gas*, for which $\alpha = 0$. In this case the occupation index shows a dependence upon ϵ much different than for Maxwell-Boltzmann systems. This will be discussed later. At all temperatures, on the other hand, the distribution approaches the Maxwell-Boltzmann form for energies ϵ large compared with kT.

A common property shared by all of these distributions is that, at sufficiently high energies, where the occupation index becomes considerably less than unity, *each distribution becomes essentially a Maxwell-Boltzmann distribution*, because the exponential factor greatly exceeds unity in this energy range. In many situations of physical interest, this fact is of great utility in deducing the relative occupation indices of two levels which both lie in this "high energy" range. Thus for such a case the relative occupation indices of two levels having energies ϵ_1 and ϵ_2 would be

$$\frac{n(\epsilon_1)}{n(\epsilon_2)} = \frac{e^{\epsilon_2/kT}}{e^{\epsilon_1/kT}} = e^{(\epsilon_2 - \epsilon_1)/kT} \tag{2}$$

In applying this result to a given system, the various energy levels which lie in this "Maxwell-Boltzmann" range are measured from some convenient zero point, which is often taken to be the ground-state energy of the system. Since for a given temperature the energy states lying more than a few kT above this ground-state energy are only very sparsely occupied, the occupation index of these higher levels is very nearly proportional to the so-called *Boltzmann factor* $e^{-\epsilon'/kT}$, where $\epsilon' = \epsilon - \epsilon_0$ and ϵ_0 is the ground-state energy.

To illustrate the range of utility of the Boltzmann factor, we shall cite a few of its common applications:

The excited states of atoms. By far the most common mode of excitation of an atom is by thermal collisions with other atoms. In this case one may regard the problem in terms of several types of systems in thermal equilibrium, one system being that of the atoms regarded as independent particles inside a large container and another being an *individual atom* regarded as an independent system having various possible energy states. (One might wish to treat all similar atoms as constituting a single system having highly degenerate discrete energy levels, but this is simply a matter of normalizing the system to correspond to the total number of atoms present.)   One can thus determine, using the Boltzmann factor, the relative equilibrium populations of the various energy states of the atom as a function of the temperature, and from this information it is possible to predict more or less quantitatively how the intensities of various spectral transitions should vary with temperature. This information can also be used in the reverse way, namely, given the relative intensities of various spectral transitions, to find the temperature of the source.

Excited states of molecules.   Information similar to that described above for the excited states of atoms can be found for various other types of systems having energy levels that can be thermally excited.   Thus the excitation of the various *vibrational and rotational states of gaseous molecules* can be analyzed by using a similar procedure.

Dissociation or ionization.   The probability of thermal dissociation or ionization of atomic or molecular systems depends, among other things, upon the Boltzmann factor.

Relative population.   The relative populations of the various quantized orientation states of certain polarized or magnetized bodies in a field can be evaluated by using the Boltzmann factors for the various states.

EXERCISES

10-6. (a) Find the relative numbers of hydrogen atoms that are in the ground state and in the first, second, and third excited states ($n = 2,3,4$) in the solar chromosphere.   Assume T to be 5000°K and remember to include the *statistical weights* of the various levels.   (b) Why does the Balmer series appear prominently in *absorption* in the solar spectrum?

10-7. For gaseous HCl in the $v = 0$ vibrational state, compare the relative numbers of molecules in the rotational states $K = 1$ and $K = 0$ at (a) $T = 300°$K and (b) $T = 3.00°$K.   *Ans.:* (a) 2.7, (b) 1.4×10^{-4}.

10-8. If the equilibrium concentration of ortho-hydrogen in hydrogen is 1 per cent at a certain temperature T_0, at what temperature will the concentration be reduced to 0.01 per cent?   *Hint:* Remember to include all statistical weights.   *Ans.:* $T_1 = 0.597T_0$.

10-9. (a) Evaluate the magnetic moment of a neutral hydrogen atom in its ground state.   (b) If atomic hydrogen gas at 100°K is subjected to a magnetic field of 10^4 gauss, what will be the percentage difference between the numbers of atoms aligned parallel to the field and opposite to it?

10-10. At what temperatures will dissociation of the following systems into their component parts become significant, that is, for what temperature does the Boltzmann-factor exponent equal -1? (*a*) Molecular

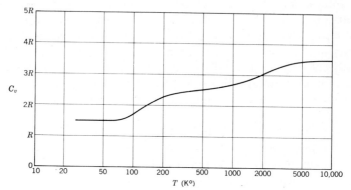

Fig. 10-3. Molar specific heat of gaseous hydrogen.

deuterium into deuterium atoms. (*b*) Atomic deuterium into ionized deuterium. (*c*) A deuteron into a proton and neutron. (Binding energy of d $\approx 2.2 \times 10^6$ eV.)

10-11. Using the values for the vibrational frequencies in the electronic ground state given in Appendix D, compute the relative numbers of molecules in the $v = 1$ and $v = 0$ vibrational states at 300 and 1000°K for the diatomic molecules given in the table below. (See Herzberg,[1] p. 123.)

Molecule	$\dfrac{n(1)}{n(0)}$ (300°K)	$\dfrac{n(1)}{n(0)}$ (1000°K)	Molecule	$\dfrac{n(1)}{n(0)}$ (300°K)	$\dfrac{n(1)}{n(0)}$ (1000°K)
H_2	2.16×10^{-9}	2.51×10^{-3}	O_2		
HCl			S_2		
N_2			Cl_2		
CO			I_2	3.60×10^{-1}	7.36×10^{-1}

10-12. Derive an expression for the relative numbers of molecules occupying the various possible rotational states at temperature T.

Ans.:
$$N_J = A(2J + 1) \exp - \frac{J(J + 1)\hbar^2}{2m_r r_J{}^2 kT} \tag{3}$$

10-13. Show that the rotational level which has the greatest population

---

[1] "Spectra of Diatomic Molecules," D. Van Nostrand Company, Inc., Princeton, N.J., 1950.

will have an approximate J value

$$J_{max} = \left(\frac{m_r r_J{}^2 kT}{\hbar^2}\right)^{1/2} - \frac{1}{2}$$

10-14. Interpret the general trend of the absorption maxima for HCl vapor shown in Fig. 9-13 as being due to the thermal distribution of the HCl molecules among their various rotational states. With this interpretation, calculate the approximate temperature of Dr. Imes' laboratory.

10-15. The specific heat of gaseous H_2 is shown in Fig. 10-3 as a function of the temperature. Interpret this function semiquantitatively, using the quantum statistics.

10-3. Applications of the Maxwell-Boltzmann Distribution Law

We shall now examine in somewhat greater detail some of the properties of two quantum-mechanical systems that are governed by Maxwell-Boltzmann statistics. These systems were treated at an early stage in the development of quantum mechanics, and they still retain a certain practical importance because of their fundamental nature. Both systems involve one-dimensional harmonic oscillators. One system is that of a solid, crystalline body in a state of thermally excited elastic vibration, and the problem is to describe the energy content of the vibrating body as a function of the temperature. The other system is that of the electromagnetic field at thermal equilibrium inside a hollow enclosure, and the problem is to obtain the distribution of energy among the normal modes of the cavity.

For brevity we shall first consider the thermal properties of a number of harmonic oscillators of the same frequency. Such oscillators may be treated as identical *distinguishable* elements whose possible energy states are $(n + \frac{1}{2})\hbar\omega$. If there are dN such oscillators, and if dN_s signifies the number of oscillators in the sth quantum state, we may write

$$\frac{dN_s}{dN} = A \exp\left(-\epsilon_s/kT\right) = A \exp\left[-(s + \tfrac{1}{2})\hbar\omega/kT\right] \tag{1}$$

To evaluate A, we have

$$1 = \sum_{s=0}^{\infty} \frac{dN_s}{dN} = A \exp\left(-\tfrac{1}{2}\hbar\omega/kT\right) \sum_{s=0}^{\infty} \exp\left(-s\hbar\omega/kT\right) \tag{2}$$

The sum may easily be evaluated if we observe that it is of the well-known form $1 + x + x^2 + \cdots = 1/(1 - x)$. Thus

$$A = \exp\left(\tfrac{1}{2}\hbar\omega/kT\right)[1 - \exp\left(-\hbar\omega/kT\right)] \tag{3}$$

Using this result, we find the average energy of the oscillators to be

$$
\begin{aligned}
E_{\mathrm{av}} &= \sum_{s=0}^{\infty} \epsilon_s \frac{dN_s}{dN} \\
&= A \sum_{s=0}^{\infty} \epsilon_s e^{-\epsilon_s/kT} \\
&= AkT^2 \frac{\partial}{\partial T} \sum_{s=0}^{\infty} e^{-\epsilon_s/kT} \\
&= \tfrac{1}{2}\hbar\omega + \frac{\hbar\omega}{e^{\hbar\omega/kT} - 1}
\end{aligned}
\tag{4}
$$

EXERCISES

10-16. Supply the missing steps in the derivation of Eq. (4).

10-17. Show that, as kT becomes very large compared to the interval $\hbar\omega$ between successive energy states of the oscillators, $E_{\mathrm{av}} \to kT$.

We are now ready to apply the above results to our two problems.

A. *The Vibrational Energy of a Crystalline Solid.* The earliest experimental results concerning the energy content of solids showed that the specific heat, expressed on an atomic basis for various chemical elements, is roughly the same from element to element, and is also roughly independent of the temperature (law of Dulong and Petit). The value of the specific heat is approximately $3R$, where $R = N_0 k$ is the *gas constant per mole*. This experimental result can readily be understood on the basis of the classical statistical mechanics, for if a solid body contains N_0 atoms per mole, the number of *degrees of freedom* required to describe its motion is $3N_0$. The law of equipartition of energy then requires that each degree of freedom possess an average energy kT, so that the specific heat of such a body should be

$$
C_v = \frac{dE}{dT} = 3N_0 k = 3R
\tag{5}
$$

With the advent of techniques for producing very low temperatures, however, it was found that the specific heats of all solid bodies *fall rapidly toward zero with decreasing temperature*, and in the lowest temperature range have values proportional to the cube of the absolute temperature. This is quite in disagreement with the above results expected upon the basis of classical statistical mechanics.

Einstein (1911) was the first to recognize that the problem is fundamentally a quantum-mechanical one. He assumed that the $3N_0$ degrees of freedom of a solid body can be represented by $3N_0$ harmonic oscillators of the same frequency, and derived Eq. (4) for the average energy of these oscillators. The derivative of this expression with respect to temperature, times $3N_0$, would then be the specific heat of such a system. Einstein's theory of specific heats agreed qualitatively with experiment at low temperatures, but it failed to show the required T^3 variation at the lowest temperatures. Nernst and Lindemann introduced several different frequencies for the oscillators, but this did not fundamentally improve the situation.

The basic problem was solved by Debye (1912), who pointed out that the thermal vibrations of a solid body should be treated as being just an excitation of the various *elastic modes of vibration* of the solid body. The frequencies of these various oscillations must be distributed more or less as are the frequencies of the modes of vibration of a three-dimensional elastic continuum, except that the actual atomic nature of a real solid leads to a *limited number* of modes of vibration $(3N_0)$ and to a *maximum frequency*, corresponding to motion of adjacent atoms in opposite directions. Debye assumed for simplicity that the frequency spectrum was of the same form as that for an elastic *continuum:*

$$dN = N(\omega)\, d\omega = A\omega^2\, d\omega \tag{6}$$

where dN is the number of modes of vibration whose frequencies lie between ω and $\omega + d\omega$. This distribution was supposed to be valid up to a maximum frequency ω_{max} such that

$$3N_0 = \int_0^{\omega_{max}} dN = \int_0^{\omega_{max}} A\omega^2\, d\omega$$

which gives
$$A = \frac{9N_0}{\omega_{max}^3} \tag{7}$$

If the Debye frequency distribution is combined with Eq. (4), one obtains for the total energy of the solid body,

$$E = \frac{9N_0 \hbar \omega_{max}}{8} + \frac{9N_0 \hbar}{\omega_{max}^3} \int_0^{\omega_{max}} \frac{\omega^3\, d\omega}{\exp\,(\hbar\omega/kT) - 1} \tag{8}$$

Differentiation of this energy with respect to the temperature then leads to the specific heat:

$$C_v = \frac{\partial E}{\partial T} = \frac{9N_0 \hbar^2}{\omega_{max}^3 kT^2} \int_0^{\omega_{max}} \omega^4 \frac{\exp\,(\hbar\omega/kT)\, d\omega}{[\exp\,(\hbar\omega/kT) - 1]^2} \tag{9}$$

The above equation can be put in a more concise form by changing the variable of integration to the dimensionless variable $x = \hbar\omega/kT$.

Thus we have

$$C_v = \frac{9N_0k^4T^3}{\hbar^3\omega_{max}^3}\int_0^{\hbar\omega_{max}/kT}\frac{x^4e^x\,dx}{(e^x-1)^2}$$
$$= 9R\left(\frac{T}{\theta}\right)^3\int_0^{\theta/T}\frac{x^4e^x\,dx}{(e^x-1)^2} \tag{10}$$

where $R = N_0k$ and $\theta = \hbar\omega_{max}/k$. θ is called the *Debye characteristic temperature*. The Debye theory of specific heats is thus a *one-parameter*

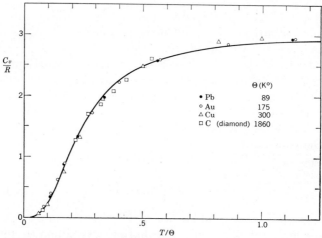

FIG. 10-4. Comparison of the Debye specific-heat curve and the observed specific heats of a number of simple substances.

theory. The agreement of this theory with experiment is quite good for many solids, as is shown in Fig. 10-4.

EXERCISES

10-18. Supply the missing steps in the derivation of Eq. (10) from Eqs. (4), (6), and (7).

10-19. Show that, at very low temperatures, the specific heat given by Eq. (10) is proportional to T^3, and at very high temperatures approaches $3R$.

Shortly after Debye put forward his theory of specific heats, Born and von Karman (1913) pointed out that the true frequency spectrum of an elastic solid body would differ markedly from the $A\omega^2$ form suggested by Debye and that an exact theory would require the use of the more complex spectrum. Various investigations have shown that there are indeed certain differences between the Debye theory and experiment and that these differences can be accounted for in terms of the expected differences between the simple Debye frequency spectrum and the actual spectrum. The Debye theory is so very simple, however, and has proved so satisfactory, that it is still used for most practical applications.

B. *The Spectral Distribution of Cavity Radiation.* As our second application of Maxwell-Boltzmann statistics to quantum-mechanical systems, we shall derive the Planck distribution law for black-body radiation. Our quantum-mechanical model of the electromagnetic field will be the same one that was used in connection with radiative transitions of atoms; that is, we assume that the electromagnetic field inside a cavity can be represented by a number of independent harmonic oscillators which correspond to the various normal modes of oscillation of the cavity. The *frequency spectrum* for this case is just the same as was used for deriving the Rayleigh-Jeans law (Sec. 2-1B):

$$N(\lambda) = \frac{8\pi}{\lambda^4} \tag{11}$$

per unit volume. Combining this with Eq. (4) and writing the entire expression in terms of λ, we have

$$dU_\lambda = \frac{4}{c} I(\lambda,T) \, d\lambda = \left[\frac{1}{2} \frac{hc}{\lambda} + \frac{hc}{\lambda(e^{hc/\lambda kT} - 1)} \right] N(\lambda) \, d\lambda$$

$$I(\lambda,T) = \frac{\pi hc^2}{\lambda^5} + \frac{2\pi hc^2}{\lambda^5} (e^{hc/\lambda kT} - 1)^{-1} \tag{12}$$

which is just the same as Eq. 2-1(13) except for the *zero-point* energy of the oscillators represented by the first term. This zero-point energy, which is infinite when integrated over the entire spectrum, can be ignored on the ground that a slightly different model of the electromagnetic field can be devised in which the zero-point energy of the field oscillators does not enter (Chap. 6). We can in any case ignore this term because the energy represented by it cannot be transferred from the electromagnetic field to the material walls of the cavity, so that it will not show up in measurements of the energy density of the field.

EXERCISES

10-20. Under the physical conditions that often hold in the laboratory and in nature, one can treat ordinary gases as being a mixture of various kinds of identical, distinguishable particles, because the occupation index of the free-particle energy states is very small over the entire energy range. Derive an expression for the distribution of molecular speeds, assuming the Maxwell-Boltzmann statistics to be valid for the gas molecules. *Ans.:* $dn = 4\pi N(m/2\pi kT)^{3/2} \exp(-\frac{1}{2}mv^2/kT)v^2 \, dv$.

10-21. Find the average speed of the molecules of a perfect gas. *Ans.:* $\bar{v} = (8kT/\pi m)^{1/2}$.

10-4. Applications of the Fermi–Dirac Distribution Law

We have already mentioned a possible application of the Boltzmann factor to the problem of finding the approximate relative populations of the various excited states of a complex atom that is in thermal equilibrium with its surroundings. In such a treatment use is made of the fact that the Fermi-Dirac distribution, which applies more exactly to this problem, approaches the pure exponential form of the Maxwell-Boltzmann distribution for sufficiently high energies. It is of course true that the Fermi-Dirac statistics in their complete form could be used in this problem if it were desired to evaluate more accurately distribution of the atoms among the various excited levels, including those near the ground state itself. This application is rarely made, however, and we shall not consider it further.

A. The Free-electron Theory of Metals. By far the most useful and fruitful application of the Fermi-Dirac statistics is to the problem of finding the energy distribution of a large number of independent Fermi particles which occupy a relatively small container. For many purposes, this is quite a good approximate model of *the conduction electrons inside a metal.* The cancellation of the electron repulsions by the atoms of the metal makes it possible to regard the electrons as being *nearly free particles* moving inside a box (i.e., inside the boundary of the piece of metal). The *density* of the electrons in a metal is so great, however, that most of the lowest energy states are occupied, and the exclusion principle is therefore very much in operation. Such a model of a metal is called a *completely degenerate Fermi-gas.* We shall consider only a few of the properties of such a system at this time, since a broader treatment of both metals and insulators is given in the next chapter.

If we take as our one-particle system a free particle that is constrained to remain inside a rectangular box, we can immediately use the results of Exercise 4-26 and write for the distribution of energy levels, allowing for a two-fold spin degeneracy for each of the one-particle levels,

$$g_s = \frac{8\pi V (2m^3)^{1/2}}{h^3} \epsilon_s^{1/2} \, \Delta\epsilon_s \tag{1}$$

Thus we have for the system of N_0 particles, from Eq. 10-1(32b),

$$n_s = \frac{[8\pi V (2m^3)^{1/2}/h^3]\epsilon_s^{1/2} \, \Delta\epsilon_s}{e^{\alpha + \beta\epsilon_s} + 1} \tag{2}$$

In order to evaluate α, we first assume that the temperature is very low, so that *all of the lowest N_0 one-particle energy states, and none of the higher energy states, are occupied.* Thus the occupation index is equal to unity for ϵ less than a certain energy ϵ_m (i.e., the *Fermi energy*) and is equal to zero for ϵ greater than this energy. We must therefore have, utilizing the discussion of Sec. 10-2, $-\alpha kT = \epsilon_m$, where ϵ_m is that kinetic energy for which

$$\sum_{\epsilon=0}^{\epsilon=\epsilon_m} g_s(\epsilon) = N_0$$

If we substitute Eq. (1) and approximate the summation by an integral, we obtain

$$N_0 = \frac{8\pi V (2m^3)^{1/2}}{h^3} \int_0^{\epsilon_m} \epsilon^{1/2} \, d\epsilon \quad \text{or} \quad \epsilon_m = \frac{h^2}{2m} \left(\frac{3N_0}{8\pi V} \right)^{2/3} \tag{3}$$

The value of ϵ_m may be expected to be roughly independent of the temperature as long as the range of ϵ over which the occupation index

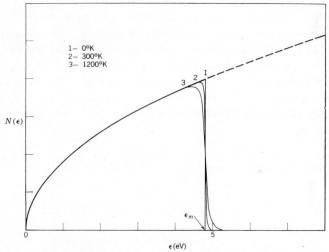

FIG. 10-5. The distribution of electron kinetic energies inside a metal according to the free-electron theory (schematic).

falls to zero is small compared to ϵ_m, that is, for temperatures such that $\epsilon_m/kT \gg 1$. This condition is satisfied for ordinary metals up to quite high temperatures (of the order of thousands of degrees) so that regarding ϵ_m or αkT as a constant is an excellent approximation.

The kinetic-energy distribution of the electrons of a metal should thus have the shape indicated in Fig. 10-5 and described by

$$N(\epsilon) = \frac{3N_0 \epsilon_m^{-3/2} \epsilon^{1/2}}{2[e^{(\epsilon - \epsilon_m)/kT} + 1]} \tag{4}$$

EXERCISES

10-22. Supply the missing steps in the derivation of Eq. (4).

10-23. Use Eq. (3) to calculate the value of the Fermi energy ϵ_m (in electron volts) for Na, Cu, and Al, assuming that these metals have one, one, and three free electrons per atom, respectively.   *Ans.:* See Table 10-1.

10-24. Over what range of temperature for Na, Cu, and Al is $\epsilon_m \geq 10kT$?

The results obtained above may be somewhat extended with the help of certain information obtained from the photoelectric effect to yield a quite satisfactory semiquantitative theory of metals.   This so-called *free-electron* theory of metals will now be described briefly, together with some of its applications.

The distribution function (4) for the electron kinetic energies inside a metal was obtained on the assumption that the electrons are free, non-interacting particles constrained to move inside a box having impenetrable walls.   Actually, of course, electrons can and do escape from inside metals under certain conditions, so that a more nearly correct potential function would be one which is approximately constant inside the metal and which undergoes a rapid increase at the boundary, attaining a new constant value at great distances outside the metal.   Furthermore, in the space immediately outside the metal the potential energy should vary as $1/r$ because of the electrostatic image forces which act upon an electron in this region.

TABLE 10-1. AVERAGE POTENTIAL ENERGY W_a, FERMI ENERGY ϵ_m, AND WORK FUNCTION $e\phi$ FOR A NUMBER OF METALS‡

Metal	Valence	W_a, eV	ϵ_m, eV	$e\phi$, eV
Li	1	6.9	4.72	2.2
Na	1	5.0	3.12	1.9
K	1	3.9	2.14	1.8
Cu	1	11.1	7.04	4.1
Ag	1	10.2	5.51	4.7
Au	1	10.3	5.54	4.8
Be	2	....	14.3	
Ca	2	7.5	4.26	3.2
Al	1	8.6	5.63	3.0
Al	3	14.7	11.7	3.0

‡ Adapted by permission from F. Seitz, "The Modern Theory of Solids," McGraw-Hill Book Company, Inc., New York, 1940.

Let us define the zero of potential energy of an electron to be the value attained when the electron is removed to infinity, and designate the average potential energy inside the metal as $-W_a$. In order to determine W_a, we make use of one of the fundamental experimental facts of the photoelectric effect, namely, the existence of an *energy threshold* for the incident quanta, below which they are unable to eject electrons from a metal (Sec. 2-1C). This effect may be interpreted to mean that there are essentially no electrons inside a metal whose total energies—kinetic plus potential—are sufficiently great that they can be elevated above

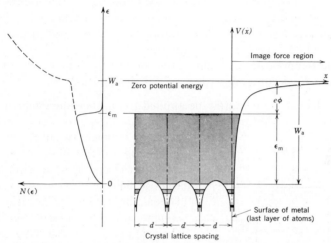

Fig. 10-6. Schematic diagram of the potential energy of an electron inside and outside a metal. The potential energy is taken to be zero at a great distance outside the metal and to vary as r^{-1} near the metal surface because of the electrostatic image forces. Inside the metal the potential energy attains a value $-W_a$ of magnitude equal to the Fermi energy ϵ_m plus the observed photoelectric threshold energy $e\phi$. The distribution of available energy states is shown at the left in a form similar to the diagram of Fig. 10-5.

zero energy by a quantum of less energy than the threshold value $h\nu_0$; and the abrupt onset of photoelectric current for quanta above this energy may further be taken to mean that there are *many* electrons which can be elevated above zero energy by quanta of energy only slightly greater than $h\nu_0$. It is clear that this abrupt drop in number of electrons with increasing energy is in complete accord with the properties of the Fermi-Dirac distribution, and we may therefore identify this limiting electron energy with the Fermi energy ϵ_m. Our modified picture of a metal is therefore as follows:

1. The electrons inside a metal are approximately free particles moving under the influence of a potential-energy function something like that shown in Fig. 10-6.
2. In the kinetic-energy range $0 < \epsilon < W_a$ the states available to each electron are approximately the same as for a particle moving in an

impenetrable box whose shape is that of the metal, distributed as described by the function (1).

3. In the range $\epsilon > W_a$ the available energy states become more dense, since for positive energy the electrons are free to move in the space outside the metal as well as inside it. (When they are outside the metal, their kinetic energies are $\epsilon - W_a$.)

4. At absolute-zero temperature the lowest N of the above levels are occupied. The highest occupied level has a kinetic energy $\epsilon = \epsilon_m$ and a total energy $-h\nu_0$. These relationships are shown in Fig. 10-6. The quantity $h\nu_0$ is called the *work function* of the metal and is usually designated $e\phi$, where ϕ is measured in *volts*.

5. At room temperature and above, the energy cutoff at $-e\phi$ is not sharp but is smeared out as described by the distribution (4).

This model can immediately be applied to another phenomenon—that of the *contact potential* of a metal. If a parallel-plate capacitor is constructed by using two different metals for the two plates and if these plates are short-circuited by a wire, a small electric field is observed to exist between the plates. The field corresponds to a difference of potential of the two plates of a volt or so, and this difference in potential is called the *contact difference of potential* of the two metals.

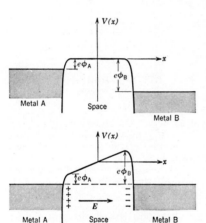

FIG. 10-7. Energy relationships of two dissimilar metals, illustrating the origin of the contact difference of potential.

It has long been recognized that the contact-potential series of metals is correlated with the photoelectric thresholds of the metals, the more strongly electropositive metals (sodium, potassium, cesium, etc.) being sensitive to light of longer wavelength. The relation can be seen diagrammatically in Fig. 10-7, which refers to the parallel-plate capacitor made of two different metals mentioned above. In Fig. 10-7a the quantities $e\phi$ are shown for the two metals. The zero of potential is at infinity, and the metals are supposed to be insulated from one another and uncharged. If these two metals are now connected by a wire, a few electrons will move from metal A into metal B, changing the relative electrostatic potential of the two metals, *until the tops of the energy distributions become equalized*. This is shown in Fig. 10-7b. In this situation an electric field, represented by the sloping connecting line, exists in the space between the two metals. From the diagram it is clear that the potential difference represented by the electric field equals $\phi_B - \phi_A$.

The above analysis also shows clearly the energetic relationships that must exist if these two metals are used as the anode and cathode of a photoelectric cell. (If an external battery is included in the circuit, the difference in height of the tops of the energy distributions is equal to the battery emf.) In order to be removed from one metal with sufficient energy to reach the other, an electron must be given enough energy to surmount the highest point of potential barrier in the space between the two metal surfaces. For the case shown in Fig. 10-7b, this point is $e\phi_B$, so that the cutoff point of current flow in this cell is at a frequency corresponding to the higher work function of the two metals, *irrespectively of whether metal A or metal B is irradiated.*

EXERCISES

10-25. Devise a means of measuring the smaller work function for the case shown in Fig. 10-7 and illustrate the resulting energy relations by a diagram similar to that of Fig. 10-7b.

10-26. The electrodes of a photoelectric cell are made of different metals. When monochromatic light of wavelength 0.2480μ falls upon electrode A, it is found that a retarding potential of 2.5 V must be furnished by an external battery to just stop the flow of current, while if this light falls upon electrode B, a retarding battery potential of 1.5 V is required. Find the work functions of the two metals.

The electronic-energy distribution given by Eq. (4) can also be used in combination with the measured work function to deduce many of the *thermionic* properties of metals. In particular, it is possible to derive equations which correctly describe the variation of thermionic emission current with temperature and with voltage under conditions of so-called *temperature-limited emission.*

The basic physical circumstances involved in thermionic emission are that, at a given temperature, there are a certain number of electrons whose thermal energies are sufficiently great that they can escape from the metal, provided they are sufficiently near one of the boundary surfaces and are moving in a suitable direction. The situation can be analyzed as follows: Suppose the surface of the metal is a plane surface perpendicular to the x-direction. In this case the y- and z-components of motion of an electron would not be affected as the electron passes through the metal surface, so that the energy required to escape must come from the x-component of the electron's motion. Thus the electron will escape

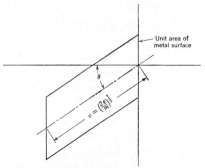

FIG. 10-8. Illustrating the incidence of an interior electron upon the surface of a metal. Electrons within a column of cross-sectional area $\cos\theta$ and length v will, in unit time, strike a unit area of the metal surface at an angle θ with the normal.

only if it arrives at the metal surface with an energy $p_x^2/2m$, due to its x-component of motion, which exceeds $\epsilon_m + e\phi$ (see Fig. 10-6) i.e.,

$$\epsilon > W_a \qquad \text{and} \qquad \cos\theta > \left(\frac{W_a}{\epsilon}\right)^{1/2} \tag{5}$$

where θ is the angle between the direction of motion of the electron and the normal to the metal surface (see Fig. 10-8).

The number of electrons of energy ϵ that will strike a unit area of the metal surface, at an angle θ per unit time and solid angle, with sufficiently high energy to escape, is

$$dN = \frac{N(\epsilon)\,d\epsilon}{V}\cos\theta\,v\,\frac{d\Omega}{4\pi} \tag{6}$$

$$= \frac{3N_0\epsilon_m^{-3/2}\epsilon^{1/2}\,d\epsilon\,\cos\theta\,2^{1/2}\epsilon^{1/2}2\pi\,\sin\theta\,d\theta}{2[e^{(\epsilon-\epsilon_m)/kT}+1]\,Vm^{1/2}4\pi}$$

The current emitted from this surface will thus be approximately

$$I = e\int dN = \frac{3eN_02^{1/2}}{\epsilon_m^{3/2}4Vm^{1/2}}\int_{W_a}^{\infty}\int_0^{\cos^{-1}(W_a/\epsilon)^{1/2}}\epsilon e^{-(\epsilon-\epsilon_m)/kT}\sin\theta\,\cos\theta\,d\theta\,d\epsilon$$

$$= \frac{3eN_02^{1/2}}{\epsilon_m^{3/2}4Vm^{1/2}}\int_{W_a}^{\infty}\tfrac{1}{2}\epsilon\left(1-\frac{W_a}{\epsilon}\right)e^{-(\epsilon-\epsilon_m)/kT}\,d\epsilon$$

if we now let $\epsilon = W_a + \Delta$, so that $\epsilon - \epsilon_m = e\phi + \Delta$, we have

$$I = \frac{3eN_02^{1/2}}{\epsilon_m^{3/2}8Vm^{1/2}}e^{-e\phi/kT}\int_0^{\infty}\Delta e^{-\Delta/kT}\,d\Delta$$

$$= \frac{4\pi me}{h^3}(kT)^2e^{-e\phi/kT} \tag{7}$$

This equation is called the *Richardson-Dushman equation of thermionic emission*. It was first derived thermodynamically, and it differs from a corresponding classical equation, derived upon the assumption that the electrons obey Maxwell-Boltzmann statistics, in the constant coefficient and in the power of T which multiplies the exponential term. This classical equation is

$$I_{cl} = \frac{eN_0}{(2\pi m)^{1/2}}(kT)^{1/2}e^{-e\phi/kT} \tag{8}$$

It is difficult to make sufficiently precise experimental measurements of thermionic emission to test the validity of the Richardson-Dushman equation because of the large effects of surface contaminations and surface irregularities, the anisotropy of the electron motion due to crystal-lattice effects, and the possibility of internal reflection of an electron which actually has sufficient energy to escape the metal.[1] In spite of these experimental difficulties and theoretical shortcomings, however, this equation is considered to be a valid description of thermionic emission when written in the form

$$I = AT^2 e^{-e\phi/kT} \tag{9}$$

where A is an experimentally determined coefficient for a given metal.

EXERCISES

10-27. Supply the missing steps in the derivation of Eq. (7).

10-28. Evaluate the theoretical value of A in Eq. (9). *Ans.: $A = 1.23 \times 10^6$ A m^{-2}°K^{-2}*

In addition to a variation with temperature, the saturation thermionic current from an emitting surface also shows some dependence upon the *electric-field strength* at the metal surface. There are here two physical effects in operation; both of them tend to increase the thermionic current with increasing field strength at a given temperature. One of the effects is a lowering of the effective work function with increasing field, so that more electrons can "get over the top" of the surface barrier. Another effect is the escape, through the surface barrier, of electrons whose energies are classically not great enough to surmount the barrier.

The first of these effects can be analyzed qualitatively as shown in Fig. 10-9. In this figure the potential-energy curve for zero applied field is the "image-force" potential that results from the attraction between an electron outside the surface and its image inside the surface. (When the image potential becomes equal to $-W_a$, it is arbitrarily cut off and set equal to a constant inside the metal.) If a uniform electric field is now applied to the metal, the net potential is the sum of the zero-field

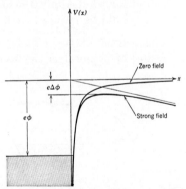

FIG. 10-9. Schematic diagram of the variation of potential energy near a negatively charged metal surface, showing the lowering of the surface barrier by the field.

---

[1] See Chap. 4.

potential and the potential due to the field. It is seen from the figure that this resultant potential has a lower maximum value than does that of the zero-field case, which corresponds to a *smaller work function* for the metal, and thus leads to a *larger thermionic current*.

EXERCISE

10-29. (*a*) Show that the analytic expression for the zero-field potential energy represented in Fig. 10-9 is

$$V(x) = -W_a \qquad x < \frac{e^2}{16\pi\varepsilon_0\, W_a}$$

$$= -\frac{e^2}{16\pi\varepsilon_0 x} \qquad x > \frac{e^2}{16\pi\varepsilon_0\, W_a}$$

where the metal "surface" is taken to be at $x = 0$. (*b*) Superimpose a uniform field E upon the above potential and show that the work function is lowered by an amount

$$e\, \Delta\phi = -\frac{e^{3/2} E^{1/2}}{(4\pi\varepsilon_0)^{1/2}}$$

(*c*) Find for what applied field strength the effective work function of a surface would be reduced by 0.1 V. *Ans.: 7 × 10⁶ V m⁻¹.*

The second of the effects of an applied field upon the thermionic emission of a metal is the so-called *barrier-penetration*, or *tunnel*, *effect*

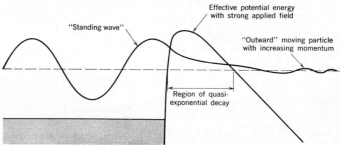

Fig. 10-10. Schematic diagram of the wave function of an electron near the boundary of a metal with a strong applied field, illustrating the phenomenon of barrier penetration leading to field emission.

which was discussed in Chap. 4. Unlike the effect described above, which can be treated classically, the barrier-penetration effect is a quantum-mechanical one. It can be understood qualitatively with the aid of Fig. 10-10.

This figure shows the region near a surface of a metal to which a very strong electric field is applied. In this case the wave functions used previously for a particle inside the metal are not quite the true energy eigenfunctions for the system, since with the application of the external field the potential function differs from the one for which the "standing-wave" solutions of Exercise 4-26 were obtained. There are *two* differences, in fact, which are of importance for our present considerations; one of them is that we have been applying to a *real metal* the results of an idealized treatment which assumed an *infinite* potential outside the metal. However, the fact that the potential energy is actually finite outside the metal introduces only very small differences into the wave functions and energy levels that represent bound states for the electrons, the main effects being a slight penetration of the particles into the classically forbidden space outside the metal and the existence of a continuum of unbound states of positive total energy, as was discussed for a one-dimensional particle in Sec. 4-7.

The second difference is that a new classically allowed region outside the metal is now present, so that the wave functions for the bound states do not diminish exponentially for all distances outside the metal but instead become oscillatory (with decreasing wavelength) within this new classically allowed region.   Now, if a particle is initially in one of the original eigenstates *inside* the metal, the effect of this new feature is to introduce a slow "leakage" of the particle through the barrier.   The wave function inside the metal is then almost a perfect standing wave, corresponding to an eigenstate of the original system, but, when analyzed in terms of a motion of the particles in directions toward or away from the metal surface, it gives a slightly higher probability that the particle inside is moving toward the surface than away from it.   The wave function in the forbidden region still retains its quasi-exponential character, but now has a small amount of an increasing "exponential" term in addition to its former pure "exponential" attenuation.   Outside the metal in the new classically allowed region, the wave function corresponds to a pure motion *away* from the metal, with a wavelength that decreases with increasing distance in accordance with the steady outward acceleration of the particle.

The above description of the penetration of a particle through the electrical potential barrier at the surface of a metal can be made quantitative, but the analysis is rather involved.   It suffices to say that the fundamental mechanism involved is as described above, and that this gives a satisfactory description of the so-called *field emission* of electrons from cold metal surfaces.   We shall also find in Chap. 15 that this same barrier-penetration phenomenon is responsible for the α-particle radioactivity of the heavy elements.

A satisfactory elementary theory of *electrical conductivity* can also be devised by using the free-electron theory of metals, following an analysis quite similar to that used by Drude, Lorentz, and others in the classical electron theory. The primary difference between the present model and the classical electron theory is that most of the electrons are prevented by the exclusion principle from contributing to the conductivity. The reason for this is that, although an electric field acting upon an electron tends to bring about a change in its momentum, this cannot happen unless a large number of nearby unoccupied energy states are available for the electron to jump to. Thus it is only those electrons whose energies are near ϵ_m that can be shifted into nearby unoccupied states by the electric field.

In spite of the fact that a small number of electrons contribute to the electrical conduction, the conductivity is still in good agreement with experimental values because of a compensating increase in the *mean free paths* of the electrons inside the metal. The equation obtained for the conductivity is

$$\sigma = \frac{e^2 N l(\epsilon_m)}{V m u(\epsilon_m)} \tag{10}$$

where $l(\epsilon_m)$ is the mean free path of electrons at the Fermi energy ϵ_m and $u(\epsilon_m) = (2\epsilon_m/m)^{1/2}$ is the speed of an electron at this energy. In order to obtain agreement with observed conductivities, $l(\epsilon_m)$ must be of the order of $1 - 6 \times 10^{-2}$ μ at ordinary temperatures. Mean free paths of this size can be deduced theoretically, using a somewhat extended theory.

EXERCISE

10-30. (a) Calculate the speed $u(\epsilon_m)$ for Cu, using the results of Exercise 10-23. (b) Using the observed resistivity of Cu (1.68×10^{-8} Ω m), calculate the mean free path $l(\epsilon_m)$ for the conduction electrons. *Ans.:* $l(\epsilon_m) = 3.9 \times 10^{-2}$ μ. (c) To how many crystal-lattice spacings does this mean free path correspond?

One of the great difficulties with the classical electron theory was that it assumed equipartition of energy between the vibrations of the metal crystal lattice and the free electrons, so that the free electrons should have made a sizable contribution to the *specific heat* of a metal. This contribution, which should have been equal to $\frac{3}{2}R$, was not observed, almost the entire specific heat being well accounted for by the lattice vibrations. This apparent anomaly is easily accounted for by the Fermi-Dirac statistics, for we can now see that, as the temperature changes, only a *small fraction* of the electrons undergoes an increase in energy. The Fermi distribution leads to a specific heat which varies linearly with T for low temperatures. Such a variation is observed experimentally.

EXERCISE

10-31. Use Eq. (4) to derive an expression for the specific heat of electrons, valid for low temperatures.

Note:
$$\int_0^\infty \frac{x\,dx}{e^x + 1} = \frac{\pi^2}{12}$$

Ans.:
$$C_e \approx \frac{\pi^2 N_0 k^2}{2\epsilon_m} T$$

The *paramagnetic susceptibility* of the alkali metals can be accounted for by considering the Fermi-Dirac distribution (4) to be a superposition of two equal distributions, one for each of the two possible spin orientations of an electron along a given axis. When a magnetic field is applied, the energy of each state in these two distributions is shifted up and down, respectively, by an amount $e\hbar B/2m$ (see Fig. 10-11). When equilibrium is attained, the tops of the two distributions must be at the same total energy, which requires that there be *more electrons oriented antiparallel to the field than parallel to it.* The difference in population between the two distributions is then, per unit volume,

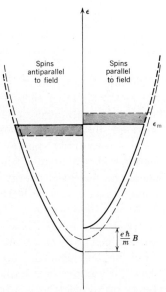

FIG. 10-11. Illustrating the origin of the paramagnetic susceptibility of the alkali metals.

$$\frac{\Delta N}{V} = \frac{N(\epsilon_m)}{V} \frac{e\hbar B}{2m} = \frac{3N_0}{2V\epsilon_m} \frac{e\hbar B}{2m} \quad (11)$$

The *magnetization* which results from this population difference is, recalling Eq. 5-5(4),

$$M = \frac{\Delta N}{V} \frac{e\hbar}{2m} = \frac{3N_0 e^2 \hbar^2 B}{8m^2 V \epsilon_m}$$

Thus the susceptibility, which agrees approximately with experiment, is

$$\chi = \frac{M}{H} \approx \frac{\mu_0 M}{B} \approx \frac{3\mu_0 N_0 e^2 \hbar^2}{8m^2 V \epsilon_m} \quad (12)$$

The foregoing description of the free-electron theory of metals using the Fermi-Dirac statistics has served to illustrate a few of the more important and distinctive features of the theory. Taken as a whole, the above applications of the Fermi-Dirac distribution provide a most satisfactory semiquantitative theory of metals which serves as an appropriate zero-order approximation to which to apply the perturbations introduced by the periodic nature of the crystal lattice. In the next chapter we shall consider some features of the extension of this theory to other solids.

B. *The Thomas-Fermi Statistical Atomic Model.* As a final example of the application of Fermi-Dirac statistics, we shall describe a useful approximate representation of the distribution of electrons about the nucleus of an atom.[1] This model, of considerable value in such problems as that of the scattering of charged particles by atoms, is based upon the following ideas: If we imagine the space surrounding an atom to be divided up into a large number of small "boxes" such that the potential energy does not change appreciably throughout any given one, the energy states available to an electron within each box will be equal to the potential energy for that box plus varying amounts of kinetic energy appropriate to the size and shape of the box. We then imagine the multitude of energy states so obtained to be occupied by the electrons of the atom, subject to the exclusion principle and also to the condition that *no states of positive total energy can be occupied*—for an electron in such a state would not be bound within the atom. Since the number of energy states available to a particle in a box is proportional to the volume of the box, it does not matter in what way the space around the atom is divided, provided only that the boxes are sufficiently small that the potential remains essentially constant throughout each one. Clearly, the lowest energy states available will be those in the boxes nearest the nucleus where the potential energy is lowest; and since there will also be more bound states available in these cells, we expect to find most of the electrons near the nucleus. There are not enough bound states close to the nucleus to accommodate all of the electrons, however, so that some of the outlying cells will also be occupied.

Let us now consider the situation analytically. Consider a small "box" of volume $d\tau$ at a distance r from the nucleus, where the potential energy of an electron is $V(r)$. The electrons within this volume should possess total energies extending from $V(r)$ up to 0, that is, kinetic energies ranging from 0 up to $-V(r)$ and distributed according to $\epsilon^{1/2}$, as are the energies of free particles. We may now make use of a result obtained in Chap. 4 in connection with the energy states available to a particle inside a rectangular box, namely, that a constant number of energy states are available per unit "volume" of phase space. In Exercise 4-28 it was found that there is in fact just one state per h^3 "volume" of phase space independent of the dimensions of the box. Applying this result to the atom, we conclude that, if p_0 is the momentum which gives an electron a kinetic energy $-V(r)$, the number of occupied states within the

---

[1] Thomas, *Proc. Cambridge Phil. Soc.*, **23**, 542 (1927); Fermi, *Z. Physik*, **48**, 73 (1928).

volume element $d\tau$ must be

$$dn = 2\,\frac{\tfrac{4}{3}\pi p_0{}^3\, d\tau}{h^3} \tag{13}$$

where the factor 2 takes account of the two spin orientations of an electron. If we substitute for p_0 its value

$$p_0 = [-2mV(r)]^{\frac{1}{2}} \tag{14}$$

and for $d\tau$ the value

$$d\tau = 4\pi r^2\, dr \tag{15}$$

we obtain

$$dn = \frac{32\pi^2 r^2 [-2mV(r)]^{\frac{3}{2}}}{3h^3}\, dr \tag{16}$$

Equation (16) describes the radial distribution of the electrons in terms of the potential-energy distribution $V(r)$. But $V(r)$ is itself produced both by the nuclear charge *and by the electron distribution*. Using Gauss's law we may evaluate the radial electric field, and from this the potential energy, in terms of the electron distribution:

$$-4\pi r^2 \varepsilon_0 \frac{dV}{dr} = -Ze^2 + e^2 \int_0^r \frac{dn}{dr}\, dr \tag{17}$$

If we differentiate this equation, we may eliminate dn/dr, using Eq. (16), and obtain an equation for $V(r)$ alone:

$$-\frac{d}{dr}\left(r^2 \frac{dV}{dr}\right) = \frac{8\pi e^2 r^2 [-2mV(r)]^{\frac{3}{2}}}{3\varepsilon_0 h^3} \tag{18}$$

The boundary conditions which determine the two arbitrary constants in the solution of this second-order differential equation are

(a)
$$V(r) \xrightarrow[r \to 0]{} -\frac{Ze^2}{4\pi\varepsilon_0 r} + \text{const}$$

(b)
$$rV(r) \xrightarrow[r \to \infty]{} 0 \tag{19}$$

These are, of course, just the requirements that the field near the nucleus must be that of the nucleus alone, and that there must be no net charge inside an infinite sphere (assuming that we are dealing with a neutral atom). Equation (18) is commonly written in dimensionless form by defining new variables χ and x to replace V and r:

(a)
$$V(r) = -\frac{Ze^2}{4\pi\varepsilon_0 r}\,\chi$$

(b)
$$r = \frac{1}{2}\left(\frac{3\pi}{4}\right)^{\frac{2}{3}} \frac{4\pi\varepsilon_0 \hbar^2}{me^2 Z^{\frac{1}{3}}}\, x = \frac{0.886}{Z^{\frac{1}{3}}}\, a_0 x \tag{20}$$

In terms of these new dimensionless variables, Eq. (18) becomes

$$x^{\frac{1}{2}} \frac{d^2\chi}{dx^2} = \chi^{\frac{3}{2}} \tag{21}$$

and the boundary conditions (19) become

(a) $\chi(0) = 1$
                                                                    (22)
(b) $\chi(\infty) = 0$

The solution of Eq. (21) is shown graphically in Fig. 10-12. Quite accurate numerical values are available in tabular form.[1]

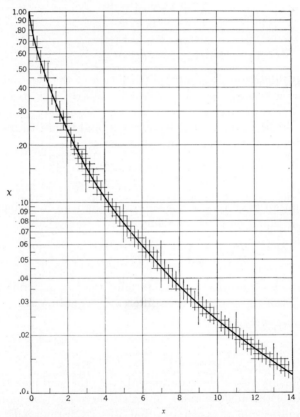

FIG. 10-12. Graph of the function $\chi(x)$ for the Thomas-Fermi atomic model.

The validity of the Thomas-Fermi atomic model depends principally upon the following two conditions:

1. There must be a sufficient number of electrons in the atom to make a statistical treatment applicable.
2. The potential $V(r)$ must not change by a large fraction of its own value within a distance of the order of the electron wavelength for the occupied states.

[1] Bush and Caldwell, *Phys. Rev.*, **38**, 1898 (1931).

Both of these conditions are more nearly satisfied for larger values of Z; the model becomes quite accurate for the heavy elements.

EXERCISES

10-32. Derive Eq. 17, using Gauss's law.
10-33. Derive Eq. (18) from Eqs. (17) and (16).
10-34. Derive Eq. (21) from Eqs. (18) and (20).
10-35. Show that Eqs. (22) and (19) are equivalent.

10-5. Applications of the Einstein-Bose Distribution Law

To conclude our discussion of the three quantum-mechanical distribution laws, we shall now cite two examples of the application of the Einstein-Bose statistics to physical systems. The first of these applications is to a *photon gas*, the application for which the distribution law was developed by Bose in 1924. The second application is to *liquid helium*, which has only recently been successfully treated, although some of the features of the condensation of a Bose system at a finite temperature were correctly deduced by Einstein soon after the development of the Bose statistics.

A. The Photon Gas. In this section we shall obtain the Planck distribution law of black-body radiation by treating the electromagnetic radiation *as a gas of particles of zero rest mass*—quanta—inside an isothermal enclosure. This point of view represents the opposite extreme to the one used in Sec. 10-3, in which the radiation was treated from the standpoint of its *wave properties*. This alternative derivation thus illustrates once more the equivalence of the wave properties and discrete-particle properties of electromagnetic radiation.

We must first consider the effect upon the Einstein-Bose distribution law that is brought about by a property of the photon gas inside an isothermal cavity that is not shared by other systems that we have so far discussed, namely, that *the number of particles is not necessarily constant*, since quanta can be introduced into and removed from the cavity by the matter which forms the cavity wall. Clearly, this situation requires the removal of the restriction 10-1(3) that the number of particles in the system remain constant. The effect upon the distribution law 10-1(21) is to make α identically zero, which is of course equivalent to having not introduced it into the analysis in the first place. Our distribution law for quanta must then be

$$n_s = \frac{g_s}{e^{\epsilon_s/kT} - 1} \tag{1}$$

We must next determine g_s.   This would appear to be difficult, because of the fact that a massless particle such as a photon is not described correctly by the Schroedinger (nonrelativistic) theory.   This difficulty can be easily sidestepped by observing that the quantization of the kinetic energy of the particles inside a rectangular container, treated in Sec. 4-10 and Exercise 4-22, amounted actually to a quantization of the square of the momentum, since an integral number of wavelengths had to be fitted into each of the rectangular dimensions of the box.   We can use this same result for the case of particles of zero rest mass because the de Broglie relation 2-2(1) is valid relativistically, so that the various possible values of the square of the momentum of a quantum should be

$$p_{lmn}^2 = \frac{h^2}{4}\left(\frac{l^2}{a^2} + \frac{m^2}{b^2} + \frac{n^2}{c^2}\right) \tag{2}$$

The relativistic relation $pc = \beta W$ between energy and momentum for a particle then gives for the various possible values of the *energy*,

$$\epsilon_{lmn}^2 = c^2\,\frac{h^2}{4}\left(\frac{l^2}{a^2} + \frac{m^2}{b^2} + \frac{n^2}{c^2}\right) \tag{3}$$

where l, m, and n are again positive integers ≥ 1.   The distribution of ϵ^2 for the present problem is therefore similar to that of ϵ in our previous application.   Thus we must have, from Eq. 4-10(5),

$$g_s = N(\epsilon^2)\,d(\epsilon^2) = \frac{2\pi V}{c^3 h^3}\,\epsilon\,d(\epsilon^2)$$

$$= \frac{2\pi V}{c^3 h^3}\,2\epsilon^2\,d\epsilon \tag{4}$$

The number of quanta whose energies lie between ϵ and $\epsilon + d\epsilon$ is therefore (with *two* polarization states per energy level)

$$n_s = \frac{8\pi V \epsilon_s^2\,d\epsilon_s}{c^3 h^3(e^{\epsilon_s/kT} - 1)} \tag{5}$$

and the energy of these quanta is

$$n_s \epsilon_s = \frac{8\pi V \epsilon_s^3\,d\epsilon_s}{c^3 h^3(e^{\epsilon_s/kT} - 1)} \tag{6}$$

This gives for the spectral-energy density inside the cavity

$$dU = \frac{n_s \epsilon_s}{V} = \frac{8\pi \epsilon_s^3\,d\epsilon_s}{c^3 h^3(e^{\epsilon_s/kT} - 1)} \tag{7}$$

The Einstein relation $\epsilon = h\nu = hc/\lambda$ and the relation 2-1(3) can be used to put this distribution into the form 2-1(13).

The above derivation of the Planck distribution law, in which electro-

magnetic radiation is treated as a *gas of particles*, is not the only successful example of such an alternative treatment.   Perhaps an even more
intriguing case which can be similarly treated is that of the elastic vibrations of solids which was discussed in Sec. 10-3.   Although most of the
applications of the particle viewpoint are in this case rather specialized,
it is perhaps desirable to discuss the nomenclature that is used in such
applications.

It was mentioned in Chap. 6 in connection with the quantization of
continuous systems that the standing-wave pattern of a vibrating continuum enters into physical problems in just the same way as do the
eigenfunctions of a particle in a container, and that this fact constitutes
the theoretical basis for the wave-particle duality of matter or of radiation.   A similar duality exists for the case of a vibrating elastic solid; in
this case the "particles" whose motions and interactions are physically
equivalent to those of the normal modes of vibration are given the name
"phonon," an acoustical analogue of the electromagnetic photon.   A
thermally excited solid is thus treated as a *gas of phonons*.   The nature
of this duality, both for elastic and electromagnetic vibrations, is made
clearer by the following table, in which various physical situations are
described from the two alternative viewpoints.   In this table, the words
"particle" and "wave" are understood to refer generally to photons,
phonons, electrons, etc., and to their wave equivalents.

"Particle" description	*"Wave" description*
A particle is in an energy eigenstate inside a container.	The wave medium inside a container is in one of its normal modes of vibration.
A particle is situated near a certain point and is moving with a given momentum.	A wave packet is moving through the medium, is instantaneously localized near a certain point, and is propagating with a given wavelength in a certain direction.
The momentum of a particle is p.	The wavelength of a wave packet is $\lambda = h/p$.
The interaction between two particles is negligibly small.	The modes of vibration of the wave medium are not coupled with one another.
There is a small interaction between two particles; that is, collisions can occur.	There are small terms which couple together the various normal modes; that is, diffractive scattering can occur.
Two particles (of the same or of different kinds) collide elastically and are scattered.	Two wave packets interfere with one another and are mutually diffractively scattered.

It is an important fact that almost any given physical phenomenon can be analyzed—although not always with equal ease—by treating it entirely from the particle viewpoint or entirely from the wave viewpoint. Thus in treating electrical resistivity one can consider either that the electrons, regarded as wave packets, are diffracted by the periodic normal modes of vibration of a crystal lattice or that the electrons (particles) collide with and are scattered by the particles of the phonon gas. In the Compton effect (Chap. 12) an electron (wave) is diffractively scattered by an electromagnetic wave; or an electron (particle) is struck by a photon and both are scattered. In the photoelectric effect an incident electromagnetic wave interacts with the periodic electronic wave function inside a crystal and excites the electron into a new unbound eigenstate of positive energy; or a photon impinges upon a crystal and ejects an electron from it. The analytical procedures by which the two viewpoints can be correlated for a given case are unfortunately too complicated to be presented here; it must suffice to say that each viewpoint is well suited for the treatment of certain kinds of problems and that a familiarity with both points of view is needed for attacking original problems.

B. Liquid Helium. As our second application of the Einstein-Bose statistics, we shall investigate the qualitative nature of the superfluid transition of liquid helium at 2.2°K. Ordinary helium consists almost entirely of neutral atoms of the isotope ^{4}He. These atoms have *total angular momentum zero* and must therefore be treated according to Einstein-Bose statistics. Helium gas at atmospheric pressure condenses into a liquid at 4.3°K. This is an ordinary condensation, associated with van der Waals' forces between the helium atoms. At 2.18°K, however, a different type of condensation occurs within the liquid. At this temperature the liquid suddenly acquires a number of extraordinary properties which are most simply interpreted in terms of a *mixture of two fluids*, one of which retains the same general properties as are exhibited above the transition temperature and the other of which possesses the new properties responsible for the remarkable behavior of the liquid at lower temperatures. These two fluids are usually referred to as the *ordinary-fluid component* and the *superfluid component*, respectively. The transition temperature is called the λ-*point*.

The two liquids of the mixture are qualitatively quite different: Whereas the ordinary-fluid component possesses finite viscosity, thermal resistivity, and entropy, the superfluid component possesses *zero values for all these quantities* and is able to flow freely through the normal-fluid component. Furthermore, many of the gross properties of liquid helium show discontinuities at the λ-point. The most spectacular of these are the sudden, complete disappearance of viscosity and resistance to heat flow, but other properties such as the specific heat, the density, and the dielectric constant also show definite but less dramatic changes. Many ingenious experiments that have been performed demonstrate this extraordinary behavior and provide a strong experimental basis for an interpretation in terms of a two-fluid mixture. Some of the more spectacular experiments will now be described briefly.

Perhaps the most obvious indication of the existence of a new transition within the liquid is the striking behavior of the liquid as its temperature is lowered by letting it evaporate. If the vapor above the liquid is pumped away, the liquid supplies more vapor to replace it, and in this process becomes cooler. Just as for most liquids, however, the vapor can be pumped away so rapidly that the liquid begins to boil; that is, heat cannot be conducted through the liquid sufficiently rapidly to keep the liquid-vapor interface as hot as the interior of the liquid.

As the boiling progresses, the liquid cools and its saturation vapor pressure diminishes in the normal manner. However, as the λ-point is passed there is a sudden, spectacular cessation of boiling throughout the entire fluid and the upper surface of the liquid soon becomes perfectly calm. Evaporation continues, however, as is evidenced by the gradual lowering of the upper meniscus, and the temperature continues to fall, as does the vapor pressure. It is apparent that what has happened at the λ-point is that the *thermal conductivity of the liquid has suddenly become so great that the entire liquid is at essentially the same temperature.* The heat conductivity has in fact become infinite, so far as our ability to measure it is concerned.

With regard to the coefficient of viscosity of the liquid above and below the λ-point, different results are obtained for different kinds of experimental measurements of the quantity. In one method the rate of flow of the liquid through a fine capillary or a porous membrane is measured. In this way one finds that *the coefficient of viscosity suddenly becomes zero at the λ-point*, and remains zero below this temperature, provided the applied pressure difference which produces the flow is not too large. One of the problems in this experiment is that of making a sufficiently fine capillary tube. This has been done by inserting a metal wire into a small glass capillary tube and then drawing the glass and metal down to a much smaller diameter, the glass being brought into contact with the wire in the process. The differential contraction between the glass and wire upon cooling then provides a very tiny annular gap through which ordinary gases or liquids effectively cannot penetrate but through which the superfluid component flows with ease—several cubic millimeters per second being a not uncommon flow rate.

In contrast with the preceding method, the viscosity of the liquid can also be measured by observing the period of oscillation of a torsion pendulum whose inertial element consists of a stack of thin, light, closely

spaced mica disks immersed in the liquid[1] (Fig. 10-13). If the liquid
has a high viscosity, the liquid between the disks is dragged along and
contributes significantly to the moment of inertia of the disks. If the
viscosity is small, the moment of inertia is more nearly equal to that
of the disks alone. It is found that *there is no discontinuity in the coefficient of viscosity η at the λ-point* as meas-
ured in this way, but that η gradually
diminishes as the temperature is lowered,
and appears to be approaching zero at
absolute-zero temperature.

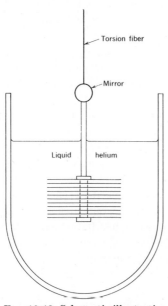

Another interesting phenomenon is that
of the so-called *fountain effect,* in which
the liquid which has flowed through a
porous plug is caused to spurt out of a
nozzle and thence fall by gravity back into
the original vessel. The apparatus which
illustrates this phenomenon is sketched in
Fig. 10-14. A porous plug connects two
vessels, one of which initially contains liquid helium. If the temperature is lowered
below the λ-point, the liquid flows through
the porous plug and attains an equilibrium
level equal to that outside. If heat is
now supplied to the second vessel, *the liq·
uid level rises and eventually overflows.* If
a constriction is placed at the top of this
vessel, the liquid spurts out and falls back
into the original vessel in a steady stream.

FIG. 10-13. Schematic illustration
of the torsion-pendulum method
of measuring the viscosity of
liquid helium.

One of the most remarkable properties exhibited by liquid helium
below the λ-point is that of the almost unattenuated propagation of *heat
pulses* at a finite speed of about 20 m s^{-1}. This phenomenon is called
"second sound," but it is actually not in any sense an acoustic phenomenon. No net mass flow is involved, nor is there any pressure change
inside the liquid—the only changes in the liquid are purely thermal in
nature. Second sound is produced if a variable heat source is immersed
in the liquid, and is detected by a temperature-sensing element. One
observes that the temperatures measured at the detector faithfully repro-
duce those "emitted" by the heater but are delayed in time in proportion
to the distance between the heater and receiver. Second sound can also
be reflected, focused, and diffractively scattered.

Perhaps the most extraordinary property of liquid helium, and one

[1] Andronikashvili, *J. Phys. U.S.S.R.,* **10**, 201 (1946).

of the least understood, is its ability to creep along the walls of the containing vessel as a thin film. If the partially filled beaker of liquid helium is suspended by thin wires inside the cryostat and is cooled below the λ-point, the liquid will flow up the inner wall of the beaker, over the top, down the outside wall, and drip from the bottom of the beaker. This surface flow is characterized by a certain small thickness and a certain linear speed. Such films are able to surmount quite high walls, on the order of several meters in height.

The various phenomena described above are not difficult to explain on the basis of the two-fluid theory:

1. The sudden cessation of boiling as the λ-point is passed is explained by the sudden appearance within the liquid of a small amount of *superfluid*, which is capable of transmitting heat through the liquid without resistance.

2. The sudden disappearance of viscosity as measured by the rate of flow through narrow capillary tubes is also explained by the sudden creation of some superfluid at the λ-point, this superfluid being able to penetrate the capillary tubes without resistance. The capillary tube experiment therefore measures the viscosity of the *superfluid component*.

3. The results of the torsion-pendulum experiment are explained on the two-fluid theory by the fact that this experiment measures the viscosity of the *normal-fluid component*, since this is the component which is dragged along by the mica disks. The result of this experiment indicates that the amount of normal fluid does not change by a large factor at the λ-point but that this component gradually disappears as the temperature is lowered toward absolute zero.

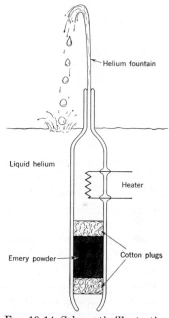

FIG. 10-14. Schematic illustration of the fountain effect.

4. The mechanism of the fountain effect is as follows: The superfluid component in the original container is able to flow through the porous plug into the second reservoir, but the normal-fluid component cannot. Within both reservoirs and at all points in the porous plug, however, the relative amounts of superfluid and normal fluid are governed *solely by the temperature of the surroundings*, and must be everywhere the

same if the system is isothermal. Thus, although only superfluid can flow through the porous plug, a certain amount of it changes into normal fluid upon arriving at the second reservoir. Furthermore, if heat is supplied to the second reservoir, the relative amount of normal fluid increases. The flow of superfluid through the porous plug is governed by the amounts (i.e., the partial pressures) in the two reservoirs, and it thus tends to flow from a region of higher concentration toward one of lower, i.e., from a region of low temperature toward one of high temperature. The normal fluid tends also to flow from a region of higher concentration toward one of lower, i.e., from a high-temperature region toward a low-temperature one. But since the normal-fluid component is blocked by the porous plug, it has no choice but to accumulate in the higher-temperature vessel and eventually overflow it. The fountain effect is thus somewhat analogous to osmotic pressure, except that the two substances involved can in the present case be transformed reversibly into one another by a temperature change.

This experiment also indicates more clearly the true nature of heat transfer within the liquid below the λ-point: in the immediate neighborhood of a heat source, some of the superfluid is transformed into normal fluid; near a heat sink, the reverse occurs. *The heat is thus carried bodily from source to sink by counterflowing streams of normal- and superfluid.* The speeds of these flows depend of course upon the relative amounts of the two fluids present, i.e., upon the temperature. Speeds of several centimeters per second are not at all uncommon.

5. The phenomenon of second sound is an extension of the preceding concept of counterflowing fluids to the transient case, in which traveling density waves of normal- and superfluid coexist in such a way that the *sum* of the densities of the two fluids remains constant but the *ratio* of densities, and therefore the local temperature and heat content, varies.

6. The surface-film effect is not well understood, but it most likely results from the fact that helium atoms have much greater attraction for other atoms—the walls of the vessel—than for one another, coupled with the absence of viscosity of the superfluid component, which permits sizable rates of mass transfer. The existence of a limited film thickness probably is connected with the range of action of the attractive force between the helium and the vessel walls, and as we will shortly see, the existence of a finite speed of travel is probably related to the excitation of helium atoms from the superfluid to the higher-energy normal-fluid states. We shall now consider the qualitative situation in liquid helium from the standpoint of quantum statistics.

In treating the properties of helium at very low temperatures, it turns out to have been possible to interpret certain aspects of the problem in a qualitative fashion by using the Einstein-Bose statistics,[1] although until the recent work of Feynman[2] it was not clear whether or not the actual interactions that exist between the helium atoms would completely invalidate the free-particle model treated by Bose and Einstein. The phenomenon of the "condensation" of an Einstein-Bose system into a superfluid form at a finite temperature can be studied qualitatively by using a free-particle model, as follows: Let the energy levels available to the particles of the system be measured from the lowest level as the zero point of energy. With the zero of energy defined in this way, it is clear that α *must be positive*, since otherwise one would have negative occupation indices for some of the levels. (That is, the denominator of the Einstein-Bose distribution would be negative for some of the energy levels of the system.) Assume first that the levels are continuously distributed as described by Eq. 4-10(5). If we then use Eq. 10-1(3) to obtain an expression for α as a function of the temperature and the parameters of the system, we may write

$$
\begin{aligned}
N &= \sum_{s=1}^{\infty} n_s = \frac{4\pi V (2M^3)^{\frac{1}{2}}}{h^3} \int_0^{\infty} \frac{\epsilon^{\frac{1}{2}}}{e^{\alpha+\beta\epsilon} - 1}\, d\epsilon \\
&= \frac{4\pi V (2M^3)^{\frac{1}{2}} k^{\frac{3}{2}} T^{\frac{3}{2}}}{h^3} \int_0^{\infty} \left[u^{\frac{1}{2}} e^{-\alpha-u} \sum_{r=0}^{\infty} e^{-r(\alpha+u)} \right] du \qquad \text{with } u = \beta\epsilon \\
&= \frac{4\pi V (2M^3)^{\frac{1}{2}} k^{\frac{3}{2}} T^{\frac{3}{2}}}{h^3} \, \Gamma\left(\frac{3}{2}\right) \sum_{p=1}^{\infty} \left(\frac{e^{-p\alpha}}{p^{\frac{3}{2}}}\right)
\end{aligned}
\tag{8}
$$

The following features of the temperature dependence of α can be seen from Eq. (8):

1. On the right we have a product of a monotonically increasing function of T and a monotonically decreasing function of α. Thus we can be sure that α *is a monotonically increasing function of* T.
2. The function of T can be made as small as we wish by letting T become small. On the other hand, there is an effective upper bound upon the function of α, namely, the value attained when α is equal to zero. *Yet the product of these two functions is a constant.*
3. The above considerations lead to the conclusion that physically acceptable values of α can be obtained *only for temperatures exceeding a certain value.*

[1] *Proc. Roy. Soc. (London)*, **153**, 576 (1936).
[2] *Phys. Rev.*, **94**, 262 (1954).

EXERCISES

10-36. Supply the missing steps in the derivation of Eq. (8).

10-37. The specific gravity of liquid helium is 0.15. Use this to find at what temperature Eq. (8) would give $\alpha = 0$ for helium at this density. *Ans.: $T_c \approx 3.2°K$.*

We have found that the straightforward application of the principles and procedures of quantum statistics to an Einstein-Bose system at low temperature leads to the conclusion that there exists, for such a system, a *critical temperature T_c* below which these procedures are apparently not valid. The questions next to be answered are what element of our treatment breaks down at this critical temperature and what feature of the *physical* situation brings about the failure? If we examine the steps taken in arriving at Eq. (8), we see that we have assumed the energy states for a particle to be *continuously distributed*. However, inasmuch as our present concern with very low temperatures involves the lowest-lying energy levels, we may expect that the actual *discrete* nature of the level distribution might play an essential role in the lowest temperature range.

Following this suggestion, let us assume that the lowest level $\epsilon_1 = 0$ is nondegenerate and that the remaining levels are continuously distributed from $\epsilon = 0$ to $\epsilon = \infty$ as described by Eq. 4-10(5). We shall now specifically exhibit the energy level at $\epsilon = 0$, the existence of which is not recognizable in the continuous distribution, and treat it separately in evaluating the summation 10-1(3):

$$
\begin{aligned}
N &= \frac{1}{e^\alpha - 1} + \frac{4\pi V (2M^3)^{1/2}}{h^3} \int_0^\infty \frac{\epsilon^{1/2}\, d\epsilon}{e^{\alpha + \beta\epsilon} - 1} \\
&= \frac{1}{e^\alpha - 1} + \left(\frac{T}{T_c}\right)^{3/2} f(\alpha) N
\end{aligned}
\tag{9}
$$

where the second quantity on the right is just the right-hand side of Eq. (8), written in terms of the critical temperature T_c. Clearly, $f(0) = 1$.

We see that it is now possible to satisfy this new equation with positive values of α for all temperatures, since the first term becomes infinite as $\alpha \to 0$. The inclusion of the lowest energy level as a separate term in our treatment has thus removed the previous difficulty of not being able to account for all of the particles at temperatures below T_c. If we now inquire into the physical situation represented by this equation, we see that, at temperatures below T_c, the parameter α will take on such values that those particles which are not included in the continuous distribution *will be found in the lowest level*. That is, a kind of *condensation* occurs; it is such that *an appreciable fraction of the particles is in the lowest energy level* at temperatures below T_c. Specifically, we find the population of the lowest level to be, approximately,

$$n_1 = N \left[1 - \left(\frac{T}{T_c} \right)^{\frac{3}{2}} \right] \tag{10}$$

and correspondingly, $\alpha = \ln \left(1 + \frac{1}{n_1} \right) \approx \frac{1}{n_1}$ (11)

The population of the lowest energy level is shown in Fig. 10-15.

According to the most recent ideas, the above rough analysis accounts qualitatively for many of the very interesting properties of liquid helium

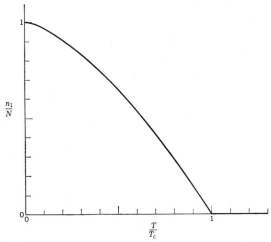

FIG. 10-15. The population of the lowest energy state of an Einstein-Bose system of free particles as a function of the temperature.

descr'bed previously. The interpretation of the properties of liquid helium below the λ-point in terms of a mixture of a superfluid component and an ordinary component, where the superfluid component corresponds to a finite part of the total sample being in the lowest energy state, represents a great triumph of the quantum statistics. The modifications in the theory which must be made in order to obtain quantitative agreement with experiment are mainly concerned with the interactions between particles; they do not fundamentally alter the conclusions reached on the basis of the simplest model. The fact that liquid ^{3}He, which is composed of Fermi particles, remains an ordinary liquid to the lowest temperatures provides a further experimental verification of the correctness of the quantum statistics.

The quantum-mechanical distribution laws described in this chapter represent an idealized limiting behavior of quantum-mechanical systems whose component elements have only a vanishingly small interaction with one another.   In spite of this idealization, the results obtained are found to apply with quite acceptable accuracy to many systems whose elements do not strictly satisfy this condition, and even to provide a reliable qualitative basis for analyzing systems with quite large internal interactions.

It should of course be recognized that the treatment given in this chapter has brought out only the most elementary aspects of quantum-statistical mechanics.   The fundamental assumptions that are involved in the most advanced treatments were scarcely mentioned here, and some of the approximations that were used in deriving the condition for maximum probability are subject to considerable suspicion, particularly in the case of the Fermi-Dirac statistics.   On the other hand, the more elaborate derivations that can be given for these distribution laws lead to the same results, which is of course why we are able to ignore the possible shortcomings of the methods we have used.   The justification of the more clear but less rigorous approach that has here been used is that it leads to correct results and at the same time is based upon fairly convincing physical grounds.

REFERENCES

Keesom, W. H.: "Helium," Elsevier Publishing Company, Amsterdam, 1942.

Kittel, C.: "Introduction to Solid State Physics," 2d ed., John Wiley & Sons, Inc., New York, 1956.

Tolman, R. C.: "The Principles of Statistical Mechanics," Oxford University Press, New York, 1938.

11

The Band Theory of Solids

The application of the Fermi-Dirac statistics to a gas of perfectly free electrons, as described in the preceding chapter, provides a quite satisfactory basis for a qualitative understanding of many of the properties of metals. There are, however, certain quantitative discrepancies between this "free-electron" model and actual metals, which can largely be attributed to the fact that the electrons of a metal are not really completely free but are acted upon by the atoms of the crystal lattice of the metal and by the other electrons. In this chapter we shall consider some of these finer details, and we shall find that, when they are taken into account, certain *entirely new qualitative features* which emerge render the theory applicable to *insulators and semiconductors* as well as to metals. This extended theory is called the *band theory of solids*, for reasons that will presently appear. The theory is as yet by no means complete, but its qualitative aspects, to which our treatment will largely be limited, are certainly correct, and these aspects alone furnish an extremely satisfying and useful qualitative picture of the nature of the solid state and of the basic physical mechanisms underlying many important phenomena.

We wish to investigate the nature of the distribution of energy levels available to an electron which moves under the influence of the atomic nuclei and the other electrons *inside a crystal*, in order to obtain a better approximation to the quantities $g_s(\epsilon_s)$ needed for an application of the Fermi-Dirac distribution law. A basic feature of the problem of the motion of an electron in a crystal is the *periodic character* of its potential-energy function. The potential energy must clearly have the *same space symmetry* and the *same periodicity* as does the crystal itself (aside from

377

small effects produced by the existence of a boundary surface in an actual case), and this can be expected to introduce a certain degree of symmetry and periodicity into the wave functions for the problem. We shall approach this problem from two essentially different points of view which lead to equivalent results. The first of these deals with the problem quite qualitatively but yields a relatively clear picture of the nature of the energy states; the second is more quantitative and leads to beautiful analytical results of a rather general character, but the qualitative features are perhaps not quite as clear. The two methods together, however, can provide us with considerable insight regarding the nature of the electronic energy states of a crystal. The two viewpoints are actually only approximately valid, both being based upon the assumption that each electron of the solid is acted upon by a fixed potential function which includes the effects of the atomic nuclei and the average effect of all of the remaining electrons. Thus we are retaining the independent-particle assumption that was utilized in developing the statistical distribution laws. This assumption, which is equivalent to the central-field approximation in the treatment of atomic wave functions, is sufficiently well satisfied that the results can be applied semiquantitatively to many problems.

11-1.   The Band Structure of the Electronic Energy States of a Crystal

A. Qualitative Features of Electronic States in a Crystal. The broad features of the distribution of energy states can be inferred by applying considerations similar to those used in the discussion of molecular binding. Thus we can imagine the crystal to be composed of a large number of identical atoms which are arranged in the same spatial pattern as in the actual crystal, but to a scale which is subject to arbitrary adjustment, and then consider the gradual transition of the system from a state of very large interatomic separation to a state of small interatomic separation.

If the interatomic separation is very large, it is clear that the possible energy levels of an electron in the crystal will be just those of an electron in an individual *atom*. These levels will, however, possess a degree of degeneracy equal to that of the atomic level *multiplied by the number of atoms in the crystal.* Thus each of the electronic levels takes on an enormous degeneracy when a large number of atoms are treated as a single crystal.

If we now consider what happens to these various degenerate energy levels as the scale of the crystal lattice is reduced from the large size considered above, we find that, just as in the case of the diatomic hydrogen molecule, the degeneracy of the atomic levels is *removed by the mutual interactions of the atoms,* with the result that *each highly degenerate level is split into a huge number of distinct levels.* These levels are so numerous, and their average separation so small, that it is permissible to regard them as being *continuously distributed* with a certain "density" $\rho(\epsilon)$ over an energy range which extends from the lowest to the uppermost level of a given group.[1] Thus each of the original atomic levels becomes a narrow *energy band* as the interatomic separation is reduced, and each band becomes broader as the mutual atomic perturbations become gradu-

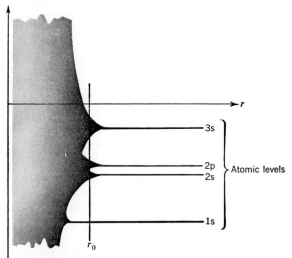

Fig. 11-1. Schematic illustration of the splitting of the atomic energy levels by the interatomic perturbations as the scale of the crystal lattice is reduced.

ally greater. This process is indicated schematically in Fig. 11-1 for some of the atomic levels.

It will later be shown by applying considerations similar to those involved in the discussion of the parity that the *wave functions* associated with the various energy levels give probability distributions $\psi_n^* \psi_n$ which have exactly the same symmetry and periodicity properties as the crystal lattice. Thus an electron in a given energy eigenstate *is equally likely to be found in the neighborhood of any of the atoms of the crystal and cannot be regarded as being associated with any individual atom.*

Figure 11-1 indicates that, for a given interatomic spacing, the "width" of the 1s band is considerably less than that of the 2s or 2p bands. This results from the fact that the effect of a perturbation upon an energy level is large only if the perturbation potential strongly "overlaps" the wave function for that level. A strong overlap occurs at greater distances of separation for wave functions which are not strongly localized near a given atom. (In more physical terms, a tightly bound electron is not as

---

[1] In terms of the notation of the previous chapter, we have $g_s = \rho(\epsilon_s) \, \Delta\epsilon_s$.

easily affected by a neighboring atom as is a less tightly bound one.) Thus the higher energy states are relatively more strongly spread into bands than are the lower ones.

Another feature indicated in this figure is the possibility of an *overlap* of bands for sufficiently small interatomic spacing. This quite commonly occurs and can have a marked effect upon the physical properties of a crystal, as will shortly be seen. Thus if the actual equilibrium inter-atomic spacing r_0 for a given crystal is one for which an overlap between two bands exists, the properties of the crystal may be quite different than if there were no overlap.

B. Criteria Which Determine the Electrical Nature of a Crystal. In order to describe the gross electrical properties of a given element, we must combine the above idea of energy bands in crystals with the *exclusion principle*, which limits the occupancy of each of the levels of a band to a single electron. In addition, we must recognize that an electric field, acting upon the electrons of a crystal, cannot change the momentum or the energy of an electron by a very large amount during the rather short time interval between successive collisions of the electron with the crystal lattice. Thus an essential requirement that must be satisfied by an electrical *conductor* is that a sufficiently large number of electrons must have energies within a range in which there are many unoccupied energy states whose energies are very close to that of the electron in question. This condition is satisfied, in the free-electron theory of the previous chapter, by those electrons lying near the top of the Fermi distribution. In the band theory now being considered, this condition is satisfied if a given band of energy levels is only *partially occupied* by electrons; it is not fulfilled if the band is *completely* occupied and if the nearest unoccupied band is separated by a sufficiently large gap from the occupied band.[1] Thus a conductor is characterized by the presence of numerous unoccupied levels adjacent to the highest occupied level, and an insulator by a *finite energy gap* between the highest occupied level and the next band of levels. For a given chemical element or compound, the popula-

---

[1] Another way of describing the situation is as follows: In order to have metallic conduction, it is necessary to have available a large number of unoccupied states which can be combined with one another to form *wave packets* which represent the motion of electrons in prescribed directions and with various momenta. The application of an electric field will then cause the phases and amplitudes of the various states contributing to a wave packet to change in a manner which corresponds to a change in direction and/or magnitude of the momentum under the action of the field, and the changes in the various possible wave packets will combine to represent an average convection of charge in the direction of the applied field. If an energy band is fully occupied, there are no remaining states whose phases and amplitudes can be changed by the field, so that there is no convection of charge in this case. (The net momentum corresponding to a completely filled band is exactly zero.)

tion of the various energy bands and thus the electrical character is, of course, determined by the number of electrons in each atom, the number of atoms in a unit cell, the exclusion principle, the conditions of separation or overlap of the various energy bands, and to a certain extent by the temperature of the crystal. In the remainder of this section we shall consider briefly the electrical nature of a few crystalline elements from the point of view of the band theory.

EXERCISE

11-1. Use the data obtained for Cu in Exercise 10-30 to calculate the increase in energy of a conduction electron which moves parallel to an applied electric field, per mean free path. Let the electric field be of such strength that the current density in the copper is 1000 A cm^{-2}. Express the energy increase in electron volts.

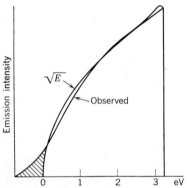

FIG. 11-2. Distribution of 3s electronic states for metallic sodium. (By permission from F. Seitz, "The Modern Theory of Solids," McGraw-Hill Book Company, Inc., New York, 1940.)

C. Metals. The metallic elements are characterized by an energy-band structure in which the uppermost filled level is contiguous to a large number of unoccupied levels. This situation occurs if a given band of levels is only partially filled or if adjacent bands near the top of the occupied levels overlap, which provides the necessary continuity in the distribution of levels. Actually, most metals have band structures in which there is a large overlap in the uppermost occupied levels, even though this overlap is not always the primary reason for the metallic character of the element.

Consider first the band structure of the alkali metals. The electron configuration of an alkali metal atom consists of a single s electron outside a completed p subshell. The energy-band structure therefore consists of a number of completely occupied bands, corresponding to the various completed subshells of the atom, with a *half-filled s band* corresponding to the half-filled s subshell. The distribution of electron states is represented schematically in Fig. 11-2 for sodium, and the band structure as a function of interatomic spacing is shown in Fig. 11-3. Figure 11-4 illustrates the energy bands of copper, which is in many respects like an alkali metal. These figures indicate that there are many unoccupied energy states near the uppermost filled level at energy ϵ_m,

which accounts for the fact that sodium and the other alkali metals are
good conductors of electricity.   As a matter of fact, the distribution of
energy states within the 3s band is very closely similar to that for *free
electrons*, if the energy scale is shifted to have its zero at ϵ_0, the lowest
energy in this band.   (The occupation index for all of the lower bands

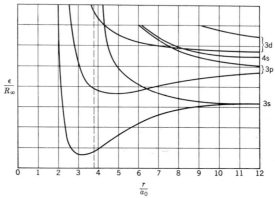

FIG. 11-3. Band structure of sodium as a function of interatomic distance.   [After
Slater, *Phys. Rev.*, **45**, 794 (1934).]

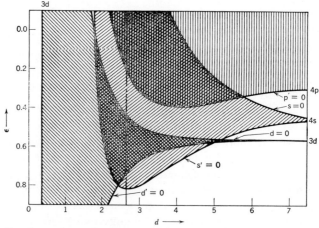

FIG. 11-4. Band structure of copper as a function of interatomic distance.   [After
Krutter, *Phys. Rev.*, **48**, 664 (1935).]

will be very near unity even for quite high temperatures, so that these
bands will not contribute to the electrical conductivity and can therefore
be ignored for these considerations.)   This is why the free-electron theory
provides a rather good description of the electrical properties of the
alkali metals.

Consider next the alkaline earths.   In the alkaline earths, an s subshell is completely occupied. This would lead to the conclusion that these elements should be *insulators,* and indeed this would be the case were it not for the fact that there is a slight overlapping of the highest occupied s band and next higher p or d band which provides the supply of unoccupied, nearby energy levels that is necessary for metallic conduction. The density of levels is shown in Fig. 11-5. According to this figure the free-electron theory should be quite inapplicable to the alkaline earths, as is in fact observed to be true. However, we shall later mention a slight modification in the theory which renders it of considerable use even in such cases.

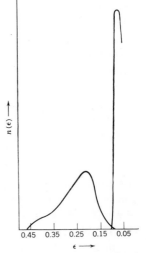

FIG. 11-5. Energy-level density in an alkaline earth. [After Manning and Krutter, *Phys. Rev.*, **51**, 761 (1937).]

D. *Insulators.* The characteristic feature of an electrically *insulating* crystal is that the uppermost occupied energy level is the highest level in some band, so that this band is completely filled, and that a *sizable energy gap* exists between this band and the next higher one. We shall discuss this situation in connection with the insulator of lowest atomic number, i.e., carbon (diamond).

The electron configuration of carbon is $1s^2\ 2s^2\ 2p^2$, so that there are two electrons in an incomplete p subshell. In view of the foregoing discussion, one might then conclude that carbon should be a metallic conductor. That this conclusion is incorrect is ascribed to the fact that the high symmetry and strong binding of the diamond crystal cause the 2p atomic levels to separate into *two* well-separated energy bands rather than one. The lower of these bands contains two states per atom, and the upper one, four states per atom. Thus the lower of these bands is completely filled by the two 2p electrons per atom, and a large energy gap (~ 6 eV) to the bottom of the upper branch of the 2p band makes diamond an insulator. Graphite, a crystalline form of carbon having different symmetry, has a much smaller separation between the energy bands, with the result that graphite is a semiconductor.

It is interesting to note that direct experimental verification of the ideas underlying the band theory of solids is provided by the use of diamond crystals as *particle counters.* In this application a diamond crystal is placed between, and in good contact with, the plates of a parallel-plate capacitor, and a quite strong electric field is applied to the crystal. Because of the fact that all of the energy states of the occupied bands are filled, this electric field is not able to change the average momentum states of the electrons, and no electric current flows. (The electric field does, however, slightly modify the electronic wave functions and energy levels, and produces a *polarization* of the crystal.)

If, now, an energetic charged particle (i.e., perhaps a cosmic-ray μ-meson) passes through the diamond crystal, a number of violent collisions will usually occur, resulting in a number of electrons in the crystal being excited to some of the levels in the higher unoccupied bands. These electrons now have energies which lie within a continuum of levels, and the electric field *can* change their momenta, resulting in a flow of current and an eventual collection of the electrons at the positive electrode of the capacitor. At the same time the vacancies in the bands from which these electrons were ejected behave somewhat like bubbles in a liquid: they are "forced" in the opposite direction to that of the electrons and are eventually "collected" at the negative electrode (that is, an electron from the negative electrode steps into each "hole" as it arrives at the electrode).

The passage of such a particle through the crystal thus results in the production of "free" electrons and holes in various energy states. These migrate rapidly toward the two electrodes, resulting in the passage of a pulse of current through the crystal. It should be emphasized that this migration of electrons and holes under the action of the electric field is similar to the movement of the electrons in a good conductor. The classical picture of insulators as consisting of bound charges, unable to leave their own atoms, is thus quite opposed to these experimental observations. The idea that the energy eigenstates extend uniformly throughout the crystal and can be compounded into wave packets which describe an almost unhindered motion of an electron through the crystal is, on the other hand, quite in accord with such observations.

EXERCISE

11-2. Crystalline sulfur is a pale yellow, transparent substance, and is one of the best known insulators. From this, deduce the approximate energy-gap width for sulfur.

E. Semiconductors. Intermediate between the good conductors of electricity and the good insulators lies a broad class of solids whose electrical conductivity is rather poor but still very much greater than that of insulators. These solids are called *semiconductors* and are of great theoretical and technical interest. Many solids are semiconductors because of the presence of impurities which interrupt the regularity of the crystal lattice and thus introduce new energy levels whose wave functions do not possess the regularity of the crystal lattice but are instead localized about the individual impurity centers. We shall briefly describe this situation shortly. Some solids, on the other hand, are semi-conductors even when they possess no imperfections in their crystal lattices; they may be called *intrinsic* semiconductors in contrast to the *impurity* semiconductors mentioned above.

The distinctive feature of an *intrinsic semiconductor* is that a *narrow energy gap* separates a filled band from an empty one. Such a substance is an insulator at sufficiently low temperatures, but at higher tempera-tures a few electrons may be excited to some of the empty levels of the higher band by *thermal excitation*. This then allows electrical conduction to occur by the migration of the holes and electrons in the nearly filled and nearly empty bands. The occupation index for this situation is given by Fig. 11-6, which shows that the Fermi level, $\epsilon_m = -\alpha kT$, lies somewhere between the two bands and is so situated that the number of vacancies in the lower band is just equal to the number of occupied states in the upper band. The Fermi level may also *shift with temperature*, as dictated by the detailed nature of the density of states $\rho(\epsilon)$ in the two bands near the Fermi level. The uppermost filled band in a semicon-ductor or insulator is called the *valence band*, and the next higher (empty) band is called the *conduction band*.

It is clear from Fig. 11-6 that the number of holes and free electrons should vary nearly exponentially with $1/T$, so that the electrical con-ductivity should also vary in this manner. This law has been well verified experimentally for several intrinsic semiconductors and is illustrated by Fig. 11-7 for silicon.

EXERCISES

11-3. Assume the density of states $\rho(\epsilon)$ in the valence and conduction bands to be symmetrical about the center of the energy gap W_g and to be the same as for *free electrons* in the conduction band. Assume also that $e^{W_g/kT} \gg 1$. Using these assumptions, show that the number of electrons per unit volume whose energies lie in the conduction band is approximately $N_e = 2(2\pi mkT/h^2)^{3/2}e^{-W_g/2kT}$.

11-4. Deduce the approximate magnitude of the energy gap in silicon from the curves of Fig. 11-7.

In contrast with the intrinsic semiconductors described above, many substances which are normally insulators may be made semiconducting by the presence of *impurity atoms* which modify the energy levels and wave functions of the solid. Furthermore, solids which are *intrinsic semiconductors* may have their properties considerably modified by the presence of impurities. Although a general theory of such *impurity*

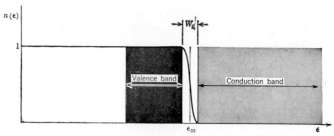

FIG. 11-6. Schematic illustration of the occupation index of the energy states of an intrinsic semiconductor at finite temperature. Some of the states in the lower band (valence band) are left unoccupied because some of the electrons are thermally excited into energy states lying in the upper band (conduction band). The Fermi energy ϵ_m does *not* in general lie midway between the two bands because the density of states in the two is usually different and the Fermi energy must be of such size that the number of vacant levels in the lower band is equal to the number of electrons in the upper band.

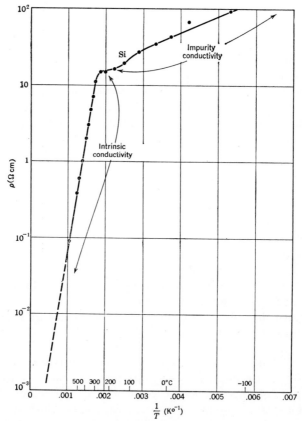

FIG. 11-7. The electrical conductivity of pure silicon as a function of the temperature, illustrating the exponential variation of conductivity with $1/T$. [After Pearson and Bardeen, *Phys. Rev.*, **75**, 865 (1949).]

semiconductors is not at present available, a theory of certain simple types of impurity semiconductors has been developed and accounts quite well for many of the phenomena associated with the technically important applications of silicon and germanium. Some of the simpler aspects of this theory will now be described.

The simplest kind of impurity semiconductor is one in which a very small fraction (10^{-5} to 10^{-10}) of the atoms of the crystal is replaced by atoms of an element having a *different valence* than the material of the solid itself. These impurity atoms are *substituted* for the atoms in the crystal; that is, they occupy the normal lattice sites in the crystal and do not produce significant distortion of the lattice. Two different kinds of such substitutional impurity atoms are of importance: (1) atoms whose valence is one or two units *greater* and (2) atoms whose valence is one or two units *less* than that of the normal atoms of the solid. Impurity atoms of type 1 are called *donor* atoms because they have *more* electrons than are needed to maintain saturated valence bonds with the surrounding normal atoms, and impurity atoms of type 2 are called *acceptor* atoms because they have *too few* electrons to maintain normal valence relationships with neighboring atoms. For example, the elements P, As, Sb, and Bi are donors with respect to Ge and Si, while B, Al, Ga, are acceptors. A semiconductor having donor impurities is called an n-*type* semiconductor because the conductivity takes place by the migration of the surplus (*negative*) electrons, while a semiconductor having acceptor impurities is called a p-*type* semiconductor because the conductivity takes place by the migration of "holes" represented by the *lack* of electrons in the solid. The behavior of these two types of semiconductor is quite similar, and the analysis of the effects of the impurities is also quite similar. We shall describe only the effects of donor impurities because this situation is more easily visualized.

If a donor impurity atom whose valence is one unit greater than that of the other atoms is introduced into the crystal lattice, all of its valence electrons except one are able to enter into normal valence bonding relations with its surrounding neighbors. The surplus electron then can be considered to be approximately free, except that it is still attracted toward the impurity atom because of the *extra unit of positive charge* of the nucleus of the atom. The electron can therefore be expected to exist in certain bound energy states around the impurity atom and in a continuum of unbound states in which it can move freely throughout the crystal. These states are analogous to the bound and free states of the electron around a free atom, except that *the electron is moving in a polarizable medium*, so that the permittivity is not ε_0, but $K\varepsilon_0$, where K is the dielectric constant of the crystal. The energy levels of the electron are thus all multiplied by a factor $1/K^2$, and the corresponding wave functions are spread out by a factor K, with respect to the energy levels and wave functions in a free atom. The one-electron approximation for the energy levels and wave functions is quite sufficiently accurate for the treatment.

EXERCISE

11-5. (*a*) Calculate the energy of the lowest bound state of an electron moving about an arsenic atom in a germanium crystal. The dielectric constant of Ge is 16. *Ans.*: $E_0 \approx 0.05$ eV. (*b*) Calculate the radius of the first Bohr orbit for (*a*) and express the result in units of the nearest neighbor spacing of the crystal lattice.

At sufficiently low temperatures, all of the excess electrons will be in the bound states near the donor atoms and will not contribute to the

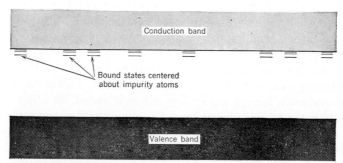

FIG. 11-8. Schematic illustration of the localized energy states introduced by the presence of donor impurities.

conductivity. As the temperature is increased, some of the excess electrons become excited thermally and are elevated into the conduction band, where they do contribute to the electrical conductivity. The ionization energy of a donor atom is usually much smaller than the energy gap between the valence band and the conduction band, so that the excitation of electrons into the conduction band from the impurity centers occurs at a much lower temperature than does the excitation of electrons from the valence band. The energetic relationships for the case of donor impurities are shown in Fig. 11-8.

EXERCISE

11-6. Describe semiquantitatively the change with temperature of the location of the Fermi level ϵ_m for the case illustrated in Fig. 11-8.

F. Rectification at a Junction. One of the most important technical uses of semiconductors is in electrical rectifiers and allied devices. The

fundamental mechanisms involved in this application are rather interesting and can be qualitatively understood on the basis of the band theory
of solids, so that we shall briefly examine the rectification phenomenon.

The features of a physical arrangement which are essential to the
phenomenon of rectification are twofold.   First, there must exist a
barrier of some kind between two conducting media over which the

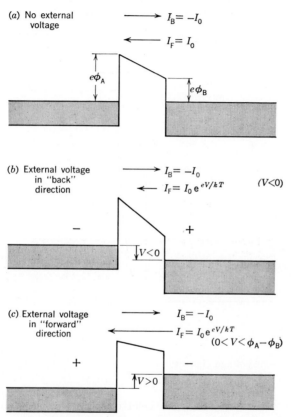

Fig. 11-9. Energy relationships in the vicinity of a rectifying junction (schematic).
The arrows indicate the direction of flow of electrons.

electrons (or holes) must pass in going from one medium to the other.
Second, the two media *must be unsymmetrical in their electronic properties;*
that is, *the work functions must be different* if the two media are metals, or,
more generally, the Fermi levels of the two media must lie at different
depths below the zero of potential outside the two media.   For simplicity,
consider an idealized case in which two metals of widely different work
function are almost in contact along a plane surface and are separated
from one another only by a very thin film of some insulating material.

If an adjustable external voltage source is then connected to the two metals, the energy relationships existing across the barrier between the two metals are as shown in Fig. 11-9.

If the externally applied voltage is zero, the Fermi levels of the two conductors attain the same level, as shown in Fig. 11-9a. In this situation there will be small equal currents in opposite directions across the

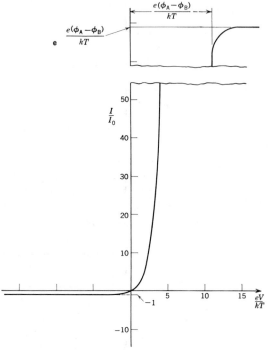

Fig. 11-10. Current-voltage characteristic curve for an ideal junction rectifier (schematic).

barrier, due to the few thermionic electrons in each metal whose energies exceed the Fermi energy plus $e\phi_A$. Call these currents I_0.

If, now, the external voltage is applied in such a direction as to lower the energy of an electron in the metal of lower work function, as in Fig. 11-9b, the electron flow from metal A to metal B is unchanged, since the barrier is unchanged in height as seen by the electrons in metal A. The electrons in metal B, however, now see a higher barrier, one of height $e\phi_A + eV$, where V is the applied voltage. The current due to the electrons going from B to A is thus changed by a factor $e^{eV/kT}$. The net current is therefore

$$
\begin{aligned}
I &= I_0(e^{eV/kT} - 1) \\
&= AT^2 e^{-e\phi_A/kT}(e^{eV/kT} - 1) \qquad V < 0
\end{aligned}
\tag{1}
$$

If the voltage is applied in the opposite direction and is less than $\phi_A - \phi_B$, a similar analysis leads to a net current described by the above equation with $V > 0$. The expected current-voltage characteristic curve is shown in Fig. 11-10.

EXERCISE

11-7. Show that, for sufficiently high voltages applied in the backward and forward directions to the rectifier of Fig. 11-9, *saturation* currents are finally reached and have the ratio

$$\frac{I_{SF}}{I_{SB}} = e^{e(\phi_A - \phi_B)/kT} \tag{2}$$

How large a work-function difference $\phi_A - \phi_B$ is necessary to yield a ratio of 100:1 at room temperature?

In actual practice, rectifiers are constructed by using a metal and an impurity semiconductor, or two impurity semiconductors, rather than two metals separated by an insulating layer. The insulating barrier that is needed for rectifying action is formed in these cases by the local removal of the excess electrons or holes from the impurity atoms near the boundary surface, or by surface treatment of the semiconductor. The variation of current with applied voltage follows a law almost identical with the one derived above.

It has also been found that the interface between a p-type and an n-type semiconductor has rectifying properties. Such an interface can be produced inside a single continuous Ge crystal in several ways, such as by changing the composition of the melt during the growth of the crystal. (Donor and acceptor impurities cancel one another's effects, atom for atom, for small concentrations.) The rectifying action of impurity semiconductors cannot be analyzed in greater detail here. It suffices to say that the mechanism is rather similar to that described above for metal-to-metal rectification and that a very similar dependence upon voltage and temperature is obtained.

It should be emphasized, before leaving this subject, that equations (1) and (2) imply a certain limitation upon the current-voltage characteristic that can be attained with any contact rectifier. That is, it is not possible to construct such a rectifier for which the current changes more abruptly with applied voltage than is indicated by the exponential term; the current cannot change more than a factor $e = 2.718$ for each kT/e change in applied voltage.

11-2. Periodicity Properties of Electronic Wave Functions

The qualitative discussion of the band theory of solids given above has served to indicate some of the basic features of the quantum-mechanical properties of electrons in solids, but of course is quite incomplete both in mathematical detail and in range of application. Our objective has been to see how the physical principles of the preceding chapters lead to the basic qualitative features of the solid state, namely, the existence of allowed and forbidden bands of energy levels for the electrons inside a solid body, the uniform extension of the corresponding wave functions

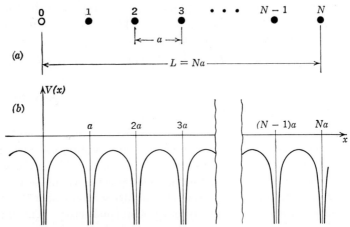

FIG. 11-11. A one-dimensional crystal: (a) arrangement of atoms. (b) potential energy.

over the entire crystal, and the situations which determine the metallic or nonmetallic character of a given element. We shall now examine somewhat more closely the nature of the wave functions of an electron which moves in a region in which its potential energy is a periodic function of its position.

A. *One Dimension.* Consider the linear array of N one-dimensional "atoms" shown in Fig. 11-11. To avoid the difficulties introduced by the existence of boundaries, we shall imagine this finite array to be part of an infinite array; however, to obtain results appropriate to a finite crystal, we shall impose upon the wave functions the *periodic boundary condition* discussed in Chap. 4:

$$\psi(x + Na) = \psi(x) \tag{1}$$

The Hamiltonian operator for a particle of mass m moving in this one-dimensional crystal is

$$\hat{H}(x) = -\frac{\hbar^2}{2m}\frac{d^2}{dx^2} + V(x) \tag{2}$$

where $V(x)$ is, by the nature of the problem, *a periodic function of x*:

$$V(x + a) = V(x) \tag{3}$$

Now, because of this property of V, the Hamiltonian operator also is periodic in x:

$$\hat{H}(x + a) = \hat{H}(x) \tag{4}$$

that is $\hat{H}(x)$ *is invariant under any primitive translation of the lattice.*

Now, we have previously seen that in such a circumstance it is usually profitable to introduce an operator whose action upon a wave function is to perform the transformation under which the Hamiltonian operator is invariant, and thence to find the eigenvalues and eigenfunctions of this operator. Let $\hat{T}(x)$ be an operator which, acting upon a function of x, yields the same function of $(x + a)$, that is,

$$\hat{T}(x)\mathrm{f}(x) = \mathrm{f}(x + a) \tag{5}$$

EXERCISES

11-8. Show that $\hat{T}(x)$ is a linear, Hermitian operator if the operator $\hat{T}^*(x)$ is so defined that $\hat{T}^*(x)\mathrm{f}(x) = \mathrm{f}(x - a)$.

11-9. Show that $\hat{T}$ commutes with $\hat{H}$ if $\hat{H}(x) = \hat{H}(x + a)$.

11-10. Show that $\hat{T}(x)$ may be written formally as $\hat{T}(x) = e^{ad/dx} = 1 + a\dfrac{d}{dx} + \cdots$ and may thus be regarded as the operator corresponding to the dynamical quantity $\exp{(iap_x/\hbar)}$.

The above properties of $\hat{T}$ assure that any nondegenerate energy eigenfunction must also be an eigenfunction of $\hat{T}$. To find these eigenfunctions, we first find the eigenvalues of $\hat{T}$. If T_0 is an eigenvalue, the eigenvalue equation is

$$\hat{T}\psi = T_0\psi \tag{6}$$

or

$$\psi(x + a) = T_0\psi(x) \tag{7}$$

If we apply the operator $\hat{T}$ a second time, we find that

$$\hat{T}^2\psi = \hat{T}T_0\psi = T_0\hat{T}\psi = T_0{}^2\psi \tag{8}$$

or

$$\psi(x + 2a) = T_0{}^2\psi(x) \tag{9}$$

Let us apply this operator N times. This yields

$$\psi(x + Na) = T_0{}^N\psi(x) \tag{10}$$

But the periodic boundary condition (1) then requires that

$$T_0{}^N = 1 \tag{11}$$

or, solving for T_0,

$$\begin{aligned} T_0 &= 1^{1/N} \\ &= e^{i2\pi l/N} \qquad l = 0, 1, \ldots, N - 1 \end{aligned} \tag{12}$$

There are thus exactly N distinct eigenvalues of $\hat{T}$—the Nth roots of unity.

We can now write down the eigenfunctions of $\hat{T}$ by inspection. These are just those functions which change only by the complex phase factor (12) when x is changed to $x + a$. Clearly, such functions must be of the form

$$\psi(x) = \pi_l(x)e^{i2\pi lx/L} = \pi_k(x)e^{ikx} \tag{13}$$

where $L = Na$, $k = 2\pi l/Na$, and $\pi_k(x)$ is *periodic in x with period a:*

$$\pi_k(x + a) = \pi_k(x) \tag{14}$$

EXERCISES

11-11. Show that functions of the form (13) satisfy the eigenvalue equation (6) with the eigenvalue e^{ika}.

11-12. Show that a particle described by a wave function of the form (13) is equally likely to be located in any of the cells of the crystal lattice.

11-13. Show that the range of variation of l can equally well be chosen to be symmetrical with respect to $l = 0$, that is,

$$l = 0, \pm 1, \pm 2, \ldots, \begin{cases} \pm \dfrac{N-1}{2} & N \text{ odd} \\[2mm] \pm \dfrac{N}{2} & N \text{ even} \end{cases}$$

A wave function having the form of Eq. (13) is called a *Bloch function.* It contains features characteristic both of free-particle wave functions and of atomic wave functions:

1. The factor $e^{i2\pi lx/L}$ has precisely the form of an energy eigenfunction of a particle inside a "box" of length L.
2. The functions $\pi_k(x)$, being periodic with the unit cell of the lattice, are analogous to the atomic eigenfunctions and, like the atomic eigenfunctions, tend to fall into well-separated discrete groups. Within each group there is, however, a certain dependence upon k.

The exponential term may be thought of qualitatively as endowing each atomic eigenstate with the character of a free-particle momentum state, in that the N eigenstates associated with a given atomic state can be compounded into wave packets similar in nature to the wave packets for a free particle. The relation between the energy and momentum is not as direct and simple as for a truly free particle, however, as will shortly be seen. Figure 11-12 illustrates schematically the nature of the Bloch functions for a one-dimensional crystal. For simplicity, the dependence of $\pi_k(x)$ upon k is ignored.

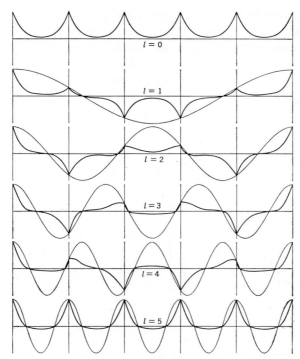

FIG. 11-12. Schematic illustration of the Bloch functions for the lowest energy band of a one-dimensional crystal having $N = 5$ atoms. The real part of ψ is shown for $l = 0, 1, 2, 3, 4,$ and 5. For $l = 0$ the function $u_0(x)$ is essentially an atomic 1s wave function. This function is gradually modified with increasing l until, at $l = 5$, the Bloch function $u_5 e^{i(2\pi x/a)}$ is again exactly periodic with period a. The energy of this state is, however, different from that of the similar function u_0^1, which corresponds to an unmodulated atomic 2s wave function.

EXERCISES

11-14. Evaluate the average momentum of a particle whose wave function is a Bloch function. (Let ψ be normalized in a unit cell.)

Ans.:

$$\langle p \rangle = \langle p_k \rangle + \hbar k$$

where

$$\langle p_k \rangle = \int_D^{D+a} \pi_k^* \frac{\hbar}{i} \frac{d\pi_k}{dx} \, dx \qquad (15)$$

11-15. Evaluate the average kinetic energy of a particle which is in an eigenstate of the form (13).

Ans.:

$$\langle T \rangle = \langle T_k \rangle + \frac{\hbar k}{m} \langle p_k \rangle + \frac{\hbar^2 k^2}{2m} \qquad (16)$$

where

$$\langle T_k \rangle = -\frac{\hbar^2}{2m} \int_D^{D+a} \pi_k^* \frac{d^2 \pi_k}{dx^2} \, dx$$

   Let us now apply the general results just derived to a specific illustrative example: Assume the potential-energy function to be a repeated array of rectangular wells as shown in Fig. 11-13. This case was first treated by Kronig and Penney.[1] Following the procedures used in Chap. 4, we may write the time-independent wave functions $u(x)$ in the form (assuming $E < V$)

(a) $u_l(x) = Ae^{\beta x} + Be^{-\beta x} \qquad -b < x < 0$

(b) $u_r(x) = Ce^{i\alpha x} + De^{-i\alpha x} \qquad 0 < x < c$ (17)

with $\alpha = (2mE)^{\frac{1}{2}}/\hbar$ and $\beta = [2m(V - E)]^{\frac{1}{2}}/\hbar$ as in Chap. 4. These

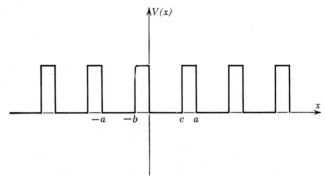

FIG. 11-13. Potential-energy function for a one-dimensional array of rectangular potential wells.

solutions must match smoothly at $x = 0$, which yields the two equations

(a) $A + B = C + D$

(b) $\beta A - \beta B = i\alpha C - i\alpha D$ (18)

Furthermore, in order to satisfy the condition that $\pi_k(x) = u(x)e^{-ikx}$ be periodic in x with period a, we must also have $\pi_k(c) = \pi_k(-b)$ and $\pi_k'(c) = \pi_k'(-b)$. Thus

(a) $e^{+ikb}(e^{-\beta b}A + e^{\beta b}B) = e^{-ikc}(e^{i\alpha c}C + e^{-i\alpha c}D)$

(b) $e^{ikb}[(\beta - ik)e^{-\beta b}A - (\beta + ik)e^{\beta b}B] = ie^{-ikc}[(\alpha - k)e^{i\alpha c}C$
$$- (\alpha + k)e^{-i\alpha c}D] \quad (19)$$

   Now ordinarily one would expect the only solutions of the above equations to be $A \equiv B \equiv C \equiv D \equiv 0$; in order that at least one of these coefficients shall be finite, the determinant of the coefficients must vanish. This gives

$$\begin{vmatrix} 1 & 1 & 1 & 1 \\ \beta & -\beta & i\alpha & -i\alpha \\ e^{ikb}e^{-\beta b} & e^{ikb}e^{\beta b} & e^{-ikc}e^{i\alpha c} & e^{-ikc}e^{-i\alpha c} \\ (\beta - ik)e^{ikb}e^{-\beta b} & -(\beta + ik)e^{ikb}e^{\beta b} & i(\alpha - k)e^{-ikc}e^{i\alpha c} & -i(\alpha + k)e^{-ikc}e^{-i\alpha c} \end{vmatrix} \equiv 0$$

[1] Proc. Roy. Soc. (London), **130**, 499 (1931).

When expanded and simplified,[1] the determinant becomes

$$\cos ka = \frac{\beta^2 - \alpha^2}{2\alpha\beta} \sinh \beta b \sin \alpha c + \cosh \beta b \cos \alpha c \qquad (20)$$

In this equation we may for the time being regard k as an independent variable which takes on the N equally spaced values $2\pi l/Na$, where l ranges in integral steps from $-N/2$ to $+N/2$ (see Exercise 11-13); thus, on the *left*, $\cos ka$ ranges through a full cycle from -1 to $+1$ and back to -1. Each of these values of k defines a *discrete sequence of energy values* in the manner described in Sec. 4-7. We shall now show that these discrete values lie in bands which are separated by finite gaps.

Now consider the *right-hand* side of Eq. (20) as a function of E, as E varies from 0 to V. If the quantities V and c are of such size that a single well of this depth and width would have several bound states, then $\sin \alpha c$ and $\cos \alpha c$ will each go through several cycles as E varies over its range.

Thus the right-hand side of Eq. (20) undergoes a quasi-sinusoidal variation with $E^{1/2}$ *whose amplitude is greater than unity*, as can be seen by rewriting the expression in the equivalent form

$$\cos ka = F(E) \cos [\alpha c - \Delta(E)] \qquad (21)$$

in which

$$F(E) = \left[1 + \frac{V^2}{4E(V - E)} \sinh^2 \beta b \right]^{1/2}$$

and

$$\tan \Delta(E) = \frac{V - 2E}{2E^{1/2}(V - E)^{1/2}} \tanh \beta b$$

We therefore conclude that *those energies for which the right-hand side of Eq. (20) or Eq. (21) exceeds unity are forbidden*. The situation is illustrated graphically in Fig. 11-14, in which the right-hand side of Eq. (21) is plotted vs. E for a lattice of rectangular wells whose width and depth are the same as for the well shown in Fig. 4-4 and whose separation b is equal to one-sixteenth of the well width:[2]

$$V = \frac{h^2}{2m} \frac{49}{16c^2} \qquad c = 16b$$

---

[1] The determinant may be treated as follows: Multiply row 3 by ik and add to row 4; multiply rows 3 and 4 by e^{-ikb} and observe that $a = b + c$; add column 1 to column 2 and column 3 to column 4; subtract half of column 2 from column 1 and half of column 4 from column 3; change to hyperbolic and trigonometric functions; expand in terms of elements of row 4.

[2] If we also denote $E/V = f$, Eq. (21) becomes for this case

$$\cos ka = \left[1 + \frac{\sinh^2 0.687(1 - f)^{1/2}}{4f(1 - f)}\right]^{1/2} \cos [11f^{1/2} - \Delta(E)]$$

where

$$\tan \Delta(E) = \frac{1 - 2f}{2f^{1/2}(1 - f)^{1/2}} \tanh 0.687(1 - f)^{1/2}$$

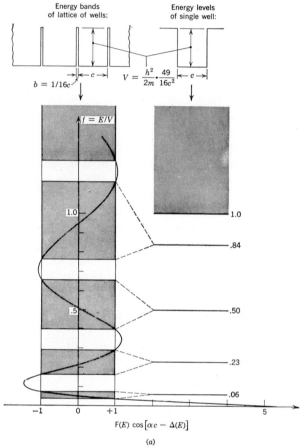

FIG. 11-14. (a) Plot of the expression $F(E) \cos [\alpha c - \Delta(E)]$ for an array of one-dimensional rectangular wells whose dimensions are given in the text.

From this figure we see that the permitted energy eigenvalues lie in bands whose positions are in close agreement with the discrete levels obtained in Exercise 4-17.

EXERCISE

11-16. Treat the case $E > V$ and show that Eq. (20) is valid for this range of E also. Thus show that energy gaps exist also in the energy range $E > V$, which is a *continuum* for the case of a single well.

Let us now look at the energy-level distribution from a slightly different viewpoint. We see from Eqs. (20) and (21) that the values of E which lie at the upper or lower boundaries of the energy bands are defined by the limiting values ± 1 of $\cos ka$. Now Exercise 11-14 shows

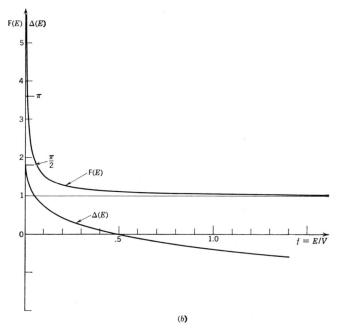

(b)

FIG. 11-14. (b) $F(E)$ and $\Delta(E)$ plotted as functions of E.

that the wave number k is closely related to the *momentum* of the particle, just as for a particle in a box, which suggests that we should regard k as being *unlimited* in its range. Cos ka is then a *cyclic* function of k, and the energy discontinuities associated with the forbidden gaps occur at the values $ka = \pm n\pi$ with n an integer. *These are just the values for which a free particle would have a wavelength equal to an integral submultiple of twice the crystal spacing a, and would be diffractively reflected by the lattice:*

$$k = \frac{2\pi}{\lambda} = \pm \frac{n\pi}{a} \qquad \text{or} \qquad \lambda = \frac{2a}{n} \tag{22}$$

The variation of E with k for a lattice of rectangular wells thus has the form indicated in Fig. 11-15, which refers to the specific case treated in Fig. 11-14. This figure indicates that the energy follows roughly the relation $E = \hbar^2 k^2/2m$ for a free particle, as indicated by the dashed curve, but that this relationship is considerably modified near the energy discontinuities at $ka = \pm n\pi$.

The curve of Fig. 11-15 also illustrates another important and fundamental property of the motion of a particle in a crystal lattice. In the neighborhood of each discontinuity the energy varies quadratically with $k - n\pi/a$:

$$E \approx E_n \pm \tfrac{1}{2}\Delta_n \pm \tfrac{1}{2}A_n \left(k - \frac{n\pi}{a} \right)^2 \tag{23}$$

Now it is found that, if a particle moving in a crystal lattice is acted upon also by an outside force, its acceleration is not F/m but is instead F/m^*, where m^* is an *effective mass* exhibited by the particle. m^* differs from m because of the interaction between the particle and the lattice; the effective mass may be greater or less than m, and may even be infinite

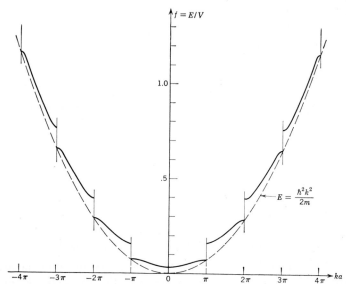

FIG. 11-15. Variation of E with k for a one-dimensional rectangular lattice.

or negative. Quantitatively one finds that m^* is given in the one-dimensional case by[1]

$$m^* = \frac{\hbar^2}{d^2E/dk^2} \tag{24}$$

Thus m^* is positive and nearly constant for states lying near the bottom of an energy band, negative for states lying near the top of a band, and infinite at some energy in the interior of a band. The concept of the effective mass is of quite broad applicability. Many of the expressions derived on the basis of the free-electron theory are valid not only for metals but also for semiconductors and insulators if m^* evaluated at the Fermi energy is used in place of m. Unfortunately, space does not permit further discussion of this important concept.

B. Three Dimensions, Brillouin Zones. We shall now summarize briefly some of the more important aspects of the generalization of the results obtained above to three dimensions. The main complicating

[1] For a derivation of this result, see C. Kittel, "Introduction to Solid State Physics," 2d ed., pp. 288 ff., John Wiley & Sons, Inc., New York, 1956.

factor encountered in three dimensions stems from the numerous different crystal-lattice structures that are encountered. We shall discuss some of the simpler aspects of crystal structure in connection with X-ray diffraction; our present discussion must accordingly be restricted mainly to the simplest case, that of a cubic lattice in the shape of a rectangular parallelepiped having N_x, N_y, and N_z unit cubes along its three edges. The periodicity condition imposed upon the wave function is then

$$\psi(x + \alpha N_x a,\ y + \beta N_y a,\ z + \gamma N_z a) = \psi(x,y,z) \tag{25}$$

where α, β, and γ may equal 1 or 0 independently. By the same line of argument as was used in deducing the form (13) for one dimension, we now obtain

$$\psi(r) = \chi_K(r)\ \exp\ (iK \cdot r) \tag{26}$$

where $\chi_K(r)$ is a function having the same periodicity as the lattice and the propagation vector K is defined as usual by

$$K = 2\pi \left(\frac{l}{N_x a}\, i + \frac{m}{N_y a}\, j + \frac{n}{N_z a}\, k \right) \tag{27}$$

l, m, and n being positive or negative integers, or l, m, $n = 0$, ± 1, ± 2, ... Thus, just as in the one-dimensional case, the Bloch form (26) combines an "atomic" type function with a plane wave characteristic of a particle in a box. There are just $N_x\, N_y\, N_z$ distinct values of $\exp\ (iK \cdot r)$, obtained by letting l, m, and n range independently over any N_x, N_y, and N_z consecutive values, respectively, but as in the one-dimensional case it is permissible and often advantageous to maintain the analogy between K and the momentum vector of a particle in a box by allowing l, m, and n to take on integral values without restriction.

In considering the dependence of the energy eigenvalues upon K, one again finds that E is a piecewise continuous function of K, with "gaps" occurring for certain values of K. Let us consider the propagation vector K as ranging over a lattice of points in "*wave number space*," also called *reciprocal space*, and imagine that surfaces of constant E are scribed in this space. What do these surfaces look like, and where do the "gaps" occur?

For small values of K the energy varies quadratically with K, as for a free particle, so that the surfaces of constant energy are ellipsoids which reduce to spheres for the case of cubic lattices. For larger values of K the energy surfaces may depart quite markedly from an ellipsoidal shape, but must of course always possess the same symmetry as does the crystal lattice.

If one follows the variation of energy with $|K|$ for a *given direction* of K, a curve similar to that shown in Fig. 11-15 for one dimension, consisting of a number of segments separated by gaps, is usually obtained. As the direction of K is changed, the curves change correspondingly in a smooth way, and the points of constant energy on these curves may be regarded as generating the constant-energy surfaces we are considering.

A feature of great importance is that of the locus of those K vectors at which a gap occurs in the E vs. $|K|$ curves just discussed. For a given energy gap this locus is a closed polyhedron having the symmetry of the

Fig. 11-16. Schematic illustration of a possible set of constant-energy surfaces for a cubic lattice.

lattice, called a *Brillouin zone*. It will be shown in the next chapter that a K vector from the origin to a point on such a polyhedron is a wave-number vector for which *Bragg reflection* occurs. Thus just as in one dimension the energy-band gaps are associated with momenta for which the crystal diffractively reflects the electrons. The energy surfaces that might be obtained for a cubic crystal are indicated schematically in Fig. 11-16.

The energy-band structure plays a decisive role in determining the macroscopic electronic properties of a crystal, as has been pointed out previously, and the concept of Brillouin zones provides an important tool for studying in some detail the influence of the crystal structure upon the electronic properties, particularly those that may be directionally sensitive. As a qualitative example, consider crystals whose band structures possess (1) a large overlap, (2) a slight overlap or narrow gap,

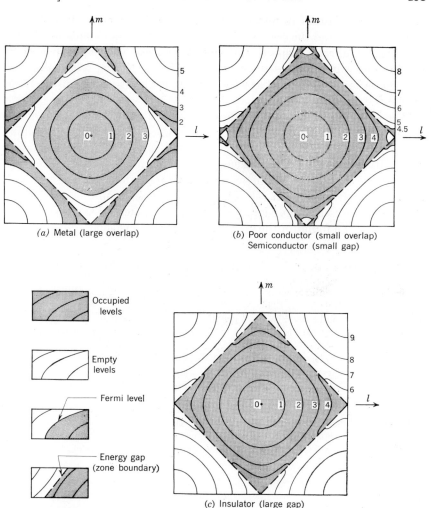

FIG. 11-17. Schematic diagrams of the energy-band structure of a cubic crystal in the l, m plane of reciprocal space. (a) A crystal in which there is a strong overlap in two neighboring bands. Such a crystal must be a good electrical conductor. (b) A crystal having but a slight overlap is a poor electrical conductor, and various properties such as the Hall effect may show strong directional dependence; a crystal having a slight energy gap will be a semiconductor. Semiconductors almost always show strong directional effects in their electrical behavior. (c) A crystal having a large band gap and an even number of electrons per unit cell of the lattice will be an insulator.

or (3) a large gap, at the Fermi level (Fig. 11-17). In the case of a large overlap there will be empty levels lying contiguous to the occupied levels for nearly all directions of K; such a crystal must be a good electrical conductor and will exhibit little directional sensitivity in its electrical properties. In the case of a slight overlap or a slight gap, the occupied states are contiguous to unoccupied states only for relatively restricted directions of K, and the crystal will accordingly be either a poor conductor or a semiconductor, respectively, whose electrical properties may show considerable directional sensitivity.[1] A crystal having a large energy gap separating the occupied and the unoccupied levels will be an electrical insulator.

The foregoing treatment, while quite incomplete, serves to indicate one of the ways by which the electronic properties of simple crystalline solids may be related to the crystal structure. A great deal more information can be deduced concerning the motions of the electrons inside a crystal lattice, both with and without the action of external electric or magnetic fields. Unfortunately, these results lie outside the scope of the present discussion. Further development of the band theory of solids can be found in the specialized texts on the solid state.

REFERENCES

Brillouin, L.: "Wave Propagation in Periodic Structures," McGraw-Hill Book Company, Inc., New York, 1946.

Dekker, A. J.: "Solid State Physics," Prentice-Hall, Inc., Englewood Cliffs, N.J., 1957.

Kittel, C.: "Introduction to Solid State Physics," 2d ed., John Wiley & Sons, Inc., New York, 1956.

Seitz, F.: "The Modern Theory of Solids," International Series in Pure and Applied Physics, McGraw-Hill Book Company, Inc., New York, 1940.

Shockley, W.: "Electrons and Holes in Semiconductors," D. Van Nostrand Company, Inc., Princeton, N.J., 1950.

[1] The directional dependence of certain properties such as the Hall effect and the magnetoresistive effect has been used experimentally to obtain information concerning the location of the Fermi surface in relation to the boundaries of the Brillouin zones.

12

X-Rays

As a final topic in atomic physics, we shall now examine some of the more elementary properties of X-rays. Quite aside from their importance in technical and medical applications, X-rays possess many properties of great physical interest and have contributed very substantially to our understanding of the basic properties of matter. Furthermore, X-rays are of great importance in many phases of nuclear-physics research, so that a consideration of their properties will serve both to clarify some aspects of atomic physics and to introduce some of the concepts that are of importance in our forthcoming discussion of nuclear physics.

12-1. Historical Introduction

X-rays were discovered quite accidentally[1] by Roentgen (1895) through their ability to induce fluorescence in certain salts of heavy metals, even after passing through several millimeters thickness of optically opaque material. Roentgen immediately dropped his investigations of electrical discharges in gases, in which he was engaged at the time of his discovery, and proceeded to study the new radiations. His first observations established many of the fundamental properties of X-rays, and

---

[1] In retrospect it seems certain that X-rays had been present in many laboratories for several years, produced by the high voltages that were often used in connection with gaseous discharge tubes; Roentgen was merely the first experimenter to *notice* them. A similar situation apparently existed in connection with the polarization of electrons in β-decay, which will be discussed in Chap. 20, and with a number of other phenomena. Experiences of this kind illustrate the importance of remaining wide awake and ready for the unexpected: An effect which appears at first to be an annoying experimental difficulty standing in the way of accurate measurement may well turn out to be of far greater importance than the quantity being measured.

the work of numerous other investigators who immediately began to study the phenomenon quickly confirmed his results and extended them in many directions.   Some of these early researches showed that:

1. X-rays are produced when cathode rays (i.e., electrons) of high energy strike the anode of a discharge tube (Roentgen, 1895).
2. Anodes made of heavy metals generate X-rays more strongly than do those made of light metals (Roentgen, 1895).
3. X-rays are uncharged, since they are undeviated by magnetic fields (Roentgen, 1895).
4. X-rays are capable of discharging electrified bodies surrounded by gases (Roentgen, 1895).   This property of rendering gases conducting by producing ions in them provides a quantitative means of measuring X-rays and is widely utilized as a means of standardizing the output of X-ray generators.   The instrument that is based upon this property is called an *ionization chamber;* it measures the charge carried by ions produced inside a certain volume of air by the X-rays.[1]
5. Substances composed wholly of the lighter elements are relatively more transparent toward X-rays than are those which contain the heavier elements (Roentgen, 1895).
6. X-rays affect photographic emulsions (Roentgen, 1895).   This property has been widely used as a means of detecting and measuring X-rays and their effects upon matter.
7. X-rays are not easily reflected or refracted by ordinary substances (Roentgen, 1895).   Both reflection and refraction of X-rays have been observed under special conditions, however.
8. The X-rays generated by the early tubes were shown by slit-diffraction experiments to be undulatory radiations of wavelength $\sim 10^{-10}$ m (~ 1 Å), and were correctly concluded to be electromagnetic in character (Haga and Wind, 1899).
9. X-rays are *polarizable* under certain conditions (Barkla, 1906).
10. X-rays can be *diffracted* by crystals, which shows both that X-rays are *waves* and that crystals are composed of regular arrays of atoms (Friedrich, Knipping, and Laue, 1912).   This experiment provides the means both for analyzing an X-ray beam into its monochromatic components and for deducing the space structure of various crystals. X-ray crystallographic investigations have contributed tremendously to our knowledge and understanding of molecular and crystal structure, and, through the X-ray crystal spectrometer, our knowledge of many basic properties of matter has been greatly enhanced.

---

[1] The unit of total dosage of X-rays is called the Roentgen unit, R.   It is the amount of X-ray total flux that produces 1 esu each of positive and negative charge in 1 cm^3 of dry air at standard temperature and pressure.   A commonly used unit of intensity of X-rays is the Roentgen per hour, R hr^{-1}.

11. The X-rays from an X-ray tube consist of a *continuous* spectrum (the "white" radiation) upon which are superimposed a number of "bright lines" which are characteristic of the anode material of the tube (W. H. Bragg, 1913).   These bright lines are called the *characteristic radiation* of a material (Fig. 12-1).

12. The wave number of the characteristic radiation increases with increasing atomic weight of the anode material (Moseley, 1913).

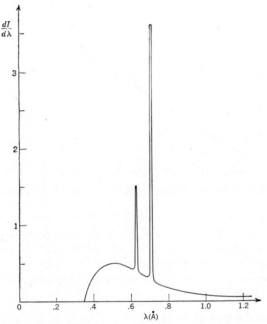

FIG. 12-1. The spectral distribution of X-rays from an X-ray tube, showing the characteristic radiation superimposed upon the "white" radiation. [Adapted from Ulrey, *Phys. Rev.*, **11**, 401 (1918).]

Moseley made use of the then recently formulated Bohr theory of one-electron atoms to identify the characteristic radiation as corresponding to transitions between the lowest-lying energy levels of an atom with a nuclear charge, Ze, in which Z is approximately equal to half the atomic weight of the material.   He was able for the first time to define a *unique sequence* of the elements upon the basis of the nuclear charge number, or *atomic number Z*, by showing that in this sequence, the square root of the wave number of a given line in the characteristic radiation varied precisely as the integer Z, as required by the Bohr formula.   This variation is shown in Fig. 12-2.

With the above brief outline of some of the more important early discoveries concerning the properties of X-rays as a guide, we shall now consider in somewhat greater detail the more elementary of their properties. Unfortunately, it will not be possible for us to treat these properties in a rigorous manner because of the great complexity of the phenomena

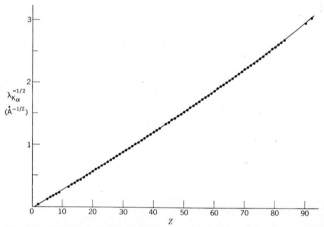

FIG. 12-2. Moseley diagram of the characteristic X-rays from a number of elements.

involved. We shall endeavor, rather, to make plausible the empirically observed behavior of X-rays and to acquire an understanding of the basic physical processes involved. In some cases a purely classical analysis will provide such an understanding, but in other cases the broader content of the quantum theory is essential to the treatment.

12-2. The Production of X-rays

The problem of the mechanism of production of X-rays was one of the earliest questions to receive a satisfactory theoretical treatment. Although certain aspects of the production mechanism intimately involve the quantum theory, certain others can be analyzed with satisfactory accuracy by using the ideas of classical electrodynamics. The question of the *angular distribution of the intensity* of the "white" X-rays is one of those for which a classical treatment is suitable, and we shall therefore begin our discussion at this point. We shall eventually consider several mechanisms of producing X-rays, but we shall first treat what may be considered to be the primary mechanism, namely, the *deceleration of a rapidly moving charge.*

A. Bremsstrahlung. Soon after their discovery, it was recognized that X-rays result from the sudden stopping, or deceleration, of a moving charge. The radiation that results from this process has come to be called *bremsstrahlung* ("braking radiation"), a term which was taken over from the early work of theoreticians, notably Sommerfeld.

It is, of course, well known from classical electromagnetic theory that radiation is emitted by an accelerating charge. The classical expressions for the fields and Poynting vector at time t at a distance r from an accelerating charge q are

$$E = \frac{q([a] \times r) \times r}{4\pi\varepsilon_0 c^2 r^3} \tag{1}$$

$$B = \frac{\mu_0 q[a] \times r}{4\pi c r^2} \tag{2}$$

$$S = e_r \frac{q^2([a] \times r)^2}{16\pi^2\varepsilon_0 c^3 r^4} \tag{3}$$

where $[a]$ is the value of the vector acceleration of the particle at an earlier time $t - r/c$ and it is assumed that the speed of the particle remains small compared with the speed of light.

If we apply the above results to the case of a particle which is moving (not too rapidly) along a line and is suddenly given an acceleration along its direction of motion, we may write

$$E = e_\theta \frac{q[a]}{4\pi\varepsilon_0 c^2 r} \sin\theta \tag{4}$$

$$B = e_\varphi \frac{\mu_0 q[a]}{4\pi c r} \sin\theta \tag{5}$$

$$S = e_r \frac{q^2[a]^2}{16\pi^2\varepsilon_0 c^3 r^2} \sin^2\theta \tag{6}$$

where θ is measured from the line of motion of the particle to the point of observation, as shown in Fig. 12-3 for a negative charge.

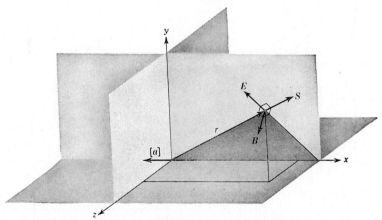

FIG. 12-3. Illustrating the relations between **E**, **B**, **S**, and [a] for a negative charge.

We thus see that the radiation from a linearly accelerating charge is polarized with its electric vector in the plane of [a] and r and that the power emission per steradian at the angle θ is

$$\Pi(\theta) = r^2|\mathbf{S}| = \frac{q^2[a]^2 \sin^2 \theta}{16\pi^2\varepsilon_0 c^3} \tag{7}$$

This expression shows that no energy is emitted either forward or backward along the direction of the acceleration, that the power emitted is a maximum at right angles to this direction, and that the radiated power is distributed symmetrically about the polar axis and about the equatorial plane. The above properties are described graphically in Fig. 12-4, which shows the so-called *radiation pattern* of an accelerated charge.

As was previously pointed out, the above results apply only to cases in which the speed of the radiating particle is at all times much less than the speed of light. This condition may be stated in an equivalent form, namely, that the *accelerating voltage* applied to the X-ray tube must be much less than mc^2/e ($mc^2 = 0.511$ MeV for electrons). Thus the X-rays from tubes operating at only a small fraction of 500 kV should substantially follow the above angular distribution.

On the other hand, there are many cases of practical interest in which the speed of the radiating particle is initially quite comparable with the speed of light. In these cases the angular distribution of the X-rays

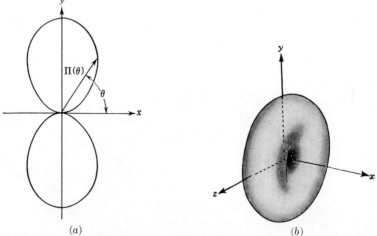

(a)   (b)

Fig. 12-4. The radiation pattern of slowly moving accelerated charge. (a) polar plot, with x-axis as polar axis; (b) perspective sketch, to illustrate the three-dimensional nature of the pattern. Note that the radiated intensity is maximum in the equatorial (y-z) plane, and is zero along the direction of the acceleration (x-axis).

is observed to be *peaked in the forward direction*. The angular distribution of the radiation of a decelerating charged particle can be found through the use of relativistically correct retarded potentials for the particle, called the Lienard-Wiechert potentials.[1] We can solve this problem by the use of a somewhat more easily visualized procedure, however, by applying a Lorentz transformation to the quantities appearing in Eq. (7).

In order to carry out the procedure suggested above, we consider two coordinate systems S and S′ moving with respect to one another with

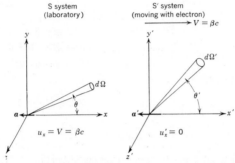

Fig. 12-5. The radiation from an accelerated charge as viewed by two observers in relative motion.

constant speed βc along their common x-axes, and imagine that in the S′ system a charge q which is initially at rest is given an acceleration a' along the negative x′-axis for a short time τ'. During this time the charge radiates power as described by Eq. (7). Our problem is to describe this process from the viewpoint of an observer in the system S. To such an observer, the charge q is initially moving with a speed $+\beta c$ and then suffers a deceleration a for a short time τ during which it radiates away a small fraction of its kinetic energy. In applying the Lorentz transformation to this situation we shall start with the energy in the S′ system which is radiated into a solid angle $d\Omega'$ at an angle θ' with respect to the x′-axis. We shall then apply the Lorentz transformation to this energy, regarded as residing in quanta, and also to the solid angle $d\Omega'$, which will be transformed into a solid angle $d\Omega$ at an angle θ with respect to the x-axis. The two situations are shown in Fig. 12-5. The various quantities which enter into the problem transform in the following ways:

[1] See J. A. Stratton, "Electromagnetic Theory," p. 475, McGraw-Hill Book Company, Inc., New York, 1941; W. R. Smythe, "Static and Dynamic Electricity," 2d ed., p. 571, McGraw-Hill Book Company, Inc., New York, 1950; or Panofsky and Phillips, "Classical Electricity and Magnetism," Addison-Wesley Publishing Company, Reading, Mass., 1955.

1. The energy and the components of momentum of the photons within the solid angle $d\Omega$. (Recall that $Pc = W$ for quanta.)

$$P'_x = P' \cos \theta' = \gamma \left(P_x - \beta \frac{W}{c} \right) = \frac{\gamma W}{c} (\cos \theta - \beta) \tag{8}$$

$$P'_y = P' \sin \theta' = P_y = \frac{W}{c} \sin \theta \tag{9}$$

$$P'_z = 0 = P_z \qquad \text{assume quantum travels in the } xy \text{ plane}$$
$$W' = \gamma(W - \beta c P_x) = \gamma W(1 - \beta \cos \theta) \tag{10}$$

2. The angle θ'. (Compare Exercise 1-15.)

$$\cos \theta' = \frac{P'_x}{P'} = \frac{cP'_x}{W'} = \frac{\cos \theta - \beta}{1 - \beta \cos \theta} \tag{11}$$

and $$\sin^2 \theta' = 1 - \cos^2 \theta' = \frac{(1 - \beta^2) \sin^2 \theta}{(1 - \beta \cos \theta)^2} \tag{12}$$

3. The solid angle $d\Omega'$.

$$d\Omega' = 2\pi \sin \theta' \, d\theta' = -2\pi \, d(\cos \theta')$$
$$= \frac{d\Omega \, (1 - \beta^2)}{(1 - \beta \cos \theta)^2} \tag{13}$$

4. The acceleration of the charge. [See Eq. 1-4(7).]

$$a' = \frac{du'_x}{dt'} = \frac{a_x}{(1 - \beta^2)^{3/2}} \qquad \text{since } u'_x = 0 \tag{14}$$

5. The duration of the radiation pulse.

$$\tau' = \frac{\tau}{\gamma} \tag{15}$$

Using the above transformation properties together with the equation (7) for the power emission per unit solid angle in the S' system, we may write for the total energy emitted into $d\Omega'$ during the time τ'

$$dW' = \Pi'(\theta')\tau' \, d\Omega'$$
$$= dW \, \gamma(1 - \beta \cos \theta) \qquad \text{from Eq. (10)}$$

But the intensity $\Pi(\theta)$ is defined by the relation $dW = \Pi(\theta)\tau \, d\Omega$ so that

$$\Pi(\theta) = \frac{\Pi'(\theta') \, d\Omega' \, \tau'}{d\Omega \, \tau\gamma(1 - \beta \cos \theta)}$$
$$= \frac{q^2[a']^2 \sin^2 \theta' \, d\Omega' \, \tau'}{16\pi^2 \varepsilon_0 c^3 \, d\Omega \, \tau\gamma(1 - \beta \cos \theta)}$$

Eliminating all primed quantities we finally obtain

$$\Pi(\theta) = \frac{q^2[a]^2 \sin^2 \theta}{16\pi^2 \varepsilon_0 c^3 (1 - \beta \cos \theta)^5} \tag{16}$$

The denominator of this expression causes the angular distribution of the radiated power from a charge which is accelerated parallel to its direction of motion to be directed predominantly in the *forward direction*, rather than being symmetrically distributed in the forward and backward directions. This equation describes rather accurately the observed distribution of the "white" X-rays from (thin-target) tubes operating at voltages up to about 1 MeV and furnishes a good qualitative guide even up to much higher voltages. At voltages of more than $\sim$1 MeV, the production of electron-positron pairs, and other effects not included in the classical model, cause the above formula to be inaccurate. Figure 12-6 shows the angular distribution of the radiation for speeds βc corresponding to kinetic energies of 0.01, 0.03, 0.10, and 0.30 times mc^2.

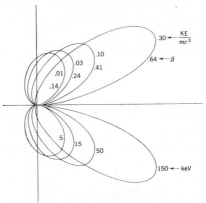

Fig. 12-6. The radiation patterns of accelerated particles moving at various speeds [Eq. 12-2(16)]. The curves are normalized to the same total power.

EXERCISES

12-1. Verify the correctness of the transformations described in (1) to (5), and carry out the substitutions which lead to Eq. (16).

12-2. Compute the angle θ_m at which the radiation from an accelerating electron is a maximum as a function of the kinetic energy of the electron.

12-3. An electron moving initially with speed βc is brought to rest with a uniform deceleration, during which time it moves only a small distance. Find the total energy emitted per unit solid angle at an angle θ with respect to the motion.

12-4. A collimated stream of high-speed, monoenergetic electrons strikes a thin aluminum foil at normal incidence and passes through it almost unchanged in energy. The intensity of the X-rays emitted by these electrons is measured at 45° and 135° with respect to the direction of the stream, and the ratio of these intensities is found to be R. Find the speed of the electrons.

B. Spectral Distribution of Bremsstrahlung. Although the classical formulas discussed above provide a rather satisfactory description of the *angular* distribution of the X-ray energy emitted by a stopping particle, the classical theory fails very badly in describing the *spectral* distribution of such X-rays. That this must be so can be seen qualitatively by thinking in terms of the conversion of kinetic energy of the particle into electromagnetic *quanta* of various energies. The maximum energy that can be given up by the particle is its kinetic energy, so that there must be a *highest frequency* that can appear in the X-ray spectrum:

$$\nu_{max} = \frac{T}{h} = \frac{e\phi}{h} \tag{17}$$

where $T = e\phi$ is the kinetic energy of the electron as it strikes the target. (This relationship is the basis of a very precise measurement of the ratio h/e, since only two quantities must be measured with high precision.)

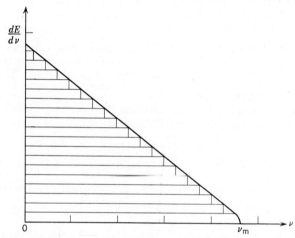

FIG. 12-7. Schematic illustration of the thick-target X-ray spectrum, regarded as composed of a number of thin-target distributions.

Several attempts have been made to derive the shape of the spectral distribution of X-radiation. All existing treatments of this problem are quite complicated, but fortunately both the theoretical and experimental results indicate that the emitted energy should be *approximately uniformly distributed in frequency* up to the high-frequency limit given by Eq. (17), provided that the X-ray target is so thin that only a negligible fraction of the electrons is absorbed. That is, for a *thin target*,

$$\frac{dE}{d\nu} \approx K \qquad 0 < \nu < e\phi/h \tag{18}$$

where K depends upon target material and thickness, the voltage of the electrons, etc., but only very little upon ν. The spectrum from a *thick* target can be thought of as being compounded from a succession of thin-target spectra, such that the average energy of the electrons that penetrate into the deeper layers is progressively diminished[1] with increasing depth. The resulting spectral distribution is then very nearly a linear function of ν, as shown in Fig. 12-7, except for a small "break" near ν_m which causes the curve to intersect the ν-axis quite steeply. An empirical formula (Kulenkampff, 1922) which closely approximates this curve for many anode materials and voltages is

$$\frac{dE}{d\nu} = C[Z(\nu_m - \nu) + 0.0025Z^2] \qquad (19)$$

where C is independent of Z and the applied voltage. This relation shows that the X-ray output is very nearly proportional to Z and to ϕ^2.

EXERCISES

12-5. Show, from Eq. (19), that the X-ray energy output of a tube is very nearly proportional to the square of the voltage applied to the tube.

12-6. The thin-target momentum distribution of the X-ray quanta from a high-energy electron accelerator follows the law

$$\frac{dN}{dp} = \frac{A}{p} \qquad 0 < p < p_{max}$$

where p is the momentum of the photon and A is a constant for given incident-electron energy. Derive this law, starting from results presented above.

C. Mechanism of Production of Bremsstrahlung, Polarization. It has been indicated in *A* that the continuous bremsstrahlung X-radiation is produced by the rapid (but classically continuous) deceleration of charged particles as they pass through matter. On the other hand, the discussion just concluded, regarding the spectral distribution of the radiation, has emphasized that the phenomenon is a quantum-mechanical one. We shall now describe a little more closely, but qualitatively, the basic mechanism involved in the bremsstrahlung process for energies less than $2mc^2$.

When a charged particle passes through matter, it interacts electrically with the nuclei of the atoms of the matter and also with the electrons surrounding the nuclei. The second of these interactions usually results only in an electronic excitation or ionization of some of the atoms which lie near the path of the incoming particle, and the particle loses numerous

---

[1] The major cause of this gradual energy loss is by ionizing collisions with the electrons of the target; X-rays are not ordinarily produced in such collisions. The detailed mechanism of collisions of this type will be considered in Chap. 14.

small amounts of energy in these encounters, gradually diminishing in energy and eventually coming to rest. This relatively slow deceleration of the particle is not, however, responsible for the X-radiation.

The emission of X-rays results, rather, from the less frequent, but individually more catastrophic, encounters with the *atomic nuclei*. In such encounters the force acting upon the incoming particle, and therefore its acceleration, is Z times greater than in the individual electronic encounters, and in addition the more massive nucleus does not recoil as readily as does a light electron. One would thus expect the radiated energy to be Z^2 times greater from these nuclear encounters than from the electronic encounters. In one manner of analyzing the system quantum-mechanically, one finds that there is a certain probability that a given encounter between the incoming particle and the atomic nucleus will result only in a deflection of the particle and a corresponding recoil of the nucleus, with *no emission of radiation*. This is called an *elastic Rutherford scattering* and is described correctly by Eq. 2-1(17). On the other hand, there is also a certain probability that a given encounter will result in the *emission of a photon* as well as in a deflection of the incoming particle and a recoil of the nucleus. This is called an *inelastic, or radiative, collision*. Now, the emission or nonemission of a photon in a given encounter is purely statistical, and it is the average behavior over a large number of encounters that follows the classical Rutherford formula and the classical radiation formula quite closely. It is found experimentally that, at tube voltages less than about 1 MeV, the "efficiency" of conversion of the electron's kinetic energy into X-rays rarely exceeds one or two per cent and is often much less.

Certain departures from the results based upon the assumption of a uniform linear deceleration can be successfully analyzed classically by allowing for the actual inverse-square nature of the interaction forces. Thus, for example, the *polarization* of the bremsstrahlung can be treated qualitatively in this way: If the assumption that the deceleration of the particle is parallel to its velocity were valid, one would conclude that X-radiation (from a thin target, at any rate) would be *completely polarized*, with its electric vector lying in the plane of *a* and *r*. This is observed not to be the case, although a sizable degree of polarization *is* observed. An explanation for this discrepancy is easy to find. The various encounters will clearly involve components of acceleration *transverse* to the direction of motion of the radiating particle, and these ought to lead to corresponding radiation components whose polarization differs from that described above. Furthermore, the relative contribution of the transverse acceleration components to the total radiation ought to be greater for the *more distant encounters*, which involve the emission of lower-energy quanta. Thus the degree of polarization ought to be greatest for quanta near the high-energy limit, i.e., for which the acceleration is nearly along the average velocity, and least for the low-energy quanta. This is observed to be the case; the polarization may be as low as 10 per cent for low-energy quanta and may approach 60 to 70 per cent for the most energetic quanta, as is indicated in Fig. 12-8.

D. *The Characteristic Radiation.*   In addition to the continuous spectrum described in the preceding sections, an X-ray tube emits certain well-defined frequencies of radiation called the *characteristic* radiation of the anode material (Fig. 12-1).   It is easy to interpret this radiation in the light of our previous discussion of the shell structure of the atom (Chap. 7).   One of the principal features of the shell model is the interpretation of the energy states of the atom in terms of the occupation, by the individual electrons, of various possible "one-electron" energy levels in the average central field of the nucleus and the remaining electrons, with the exclusion principle governing the occupation of these individual levels.   Thus an atom is pictured as having two electrons in the 1s state, two in the 2s state, six in the 2p state, etc., as was described in connection with the periodic table.   If one or more of these electrons is removed from its "normal" energy state and occupies some other energy level,

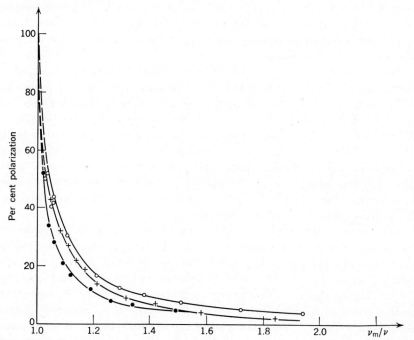

Fig. 12-8. Polarization of white X-radiation as a function of quantum energy. [Adapted from Cheng, *Phys. Rev.*, **46**, 243 (1934).]

the atom is in an *excited state* and will proceed to attain the lowest possible energy state corresponding to the number of electrons it possesses, by the emission of quanta. It should of course be remembered that all excited states of an atom, including those now under discussion, are a property of the *atom as a whole,* and can only be described in terms of the possible energy levels of the individual constituent electrons as an approximation.

One of the principal mechanisms of energy loss by the incoming particle is through the excitation or ionization of the atoms of the anode material. This excitation can be of various degrees, depending upon which one of the Z electrons of an atom is removed by the incoming particle. The least excitation results if an electron is removed from the outermost shell, leaving the atom ionized but otherwise unexcited. On the other hand, the greatest excitation results if an electron is removed from the 1s subshell of the atom and is completely ejected from the atom. (The exclusion principle of course prevents an electron being raised from the 1s subshell to any higher shell which is already fully occupied by electrons, so that such an electron must either be raised to one of the ordinary optical levels or, much more probably, be completely ejected from the atom.) The passage of a fast particle through the anode material thus results in many atoms being excited, each by the removal of an electron from one of the atomic subshells. *The radiation which results when the atoms return to an unexcited state is just the characteristic radiation under discussion.*

The conventional description of the various transitions involved in the emission of the characteristic X-radiation makes use of a somewhat different terminology than is used in describing optical transitions, but the fundamental mechanisms involved are of course precisely similar in X-ray and optical spectra. The various shells are now called the K, L, M, N, ... , using a notation introduced by Barkla; the K shell corresponds to the lowest one-electron energy state (that is, $n = 1$). The electrons which occupy the energy states in the K shell are called K electrons; those in the L shell, L electrons; etc. There are thus two K electrons, eight L electrons, eighteen M electrons, etc.

The various substates of a given shell are labeled in the order of increasing energy by a roman subscript, I, II, III, etc. Thus the K shell has only one level, the L shell *three* levels called L_I, L_{II}, L_{III} (corresponding to the $^2S_{1/2}$, $^2P_{1/2}$, and $^2P_{3/2}$ states, respectively) and similarly for the higher shells. The state of excitation of an atom is then given in this notation by identifying *from which of the above states an electron is missing.* Thus L_{II} refers to a state of excitation in which an electron is missing from the one-electron state 2p $^2P_{1/2}$, the other states being normally occupied by electrons.[1]   The X-ray term diagram corresponding to these various states of excitation appears *inverted* with respect to the above ordering of subshell energies, because of the above-described correlation of a state of excitation with the *absence* of an electron from one of the one-electron levels.   In the X-ray term diagram the K level therefore lies highest, since it requires the most energy to remove a K electron from an atom; the L_I level is next highest; then L_{II}, L_{III}, M_I, M_{II}, M_{III}, M_{IV}, M_V; and so on.   This is illustrated in Fig. 12-9.

If an atom is in a given state of excitation by virtue of the absence of an electron from one of its normally occupied shells, it may reduce its degree of excitation by supplying an electron from a higher shell to fill the unoccupied state and emitting a quantum corresponding to the energy difference between the two states.

The customary notation for the various spectral series that result from such transitions is as follows: The series of lines which result from an electron from higher shells dropping into an unoccupied K level is called the K *series of X-ray lines*, and the various lines of the series are called $K\alpha$, $K\beta$, $K\gamma$, $K\delta$, etc., in the order of increasing frequency.   (Thus the K series of X-ray lines of hydrogen is identical with the Lyman series.) The series which results from transitions into an unoccupied L level is called the L series, and so on.   The usual electric-dipole selection rules govern the transitions which lead to the brightest of the observed lines.

We thus see that the characteristic X-ray spectrum of a given material is like that of a one-electron atom, in which the effective potential function which governs the energy levels is that due to all electrons except the one which is missing in the excited atom.   The variation in the frequencies of the characteristic X-ray lines with atomic number therefore follows roughly a Z^2 law.   Although the dependence of the energy levels upon n and l differs considerably from the form of the Bohr formula, the Z^2 dependence of the energy of a given level is rather accurate, especially for the K radiation.[2]   This is the basis of Moseley's important work in identifying the correct ordering of the elements upon the basis of the frequencies of the characteristic K series X-ray lines.[3]   His work not only established the ordering of elements but also served as an important confirmation of the correctness of the Bohr-Rutherford nuclear atomic model.   Figure 12-2 illustrates Moseley's law: $\nu^{1/2} = K(Z - \sigma)$ for the

---

[1] The notation introduced in this section seems to be in common use, but one occasionally finds other notation, similar in appearance but actually different, in textbooks and in the literature.

[2] More precisely, the K radiation follows a $(Z - \sigma)^2$ law with $\sigma \approx 3$ because of the shielding effect of the remaining electrons, the L radiation a $(Z - \sigma')^2$ law, etc.

[3] *Phil. Mag.*, **26**, 1024 (1913); **27**, 703 (1914).

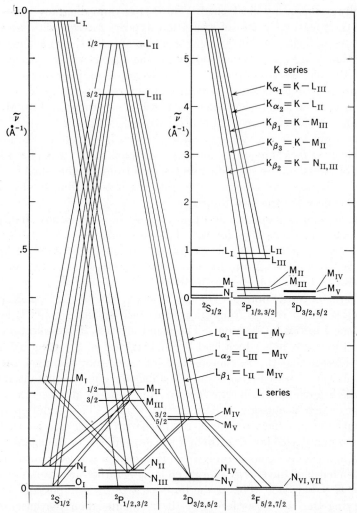

FIG. 12-9. X-ray term diagram for tungsten.   Note the large magnitudes of the fine-structure effects discussed in Chap. 5 and the screening effects of Chap. 7.   The notation for designating the levels is that of Sommerfeld; that for the transitions ($K\alpha_1$, $K\alpha_2$, etc.) is that of Siegbahn.   Other notations are often encountered, however, so that caution should be used in interpreting the literature.

K radiation, and Table 12-1 gives the X-ray term values for a number of elements.

EXERCISE

12-7. An X-ray tube having a copper target is to be operated at the highest voltage for which the K series of characteristic lines does not appear.   What is this voltage?   *Ans.:* 8.9 kV.

TABLE 12-1. X-RAY TERMS FOR VARIOUS ELEMENTS

The various levels are given in kilo electron volts; 1 keV $\leftrightarrow$ 12.398Å.

Term	13 Al	29 Cu	42 Mo	47 Ag	74 W	82 Pb
K	1.560	8.990	20.03	25.50	69.55	88.10
L_I	0.1152	1.098	2.870	3.815	12.10	15.89
L_{II}	0.0730	0.955	2.630	3.525	11.53	15.21
L_{III}	0.0725	0.934	2.521	3.355	10.21	13.07
M_I	......	0.123	0.509	0.720	2.815	3.867
M_{II}	0.005	0.0753	0.412	0.604	2.575	3.570
M_{III}	......	......	0.395	0.573	2.278	3.091
M_{IV}	......	0.0038	0.234	0.374	1.865	2.598
M_V	......	......	0.2302	0.368	1.803	2.495
N_I	......	......	0.0655	0.0971	0.590	0.905
N_{II}	......	......	0.0371	0.0582	0.485	0.778
N_{III}	......	......	....	......	0.418	0.658
N_{IV}	......	......	....	0.0061	0.252	0.447
N_V	......	......	0.007	0.0049	0.239	0.425
N_{VI}	......	......	....	......	0.027	0.143
N_{VII}	......	......	....	......	......	0.136

12-3. The Interactions of X-rays with Matter

We shall now consider briefly the more important phenomena which result from the passage of X-rays through matter. If a beam of X-rays

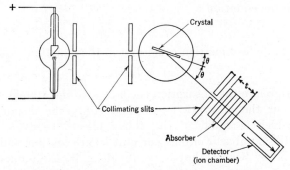

FIG. 12-10. Experimental arrangement for studying the absorption of monochromatic X-rays.

is passed through a crystal monochromator and thence through various thicknesses of matter, it is observed that the beam is *attenuated* with increasing absorber thickness. This attenuation, or "absorption," is caused by many mechanisms, some of which will be treated in this section. The most important of the mechanisms, for X-ray photon energies below about 1 MeV, are ionization (photoelectric absorption), scattering by the electrons of the absorbing material (unmodified and modified, or Compton, scattering), and diffractive scattering by a crystal lattice. We shall now briefly discuss the observed features of the absorption of X-rays and the mechanisms which are important in the absorption process.

A. *The Absorption of Monochromatic X-rays.* If a collimated beam (Fig. 12-10) of monochromatic X-rays is passed through an adjustable thickness of absorbing material and the intensity of the transmitted beam is measured, it is found that the transmitted beam *decreases exponentially in intensity with increasing absorber thickness.*

EXERCISE

12-8. Show that an exponential attenuation results if the decrease in intensity in a differential thickness of absorber is proportional to the intensity reaching that layer of absorber.

We shall shortly relate this attenuation to the mechanisms which cause it. Some of the mechanisms can best be studied by means of absorption experiments, however, so we shall now describe the empirically observed results of such experiments. Some most important and interesting information comes out of a study of the variation of the *monochromatic linear absorption coefficient* σ_L with the wavelength of the X-rays. It is found[1] that, at very short wavelengths ($\lambda < \sim 10^{-11}$ m or $h\nu > \sim 80$ keV), the absorption coefficient for a given material varies according to the equation

$$\sigma_L(\lambda) = a\lambda^3 + b \tag{1}$$

where $\sigma_L(\lambda)$ is the attenuation coefficient in the equation

$$I(x) = I_0 e^{-\sigma_L x} \tag{2}$$

In order to relate the absorption coefficient to the atomic mechanisms responsible for the absorption, we shall study the *atomic absorption coefficient*, which is equal to the above linear absorption coefficient divided by the number of atoms per unit length and per unit cross section of the X-ray beam[2] (that is, we divide σ_L by the number of atoms

---

[1] Siegbahn, *Physik. Z.*, **15** (1914).

[2] Another very commonly used absorption coefficient is the so-called *mass absorption coefficient.* This is obtained by dividing the linear absorption coefficient by the *density* of the material. In this case the variable x is replaced by the variable ρx in the absorption law (2), so that the "thickness" of an absorber is measured in *mass of absorber per unit area of beam.* In MKS units, the unit of absorber thickness would thus be *kilograms per square meter.* A more commonly used unit is the gram per square centimeter, which is equal to 10 kg m⁻².

per unit volume). Thus if A is the chemical atomic weight,

$$\sigma_a = \frac{\sigma_L}{\rho N_0/A} = \frac{\sigma_L A}{\rho N_0} \qquad (3)$$

On this basis it is found that the coefficients in the law corresponding to Eq. (1) depend upon the atomic number of the absorbing material in a relatively simple way. Empirically[1] one finds

$$\sigma_a = C_{0K} Z^4 \lambda^3 + B \qquad (4)$$

for sufficiently short wavelengths, with $C_{0K} = 2.25$ m^{-1}. B is negligible except for the lightest elements and wavelengths $< \sim 10^{-10}$ m.

EXERCISES

12-9. Show from Eq. (3) that the atomic absorption coefficient has the dimensions of *area* per atom. One can think of σ_a as the equivalent *cross-sectional area* of a perfectly absorbing surface which would remove as much radiation from a beam as does an atom. In the terminology of nuclear physics, σ_a is the *total interaction cross section* for the absorption of X-rays of wavelength λ.

12-10. What thickness of lead is required to decrease the intensity of 100-keV X-rays by a factor of 400? What thickness of iron would be necessary? Neglect B for this estimate. (Compare Fig. 12-12.)

As the wavelength of the X-rays under consideration is gradually increased, the absorption cross section σ_a increases as described by the above equation, until a certain wavelength is reached. At this wavelength the absorption coefficient *decreases abruptly by a factor of several-fold*, and then again follows a variation with λ similar to Eq. (4), except with a smaller coefficient C. Further increase in λ eventually reveals *a series of three abrupt decreases* similar to the above one, followed by another range governed by an equation of the previous form. This behavior is represented schematically in Fig. 12-11, which shows σ_m plotted vs. λ. The nature of the discontinuity at K is better illustrated in Fig. 12-12, which illustrates $\log \sigma_m$ vs. $\log \lambda$. The existence of these "absorption edges" was first discovered by de Broglie.[2] Between K and L$_1$, the variation of σ_a with λ is given by

$$\sigma_a = C_{KL} Z^4 \lambda^3 + B \qquad \text{with } C_{KL} = 0.33 \text{ m}^{-1} \qquad (5)$$

The above results can be interpreted quite simply in terms of our previous discussion of the characteristic radiation. It is found that the

---

[1] Bragg and Pierce, *Phil. Mag.*, **28**, 626 (1914); Hull and Rice, *Phys. Rev.*, **8**, 326 (1916).

[2] *Compt. rend.*, **158**, 1493 (1914); **163**, 87 (1916); **163**, 352 (1916).

wavelengths at which the discontinuities in the absorption cross section occur *are precisely the shortest wavelengths appearing in each of the series of characteristic lines of the absorbing material* when it is used as the anode (target) of an X-ray tube. It thus appears that *the major part of the absorption cross section is due to the photoelectric ejection of electrons from the atoms of the absorbing material.* The discontinuities in the absorption curves correspond to the critical quantum energies, below which the quantum is unable to eject an electron from a given level. Thus the first discontinuity described above is called the K-*absorption edge* because it represents the critical wavelength for the ejection or non-ejection of a K electron. Similarly, the next three absorption edges are called the L-absorption edges, and are subdivided L_I, L_{II}, and L_{III}. There are five M-absorption edges, seven N-edges, and so on.

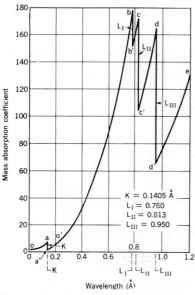

FIG. 12-11. The variation of the monochromatic mass absorption coefficient of lead as a function of λ. (By permission from F. K. Richtmyer, E. H. Kennard, and T. Lauritsen, "Introduction to Modern Physics," 5th ed., McGraw-Hill Book Company, Inc., New York, 1955.)

A careful study of the shape of the absorption curve near an absorption edge reveals that an edge is not perfectly sharp and abrupt, but shows a certain amount of "*fine structure*," corresponding to the elevation of an electron into one of the ordinary optical levels above the uppermost occupied levels rather than into the continuum. This fine structure is only a few electron volts in extent and can usually be neglected.

B. Photoelectric Absorption, Fluorescent Radiation, the Auger Effect. We have just seen that the greater part of the absorption of X-rays of energy less than ~1 MeV can be attributed to the photoelectric absorption of the X-ray quanta and the consequent ejection of electrons from the atom. Let us now examine this process a little more carefully.

The analytic form (4) and (5) of the absorption cross section has three distinct features of interest—the power-law variation with Z and λ ($Z^4\lambda^3$), the coefficients of this power law (C_{0K}, C_{KL}, C_{LM}, etc.), and the additive term B. The power law $Z^4\lambda^3$ was first derived on the basis of the old quantum theory by Kramers.[1] His analysis has since been sup-

[1] *Phil. Mag.,* **46**, 836 (1923).

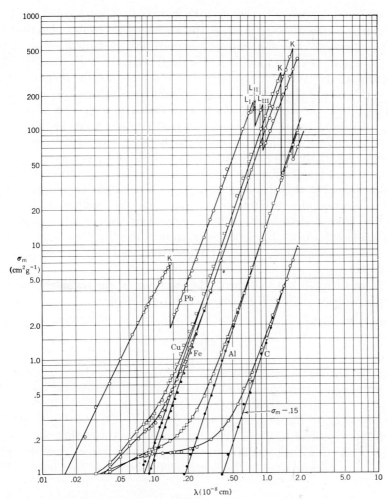

FIG. 12-12. The mass absorption coefficient for several elements. [From data of Allen, *Phys. Rev.*, **27**, 266, and **28**, 907 (1926).]

planted by a more exact one based upon the modern quantum theory, but the $Z^4\lambda^3$ result is obtained also in this theory as a good first approximation. (Some small, systematic variations in the exponents with Z and λ are observed, but these are usually ignored.)   We cannot go into this matter very deeply here, but it can be said that the reason that Z and λ both appear in this expression is that an interaction of the photoelectron with the *nucleus* is essential to the photoelectric process in order to conserve linear momentum, and the likelihood that a given electron will be ejected by a photon of given energy is smaller as the momentum that must be transmitted to the nucleus is greater.   Thus the increasing strength of the interaction with the nucleus with increasing Z for a given shell permits large momenta to be given to the nucleus more easily. On the other hand, for given Z, the momentum that must be transmitted to the nucleus in ejecting an electron increases with decreasing λ. Such considerations account satisfactorily for the signs, but of course not the magnitudes, of the exponents of Z and λ in the absorption cross section.[1]

The above considerations, that the probability of ejection of a photoelectron diminishes as the momentum that must be given to the nucleus increases and as the coupling with the nucleus decreases, account also for the apparently anomalous fact that a photon of given energy is more likely to eject an electron that is *tightly bound* in the atom than one that is weakly bound if this is energetically possible. Thus the coefficient C_K for the ejection of a K electron is greater than the coefficient C_L for the ejection of an L electron, as is demonstrated by the fact that the absorption coefficient drops by a factor of from 3 to 12 at the K absorption edge, even though there are eight L electrons, and only two K electrons, in an atom.

The quantity B has to do with the *scattering* of the radiation from the original beam. This will be considered in C, below.

EXERCISES

12-11. Evaluate the coefficients C_{0K}, C_{KL}, etc., which describe the absorption cross section in formulas (1) and (5) in terms of the coefficients C_K, C_L, etc., described above. *Ans.:* $C_{0K} = C_K + C_L + C_M + \cdot \cdot \cdot \cdot$ and $C_{KL} = C_L + C_M + \cdot \cdot \cdot \cdot$.

12-12. A general result in quantum mechanics states that the integral of the absorption cross section associated with a given level over the entire spectrum is proportional to the number of electrons in that level. That is,

$$\int_0^\infty \sigma_{ai}(\lambda) \, d\lambda = \alpha n_i$$

where $i = $ K, L, ... and α is a constant coefficient.

Using the form $\sigma_{ai} = C_i Z^4 \lambda^3$ up to an absorption edge, evaluate the relative magnitudes of C_K and C_L in terms of the K- and L-shell energies E_K and E_L. If these energies are approximately given by the Bohr formula, what is the approximate numerical relation between C_K and C_L?

---

[1] The Z^4 dependence can be most easily understood by recognizing that the photoelectric process is essentially an electric-dipole transition of an electron from a bound state to an unbound state by the absorption of an incident quantum. The matrix element for such a transition contains a factor Z^2, and the square of the matrix element, a factor Z^4.

We shall now consider two phenomena that are commonly associated with the photoelectric absorption of X-rays, although both are actually quite independent of the photoelectric process as such.   One of these is the phenomenon of *fluorescent radiation* accompanying the excitation of the atoms by the photoelectric process.   This radiation is of course just the characteristic radiation of the absorber.   For a given wavelength of incident radiation, those series of the characteristic spectrum will appear whose initial levels can be excited by this radiation.   This process is called *X-ray fluorescence* because the reradiated quanta are characteristic of the absorbing material rather than of the incident radiation, just as in ordinary optical fluorescence.

The other phenomenon associated with photoelectric absorption is the so-called *Auger effect.*[1]   In this effect, an atom which is in an excited state reduces its excitation by simultaneously dropping an electron from a higher shell into the vacant electronic state and *ejecting* another electron, usually from this same higher shell, from the atom.   This process thus differs from the familiar one of radiative deexcitation in that the energy which ordinarily emerges as a quantum is used to remove another electron from the atom.   It should be emphasized, however, that the second electron is *not* ejected by the photoelectric absorption of a photon emitted by the atom as the first electron drops into the vacant level, but emerges *directly* in the process of readjustment of the atom.   Thus, for example, the initial state of the atom might correspond to the absence of a K electron, and the final state to the absence of an L_I and an L_{III} electron. The difference in energy between these states then appears as *kinetic energy* of the ejected electron.   This form of readjustment of an excited atom is called an *Auger transition*.   Auger transitions occur with quite high probability, and they compete strongly with, or even overshadow, radiative deexcitation of an atom.   Indeed, an atom may emit two or more Auger electrons in a sequence of Auger transitions from an excited state.   In optical spectroscopy this effect is called *autoionization*.

EXERCISE

12-13. If X-rays of wavelength corresponding to the $K\alpha_1$ line of Mo irradiate copper, (*a*) with what energy, in electron volts, will K-photoelectrons be ejected from the copper atoms?   (*b*) If a copper atom subsequently undergoes an Auger transition to the $L_I L_{III}$ excited state, with what kinetic energy will the Auger electron be ejected?   (Assume that the excitation energy of a doubly ionized atom is the sum of the excitation energies of the two corresponding singly ionized levels.)

[1] Auger, *Compt. rend.*, **180**, 65 (1925); **182**, 773, 1215 (1926); *J. phys.*, **6**, 205 (1925).

C. Classical Theory of the Unmodified Scattering. In the earliest investigations of the effects of X-rays upon matter, the scattering of the X-rays by the individual electrons of the matter was considered to play a prominent role. Such an effect does in fact occur, but, as we have seen, it is by no means necessarily the dominant mechanism for the removal of X-rays from a collimated beam. We shall consider here the case of *incoherent* scattering by the individual electrons—the so-called *Thomson scattering*—and shall treat later the case of X-ray *diffraction* (coherent scattering). We shall follow the early classical treatment of this scattering.

The classical interaction of an electromagnetic wave with an electron which is situated in a stationary atom is primarily through the *electric field* of the wave. (The magnetic field acts only after the electric field has given the electron a velocity, and the magnetic force is then such as to give the electron a velocity *in the direction of propagation of the wave.* This is one way of visualizing the origin of *radiation pressure*.) The electric field of the incident plane wave may be visualized as "shaking" the electron, and the vibration of the electron then results in a radiation of electromagnetic energy in a *dipole pattern* at the same frequency as the incident radiation. This scattering without change in frequency is called *unmodified scattering*.

Let the intensity of plane-polarized radiation incident upon an electron be

$$S_{av} = \left(\frac{\varepsilon_0}{\mu_0}\right)^{\frac{1}{2}} (E^2)_{av} \tag{6}$$

The electric field will accelerate the electron with an acceleration whose instantaneous value is

$$a = -\frac{eE}{m} \tag{7}$$

The accelerated electron will thereupon *radiate energy* as described by Eq. 12-2(7). In terms of the incident intensity, we have for the radiated power per unit solid angle

$$\Pi(\theta)_{av} = \frac{e^4(E^2)_{av} \sin^2 \theta}{16\pi^2 \varepsilon_0 m^2 c^3}$$
$$= \frac{\mu_0^{\frac{1}{2}} e^4 S_{av} \sin^2 \theta}{16\pi^2 \varepsilon_0^{\frac{3}{2}} m^2 c^3} \tag{8}$$

where θ is measured from an axis parallel to the electric vector of the incident wave. The *total scattered power* is

$$\Pi_{av} = \frac{e^4 S_{av}}{6\pi \varepsilon_0^2 m^2 c^4} \tag{9}$$

The ratio Π_{av}/S_{av} represents the fraction of the incident intensity that is scattered by a *single electron*. It is

$$\sigma_0 = \frac{e^4}{6\pi\varepsilon_0{}^2m^2c^4} \tag{10}$$

EXERCISES

12-14. Supply the missing steps in the derivation of Eq. (10).

12-15. Show that σ_0 has the dimensions of a *cross-sectional area*. This quantity can be thought of as the cross-sectional area of the incident beam which is removed by the electron and scattered in a directional pattern given by Eq. 12-3(8). σ_0 is called the *Thomson scattering cross section* of an electron.

12-16. Evaluate the Thomson cross section in terms of the projected area of the classical electron. What is its numerical magnitude? *Ans.:* $\sigma_0 = \frac{8}{3}\pi r_0{}^2 = 6.66 \times 10^{-29}\ m^2 = 0.666$ barn.

In comparing the actual scattering with that expected upon the basis of the above classical analysis, it is of interest to consider not only the *total* scattered intensity but also the *angular distribution* of the intensity with respect to the direction of propagation and the direction of polarization of the incident beam. In order to do so, it is most convenient to change to a new system of polar coordinates whose polar axis z' is along the direction of propagation of the original radiation and whose azimuth angle ϕ' is measured from the direction of polarization (electric vector) of the radiation considered above. Thus we have the following connection between the old and new coordinates:

Physical direction	Coordinate axis	
	Old	New
S (propagation direction)	$y = r \sin\theta \sin\varphi$	$z' = r \cos\theta'$
E (polarization direction)	$z = r \cos\theta$	$x' = r \sin\theta' \cos\varphi'$
H	$x = r \sin\theta \cos\varphi$	$y' = r \sin\theta' \sin\varphi'$

We therefore may write Eq. (8) in the equivalent form

$$\Pi(\theta',\varphi') = r_0{}^2 S_{av}(1 - \sin^2\theta' \cos^2\varphi') \tag{11}$$

Finally, if, as is usually the case, the incident beam is only *partially polarized*, or is *unpolarized*, we must add to the above scattered intensity another contribution due to the radiation polarized in the x-direction. If the intensity of this latter component is equal to α times that of the component treated above, we have

$$\Pi(\theta',\varphi') = \frac{r_0{}^2 S_{av}}{(1 + \alpha)} [1 + \alpha - \sin^2 \theta'(\cos^2 \varphi' + \alpha \sin^2 \varphi')] \quad (12)$$

where S_{av} refers to the *total* beam intensity.

The ratio $\Pi(\theta',\varphi')/S_{av}$ has the dimensions of an *area* and may be called the *differential scattering cross section* $d\sigma_0/d\Omega$ for scattering X-rays into a unit solid angle in the direction (θ',φ').

EXERCISES

12-17. Supply the missing steps in the derivation of Eq. (12).

12-18. Show that, if the incident beam is unpolarized and of intensity S_{av}, the scattered intensity is

$$\Pi(\theta',\varphi') = \tfrac{1}{2} r_0{}^2 S_{av}(1 + \cos^2 \theta') \quad (13)$$

12-19. Show from Eq. 12-2(1) that the radiation scattered at right angles to the original beam ($\theta' = \pi/2$) should be *plane-polarized* with its electric vector perpendicular to the plane containing the beam and the point of observation, *irrespective of the condition of polarization of the original beam.* This fact was used by Barkla (1906) to show that the radiation from an X-ray tube is *not* polarized entirely in the direction of the electron beam (see Sec. 12-2C).

Actual measurements of the magnitude and angular distribution of the scattered X-rays agree only roughly with the above analysis. The observed total scattering cross section σ_s is in fair agreement with the Thomson cross section σ_0 for the light elements and for wavelengths between 2×10^{-11} and 10^{-10} m, but differs considerably from this value for the heavy elements and for wavelengths outside this range. The angular distribution of the scattered intensity similarly shows some agreement with the $1 + \cos^2 \theta'$ form for light elements and wavelengths in the above range, and disagreement for heavy elements. This is shown in Figs. 12-13 and 12-14.

The discrepancies outlined in Fig. 12-13 can be partially interpreted in the long-wavelength region in terms of a tendency for the electrons associated with a given atom to scatter *in phase with one another.* In such a case the scattered power varies as the *square* of the number of electrons in the atom (since the amplitudes add in this case) and as the number of atoms present. A complete quantum-mechanical treatment of the problem gives quite good agreement with experimental results. The present classical treatment is of importance mainly in order to bring out the *order of magnitude* of the scattering effect and to illustrate again

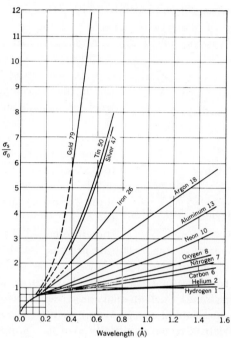

FIG. 12-13. The total scattering cross section as a function of wavelength. (By permission from A. H. Compton and S. K. Allison, "X-rays in Theory and Experiment," copyright D. Van Nostrand Company, Inc., Princeton, N.J., 1935.)

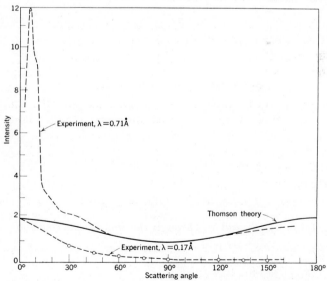

FIG. 12-14. Angular distribution of scattered radiation for carbon and iron. (By permission from A. H. Compton and S. K. Allison, "X-rays in Theory and Experiment," copyright D. Van Nostrand Company, Inc., Princeton, N.J., 1935.)

the concept of a *cross section* associated with a physical process involving the bombardment of one type of particle by another.

D. *Compton Scattering or Modified Scattering.* In the study of the scattering of X-rays by matter, a type of scattering was observed[1] in which the wavelength of the scattered radiation is not the same as that of the incident radiation, but still is related to the incident radiation rather than to the absorber. Compton[2] was the first to suggest an explanation of these observations and show by direct measurements that the explanation is valid; the effect is therefore called the Compton effect.

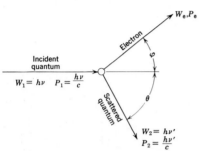

FIG. 12-15. Diagrammatic sketch of a Compton collision.

The distinctive feature of this scattering process, which made Compton's explanation a bold one in its day, is that it can be analyzed by applying the energy and momentum conservation laws to a collision *between a quantum and an electron.* Such an analysis proceeds as follows:

Assume that a quantum of energy $h\nu$ moves along the x-axis and strikes a perfectly free electron which is initially at rest. Suppose the quantum departs at an angle θ with respect to its initial direction, and that the electron recoils at an angle φ, as shown in Fig. 12-15.

The conservation laws may be written in the form

(a) $$\frac{h\nu}{c} = \frac{h\nu'}{c} \cos \theta + P_e \cos \varphi$$

(b) $$0 = \frac{h\nu'}{c} \sin \theta - P_e \sin \varphi \qquad (14)$$

(c) $$h\nu + m_0c^2 = h\nu' + W_e$$

We eliminate φ by rearranging (a) and (b), squaring, and adding:

$$\frac{h^2\nu^2}{c^2} - \frac{2h^2\nu\nu'}{c^2} \cos \theta + \frac{h^2\nu'^2}{c^2} = P_e{}^2$$

or $$h^2(\nu - \nu')^2 + 2h^2\nu\nu'(1 - \cos \theta) = P_e{}^2c^2$$

Solving (c) for $W_e{}^2$,

$$h^2(\nu - \nu')^2 + 2h(\nu - \nu')m_0c^2 + m_0{}^2c^4 = W_e{}^2$$

We then eliminate W_e and P_e through the relativistic relation

$$W_e{}^2 - P_e{}^2c^2 = m_0{}^2c^4$$

[1] Gray, *J. Franklin Inst.*, Nov., 1920.
[2] *Bull. Nat. Research Council (U.S.)*, **20**, 16 (1922); *Phys. Rev.*, **21**, 715 (1923); **22**, 409 (1923).

with the result

$$2h(\nu - \nu')m_0c^2 = 2h^2\nu\nu'(1 - \cos \theta)$$

from which we finally obtain

$$\frac{\nu - \nu'}{\nu\nu'} c = \lambda' - \lambda = \frac{h}{m_0c} (1 - \cos \theta) \qquad (15)$$

This equation asserts that the change in wavelength of a photon in a Compton collision is *independent of the initial wavelength* and depends only upon the *angle of scattering*. The maximum change in wavelength occurs for backward scattering and is then equal to $2h/m_0c$. The quantity h/m_0c is called the *Compton wavelength of the electron*. It is equal to

$$\frac{h}{m_0c} = 24.26 \times 10^{-13} \text{ m} = 0.02426 \text{ Å} \qquad (16)$$

EXERCISES

12-20. Supply the missing steps in the derivation of Eq. (15).

12-21. At what tube voltage does the wavelength of the most energetic X-ray equal the Compton wavelength? *Ans.:* $\phi = m_0c^2/e = 0.511$ MV.

In most X-ray applications, the Compton shift is quite small and requires rather high resolution for quantitative study. In practice, the Compton-scattered radiation is not as sharply defined in wavelength as is the incident radiation, because of the *Doppler shifts* due to the motions of the electrons within the solid absorber.[1]

We are not able here to derive the *cross section* for Compton scattering. A quantum-mechanical treatment of the problem of the scattering of radiation by perfectly free electrons leads to the equations, called the *Klein-Nishina formula* (1929), for the differential cross section

$$\frac{d\sigma_s}{d\Omega} = \frac{\frac{1}{2}r_0^2(1 + \cos^2 \theta')}{(1 + 2\epsilon \sin^2 \frac{1}{2}\theta')^2} \left[1 + \frac{4\epsilon^2 \sin^4 \frac{1}{2}\theta'}{(1 + \cos^2 \theta')(1 + 2\epsilon \sin^2 \frac{1}{2}\theta')} \right] \qquad (17)$$

and for the total cross section (Fig. 12-16)

$$\sigma_s = 2\pi r_0^2 \left\{ \frac{1 + \epsilon}{\epsilon^2} \left[\frac{2 + 2\epsilon}{1 + 2\epsilon} - \frac{\ln (1 + 2\epsilon)}{\epsilon} \right] + \frac{\ln (1 + 2\epsilon)}{2\epsilon} - \frac{1 + 3\epsilon}{(1 + 2\epsilon)^2} \right\} \qquad (18)$$

where r_0 is the classical electron radius and ϵ is the photon energy in units of m_0c^2 (that is, $\epsilon = h\nu/m_0c^2$). This equation does not include

---

[1] This effect provides a possible means of studying the motions of electrons in solids.

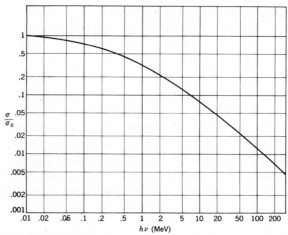

FIG. 12-16. Graph of total scattering cross section for free electrons [Eq. 12-3(18)]. Extensive graphs relating to Compton scattering can be found in *Natl. Bur. Standards Circ.* 542, 1953.

any of the effects of coherence in the scattering of the electrons in a given atom, since all electrons are supposed to be free, independent, and distributed at random.

EXERCISE

12-22. Show that, as $\epsilon \to 0$, $\sigma_s \to \sigma_0$. Show also that, as $\epsilon \to \infty$, $\sigma_s \to \frac{3}{8}\sigma_0(1/\epsilon)(\frac{1}{2} + \ln 2\epsilon)$.

In the observations of X-ray scattering at a given angle θ', each monochromatic "line" is observed to be scattered into *two* lines, one *unmodified* in frequency and the other *modified* according to the Compton formula (15). These are called *unmodified scattering* and *modified, or Compton, scattering* and are attributable in an approximate way to those electrons in the atom whose binding energies are respectively greater than or less than that energy represented by the change in frequency of the modified line. Thus, if the energy that would be given to a free electron in a Compton scattering is sufficient for that electron to escape from the atom, the electron will probably be ejected and the scattered radiation will be shifted in frequency as re-

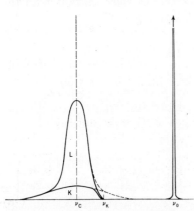

FIG. 12-17. Line profile of scattered monochromatic X-rays, showing the unmodified and modified lines (schematic). For measurements see Kirkpatrick and DuMond, *Phys. Rev.*, **54**, 802 (1938).

quired by the Compton formula. On the other hand, if this energy is insufficient for the electron to escape, the atom will be left in its initial state and the scattered radiation will be *unmodified* in frequency.

If the line profiles of the unmodified and modified lines are studied by using a crystal spectrometer of high resolution, a plot similar to Fig. 12-17 is obtained for a given frequency ν_0 and angle of observation θ. This figure indicates schematically that only a *very* slight broadening of the unmodified line due to atomic thermal motions and a *very* slight shift due to atomic Compton collisions would be present, whereas one should observe a considerable broadening of the Compton line by the Doppler effect of the *electronic motions within the atom*. The total line profile of the Compton line would be the sum of contributions due to the various shells of electrons. In the figure, a case in which even the K electrons could be removed from the atom has been assumed.

EXERCISE

12-23. Interpret the "step" in the Compton line in Fig. 12-17 at ν_K.

12-4. X-ray Diffraction

A most important X-ray phenomenon, used in both the study of X-rays and the application of X-rays to the study of matter, is that of X-ray diffraction. The *crystal monochrometer* is used to produce monochromatic X-rays; the *crystal spectrometer* is used to study the spectral distribution of X-rays in many kinds of experiments. Monochromatic X-rays are used to study the structure of crystals and complex molecules through the diffractive properties of those bodies. Few experimental techniques have enjoyed as wide and as fruitful an application as has X-ray diffraction. In view of the importance of the phenomenon, we shall now consider some of its more elementary aspects.

The earliest attempts to detect diffraction of X-rays made use of very narrow slits (Haga and Wind, 1903; Walter and Pohl, 1909.) These experiments were on the ragged borderline of success, and only estimates of the order of magnitude of X-ray wavelengths could be made from them. The experiments suggested that X-rays possess wavelengths of about an angstrom unit. The theorist M. von Laue (1912) was the first to conceive of the possibility of using the *regularly spaced atoms of a crystal* as a three-dimensional diffraction grating, and in cooperation with Friedrich and Knipping he succeeded almost at once in obtaining a diffraction pattern from a single crystal of rock salt. This experiment immediately provided a means of making rather accurate *absolute measurements*, and extremely precise *comparisons*, of X-ray wavelengths.

The conditions that must be satisfied in order to obtain a diffracted beam of X-rays from a crystal may be visualized as follows: If the wave-fronts of a monochromatic plane wave are traced (using Huygens' construction) as the wave travels through a crystal (treating each atom as a scatterer which vibrates in phase with the impinging wave), the condition for constructive interference is that the path lengths along the incident beam from a given wavefront to any two atoms in the crystal, and then in the direction of the diffracted beam to any other given wavefront, *must differ by an integral number of wavelengths* (Fig. 12-18). This condition can be treated in a mathematically rigorous manner to deduce the necessary and sufficient conditions that diffraction may occur. For the moment, however, we shall instead make use of a geometrical approach.

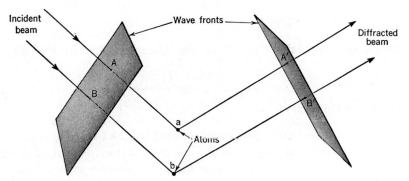

FIG. 12-18. Illustrating the condition leading to reinforcement of the X-ray waves scattered from two atoms in a crystal: The path lengths BbB' and AaA' must differ by an integral number of wavelengths.

The atoms of a simple crystal can be visualized as being arranged in a series of *parallel planes* which are equally spaced from one another (Fig. 12-19). These planes may be chosen in a great many possible ways by selecting *any three noncollinear atoms* in the crystal and passing a plane through them. Other atoms than the chosen ones will also lie in this plane (if the crystal is sufficiently large), and if the plane is moved perpendicularly to itself, it will eventually reach a position such that at least one, and therefore many, atoms will again lie in it. If this required minimum distance is d, then a further translation d will again leave it in a position such that a large number of atoms lie in the plane. In this way each atom of the crystal can be shown to lie on some plane in the family of equally spaced planes. Such a family of planes has the following properties:

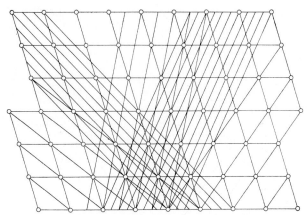

FIG. 12-19. Illustration of the various families of atomic planes of a two-dimensional lattice.

1. The arrangement of atoms in all planes follows the same geometrical pattern, because of the invariance of the crystal lattice with respect to any parallel translation which moves one atom into another.

2. The surface density of atoms in a given plane, divided by the distance between planes, is a constant equal to the number of atoms per unit volume in the crystal.

Now consider a single one of the planes of any given family of planes. It is clear that the stated condition of integrally related path differences would be satisfied for this plane if the incident and diffracted rays were to satisfy the condition for *specular reflection*, since in this case the path difference for any pair of atoms in the plane would be zero. (For this one plane, there would also be other possible directions of reinforcement of the incident and diffracted beams that would not satisfy this condition for specular reflection. These will be mentioned again shortly.) Now if the other, parallel, planes are introduced, it is clear that, for *certain angles* of specular reflection, the diffracted beams *from successive planes* will also reinforce one another, and thus will satisfy the above-stated condition. If θ is the angle of incidence and diffraction measured from the plane, we must therefore have (Fig. 12-20)

$$\Delta L = 2d \sin \theta = n\lambda \qquad (1)$$

The diffraction of X-rays from a crystal is thus dependent *only upon the distance between successive planes in a given family;* it is independent of the detailed arrangement of the atoms within each plane. The condition (1) is called the *Bragg condition,* and the diffraction process itself is called *Bragg reflection.* The term *reflection* is used to express the fact that diffraction occurs only under conditions of "specular reflection" from some set of planes within the crystal.

Although the above elementary derivation of the Bragg condition brings out many of the essential features of X-ray diffraction, it is instructive to approach the problem from a more general point of view which exhibits several important properties of crystal lattices and which serves also to introduce some useful terminology. We shall now examine X-ray diffraction from this more general viewpoint, beginning with a brief description of crystal structure.

A physical crystal consists of a regular three-dimensional array of identical groups of atoms, in which corresponding points of each group

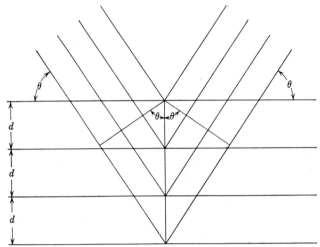

Fig. 12-20. Illustration of the Bragg condition for X-ray reflection from successive atomic planes.

fall on what is called a *space lattice*. The identical groups may contain any number of atoms from one, as for some crystalline pure elements, to thousands, as for many organic crystals. Regardless of the complexity of the atomic clusters that compose the crystal, however, the space lattice upon which these groups fall must be one of *fourteen possible types*, which are illustrated in Fig. 12-21.

A crystal lattice may conveniently be thought of as being built up of a solid stack of parallelepipedal blocks whose corners define the lattice. One such block is called a *unit cell* of the lattice, provided it is the smallest unit of the given shape which is needed to define all of the lattice points. The unit cell of a given lattice can be chosen in an infinite number of ways, but usually one of the ways is especially simple and is accepted conventionally for this purpose. Inside the unit cell of a given crystal there will, of course, appear exactly one of the groups of atoms whose repeated array constitutes the crystal. The arrangement of the atoms within the unit cells of a few simple crystals is shown in Fig. 12-22. The

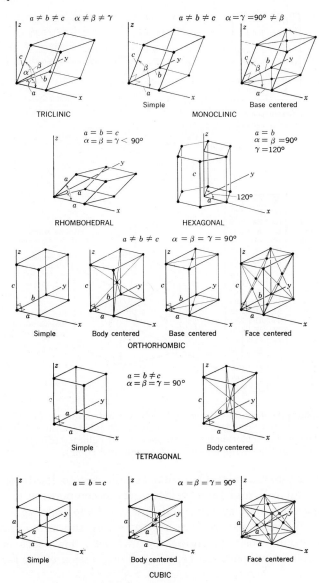

FIG. 12-21. The fourteen space lattices.

crystalline forms of a number of elements and simple compounds are given in Table 12-2.

For treating crystal problems analytically, one needs an appropriate coordinate system with which to define the locations of the atoms within the unit cell and the locations of the space-lattice points at which the corners of the unit cell lie. For many purposes this is most conveniently

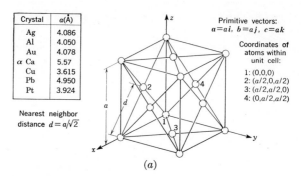

Crystal	a(Å)
Ag	4.086
Al	4.050
Au	4.078
α Ca	5.57
Cu	3.615
Pb	4.950
Pt	3.924

Nearest neighbor distance $d = a/\sqrt{2}$

Primitive vectors:
$a = ai, \; b = aj, \; c = ak$

Coordinates of atoms within unit cell:

1: $(0,0,0)$
2: $(a/2,0,a/2)$
3: $(a/2,a/2,0)$
4: $(0,a/2,a/2)$

(a)

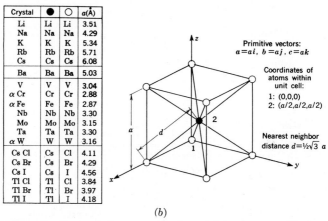

Crystal	●	○	a(Å)
Li	Li	Li	3.51
Na	Na	Na	4.29
K	K	K	5.34
Rb	Rb	Rb	5.71
Cs	Cs	Cs	6.08
Ba	Ba	Ba	5.03
V	V	V	3.04
α Cr	Cr	Cr	2.88
α Fe	Fe	Fe	2.87
Nb	Nb	Nb	3.30
Mo	Mo	Mo	3.15
Ta	Ta	Ta	3.30
α W	W	W	3.16
Cs Cl	Cs	Cl	4.11
Cs Br	Cs	Br	4.29
Cs I	Cs	I	4.56
Tl Cl	Tl	Cl	3.84
Tl Br	Tl	Br	3.97
Tl I	Tl	I	4.18

Primitive vectors:
$a = ai, \; b = aj, \; c = ak$

Coordinates of atoms within unit cell:

1: $(0,0,0)$
2: $(a/2,a/2,a/2)$

Nearest neighbor distance $d = \frac{1}{2}\sqrt{3}\, a$

(b)

FIG. 12-22. Examples of commonly encountered crystal structures. (a) Face-centered cubic lattice. (b) Body-centered cubic lattice (simple cubic for the compounds CsCl, etc.). (c) The sodium chloride lattice (face-centered cubic). (d) Hexagonal close-packed lattice. (e) The diamond lattice (face-centered cubic).

done by using rectangular Cartesian coordinates to locate the atoms within the unit cell, and three vectors, whose components are referred to this same coordinate system, to define the shape and size of the unit cell itself. While it may require a great many sets of coordinates to define the atomic positions inside the unit cell, one needs only *three* vectors to define the space lattice, using an expression of the form

$$\boldsymbol{r} = n_1\boldsymbol{a} + n_2\boldsymbol{b} + n_3\boldsymbol{c} \tag{2}$$

where $\boldsymbol{r}$ is a vector from one lattice point to another; $\boldsymbol{a}$, $\boldsymbol{b}$, and $\boldsymbol{c}$ are three linearly independent vectors called the *primitive vectors* of the lattice, and n_1, n_2, and n_3 are numbers which may independently take on any positive or negative integral value. The primitive vectors define the edges of the unit cell.

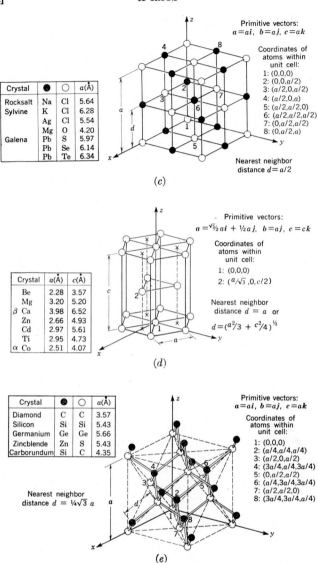

Primitive vectors:
$a = ai, \; b = aj, \; c = ak$

Coordinates of atoms within unit cell:
1: $(0,0,0)$
2: $(0,0,a/2)$
3: $(a/2,0,a/2)$
4: $(a/2,0,a)$
5: $(a/2,a/2,0)$
6: $(a/2,a/2,a/2)$
7: $(0,a/2,a/2)$
8: $(0,a/2,a)$

Crystal	●	○	a(Å)
Rocksalt	Na	Cl	5.64
Sylvine	K	Cl	6.28
	Ag	Cl	5.54
	Mg	O	4.20
Galena	Pb	S	5.97
	Pb	Se	6.14
	Pb	Te	6.34

Nearest neighbor distance $d = a/2$

(c)

Primitive vectors:
$a = \tfrac{\sqrt{3}}{2}ai + \tfrac{1}{2}aj, \; b = aj, \; c = ck$

Coordinates of atoms within unit cell:
1: $(0,0,0)$
2: $(a/\sqrt{3}, 0, c/2)$

Nearest neighbor distance $d = a$ or
$d = (a^2/3 + c^2/4)^{1/2}$

Crystal	a(Å)	c(Å)
Be	2.28	3.57
Mg	3.20	5.20
β Ca	3.98	6.52
Zn	2.66	4.93
Cd	2.97	5.61
Ti	2.95	4.73
α Co	2.51	4.07

(d)

Primitive vectors:
$a = ai, \; b = aj, \; c = ak$

Coordinates of atoms within unit cell:
1: $(0,0,0)$
2: $(a/4,a/4,a/4)$
3: $(a/2,0,a/2)$
4: $(3a/4,a/4,3a/4)$
5: $(0,a/2,a/2)$
6: $(a/4,3a/4,3a/4)$
7: $(a/2,a/2,0)$
8: $(3a/4,3a/4,a/4)$

Crystal	●	○	a(Å)
Diamond	C	C	3.57
Silicon	Si	Si	5.43
Germanium	Ge	Ge	5.66
Zincblende	Zn	S	5.43
Carborundum	Si	C	4.35

Nearest neighbor distance $d = \tfrac{1}{4}\sqrt{3}\,a$

(e)

Fig. 12-22. (Continued)

EXERCISES

12-24. Find suitable unit cells, primitive vectors, and atomic coordinates within the unit cell for the following lattices: (a) face-centered cubic, (b) body-centered cubic, (c) hexagonal close-packed, (d) NaCl, (e) CsCl, (f) C (diamond), and (g) C (graphite).

TABLE 12-2. CRYSTALLINE FORMS OF SOME ELEMENTS AND SIMPLE COMPOUNDS
AT OR NEAR ROOM TEMPERATURE

Substance	System	Type	Lattice const ($\dot{A}$)		
			a	b	c
Ag	Cub.f-c	Cu	4.086		
Al	Cub.f-c	Cu	4.050		
Au	Cub.f-c	Cu	4.0781		
Ba	Cub.b-c	W	5.025		
C (dia.)	Cub.f-c	Fig. 12-22e	3.567		
C (graph.)	Hex.	. . . .	2.4612	. . . . . .	6.7079
α Ca	Cub.f-c	Cu	5.57		
Cd	Hex.c-p	Mg	2.9736	. . . . . .	5.6058
Cr	Cub.b-c	W	2.884		
Cu	Cub.f-c	Fig. 12-22a	3.615		
α Fe	Cub.b-c	W	2.8665		
Ge	Cub.f-c	Diamond	5.6575		
I$_2$	Orth.	Fig. 12-21	7.25	**9.77**	4.77
In	Rhomb.	Fig. 12-21	3.24	. . . . . .	4.94
K	Cub.b-c	W	5.344		
Li	Cub.b-c	W	3.509		
Mg	Hex.c-p	Fig. 12-22d	3.2028	. . . . . .	5.1998
Mo	Cub.b-c	W	3.150		
Na	Cub.b-c	W	4.291		
Ni	Cub.f-c	Cu	3.524		
Pb	Cub.f-c	Cu	4.950		
Pt	Cub.f-c	Cu	3.924		
S	Orth.	. . . .	12.92	24.55	10.48
Si	Cub.f-c	Diamond	5.4306		
α Ti	Hex.c-p	Mg	2.953	. . . . . .	4.729
U	Orth.	. . . .	4.945	5.865	2.852
V	Cub.b-c	W	3.040		
α W	Cub.b-c	Fig. 12-22b	3.1648		
Zn	Hex.c-p	Mg	2.6590	. . . . . .	4.9351
AgBr	Cub.f-c	NaCl	5.755		
AgCl	Cub.f-c	NaCl	5.545		
BN	Hex.	Graphite	2.51	. . . . . .	6.69
CsCl	Cub.	Fig. 12-22b	4.110		
CsI	Cub.	CsCl	4.562		
CuCl	Cub.	ZnS	5.407		
KBr	Cub.	NaCl	6.578		
KCl	Cub.	NaCl	6.28		
MgO	Cub.	NaCl	4.203		
NaBr	Cub.	NaCl	5.94		
NaCl	Cub.	Fig. 12-22c	5.62737		
PbS	Cub.	NaCl	5.97		
PbSe	Cub.	NaCl	6.14		
PbTe	Cub.	NaCl	6.34		
SiC	Cub.	ZnS	4.348		
α SiO$_2$	Hex.	. . . .	4.903	. . . . . .	5.393
ZnS	Cub.	Fig. 12-22e	5.43		

12-25. Find the volume of a unit cell of a crystal lattice in terms of a, b, and c.   *Ans.:* $V = a \cdot b \times c$.

The diffractive scattering of X-rays by a crystal is dependent quantitatively both upon the arrangement of atoms within the unit cell and upon the space-lattice parameters, but these two effects can be treated separately on the basis of well-known principles of physical optics.   Thus, the possible *directions of incidence* for which diffraction can occur depend upon the space lattice, while the *intensities* of the radiation scattered in these various possible directions depend upon the scattering properties of the atoms in the unit cell.   Both of these effects are of the greatest importance in analyzing crystal structures by X-ray diffraction.   We shall first treat the diffractive effect by considering a hypothetical problem of the scattering of X-rays by point scatterers situated at the lattice sites of a point lattice.

Consider a collimated beam of monochromatic X-rays of wavelength λ which is incident upon a crystal, and suppose there is a diffracted beam leaving the crystal in a certain direction.   Designate the directions of the incident and diffracted beams by the unit vectors e_i and e_d.   The condition that must be satisfied if the incident beam is to be diffracted into the new direction is, of course, that the optical paths of any two scattered rays, measured between fixed wavefronts, *must differ by an integral number of wavelengths*.   This condition assures that all scattered rays will reinforce one another in the diffracted beam.

EXERCISE

12-26. Show that the ratio of intensity of two scattered beams, one of which satisfies the above condition and the other of which does not, should be of the order of the number of unit cells that are present in the crystal.

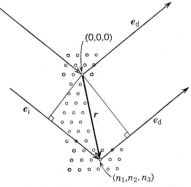

To express the above condition analytically, consider the path difference between the ray which scatters from the origin and the ray which scatters from the general lattice point r given by Eq. (2).   From Fig. 12-23 this path difference is easily seen to be

$$\Delta_r = r \cdot e_i - r \cdot e_d$$
$$= r \cdot (e_i - e_d) \qquad (3)$$

FIG. 12-23. Illustrating the path differences for X-rays scattered from two lattice sites in a crystal.

Therefore, the necessary condition for reinforcement is that

$$\mathbf{r} \cdot (\mathbf{e}_i - \mathbf{e}_d) = n_r \lambda \tag{4}$$

where n_r is an integer which depends upon $\mathbf{r}$. By considering the dependence of n_r upon $\mathbf{r}$ under conditions wherein n_1, n_2, and n_3 are separately varied, one finds that n_r must be specifically of the form

$$n_r = m_1 n_1 + m_2 n_2 + m_3 n_3 \tag{5}$$

where m_1, m_2, and m_3 are fixed integers for all $\mathbf{r}$ for *given* directions $\mathbf{e}_i$ and $\mathbf{e}_d$, but which, when allowed to take on various values, *serve to fix*

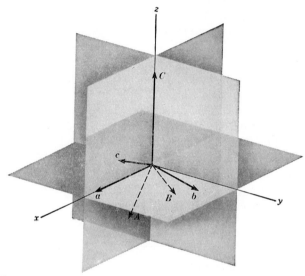

FIG. 12-24. Illustration of the reciprocal vectors $\mathbf{A}$, $\mathbf{B}$, and $\mathbf{C}$ which correspond to given lattice vectors $\mathbf{a}$, $\mathbf{b}$, and $\mathbf{c}$.

the possible directions $\mathbf{e}_i$ *and* $\mathbf{e}_d$ *at which diffractive scattering can occur.* Our problem is therefore to find what values of $\mathbf{e}_i - \mathbf{e}_d$ satisfy Eq. (4) as a function of m_1, m_2, and m_3.

If we insert the forms (2) and (5) for $\mathbf{r}$ and n_r into Eq. (4), we obtain

$$(n_1 \mathbf{a} + n_2 \mathbf{b} + n_3 \mathbf{c}) \cdot (\mathbf{e}_i - \mathbf{e}_d) = (m_1 n_1 + m_2 n_2 + m_3 n_3)\lambda \tag{6}$$

which must hold true for all values of n_1, n_2, and n_3 for given m_1, m_2, and m_3. This requires that the coefficients of n_1, n_2, and n_3 on the two sides separately be equal:

(a) $\mathbf{a} \cdot (\mathbf{e}_i - \mathbf{e}_d) = m_1 \lambda$

(b) $\mathbf{b} \cdot (\mathbf{e}_i - \mathbf{e}_d) = m_2 \lambda$ (7)

(c) $\mathbf{c} \cdot (\mathbf{e}_i - \mathbf{e}_d) = m_3 \lambda$

These are called the *Laue equations.*

To interpret these conditions most simply it is advantageous to introduce three new vectors A, B, and C, which are defined in terms of the primitive lattice vectors by the relations

$$A = \frac{b \times c}{a \cdot b \times c} \qquad B = \frac{c \times a}{a \cdot b \times c} \qquad C = \frac{a \times b}{a \cdot b \times c} \qquad (8)$$

EXERCISE

12-27. Show that $a \cdot A = b \cdot B = c \cdot C = 1$ and $a \cdot B = a \cdot C = b \cdot A = b \cdot C = c \cdot A = c \cdot B = 0$.

The vectors A, B, and C are called the *reciprocal vectors* to a, b, and c, and are, of course, a valid set of linearly independent vectors in terms of which any vector can be expressed (Fig. 12-24). In particular, if we express $e_i - e_d$ in terms of these reciprocal vectors in the form

$$e_i - e_d = \alpha A + \beta B + \gamma C \qquad (9)$$

then, with the aid of Exercise 12-27, Eq. (7) may be written in the especially simple form

(a) $\qquad\qquad\qquad\qquad \alpha = m_1\lambda$
(b) $\qquad\qquad\qquad\qquad \beta = m_2\lambda \qquad\qquad\qquad (10)$
(c) $\qquad\qquad\qquad\qquad \gamma = m_3\lambda$

so that $\qquad\qquad e_i - e_d = \lambda(m_1 A + m_2 B + m_3 C) = \lambda R \qquad (11)$

where R *is a vector to some point of a lattice whose primitive vectors are* A, B, *and* C. This lattice is called the *reciprocal lattice* corresponding to the original lattice. The original lattice is called the *direct lattice*.

EXERCISES

12-28. Find the forms of the reciprocal lattices which correspond to the simple, face-centered, and body-centered cubic lattices. *Ans.:* Simple, body-centered, and face-centered cubic.

12-29. Show that the reciprocal lattice of the reciprocal lattice is the direct lattice.

12-30. Find the "volume" of a unit cell of the reciprocal lattice. *Ans.:* $V' = 1/a \cdot b \times c$.

The significance of Eq. (11) can be analyzed as follows:

1. The vector $e_i - e_d$ has a length $2 \sin \theta$, lies in the plane defined by e_i and e_d, and makes equal angles $\pi/2 - \theta$ with vectors e_i and e_d (Fig. 12-25).
2. Every vector R of the reciprocal lattice *is normal to a certain set of*

atomic planes within the crystal and has a length inversely proportional to the spacing between successive planes, if m_1, m_2, and m_3 are relatively prime. To show this, consider the equation

$$\frac{R \cdot r}{|R|} = p \tag{12}$$

in which $R = m_1 A + m_2 B + m_3 C$ is fixed, r is allowed to vary, and p is an adjustable parameter. This is the equation of a plane perpen-

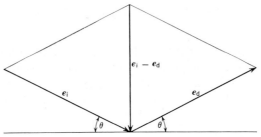

FIG. 12-25. The relation between $e_i - e_d$, e_i, and e_d.

dicular to R and passing a distance p from the origin of r. A lattice site $n_1 a + n_2 b + n_3 c$ will evidently lie in this plane if

$$R \cdot r = m_1 n_1 + m_2 n_2 + m_3 n_3 = p|R| \tag{13}$$

We first show that some of the atoms of the crystal certainly lie in such a plane for some value of p. Indeed, the atom at the origin and the three atoms whose lattice coordinates are

(a) $(n_1 = 2m_2 m_3, \quad n_2 = -m_1 m_3, \quad n_3 = -m_1 m_2)$

(b) $(n_1 = -m_2 m_3, \quad n_2 = 2m_1 m_3, \quad n_3 = -m_1 m_2)$

(c) $(n_1 = -m_2 m_3, \quad n_2 = -m_1 m_3, \quad n_3 = 2m_1 m_2)$

clearly lie in the plane (12) with $p = 0$. Therefore, because of the translational invariance properties of the lattice, *a multitude of other atoms lie in the same plane, such a plane must pass through every lattice site in the crystal, and these planes must be uniformly spaced.*

To find the spacing between planes, we must find the smallest positive value of p for which another such plane exists. Since the quantity $m_1 n_1 + m_2 n_2 + m_3 n_3$ is an integer, the value of p we seek will be the one corresponding to setting $m_1 n_1 + m_2 n_2 + m_3 n_3$ equal to the smallest possible positive integer. Thus, we must find what values of n_1, n_2, and n_3 will make

$$m_1 n_1 + m_2 n_2 + m_3 n_3 = n > 0 \tag{14}$$

as small as possible for given m_1, m_2, m_3. It is easy to prove that n must be a divisor of all three integers m_1, m_2, and m_3. For, suppose

n does *not* divide m_1. Then we may set $m_1 = sn - t$, where s and t are integers, and $t < n$, so that Eq. (14) may be written

$$m_1 n_1 + m_2 n_2 + m_3 n_3 = \frac{m_1 + t}{s} \tag{15}$$

or, rearranging,

$$m_1(sn_1 - 1) + sm_2 n_2 + sm_3 n_3 = t < n \tag{16}$$

which shows that there is another combination n_1', n_2', n_3' which gives an integral value *less than* n. But since n is already supposed to be the smallest such integer, this is impossible. Thus, by contradiction we find that n is a divisor of m_1. Similarly it must also be a divisor of m_2 and m_3. Thus, we take n as the greatest common divisor of m_1, m_2, and m_3, and we may then write Eq. (12) in the form

$$p_{\min} = \frac{n}{|R|} = d \tag{17}$$

as the distance between successive planes of atoms. The condition (11) can then be written in the form

$$e_i - e_d = \lambda|R|e_R = \frac{\lambda n}{d} e_R \tag{18}$$

or, finally,

$$2 \sin \theta \, e_R = \frac{n\lambda}{d} e_R \qquad \text{or} \qquad n\lambda = 2d \sin \theta \tag{19}$$

which is again the Bragg condition. The integer n evidently defines the *order* of the diffractive scattering; if m_1, m_2, and m_3 are relatively prime, then $n = 1$ and the vector R is equal in magnitude to the reciprocal of the interplanar spacing.

3. It should be noted that the condition (18) defines only the direction of a line in the plane of incidence and diffraction, but does not specify the azimuthal orientation of this plane about the line. Indeed, the azimuthal orientation of the plane is arbitrary, just as was found in the simpler derivation of the Bragg condition given previously. This freedom is, of course, a consequence of the basic requirement that $\theta_i = \theta_d$ since, for this condition, all points in a given atomic plane will scatter in phase, just as for specular reflection.

EXERCISE

12-31. Show that the density of atoms in a given plane is proportional to the spacing between successive planes.

We have just found that for every possible set of directions for the incident and diffracted beams there exists a family of planes perpendicular to the bisector of the angle between the incident and diffracted directions. Thus, all possible conditions of diffraction from a given crystal can be expressed in the form of the Bragg-reflection condition applied to all possible families of planes that can be drawn within the crystal.

The numbers m_1, m_2, and m_3, which define the orientation and spacing of the atomic planes from which the X-rays are reflected, are not neces-

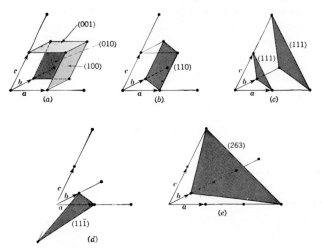

FIG. 12-26. Illustration of the Miller indices of some atomic planes.

sarily relatively prime. As we have seen above, their G.C.D. is equal to the order of the diffracted beam. If these numbers are reduced to relative primality by factoring out their G.C.D., they are called the *Miller indices* of the planes in question. Thus if we write

$$m_1 = nh \qquad m_2 = nk \qquad m_3 = nl \tag{20}$$

where n is the G.C.D. of m_1, m_2, and m_3, then the Miller indices h, k, and l are relatively prime and are respectively inversely proportional to the intercepts of the corresponding plane on the primitive vectors $\boldsymbol{a}$, $\boldsymbol{b}$, and $\boldsymbol{c}$. Figure 12-26 illustrates this property of the Miller indices.

EXERCISES

12-32. Prove that the Miller indices are inversely proportional to the intercepts of the corresponding plane on the three primitive vectors, each intercept being measured in terms of the length of the corresponding primitive vector.

12-33. The specific gravity of rock salt is 2.1632. Find the inter-

atomic spacing between neighboring Na and Cl atoms in the crystal.
Ans.: $d = 2.820 \times 10^{-4} \mu$.

12-34. Find the spacing between successive (110), (111), and (210) planes in rock salt.

12-35. At what angle will the Kα radiation of copper be reflected in first order from the (100) planes of rock salt? *Ans.*: $\theta = 7°50'$.

12-36. In one form of X-ray spectrometer, first introduced by Bragg, a narrow beam of heterochromatic X-rays is allowed to fall upon a cleavage face of a crystal, and the reflected X-rays are detected by an ioniza-

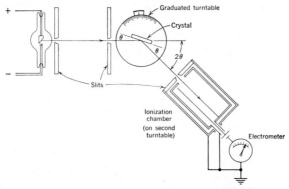

FIG. 12-27. A Bragg X-ray spectrometer. Calcite ($d = 3.02945$ Å) is often used in this kind of spectrometer.

tion chamber or a photographic plate. The crystal is mounted upon a pivoted table so that the angle of incidence of the X-rays can be adjusted. The intensity of the reflected beam can thus be measured as a function of the angle of tilt of the crystal (Fig. 12-27).

The Kα radiation from a molybdenum-target X-ray tube is scattered at 90° from a block of carbon and is then analyzed with a Bragg spectrometer using a calcite crystal cut along its cleavage plane. (*a*) Calculate the change in wavelength of the X-rays scattered by Compton scattering. (*b*) Determine the angular separation in the first order between the modified and unmodified lines; that is, through what angle must the crystal be rotated to detect these two lines? (*c*) Find the angle between the direction of the incident photon beam and the recoiling electron. (*d*) Determine the kinetic energy of the recoil electron. *Ans.*: (*a*) $\Delta\lambda = 0.024$ Å. (*b*) $\theta = 13'.8$. (*c*) $\phi = 44°2'$. (*d*) $T = 576$ eV.

12-37. In the rotating-crystal method of X-ray crystallography, a small crystal is mounted on a spindle and is rotated while being irradiated by a narrow beam of monochromatic X-rays. As the crystal passes through the various orientations which satisfy the Bragg-reflection conditions, the X-ray beam is diffracted in the corresponding spatial directions and strikes a strip of film which is held in a cylindrical shape with the spindle as an axis (Fig. 12-28). The result of this process is that a number of spots appear on the film; these spots are characteristic of the structure and orientation of the crystal, and may be used as a quantitative measure of the crystal parameters. If a rock salt crystal is placed upon the spindle with one cubic axis oriented parallel to the spindle, at what angles θ will spots appear in the equatorial plane of the apparatus if X-rays of wavelength 0.40 Å are used?

The technique of X-ray crystallography involves finding the orientations and separations of the various atomic planes of a crystal by meas-

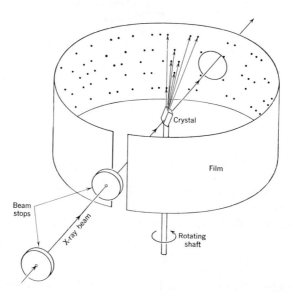

FIG. 12-28. The rotating-crystal method of X-ray crystallography.

uring the locations of the "spots" formed by the diffracted beams as they impinge upon a photographic emulsion. Such measurements not only serve to measure the crystal type and the size of the unit cell but also provide, under favorable conditions, an extremely accurate means of measuring certain other properties of the crystal such as the minute distortions of the lattice which result from certain types of impurity and the amplitudes of the thermal vibrations of the atoms in the crystal.

Many X-ray instruments utilize the Bragg-reflection condition, and this condition also appears in many analytical procedures such as are used in the determination of crystal structure. Unfortunately we are not able to investigate these applications in detail, owing to lack of space. Many X-ray instruments and techniques are described in textbooks and in the literature.

12-5. Atomic and Crystal Structure Factors

Many crystals of present-day interest are quite complicated, and their unit cells may contain hundreds or thousands of atoms of various kinds. In such cases the real problem which confronts the X-ray crystallographer is not that of determining the parameters of the crystal lattice, but rather that of determining the arrangement of the atoms within the unit cell.   For this purpose one makes use of the *intensities* of the diffracted beams rather than merely their spatial directions, as was mentioned previously.   In many cases a certain amount of information may already be known about this arrangement, and the arrangement may even be restricted to one of a very few possibilities.   In such cases the X-ray technique may serve to eliminate some of the possibilities and may even definitely establish the correctness of one of them.   (In principle, an exact measurement of the intensities of all of the scattered beams would serve to evaluate the coefficients of a Fourier expansion of the charge density throughout the unit cell.)

We shall now consider briefly the *directional distribution* of the radiation scattered from the atoms which occupy a unit cell of the crystal. Within an accuracy that is quite sufficient for most purposes, one can regard this scattering as resulting from the "shaking" of each tiny volume element of the electronic charge distribution under the influence of the incoming wave, the scattered radiation being just the combined electric-dipole radiation of all of these charge elements, each vibrating with a phase that is appropriate to its location within the unit cell.   The angular distribution of the scattered radiation is thus

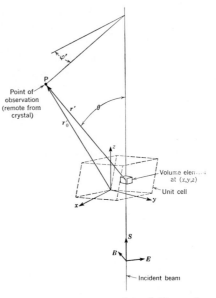

FIG. 12-29. The scattering of X-rays by an element of charge in a unit cell.

analogous to the radiation pattern of an antenna array, where the antennas are fed in certain phase relationships to one another.   The situation is shown in Fig. 12-29.

To treat the problem analytically, let $\rho(\mathbf{r})$ be the charge-density distribution throughout the unit cell.   In terms of the physical probability density discussed in Chap. 7 we may write

$$\rho(\mathbf{r}) = -e\psi^*\psi \tag{1}$$

where, for an atomic crystal,

$$\int_{\text{unit cell}} \psi^*\psi \, dV = Z \qquad \text{the atomic number}$$

Let the incoming X-ray beam have the form

$$E = E_0 \exp\left[i\left(2\pi\frac{z}{\lambda} - \omega t\right)\right] \qquad \text{real part} \qquad (2)$$

This field will produce an acceleration of each electron of the atomic charge cloud, equal to

$$a = -\frac{e}{m}E = -\frac{e}{m}E_0 \exp\left[i\left(2\pi\frac{z}{\lambda} - \omega t\right)\right] \qquad (3)$$

A charge element dq, moving with this acceleration, will radiate an electric field equal to [see Eqs. 12-2(1) and (4)]

$$dE_s = \frac{dq([a] \times r') \times r'}{4\pi\varepsilon_0 c^2 r'^3} \qquad (4)$$

Because of the assumed sinusoidal form of the original X-ray beam, the retarded acceleration $[a]$ is equal to

$$[a] = -\frac{e}{m}E_0 \exp\left[i\left(2\pi\frac{z}{\lambda} + 2\pi\frac{r'}{\lambda} - \omega\tau\right)\right] \qquad (5)$$

where τ is the instant of observation at P. And since r' is very large compared to an atom or a unit cell, we may approximate

$$r' = r_0' - x\sin\theta'\cos\varphi' - y\sin\theta'\sin\varphi' - z\cos\theta' \qquad (6)$$

where (θ',φ') are the polar angular coordinates of the point of observation as seen from the scatterer, the polar axis being taken in the direction of the incident X-ray beam. Finally, inserting the value $dq = \rho \, dV = -e\psi^*\psi \, dV$ together with Eqs. (5) and (6) into Eq. (4), we obtain

$$E_s = \frac{[(E_0 \times r') \times r']e^2}{4\pi\varepsilon_0 mc^2 r'^3} \exp\left[i\left(2\pi\frac{r_0'}{\lambda} - \omega\tau\right)\right]$$
$$\times \int_{\text{unit cell}} \psi^*\psi \exp\left\{\frac{2\pi i}{\lambda}[-x\sin\theta'\cos\varphi' - y\sin\theta'\sin\varphi'\right.$$
$$\left. + z(1 - \cos\theta')]\right\} dx\, dy\, dz \qquad (7)$$

for the electric vector of the radiation scattered in the direction (θ',φ'). The coefficient before the integral is equal to the electric vector of the radiation scattered by a *single electron*, so that the integral can be regarded as a geometrical factor which compares the amplitudes of the scattered radiation with that due to a single electron in the given direction (θ',φ').

It is called the *crystal structure factor* of the electronic charge distribution within the unit cell:

$$F(\lambda,\theta',\varphi') = \int_{\text{unit cell}} \psi^*\psi \, \exp\left\{-\frac{2\pi i}{\lambda}[x \sin\theta' \cos\varphi' + y \sin\theta' \sin\varphi' - z(1 - \cos\theta')]\right\} dx \, dy \, dz \quad (8)$$

Although the precise evaluation of the structure factor is hopelessly complicated for even the simplest cases, since one must know the wave function ψ for the electrons inside the unit cell, one can obtain quite useful approximations to this quantity by using the charge distributions found from the Hartree or Thomas-Fermi atomic models and combining these *atomic structure factors* with appropriate amplitude and phase for the various atoms within the unit cell.

EXERCISES

12-38. Carry out in detail the steps leading to Eq. (7).

12-39. The atomic structure factor is usually expressed as a function of the Bragg-reflection angle θ rather than θ' and φ', under the assumption that the atomic electronic charge distribution is spherically symmetric. Express Eq. (8) in terms of the Bragg angle, using a rectangular-coordinate system whose ξ and η axes lie parallel to the Bragg planes and whose ζ axis is perpendicular to them.

Ans.: $$f(\lambda,\theta) = \int_{\text{atom}} \psi^*\psi \, \exp\frac{4\pi i\zeta \sin\theta}{\lambda} \, d\xi \, d\eta \, d\zeta \quad (9)$$

[Note that $f(\lambda,\theta)$ is a function of $\sin\theta/\lambda$ in this case.]

12-40. Show that the atomic structure factor is equal to Z for radiation scattered in the forward direction ($\theta = 0$ or $\theta' = 0$).

12-41. Show that the crystal structure factor is equal to the number of electrons in a unit cell for scattering in the forward direction.

12-42. Obtain an expression for the crystal structure factor in terms of the atomic structure factors f_j of the various atoms in the unit cell.

Ans.: $$F(\lambda,\theta) = \sum_j f_j \exp\left(4\pi i\zeta_j \frac{\sin\theta}{\lambda}\right)$$

$$= \sum_j f_j \exp[2\pi ni(hx_j + ky_j + lz_j)]$$

(ζ_j is the ζ-coordinate of the jth atom, and $\mathbf{r}_j = x_j\mathbf{a} + y_j\mathbf{b} + z_j\mathbf{c}$.)

12-43. Use the expression of Exercise 12-42 to evaluate the relative intensities of the first-order Bragg reflections from the (100), (110),

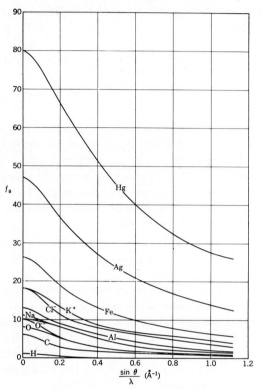

Fig. 12-30. Some atomic structure factors.

(111), and (221) planes, and the second-order reflections from the (100) and (110) planes of rock salt. Note that the structure factor refers to the scattered *amplitude*, and that $\sin \theta/\lambda = n/2d$.

Partial ans.:

$$I_{100} = 0 \qquad I_{111} \sim 16(f_{Na} - f_{Cl})^2 \qquad \text{first order}$$
$$I_{100} \sim 16(f_{Na} + f_{Cl})^2 \qquad\qquad\quad \text{second order}$$

12-44. The first-order Bragg reflection from the (111) planes of rock salt is quite strong but is almost entirely missing for the analogous case of sylvine (KCl). Explain this fact.

The structure factors of many atoms and ions have been worked out by James and Brindley.[1] Some of these are given as a function of $\sin \theta/\lambda$ in Fig. 12-30.

12-6. Refraction and Reflection of X-rays

In his first investigations of X-rays, Roentgen found that they undergo exceedingly little if any refraction or reflection when they go from one

[1] *Z. Krist.*, **78**, 470 (1931); *Phil. Mag.*, **12**, 81 (1931).

medium into another. This result is in direct agreement with the classical dispersion theory, according to which indices of refraction differing from unity by only a few parts per million would be expected for ordinary solids. However, with refined techniques, even the tiny deviations of an X-ray beam that result from refraction indices of this size have been measured and must be taken into account in precise work.

According to the classical theory of dispersion, the refraction of light is supposed to result from the interaction between the radiation and the atomic oscillators which give rise to the spectral radiation characteristic of the medium in question. If there are N_i oscillators of frequency ν_i per unit volume of the medium, the classical dispersion theory leads to the relation[1]

$$\mu^2 - 1 = \frac{e^2}{\pi m} \sum_i \frac{N_i}{\nu_i^2 - \nu^2} \tag{1}$$

where e is the charge and m the mass of each oscillator and μ is the index of refraction. This formula gives a fairly good description of the refractive indices of many substances throughout the visible range if one assumes that the characteristic frequencies ν_i lie generally in the near ultraviolet. If this same formula is applied to X-rays, for which the frequency ν can be assumed to be much greater than the oscillator frequencies ν_i, one can neglect the ν_i in Eq. (1), so that a much simpler relation is obtained:

$$\mu^2 - 1 = \frac{-e^2}{\pi m \nu^2} \sum_i N_i = \frac{-e^2 N}{\pi m \nu^2} \tag{2}$$

where $N = \sum_i N_i$ is the total number of oscillators, i.e., electrons, per unit volume. Thus in the X-ray spectral region, the refractive index of material media should be *less than unity* by a small amount. This means, for instance, that a prism should bend an X-ray beam *away* from its broad end, and total reflection should occur within the *less-dense* medium for grazing angles of incidence.

EXERCISES

12-45. The refractive index of Al is

$$1 - \mu = 1.68 \times 10^{-6} \qquad \text{for Mo K}\alpha \text{ radiation}$$
$$1 - \mu = 8.4 \times 10^{-6} \qquad \text{for Cu K}\alpha \text{ radiation}$$

Compare these measured values[2] with those given by Eq. (2).

[1] For a derivation of this formula, see J. C. Slater and N. H. Frank, "Electromagnetism," chap. 9, International Series in Pure and Applied Physics, McGraw-Hill Book Company, Inc., New York, 1947.

[2] Davis and Slack, *Phys. Rev.*, **27**, 18 (1926).

12-46. Compute at what angle of incidence one should observe total reflection for X-rays of wavelength 1.279 Å (tungsten L-radiation) striking silver.[1]   *Ans.: $\theta_t = 21.6'$ from plane of surface.*

When the dispersion problem is treated quantum-mechanically, one obtains a formula quite similar to Eq. (1), in which the various frequencies ν_i are those which can induce electric-dipole quantum transitions in the atoms of the medium and the coefficients N_i are related to the statistical weights and transition probabilities for the respective transitions. (These are usually referred to as the *oscillator strengths* of the transitions.) Thus one expects that the refractive index of a substance should follow Eq. (2) *except for frequencies near an absorption edge.*   This is observed to be the case.

In ordinary work the refraction of X-rays is of no practical consequence, but in the most precise work, especially in X-ray diffraction, it must sometimes be taken into account.   The effect of refraction upon X-ray diffraction can best be visualized by observing that (1) the Bragg condition 12-4(1) refers specifically to the angles of incidence and reflection and the wavelength of the rays *inside the crystal* and (2) these rays will be bent slightly both on entering and on leaving the crystal and will have a slightly greater wavelength inside than outside the crystal.   These effects will then lead to an erroneous value for the interplanar spacing d if the externally measured angles and wavelength are used in the Bragg formula.

EXERCISES

12-47. Referring to Fig. 12-31, derive an approximate equation connecting the externally measured angles and wavelength with the inter-

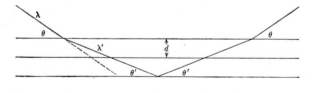

FIG. 12-31. The refraction of X-rays upon entering and leaving a crystal.

planar spacing for the case of Bragg reflection from a cleavage face of a crystal.

Ans.:
$$n\lambda = 2d \sin \theta \left[1 + \frac{4(\mu - 1)d^2}{n^2\lambda^2} \right] \tag{3}$$

12-48. Insert the value of $\mu - 1$ obtained from Eq. (2) into Eq. (3)

---

[1] This measurement was made by Compton, *Phil. Mag.*, **45**, 1121 (1923).

and thus evaluate the effective grating spacing of calcite as a function of n.

Ans.: $$d_n = 3.02945 \left(1 - \frac{1.4 \times 10^{-4}}{n^2}\right) \times 10^{-10} \text{ m} \tag{4}$$

REFERENCES

Compton, A. H., and S. K. Allison: "X-rays in Theory and Experiment," D. Van Nostrand Company, Inc., Princeton, N.J., 1935.

Richtmyer, F. K., E. H. Kennard, and T. Lauritsen: "Introduction to Modern Physics," 5th ed., International Series in Pure and Applied Physics, McGraw-Hill Book Company, Inc., New York, 1955.

Sproull, W. T.: "X-rays in Practice," McGraw-Hill Book Company, Inc., New York, 1946.

13

Basic Properties of Nuclei

In our study of atomic and molecular structure we have so far treated atomic *nuclei* as negligibly small bodies which possess mass, charge, and spin, but which are otherwise quite inert.   Actually, of course, nuclei are known to possess a host of properties in addition to those mentioned, and under certain circumstances are anything *but* inert!   In the ensuing chapters we shall examine some of the properties of nuclei and some of the more fundamental relationships between nuclei and also between the particles of which nuclei are composed.

Our knowledge of nuclear properties has accumulated side by side with our knowledge concerning atomic structure.   Indeed, nuclear physics and atomic physics both received their greatest forward impetus in a single step: the discovery of the nuclear theory of atomic structure by Lord Rutherford in 1910.   Since that great epoch, however, investigations of atomic structure and nuclear structure have proceeded along rather different paths.   To be sure, these paths have often joined briefly, as in hyperfine structure and molecular rotational energy states; but in a broad sense the phenomena and experimental methods through which these two aspects of the structure of matter have been studied have been quite different.   In the preceding chapters we have seen something of the extent to which the electronic structure of the atom and the relationships of atoms to one another result from the operation of a few fundamental principles, collectively called the *quantum theory of matter*.   Unfortunately, no comparable degree of success has yet been attained in the analogous problem of describing the properties of nuclei in terms of a few basic principles, although great strides have been made in this direction within recent years.

458

In the interests of efficiency and clarity in our later study of the various properties of nuclei, it is desirable to draw upon the substantial body of empirical information that is now available concerning nuclei and nuclear particles. In the present chapter we shall therefore examine a part of this empirical knowledge, beginning with a brief outline of some of the early work in which many of the basic properties of nuclei were discovered.

The study of atomic nuclei had its beginning with the discovery by Becquerel (1896) that certain fluorescent salts of uranium have the property of blackening a photographic emulsion through a layer of optically opaque paper and a thin plate of silver, even after several days have passed since the last exposure of the salts to sunlight. Like the discovery of X-rays in the previous year, Becquerel's discovery was quite accidental in all respects save that his training and interests happened to lead him to study the phenomenon of fluorescence and its possible relationship to the emission of X-rays, and that he had sufficiently acute perception to recognize the significance of his discovery and pursue it to a definite conclusion.

Immediately following Becquerel's discovery, many researchers took up the study of the new phenomenon, which was called *radioactivity*, and information was rapidly accumulated concerning the properties of radioactivity and radioactive substances. The nature of these researches and some of their interdependence is brought out by the following chronological tabulation of some of the more important early results:

1896   *Radioactivity* of uranium was discovered and was shown to be *independent of the state of chemical combination or the physical condition* of the uranium with which the phenomenon was associated. It was also found that radioactivity involves the emission of certain radiations which have the property of *ionizing air*, in a manner similar to X-radiation (Becquerel).

1898   *Thorium* compounds were found also to show radioactivity, and *polonium and radium* were discovered by chemical decomposition and fractionation of uranium ores (Curie and Curie).

1899   *Radon gas* ("radium emanation") was discovered as the result of tracking down the source of a troublesome erratic background effect in certain delicate measurements of the ionization due to radioactive substances (Rutherford).

     As a result of the measurement of the total ionization current produced in the air between two plates by a radioactive coating on one of the plates, *two radiations were distinguished* and named on the basis of their ability to penetrate absorbing material: The *less penetrating* of these rays, but those causing the greater part of the ionization, were called *α-rays*. They are completely absorbed by

a few thousandths of a centimeter of aluminum, or a few centimeters of air. The *more penetrating* rays, which were found
responsible for only a small part of the ionization current, were
called *β-rays*. These rays are able to penetrate several millimeters
of aluminum, or several meters of air (Rutherford).

The β-rays were deflected magnetically and were shown to consist
of high-speed *negative particles* of various velocities (Rutherford).
Actinium was discovered in pitchblende (Debierne).

1900 *Chemical-separation procedures* were used to isolate and study the
activities of the various products of radioactive decay (Rutherford
and Soddy, Crookes).

With the isolation of radon and some of the other short-lived
products of radioactive decay, it was recognized for the first time
that radioactivity does not go on forever, but decreases in intensity
at a rate characteristic of the particular substance.

An even more penetrating radiation than the β-rays was found to
be present in the decay of radium, and it was shown that this
radiation is not deviated by a magnetic field. These rays were
named γ-rays (Villard).

1903 The *transmutation properties* of radioactivity were now quite
clearly recognized, and the relation of the chemical change of an
atom to the type of radiation emitted in a given transformation
was correctly stated. Expressed in modern terminology, these
relationships are (Rutherford and Soddy):

1. If an atom undergoes α-particle decay, its atomic weight
   decreases by almost exactly four units and its atomic number
   decreases by two units.
2. If an atom undergoes β-particle decay, its atomic weight
   decreases by only a very small fraction of a unit but its atomic
   number *increases by one unit*.
3. If an atom undergoes γ-ray decay, neither its atomic weight
   nor its atomic number undergoes any appreciable change.

The term "isotope" was introduced to distinguish atoms having
identical chemical properties but different masses and radioactive
decay properties (Soddy).

α-rays were successfully deflected in electric and magnetic fields,
and an e/m value about half that for H was obtained. Speeds of
about 2.5×10^9 cm s^{-1} were deduced from these measurements
(Rutherford).

The regular occurrence of helium gas in radioactive-ore deposits
led to the conjecture that *the α-rays might actually be helium atoms*.
Individual α-rays were detected for the first time by their ability to

produce a scintillation on a ZnS screen (Crookes, Elster and Geitel). The *heating effect* of radium was measured and found to be about 400 J h^{-1} g^{-1}. This was recognized as corresponding to an enormous amount of energy per atom—*millions of times* the energy available in ordinary chemical reactions (Curie and Laborde).

1904 The idea that *α-rays are particles having definite ranges* in matter (rather than a radiation having an exponential attenuation) was proposed and experimentally demonstrated. The characteristic ranges in air of the α-particles from various radioactive substances were measured. They were found to lie between about 3 cm and about 9 cm at normal temperature and pressure and to correspond to kinetic energies of from 4 *to* 9 *million electron volts* for the α-particles (Bragg).

1905 A new description of radioactive decay as a *statistical process* involving a certain *probability* of disintegration per unit time for each atom was advanced (Schweidler).

1906 Measurements of α-particle scattering in thin foils of gold, alu-
to minum, etc., were made during this period (Rutherford, Geiger,
1908 Marsden).

1908 In this year the first application of the *proportional counter* was made as a detection device for studying *individual* α-particles. With this method the specific activity of radium was measured and found to be 3.5 × 10^{10} disintegrations s^{-1} g^{-1} (Rutherford and Geiger).

The charge on an α-particle was measured by measuring the charge carried by a known number of α-particles (from a source of known specific activity) using a sensitive electrometer. The α-particle was by this time generally considered to be *doubly charged*, so that the electronic charge was estimated to be 4.65 × 10^{-10} esu. This was for some years the most accurate value of *e*; the modern value is 4.80 × 10^{-10} esu, see Appendix A. (Rutherford and Geiger).

1909 Although it was strongly suspected that α-particles are helium ions, this had not yet been proved. α-particles were collected inside an evacuated vessel by allowing them to pass through an extremely thin glass wall for a few days, and then the spectrum of a gaseous discharge inside this tube was observed. The definite presence of the helium spectrum proved that α-particles are in fact *identical* with helium (Rutherford and Royds).

The *total number of ions* produced by an α-particle traversing hydrogen gas at normal temperature and pressure was measured and found to be about 2.4 × 10^5 from the α-particles from RaC (Geiger).

1910 *α-particle "tracks"* were observed in a photographic emulsion (Kinoshita).

1911 The nuclear model of the atom was proposed (Rutherford).

1912 α- and β-particle tracks were first observed in a *cloud chamber* (C. T. R. Wilson).

1913 Magnetic-deflection methods were used to demonstrate that ordinary neon consists of two isotopes, of mass numbers 20 and 22 (Thomson).

1919 The first *artificial nuclear disintegration* was detected by observing scintillations on a ZnS screen produced by particles whose range through the air was *much longer* than that of the α-particles emerging from the source. These long-range particles were *protons* produced in the bombardment of nitrogen by the α-particles (Rutherford).

1920 The *radii* of certain heavier nuclei (copper, silver, gold) were shown to be less than 1.2 to 3.4 × 10⁻¹² cm, by observing that the Rutherford scattering law is valid for distances of approach of this magnitude (Chadwick).

1922 Definite evidence of the artificial disintegration by α-particles
to   of all elements up to potassium (with the exception of Li, Be, C,
1924 and O) in which *protons* are emitted was obtained (Rutherford and Chadwick).

1925 The first study of the artificial disintegration of nitrogen nuclei using the Wilson cloud chamber as a detection device was carried out. This permitted simultaneous observation of the incident α-particle, the emitted proton, and the recoil nucleus, and made possible a determination of the *energy balance* in the reaction (Blackett).

1926 Definite departures from the Coulomb law were detected in
to   α-particle scattering from hydrogen, helium, magnesium, and
1927 aluminum.   The radii at which these departures were first noticeable were about 2.5 to 12 × 10⁻¹³ cm.   These radii may be taken as the radii of the *nuclei* (Rutherford and Chadwick, Bieler).

1928 The correct theoretical explanation of α-radioactivity as a quantum-mechanical problem involving the *penetration of a barrier* was advanced (Gamow, Gurney and Condon).

1932 The *neutron* was discovered, as a result of discrepancies in the interpretation of certain experiments involving the products of the bombardment of light elements with α-particles (Chadwick, Curie and Joliot).

The first nuclear reaction using *machine-accelerated particles* was studied (Cockroft and Walton).

Nuclear physics in the modern sense dates from these last two events, since the discovery of the neutron provided the missing link in the problem of the *constituents* of nuclei, while the successful operation of a high-voltage accelerator capable of inducing nuclear reactions provided a more controllable means of studying nuclear structure than is possible with natural radioactivity alone.

It is interesting to note that, within 15 years of the discovery of radioactivity, virtually all of the basic techniques of detection and isolation of radioactive particles and nuclei were used. Such detection devices as the ionization chamber, the scintillation counter, the photographic emulsion, the proportional counter, and the cloud chamber, and such techniques as those of chemical separation and magnetic deflection, were in regular use. Modern nuclear research makes use of these same devices and techniques, refined with respect to their sensitivity and precision, but still basically similar to those used in the early days of the study of radioactivity.

Let us now examine in somewhat greater detail some of the more basic properties of nuclei.

13-1. Charge and Mass

The properties of nuclei that are most directly familiar from our experience with atomic physics are *charge and mass*. As we saw in Chap. 7, it is the nuclear charge which fundamentally defines the gross physical and chemical properties of an atom, all atoms of given nuclear charge being chemically almost indistinguishable. The charge carried by a nucleus is an integer Z times the charge of an electron and is of opposite sign. Experimentally, Z is known to be an integer within about one part in a thousand from mass-spectroscopic data, but by indirect means it is limited much more closely, perhaps to within a part in 10^{15}, because of the observed absence of a number of effects that would be expected if matter were not precisely electrically neutral when not ionized.[1] It is universally assumed that Z is exactly an integer.

The masses of nuclei are known roughly from chemical atomic weights, the conversion from gram-atomic weights to atomic masses being made through Avogadro's number N_0. Atomic masses determined from chemical data are generally subject to considerable uncertainty because of the fact that ordinary chemical elements are usually mixtures of several isotopes. The masses of nuclei and other particles are commonly expressed in terms of the following systems of units:

[1] Hughes, *Phys. Rev.*, **105**, 170 (1957).

1. In ordinary metric cgs or MKS units (g or kg).
2. In atomic mass units (amu). In this system of units the atomic mass of the isotope $^{16}_8O$ is taken to be exactly 16. This is the basis of the so-called *physical scale* of atomic weights. It differs slightly from the *chemical scale* of atomic weights in which the atomic weight of *ordinary oxygen* (i.e., of a mixture of 99.76 per cent ^{16}O, 0.039 per cent ^{17}O, and 0.204 per cent ^{18}O) is taken to be exactly 16.
3. In units of the electron mass (m_e). This unit is most commonly used in connection with the so-called "elementary particles," to be discussed later (Chap. 20).
4. In "energy units" of a million electron volts (MeV). In this system the rest-mass energy M_0c^2 is what is actually given. Thus the mass of an electron is often said to be 0.511 MeV. What is meant is $m_ec^2 = 0.511$ MeV.

The relationships between these systems of units and the masses of the proton, neutron, and hydrogen atom in the various systems are given in Table 13-1.

TABLE 13-1. NUCLEAR MASS UNITS

Quantity	kg	amu	m_e	MeV/c^2
1 kg....................	....	6.02486×10^{26}	1.09790×10^{30}	5.61000×10^{29}
1 amu..................	1.65979×10^{-27}	....	1822.28	931.141
1 m_e	9.1083×10^{-31}	0.548763×10^{-3}	....	0.510076
1 MeV/c^2..............	1.78253×10^{-30}	1.07395×10^{-3}	1.957039	
Proton rest mass M_p.......	1.67239×10^{-27}	1.007593	1836.12	938.211
^{1}H rest mass M_H..........	1.67330×10^{-27}	1.008142	1837.12	938.722
Neutron rest mass M_n......	1.67470×10^{-27}	1.008983	1838.65	939.505

The most accurate "direct" measurements of the masses of nuclei have been made with devices called *mass spectrometers*. These devices measure the ratio of mass to charge for an ion by the use of a suitable combination of electric and magnetic fields, just as was done in the original determination of e/m for electrons. With modern techniques, a precision of about one part in 10^6 can be attained for atoms whose atomic weight is less than about 40. The principal features of high-precision mass-spectrometer design and operation are:

1. Singly or multiply charged ions of the element or compound being measured are formed by some means, such as by an electrical discharge in a gaseous mixture containing the material or by evaporation from a solid deposit of the compound on a hot filament.
2. The ions are accelerated by an electric field and are emitted through a narrow slit with approximately homogeneous energy.

3. The beam of ions emerging from the slit is somewhat inhomogeneous in speed and somewhat divergent in angle. A critical combination of electrostatic and magnetic deflection is then applied to this beam, with the result that all ions with a given value of M/q are accurately focused at a single line in space (the "image" of the slit), irrespective of the velocity and directional inhomogeneities originally present among the ions. This action is called *velocity focusing* and *directional focusing* and is an important factor in the design of a high-resolution instrument.

4. The arrival of the ions at their various characteristic points is registered either photographically (in which case the device is called a *mass spectrograph*) or electrically (by measuring the current of the beam as it strikes a collecting electrode).

5. In actual practice, the most precise results are obtained when the masses of two compounds of almost equal M/q values are *compared* and the *difference* of their M/q values determined. Such a procedure is relatively insensitive to calibration errors and nonlinearities in the instrument. It is called the *matched-doublet* method.

Examples of the precision attainable with the matched-doublet method are given in Table 13-2. In this table, A/n is the value of the ratio

TABLE 13-2

Doublet	A/n	Nominal mass difference, mmu
$(^{12}C\ ^1H_4)^+ - {}^{16}O^+$	16	36.369 ± 0.021
$^2H_3^+ - {}^{12}C^{++}$	6	42.230 ± 0.019
$^1H_2^+ - {}^2H^+$	2	1.539 ± 0.0021

of the molecular weight to the degree of ionization of the compounds being compared. Thus doubly ionized carbon has a weight of 12 and a degree of ionization of 2, and is compared with a *singly ionized triatomic molecule of deuterium*. This doublet difference is sometimes written $^2H_3 - \frac{1}{2}\ ^{12}C$, since the M/q ratio of doubly ionized ^{12}C is the same as for a singly ionized atom of half of the mass of ^{12}C.

It should be pointed out that a mass spectrometer does not measure the masses of *nuclei*, but of ionized *atoms and molecules*. Thus the masses of the electrons and the mass equivalent of the electronic binding energy are included in mass-spectroscopic measurements. It turns out that this is not a disadvantage but an asset, because it is *atomic* masses that enter into considerations of the stability of nuclei and the energy balance of nuclear reactions.

EXERCISES

13-1. Use the experimental values in Table 13-2 to derive the masses of neutral ^1H, ^2H, and ^{12}C atoms, in terms of that of ^{16}O.

13-2. Approximately what would the sensitivity of the matched-doublet method have to be in order to detect the difference in ionization energy between hydrogen and carbon? Take the first ionization potential of carbon to be 11.2 eV and the next ionization potential to be 24.3 eV. Take the ionization potential of hydrogen to be 15.6 eV in the $^2H_3{}^+$ molecule.

Mass spectrometers are used not only for the measurement of atomic masses, but also for quantitatively determining the *relative abundances* of the stable isotopes of an element and carrying out physical separations of the isotopes. During World War II, kilogram quantities of the rare isotope ^{235}U were obtained through the use of numerous large mass spectrometers.

As an illustration of the design layout of a precision mass spectrometer, Fig. 13-1 shows an instrument used by Mattauch (1938). Other such instruments have been devised and used by Aston, Bainbridge, Dempster, Nier, and others.[1] In this instrument, the slit S_2 defines a narrow beam

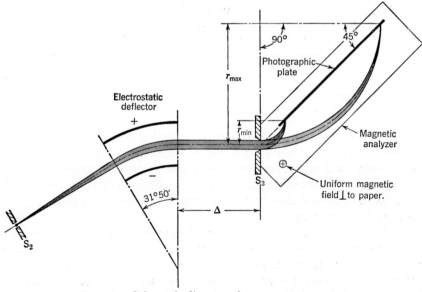

FIG. 13-1. Schematic diagram of a mass spectrometer.

[1] For a discussion of several types of mass spectrometers, see Mass Spectroscopy in Physics Research, *Natl. Bur. Standards Circ.* 522, 1953.

of ions having a small angular divergence and a small energy inhomogeneity, the electrostatic deflector is designed to render the emergent beam parallel for a certain energy, and there is a certain "velocity dispersion" of the "focal length" of this part of the system. The wide slit S_3 accepts a certain part of this beam, and the magnetic field focuses the parallel beam on a photographic plate. The radius of curvature and angular deflection in the magnetic field are designed to give a velocity dispersion of focal length which *exactly cancels* that of the electrostatic deflector. There are many possible combinations of electrostatic and magnetic deflection for which focusing can occur both with respect to direction and velocity for a *given* mass; this instrument has the added feature of satisfying this condition for a wide range of masses.

Various investigators have measured the masses of many atoms by the matched-doublet method, and quite accurate mass values are available for most elements. These measurements have a great importance in nuclear physics; for, as we shall see later, it is through a comparison of the measured mass with that to be expected for the known number of "hydrogen atoms" and neutrons comprising a given atom that the *binding energy* of a nucleus can be determined. The binding energies of nuclei are commonly measured in MeV or keV (thousands of electron volts), or in amu or mmu (1 milli–mass unit $\equiv 10^{-3}$ amu).

Within recent years an indirect method of determining atomic masses has been developed to a degree of precision which surpasses that of even the most precise mass-spectrometer values. In this method, mass differences between the ingoing and outgoing particles of a nuclear reaction are evaluated by measuring the *energy balance* of the nuclear reaction. Internal checks upon the accuracy of a given mass value are provided by utilizing several nuclear reactions to connect the mass in question with various other masses. The network of relationships between masses so obtained is sufficiently overdetermined that errors of measurement can often be brought to light and new measurements made. The atomic mass values which are given in Appendix G were largely obtained through this method. It is, incidentally, of considerable interest to observe that the agreement of the mass-spectrometer mass values with those obtained from the energetics of nuclear reactions provides an exceedingly strong experimental confirmation of the correctness of Einstein's law of conservation of mass energy.

13-2. Nuclear Angular Momentum

Experimental evidence that nuclei possess *angular momentum, or spin,* comes from various sources:

1. The so-called hyperfine structure of atomic spectral lines, as revealed

by interferometers or very high-resolution spectrometers, is in part attributable to a slight splitting of the atomic energy levels by a mechanism similar to that which gives the familiar LS fine structure, but of much weaker strength (see Sec. 5-12).[1] The multiplicities of the various hyperfine levels are in all measurable cases just what would result from the quantized addition of the electronic total angular momentum J with another angular momentum I, where I is the same for all energy levels of a given atom. Furthermore, the Zeeman splitting of the hyperfine levels is also what would be expected from an additional angular momentum of the same magnitude I for each level.

2. The alternation of intensities in the rotational fine structure of the electronic and vibrational transitions of homonuclear diatomic molecules (Sec. 9-6), which results from the different statistical weights of odd and even rotational states of the molecule, affords a quite direct means of measuring the spins of some nuclei. It is of interest to observe that, whereas most nuclear spin effects depend quantitatively upon the interaction of the nuclear magnetic-dipole moment or electric-quadripole moment with external magnetic or electric fields, the present effect is brought about by the operation of the exclusion principle, and is therefore not basically dependent upon any property other than the spin itself. The spins of several nuclei have been measured successfully by this means.

3. The existence of nuclear angular momentum is strongly supported by a great body of evidence obtained from studies of β- and γ-transitions of excited nuclei and studies of nuclear magnetic resonance, and many of the known nuclear spins have been obtained from, or checked against, such evidence.

13-3. Magnetic Moments of Nuclei; Nuclear Magnetic Resonance

The existence of magnetic moments in nuclei was first established in connection with the hyperfine splitting of atomic energy levels (Sec. 5-12). The effects extend much farther than this, however, and nuclear magnetism plays a prominent role in many quite diverse physical phenomena.

By far the most precise measurements of the magnetic moments of nuclei have been obtained by using the phenomenon of *nuclear magnetic resonance,* in which nuclei situated in a homogeneous magnetic field are caused to jump from one quantized orientation to another by the applica-

---

[1] In some cases, part of the apparent hyperfine structure of a spectral line is in reality an isotope effect, traceable to the differing nuclear properties of the various isotopes of a given atom that are present in the radiation source.

tion of a weak oscillating field whose frequency is equal to the Larmor precessional frequency of a nucleus in the homogeneous field.

Nuclear resonance effects were discovered independently by Purcell, Torrey, and Pound[1] and by Bloch, Hansen, and Packard.[2] The former investigators detected the absorption of energy by the sample as the applied frequency was swept slowly through the resonance value, while the latter group detected a signal induced in a pickup coil by the precessing nuclei in the sample. The second method is called *nuclear magnetic induction*.

EXERCISE

13-3. (*a*) Calculate the angular frequency of the Larmor precession of a classical symmetric top whose angular momentum is I and whose magnetic moment is $\mu = g_N(e/2M)I$ in a magnetic field whose induction is B.

Ans.:
$$\omega_L = g_N \frac{eB}{2M_p} \tag{1}$$

(*b*) Calculate the energy levels of a nucleus whose spin quantum number is I and whose magnetic moment is $\mu = g_N(e/2M_p)I$ in a magnetic field B. Using the selection rule $\Delta M_I = \pm 1$, evaluate the angular frequency of an electromagnetic quantum which will induce the system to jump from one orientation to another.

Ans.:
$$E_{M_I} = g_N \frac{e}{2M_p} M_I \hbar B \qquad \omega = g_N \frac{e}{2M_p} B = \omega_L \tag{2}$$

All magnetic-resonance experiments involve the transition of the nuclear spins from a given state to one of higher energy. If one is to obtain a detectable effect, two conditions must be satisfied: First, there must be more nuclei in the lower states than in the upper ones, so that the upward absorptive transitions outnumber the downward induced emissive transitions which return energy to the oscillator. Second, the nuclei that are in the upper state must have some means of returning to a lower state other than by emitting radiation. The first of these conditions is satisfied if the populations of the various energy states are in statistical equilibrium at a sufficiently low temperature in a sufficiently strong magnetic field; the second is satisfied if there is sufficiently strong coupling between the nuclear moment and the solid or liquid system in which it is situated. This coupling leads to nonradiative transitions

[1] *Phys. Rev.*, **69**, 37 (1946).
[2] *Phys. Rev.*, **69**, 127 (1946); **70**, 474 (1947).

of the nuclei, the energy being transferred to the crystal lattice where it appears ultimately as heat. This energy transfer, which permits the nuclear spins to return toward statistical equilibrium, is called spin-lattice relaxation.

EXERCISES

13-4. Calculate the approximate difference in population of two adjacent nuclear energy states at a temperature T in a magnetic field B, if there are N_0 nuclei in the sample.

$Ans.$:
$$\Delta N \approx \frac{N_0}{2I + 1} g_N \frac{e}{2M_p} \frac{B\hbar}{kT} \tag{3}$$

13-5. If the relaxation time (the time required for a nucleus to return from the upper to the lower state by spin-lattice relaxation) is approximately 10 μs, compute the approximate maximum power absorbed by the protons in 1 cm.^{3} of ice in a field of 10^4 gauss. Assume the nuclear g-factor for protons to be $g_p = 2.8.$ $Ans.$: $P \approx 2 \times 10^{-5}$ W.

13-6. Two nuclear magnetic moments can be compared with high precision by measuring the frequencies at which they precess in the *same magnetic field*. In this way one avoids the relatively difficult problem of measuring a magnetic field with high precision. Show that $g_1/g_2 = \omega_1/\omega_2$.

13-7. The angular velocity with which a charged particle executes a circular orbit in a magnetic field is called the *cyclotron frequency* of the particle in the given field. The magnetic moment of the proton has been measured with great precision by comparing its Larmor and cyclotron frequencies in the same magnetic field.[1] Show that, in terms of these frequencies and the charge and mass of the proton,

$$\mu_p = \frac{e\hbar}{2M_p} \frac{\omega_L}{\omega_c} = \frac{\omega_L}{\omega_c} \mu_N \tag{4}$$

where $\mu_N = e\hbar/2M_p$ is the *nuclear magneton*.

13-4. Electric-quadripole Moments

In addition to electric-charge and magnetic-dipole moments, certain nuclei are known also to possess an *electric-quadripole moment*. While these quadripole moments appear in the same general physical contexts as do the magnetic-dipole moments—e.g., in optical and microwave hyperfine structure of atomic and molecular energy levels and in nuclear magnetic-resonance effects—the behavior of an electric quadrupole is rather different, both qualitatively and quantitatively, from that of a magnetic dipole. The basis for the detection and measurement of the

---

[1] Hipple, Sommer, and Thomas, *Phys. Rev.*, **76**, 1877 (1949); **80**, 487 (1950).

electric-quadripole moment can be visualized most simply in terms of classical concepts. Consider a small, localized distribution of electric charge subjected to an electrostatic potential which originates from charges external to the localized system. The energy of the charge system is given by the familiar expression

$$W = \int_v \rho\phi \, dv \tag{1}$$

where ρ is the charge density, ϕ is the electrostatic potential, and the integral extends over the localized distribution. To express this energy in terms of the electric moments of the distribution, we expand the potential $\phi(x,y,z)$ in a Taylor's series about an origin situated at the center of charge. In tensor language, such an expression is

$$\phi(x,y,z) = \phi_0 + \frac{\partial\phi}{\partial x_j}\bigg|_0 x_j + \frac{1}{2!}\frac{\partial^2\phi}{\partial x_j \, \partial x_k}\bigg|_0 x_j x_k + \cdots \tag{2}$$

Each of the derivatives appearing in this expression is to be evaluated at the origin, and is of course constant with respect to the variables x_j of integration. Insertion of this expression into Eq. (1) gives

$$W = \phi_0 \int_v \rho \, dv + \frac{\partial\phi}{\partial x_j}\bigg|_0 \int_v \rho x_j \, dv + \frac{1}{2!}\frac{\partial^2\phi}{\partial x_j \, \partial x_k}\bigg|_0 \int_v \rho x_j x_k \, dv + \cdots \tag{3}$$

The integrals which appear in this equation are the various *moments* of the distribution with respect to the given origin. These moments are tensor quantities. For example, the first integral

$$\int_v \rho \, dv = Q \tag{4}$$

is just the *net charge*—a scalar quantity; the three integrals of the second term

$$\int_v \rho x_j \, dv = p_j \tag{5}$$

are the three components of the first-rank tensor *electric-dipole moment;* and the nine integrals of the third term

$$\int_v \rho x_j x_k \, dv = Q_{jk} \tag{6}$$

are the components of the second-rank tensor *electric-quadripole moment* of the distribution. In general, the integrals

$$\int_v \rho x_j x_k \cdots x_s \, dv = Q_{jk\ldots s} \tag{7}$$

are the components of a tensor of rank equal to the number N of x's appearing in the integrand and are called the *electric 2^N-pole moments* of the charge distribution. In terms of these moments, the energy (1) may be expressed

$$W = Q\phi_0 + p_j \frac{\partial \phi}{\partial x_j}\bigg|_0 + \frac{1}{2!} Q_{jk} \frac{\partial^2 \phi}{\partial x_j\, \partial x_k}\bigg|_0 + \cdots$$
$$+ \frac{1}{l!} Q_{jk\cdots s} \frac{\partial^l \phi}{\partial x_j\, \partial x_k \cdots \partial x_s}\bigg|_0 + \cdots \quad (8)$$

The first two terms of this expression are recognized at once as the net charge times the potential and as the scalar product of the electric-dipole moment with the gradient of the potential, respectively. The latter is of course just the familiar expression for the orientation energy of an electric dipole in an electric field:

$$p_j \frac{\partial \phi}{\partial x_j}\bigg|_0 = -\boldsymbol{p} \cdot \left(-\nabla\phi\bigg|_0\right) = -\boldsymbol{p} \cdot \boldsymbol{E}_0 \quad (9)$$

Considering now the electric-quadripole term, we see that the various components of Q_{jk} are "acted upon" by corresponding gradient components of the electric field. Thus a uniform field does not produce any torques or forces upon an electric quadrupole: an electric-quadripole moment can interact only with an *inhomogeneous field*.

For simplicity, let us now restrict our attention to charge distributions which are cylindrically symmetric about some symmetry axis; this case is the one encountered in nuclear physics, and the symmetry axis is conventionally taken as the z-axis. For this case the quadripole tensor is diagonal:

$$Q_{jk} = \begin{pmatrix} Q_{xx} & 0 & 0 \\ 0 & Q_{yy} & 0 \\ 0 & 0 & Q_{zz} \end{pmatrix} = \begin{pmatrix} \int \rho x^2\, dv & 0 & 0 \\ 0 & \int \rho y^2\, dv & 0 \\ 0 & 0 & \int \rho z^2\, dv \end{pmatrix} \quad (10)$$

The quadripole energy is correspondingly equal to

$$W_Q = \tfrac{1}{2} Q_{xx} \frac{\partial^2 \phi}{\partial x^2} + \tfrac{1}{2} Q_{yy} \frac{\partial^2 \phi}{\partial y^2} + \tfrac{1}{2} Q_{zz} \frac{\partial^2 \phi}{\partial z^2} \quad (11)$$

where, because of the cylindrical symmetry, $Q_{xx} = Q_{yy}$. Furthermore, Laplace's equation must be satisfied by the three partial derivatives. Equation (11) may therefore be written in the equivalent form

$$W_Q = \tfrac{1}{2}(Q_{zz} - Q_{xx}) \frac{\partial^2 \phi}{\partial z^2} \quad (12)$$

$$= \tfrac{1}{4} eQ \frac{\partial^2 \phi}{\partial z^2} \quad (13)$$

The quantity Q, which completely defines the interaction energy of a cylindrically symmetric charge distribution, is conventionally referred to as "the" *electric-quadripole moment* of the distribution, although from the above analysis we see that it is in reality a combination of the components of the second-rank quadripole-moment tensor referred to its principal axes.

EXERCISES

13-8. Complete the derivation of Eq. (12) from Eq. (11).

13-9. Show that

$$eQ = \int \rho(3z^2 - r^2) \, dv \tag{14}$$

13-10. Show also that

$$eQ = 2 \int \rho r^2 \, P_2(\cos \theta) \, dv \tag{15}$$

13-11. Show that $Q \equiv 0$ for a spherically symmetric charge distribution, $Q > 0$ for a prolate, $Q < 0$ for an oblate spheroidal distribution.

13-12. Show that the dimensions of Q are those of area.

13-13. Evaluate the electric charge, dipole moment, and quadripole moments of the axially symmetric charge distributions shown in Fig. 13-2.

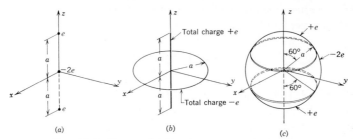

Fig. 13-2. Charge distributions for Exercise 13-13.

13-14. Suppose an axially symmetric charge distribution is not aligned exactly with the z-axis, but is precessing about it at an angle α. Evaluate W_Q for this case, in terms of Q and $\partial^2 \phi / \partial z^2$. For simplicity, assume the field to be axially symmetric about the z-axis—this condition actually does not affect the result if the precession is uniform. *Hint:* Transform the diagonal tensor (10) to new axes tilted at angle α, and observe that $\partial^2 \phi / \partial x^2 = \partial^2 \phi / \partial y^2 = -\frac{1}{2} \partial^2 \phi / \partial z^2$ are the only nonvanishing second derivatives for an axially symmetric field.

Ans.:
$$W_Q = \frac{1}{8} eQ \frac{\partial^2 \phi}{\partial z^2} (3 \cos^2 \alpha - 1) \tag{16}$$

The above classical discussion of electric quadripoles can be extended to quantum mechanics by finding an appropriate quantum-mechanical operator to represent W_Q. Thus from the results of Sec. 2-5C, the perturbation of a given energy state brought about by the electric-quadripole interaction should be

$$\Delta E = \langle W_Q \rangle = \tfrac{1}{8}eQ \frac{\partial^2 \phi}{\partial z^2} \langle 3 \cos^2 \alpha - 1 \rangle \tag{17}$$

where Q is regarded as fixed, as for a rigid body. Our previous experience with fine-structure effects (Sec. 5-8) then leads us to evaluate the term in brackets by using unperturbed wave functions for which $|I|^2$ and I_z are precisely defined. Thus we may write

$$\langle W_Q \rangle = \tfrac{1}{8}eQ \frac{\partial^2 \phi}{\partial z^2} \left\langle \frac{3 I_z^{\,2}}{|I|^2} - 1 \right\rangle$$
$$= \tfrac{1}{8}eQ \frac{\partial^2 \phi}{\partial z^2} \left[\frac{3M_I^2}{I(I+1)} - 1 \right] \tag{18}$$

EXERCISES

13-15. Supply the missing steps in the derivation of Eq. (18) from Eq. (17).

13-16. Evaluate the maximum energy shift that can be observed for a body whose quadripole moment is Q.

Ans.:
$$\langle W_Q \rangle_{\max} = \tfrac{1}{8}eQ \frac{\partial^2 \phi}{\partial z^2} \frac{2I-1}{I+1} \tag{19}$$

13-17. Show that energy-level shifts due to an electric-quadripole moment can be observed experimentally only if the angular-momentum quantum number $I \geq 1$. Thus nuclei (or other systems) of spin 0 or $\tfrac{1}{2}$ cannot exhibit electric-quadripole effects.

13-18. In most situations in which the electric-quadripole moment produces measurable effects, the inhomogeneous electric field is not applied externally by the experimenter, but exists as a result of the particular distribution of the electrons throughout the atom or molecule. To obtain a rough estimate of the magnitudes to be expected, evaluate $\partial^2 \phi / \partial z^2$ at the nucleus of a hydrogen atom which is in the state u_{210}. *Hint:* First evaluate the potential inside a spherical shell of radius R on which a surface charge $\sigma = F(R) \cos^2 \theta$ is situated, and then add together the field gradients, at the origin, due to the various shells from $R = 0$ to ∞. Use the relation $\sigma = -e\psi^*\psi \, dR$.

Ans.:
$$\phi_{\text{shell}} = \frac{RF(R)}{3\varepsilon_0} + \frac{2F(R)r^2 P_2(\cos \theta)}{15\varepsilon_0 R} \qquad r < R$$
$$\left. \frac{\partial^2 \phi}{\partial z^2} \right|_0 = \left(\frac{-e}{4\pi\varepsilon_0} \right) \left(\frac{1}{30 a_0'^{\,3}} \right)$$

13-19. Compare the perturbation energy caused by the quadripole moment of the deuteron ($I = 1$, $Q = 10.9 \times 10^{-27}$ cm^2) in the u_{210} state of atomic deuterium with that due to the deuteron magnetic moment $\mu_D = 0.857\mu_N$ in a magnetic field of 10^4 gauss. *Ans.:* $\Delta E_Q / \Delta E_\mu \approx 10^{-3}$.

The preceding extension of classical quadripole effects to quantum mechanics is only partially correct, since it was assumed that Q is a fixed parameter of the system. Actually, of course, the charge distribution throughout a nucleus must be described in terms of a probability distribution, just as for the electronic-charge cloud of an atom or molecule. This feature adds to the complication of the problem but does not alter the essential results as described above. For our present purposes it suffices to say that effects which are most readily explained in terms of an electric-quadripole moment have been found in certain nuclei. As is expected from the result of Exercise 13-17, these effects are found only in nuclei whose I $\geq$ 1. The quadripole moments which have so far been observed range between the values

$$-1.7 \times 10^{-24} \text{ cm}^2 \leq Q_m \leq + 10 \times 10^{-24} \text{ cm}^2 \qquad (20)$$

The measured quadripole moments Q_m given in Appendix G are defined by $Q = (2I + 2)Q_m/(2I - 1)$.

13-5. Radioactivity

One of the first known properties of nuclei, and one which preceded by more than a decade the nuclear theory of atomic structure itself, is that of the instability of nuclei with respect to decomposition into smaller pieces. This property is called *radioactivity*, and the decomposition process is called *radioactive decay*. Radioactivity occurs in several forms and is of great theoretical and practical interest.

In the early studies of radioactivity it was soon recognized that the mixture of radioactive elements that occur together in an ore deposit constitute a chain, or series, of radioactive atoms produced from the decay of a "parent" atom present in the ore. The identification and determination of the properties of the various members of a series became one of the most active branches of nuclear research, and has remained quite active up until the present time. In Chap. 15 we shall examine these series in more detail.

Although our description of radioactivity has so far been limited mainly to the radioactive nuclear species found in nature, it is a familiar fact that radioactivity can be induced artificially, and it is now true that the artificially produced radioactive species outnumber those occurring naturally by a quite large factor. Among the artificially radioactive nuclei, both α- and β-activities have been observed, and in addition many cases are known whose decay is unlike that of any naturally occurring nucleus. These other types of decay have the following gross properties:

1. In a type of β-activity which commonly occurs, a *positive* electron is emitted rather than a negative one. This is called β^+-decay, or *positron decay*. In this type, the nuclear charge *decreases by one unit* and the mass remains approximately unchanged.

2. Another type of process which occurs along with β^+-decay is the nuclear "capture" of one of the orbital electrons of the atom. This process is called *electron capture*. If the electron comes from the K-shell, the process is called K-capture, and so on. The changes in mass and charge are the same as for β^+-decay.

3. It is sometimes observed that *neutrons* are emitted from a radioactive nucleus. These neutrons are called *delayed neutrons*, as distinguished from the so-called *prompt* neutrons which emerge directly from many nuclear reactions. Although delayed neutrons give the superficial appearance of a radioactive decay with a characteristic lifetime of perhaps several seconds or longer, they are known to be emitted as a *secondary process*, subsequent to a β-decay.

Further features of the above processes will be described later in connection with nuclear stability and the different types of nuclear reactions.

13-6. The Constituents of Nuclei

From the earliest investigations of the atomic weights of the elements it was recognized that there is a definite tendency for atomic weights to be integral multiples of the atomic weight of hydrogen. There are, of course, several clear exceptions to this rule; one of the most striking is chlorine, whose atomic weight is almost exactly 35.5. The discovery of *isotopes* provided a satisfactory explanation of the exceptions, and it became an accepted fact that the individual isotopes of the elements tend to have masses which differ, from one to another, by small integral multiples of the mass of a hydrogen atom. This strongly suggested that all atoms are in some way constructed out of hydrogen atoms.

With the advent of the nuclear theory of the atom, the above picture of atomic structure took a more specific form, in which the nucleus of an atom was supposed to be compounded from *nuclei of hydrogen* and the atom was rendered electrically neutral by a corresponding number of *electrons* circulating about it in some manner.

The early work of Barkla on the scattering of X-rays from the light elements, and later, that of Moseley on the characteristic X-radiation of the elements, brought a new feature into the problem.   These researches indicated quite clearly that the number of electrons in the outer parts of an atom, and the nuclear charge number Z, are each only about *half the atomic weight* in units of the weight of hydrogen.   This result indicated that the nucleus of an atom cannot be composed wholly of protons, but must contain something else as well.   However, it was well known that those radioactive elements which undergo β-decay eject a *negative electron* from the nucleus, so that it seemed safe to infer that the nucleus *must have contained the electron before emitting it*.   It was thus concluded that the *mass* of a nucleus is contributed almost wholly by protons, but that the electrical effects of about half of these are canceled by a number of electrons *inside the nucleus*.   This view of nuclear structure was widely held, but not entirely without misgivings, until the discovery of the neutron in 1932.   Difficulties with the proton-electron theory of nuclear structure arose because of the great success of the new quantum theory in describing the electronic properties of matter.   These difficulties may be summarized as follows:

1. It was recognized that, because of the known *small size* of nuclei, it would require a tremendous binding energy to hold an electron inside such a small space.

     From the analysis of the energy levels of a relativistic particle in a box it follows that the minimum kinetic energy of an electron inside a nucleus of diameter $\sim 10^{-14}$ m would be about 60 MeV.   The energy with which β-particles emerge in β-decay is only 2 or 3 MeV, however.   These two numbers are difficult to reconcile with one another.

2. An even more serious objection to the proton-electron theory of nuclei comes from a consideration of *nuclear spin*.   It is easy to prove that a compound "particle" which is composed of a number of sub-particles must have integral spin if it contains an even number, and half-integral spin if it contains an odd number, of Fermi particles. Now it is known that *protons and electrons are both Fermi particles*. Thus the spin character of a nucleus composed of protons and electrons would be governed by the number of unneutralized protons (that is, by Z), since the remaining protons plus the neutralizing electrons would always constitute an even number of Fermi particles.   Thus all nuclei of odd atomic number should have half-integral spin, and all nuclei of even atomic number should have integral spin.

     Actually, some clear violations of this requirement are observed. One of the clearest is that of ordinary nitrogen.   This nucleus should have half-integral spin $(Z = 7)$, whereas the spin deduced from the band spectrum of the homonuclear N_2 molecule is *one* unit. This is an extremely serious objection to the proton-electron hypothesis.

3. When the *magnetic moments* of nuclei were studied by using the hyperfine splitting of certain spectral lines, it was found that nuclei have magnetic moments much smaller than the magnetic moment of' an electron. It thus became necessary to assume that the electrons that are bound within a nucleus do not possess their full magnetic moment, but only a small part of it. This suppression of the electronic magnetic moment is difficult to accept.

4. It is very difficult to see how electrons can play the dual role in atomic structure that is required of them under the proton-electron hypothesis. One would think, if *any* electrons could be bound *inside* nuclei, that *all* the electrons of an atom would be bound in this way, rather than only about half of them.

The experimental discovery of the *neutron* led Heisenberg (1932) to suggest that the nuclei might be composed of *protons and neutrons* rather than protons and electrons. This hypothesis has led to a satisfactory and complete interpretation of many of the properties of nuclei. We shall postpone a description of the experiments by which the detailed properties of the neutron have been determined, since these are somewhat indirect and require the use of concepts that we have not yet discussed. It suffices to say for the present that the neutron has a mass slightly greater than that of the proton, and spin of $\frac{1}{2}$. All nuclei are considered to be composed of a certain number Z of protons and a certain number N of neutrons. Thus the spin character of a nucleus is determined by $Z + N$, rather than by Z. The integral spin of nitrogen, described above, is thus explained by the fact that the nitrogen nucleus contains a total of 14 Fermi particles. This and all other spectroscopic evidence agrees with the proton-neutron hypothesis of nuclear structure.

In accepting the proton-neutron hypothesis of nuclear structure, one must then explain the origin of the β-particles ejected in β-decay. As we shall see in Chap. 15, the modern explanation of β-decay is based upon the idea that protons and neutrons are not immutable particles, but may under certain circumstances be transformed one into the other.

13-7. Symbols and Nomenclature for Nuclear Structure

We now list some of the nomenclature and symbols that are universally used to describe nuclei. For a given nucleus

Z = number of protons (*atomic number*)

N = number of neutrons (*neutron number*)

$A = N + Z$ = total number of protons and neutrons, referred to generically as *nucleons* (*mass number*)

$N - Z = A - 2Z$ = excess of neutrons over protons (*isotopic number*)

The atomic number and mass number are herein denoted by a preceding subscript and superscript, respectively, to the chemical symbol. For example, $_{82}^{206}$Pb denotes an isotope of lead ($Z = 82$) containing 124 neutrons. Any particular nuclear species is referred to as a *nuclide*.

Special names are used to describe classes of nuclei having the same values of one or more of the above quantities:

Characteristics of nuclei	Name
Same Z, different N.......	Isotopes
Same N, different Z.......	Isotones
Same A, different N, Z....	Isobars
Same A, same Z..........	Isomers

The last-named situation is exemplified by $_{91}^{234}$Pa, the third member of the uranium series. This element is formed in an excited state, called UX$_2$ (uranium X$_2$), from which it decays with a half-life of 1.14 min in either of two ways: 99.85 per cent of the $_{91}^{234}$Pa atoms decay by a β-transition into $_{92}^{234}$U; the remaining 0.15 per cent, however, decay with the emission of a γ-ray to $_{91}^{234}$Pa in its ground state, which in turn decays by β-decay to $_{92}^{234}$U, with a half-life of 6.7 h. The unexcited form of $_{91}^{234}$Pa is called UZ (uranium Z). The γ-ray transition between these two isomeric states is called an *isomeric transition*.

We shall sometimes refer to a general nuclide by using the notation (Z,A).

13-8. Excited States of Nuclei, Nuclear Reactions

As we have just seen, an atomic nucleus is now known to be an aggregate of protons and neutrons, presumably held together by forces whose range of action is sharply limited to distances less than about 1.5×10^{-15} m. It is assumed that the laws of quantum mechanics govern the motions of the particles comprising the nucleus—an assumption which is currently thought to be valid provided certain extensions of the theory and our ideas of what kinds of particle ultimately compose nuclei are made. In the present discussion we shall be concerned with properties of nuclei which apparently *can* be satisfactorily described in the language of the familiar atomic quantum mechanics, and we shall set aside for later examination the important question of whether a nuclear aggregate of protons and neutrons does or does not follow quantum mechanics in exact detail.

We visualize a nucleus, then, as consisting of two kinds of indistinguishable Fermi particles—protons and neutrons—whose motions are described by ordinary nonrelativistic quantum mechanics. One would expect such a system to have certain *energy eigenstates* in which it can exist, and this is observed experimentally to be the case. The lowest of these energy states is the *ground state*, and the higher states represent *states of excitation* of the nucleus.[1] These states are, as in the atomic case, eigenstates for other dynamical quantities in addition to the energy. The total angular momentum (orbital plus spin), the z-component of this angular momentum, and the parity are examples of such other quantities which have quantized values for the various nuclear energy states.

If a nucleus happens to be in one of its states of excitation, we find that it may decay to a lower energy state by one of a number of possible processes:

1. *By the emission of a γ-ray.* This method of deexcitation is closely analogous to the familiar atomic case, in which an excited atom emits a quantum as it reduces its state of excitation. The large energy differences in the nuclear case cause the emitted radiation to fall in the X-ray range of wavelengths, photon energies of a few million electron volts being not uncommon. It is found that electric-dipole radiation is quite commonly emitted in nuclear transitions, but that magnetic-dipole, electric- and magnetic-quadripole, and even higher orders of radiation are also often emitted. As in the atomic case, the lifetime of an excited state against γ-ray emission tends to be longer, the higher the multipole order of the radiation.

2. *By the emission of an electron in an Auger transition* (internal conversion). Rather than emit a γ-ray in the process of deexcitation, an atom may change its state of nuclear excitation and eject one of its electrons, usually K-shell, with a kinetic energy equal to the energy change of the nuclear state minus the binding energy of that electron in the atom. Such a transition is called *internal conversion*, and the emitted electron is called an *internal-conversion electron*. The process is sometimes described as the photoelectric ejection of an electron by a γ-ray emitted from the nucleus, but a more correct view is that the nuclear deexcitation and the electron ejection are a *single process*, with no intermediate γ-ray involved. That this is true is evidenced by the fact that internal-conversion electrons are sometimes observed in connection with energy transitions in which *γ-rays are not present*. The internal-conversion process is thus an example of an *Auger transition* and emphasizes again the fact that each energy level of an atom is really a property of the atom *as a whole*, and not its individual parts.

---

[1] It is convenient to have a uniform terminology with which to describe all nuclei that exist as distinct entities, even though this existence might be of rather short duration. It is thus customary to refer to the lowest energy state of a given nuclide as the ground state of that nuclide even for nuclides which are unstable toward decay. Strictly speaking, such a state is not the lowest energy state of the system but is a relatively long-lived excited state.

3. *By the emission of a nuclear particle.* If the level of excitation of a nucleus is great enough, the total energy of the system may exceed that of a nucleus of lower mass number plus a free nucleon, in which case the emission of a nuclear particle becomes probable. The particles most commonly emitted are neutrons and protons, but simple composite particles of high stability such as α-particles are also frequently observed. The excitation energies above which particle emission becomes probable are generally in the neighborhood of the binding energy of a nucleon, namely, about 8 MeV.

All excited states of nuclei are observed to have certain "widths" which depend upon the *decay lifetimes* of the states, just as in the atomic case. These widths vary over a great range, being generally narrowest for states involving γ-ray emission and broadest for states involving particle emission. If the only mode of decay of a given state is by γ-ray emission, for which the characteristic lifetimes vary upward from about 10^{-14} s, the level will be quite narrow compared to the interval between levels; whereas if a state can decay by particle emission, it is usually found that the width of the level is much greater than for γ-ray decay, and for states of high excitation may become so great that neighboring states overlap strongly. In this case it is difficult to utilize the notion of discrete, independent energy levels at all, and one must recognize that each possible value that the energy of the system may have actually corresponds to the nucleus being in a *composite* energy state consisting of a linear combination of possibly several broad, overlapping, rapidly decaying "eigenstates."

EXERCISES

13-20. Calculate the approximate width, in electron volts, of a nuclear energy eigenstate which decays by γ-ray emission with a mean decay lifetime of 10^{-14} s. This represents a typical decay lifetime for electric dipole transitions.

13-21. The "profile" of a nuclear energy level involving particle emission can often be measured experimentally by studying the *formation* of this energy state from high-energy particles bombarding appropriate nuclei. This profile is called the *excitation function* of the level; and if the level is rather narrow, it is said to be a *resonance* level. Experimental measurements of the widths of such levels often give values ranging from about 10^3 to about 10^5 eV in the lighter elements. Calculate the approximate range of *decay lifetimes* corresponding to this range of widths.

The measurement of the energy values, widths, and decay properties

of the various excited states of nuclei, and their classification according
to their quantum properties (spin, parity, etc.), is exactly analogous to the
similar study that was made of the electronic states of excitation of
atoms in the early studies of atomic structure.   The phenomena that are
available and the experimental techniques that are used are rather
different in the two cases, but the *objective* is exactly the same, namely, to
obtain quantitative information regarding the fundamental interactions
between the particles comprising the system.

The earliest studies of the excited states of nuclei were made by
measuring the energies with which α-particles were emitted from a radioactive nucleus, and comparing these with the energies of the γ-rays
emitted from the daughter nucleus.   The α-particle energies were found
by measuring the ranges of the α-particles in air, and γ-ray energies
were found by measuring the momenta of the internal-conversion electrons accompanying them, utilizing the curvatures of the electron trajectories in a magnetic field.   By such quantitative measurements it was
established that the α-particles often fell into two or more well-defined
energy groups, and that *the difference in energy between two such groups
corresponded exactly to the observed γ-ray energies*.   Thus it was concluded
that the daughter nucleus was sometimes formed in an excited state
from which it subsequently decayed by γ-ray emission.   The existence
of such excited states in several nuclei was established in this way.

The above procedure is of course limited to the few naturally radioactive nuclei.   The systematic, quantitative study of the excited states
of nuclei in all ranges of the mass number dates from the first artificial
nuclear disintegration using artificially accelerated particles (Cockroft
and Walton, 1932), and today virtually all investigations of the excited
states of nuclei make use of *nuclear reactions* between a stationary nucleus
and bombarding particle of accurately known energy.

Much of the nomenclature of nuclear reactions is parallel to that used
in describing simple *chemical* reactions.   Thus, a nuclear reaction is often
written in the form $A + B \rightarrow C + D$; for example, $^{14}N + \alpha \rightarrow ^{17}F + n$,
where A and B refer to the initial, or "reacting," nuclei and C and D
refer to the final, or "product," nuclei.   Furthermore, the *energy balance*
of such a reaction is often expressed in a form analogous to that used in
chemical reactions: if energy must be supplied to A and B in order to
match the rest-mass energy of $C + D$, the reaction is said to be *endoergic*,
and if energy is released, *exoergic*.   This energy input or output is referred
to as the Q-value of the reaction, and is taken to be *positive if the reaction
is exoergic*.   It is usually expressed in MeV, and it is sometimes written
as one term in the nuclear-reaction equation.   Thus, $A + B \rightarrow C + D
+ Q$.   Another commonly used form in which a nuclear reaction is
expressed is $A(B,D)C$, where A refers to the stationary, or *target*, nucleus;

B to the incident, or *bombarding*, particle; D to the less massive of the product particles; and C to the more massive product, or *residual nucleus*. For example, $^{14}N(\alpha,n)^{17}F$.

13-9. Particles

During the past century the research efforts of chemists and physicists have turned steadily from investigation of the large-scale properties of materials toward the elucidation of the atomic and subatomic structure of matter. The course of modern physics has been marked by the discovery of one, then another, then many so-called "elementary particles" which seem to constitute the basic "building blocks" from which all things are made.

In the preceding chapters we have seen how atoms are constructed out of electrons and nuclei, and in the present chapter we have seen that atomic nuclei are in turn made out of protons and neutrons. It might thus appear that all matter could be composed solely of these elementary ingredients. But this picture, simple as it seems, does not suffice completely to describe the structure of matter, for experiments with high-energy particles have brought to light numerous subnuclear particles whose role in nuclear structure is by no means clear. Most of these new particles have masses lying between those of the electron and proton and are called *mesons*, but several particles called *hyperons*, which are more massive than a proton, are also known. The elucidation of the properties of these particles and their role in the scheme of things is a matter of great interest and importance, and it is little wonder that a large part of the efforts of physicists have been expended in this direction.

In Chap. 20 we shall describe the more important properties of all known particles insofar as they are available. For the present it will suffice to observe that the existence of these particles must be taken into account in any exact treatment of nuclear structure, and that there is good reason to believe that the forces which hold nuclei together are transmitted via an exchange of mesons between nucleons.

13-10. Conclusion

The properties of nuclei which have just been described may be regarded as being directly established by experiment, and thus as being independent of any particular model by which we may wish to describe nuclear structure. One of the basic tasks now before us is just that of interpreting, or as we say, "explaining," these measured properties in terms of the intrinsic properties of nucleons and certain quantitative rules of behavior. In the following chapters we shall try to piece together a qualitative picture of nuclear structure in which the known properties of nuclei are interrelated as simply and clearly as possible; this will lead naturally to an examination of the several nuclear models which endeavor to provide a quantitative basis for these properties.

We shall begin our treatment with a brief introduction to the theory of particle scattering, treating for the most part the Rutherford law and its application to the physics of the passage of charged particles through matter. Next, the phenomenological relationships which govern radioactive decay, both for naturally occurring and artificially produced radioactive species, will be examined. This will lead naturally to a consideration of some of the simpler aspects of the mechanisms responsible for α- and β-decay of nuclei, and thence to a consideration of those gross properties of nuclei which seem to render some nuclei stable and others unstable. We shall then consider the dynamical relationships which exist in nuclear reactions and the gross properties of nuclear reactions and excited states. This will lead us to examine the various nuclear models which strive to describe some of the details of nuclear structure. The nature of the binding forces which hold nuclei together, the character of the elementary nucleon-nucleon interaction, and the properties of nuclear and subnuclear particles will then be examined in turn. Finally, some of the phenomena in the universe-at-large in which nuclear physics appears to play a prominent role will be briefly considered.

REFERENCES

Jauncey, G. E. M.: The Early Years of Radioactivity, *Am. J. Phys.*, **14**, 226 (1946).

Richtmyer, F. K., E. H. Kennard, and T. Lauritsen: "Introduction to Modern Physics," 5th ed., International Series in Pure and Applied Physics, McGraw-Hill Book Company, Inc., New York, 1955.

Rutherford, Chadwick, and Ellis: "Radiations from Radioactive Substances," The Macmillan Company, New York, 1930.

14

Particle Scattering

Before beginning our more detailed examination of the properties of nuclei we shall consider some of the important aspects of the *scattering of particles*, because quantitative measurement of nuclear properties often depends directly or indirectly upon this phenomenon. We shall first consider in some detail the classical theory of Rutherford scattering and the effects of this type of scattering upon particles which pass through matter, and then we shall treat briefly a few elementary features of particle scattering in general.

14-1. The Classical Theory of Rutherford Scattering

By far the most common result of a collision of two charged particles is that the particles are deflected or *scattered* because of their mutual electrostatic attraction or repulsion. These scatterings were studied exhaustively by Rutherford and his students, and after numerous careful measurements Rutherford was led to adopt the nuclear atomic model to explain the observed results. In these famous experiments, energetic α-particles from the decay of radioactive atoms were allowed to pass through very thin metal foils, and the deflections of the individual α-particles from the original direction of motion were measured by observing the scintillations produced by their impacts upon a zinc sulfide screen, as was described in Sec. 2-1D.

When an α-particle penetrates an atom, large angular deflections will result if the α-particle passes close to the positive nucleus, and most of this deflection will take place while the α-particle is relatively near the nucleus. For this reason the shielding effect of the electrons can be neglected as a first approximation, and the orbit followed by the α-particle should therefore be essentially that followed by a particle moving in an inverse-square repulsive force field.   We shall now briefly review this classical problem.

The dynamical system is shown in Fig. 14-1, which depicts the trajectory followed by a particle of mass m_1 and charge $Z_1 e$ which is initially

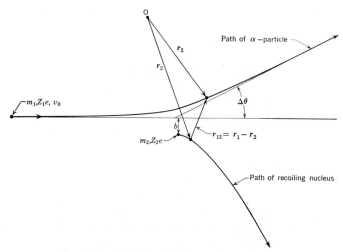

FIG. 14-1. A Rutherford scattering encounter.

moving with speed v_0 in such a direction that it would pass, were it not deflected, a distance b from a second particle of mass m_2 and charge $Z_2 e$ which is initially at rest with respect to the observer. b is called the *impact parameter* of the collision.   Actually, of course, the incident particle is deflected and the target particle is ejected, as is indicated in the figure.

The equations of motion (nonrelativistic) of the two particles are

(a)
$$m_1 \ddot{r}_1 = \frac{Z_1 Z_2 e^2 (r_1 - r_2)}{4\pi \varepsilon_0 |r_1 - r_2|^3}$$

(b)
$$m_2 \ddot{r}_2 = - \frac{Z_1 Z_2 e^2 (r_1 - r_2)}{4\pi \varepsilon_0 |r_1 - r_2|^3}$$

$$(1)$$

These equations may be combined to yield

(a) $\qquad m_1 \ddot{r}_1 + m_2 \ddot{r}_2 = (m_1 + m_2)\ddot{r}_{\text{CM}} = 0$

(b) $\qquad \ddot{r}_1 - \ddot{r}_2 = \ddot{r}_{12}$

$$(2)$$

$$= \frac{m_1 + m_2}{m_1 m_2} \frac{Z_1 Z_2 e^2}{4\pi \varepsilon_0 r_{12}^3} r_{12}$$

which show that *the center of mass r_{CM} moves with constant velocity, and the relative-position vector r_{12} varies as if a single particle, of mass $m_1 m_2/(m_1 +$*

m_2), *were moving around a FIXED force center.* $m_1 m_2/(m_1 + m_2)$ is called the *reduced mass* m_r. We shall for the time being ignore the CM motion and fix our attention on Eq. (2b), which may be written in the abbreviated form

$$\ddot{\mathbf{r}}_{12} = \frac{K}{r_{12}^3} \, \mathbf{r}_{12} \tag{3}$$

where $K = \dfrac{Z_1 Z_2 e^2}{4\pi\varepsilon_0 m_r}$

To study the shape of the trajectory, we introduce polar coordinates (r,θ) in the plane of the motion. Then, if $\mathbf{e}_r$ and $\mathbf{e}_\theta$ are unit vectors along and perpendicular to $\mathbf{r}$, we may write

$$\begin{aligned} \mathbf{r}_{12} &= r\mathbf{e}_r \\ \dot{\mathbf{r}}_{12} &= \dot{r}\mathbf{e}_r + r\dot{\mathbf{e}}_r = \dot{r}\mathbf{e}_r + r\dot{\theta}\mathbf{e}_\theta \\ \ddot{\mathbf{r}}_{12} &= \ddot{r}\mathbf{e}_r + 2\dot{r}\dot{\theta}\mathbf{e}_\theta + r\ddot{\theta}\mathbf{e}_\theta - r\dot{\theta}^2\mathbf{e}_r \end{aligned} \tag{4}$$

and

EXERCISE

14-1. Show that $\dot{\mathbf{e}}_r = \dot{\theta}\mathbf{e}_\theta$ and that $\dot{\mathbf{e}}_\theta = -\dot{\theta}\mathbf{e}_r$.

Using the above relationships, Eq. (3) may be written in component form:

(a)

$$\ddot{r} - r\dot{\theta}^2 = \frac{K}{r^2}$$

(b)

$$r\ddot{\theta} + 2\dot{r}\dot{\theta} = 0 \tag{5}$$

If the second of these equations is multiplied by r, it may be integrated:

$$r^2\dot{\theta} = \text{const} = -v_0 b \tag{6}$$

This equation expresses the conservation of angular momentum.

EXERCISE

14-2. Verify that the initial value of $r^2\dot{\theta}$ is $-v_0 b$.

In order to find the relative deflection angle θ_c, we must find the dependence of r upon θ. To do this, we change to θ as independent variable, and for ease in integrating the equations, to $u = 1/r$ as the dependent variable:

(a)

$$r = \frac{1}{u}$$

(b)

$$\dot{r} = -\frac{\dot{u}}{u^2} = -\frac{\dot{\theta}(du/d\theta)}{u^2} = +v_0 b \frac{du}{d\theta} \tag{7}$$

(c)

$$\ddot{r} = +v_0 b\dot{\theta}\frac{d^2 u}{d\theta^2} = -v_0^2 b^2 u^2 \frac{d^2 u}{d\theta^2}$$

Thus Eq. (5a) becomes

$$-v_0^2 b^2 u^2 \frac{d^2 u}{d\theta^2} - v_0^2 b^2 u^3 = K u^2$$

or
$$\frac{d^2 u}{d\theta^2} + u = -\frac{K}{v_0^2 b^2} \tag{8}$$

which may be integrated to give

$$u = \frac{1}{r} = A \cos (\theta + \delta) - \frac{K}{v_0^2 b^2} \tag{9}$$

This equation is the polar-coordinate equation of a *conic section with the pole at one focus*. The eccentricity is given by e in the equation

$$r = \frac{ef}{1 - e \cos (\theta + \delta)} \tag{10}$$

EXERCISES

14-3. Supply the missing steps in the derivation of Eq. (9).

14-4. Evaluate the constants A and δ in terms of the initial conditions $\theta_0 = \pi$, $r_0 = \infty$, $\dot{r}_0 = -v_0$, $r_0^2 \dot{\theta}_0 = -v_0 b$.

Ans.: (a)
$$A = \frac{(1 + K^2/v_0^4 b^2)^{1/2}}{b}$$

(b)
$$\tan \delta = \frac{v_0^2 b}{K} \tag{11}$$

14-5. Find the eccentricity e of the orbit and show that the orbit is a hyperbola. *Ans.:* $e = (1 + v_0^4 b^2/K^2)^{1/2} > 1$.

The angular deflection of the incoming particle is now easily found. First, the angle θ_c, corresponding to $r = \infty$, is found from Eq. (9) by setting

$$A \cos (\theta + \delta) - \frac{K}{v_0^2 b^2} = 0$$

One solution of this equation is known to be $\theta = \pi$, since this is the initial condition. The other must be $\pi - 2\delta$, since $\cos (\pi - \delta) = \cos (\pi + \delta)$, so that one obtains from Exercise 14-4

$$\tan \tfrac{1}{2}\theta_c = \cot \delta = \frac{K}{v_0^2 b} \tag{12}$$

However, the angle θ_c is the angular deflection of the particle in the CM system, since we have been working with the *relative* displacement r_{12}. To find the angular deflection in the *lab* system, the motion of the CM system must be added, and we shall soon consider this matter further. Fortunately, in many cases of practical interest the ratio m_1/m_2 is sufficiently small that the CM may be considered stationary. For the analysis of α-particle scattering in heavy-metal foils, this approximation is justified and will be used in what follows.

The expression (12) shows that there is no restriction upon the size of the deflection angle θ_c that can be obtained in a single encounter since, contrary to the case of the Thomson atom (Sec. 2-1D), the large deflections now are associated with small impact parameters. One can

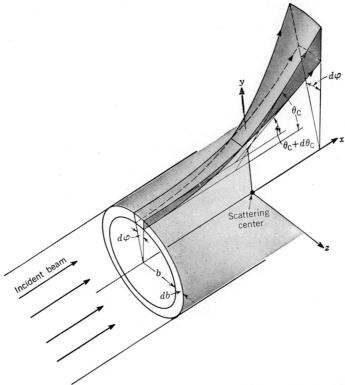

FIG. 14-2. Those particles which are aimed toward the annulus of area $2\pi b\,db$ will be scattered through angles between θ_c and $\theta_c + d\theta_c$.

thus account, at least qualitatively, for *any* size of deflection angle on the basis of the Rutherford nuclear atomic model. However, to test this model quantitatively, the *distribution of the angular deflections* must be deduced from the theory and compared with experiment. Since the main question at issue involves large deflections, or small impact parameters, one can assume that only a single close encounter is at all likely to be experienced by a given α-particle. To evaluate the distribution function $N(\theta_c)$, one must find how many particles will be aimed with impact parameters between b and $b + db$, such that they will suffer deflections lying between θ_c and $\theta_c + d\theta_c$. From Fig. 14-2, we see that

$$dN = N(\theta_c)\, d\theta_c = N_0 nt\, 2\pi b\, db$$

$$N(\theta_c) = 2\pi N_0 nt\, \frac{b\, db}{d\theta_c}$$

$$= \pi N_0 nt K^2\, \frac{\cos \frac{1}{2}\theta_c}{v_0^4 \sin^3 \frac{1}{2}\theta_c} \tag{13}$$

where N_0 is the number of incident α-particles, t is the foil thickness, and n is the particle density of the foil, in particles per unit volume.

EXERCISE

14-6. Supply the missing steps in the derivation of Eq. (13).

It is desirable to evaluate the number of particles scattered into a unit solid angle at a given angle θ_c with respect to the beam. Thus we use the relation connecting the solid angle Ω with θ_c:

$$\Omega = 2\pi(1 - \cos \theta_c)$$

so that

$$d\Omega = 2\pi \sin \theta_c\, d\theta_c$$

and thence

$$\frac{dN}{d\Omega} = \frac{N(\theta_c)}{2\pi \sin \theta_c}$$

$$= \frac{N_0 nt K^2}{4 v_0^4 \sin^4 \frac{1}{2}\theta_c}$$

$$= \frac{N_0 nt Z_1^2 Z_2^2\, e^4}{64\pi^2 \varepsilon_0^2 m_1^2 v_0^4 \sin^4 \frac{1}{2}\theta_c} \tag{14}$$

The above law is called the *Rutherford law of single scattering*. We see that the number of scattered α-particles per steradian should be

1. Directly proportional to the thickness t of the foil
2. Directly proportional to the *square of the atomic number* Z_2 of the scattering nuclei in the foil
3. Inversely proportional to the *square of the kinetic energy* $\frac{1}{2}m_1 v_0^2$ of the α-particles
4. Inversely proportional to the *fourth power of the sine of half the scattering angle* θ_c

As was mentioned in Chap. 2, these features were most carefully tested by Geiger and Marsden (1913) by varying the thickness of the foil, its composition, the energy of the incident α-particles, and the angle of the screen with respect to the incident beam. The results of these experiments were in excellent agreement with the theoretical predictions, and the Rutherford nuclear atom was thereby established beyond any

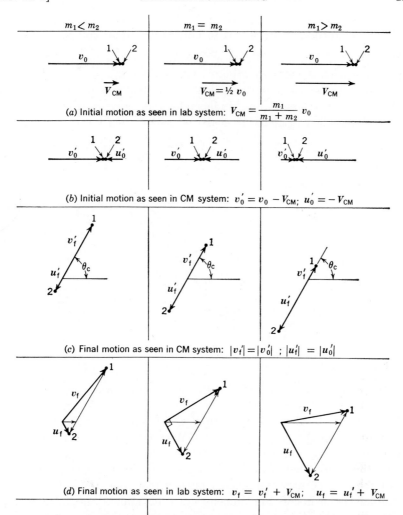

$m_1 < m_2$ | $m_1 = m_2$ | $m_1 > m_2$

(a) Initial motion as seen in lab system: $V_{CM} = \dfrac{m_1}{m_1 + m_2} v_0$

(b) Initial motion as seen in CM system: $v_0' = v_0 - V_{CM};\ u_0' = -V_{CM}$

(c) Final motion as seen in CM system: $|v_f'| = |v_0'|\ ;\ |u_f'| = |u_0'|$

(d) Final motion as seen in lab system: $v_f = v_f' + V_{CM};\ u_f = u_f' + V_{CM}$

FIG. 14-3. Illustrating the kinematic relationships in transforming from the CM to the lab system for three cases: $m_1 < m_2$, $m_1 = m_2$, and $m_1 > m_2$. The transformations are carried out in three steps: The initial motion in the lab system is first transformed to the CM system, and the scattering is then analyzed in the CM system (the line of motion is rotated through an angle θ_c and the speeds of the two particles are unchanged). Finally, the resulting motion is transformed back to the lab system.

possible doubt as being quantitatively as well as qualitatively correct. Later theoretical treatment of the problem, based upon the quantum theory, also led to the Rutherford scattering law (14), although the simple classical picture of localized particles executing hyperbolic orbits is of course not valid. The quantum-mechanical derivation of Eq. (14) is somewhat complicated, and for this reason we shall not give it.

14-2. Transformation from CM to Laboratory Coordinates

We now return to the problem of transforming the result of a scattering experiment from the CM coordinate system to the lab system. Although our present concern is with the special case of Rutherford scattering, this problem is of course quite independent of any particular law of force and must be considered in nearly all particle-scattering experiments.

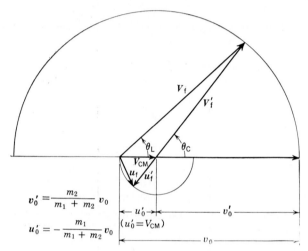

$$v_0' = \frac{m_2}{m_1 + m_2}\, v_0$$

$$u_0' = -\frac{m_1}{m_1 + m_2}\, v_0 \qquad (u_0' = V_{CM})$$

FIG. 14-4. Geometrical representation of the kinematics of elastic scattering of a particle by a more massive target particle.

It is almost always the case that a scattering experiment is most simply analyzed theoretically in the CM system, in which the center of mass is stationary, but is actually *performed* in such a system that one of the particles involved (called the target particle) is initially at rest or nearly so. (In certain high-energy nuclear reactions, the initial motions of the nucleons inside the nucleus must be allowed for.) One great advantage of the CM system for analyzing the motion is that, for elastic scattering, the incident particle and target particle *recede from the collision with the same speeds as they possessed initially*, even though their line of motion has been rotated through an angle θ_c. The kinematic relationships connecting the CM and lab systems are indicated in Fig. 14-3 for three important cases: (1) $m_1 < m_2$, (2) $m_1 = m_2$, and (3) $m_1 > m_2$. It is assumed that all speeds are sufficiently slow that nonrelativistic mechanics can be used.

A simple geometrical representation of the above results is indicated in Fig. 14-4. From this figure it is easily seen that the target particle cannot have a component of motion opposite the incident direction, and that the incident particle cannot be scattered backward if $m_1 > m_2$.

EXERCISES

14-7. Show that

$$\cot \theta_L = \cot \theta_c + (m_1/m_2) \csc \theta_c \qquad (1)$$

14-8. In order to transform the distribution function $dN/d\Omega$ from CM to lab coordinates, one must know the transformation law for solid angles. Show that

$$\frac{d\Omega_c}{d\Omega_L} = \frac{[(m_1/m_2)^2 + 2(m_1/m_2) \cos \theta_c + 1]^{3/2}}{1 + (m_1/m_2) \cos \theta_c} \qquad (2)$$

if the scattering is independent of the azimuth angle ϕ.

14-9. Use a diagram analogous to Fig. 14-4 to find the maximum possible deflection angle, in the lab system, of an incident particle which is more massive than the target particle.

$$\text{Ans.:} \qquad \sin \theta_{L\,\text{max}} = \frac{m_2}{m_1} \qquad (3)$$

14-10. Show that, if $m_1 = m_2$, the incident and target particles move at right angles to one another in the lab system after impact.

14-11. Derive a relativistically correct equation relating θ_L and θ_c.

$$\text{Ans.:} \qquad \cot \theta_L = \gamma_{CM} \left(\cot \theta_c + \frac{W_{1c}}{W_{2c}} \csc \theta_c \right) \qquad (4)$$

where $\gamma_{CM} = (1 - \beta_{CM}^2)^{-1/2}$, $\beta_{CM} = P_1/(W_1 + m_2)$, W_{1c} and W_{2c} are the total energies of particles 1 and 2 in the CM system, and P_1 and W_1 are the momentum and total energy of the incident particle in the lab system.

14-12. A certain fragment from the fission of ^{235}U is moving through the gas of a cloud chamber when it is scattered by one of the atoms in the gas. Its angle of deflection is observed to be 12.2°, and the atom with which it collides recoils at an angle of 60° from the original direction of the fission fragment. The cloud-chamber gas is a mixture of ^{40}A and a saturated vapor of C_2H_5OH and H_2O. Find the atomic weight of the fission fragment. *Note:* When a nucleus undergoes fission, the two daughter nuclei have very roughly equal mass. *Ans.: A = 140.*

14-13. A beam of α-particles of kinetic energy 4.5 MeV passes through a thin foil of ^9Be. The number of α-particles scattered between 30° and 90° and between 90° and 150° is measured. What should be the ratio of these numbers? *Ans.: R = 16.0.*

14-14. Show that the scattering of smooth, hard, perfectly elastic spheres of radii a and A is isotropic in the CM system, according to classical mechanics. *Ans.: $dN/d\Omega_c = \frac{1}{4} N_0 nt(a + A)^2$.*

14-15. A target of beryllium having 10^{19} atoms cm^{-2} normal to the direction of incidence is bombarded by a beam of 1.00-MeV protons. What is the probability that a given proton will be scattered by the Coulomb field of a beryllium nucleus through an angle $\theta_L = 138°$ into a solid angle $d\Omega_L = 10^{-3}$ sr? *Useful information:*

$$\frac{\text{Mass of beryllium}}{\text{mass of proton}} = 8.96$$

$m_e c^2 = 0.511$ MeV, $e^2/4\pi\epsilon_0 m_e c^2 = 2.82 \times 10^{-13}$ cm. *Ans.:* $P = 2.17 \times 10^{-10}$.

14-16. Use conservation of energy and angular momentum to evaluate the distance of closest approach of two charged particles. *Ans.:* $r_{\text{min}} = (K/v_0^2)(1 + \csc \frac{1}{2}\theta_c)$.

14-17. When α-particles from RaC′(^{214}Po) bombard aluminum foil, it is found that the Rutherford scattering law begins to break down at a lab scattering angle of about 60°. What is the approximate effective size of an aluminum nucleus in this experiment if an α-particle has a radius of about 2.0×10^{-15} m? *Ans.:* $R \approx 6 \times 10^{-15}$ m.

14-3. Multiple Coulomb Scattering and Energy Loss by Ionization

The Rutherford law of single scattering provides an accurate description of the large-angle scatterings which charged particles undergo in traversing matter. But these so-called *single Coulomb scatterings* do not by any means account for all that happens to the charged particle; for in its path through the target material, it passes nearby multitudes of atoms, and while each of these encounters affects the onrushing particle but little, their cumulative effect can be, and is, quite significant. There are here two principal effects which are of great importance in the study of particle physics: *multiple Coulomb scattering* and *energy loss by ionization.*

As a fast-moving charged particle passes near or through an atom, it interacts with the *circumnuclear electrons* as well as with the *atomic nucleus.* In this process an electron is occasionally given sufficient energy to escape from the atom to which it is bound. This energy is of course acquired at the expense of the fast-moving incident particle, with the result that the moving particle is decelerated and may ultimately be brought to rest. As we shall now see, almost all of the *energy loss* of the incident particle results from the electronic encounters, while almost all of the *scattering deflection* results from the *nuclear* encounters.

Consider a fast-moving particle of mass m_1, charge Z_1e, and speed v_0 which passes through a stationary Rutherford atom of mass m_2 and charge Z_2e. If this particle does not pass too near the atomic nucleus or any of the electrons, its direction of motion and speed are essentially unchanged by the collision, and further, neither the atomic nucleus nor its surrounding electrons move appreciably during the time of traversal of the atom by the incident particle. In this approximation it is easy to evaluate the magnitude of the transverse impulse $P_\perp$ suffered by the nucleus and each electron. Taking first the nucleus, we may write

$$
\begin{aligned}
P_{\perp n} &= \int_{-\infty}^{\infty} q_1 E_\perp \, dt = q_1 \int_{-\infty}^{\infty} E_\perp \frac{ds}{v_0} \\
&= \frac{q_1}{2\pi\varepsilon_0 b_n v_0} \int_{-\infty}^{\infty} 2\pi b_n \varepsilon_0 E_\perp \, ds = \frac{q_1 q_2}{2\pi\varepsilon_0 b_n v_0} = \frac{Z_1 Z_2 e^2}{2\pi\varepsilon_0 b_n v_0}
\end{aligned}
\tag{1}
$$

where the integral was evaluated by using Gauss's law and b_n is the impact parameter (distance of closest approach) for the nucleus. A similar procedure leads to the corresponding quantity for an electronic encounter:

$$
P_{\perp e} = \frac{Z_1 e^2}{2\pi\varepsilon_0 b_e v_0}
\tag{2}
$$

where b_e is the impact parameter for the electron.

The above expressions provide a means for comparing the relative magnitudes of the effects of the electrons and the nucleus upon the motion of the charged particle. Figure 14-5 shows this diagrammatically; it represents a view of the atom being traversed, as seen along the line of motion of the charged particle, the individual transverse impulses being represented by vectors. It is quite clear that the impulses of the electrons are individually much smaller, and furthermore tend to cancel one another, while the impulse of the nucleus is relatively large (by the factor Z_2) and is not canceled. Thus we conclude that the net transverse impulse upon the passing particle is overwhelmingly due to its interaction

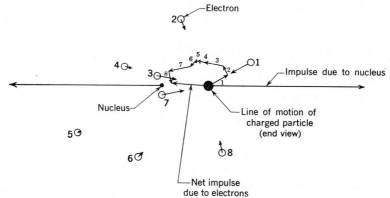

FIG. 14-5. Comparison of the impulse transferred to the electrons of an atom by a fast-moving positive particle with the impulse transferred to the atomic nucleus.

with the nucleus of the atom. It is this impulse, of course, which causes the *scattering* of the incident particle.

When we consider the *kinetic energy* imparted to the nucleus and electrons, on the other hand, the situation is quite different. Each particle carries away an energy approximately equal to $P_\perp^2/2M$, and since the electrons are far less massive, and also much more numerous, than the nucleus, it is the *electrons* which receive the lion's share of the *energy* lost by the charged particle.

EXERCISES

14-18. Justify each step in the derivation of Eq. (1).

14-19. Evaluate the kinetic energy transmitted to the nucleus and to the electrons, and thus verify that nearly all of the energy lost by the incoming particle is transferred to the electrons. Assume $A_n \approx 2Z_n$ and $M_H \approx 2000m_e$. *Ans.:* $E_e/E_n \approx 4000b_n^2(1/b_e^2)_{av}$.

14-20. Show that the average energy lost by the incident particle in traversing a given thickness of matter should be approximately inversely proportional to the square of its speed.

Although the above treatment provides a simple and qualitatively correct basis for evaluating the energy loss and scattering of charged particles, it is desirable to investigate these matters somewhat more fully. Let us therefore examine in greater detail what should happen to the electrons as the charged particle passes through the atom, in order to obtain a more nearly correct value for the energy loss than is given by the approximation used above. The basic factor that we must now introduce is the quantum character of the energy states available to the atom. Thus if the energy that would be imparted to an electron by the passing particle is very small, the electron may not be left with sufficient energy to escape from the atom. Of course the atom may be raised to an excited state even if no electron escapes from it, but this possibility is quite negligible compared to that of outright ionization of the atom. Actually the problem is tremendously complex and has not yet been treated with satisfactory accuracy, but one can still obtain useful results even though many minor effects are neglected. We shall therefore assume that:

1. The energy transferred to an electron is to be evaluated by using the classical Rutherford theory.
2. If this energy is greater than its ionization energy, the electron will actually escape from the atom.
3. If this energy is less than its ionization energy, the electron will remain bound and no energy whatever will be transferred.

To find the average energy loss of the charged particle, we must first find what energy would be transferred to a stationary, free electron for a given impact parameter b.   From Fig. 14-6 we see that the energy given the electron is

$$E_e = \tfrac{1}{2}m_e u_f{}^2 = \tfrac{1}{2}m_e(2V_{CM}\sin\tfrac{1}{2}\theta_c)^2 = 2m_e\left(\frac{m_1 v_0}{m_1 + m_e}\right)^2 \sin^2 \tfrac{1}{2}\theta_c \quad (3)$$

$$= \frac{2m_e m_1{}^2}{(m_1 + m_e)^2}\, v_0{}^2\,\frac{1}{1 + \cot^2 \tfrac{1}{2}\theta_c} = \frac{2m_r{}^2 v_0{}^2}{m_e(1 + v_0{}^4 b^2/K^2)} \quad (4)$$

where we have used Eq. 14-1(12) in the final step.   Next we must evaluate the probability $d\sigma$ that a given energy transfer will take place.   In a

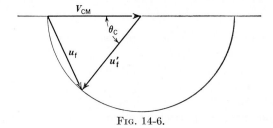

FIG. 14-6.

thickness dx of target material whose atom density is N per unit volume, this is, for the ith electron in an atom $(1 \le i \le Z)$,

$$d\sigma_i = N\, dx 2\pi b_i\, db_i = N\, dx 2\pi b_i \frac{db_i}{dE_e}\, dE_e \quad (5)$$

But from Eq. (4) we find

$$\frac{2b_i\, db_i}{dE_e} = \frac{d(b_i{}^2)}{dE_e} = \frac{2m_r{}^2 K^2}{m_e v_0{}^2 E_e{}^2} \quad (6)$$

So that
$$d\sigma_i = N\, dx \frac{2\pi m_r{}^2 K^2\, dE_e}{m_e v_0{}^2 E_e{}^2} = N\, dx \frac{Z_1{}^2 e^4\, dE_e}{8\pi\varepsilon_0{}^2 v_0{}^2 E_e{}^2 m_e} \quad (7)$$

Finally, to obtain the total average energy transfer in dx, we must multiply $d\sigma_i$ by E_e, integrate over all possible energy transfers for a given electron, and sum over all electrons in an atom for which the maximum possible energy transfer $E_{\max}$ exceeds the ionization energy I_i [see assumption (3) above]

$$dE = \sum_{i=Z'}^{Z} \int_{I_i}^{E_{\max}} E_e\, d\sigma_i = \frac{NZ_1{}^2 e^4}{8\pi\varepsilon_0{}^2 m_e v_0{}^2}\, dx \sum_{i=Z'}^{Z} \ln \frac{E_{\max}}{I_i} \quad (8)$$

(Z' is that value of i above which $E_{\max} > I_i$.)

From the dynamics of the collisions we easily find the maximum possible energy transfer to be (nonrelativistic)

$$E_{\max} = \tfrac{1}{2}m_e(2V_{CM})^2 = \frac{2m_r^2}{m_e}v_0^2 \tag{9}$$

and the minimum energy transfer is taken as I_i, the ionization energy of the ith electron. Again, *electrons for which $I_i > E_{\max}$ are omitted from the summation.* We thus find the average rate of energy loss of a charged particle to be

$$\frac{dE}{dx} = \frac{NZ_1^2e^4}{8\pi\varepsilon_0^2 m_e v_0^2} \sum_{i=Z'}^{Z} \ln \frac{2m_r^2 v_0^2}{m_e I_i} \tag{10}$$

The above equation exhibits the main observed features of the ionization energy loss; they are as follows:

1. The rate of energy loss should be precisely proportional to the square of the charge of the incident particle.
2. The rate of energy loss should be approximately inversely proportional to the square of its speed v_0.
3. The proportionality "constant" should increase slowly with increasing speed of the incident particle, both because the maximum energy transfer increases and because, as a consequence of this, the number of terms in the summation increases.

The observed rate of energy loss of charged particles is in rather good agreement, both qualitatively and quantitatively, with the above formula. Of course, perfect agreement under all conditions is not to be expected, for several important reasons:

1. The actual behavior of atomic electrons whose energy is calculated to be near the ionization energy is not as simple as has been assumed above. There is actually a *smooth transition* of the probability of ejection from zero toward unity as the calculated electron energy grows from zero upward, rather than a *sudden jump* just as the electron energy passes the ionization value, because of the statistical character of quantum-mechanical systems. This effect is quite difficult to calculate for any except the simplest atoms.
2. If the speed of the incident particle is comparable with c, the nonrelativistic treatment used here breaks down. Here there are at least two distinct effects at work; one of them is that the maximum energy transfer, and in fact all dynamical quantities, must be calculated relativistically by using the principles of Chap. 1; and the other is that the Coulomb force law used in the derivation of the Rutherford scattering formula is not correct at high speeds. As we saw in Chap. 1, the electric field surrounding a fast-moving charged particle becomes

"squashed" into the equatorial plane as the speed approaches c, and the field "reaches out" farther in the equatorial direction. This effect leads to an *increase* in the rate of energy loss for very high-energy particles, and the rate should in fact become infinite as $v_0 \to c$.

3. The above-mentioned divergence does not in fact appear experimentally, because of another effect which we have ignored: The nearby atoms become polarized by the passing particle and shield the more distant atoms from its field, reducing the energy loss. This

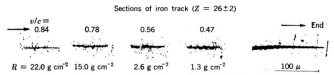

Sections of iron track ($Z = 26 \pm 2$)

$v/c =$
0.84   0.78   0.56   0.47   End

$R = 22.0$ g cm^{-2} 15.0 g cm^{-2} 2.6 g cm^{-2} 1.3 g cm^{-2} 100 μ

FIG. 14-7. Illustration of the reduction of ionizing power of a multiply charged particle as it nears the end of its range. This is caused by the fact that the incident particle picks up electrons from the medium, thus reducing the effective charge of the moving particle. (Courtesy of Prof. M. F. Kaplon, University of Rochester.)

polarization effect is also responsible for the phenomenon of *Cerenkov radiation*, which arises if a charged particle travels through a medium faster than the speed of light in the medium.

4. As the incident particle slows down, it will eventually pick up and carry along some of the less tightly bound electrons of the resisting medium. This reduces the effective charge of the particle and therefore its energy-loss rate. This effect is most important for particles of high atomic number and is quite easily seen in the tracks left by heavy cosmic-ray particles in photographic emulsions (Fig. 14-7).

All in all, the ionization loss is a tremendously complex phenomenon, and for the most precise work it has proved necessary to resort to an experimental measurement for each stopping material of interest, using particles, usually protons, of accurately known energy. The curves shown in Fig. 14-8 have been obtained in this way and are believed to be accurate within a few per cent. The curves show essentially the features that are expected on the basis of the preceding discussion. One very important feature which has been discussed only qualitatively above is that *the energy-loss rate of any particle goes through a broad minimum at a speed somewhat below that of light, and for higher speeds rises only slightly above this minimum value.* In terms of the energy or momentum of the incoming particle, this minimum occurs at a kinetic energy or momentum of about M_0c^2 or M_0c, respectively, and for higher energy or momentum the loss rate approaches a value called the *plateau value*. This behavior provides the basis for the measurement of the charge of a fast-moving particle, for the Z_1^2 dependence far overshadows any small

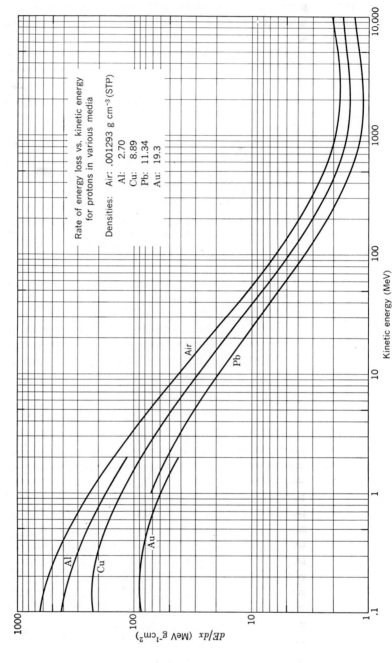

Fig. 14-8. Energy-loss rate of protons traversing air, aluminum, copper, lead, and gold. (Plotted from data given in "American Institute of Physics Handbook," McGraw-Hill Book Company, Inc., New York, 1957.)

uncertainty in the factor by which the ionization increases from the minimum to the plateau. Thus a fast proton or other singly charged particle should have an energy-loss rate equal to the minimum value I_0; a fast α-particle, a rate near $4I_0$; a fast lithium nucleus, near $9I_0$; etc.

As a result of the above-described property, most of the tracks observed in a cloud chamber, bubble chamber, or nuclear emulsion will appear to have the same density of ions; this energy-loss rate is called *minimum ionization*, and it is used as a convenient reference value for the measurement of the energy-loss rate of heavily ionizing particles whose tracks also appear in the chamber or emulsion.

14-4. Primary and Secondary Ionization; Knock-on Electrons; Delta Rays

We have so far considered only the direct effects of the incident particle, namely, the ejection of electrons from the atoms of the stopping material through which the particle travels. But there are also various secondary effects that must sometimes be considered. One of the most important of these is the further ionization of the stopping material by the more energetic of the ejected electrons. The directly ejected electrons and the ionized atoms they leave are referred to as *primary* ions. If one of these primary electrons possesses sufficient energy, it may subsequently eject further electrons from nearby atoms before being brought to rest. These latter electrons and ionized atoms are called *secondary* ions. The energetic electron is called a *knock-on electron*, and if its track in a cloud chamber, photographic emulsion, or bubble chamber is relatively short but still visible, the track is called a *delta ray*. Delta rays are most prominently visible along the track of a particle which is moving at a speed of a few tenths c, because in this speed range the density of primary ions along the track is large, and their energy is high. Many delta rays can be seen along the track shown in Fig. 14-7.

For most materials, the energy removed from the incident particle per *primary*-ion pair is about 30 eV, and the average number of secondary-ion pairs made by each primary electron is between two and three.

14-5. Range; Straggling

In many experiments one wishes to know *how far* a given particle having given energy will penetrate into a given absorbing material, or how much of its energy will be lost in traversing a given thickness of the absorber. To answer such questions, one needs to know the so-called *range-energy relation* for the given particle and absorbing material. An approximate form for this function can be found by integrating Eq. 14-3(10): ignoring the slowly varying logarithmic terms, one finds that the range (penetration distance) should be approximately

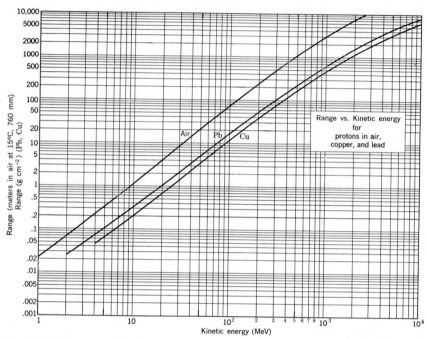

FIG. 14-9. Range vs. kinetic energy for protons in air, copper, and lead. (Drawn from data given in "American Institute of Physics Handbook," McGraw-Hill Book Company, Inc., New York, 1957. For more extensive graphs, see Appendix F.)

$$R \approx \frac{\text{const.}}{Z_1^2 m_1} E^2 \tag{1}$$

The curves shown in Fig. 14-9 were obtained experimentally for protons; they may be extended to other particles by using the scaling relation

$$R(M',Z',E') = \frac{M'Z^2}{MZ'^2} R\left(M,Z,\frac{M}{M'} E'\right) \tag{2}$$

It will be seen from these curves that, over an appreciable span of energies, the range is approximately proportional to the square of the energy, as suggested by Eq. (1). At very high (relativistic) energies, the range becomes proportional to the energy, as it should since, in this region, dE/dx is essentially constant. In these curves the range is expressed in units of grams per square centimeter. Except for relatively subtle effects such as polarization, the range is dependent only upon the number of atoms of the stopping material that have been traversed, and not upon the density with which these atoms are packed. Thus the quantity $N\,dx$ which first appears in Eq. 14-3(5) may be transformed into grams per square centimeter units by using the relations

$$N = N_0 (\rho/A) \qquad \text{and} \qquad \rho \, dx = dR$$

where N_0 is Avogadro's number, ρ the density, A the atomic weight, all in cgs units, and R the range in grams per square centimeter. Thus

$$N \, dx = (N_0/A) \, dR \tag{3}$$

When R is expressed in grams per square centimeter units, it is observed that the range-energy relation is *roughly independent of the stopping material*. In these units the energy-loss rate corresponding to minimum ionization for singly charged particles varies from about 2 MeV g^{-1} cm^2 for the light elements down to about 1 MeV g^{-1} cm^2 for the heaviest elements. These numbers are quite useful for making rough calculations of energy loss.

When the ranges of particles of a given kind, known to have equal energy, are measured, it is found that the ranges are not exactly equal but are distributed over a small interval. This phenomenon is called *range straggling* and is traceable to several effects:

1. In computing the range from the energy-loss curve, it is assumed that the particle follows a straight path. Actually, of course, the particle is scattered by its encounters with the atomic nuclei and therefore follows a zigzag path which carries it only more or less along a fixed direction. Clearly, the measured range will be less than the actual range because of this effect.

2. While the computed or the measured energy-loss curves represent accurately the *average* behavior of the incident particle, there are finite *fluctuations* about this average because of the statistical nature of the scattering process. These fluctuations are of course the more violent, the fewer the encounters, and can be quite important if an accurate measurement of ΔE is to be made over a small interval ΔR.

3. Density inhomogeneities in the absorbing material can lead to straggling if different particles traverse the absorber in different places.

4. As a multiply charged particle slows down it picks up electrons from the absorber, as previously described. This process is of course also a statistical one, and the particle will gain or lose electrons in a more or less random manner depending upon the history of the particular particle. This fluctuation in effective charge in turn leads to fluctuations in energy-loss rate and therefore in range. Even a singly charged particle can pick up electrons near the end of its range, but this is important only for very low-energy particles which come to rest in a gaseous medium.

The resulting distribution of the observed range is quite complicated but can be determined approximately for many cases of practical interest:

In work involving particles which actually stop, it is often assumed that the differential range distribution is Gaussian; while in work with fast, lightly ionizing particles, the energy loss in a given small layer is assumed to be distributed according to a distribution which is obtained by summing the direct energy loss of the particle and that of its accompanying knock-on electrons in the layer, using a Poisson-like distribution to describe both the knock-on population and the fluctuations in direct energy loss in the layer.[1]

EXERCISES

14-21. Using the fact that dE/dx in a given material is accurately of the form $Z_1^2 f(v_0)$, derive the scaling relation (2).

14-22. A proton whose path has a radius of curvature of 250 cm in a magnetic field of 10^4 gauss traverses a lead plate whose thickness is 40 g cm^{-2} along the path of the proton. What should the radius of curvature be after it emerges from the plate? *Ans.:* $\sim$200 cm.

14-23. A cosmic-ray particle whose momentum is 380 MeV/c is observed to produce an ionization density of approximately 2.0 times minimum in the argon gas of a cloud chamber. What is the mass of the particle? *Ans.:* $M \approx 950 m_e$.

14-6. Some General Properties of Particle Scattering

Much of the information that has been obtained about nuclear forces and nuclear structure has come from scattering experiments analogous to those pioneered by Rutherford. However, the forces which hold nuclei together are quite different from the inverse-square law of electrostatics, for nuclear forces are both immensely stronger and of immensely shorter range than electrical forces. Furthermore, while a classical treatment of the problem of Coulomb scattering happens to lead to the correct angular distribution for the scattered particles, this is not commonly the case, and therefore one must generally make use of quantum mechanics in order to interpret scattering data correctly. We shall now briefly consider some of the more elementary aspects of one quantum-mechanical treatment of particle scattering.

In Chap. 4 we considered several one-dimensional examples of the reflection and transmission of particles at potential-energy discontinuities. In that treatment certain characteristic features of particle scattering were apparent, such as the backward scattering, or reflection, of particles whose energies are great enough classically to surmount a

---

[1] For an extensive discussion of energy loss, scattering, and straggling, see Allison and Warshaw, *Revs. Modern Phys.*, **25**, 779 (1953).

barrier and the resonance transmission of particles whose energies bear just the right relationship to the barrier height and width. These same phenomena are found in somewhat modified form in the three-dimensional case also.

In one dimension, there are but two possible alternatives for the scattered particles: to be transmitted or reflected. The entire result of the scattering from a given barrier or well, for particles of given energy, is thus that the particles are *reflected* with a certain amplitude and phase, or are *transmitted* with a certain amplitude and phase, with respect to the amplitude and phase of the incident particles. Given the "shape" of the barrier or well, one can predict quantitatively what the amplitudes and phases of reflected and transmitted beams should be for given particle energy. However, we are actually faced with the reverse problem: *given the scattering properties,* i.e., amplitudes and phases of the reflected and transmitted waves as a function of energy for a beam of particles imping-ing upon a scattering center, *to find the shape of the barrier or well.*

In three dimensions there are of course more alternatives for the incident particles: They may not be scattered at all but instead proceed with unchanged amplitude and phase; they may, on the other hand, be scattered *in any direction* with amplitudes and phases which depend upon the shape of the scattering potential and upon the energy.

We shall now examine certain aspects of the scattering of relatively low-energy particles by a *spherically symmetric* force field, using a simpli-fied version of the so-called *method of partial waves.* Let us consider the scattering of particles of mass m by particles of mass M, and suppose that these particles interact via a spherically symmetric force field whose potential function is $V(r)$, where r is the distance between the particles. By the procedure outlined in Chap. 5, this problem may easily be sep-arated into one involving the motion of the center of mass, which moves like a free particle of mass $m + M$, and one involving the relative motion of the two particles, which is equivalent to the motion of a particle of mass m_r moving in a *fixed* potential field $V(r)$. (m_r is the reduced mass.) We shall consider only the latter problem; the transformation to the lab system is easily made by using the results of Exercises 14-7 and 14-8.

The Schroedinger equation for the relative motion can be separated by using the same procedure as in Chap. 5, and the resulting wave functions can be expressed in the form

$$\psi(r,\theta,\varphi) = \sum_{l,m} A_{lm} R_l(r) P_l^m(\cos\theta) e^{im\varphi} \qquad (1)$$

in which $R_l(r)$ is a solution of the radial equation

$$\frac{1}{r^2}\frac{d}{dr}\left(r^2\frac{dR_l}{dr}\right) + \frac{2m}{\hbar^2}\left[E - V(r) - \frac{\hbar^2 l(l+1)}{2m_r r^2}\right]R_l = 0 \qquad (2)$$

where E is one of the continuum of possible energies of the unbound system and is to be regarded as *precisely specified in advance*.

By this procedure one could evidently evaluate the wave function for any given values of l and m, if the potential $V(r)$ were known. But what we actually must do is try to determine $V(r)$ from scattering data by using particles whose angular momentum is *not precisely specified in advance*. Indeed, in an actual experiment almost *all possible* angular momenta are present, since it is not feasible experimentally to restrict the paths of the incident particles to atomic dimensions, let alone to nuclear ones.

Guiding our procedure by the experimental situation, therefore, we shall seek wave functions of the form (1) which describe, at points greatly distant[1] from the scattering center, a superposition of a *plane wave* moving in the positive z-direction, representing both the incoming and nonscattered outgoing particle, and a spherical wave, representing the scattered particles:

$$\psi(r,\theta,\varphi) \approx e^{ikz} + \frac{1}{r} f(\theta) e^{ikr} \tag{3}$$

where $k = 2\pi/\lambda = p/\hbar = (2m_r E/\hbar^2)^{1/2}$, and the choice of the z-axis as the direction of the incident beam assures that the scattering will be independent of φ. Now we can measure $|f(\theta)|^2$ experimentally, since this quantity is proportional to the number of particles scattered into a unit solid angle at an angle θ away from the beam.[2] Thus if we express $f(\theta)$ also in terms of the *properties of the scattering potential*, we will have a connection between experimental measurements and the "shape" of the potential function. What we must now do, therefore, is express the wave function (3) in terms of the solutions $R_l(r)$ of Eq. (2) in a form

$$\psi(r,\theta,\varphi) = \sum_l R_l(r) P_l(\cos \theta) \tag{4}$$

which is to be valid in the region far removed from the origin.

In the *absence* of a scattering potential, $R_l(r)$ satisfies the equation

$$\frac{d^2 R_{l0}}{dr^2} + \frac{2}{r} \frac{dR_{l0}}{dr} + \frac{2m}{\hbar^2} \left[E - \frac{\hbar^2 l(l+1)}{2m_r r^2} \right] R_{l0} = 0 \tag{5}$$

---

[1] By "points greatly distant from the scattering center" we mean many atomic diameters. At such distances the incoming and the scattered beams are both present.

[2] For experimental measurement of $|f(\theta)|^2$ the particle detector is ordinarily placed at least several centimeters distant from the scatterer so it will be *outside* the incoming beam. If one were to carry out measurements within the incident beam, there would of course appear *interference terms* between the incoming and scattered particles.

or, changing to the dimensionless variable $\rho = kr$, the equation

$$R''_{l0} + \frac{2}{\rho} R'_{l0} + \left[1 - \frac{l(l+1)}{\rho^2} \right] R_{l0} = 0 \tag{6}$$

The solution of this equation is a so-called *spherical Bessel function:*

$$R_{l0}(\rho) = \left(\frac{\pi}{2\rho} \right)^{1/2} J_{\pm(l+1/2)}(\rho) \tag{7}$$

where $J_{\pm(l+1/2)}(\rho)$ is an ordinary Bessel function of order $\pm(l + \frac{1}{2})$. A few of these functions are

$$
\begin{array}{ll}
(a)\ \ R_{00} = \dfrac{\sin kr}{kr} & S_{00} = -\dfrac{\cos kr}{kr} \\[2mm]
(b)\ \ R_{10} = \dfrac{\sin kr}{k^2 r^2} - \dfrac{\cos kr}{kr} & S_{10} = -\dfrac{\cos kr}{k^2 r^2} - \dfrac{\sin kr}{kr} \\[2mm]
(c)\ \ R_{20} = \left(\dfrac{3}{k^3 r^3} - \dfrac{1}{kr} \right) \sin kr & S_{20} = -\left(\dfrac{3}{k^3 r^3} - \dfrac{1}{kr} \right) \cos kr \\[4mm]
\qquad\quad - \dfrac{3}{k^2 r^2} \cos kr & \qquad\quad - \dfrac{3}{k^2 r^2} \sin kr
\end{array}
\tag{8}
$$

We can use only R_{l0}, which is finite for all values of ρ. In the region distant from the origin these functions have the form

$$R_{l0} \approx \frac{\sin (kr - \frac{1}{2}l\pi)}{kr} \qquad kr \gg 1 \tag{9}$$

In the *presence* of the scattering potential but in the region far outside its range, we shall therefore use as an asymptotic solution

$$R_l \approx A_l \frac{\sin (kr - \frac{1}{2}l\pi + \delta_l)}{kr} \tag{10}$$

This differs from Eq. (9) only by the inclusion of an additional *phase shift* δ_l which arises from the scattering potential.

In order to evaluate $f(\theta)$, we must also expand the incident-particle wave function as a series of Legendre functions:

$$e^{ikz} = e^{ikr \cos \theta} = \sum_l B_l(r) P_l(\cos \theta) \tag{11}$$

We evaluate B_l in the usual manner by multiplying both sides of Eq. (11) by $P_s(\cos \theta) \sin \theta \, d\theta$ and integrating from 0 to π. In terms of the variable $\mu = \cos \theta$ we thus find, using the orthogonality of the Legendre functions,

$$B_s(r) = \frac{2s+1}{2} \int_{-1}^{+1} P_s(\mu) e^{ikr\mu} \, d\mu \tag{12}$$

where $\mu = \cos \theta$. The first three such coefficients are

$$(a) \qquad B_0(r) = \tfrac{1}{2} \int_{-1}^{+1} e^{ikr\mu} \, d\mu = \frac{\sin kr}{kr}$$

$$(b) \qquad B_1(r) = \tfrac{3}{2} \int_{-1}^{+1} \mu e^{ikr\mu} \, d\mu = \frac{3}{ikr} \cos kr - \frac{3}{ik^2r^2} \sin kr \qquad (13)$$

$$(c) \qquad B_2(r) = \tfrac{5}{2} \int_{-1}^{+1} \tfrac{1}{2}(3\mu^2 - 1)e^{ikr\mu} \, d\mu$$

$$= \left(\frac{5}{kr} - \frac{15}{k^3r^3}\right) \sin kr + \frac{15 \cos kr}{k^2r^2}$$

We are now ready to evaluate $f(\theta)$. This we do by equating Eqs. (3) and (4), using the forms (10) and (11) for $R_l(r)$ and $e^{ikr\cos\theta}$. Since we are matching solutions in the range $kr \gg 1$, we neglect in $B_l(r)$ all terms involving higher inverse powers of r than the first, with the result

$$f(\theta)e^{ikr} = \sum_l \frac{1}{k}[A_l \sin (kr - \tfrac{1}{2}l\pi + \delta_l)$$

$$- (2l + 1)i^l \sin (kr - \tfrac{1}{2}l\pi)]P_l(\cos \theta) \quad (14)$$

If we now make use of the fact that the coefficients of e^{ikr} and e^{-ikr} on the two sides of the equation must separately be equal, we obtain the two equations

$$(a) \qquad f(\theta) = \sum_l \frac{1}{2ik}[A_l e^{-\frac{1}{2}il\pi+i\delta_l} - (2l + 1)i^l e^{-\frac{1}{2}il\pi}]P_l(\cos \theta)$$

$$(b) \qquad 0 = \sum_l \frac{1}{2ik}[A_l e^{-\frac{1}{2}il\pi-i\delta_l} - (2l + 1)i^l e^{\frac{1}{2}il\pi}]P_l(\cos \theta) \qquad (15)$$

Equation (15b) may be solved for A_l, and this coefficient then eliminated in Eq. (15a), with the result

$$f(\theta) = \frac{1}{2ik} \sum_l (2l + 1)(e^{2i\delta_l} - 1)P_l(\cos \theta) \qquad (16)$$

This is the basic result of the method of partial waves. According to this equation the scattering effect caused by a given scattering potential may be regarded as a *summation of the effects of all possible angular momenta of the incident particle*, the wave train associated with a particle of given angular momentum being scattered with a certain amplitude and its phase shifted by an amount δ_l with respect to the incident wave train. The phase shifts are uniquely determined by the scattering potential, and presumably, vice versa. The experimental problem is therefore to obtain sufficiently precise scattering data to permit an accurate evaluation of the phase shifts.

The quantity actually measured experimentally is not $f(\theta)$ itself but

$|f(\theta)|^2$, the probability that a particle will be found moving in the direction θ per unit solid angle per incident particle per unit area. This is just the definition of the differential cross section $d\sigma/d\Omega$ for the scattering process, so that we have

$$\frac{d\sigma}{d\Omega} = |f(\theta)|^2 = \frac{1}{k^2} \left| \sum_l (2l + 1)e^{i\delta_l} \sin \delta_l P_l(\cos \theta) \right|^2 \tag{17}$$

$$= \frac{1}{k^2} \sum_l \sum_{l'} (2l + 1)(2l' + 1)e^{i(\delta_l - \delta_{l'})} \sin \delta_l \sin \delta_{l'} P_l(\cos \theta) P_{l'}(\cos \theta)$$

$$\tag{18}$$

The total cross section is obtained by integrating Eq. (18) over all solid angles:

$$\sigma = 2\pi \int_0^\pi \frac{d\sigma}{d\Omega} \sin \theta \, d\theta = \frac{4\pi}{k^2} \sum_l (2l + 1) \sin^2 \delta_l \tag{19}$$

In practice, the above analysis is best applied to cases in which the scattering potential is of quite limited range, outside which it can be neglected. The phase shifts which characterize such potentials show a dependence upon energy such that, at the lowest energies, only δ_0 is of significant size and all others are negligible; at somewhat higher energies, δ_0 and δ_1 are of comparable size but all others are still negligible; and for higher and higher energies, the phase shifts associated with larger and larger l values become important. This behavior corresponds to the classical result that a particle of momentum $p = (2mE)^{1/2}$ which passes within given distance b of the origin cannot possess an angular momentum greater than pb; particles having angular momenta greater than this would pass outside the radius a and would not be scattered by a potential limited to this range. In the partial wave analysis given above, it is found that the limiting l value above which the phase shifts rapidly become negligible is approximately

$$l = \frac{b \sqrt{2mE}}{\hbar} = bk \tag{20}$$

According to the above results, therefore, the scattering of very low-energy particles by a potential of sharply limited range should be *isotropic*, since only the $P_0(\cos \theta)$ term would enter into Eq. (16) or Eq. (18). Since only particles having zero angular momentum would be scattered, this case is called "pure S-wave scattering." Other l values for the scattered particles lead to P-wave scattering ($l = 1$), D-wave scattering ($l = 2$), and so on. Equation (18) shows that, if more than one l-value enters into the scattering, there are *interference effects* between the various l-values.

Although the behavior just described greatly simplifies the evaluation of phase shifts from experimental data, it of course introduces the drawback that one also cannot obtain very much information concerning the "shape" of the potential function from such limited data. Ultimately, the values of all phase shifts at all energies would be needed to specify the potential completely, and if taken literally, this would be experimentally unattainable. Our state of knowledge concerning nuclear forces is as yet sufficiently incomplete that even such restricted information as is provided by a phase-shift analysis can be of considerable use, and considerable effort has been directed toward the evaluation of phase shifts in several nuclear scattering experiments.

Before leaving the subject of partial waves and phase shifts, it is of some interest to observe that a simple explanation of the well-known Ramsauer-Townsend effect is provided by the method of partial waves. In this phenomenon, electrons of controlled energy which collide with atoms of a heavy rare gas are observed to be only very weakly scattered if their energy is equal to some certain value, but to be strongly scattered both at lower and at higher energies. According to the method of partial waves, one would expect such a result if the scattering potential were such as to give a large phase shift δ_0 at low energy where higher l-values have not yet become important. In particular, if δ_0 passes through 180° and all other phase shifts are negligible, we see from Eq. (19) that the total scattering cross section would vanish. This is just what happens in the cases where the phenomenon is observed.

EXERCISES

14-24. Verify that wave functions having the form (7) satisfy Eq. (6) if $J_{\pm(l+\frac{1}{2})}(\rho)$ satisfies the ordinary Bessel differential equation

$$J'' + \frac{1}{\rho} J' + \left[1 - \frac{(l + \frac{1}{2})^2}{\rho^2} \right] J = 0$$

14-25. Verify that Eq. (8a) satisfies Eq. (5).

14-26. Carry through the steps leading to Eq. (12).

14-27. By writing $P_s(\mu) = \sum_{r=0}^{s} a_r \mu^r$ and integrating once by parts, or otherwise, show that the leading term (lowest inverse power of r) in $B_s(r)$ is

$$B_s(r) = \frac{2s + 1}{2ikr} [P_s(1)e^{ikr} - P_s(-1)e^{-ikr}]$$

$$= \frac{2s + 1}{kr} \sin kr \qquad s \text{ even}$$

$$= \frac{2s + 1}{ikr} \cos kr \qquad s \text{ odd}$$

14-28. Supply the missing steps in deriving Eq. (16) from Eq. (14).

14-29. Derive Eq. (19) from Eq. (18).

14-30. Evaluate the phase shifts δ_0 and δ_1 as a function of energy for particles scattered from a fixed, impenetrable sphere of radius a. *Hint:* Since $r = 0$ is excluded, the wave function may contain both R_{l0} and S_{l0} functions. Also, the wave function must be zero at $r = a$. *Ans.:* $\delta_0 = ka$, $\delta_1 = ka - \tan^{-1} ka$.

14-31. Calculate the differential scattering cross section for the above case for $ka \ll 1$ and $ka = 0.5$, assuming that only δ_0 and δ_1 are important in this energy range. *Ans.:* $d\sigma/d\Omega = a^2$, $d\sigma/d\Omega = a^2(0.920 + 0.124 \cos \theta + 0.053 \cos^2 \theta)$.

14-32. Calculate the total scattering cross section for the system of Exercise 14-30 in terms of the geometrical cross section $\sigma_g = \pi a^2$, for $ka \ll 1$, $ka = 0.5$, and $ka = 1.0$, assuming as above that δ_0 and δ_1 are the only important phase shifts. *Ans.:* $\sigma = 4\sigma_g$, $\sigma = 3.73\sigma_g$, and $\sigma = 3.35\sigma_g$.

The last three exercises illustrate some important features of scattering by potentials of finite range:

1. The phase shifts due to a purely repulsive potential are all positive—the wave trains are "pushed out" by such a potential. Conversely, the phase shifts of a purely attractive potential are all negative—the wave trains are "pulled in" in this case.

2. The scattering is isotropic for the lowest energies; the first anisotropy that enters is proportional to $\cos \theta$, with amplitude approximately $2\delta_1/\delta_0$ relative to the amplitude of the isotropic term. This term appears as a forward-backward anisotropy in the CM system.

3. The total scattering cross section at the lowest energies is 4 times the so-called geometrical cross section, $\sigma_g = \pi a^2$, of the scatterer. Such a result is of course not possible classically, since only those incident particles striking within a circle of area πa^2 could possibly be scattered. However, this result is exactly what is observed in the optical case of scattering of light waves or radio waves by scatterers much smaller than a wavelength in size.

At higher energies the scattering cross section approaches $2\pi a^2$, which corresponds to the optical case of scattering by objects much *larger* than a wavelength in size. Thus even in this range the scattering cross section is greater than σ_g.

REFERENCES

Mott, N. F., and H. S. W. Massey: "The Theory of Atomic Collisions," Oxford University Press, New York, 1933.

Schiff, L. I.: "Quantum Mechanics," 2d ed., International Series in Pure and Applied Physics, McGraw-Hill Book Company, Inc., New York, 1955.

15

Radioactivity

No nuclear property is more characteristic than that of *radioactivity*, the spontaneous decomposition of a nucleus into lighter fragments. This phenomenon has been of tremendous utility in the study of nuclear structure, and in recent years it has also played an increasingly important role in many other fields of science not directly related to nuclear physics. Numerous techniques involving radioactivity have been applied to a range of physical and chemical problems which covers virtually all fields of scientific endeavor from anthropology to cosmology. Because of the importance of radioactivity both as a phenomenon and as a research tool, we shall treat in the present chapter some of the more important phenomenological properties of natural and artificially induced radioactivity, and also some of the elementary theoretical aspects of α- and β-decay.

15-1. The Four Radioactive Series

As was described in Chap. 13, one finds in nature three types of radioactivity, called α-, β-, and γ-activity, which are characterized as follows:

1. In an α-decay, a helium nucleus is emitted from the radioactive nuclide, leaving the latter with two units less charge and four units less mass:

$$(Z,A) \to (Z - 2, A - 4) + {}^{4}_{2}\text{He} \tag{1}$$

2. In a β-decay, a negative electron is emitted, leaving the nucleus with one unit *more* charge and the same mass number:

$$(Z,A) \to (Z + 1, A) + \beta^{-} \tag{2}$$

3. In a γ-decay, an electromagnetic quantum is emitted, leaving the charge and mass number of the nucleus unchanged:

$$(Z,A)^* \rightarrow (Z,A) + h\nu \tag{3}$$

It was also mentioned in Chap. 13 that naturally occurring radioelements are found to form long series, the members of a series each being the "daughter" of some "parent" nuclide. The presence of γ-*rays* in radioactive series results from the fact that an atom is not necessarily "born" in its lowest energy state when it is formed in the process of decay of its immediate progenitor in a series, but can be formed in an *excited state* from which it decays by γ-ray emission. The γ-rays from radioactive decay are analogous to the optical and X-ray line spectra connecting the various electronic excited states of atoms.

From the relationships listed above, we see that the mass number and the atomic number of a given member of a radioactive series are related to those of an ancestor by the equations

$$A = A_0 - 4N_\alpha \tag{4}$$
$$Z = Z_0 - 2N_\alpha + N_\beta \tag{5}$$

where N_α is the number of α-particles and N_β the number of β particles that are emitted in arriving at the given member.

Equations (4) and (5) suggest that there might exist *four different series* of radioactive elements, each characterized by a different value of m in the following expression for the mass numbers of its members:

$$A = 4n + m \tag{6}$$

These are called the $4n$, $4n + 1$, $4n + 2$, and $4n + 3$ series. Only three of these four series are found in nature; the fourth series, corresponding to $4n + 1$, has been found and studied only rather recently in connection with the artificial production of the transuranic elements. Some of the properties of the four radioactive series are indicated in Table 15-1 and in Figs. 15-1 to 15-4.

15-2.   The Growth and Decay of Radioactivity

In the study of the chemical relationships between the various radioactive elements which are found mixed together in a given ore deposit, it is found that each member of a radioactive family is characterized by a certain *half-life* $T_{1/2}$ which governs the rate of its disappearance after it is isolated from the other members of the family. This disappearance is governed by the so-called *radioactive-decay law:*

$$N(t) = N_0 2^{-t/T_{1/2}} \tag{1}$$

in which N is the number of atoms still present at time t and N_0 the initial number at time $t = 0$.   Experimentally it is observed that the rate of disintegration of a given type of radioactive atom is quite independent of the physical or chemical state of the atom, and is in fact basically an intrinsic property of the nucleus itself.

Schweidler (1905) advanced the hypothesis, now known to be quite correct, that radioactive-decay processes are *statistical* in character, each radioactive atom of given type that is "alive" at time t having the same

TABLE 15-1. GROSS PROPERTIES OF THE FOUR RADIOACTIVE SERIES

| Series | Name | Parent | | Stable end product |
		Symbol	Half-life	
$4n$	Thorium	^{232}Th	1.39×10^{10} y	^{208}Pb
$4n + 1$	Neptunium	^{237}Np	2.25×10^{6} y	^{209}Bi
$4n + 2$	Uranium	^{238}U	4.51×10^{9} y	^{206}Pb
$4n + 3$	Actinium	^{235}U	7.07×10^{8} y	^{207}Pb

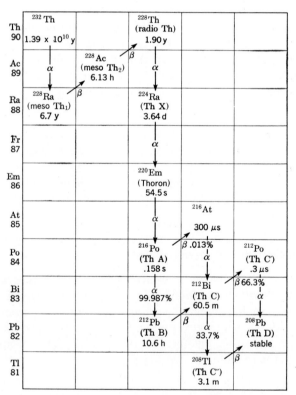

FIG. 15-1. The thorium ($4n$) radioactive series.

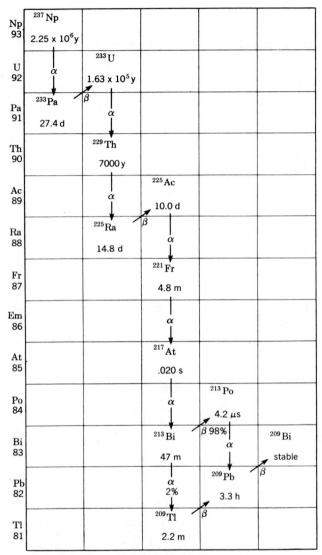

FIG. 15-2. The neptunium $(4n + 1)$ radioactive series.

probability of decay in a short time interval dt. Those atoms which do not then decay have again the same probability of decaying in the next interval dt, and so on. In terms of the *disintegration constant* λ, defined as the fraction of the atoms present that decay per unit time, we may write

$$\frac{dN}{N} = -\lambda \, dt \qquad (2)$$

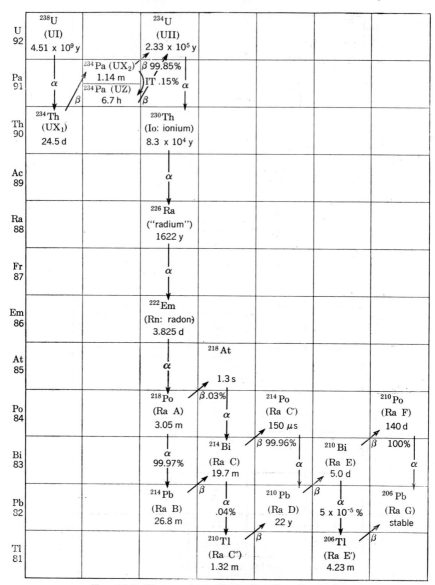

FIG. 15-3. The uranium $(4n + 2)$ radioactive series.

This equation may be readily integrated to yield the familiar exponential decay law

$$N = N_0 e^{-\lambda t} \tag{3}$$

Comparing with Eq. (1), we see that the disintegration constant is equal

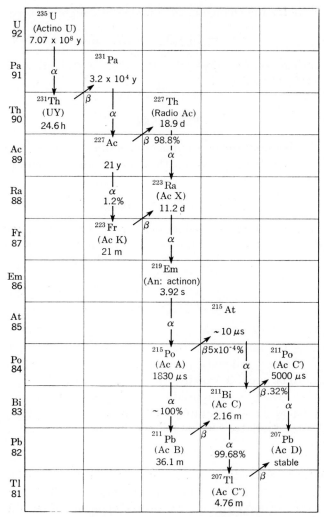

FIG. 15-4. The actinium $(4n + 3)$ radioactive series.

to

$$\lambda = \frac{\ln 2}{T_{1/2}} = \frac{0.693}{T_{1/2}} \tag{4}$$

In many cases the situation is made more complicated by the fact that a given radioelement may be one of a series, its immediate parent having been formed by the disintegration of some other radioelement. The time variation of the amount of a given nuclide is not necessarily governed by the above equations in such a case. If we designate the number of parent atoms by N_p and the number of daughter atoms by N_d, we may evidently write for the change in the number of daughter atoms

$$dN_d = (+\lambda_p N_p - \lambda_d N_d)\, dt \qquad (5)$$

The following exercises deal with this and several other important aspects of radioactive decay.

EXERCISES

15-1. Find the *mean life* of a radioactive atom.   *Ans.:* $\tau_m = 1/\lambda$.   (6)

15-2. Find the relation between the half-life and the mean life of a radioactive atom.

Ans.: $$T_{\frac{1}{2}} = \tau_m \ln 2 = 0.693\tau_m \qquad (7)$$

15-3. If a given atom can undergo either of two kinds of radioactive decay, with disintegration constants λ_1 and λ_2 for the two modes, show that

$$N(t) = N_0 e^{-(\lambda_1+\lambda_2)t} \qquad (8)$$

and that $$\tau_m = \frac{1}{\lambda_1 + \lambda_2} \qquad (9)$$

15-4. The *activity R* of a radioactive sample is defined as the number of disintegrations per second from the sample. Two units of activity are in use: One of these, redefined from an older usage, is the *curie*, which is now taken to be equal to exactly 3.7×10^{10} disintegrations per second. A smaller unit, which is appropriate to weaker activities, is the Rutherford (Rd), equal to 10^6 disintegrations per second. Show that the activity of a pure isolated nuclide follows the law

$$R = \lambda N = R_0 e^{-\lambda t} \qquad (10)$$

15-5. A given radioactive sample may contain a chemically inseparable mixture of isotopes of a given element having different half-lives. Show that the ratio f of activities of any two components of such a mixture varies with time according to the law

$$f = f_0 e^{-(\lambda_2-\lambda_1)t} \qquad (11)$$

so that after a sufficiently long time *only the longest-lived component will be present in appreciable concentration.* This often simplifies the analysis of the composition of such a mixture, since the effect of the longest-lived component can first be determined and subtracted from the total activity, and the remaining activity can then be treated similarly to find the next-longest-lived component, etc.

15-6. A given radioactive sample may be initially pure, but its successive decay products may themselves be radioactive with various half-

lives. The amount of a given product present in the sample will therefore tend to grow because of the decay of its immediate progenitor, and
will tend to decay because of its own instability, as expressed by Eq. (5).
Write the set of differential equations which describe this situation, and
show that a stepwise solution of these equations is possible, leading to the
equations

(a) $\quad N_1(t) = N_0 e^{-\lambda_1 t}$

(b) $\quad N_2(t) = \dfrac{N_0 \lambda_1}{(\lambda_2 - \lambda_1)} (e^{-\lambda_1 t} - e^{-\lambda_2 t})$ $\hspace{3cm}$ (12)

(c) $\quad N_3(t) = N_0 \lambda_1 \lambda_2 \left[\dfrac{e^{-\lambda_1 t}}{(\lambda_2 - \lambda_1)(\lambda_3 - \lambda_1)} + \dfrac{e^{-\lambda_2 t}}{(\lambda_1 - \lambda_2)(\lambda_3 - \lambda_2)} \right.$

$\left. \hspace{5cm} + \dfrac{e^{-\lambda_3 t}}{(\lambda_1 - \lambda_3)(\lambda_2 - \lambda_3)} \right]$

15-7. If the initial member of a given series of radioactive atoms is
formed at a *constant rate*, the various members of the series will eventually
attain *equilibrium* concentrations in the sample, dependent upon their
respective mean lives, such that all members have equal activity. This
kind of equilibrium is called *secular equilibrium*, and it is approximately
satisfied for the natural radioactive series, whose initial members are so
much longer-lived than any of the decay products that they can be
considered to be decaying at a constant rate.

If the rate of production of a certain member of a radioactive series
is not constant but changes very little during the lifetime of any of the
subsequent members, a condition called *transient equilibrium* is eventually
attained. This situation also implies an equal activity for all members
of the series at any given time, but this activity varies slowly with time,
in step with the rate of production of the initial member.

Show that the condition of secular equilibrium implies *equal activity*
for all members of a nonbranching radioactive series, and thus find how
many kilograms of ^{226}Ra should be present per kilogram of ^{238}U in a given
ore. *Ans.:* $W_{\text{Ra}} = 0.342 \times 10^{-6} W_{\text{U}}$.

15-8. (a) A 100-g pulverized sample of thorium-bearing rock is placed
around a geiger counter which is sensitive to β-particles. The counter
efficiency, determined by calibration with an artificially prepared sample
of known β-activity, is 0.11. The measured counting rate for the test
sample is 121 counts s^{-1}. How many grams of thorium are present in
the sample? (b) The rock sample of (a) is analyzed chemically for lead,
and the lead so obtained is mixed with 100 g of inert rock and replaced
around the geiger counter. The elapsed time, from the separation of the
lead from the rock sample to its replacement around the geiger counter, is
exactly 30 min. What should the counting rate be, just as the new
sample is placed around the counter? *Ans.:* (a) 0.067 g. (b) 37.5 s^{-1}.

15-3. The Ages of Minerals and of the Earth

One of the most fascinating aspects of the naturally occurring radioactive elements is that they provide a means, under suitable conditions, of determining the age of a given sample of rock, and indirectly the age of the earth and perhaps of the universe itself. Even a rough inspection of the abundances of radioactive elements and of their stable decay products reveals qualitative information of great interest:

1. The fact that radioactive elements exist at all in nature implies that the earth has not existed for an infinite length of time and suggests that the elements as we know them might have been created at a time when the physical state of the matter was considerably different than it is today. The time that has elapsed since this epoch may be called the *age of the elements*.

2. The absence of the neptunium radioactive series implies that the earth is many times 2×10^6 years old, since there is good reason to believe that this series was initially created along with the other three.

3. The present relative abundances of ^{235}U and ^{238}U ($\sim 1:140$) suggest that the elements are perhaps not much older than $7 \times 10^8 \times \log_2 140$ years (that is, 5×10^9 years), because at this time in the past the relative abundances of the two isotopes would have been about the same, a condition which is observed to be roughly satisfied for numerous *stable* nuclei.

Many attempts have been made to arrive at a quantitative measurement of the age of the earth and of the elements. Some of these investigations have made use of the following kinds of measurements:

1. The ratio of uranium to helium. In the course of decay of ^{238}U to ^{206}Pb and of ^{235}U to ^{207}Pb, eight and seven α-particles are emitted, respectively. These α-particles have quite small ranges in solid matter and may become trapped inside a rock specimen if the rock is quite fine grained and impervious to penetration by gas or water. By measuring the amount of uranium present in a rock and also the amount of helium, it is possible to arrive at an "age" for the rock which represents the time that has elapsed since the rock last solidified. In this way ages up to about 2×10^9 years have been found for granite samples. The ages of *iron meteorites* have also been measured by using this method, with similar results. In the case of the meteorites, however, there is some reason to believe that α-particles can be introduced into the iron by other means than by the decay of radioactive atoms, namely, by *cosmic-ray bombardment*. This method is not considered reliable because of the possibility of loss of helium over long periods of time even from quite homogeneous rocks.

2. The ratio of ^{238}U to ^{206}Pb, or ^{232}Th to ^{208}Pb. In a mineral in which all of the lead that is present can be assumed to have been produced from radioactive parents originally present in the sample, it is possible to deduce the age of the specimen by measuring the amounts of a given radioactive parent and stable end product that it contains. The absence of the nonradiogenic lead isotope ^{204}Pb is usually taken as an indication that the lead is purely radiogenic. This method is subject to errors due to differential leaching or introduction of one or the other of the elements of interest.

3. The ratio of ^{206}Pb to ^{207}Pb. It is found experimentally that *uranium from all available sources has the same relative isotopic abundances of* ^{238}U *and* ^{235}U, *namely*, 138.5:1. This implies that all uranium was originally formed with a certain "universal" relative isotopic constitution and has not subsequently been subject to local changes of a chemical or radiogenic nature. (That is, none of the uranium has been formed from a now extinct progenitor since it was deposited in a solid form.) This means that the lead that has been formed from the decay of the uranium in a sample will possess an isotopic ratio ^{206}Pb/^{207}Pb which is dependent upon the age of the sample because of the different decay rates of ^{238}U and ^{235}U. This method is not subject to as much error as the previous one because of differential leaching or introduction of lead and uranium in the rock sample, since the two types of uranium and the two types of lead involved are subject to the same chemical factors. For accurate results, however, it is still necessary that the rock be a "closed" system with respect to both the uranium and the lead. This can sometimes be checked by intercomparing the ages as determined from ^{238}U/^{206}Pb, ^{235}U/^{207}Pb, and ^{206}Pb/^{207}Pb.

In practice, one often desires to know the age of a rock specimen which may already have contained some lead at the time of its formation, so that not all of the lead now present is the product of the uranium or thorium in the sample. In this case one could still use the method described above if the isotopic constitution of the nonradiogenic lead were known. Now the isotopic constitution of *all* lead, taken as a whole, *has been changing systematically since the elements were formed*, because of the addition of ^{206}Pb, ^{207}Pb, and ^{208}Pb to the "primordial" lead through the decay of ^{238}U, ^{235}U, and ^{232}Th. Thus, if the relative cosmic abundances of uranium, thorium, and lead were known, and if the isotopic constitution of *primordial lead* were known, it would be possible to evaluate these changes in the average isotopic constitution of lead with time.

Measurements of the isotopic constitution of lead taken from ore deposits of various ages and from other sources have been made, and the isotopic constitution of average lead as a function of time has been found approximately;[1] the results are shown in Table 15-2. This table shows a quite significant change in the relative amounts of all radiogenic lead species over the period represented.

TABLE 15-2. AVERAGE ISOTOPIC CONSTITUTION OF LEAD AT
VARIOUS PAST TIMES

Time, 10^6 y	^{206}Pb/^{204}Pb (^{238}U series)	^{207}Pb/^{204}Pb (^{235}U series)	^{208}Pb/^{204}Pb (^{232}Th series)
0–50	18.5	15.6	38.4
1060	16.7	15.5	36.3
3500	11.4	13.5	31.1
Meteoritic (primordial?)	9.4	10.3	29.2

On the basis of this method of analysis, it has been estimated that the age of the meteorite material, since the time of its isolation from uranium- or thorium-containing matter, is about 4.5×10^9 years. This figure may be taken as a lower limit to the age of the universe itself, by which we mean the time span since the genesis of the elements as we now know them. It is most interesting and perhaps quite significant that this figure is in fair agreement with measurements of the age of the universe as defined by the rates of recession of the galaxies. This latter "age" is obtained by extrapolating the observed linear velocity vs. distance relation of the galaxies backward in time to an epoch at which all of the galaxies would have been in the same general neighborhood. The agreement of the ages of the universe as defined by these two quite different kinds of observation is often taken as strong support for the hypothesis that the elements as we now observe them were formed in a cataclysmic event about 5 to 10 billion years ago. More will be said about the formation of the elements in Chap. 21.

EXERCISES

15-9. A certain rock solidified at a certain time in the past, and at that time contained no lead but did contain some uranium. Assuming no interchange with the outside, find expressions, as a function of the age of the rock, for the atomic ratios ^{206}Pb/^{238}U, ^{207}Pb/^{235}U, and ^{206}Pb/^{207}Pb. *Ans.*: ^{206}Pb/^{238}U $= (2^{t/T_{1/2}} - 1)$.

15-10. From the data given in Table 15-2, compute the approximate present relative abundances of uranium, thorium, and lead in the earth's outer crust.

[1] Tilton et al., *Bull. Geol. Soc. Am.*, **66**, 1131 (1955).

15-4.   Other Naturally Occurring Radionuclides

The three radioactive series of elements which occur in nature do not quite account for all naturally occurring radioactive elements.   A few other elements which are definitely radioactive with half-lives of the order of billions of years have been found.   It is virtually certain that there exist still other elements which are unstable, but whose lifetimes are so long that their activity has not yet been detected.   A list of the known unstable elements which are not members of one of the three radioactive series and which are not being currently produced by cosmic rays is given in Table 15-3.

TABLE 15-3. NATURALLY OCCURRING RADIONUCLIDES WITH $Z < 82$

Element	Abundance, %	Type of activity	Half-life, y
$^{40}_{19}\text{K}$	0.0119	β^-, EC	1.2×10^9
$^{50}_{23}\text{V}$	0.25	EC	4×10^{14}
$^{87}_{37}\text{Rb}$	27.85	β^-	6.2×10^{10}
$^{115}_{49}\text{In}$	95.77	β^-	6×10^{14}
$^{138}_{57}\text{La}$	0.089	β^-, EC	1.0×10^{11}
$^{142}_{58}\text{Ce}$	11.07	α	5×10^{15}
$^{144}_{60}\text{Nd}$	23.87	α	3×10^{15}
$^{147}_{62}\text{Sm}$	15.07	α	1.2×10^{11}
$^{176}_{71}\text{Lu}$	2.60	β^-	5×10^{10}
$^{187}_{75}\text{Re}$	62.93	β^-	4×10^{12}
$^{192}_{78}\text{Pt}$	0.78	α	$\sim 10^{15}$

In addition to the above elements and others that may yet be found to be radioactive, a few elements of rather short half-life are constantly being formed in measurable quantities in the atmosphere and in the top layers of the earth's surface by the bombardment of cosmic rays.   For example, the isotopes ^{14}C and ^3H (tritium) are formed by the action of neutrons on the nitrogen in the atmosphere in the reactions

$$^{14}\text{N} + \text{n} \rightarrow {}^{14}\text{C} + \text{p} \qquad \text{and} \qquad {}^{14}\text{N} + \text{n} \rightarrow {}^{12}\text{C} + {}^3\text{H}$$

These two elements undergo β-decay with half-lives of 5568 and 12.4 years, respectively.   They are currently of interest because of their application to the determination of the ages of objects which contain them.   Some of these applications will be described in Chap. 21.

15-5.   Artificial Radioactivity

For many years the only known radioactive nuclides were those which occur in nature.   But in 1934, Curie and Joliot discovered that positron activity could be induced in an aluminum foil by exposing the foil to the α-radiations from polonium. In the time that has since elapsed, almost a thousand radioactive nuclides have been produced artificially, both by the use of high-energy particle accelerators and in nuclear reactors.

Artificial radioactivity is in almost all respects similar to the natural kind and is governed by the same decay laws and series relationships. But there are two respects in which some differences have been observed: (1) in the existence of *positron decay* among artificially radioactive nuclides and (2) in the presence of a certain sensitivity of decay rate to physical or chemical state on the part of those nuclides which are unstable toward electron capture. In the next chapter we shall learn why no β^+-active radionuclides occur in nature, and later in the present chapter we shall consider some of the more important aspects of electron capture. In what follows, no distinction will be made between natural and artificial radioactivity.

15-6. Elementary Theory of α-particle Decay

The first successful application of quantum mechanics to the problem of nuclear structure was made by Gamow and by Gurney and Condon (1928), who showed that α-radioactivity—the escape of α-particles from a radioactive nucleus—is a quantum-mechanical *barrier-penetration phenomenon*. This can be seen by considering the energetics of α-particle decay together with the sizes of nuclei as determined from the Rutherford scattering experiments: Thus, whereas α-particles are observed to be ejected with energies ranging from 4.0 to 9.0 MeV, their electrostatic potential energies *alone* are approximately 15 MeV even when they are *well outside* the range of nuclear forces as determined by the distances of closest approach deduced from the Rutherford experiments. We therefore conclude that, in order to escape from the nucleus, *the α-particle must traverse a region where its potential energy exceeds its total energy.* This is indicated schematically in Fig. 15-5, which depicts the potential energy of an α-particle as a function of its distance from the nucleus. At large distances outside the nucleus, the potential-energy function is just the electrostatic potential energy, but at some radius R the nuclear forces become appreciable; inside this radius, which is about 10^{-14} m, the potential energy must drop sharply.

The emergence of α-particles having kinetic energy E when they are essentially at infinity implies that the system would have an energy eigenstate of this *total* energy if the escape of α-particles were neglected. According to the results obtained in Chaps. 4 and 10, therefore, we expect the wave function to have qualitatively the form of a standing wave inside the nucleus, a rapidly decaying quasiexponential in the region from R to R_1, and an outward-moving wave train at great distances outside

R_1. Actually, this picture is only roughly correct quantitatively, for several reasons:

1. The three-dimensional character of the problem requires that the effects of angular momentum be included. This is readily done by making use of the procedure outlined in Chap. 9 in connection with molecular binding. In this way the actual problem can be treated as a one-dimensional problem whose potential-energy function is

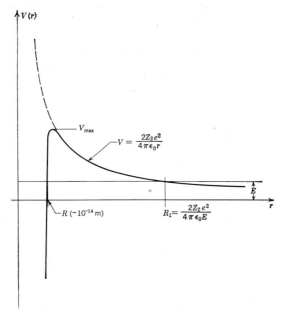

FIG. 15-5. The potential energy of an α-particle near a nucleus (schematic).

equal to the ordinary potential energy plus a "centrifugal potential." The latter quantity has the effect of increasing the height of the barrier and thus reducing the decay rate. This accounts for the observed fact that the orbital angular momentum carried off by the α-particle seldom exceeds three or four units. We shall treat only the case $l = 0$.

2. We have so far assumed implicitly that, since the α-particle emerges as a unit from the nucleus, it must have existed as a unit inside the nucleus. Actually, nuclei are composed of *protons and neutrons* whose association with one another may be so loose that only rarely would two protons and two neutrons find themselves moving together as an α-particle. We must therefore somehow take into account the fact that any two of the Z protons and any two of the N neutrons in a nucleus are potentially capable of acting in a concerted group

as an α-particle, but that this might occur only rarely for given specified pairs of protons and neutrons.

3. In the light of the preceding argument, we must also call into question the basic premise that an α-particle *has* a well-defined potential energy when *inside* the nucleus.

While the foregoing qualifications do introduce some difficulties, it is still possible to obtain significant results without unduly complicating the theory. With the above points in mind, let us represent the probability of decay—i.e., the disintegration constant λ—as a product of the number ν of times per second an α-particle collides with the barrier and the probability of transmission T per collision:

$$\lambda = \nu T \tag{1}$$

From a consideration of the elementary one-dimensional case, we may evidently express ν itself as the number N of α-particle-like associations of protons and neutrons that are likely to exist inside the nucleus at any given instant multiplied by the number of times per second that a particle, moving at some characteristic speed v, would "bounce" back and forth along a diameter $2R$ of the nucleus

$$\nu = N \frac{v}{2R} \tag{2}$$

The transmission coefficient T can be evaluated with satisfactory precision by means of a simple extension of the one-dimensional problem of transmission through a rectangular barrier (Sec. 4-8). For sufficiently low particle energy the exponential attenuation coefficient β' of Eq. 4-6 (4) will be so great that the barrier will be many "attenuation lengths" in extent; that is, if the barrier length is L, then $\beta'L \gg 1$ for low particle energies. In this case the transmission factor of a rectangular barrier would be equal to (compare Exercise 4-19)

$$T \approx e^{-2\beta'L} = \exp - 2 \int_0^L \beta' dx \tag{3}$$

If the barrier is not of constant height but changes only slowly with x, we may still approximate the transmission factor by the above expression, if we let β' *vary with* x. For our case we then have

$$T \approx \exp - 2 \int_R^{R_1} \beta'(x)\, dx \tag{4}$$

where, as in the one-dimensional rectangular barrier,

$$\beta'(x) = \left(\frac{2m}{\hbar^2}\right)^{1/2} [V(x) - E]^{1/2} \tag{5}$$

and the limits R and R_1 are those values of x for which $\beta'(x) = 0$. If we assume that $V(x)$ is exactly equal to the electrostatic potential energy at all points outside R, we may evaluate T in the form

$$\ln T = -2 \int_R^{R_1} \left(\frac{2m}{\hbar^2}\right)^{1/2} \left(\frac{2Ze^2}{4\pi\varepsilon_0 x} - E\right)^{1/2} dx \tag{6}$$

$$= -2 \left(\frac{2mE}{\hbar^2}\right)^{1/2} \int_R^{R_1} \left(\frac{R_1}{x} - 1\right)^{1/2} dx$$

since $E = 2Ze^2/4\pi\varepsilon_0 R_1$.

If we let $x = R_1 \cos^2 u$ and $R = R_1 \cos^2 u_0$, the integral becomes

$$\ln T = +4 \left(\frac{2mE}{\hbar^2}\right)^{1/2} R_1 \int_{u_0}^0 \sin^2 u\, du$$

$$= 4 \left(\frac{2mE}{\hbar^2}\right)^{1/2} R_1 (\tfrac{1}{2}u - \tfrac{1}{4}\sin 2u)_{u_0}^0$$

$$= 2 \left(\frac{2mE}{\hbar^2}\right)^{1/2} R_1 (-u_0 + \sin u_0 \cos u_0)$$

$$= -2 \left(\frac{2mE}{\hbar^2}\right)^{1/2} R_1 \left[\cos^{-1}\left(\frac{R}{R_1}\right)^{1/2} - \left(\frac{R}{R_1}\right)^{1/2}\left(1 - \frac{R}{R_1}\right)^{1/2}\right] \tag{7}$$

Now, if we make use of the fact, known from the Rutherford scattering experiment, that $R \ll R_1$, we may expand

(a)
$$\cos^{-1}\left(\frac{R}{R_1}\right)^{1/2} \approx \frac{\pi}{2} - \left(\frac{R}{R_1}\right)^{1/2}$$

(b)
$$\left(1 - \frac{R}{R_1}\right)^{1/2} \approx 1 \tag{8}$$

so that $\ln T$ becomes

$$\ln T \approx -2 \left(\frac{2mE}{\hbar^2}\right)^{1/2} R_1 \left[\frac{\pi}{2} - 2\left(\frac{R}{R_1}\right)^{1/2}\right]$$

$$= \frac{4e}{\hbar}\left(\frac{m}{\pi\varepsilon_0}\right)^{1/2} Z^{1/2}R^{1/2} - \frac{e^2}{\hbar\varepsilon_0}\left(\frac{m}{2}\right)^{1/2} ZE^{-1/2}$$

$$= 2.97 Z^{1/2}R^{1/2} - 3.95 ZE^{-1/2} \tag{9}$$

where E is in MeV, R is measured in units of 10^{-15} m, and Z is the atomic number of the *residual nucleus*.

Returning now to Eqs. (1) and (2), we may write

$$\log_{10} \lambda \approx \log_{10} \frac{Nv}{2R} + 1.28 Z^{1/2}R^{1/2} - 1.71 ZE^{-1/2} \tag{10}$$

where we have taken logarithms to the base 10 in order to compare the theory with observation more easily.

Figure 15-6 shows $\log_{10} \lambda$ plotted vs. $ZE^{-1/2}$ for a number of representative α-active nuclides from each of the four decay series. The straight

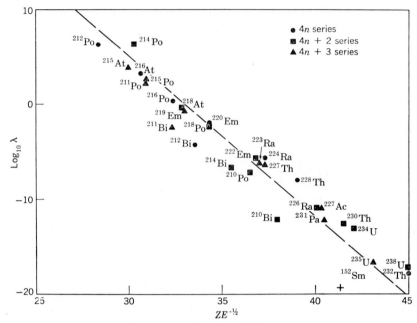

Fɪɢ. 15-6. Comparison of Eq. 15-6(10) with observation.

line drawn through the points has a slope of -1.71 as required by the theory, but otherwise it has been adjusted to fit as well as possible. Considering the tremendous range of λ that is represented—corresponding to half-lives ranging from 10^{-6} *second* to 10^{10} *years*—and the fact that the *slope* of the line is not subject to adjustment, the agreement of the barrier-penetration theory with observation must be considered to be most satisfactory indeed.

Having established the essential validity of the barrier-penetration theory, we may now use the *position* of the line which best represents the data in Fig. 15-6 to evaluate the approximate magnitude of the remaining quantities in Eq. (10). In this way we find

$$\log_{10} \frac{Nv}{2R} + 1.28 Z^{\frac{1}{2}} R^{\frac{1}{2}} \approx 55.5 \tag{11}$$

Because of the quite weak dependence of the logarithmic term upon its argument, we should now be able to obtain a reasonably accurate estimate of R from this equation. Thus, if we estimate

$N \approx 1$ α-particle inside the nucleus at any given time

$v \approx 2 \times 10^7$ m s^{-1}, about the same as the speed with which α-particles emerge from nuclei

$R \approx 10^{-14}$ m from Rutherford-scattering data

$Z \approx 85$ for representative α-active nuclei

we obtain

$$R = \left(\frac{55.5 - 21.0}{1.28 \times 9.2}\right)^2 = 8.5 \ (\times 10^{-15} \ \text{m}) \tag{12}$$

This value provides a general corroboration of the evidence provided by scattering experiments concerning the radii of nuclei and further demonstrates the essential correctness of the present model of α-decay. Of course, the value of R defined above should not be exactly equal to the "actual" radius of a nucleus for the following reasons:

1. R ought to be the *sum* of the nuclear radius and the α-particle radius.
2. If the quantity N is in error by five or six powers of ten, the value obtained for R will change by a factor of about 2.
3. The value of R which appears in the theory is in any case a fictitious one, corresponding to an effective radius where the Coulomb force is assumed to cut off sharply. Actually, the nuclear forces will first begin to be felt somewhat *outside* R, and the potential energy will be equal to E somewhat *inside* R.

A relationship similar to Eq. (10) was obtained empirically quite early in the study of the systematics of radioactive decay. When $\log_{10} \lambda$ is plotted vs. the *range in air* of the α-particles, a smooth curve is obtained, the analytical expression of which is called the *Geiger-Nuttall law*. The semiquantitative interpretation of this law by Gamow and others was quite properly considered a major triumph of the newly discovered quantum mechanics.

EXERCISE

15-11. It has been established that $^{152}_{62}\text{Sm}$ is α-active with a half-life of 1.0×10^{12} year. What energy should the α-particles from this nucleus have? (Their measured energy is 2.1 MeV.)

15-7. Beta Decay; the Neutrino Hypothesis

We now turn to the second of the three classical types of radioactivity, β-decay. The β-decay process is characterized phenomenologically by the emission of a negative or positive electron from the nucleus, and we infer that this is accompanied by a transformation of a neutron into a proton or a proton into a neutron, respectively. β-decay may be compared with α-decay as follows:

1. The energies of the ejected electrons range from a few keV to more than 15 MeV, whereas α-particles are seldom ejected with less than 4 or more than 9 MeV.

2. The number of known β-active nuclei is many times the number of α-active ones.   Several hundred β-emitters are known.

3. The half-lives of β-active nuclei range upward from about 1 s, whereas several α-active nuclei have lifetimes much shorter than 1 s.

4. α-particles emerge from nuclei with discrete, well-defined energies, indicating that such transitions take place between well-defined energy states in the parent and daughter nuclei.   In strong contrast with this, *β-particles are always observed to possess a continuous spectrum of energies extending from essentially zero up to a maximum.*[1]

5. In α-decay one is dealing with the escape from the nucleus of a type of particle which, although it may not have existed as such inside the nucleus, is at least composed of particles which *did* exist previously inside the nucleus.   In β-decay, on the other hand, we find emerging a particle which we have quite good reason to believe *did not exist inside the nucleus prior to the decay.*   This is thus analogous to the emission of electromagnetic quanta from excited systems, in which we think of the quanta as being *created* as part of the over-all decay process.

6. α-particles emerge from nuclei at speeds quite low compared with the speed of light and may therefore be treated nonrelativistically.   However, β-decay electrons quite commonly possess energies comparable with m_0c^2 and must consequently be treated relativistically.

Upon close examination one finds several puzzling discrepancies in the observed features of β-decay, some of which were historically of considerable importance in establishing our current ideas of the nature of the phenomenon.   They are summarized below:

1. The continuous distribution of electron energies is perhaps the most curious feature of β-decay.   How could this come about?   There might appear to be several possibilities.   Perhaps the nuclear energy states involved are not sharp but are so broad as to be essentially a continuum.   This is in contradiction to a great body of evidence from studies of nuclear reactions, however, which clearly requires that these nuclear states be discrete.

   Or, perhaps the electron carries off only a part of the energy, leaving the nucleus in an excited state from which it decays by γ-ray emission. Many cases are known, however, in which β-rays having a continuous energy spectrum are not accompanied by γ-activity.

   Another possibility is that, in emerging from the nucleus, a β-par-

---

[1] Sometimes one also observes *sharp lines* superimposed upon the continuous spectrum.   However, these are known to be ordinary atomic electrons which have been ejected in an *Auger transition* of the excited nucleus, and are thus quite distinct from true β-decay electrons.

ticle might collide with one or more of the atomic electrons, which would result in *more electrons* having *less energy per electron*. Two important objections can be advanced against this hypothesis: first, β^+-decays which are completely free of β^--particles are often observed and, second, calorimetric experiments, in which all of the *energy* contained in the emerging electrons is captured by stopping the particles inside the calorimeter, show a distinct discrepancy with the energy balance expected on the basis of the known masses of the nuclei involved in the decay. The emergent energy that is measured in these experiments is in fact found to correspond to the *average* electron energy as determined from the observed energy spectrum.

A strong clue is provided by the fact that no electron is ever observed to have *more* energy than is permitted by the conservation of energy. In fact, if the *maximum* observed electron energy is used in the calculations of the energy balance, there is then no departure from conservation of energy. Thus, the discrete character of the initial and final states, together with the observed continuum of electron energies, clearly poses the question of how (or perhaps *whether*) energy can be conserved in the β-decay process.

2. In the previous chapter we saw that there is a great amount of experimental evidence which strongly indicates the nuclei are composed of protons and neutrons, both of which particles are supposed to have *half-integral* spin. However, the β-decay process is known to result in the replacement of a neutron by a proton, or vice versa. If we try to write a reaction of this kind, e.g.,

$$p \rightarrow n + e^+ \qquad \text{or} \qquad n \rightarrow p + e^- \tag{1}$$

we see that *one* particle of half-integral spin has been transformed into *two* particles of half-integral spin. But, inasmuch as the two emergent particles can only have an *integral* relative orbital angular-momentum quantum number, we are faced with a clear violation of the law of combination of quantized angular momenta, according to which an even number of half-integral spins can combine only to form an *integral* spin.

3. In the β-decay of certain nuclei it is possible to measure both the electron and the recoiling nucleus. In these cases one finds that the electron seldom is moving exactly opposite the nucleus, as is required if linear momentum is conserved. Hence it appears that the β-decay process also violates the principle of conservation of *momentum*.

Pauli (1930) first suggested that the apparent violation of conservation of energy in β-decay might be explained if *another particle* were emitted together with the electron. This particle must be neutral to conserve charge and must have a rest mass small compared with that of an electron because energy *is* conserved if the most energetic electrons are used (that is, not much can be left over to be carried off as mass energy by the neutral particle). This particle must also have much weaker interaction with matter than does a γ-ray, since we find from the calorimeter experiments that it can apparently pass through large masses of solid matter without losing much of its kinetic energy. This hypothetical particle is called a *neutrino* (little neutral one). All currently known features of β-decay are properly accounted for if the neutrino possesses the following properties:

$$\text{mass } m = 0 \qquad \text{charge} = 0 \qquad \text{spin} = \tfrac{1}{2}\hbar \tag{2}$$

If the neutrino exists, all of the apparent violations of conservation of energy, momentum, and angular momentum disappear; if it does not exist, we can still say on the basis of experimental evidence that the missing energy, momentum, and spin are related to one another as they would be *if they were* carried away by a single particle having the above properties, and further, a quantitative theory of β-decay, based upon the participation of the neutrino, is able to account for all known features of β-decay. However, it would still be desirable to have more direct evidence that neutrinos exist. Such evidence has been sought in connection with the reverse process called *inverse β-decay*, wherein energetic neutrinos from a fission reactor are allowed to bombard a material in which they should induce a reaction of the basic form

$$_{1}^{1}\text{H} + \nu \rightarrow {}_{0}^{1}\text{n} + \text{e}^{+} \tag{3}$$

If this experiment gave positive results when carried out near a fission pile (carefully shielded from neutrons) but negative results otherwise, it would constitute conclusive evidence that the neutrino exists. Present results in such experiments are affirmative but still of rather limited statistical reliability.[1]

15-8. Elementary Theory of β-Decay

Let us now examine some of the more elementary theoretical aspects of β-decay. We begin by investigating those features which are governed by the conservation laws. Suppose an atom of mass M' decays at rest into another atom of mass M, an electron, and a neutrino, whose respective momenta are $\boldsymbol{p}_{\text{N}}$, $\boldsymbol{p}_{\text{e}}$, and $\boldsymbol{p}_{\nu}$ as shown in Fig. 15-7. The law of con-

---

[1] Cowan and Reines, *Phys. Rev.*, **92**, 830 (1953), report a probable positive result using reaction (3). Their experiment involved the detection of delayed coincidences between the positron and a γ-ray emitted from a Cd nucleus which captures the neutron.

servation of energy-momentum requires that

$$\boldsymbol{p}_N + \boldsymbol{p}_e + \boldsymbol{p}_\nu = 0 \tag{1}$$

and
$$M + T + w_e + w_\nu = M' \tag{2}$$

where T is the kinetic energy of the residual atom and w_e and w_ν are the total energies of the electron and neutrino. (All quantities are expressed in energy units.)

Now, inasmuch as the available energy is only a few MeV, the residual atom may be treated nonrelativistically; and since $M' \gg m_e$ while $|\boldsymbol{p}_N| \approx |\boldsymbol{p}_e|$, *the kinetic energy carried away by the residual atom is quite negligible.* Thus, the β-decay process may for practical purposes be regarded as one in which the available energy is divided in some way between the electron and neutrino, their motion being otherwise independent: whatever their directions of motion and

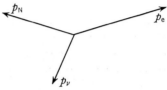

FIG. 15-7. A β-decay.

momenta, the massive nucleus recoils as necessary to satisfy momentum balance. In this case we may write with good accuracy

$$w_e + w_\nu = M' - M = E_m + m_e \tag{3}$$

where E_m is the maximum kinetic energy present in the β-decay spectrum.

EXERCISES

15-12. ^{12}N β-decays into ^{12}C with an energy release of 17 MeV. What is the maximum kinetic energy with which the ^{12}C can recoil? *Ans.:* 0.013 MeV.

15-13. In computing the energy release using tabulated atomic masses for neutral atoms, allowance must be made for the fact that, since an electron is observed to be ejected from an atom in β-decay and the nuclear charge is increased or decreased by one unit, the resulting atom is *singly ionized* positively or negatively. Denoting the masses of the parent and daughter *neutral atoms* by M_0' and M_0, show that the maximum kinetic energy appearing in the observed β-decay spectrum is

(a)
$$E_m = M_0' - M_0$$
(b)
$$E_m = M_0' - M_0 - 2m_e \tag{4}$$

for β^-- and β^+-decay, respectively.

We next consider those factors that affect the observed division of energy between the electron and neutrino. Clearly one of these factors will be some kind of a *transition probability* connecting the states describing the parent atom M' and the daughter atom M, electron, and neutrino. Another factor is a statistical one which depends upon the number of possible final states per unit energy range available to the electron and neutrino. Recalling the discussion of time-dependent transitions given in Chap. 2, the matrix element which governs the transition from the initial state ψ_i to the final state ψ_f will be of the form

$$(f|V|i) = \int \psi_f^* \hat{V} \psi_i \, d\tau \tag{5}$$

where $\hat{V}$ is the operator which describes the interaction between the nucleus, the electron, and the neutrino.

Now, clearly ψ_i will be an eigenfunction u_i of the system corresponding to the parent nucleus at rest. On the other hand, ψ_f will be an eigenfunction of the system comprising daughter nucleus, electron, and neutrino. Because of the weakness of the interaction between these latter three particles, ψ_f will be of the form

$$\psi_f = u_f \exp\left(i\boldsymbol{p}_e \cdot \boldsymbol{r}/\hbar\right) \exp\left(i\boldsymbol{p}_\nu \cdot \boldsymbol{r}/\hbar\right) \tag{6}$$

where u_f is the appropriate nuclear eigenfunction for the daughter atom, and the exponential factors are the wave functions of a *free electron and neutrino*. The wave functions u_i and u_f will, of course, differ from zero only inside the nucleus. Since the wavelengths of the electron and neutrino are considerably greater than a nuclear radius, it is convenient to use the expanded form for the exponentials:

$$\exp\frac{i}{\hbar}\left(\boldsymbol{p}_e + \boldsymbol{p}_\nu\right) \cdot \boldsymbol{r} = 1 + \frac{i}{\hbar}\left(\boldsymbol{p}_e + \boldsymbol{p}_\nu\right) \cdot \boldsymbol{r} - \frac{[(\boldsymbol{p}_e + \boldsymbol{p}_\nu) \cdot \boldsymbol{r}]^2}{2\hbar^2} + \cdots \tag{7}$$

Concerning the interaction $\hat{V}$, we may expect this to be a "contact" (delta function) interaction whose "strength" is described by a dimensional constant G, whose dimensions are energy times volume; there could also appear certain "angular correlation" terms which describe a tendency for the electron and the neutrino to emerge in certain preferred relative directions, perhaps with certain relative spin orientations as well. Also, the Coulomb interaction between the electron and the nucleus will appear in $\hat{V}$. The actual form of $\hat{V}$ has been a subject of considerable debate, partly because the observed energy spectrum of the electrons is relatively insensitive to the precise form taken for $\hat{V}$. We shall follow Fermi who, in his first treatment of β-decay (1934), assumed that $\hat{V}$ is essentially a constant, independent of such things as the electron energy and the relative spin orientations of the electron and neutrino. In this approximation the integral (5) assumes the simple form

$$(f|V|i) \approx G \int u_f^* u_i \, d\tau \tag{8}$$

where we have used only the first term of Eq. (7) and G is the (real) dimensional factor which defines the strength of the interaction.

The matrix element (8) refers to a transition in which the electron and neutrino each occupy one of the continuum of possible states available to them.   In order to evaluate the relative probabilities of these various energy values, however, we must multiply the transition probability per state by the *density of energy levels* in the continuum.   In our discussion of the free particle (Chap. 4) we found that the states available to a free particle are distributed uniformly in phase space with a density h^{-3}. Thus, we may write

$$dn = \frac{dx\, dy\, dz\, dp_x\, dp_y\, dp_z}{h^3} \tag{9}$$

as the number of states available within the phase-space volume $dx\, dy\, dz\, dp_x\, dp_y\, dp_z$.   If the β-decay takes place inside a box of unit volume, and if we transform from rectangular to spherical polar coordinates in momentum space, Eq. (9) becomes

$$dn = \frac{4\pi p^2\, dp}{h^3} \tag{10}$$

where p is the magnitude of the momentum of the particle.   If the electron and neutrino were *independent* of one another, the number of states available to them jointly would be

$$d^2n = \frac{16\pi^2 p_e^2\, dp_e\, p_\nu^2\, dp_\nu}{h^6} \tag{11}$$

Therefore, the total transition probability to states in the neighborhood of p_e and p_ν is [see Eq. 2-5(28)]

$$d^2\Pi = 4|(\mathrm{f}|V|\mathrm{i})|^2 \frac{\sin^2\, (E_\mathrm{f} - E_\mathrm{i})t/2\hbar}{(E_\mathrm{f} - E_\mathrm{i})^2} \frac{16\pi^2}{h^6}\, p_e^2 p_\nu^2\, dp_e\, dp_\nu \tag{12}$$

Now, experimentally we can measure only the *electron* momentum, since the neutrino is essentially devoid of interaction.   Thus, to eliminate the neutrino from (12) we integrate over p_ν, remembering that (12) does not require that energy be conserved.   Now $E_\mathrm{f} - E_\mathrm{i} = \Delta E = E_e + E_\nu - E_m$ and $dp_\nu = dE_\nu$ (assuming the neutrino to have zero rest mass).   We then let $\Delta E t/2\hbar = z$ and $dE_\nu t/2\hbar = dz$ (as in Sec. 6-5), so that

$$d^2\Pi = 4|(\mathrm{f}|V|\mathrm{i})|^2 \frac{\sin^2 z}{z^2} \frac{16\pi^2}{h^6}\, p_e^2 p_\nu^2\, dp_e \frac{t}{2\hbar}\, dz \tag{13}$$

The integral of (13) over z will yield the differential electron momentum spectrum:

$$d\Pi = I(p_e)\, dp_e = \frac{64\pi^3 t}{h^7}\, p_e^2\, dp_e \int_{-\infty}^{\infty} |(\mathrm{f}|V|\mathrm{i})|^2 \frac{\sin^2 z}{z^2}\, p_\nu^2\, dz \tag{14}$$

where we have removed from the integral all quantities that are independent of z. Now, the resonance denominator (z^{-2}) assures that only values of z near $z = 0$ (that is, values of the neutrino energy corresponding to energy conservation) will contribute significantly to the integral. Therefore, we may remove from the integral both the matrix element and the neutrino momentum, each evaluated at $z = 0$, since these terms vary slowly with z. This yields

$$I(p_e) \, dp_e = \frac{64\pi^3 t}{h^7} p_e^2 p_{\nu 0}^2 \, dp_e |(\mathrm{f}|V|\mathrm{i})|^2 \int_{-\infty}^{\infty} \frac{\sin^2 z}{z^2} \, dz \qquad (15)$$

The neutrino momentum at $z = 0$ may be evaluated from (3):

$$w_e + p_{\nu 0} = E_m + m_e$$

or

$$p_{\nu 0} = E_m - E_e \qquad (16)$$

Finally, we insert the approximate form (8) for the matrix element and note that $\int_{-\infty}^{\infty} z^{-2} \sin^2 z \, dz = \pi$, with the result

$$\frac{I(p_e) \, dp_e}{t} = \frac{64\pi^4}{h^7} G^2 |M|^2 p_e^2 (E_m - E_e)^2 \, dp_e \qquad (17)$$

where

$$M \approx \int u_f^* u_i \, d\tau \qquad (18)$$

and $I(p_e)/t$ is the differential momentum distribution per unit time:

$$\frac{I(p_e)}{t} = \frac{dN}{dp_e} \qquad (19)$$

In the above theoretical equations we have a result which can be checked against experimental observations, namely, *the shape of the electron momentum spectrum should be governed almost entirely by the statistical factor which describes the density of states available to the electron and the neutrino.* The test of this prediction is most conveniently carried out by means of a so-called *Kurie plot*,[1] in which $[I(p_e)/p_e^2]$ is plotted vs. E_e:

$$\left[\frac{I(p_e)}{p_e^2 t} \right]^{1/2} = G|M| \frac{8\pi^2}{h^{7/2}} (E_m - E_e) \qquad (20)$$

Figure 15-8 shows Kurie plots for several β-emitters. The expected linear relationship is well satisfied, particularly near the upper limit of the spectrum. (This incidentally provides a convenient and accurate means of determining this upper energy limit.) The main discrepancy that one ordinarily finds is a surplus of low-energy electrons in the case of β^--decay, and too few of these in β^+-decay. This is a result of the neglect of the

[1] Kurie, Richardson, and Paxton, *Phys. Rev.*, **49**, 368 (1936).

Coulomb interaction between the electron and the nucleus.   The effect can be visualized qualitatively in the following simple terms: the momentum spectrum of the electrons as they leave the nucleus is essentially that described by Eq. (17).   In the case of β^--decay, however, the electrons are attracted to the nucleus and hence lose some energy in emerging

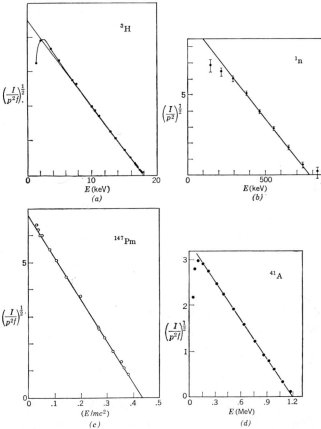

Fig. 15-8. Kurie plots for several β-emitters.   (a) ^{3}H.   [After Langer and Moffatt, *Phys. Rev.*, **88**, 689 (1952).]   (b) The neutron.   [After Robson, *Phys. Rev.*, **83**, 349 (1951).]   (c) ^{147}Pm.   [After Langer, Motz, and Price, Jr., *Phys. Rev.*, **77**, 798 (1950).] (d) ^{41}A.   [After Schwarzschild, Rustad, and Wu, *Phys. Rev.*, **103**, 1796 (1956).]

from the atom; this gives a surplus of low-energy electrons.   In β^+-decay, the repulsion of the nucleus gives the positrons more energy, resulting in a dearth of low-energy positrons.   When this effect is computed quantitatively and the distribution (17) correspondingly corrected, the agreement with experiment is quite good, as shown in Fig. 15-8a, c, and d.

While the *shape* of the β-decay spectrum is well accounted for on the basis of the statistical factor and the Coulomb effects, we must bring in also the magnitudes of the coupling constant G and the matrix element M in order to account for the observed *lifetimes* of β-decay. Equation (17) gives the differential decay probability per unit time and per unit momentum interval. To obtain the disintegration constant λ, therefore, we must integrate Eq. (17) over all momenta from 0 to p_m:

$$\frac{1}{\tau_m} = \lambda = \int_0^{p_m} \frac{I(E_e)}{t}\, dp_e = G^2|M|^2 \frac{64\pi^4}{h^7} \int_0^{p_m} p_e^2(E_m - E_e)^2\, dp_e$$

$$= G^2|M|^2 \frac{64\pi^4}{h^7} \int_0^{E_m} (E_m - E_e)^2(E_e + m_e)(2m_e E_e + E_e^2)^{1/2}\, dE_e \qquad (21)$$

$$= \frac{G^2|M|^2 64\pi^4}{60h^7} E_m^5 \left\{ \left[2 + 8\frac{m_e}{E_m} + 3\frac{m_e^2}{E_m^2} - 10\frac{m_e^3}{E_m^3} - 15\frac{m_e^4}{E_m^4} \right] \right.$$
$$\left(1 + \frac{2m_e}{E_m}\right)^{1/2} + 15\left(1 + \frac{m_e}{E_m}\right)\frac{m_e^4}{E_m^4} \ln\left[\frac{E_m}{m_e} + 1 \right.$$
$$\left. \left. + \frac{E_m}{m_e}\left(1 + 2\frac{m_e}{E_m}\right)^{1/2} \right] \right\} \qquad (22)$$

$$\approx \frac{G^2|M|^2 64\pi^4}{30h^7} E_m^5 \qquad E_m \gg m_e \qquad (23)$$

Thus, if the decay energy is much greater than the rest-mass energy of the electron, the disintegration constant ought to be approximately proportional to the fifth power of E_m. This is observed to be roughly true for some of the light elements, but the variation of $|M|^2$ from one case to another prevents any quantitative test of Eq. (23).

A more informative approach is to use the observed half-lives and the theoretical equation (22) (modified somewhat to allow for the Coulomb effects mentioned previously) to deduce information concerning the magnitude of the matrix element $G^2|M|^2$:

$$\frac{1}{T_{1/2}} = G^2|M|^2 f(Z, E_m) \qquad (24)$$

where $f(Z, E_m)$ is the function appearing in Eq. (22), corrected for Coulomb effects.[1] Thus we may use the product $fT_{1/2}$ as a *measure* of the magnitude of $G^2|M|^2$:

$$fT_{1/2} = \frac{1}{G^2|M|^2} \qquad (25)$$

The $fT_{1/2}$ values of several hundred β-emitters have been evaluated, and these are found to vary over a tremendous range—from 10^3 to 10^{18} s. As we shall now see, this range can be interpreted quantitatively in terms of *selection rules* which govern β-decay.

The results obtained so far have been based upon the approximation

[1] Feenberg and Trigg, *Revs. Modern Phys.*, **22**, 399 (1950).

that only the first term of the expansion (7) is of importance in determining the magnitude of the matrix element of the transition. In this approximation the form of M is simply that of Eq. (18), where u_f and u_i are the wave functions of the daughter and parent nuclei, respectively. Now, although we do not know the exact form of the nuclear wave functions, we *can* say with certainty that these will be eigenfunctions of the nuclear angular momentum, and that *the integral* (18) *will therefore be zero unless the angular momenta of initial and final nuclear states are equal.* This situation is, of course, familiar to us from our experience with spectroscopy and is described by saying that β-decays must obey the selection rule

$$\Delta I = 0 \tag{26}$$

Now, this result does *not* mean that *no transitions whatever* can occur except those which satisfy this selection rule, but only that such transitions should account for the *most intense* β-decays. For if the approximation (18) yields a zero value for M, we must see whether the next term in the expansion (17) will yield a nonzero result:

$$M_1 = \frac{i}{\hbar} (\boldsymbol{p}_e + \boldsymbol{p}_\nu) \cdot \int u_f^* \boldsymbol{r} u_i \, d\tau \tag{27}$$

The expression will be recognized as being quite similar to that for electric-dipole radiation, and indeed by the same analysis we find that M_1 is zero unless

$$\Delta I = 0, \pm 1 \tag{28}$$

In a similar manner we find that the successive terms in the expansion (7) correspond to transitions which permit successively greater and greater angular-momentum changes.

We can obtain a rough estimate of the factor by which each successive term in the expansion reduces the matrix element by observing that Eq. (27) is surely less than

$$|M_1| < |M| \frac{R}{\lambda} \tag{29}$$

where R is the nuclear radius and

$$\lambda = \frac{\hbar}{|\boldsymbol{p}_e + \boldsymbol{p}_\nu|} \tag{30}$$

For the usual values of E_m, we thus find

$$\frac{M_1}{M} \approx \frac{1}{100} \tag{31}$$

so that a transition for which $\Delta I = \pm 1$ should be reduced in intensity by a factor $\sim 10^4$ with respect to one for which $\Delta I = 0$.

It is now clear that the great range found for the $fT_{\frac{1}{2}}$ values should correspond to a range of ΔI values, and this is indeed found to be the case: the smallest $fT_{\frac{1}{2}}$ values correspond to $\Delta I = 0$ and the largest ones to $\Delta I = \pm 3$.

Let us now examine the correlation of the $fT_{\frac{1}{2}}$ values with the above selection rules. The smallest $fT_{\frac{1}{2}}$ values are observed to be those involving so-called *mirror nuclei*. [Two nuclei (Z',N') and (Z,N) are called mirror nuclei if $Z' = N$ and $N' = Z$.] Since the β-decay process results in the replacement of a neutron by a proton or vice versa, the mirror nuclei of interest to us are those for which

$$Z' = N' + 1 \quad \text{and} \quad N' = Z' + 1 \tag{32}$$

or
$$A' = 2Z' + 1 \quad \text{and} \quad A' = 2Z' - 1 \tag{33}$$

Examples of such nuclei are $({}^{1}_{0}n, {}^{1}_{1}H)$, $({}^{3}_{1}H, {}^{3}_{2}He)$, $({}^{7}_{3}Li, {}^{7}_{4}Be)$, $({}^{9}_{4}Be, {}^{9}_{5}B)$, $({}^{11}_{5}B, {}^{11}_{6}C)$, etc. The $fT_{\frac{1}{2}}$ values for some of these are shown in Table 15-4.

TABLE 15-4. $\text{LOG}_{10}\ fT_{\frac{1}{2}}$ VALUES FOR SOME β-TRANSITIONS INVOLVING MIRROR NUCLIDES
(Allowed and Favored Transitions)

Reaction	Half-life, s	Energy, MeV	Spin		$\log_{10} fT_{\frac{1}{2}}$
			Parent	Daughter	
${}^{1}_{0}n \to {}^{1}_{1}H + \beta^-$	750	0.78	$\frac{1}{2}$	$\frac{1}{2}$	3.0
${}^{3}_{1}H \to {}^{3}_{2}He + \beta^-$	3.9×10^8	0.02	$\frac{1}{2}$	$\frac{1}{2}$	3.0
${}^{11}_{6}C \to {}^{11}_{5}B + \beta^+$	1230	0.96	$\frac{3}{2}$	$\frac{3}{2}$	3.6
${}^{23}_{12}Mg \to {}^{23}_{11}Na + \beta^+$	11.6	3.0	$\frac{3}{2}$	$\frac{3}{2}$	3.5
${}^{41}_{21}Sc \to {}^{41}_{20}Ca + \beta^+$	0.87	5.0	$\ldots$	$\ldots$	3.4

In a later chapter we shall find that there are quite strong reasons for believing that the ground-state wave functions of the two nuclei of a mirror pair are very similar to one another, which would of course make M nearly as large as it can be: unity. Transitions involving mirror nuclei are called *allowed and favored* transitions for this reason.

Next after the allowed and favored transitions come those which still satisfy the selection rule $\Delta I = 0$, but for which u_f and u_i, while not strictly orthogonal, are sufficiently unlike one another that M is considerably less than unity. These transitions are said to be *allowed but not favored*. Examples of such transitions are given in Table 15-5.

If a transition which satisfies the selection rule $\Delta I = 0$ cannot take place (that is, if there are no states in the daughter nucleus having the

TABLE 15-5. $\log_{10} fT_{1/2}$ VALUES FOR SOME ALLOWED BUT NOT
FAVORED β-TRANSITIONS

| Reaction | Half-life, s | Energy, MeV | Spin | | $\log_{10} fT_{1/2}$ |
			Parent	Daughter	
$^{24}Na \rightarrow {}^{24}Mg + \beta^-$	5.4×10^4	5.511	4	4	6.1
$^{35}S \rightarrow {}^{35}Cl + \beta^-$	7.51×10^6	0.167	$3/2$	$3/2$	5.0
$^{45}Ca \rightarrow {}^{45}Sc + \beta^-$	1.42×10^7	0.254	$7/2$	$7/2$	5.6
$^{94}Tc \rightarrow {}^{94}Mo + \beta^+$	315	4.32	2	2	5.6
$^{169}Er \rightarrow {}^{169}Tm + \beta^-$	8.1×10^5	0.33	$(1/2)$	$(1/2)$	6.1

TABLE 15-6. $\log_{10} fT_{1/2}$ VALUES FOR SOME FORBIDDEN TRANSITIONS

| Reaction | $T_{1/2}$, s | Energy, MeV | Spin | | $\log_{10} fT_{1/2}$ |
			Parent	Daughter	
First Forbidden ($\Delta I = 0$ or 1, *Parity Change*)					
$^{87}Kr \rightarrow {}^{87}Rb + \beta^-$	4670	3.2	$(5/2)$	$3/2$	7.0
$^{111}Ag \rightarrow {}^{111}Cd + \beta^-$	6.5×10^5	1.0	$(1/2)$	$1/2$	7.2
$^{137}Xe \rightarrow {}^{137}Cs + \beta^-$	228	4.0	$(7/2)$	$7/2$	6.3
$^{141}Ce \rightarrow {}^{141}Pr + \beta^-$	2.4×10^6	0.56	$(7/2)$	$5/2$	7.7
$^{185}W \rightarrow {}^{185}Re + \beta^-$	6.3×10^6	0.43	$(3/2)$	$(5/2)$	7.5
$\Delta I = 2$, *Parity Change*					
$^{37}S \rightarrow {}^{37}Cl + \beta^-$	300	4.3	$(7/2)$	$3/2$	7.1
$^{85}Kr \rightarrow {}^{85}Rb + \beta^-$	3×10^8	0.74	$(9/2)$	$5/2$	9.2
$^{137}Cs \rightarrow ({}^{137}Ba)^* + \beta^-$	1.0×10^9	0.53	$7/2$	$(11/2)$	9.6
Second and Higher Forbidden ($\Delta I = 2$, *No Parity Change*, and $\Delta I > 2$)					
$^{87}Rb \rightarrow {}^{87}Sr + \beta^-$	1.8×10^{18}	0.13	$3/2$	$9/2$	16.5
$^{115}In \rightarrow {}^{115}Sn + \beta^-$	1.8×10^{22}	0.63	$9/2$	$1/2$	23.2
$^{129}I \rightarrow {}^{129}Xe + \beta^-$	1.0×10^{15}	0.12	$7/2$	$1/2$	13.5
$^{135}Cs \rightarrow {}^{135}Ba + \beta^-$	6.6×10^{13}	0.21	$7/2$	$3/2$	13.1
$^{187}Re \rightarrow {}^{187}Os + \beta^-$	1.2×10^{20}	0.043	$5/2$	$(9/2)$	17.7

same spin as and lower energy than the parent), it is called a *forbidden* transition. If the observed transition satisfies the condition $\Delta I = \pm 1$, it is called *first-forbidden;* if $\Delta I = \pm 2$, it is called *second-forbidden;* and so on. Examples of forbidden transitions are given in Table 15-6. Note that $\log_{10} fT_{1/2}$ increases by roughly four units for each additional degree of forbiddenness.

The above picture, while in good agreement with much of the experimental data, is not without some striking exceptions. For example, the reaction

$$\ce{^6_2He} \rightarrow \ce{^6_3Li} + \beta^- + \nu \qquad (34)$$

has a $\log_{10} fT_{\frac{1}{2}}$ value of only 2.91, yet $\ce{^6_2He}$ has a nuclear spin $I = 0$ and $\ce{^6_3Li}$, $I = 1$. This evidently represents a breakdown of the selection rules based upon the Fermi approximation in which the spin dependence of the interaction operator $\hat{V}$ is neglected. Gamow and Teller[1] showed that there are five possible types of interaction which are properly covariant under the Lorentz transformation. These are called scalar (S), 4-vector (V), 4-tensor (T), axial 4-vector (A), and pseudoscalar (P). These lead to the selection rules

(a)	S, V$: \Delta I = 0$	no parity change
(b)	T, A$: \Delta I = 0, \pm 1$	no parity change, $I = 0 \nleftrightarrow I = 0$ (35)
(c)	P$: \Delta I = 0$	parity must change

The existence of strongly allowed transitions with $\Delta I = \pm 1$, therefore, requires the presence of either the T or A interactions, or both. On the other hand, the fact that most allowed transitions are those having $\Delta I = 0$ and no parity change appears to require the S and/or V interaction. This subject is quite complex, and we shall not discuss it here, except to say that there is considerable evidence that only V and A are present. We shall discuss this matter further in Chap. 20. The student should refer to the current literature for the most up-to-date information in this field.

The transitions which do not follow the simple selection rules also yield momentum distributions whose Kurie plots do not quite have the usual linear form. The reason for this is simply that the $\boldsymbol{p}_e + \boldsymbol{p}_\nu$ factor in Eq. (27) introduces an additional energy dependence which is not present in the allowed transitions having $\Delta I = 0$. The effect of this additional term is not difficult to estimate, since we may simply multiply the energy-independent integral by $|\boldsymbol{p}_e + \boldsymbol{p}_\nu|^2$, suitably averaged over all directions of emission of the electron and neutrino. If we assume these to be independent, this average is

$$|\boldsymbol{p}_e + \boldsymbol{p}_\nu|^2_{\text{av}} = p_e{}^2 + p_\nu{}^2 \qquad (36)$$
$$= p_e{}^2 + (E_m - E_e)^2 \qquad (37)$$

Thus if we write the momentum distribution in the modified form

$$I_1(p_e) = G^2 |M_1|^2 \frac{64\pi^4}{h^9} p_e{}^2 (E_m - E_e)^2 [p_e{}^2 + (E_m - E_e)^2] \qquad (38)$$

[1] *Phys. Rev.*, **49**, 895 (1936).

we would expect to find

$$\left(\frac{I_1(p_e)}{p_e^2[p_e^2 + (E_m - E_e)^2]}\right)^{1/2}$$

$$= G|M_1| \frac{8\pi^2}{h^{9/2}} (E_m - E_e) \quad (39)$$

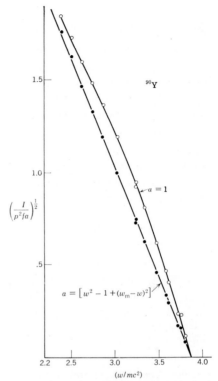

FIG. 15-9. Kurie plot for ^{90}Y, with and without the correction factor $a = [p_e^2 + (E_m - E_e)^2]$. [After Langer and Price, *Phys. Rev.*, **76**, 641 (1949).]

to be again a linear relation, as it is indeed found to be for first-forbidden spectra (Fig. 15-9). Higher-order spectra can be correspondingly corrected.

Finally, it is of interest to investigate the *strength* of the β-decay interaction. This can be evaluated for cases in which we have good reason to expect that the value of the matrix element M is known, as for the mirror nuclides where it should be nearly unity. From such cases it is found that

$$G = 1.41 \times 10^{-49} \text{ erg cm}^3 \quad (40)$$

which makes the β-decay interaction the weakest known except for that of gravity. Thus, while β-decay is, like gravitational attraction, a property of nuclei, it too appears to be essentially unrelated to the forces which govern the structure of nuclei, and its function in the over-all scheme of the universe is by no means clear as yet.

EXERCISES

15-14. Justify the use of only the first term of the expansion (7) by showing that $|\mathbf{p}_e + \mathbf{p}_\nu|R/\hbar \approx 10^{-2}$ for ordinary β-transitions.

15-15. Verify Eq. (23) by integrating Eq. (21) for the case $E_m \gg m_e$.

15-16. Suppose the neutrino possessed a small rest mass m_ν, perhaps 1 per cent or so of an electron mass. Modify the derivation of the statistical factor (15) to take such a rest mass into account. How would the expected shape of the Kurie plot be modified if $[I(p_e)/p_e^2]^{1/2}$ were still plotted vs. E_e?

15-9. Electron Capture

In nearly all cases in which a nucleus is unstable toward β^+-decay, it is also found to undergo another type of decay called *electron capture* which competes with the β^+-decay. In this process one observes that one of the orbital electrons of the atom is absorbed by the nucleus, resulting in a decrease of the nuclear charge by one unit. The elementary process is inferred to be

$$p + e^- \rightarrow n + \nu \tag{1}$$

Unlike ordinary β^+-decay, wherein the emerging electron possesses a sizable fraction of the released energy, the absorbed electron, being one of the ordinary atomic electrons, has *definite energy*, and so, therefore, has the neutrino. In fact, since the binding energy of an electron in an atom seldom exceeds a few tens of kilo electron volts, the neutrino carries away virtually all of the available energy.

The energy balance in electron capture also differs from that of β^+-decay in the important respect that β^+-decay cannot occur unless at least $2m_e$ of energy is available, as is indicated by Eq. 15-8(4b), whereas electron capture often occurs when β^+-decay cannot.

If the captured electron comes from the atomic K-shell, the process is called K-capture; if from the L-shell, L capture; and so on. K-capture is the most common type of electron capture observed, primarily because K-electrons have so much more probability of being found inside the nucleus than have the other electrons.

A theory of electron capture can be constructed along the same lines as that of β-decay, with the following differences:

1. The initial wave function ψ_i is now equal to the parent nuclear wave function u_i, multiplied by the total electronic wave function of the atom, rather than being u_i alone.
2. The final wave function is equal to the daughter nuclear wave function u_f, multiplied by the same electronic wave function as above, except with one electron removed (usually a K-electron), and multiplied also by a *plane wave* to represent the neutrino.
3. The statistical factor differs markedly from that used in β-decay because the momentum distribution of the electron is given in advance (being that appropriate to the atomic eigenstate in question) and that of the neutrino is fixed through the conservation of energy.

The probability of electron capture can be computed quite accurately relative to that of β^+-decay because both processes involve the same nuclear matrix element, which therefore drops out. The observed branching ratios agree very satisfactorily with those predicted theoretically.

REFERENCES

Allen, J. S.: "The Neutrino," Princeton University Press, Princeton, N.J., 1958.

Bethe, H. A., and P. Morrison: "Elementary Nuclear Theory," 2d ed., John Wiley & Sons, Inc., New York, 1956.

Blatt, J. M., and V. W. Weisskopf: "Theoretical Nuclear Physics," John Wiley & Sons, Inc., New York, 1952.

Friedlander, G., and J. W. Kennedy: "Introduction to Radiochemistry," John Wiley & Sons, Inc., New York, 1949.

Green, A. E. S.: "Nuclear Physics," International Series in Pure and Applied Physics, McGraw-Hill Book Company, Inc., New York, 1955.

Kaplan, I.: "Nuclear Physics," Addison-Wesley Publishing Company, Reading, Mass., 1955.

16

Systematics of Nuclear Stability

In this chapter we shall examine certain aspects of the *stability* of nuclei with respect to radioactive decay. We shall find it possible to describe the binding energy of the heavier nuclei semiquantitatively in terms of the interplay of several rather elementary physical effects, with the result that the observed pattern of isotope stability is given a rather satisfactory physical foundation. Indeed, upon the basis of this analysis we shall also be able to predict with fair accuracy the *type of decay* and the approximate energy release of a nucleus which is *not* stable. Thus what at first appears to be a chaotic and arbitrary distribution of stability among the various possible nuclei will be seen instead actually to be an *orderly arrangement* following definite and simple rules.

It should be emphasized, of course, that the discussion of the present chapter represents only a crude first approximation to an *exact* description of the properties of nuclei. The nuclear model of this chapter must be considerably refined and supplemented in order to describe the many properties of nuclei that do not find complete or satisfactory interpretations at the present level of discussion.

As an indication of the kind of observations that we shall endeavor to fit into a logical framework, we list the following empirically observed properties of nuclei:

1. Light nuclei tend to have atomic weights that are equal to about twice their atomic numbers. The heavier nuclei, on the other hand, tend to have atomic weights *somewhat greater* than twice their atomic numbers. This average ratio is a rather smoothly increasing function of the atomic number.

2. Elements of odd Z have only one or two stable isotopes, and both

of these are of odd A. Only six odd-Z nuclides violate this rule: ^{2}H, ^{6}Li, ^{10}B, ^{14}N, ^{50}V, and ^{180}Ta.

3. Elements of even Z have only one or two stable isotopes of odd A but may have *several* stable isotopes of even A.

4. Except for the special case of hydrogen, elements of even atomic number are *several times as abundant* as are elements of odd atomic number.

We shall find it possible to interpret these striking regularities in terms of a quite simple theory of nuclear structure.

16-1.   Binding Energy of Nuclei

One of the most important clues to the structure of nuclei comes from a study of their binding energies, that is, the surplus energy which nucleons give up by virtue of their mutual attractions when they become bound together as a nucleus. As was pointed out in Chap. 13, this binding energy can be found by comparing the measured mass of a nucleus with the sum of the masses of the corresponding free nucleons. Now let

$M(Z,A)$ = the measured mass of the neutral atom (Z,A)

M_H = the mass of neutral atomic ^{1}H (1.008142 amu)

M_n = the mass of a neutron ^{1}n (1.008983 amu)

Then if the nuclear and electronic binding energy were negligible, the neutral atom would have a mass

$$M'(Z,A) = ZM_H + (A - Z)M_n \tag{1}$$

The actual mass of the atom is, however, *less* than this.   The difference may be attributed to the binding energy E_B of the nucleus and of the electrons, which would have to be supplied to separate the atom into hydrogen atoms and neutrons.   According to Einstein's celebrated equation, we may evidently write

$$E_B = c^2[M'(Z,A) - M(Z,A)]$$
$$= c^2[ZM_H + (A - Z)M_n - M(Z,A)] \tag{2}$$

(It may be remarked that the universal agreement of mass-energy relations of this kind with experimental results constitutes an exceedingly strong confirmation of the correctness of the special theory of relativity.)

It is found experimentally that the *binding energy per particle*

$$\frac{E_B}{A} = \frac{c^2}{A}[ZM_H + (A - Z)M_n - M(Z,A)] \tag{3}$$

*is roughly a constant for $A > 25$, as shown in Fig. 16-1.

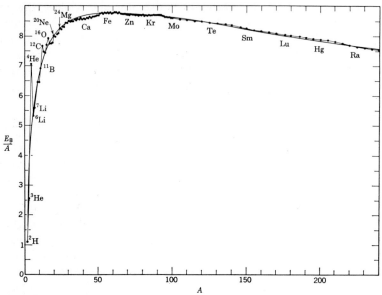

FIG. 16-1. The binding energy per particle as a function of the mass number A. The solid curve represents a semiempirical binding-energy formula [Eq. 16-3(2)].

A commonly used notation for expressing nuclear binding energies is by means of the so-called *packing fraction*. This quantity is defined by the equation

$$f = \frac{M(Z,A) - A}{A} \qquad (4)$$

where $M(Z,A)$ is expressed in atomic mass units. In this notation the packing fraction of ^{16}O is exactly zero, and may be either positive or negative for other atoms. The quantity

$$\Delta M = M(Z,A) - A \qquad (5)$$

is called the *mass excess*.

EXERCISE

16-1. Obtain an expression analogous to Eq. (3) which involves the packing fractions, f_H, f_n, and f rather than M_H, M_n, and M.

Figure 16-1 shows that those atoms in the neighborhood of $A = 55$ ($\sim$Fe) are the most stable, since they possess the maximum binding energy per nuclear particle. This means that energy would be released either in the *fusion* of two light elements to form a heavier element or in the *fission* of a heavy element to form two lighter elements. Both of these processes have been utilized in nuclear explosions, and both may be of considerable practical importance in the controlled production of nuclear energy.

The present chapter will be concerned primarily with that region of Fig. 16-1 in which $A > {\sim}40$, where the binding energy per particle shows a rather smooth variation with A. Some of the features exhibited by the lighter elements will be treated in another chapter.

16-2. The Sizes of Nuclei

We shall now briefly examine the evidence regarding the *sizes* of nuclei, since this enters critically into our further considerations of binding energy and the stability of nuclei. Experimental evidence is available from several sources:[1]

THE CROSS SECTION OF THE NUCLEUS FOR FAST NEUTRONS. Perhaps one of the most direct measurements of the radius of a nucleus is by means of the cross section it presents toward a beam of fast neutrons. The idea here is that a fast neutron—one whose wavelength is sufficiently small compared to the size of the nucleus—should cause a nuclear disintegration if it passes within the projected area of the nucleus and should not interact if it passes outside that area. (Neutrons, rather than protons, must be used in order to eliminate the effects of the purely electrical interaction.) By this means it is found that the radius of a heavy nucleus such as lead or uranium is about 10^{-14} m and that the radius of an intermediate nucleus such as iron is about 6×10^{-15} m.

THE LIFETIMES FOR α-PARTICLE DECAY. As was described in the previous chapter, the theory of α-particle decay of heavy radioactive nuclei assumes that inside a nucleus there is an energy state for an α-particle which lies above the zero of energy represented by a free α-particle and the residual nucleus. The α-particle is hindered in leaving the nucleus, however, by the existence of a potential barrier over which it cannot go because its energy is insufficiently great. The α-particle escapes from the nucleus by "leaking through" the barrier, and the probability per unit time that this will occur is related to the energy available, the height and width of the barrier, and *its radius*. When the calculations are carried out, it is found that the effective lifetime for α-particle decay is a sensitive function of the radius assumed for the nucleus, and the observed α-decay lifetimes can then be used as a means of *defining* this radius. The radii of the heavy elements, calculated in this way, lie between 8.4 and 9.8×10^{-15} m.

---

[1] For an extensive discussion of the measurement of nuclear size by various methods, see: Papers from the International Congress of Nuclear Sizes and Density Distribution held at Stanford University, *Revs. Modern Phys.* **30**, 412 (1958).

THE ELECTROSTATIC-ENERGY RELATIONSHIPS IN LIGHT NUCLEI.   In the previous chapter we saw that there are certain pairs of so-called *mirror nuclei*, which have n protons and $n + 1$ neutrons, and $n + 1$ protons and n neutrons, respectively.   Such nuclei are of great theoretical interest because the number of neutron-proton interactions is precisely the same in each nucleus of a given pair, and the numbers of neutron-neutron and proton-proton interactions are just interchanged.   It turns out that the difference in binding energy of such a pair of nuclei can be mostly accounted for in terms of the observed mass difference of a proton and a neutron and *the additional Coulomb energy of the nucleus of higher Z*. This Coulomb energy can be expressed in terms of the radius of the "sphere" throughout which the protons are assumed to be (uniformly) distributed, and agreement with observation is obtained if the radius of the sphere is taken to be about $1.3A^{1/3} \times 10^{-15}$ m.

FINE-STRUCTURE SPLITTING OF ATOMIC ENERGY LEVELS OF μ-MESONS (μ-MESIC ATOMS).   Another means of measuring the sizes of nuclei makes use of the "atomic" spectroscopy of artificially produced negative μ-mesons which are captured into "Bohr orbits" around nuclei.   The properties of μ-mesons will be described in a later chapter; it suffices here to say that a μ-meson is very much like an electron except that its mass is 207 times as great and that it has very little, if any, interaction with nuclei other than through its electric charge.

Because of the much larger mass of the μ-meson, the energy levels are 207 times greater, and the scale of the wave functions is reduced by a factor of 207 [Eqs. 5-3(38) and (39)].   This means that a negative μ-meson, moving near a neutral atom, will be captured into bound states which lie deep inside the electron cloud, and the lowest states will penetrate significantly *inside the nucleus itself*.   (The μ-meson has so small an interaction with the individual nuclear particles that it is able to penetrate the nucleus many thousands of times without "hitting" an individual proton or neutron.)

It is found that the μ-meson in its lowest states penetrates the nucleus to such an extent that the wave functions and energy levels depend in a measurable way upon the *size* of the nucleus and upon the *distribution of charge* throughout the nuclear volume.   Transitions between these lowest states provide a quantitative measure of this effect of the finite size of the nucleus, and they yield the result $R = 1.2A^{1/3} \times 10^{-15}$ m for nuclear radii in the range $\sim\!13 < Z < \sim\!80$.[1]   This use of the μ-meson as a "nuclear-probe" particle has many possibilities, only a few of which have as yet been successfully exploited.

---

[1] Fitch and Rainwater, *Phys. Rev.*, **92**, 789 (1953); Cooper and Henley, *Phys. Rev.*, **92**, 801 (1953); Wheeler, *Phys. Rev.*, **92**, 812 (1953).

SCATTERING OF HIGH-ENERGY ELECTRONS FROM NUCLEI. Some of the most precise measurements of the charge distribution throughout nuclei have been made by observing the angular distribution of high-energy (~500 Mev) electrons that have been scattered by their interaction with the nuclear electric and magnetic fields. In an impressive series of experiments, Hofstadter and his coworkers have measured the sizes of several nuclei from hydrogen to gold.[1]

All of the preceding methods of measuring the sizes of nuclei are in agreement with the hypothesis that the nuclear volume is proportional to the number of particles in it. That is, $\tfrac{4}{3}\pi R^3 \sim A$. This experimental relationship is usually expressed in the form

$$R = R_0 A^{\frac{1}{3}} \tag{1}$$

with
$$R_0 \approx 1.2 \times 10^{-15}\ \text{m} \tag{2}$$

where the given value of R_0 is accurate to within about 10 per cent.

EXERCISES

16-2. Evaluate the probability that a μ-meson in the u_{100} state will be found inside a sphere of 10^{-14} m radius scribed about a "point" nucleus of $Z = 82$. This may be taken as a rough estimate of the fraction of the time the μ-meson would be found inside a lead nucleus.

16-3. Equation (1) asserts that all nuclei have the *same density*. Evaluate this density, and express the result in grams per cubic centimeter.

16-3. The Semiempirical Mass Formula of Weizsacker

We are now in a position to combine some of the observed facts of nuclear structure into a theoretical form which satisfactorily describes many of the main features of the masses of nuclei. The mass formula to be obtained is, however, semiempirical, in that the numerical values of the parameters entering into it are evaluated as to agree with experiment, rather than being deduced from first principles.

We begin by considering the most important observed properties of nuclei that are relevant to our discussion, and the bearing these properties have upon the question of the binding energies of nuclei.

1. Nuclei have a constant density and an almost constant binding energy per particle. These properties are similar to the familiar situation in *chemical* binding and indicate a similar character for the interparticle forces in the two cases with respect to their *saturation* properties.

[1] Hahn, Ravenhall, and Hofstadter, *Phys. Rev.*, **101**, 1131 (1956); Hofstadter, *Revs. Modern Phys.*, **28**, 214 (1956).

Thus it appears that a given nucleon cannot interact strongly with all of the other nucleons in a nucleus, but enters into bonding relationships with only a *limited number of its nearest neighbors.*

2. Following the analogy with chemical forces, we may expect that, while the greatest contribution to the binding energy of a nucleus would be a term proportional to the volume of the nucleus, the contribution made by those nucleons which reside at the surface of the nucleus would not be as great as by those inside, because the surface nucleons are not completely surrounded by other nucleons with which they can form "bonds." Thus there should be *surface-tension effects,* i.e., effects tending to make the surface area of the nucleus a minimum. The binding energy associated with this effect should be proportional to the number of nucleons in the surface of the nucleus, that is, to $A^{2/3}$, and should be negative.

3. It is observed that, in the lighter elements, those nuclei having the greatest stability have *equal numbers of protons and neutrons,* and that nuclei having *even* numbers of protons *and even* numbers of neutrons are especially stable. (The prominent peaks in the binding-energy curve of Fig. 16-1 correspond to ^{4}He, ^{12}C, ^{16}O, and ^{20}Ne.)

This suggests that the distributions of energy states for protons and for neutrons in the nucleus are quite similar, and that there is a strong tendency for *pairs* of protons or neutrons to go into each energy state with opposed spins. As a matter of fact, it is believed that the specifically *nuclear* properties of neutrons and protons are identical and that the only differences between the two particles are the result of their different electrical charge distributions. We shall later

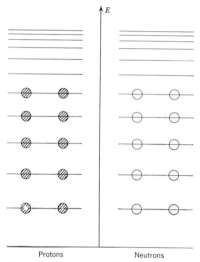

Protons   Neutrons

FIG. 16-2. Schematic illustration of the energy states available to protons and neutrons in a nucleus.

consider this matter at greater length, but for the present we shall proceed by using the hypothesis that the nuclear properties of neutrons and protons are the same.

If we represent the energy states available to protons and to neutrons schematically as in Fig. 16-2, we can deduce that the binding energy for a given A is a maximum if $N = Z$ and varies approximately quadratically with $N - Z$ if $N \neq Z$.

From considerations such as the above, one expects a term of the form $f(A)(N - Z)^2$ to appear in the binding-energy expression.

EXERCISE

16-4. Consider a nucleus with an even number of nucleons. Assume that the energy states available to protons and neutrons are spaced an amount ϵ apart in the neighborhood of the uppermost filled level when $N = Z$. Show that, if $N \neq Z$, the decrease in the binding energy is approximately equal to $\frac{1}{4}\epsilon(N - Z)^2$, assuming that only one particle of each type can occupy each level.

4. The mutual electrostatic energy of the protons must be taken into account in evaluating the binding energy. It is usually assumed that the Z protons are distributed uniformly throughout a sphere of radius $R = R_0 A^{1/3}$. Experiments indicate that this is only roughly true, but it is still a useful approximation and we shall use it in our treatment.

EXERCISE

16-5. Show that the work required to bring into superposition Z uniformly charged spheres of radius R with a charge $+e$ on each sphere is

$$W = \frac{3}{5}\frac{Z(Z - 1)e^2}{4\pi\varepsilon_0 R} = 0.71Z(Z - 1)A^{-1/3} \quad \text{MeV} \quad (1)$$

(If protons are brought, one by one, from infinity and inserted into a nucleus, each proton in turn becomes "smeared out" more or less uniformly over the nuclear volume. This expression is a fairly good approximation to the electrical work that would be done in the process.)

5. Finally, there is a small but qualitatively important effect called the *odd-even effect* which applies to cases in which there is an odd number of nucleons of one kind or the other in a nucleus. It expresses the fact that a single nucleon in a given state is not as strongly bound as are two similar nucleons in this same state. This effect is usually introduced in a form such that, if N and Z are both even, a small amount is *added* to the binding energy, and if N and Z are both odd, this same amount is *subtracted* from the binding energy. If N is even and Z is odd, or vice versa, the term is omitted.

Combining the various physical effects described above, we are led to write the binding energy of a nucleus in the form (Weizsacker, 1935)

$$
\begin{aligned}
E_{\text{B}} = {}& \alpha A && \text{volume effect} \\
& - \beta A^{2/3} && \text{surface effect} \\
& - (N - Z)^2 f(A) && \text{isotope effect} \\
& - \gamma Z(Z - 1)A^{-1/3} && \text{Coulomb effect} && (2) \\
& + \delta(A) && N \text{ and } Z \text{ both even} && \\
& - \delta(A) && N \text{ and } Z \text{ both odd} && \\
& + 0 && A \text{ odd} &&
\end{aligned}
$$

A even   odd-even effect

The coefficients α, β, and γ and the functions $f(A)$ and $\delta(A)$ are evaluated by a combination of theoretical computation and matching with experimental atomic masses. In this way the following numerical values are obtained, where E_B is expressed in MeV:‡

$$\alpha = 15.7 \text{ MeV} \qquad f(A) = (23.6/A) \text{ MeV}$$
$$\beta = 17.8 \text{ MeV} \qquad \delta(A) = (132/A) \text{ MeV}$$
$$\gamma = 0.712 \text{ MeV}$$

Finally, we may write the following equation for the *mass* of a given atom

$$
\begin{aligned}
M(Z,A) &= ZM_H + NM_n - E_B \\
&= AM_n - Z(M_n - M_H) - E_B \\
&= 939.526A - 0.784Z - 15.7A + 17.8A^{2/3} + 23.6(A - 2Z)^2 A^{-1} \\
&\qquad + 0.712Z(Z - 1)A^{-1/3} \pm 132A^{-1} \begin{cases} \text{odd-odd} \\ \text{even-even} \end{cases}
\end{aligned}
\qquad (3)
$$

This mass formula is found to describe the main features of the variation of atomic mass with A and Z sufficiently accurately for many practical purposes, for $A > \sim 5$. The general agreement of the mass formula with observed masses is shown by the solid curve in Fig. 16-1. We shall make a more detailed comparison in our forthcoming discussion of β-stability.

16-4. The Stability of Nuclei, Stability Rules

One of the most illuminating applications of the Weizsacker mass formula is to the problem of accounting for the observed pattern of stability and instability among the various possible nuclei that might be imagined to exist. We shall find that this formula is able to account surprisingly well for the so-called "nuclear stability rules" that have long been known. In order to carry out this application, we must first consider what criteria a stable nucleus must satisfy.

One might consider the most general condition for stability to be: If the rest mass of a given atom is smaller than, or equal to, the total rest mass of any combination of components into which it may be (conceptually) divided, the atom should be stable; if not, it should be unstable with respect to decay into such combinations of component parts as have a smaller total rest mass.

In applying such a criterion, however, it must be borne in mind that the existence of a mass inequality indicating instability does not in itself *guarantee* that a nucleus will have a detectable radioactivity. The

‡ Green, *Phys. Rev.*, **95**, 1006 (1954).

possible existence of *barriers*, as in α-particle decay, or of *selection rules*, or both, may make a nominally unstable atom be actually stable for practical purposes. As a matter of fact, there are a great many known nuclear species which should be unstable by the above criterion, but whose decay half-lives are unmeasurably long. For definiteness, therefore, we list briefly the kinds of instability a nucleus in its ground state most commonly exhibits:

1. Instability against heavy-particle emission. If $M(Z,A) > M(Z - 1, A - 1) + M_H$, we may conclude that the atom $M(Z,A)$ will spontaneously decay *by emitting a proton*. Similarly, for other simple heavy particles $M(Z,A) > M(Z, A - 1) + M_n$ indicates *neutron instability*, $M(Z,A) > M(Z - 2, A - 4) + M_{He}$ indicates α-particle instability, and so on.

EXERCISE

16-6. Several naturally occurring α-emitters, but no proton- or neutron-emitters, are observed. Explain this fact in terms of the mass formula and the stability criteria just given.

2. Instability against β-decay or electron capture. One of the most common types of instability of a nucleus is that leading to β-decay or electron capture. An atom is β^--unstable if $M(Z,A) > M(Z + 1, A)$, and unstable against electron capture if $M(Z,A) > M(Z - 1, A)$. If $M(Z,A) > M(Z - 1, A) + 2m_e$, it is also unstable against the emission of positrons.

3. Instability against spontaneous fission. An extension of the conditions of instability against heavy-particle emission to the extreme limit in which the masses of the two resultant fragments are *about equal*, that is, $M(Z,A) > M(Z',A') + M(Z - Z', A - A')$, where $Z' \approx \frac{1}{2}Z$, $A' \approx \frac{1}{2}A$, leads to the concept of *spontaneous fission*. It is found that at least one heavy nucleus, ^{238}U, is unstable against decay of this kind, with a half-life of about 10^{16} years. The mass formula is capable of accounting qualitatively for this process, but some refinement is necessary for a quantitative treatment. Fission will be discussed in a later chapter.

Of the above-listed types of instability an atom may possess, the instabilities toward β-decay and electron capture are by far the most important because the 8-MeV binding energy of a nucleon inside the nucleus is almost always sufficient to prevent the emission of a heavy particle. Thus, in order to investigate the stability pattern of nuclei, only β-decay and electron capture need be considered.

Let us first discuss the problem for an atom with an odd value of A

so that the $\delta(A)$ term does not enter.   Equation 16-3(3) shows that the mass of an atom of given A varies quadratically with Z on either side of a certain minimum point.   The mass may thus be written

$$M(Z,A) = C_A + \tfrac{1}{2}B_A(Z - Z_A)^2 \tag{1}$$

The coefficients may be determined in this equation by first determining Z_A (regarding Z as a continuous variable for this purpose) using the condition

$$\left.\frac{\partial M}{\partial Z}\right|_{Z = Z_A} = 0$$

That is,

$$-0.784 - 94.4(A - 2Z_A)A^{-1} + 0.712(2Z_A - 1)A^{-\frac{1}{3}} = 0$$

The second term in this expression, which originates from the isotope effect, thus tends to make $Z_A = \tfrac{1}{2}A$, whereas the first and third terms,

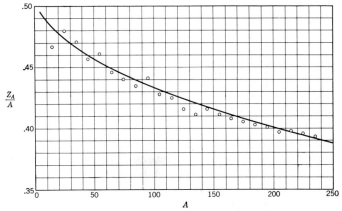

FIG. 16-3. The ratio Z_A/A vs. A, as given by Eq. 16-4(2).   The points represent the observed values of Z/A for some odd-A nuclides.

which originate from the proton-neutron mass difference and the Coulomb effect, respectively, tend to make Z_A *differ* from $\tfrac{1}{2}A$.   The result of these three effects is that

$$Z_A = \frac{95.2 + 0.712A^{-\frac{1}{3}}}{188.8 + 1.424A^{\frac{2}{3}}} A \tag{2}$$

which shows that Z_A is always less than $\tfrac{1}{2}A$, in the range of validity of the mass formula.   The ratio Z_A/A given by this equation is shown graphically in Fig. 16-3.

Inspection of Eq. 16-3(3) reveals that

$$B_A = 188.8A^{-1} + 1.424A^{-\frac{1}{3}} \tag{3}$$

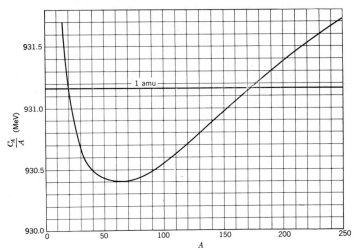

FIG. 16-4. Graph of C_A/A vs. A.

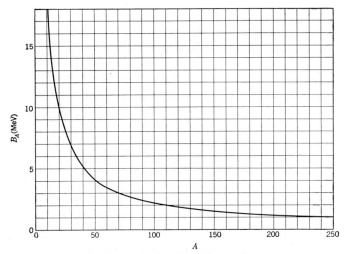

FIG. 16-5. Graph of B_A vs. A.

C_A can then be determined and is found to be

$$C_A = 947.4A + 17.8A^{2/3} - \frac{1}{2}\frac{(95.2 + 0.712A^{-1/3})^2}{188.8 + 1.424A^{2/3}}A \tag{4}$$

C_A/A and B_A are shown graphically in Figs. 16-4 and 16-5. The variation of mass with Z for odd A is shown in Fig. 16-6 for $A = 65$.

Recalling the criteria for β-stability, we see from Fig. 16-6 that *there should be only one stable isobar for a given odd value of A,* the neighboring isobars of lower Z being unstable with respect to β-decay, and of higher

Z, with respect to electron capture and possibly β^+-decay also. *The stable isobar should be that one whose atomic number lies nearest to Z_A.* It may rarely occur that Z_A lies midway between two integers, so that there might be a few cases in which two isobars of a given odd A exist.

If we compare the above result with observation, we find that it is indeed true that there is generally only one stable isobar for each odd value of A, the only exceptions being $A = 113$ (^{113}Cd, ^{113}In), $A = 115$ (^{115}In, ^{115}Sn), and $A = 123$ (^{123}Sb, ^{123}Te).

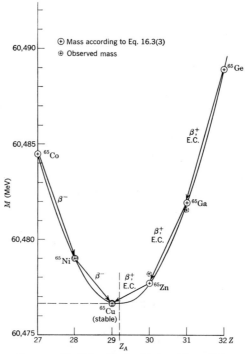

FIG. 16-6. Variation of mass with Z for fixed odd A. ($A = 65$ is used as an example.)

Taking now the case in which A is even, we proceed exactly as above and obtain the same equations for Z_A, B_A, and C_A, *but we must now include the $\delta(A)$ term.* The situation for this case is indicated in Fig. 16-7. This figure shows how the inclusion of the $\pm \delta(A)$ term gives *two* parabolas, displaced up and down from the one described by Eq. (1) by equal amounts $\pm \delta(A)$, odd-Z values (corresponding to the so-called "odd-odd" nuclei) being associated with the upper, and even-Z values ("even-even" nuclei) with the lower.

Figure 16-7 shows cases for which one or two stable isobars exist. When actual numerical values are used in the analysis, it turns out that most even mass numbers have two stable isobars, somewhat fewer have one stable isobar, and a few have three stable isobars. A further quite important result of the numerical analysis is that *no stable odd-odd isobars should exist in the range of A covered by the mass formula.* The above results are in striking agreement with observation. Only two apparently stable odd-odd nuclei, $^{50}_{23}$V and $^{180}_{73}$Ta, are observed with $A > 14$, and one, two, or three stable even-even isobars are commonly observed.

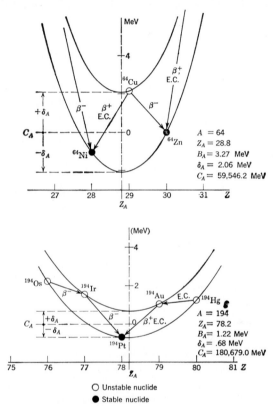

$A = 64$
$Z_A = 28.8$
$B_A = 3.27$ MeV
$\delta_A = 2.06$ MeV
$C_A = 59,546.2$ MeV

$A = 194$
$Z_A = 78.2$
$B_A = 1.22$ MeV
$\delta_A = .68$ MeV
$C_A = 180,679.0$ MeV

○ Unstable nuclide
● Stable nuclide

FIG. 16-7. Variation of mass with Z for fixed even A.

Our analysis of nuclear stability up to this point has been concerned with the stability of various *isobars* of given A. One may wish to express the information thus obtained in other ways, however. For example one may wish to know how many stable *isotopes* exist for given Z. We can analyze this situation by dividing the discussion into two parts, as before:

1. Z odd. The instability of odd-odd isobars leads us to conclude that *an element of odd Z will have no isotopes of even mass number.* (The only exceptions to this rule are the light nuclei ^{2}H, ^{6}Li, ^{10}B, ^{14}N, and the nuclides ^{50}V and ^{180}Ta.)

On the other hand, an odd-Z element must have *at least one stable isotope*, and *may have, at most, two stable isotopes, of odd mass number*. The situation leading to two stable isotopes for odd Z is as follows: The integral Z nearest to each Z_A represents the stable isobar for given odd A; but because the successive values of Z_A are spaced not at exactly half-integral intervals but rather at *smaller* intervals, there will every so often be a value of Z which lies nearest to *two successive* such Z_A values. These values of Z will then correspond to situations in which the given element will have *two* odd-A isotopes.

2. Z even. Since the situation just described might occur at an even value as well as at an odd value of Z, it is clear that the same rule regarding odd-A isotopes should hold for even Z also. That is, an element of even atomic number will have *at least one* stable isotope, and *at most two* stable isotopes, of odd mass number.

On the other hand, since all even-A isobars lead to *even* atomic numbers, we conclude that an element of even-Z number will have *at least one, and perhaps several, isotopes of even* A.

The analysis we have outlined in this section is remarkably successful in its ability to account for the observed pattern of stability of the various isotopes of the elements, and it appears to provide a reasonable physical basis for the observations. Thus, while there are some exceptions to the qualitative conclusions reached upon the basis of this simplified theory, and a large number of minor numerical inaccuracies in its detailed results, it must still be considered quite remarkable that so elementary an approach can provide so complete an interpretation of the gross systematics of both stable and unstable nuclides.

Finally, we emphasize once again that the procedure outlined in this chapter also permits one to predict the approximate *energy release* to be expected in the β-decay of a given unstable nucleus by evaluating the numerical coefficients B_A, Z_A, and $\delta(A)$ for the given value of A and inserting these into Eq. 16-4(1) for the initial and final values of Z.

EXERCISES

16-7. Verify Eqs. 16-4(2), (3), and (4).

16-8. Derive expressions for the energy release in the β-decay of an atom (Z,A) (a) of odd A and (b) of even A.

Ans.: (a) $\Delta E = B_A(Z_A - Z - \frac{1}{2})$ $\qquad\qquad\qquad\qquad$ A odd

(b) $\qquad \Delta E = B_A(Z_A - Z - \frac{1}{2}) \begin{cases} +2\delta(A) & Z \text{ odd} \\ -2\delta(A) & Z \text{ even} \end{cases}$ A even

16-9. For what elements should stable isobars exist for (a) $A = 80$ and (b) $A = 97$?

16-10. For what elements should stable isobars exist for (a) $A = 169$ and (b) $A = 166$?

16-11. Investigate the stability of the following atoms and give the types of decay and approximate energy releases for those which you find to be unstable: (a) ^{105}Pd, (b) ^{105}Ag, and (c) ^{180}Ta.

16-12. Derive an expression for the approximate energy release in the α-decay of a nucleus $(A + 2, Z + 1)$.

$$Ans.: \quad E = \frac{1.424(2Z - 1)}{A^{1/3}} + \frac{47.5}{A^{1/3}} - \frac{0.95Z(Z - 1)}{A^{4/3}} - \frac{94.5(A - 2Z)^2}{A^2}$$
$$- 34.8$$

16-13. Find the approximate energy release of the α-active nucleus $^{229}_{91}Pa$.

16-14. When ^{235}U absorbs a slow neutron, it undergoes fission. If one of the fission products is ^{141}Cs, (a) what is the other product? (Assume no neutron emission.) (b) Approximately how much energy should be released in this process? (c) If both products undergo β-decay until they become stable, how many will each undergo? *Ans.:* (a) ^{95}Rb. (b) 180 MeV. (c) 4,5.

REFERENCES

Blatt, J. M., and V. F. Weisskopf: "Theoretical Nuclear Physics," John Wiley & Sons, Inc., New York, 1952.

Feenberg, E.: Semi-empirical Theory of the Nuclear Energy Surface, *Revs. Modern Phys.*, **19**, 239 (1947).

Friedlander, G., and J. W. Kennedy: "Introduction to Radiochemistry," John Wiley & Sons, Inc., New York, 1949.

Green, A. E. S.: "Nuclear Physics," International Series in Pure and Applied Physics, McGraw-Hill Book Company, Inc., New York, 1955.

17

Dynamics of Nuclear Reactions

We have so far discussed certain aspects of nuclear structure which relate to the spontaneous transmutation of one nuclear species into another through α-decay or β-decay. As is well known, however, these processes by no means account for all of the types of transmutation a nucleus can undergo. In particular, a wide variety of *nuclear reactions* can take place, and many of these provide a powerful means of studying the finer details of nuclear structure. Indeed, by far the greater part of the quantitative information regarding nuclear structure and excited states that we now possess has been obtained through the study of nuclear reactions under well controlled experimental conditions. In such experiments a beam of particles, preferably homogeneous in energy, is allowed to impinge upon a *target* which contains atoms of a certain nuclide, and the consequences of the resulting impacts are studied quantitatively by measuring the directions of motion and speeds of the various types of particles that are produced. In the present chapter we shall examine some of the more general properties of nuclear reactions and excited states.

17-1. Kinds of Nuclear Reactions

Before considering the quantitative aspects of nuclear reactions, we shall consider briefly some of the more important mechanisms by which such reactions commonly occur, in an approximate order of increasing "violence" of interaction.

A. Rutherford Scattering and Coulomb Excitation. If a target nucleus is bombarded by charged particles, the mutual interaction of the electric charges leads to the familiar phenomenon of elastic *Rutherford scattering*,

562

the main features of which were presented in Chap. 14.   In addition to such elastic scattering, an inelastic process can also occur, in which the electric field of the incident particle, interacting with the electric moments (dipole, quadripole, etc.) of the target particle, excites the latter into a higher energy state, from which it subsequently decays by γ-ray emission. This process, which is called *Coulomb excitation*, is commonly used in the study of the low-lying excited states of the heavier nuclei.

B. Nuclear Potential Scattering.   The processes just described may occur even if the nuclear encounter is so distant as to be completely outside the range of nuclear forces.   For closer encounters, however, another type of elastic scattering occurs, in which the incident particle is scattered by the *nuclear* forces between it and the target particle, but in a relatively coherent way.   In this type of scattering, which is called *nuclear potential scattering*, the incident particle responds to an average potential field due to the nucleons of the target nucleus, and neither the incident particle nor the target nucleus is left in an excited state.   Because of the coherent nature of nuclear potential scattering, one observes characteristic quantum-mechanical interference effects between this process and the elastic Rutherford scattering.   Nuclear potential scattering was first studied intensively using neutrons as the bombarding particles, but it has also been observed quantitatively with charged particles, notably protons.

C. Surface Scattering.   A somewhat more violent interaction than that leading to nuclear potential scattering may occur, wherein the incident particle acts especially strongly upon one or more of the nucleons at the surface of the target nucleus, and is thereby scattered inelastically, leaving the target nucleus in an excited state.   This process, called *nuclear surface scattering*, is somewhat analogous to Coulomb excitation, except that it takes place through the nuclear force interaction rather than the electrical interaction between nuclei.

D. Surface Transmutation.   In an encounter similar to the above, the incident particle may actually directly eject one or more particles from the target nucleus, and itself be scattered or captured.   A reaction of this kind is called a *surface transmutation*.

E. Compound Nucleus Formation.   In contrast with the above *direct* nuclear reactions, a great body of experimental evidence indicates that many nuclear reactions take place in two rather well defined steps, if the kinetic energy of the bombarding particles does not exceed about 30 MeV. In the first step, the incident particle interacts with the stationary, or target, nucleus and forms with it a *compound nucleus* whose mass number and charge are, respectively, the sum of the mass numbers and charges of its two component parts.   The nuclei which combine to form this compound nucleus become mixed together and *soon lose their individual identities.*

The second step in the nuclear reaction occurs when this excited compound nucleus *decays into the product particles*. An important feature of this decay process is that it is *independent of the particular mode of formation of the compound nucleus, and depends only upon its state of excitation*. This separability of the formation and decay of the compound nucleus greatly aids in the study of the states of excitation of nuclei, since one can produce a given nucleus as an excited compound nucleus by using various combinations of bombarding and target particles.

One useful view of the physical process leading to the formation and subsequent decay of a compound nucleus is that the nucleons of the system are, at the instant of impact, in a relatively well-ordered state of motion—a state that is *statistically unlikely* for the composite nucleons. The nucleons then proceed to attain a more probable state of motion in which no single nucleon or small group of nucleons has a very large share of the total excitation energy, and the nucleus may remain in such a state for some time. There will, however, be *fluctuations* about this statistically probable state, and it may eventually happen that a single nucleon or some small group of nucleons will again acquire sufficient energy to escape from the nucleus, and the nucleus may then decay. According to this view the decay of an excited nucleus is closely analogous to the *evaporation of a molecule from a small drop of liquid*, with a resultant cooling (deexcitation) of the drop. If the nucleus is in a high state of excitation, several particles may "evaporate" from it in this way. These particles are often called *evaporation particles*, because of this analogy.

The concept of the compound nucleus was first introduced by Bohr in 1936 and has been applied quite fruitfully in many ways in nuclear physics. Some of these applications, and some of the limitations of the concept, will be described from time to time in our treatment of specific phenomena.

F. Stripping and Pickup Reactions.   There also exists a broad class of nuclear reactions called *stripping reactions*, in which a composite nucleus, in passing near a target nucleus, loses one or more nucleons to the target nucleus. Such a reaction is *direct*, in that a compound nucleus is not formed; its mechanism may be pictured qualitatively as follows: At a given energy and orbital angular momentum, there is a certain probability that one or more of the nucleons will be moving within the *incident* nucleus in such a way that, as the latter passes close by the target nucleus, these nucleons are also moving in just the right way to be captured into a certain "orbit" about the *target* nucleus. The stripped nucleus recoils from the encounter with a momentum change approximately equal to and opposite from that carried away by the missing nucleons. Because of the direct nature of the stripping process, the CM angular distribution of the outgoing particles is not isotropic but shows characteristic diffractive maxima and minima, predominantly in the forward direction.

The inverse of the stripping reaction is the so-called *pickup* reaction, in which one or more of the target nucleons is picked up by the passing particle. In this case the CM angular distribution also shows a diffractive pattern. Indeed, essentially the same theory may be applied to the inverse cases of (d,p) and (p,d) or (d,n) and (n,d) reactions.

G. Spallation Reactions. The reactions so far described are characteristic of relatively low bombarding energies, from essentially zero up to a few times the average nucleon binding energy of about 8 MeV. At higher energies, say above 50 MeV or so, nuclear reactions differ considerably from those observed at lower energy. The main feature of the higher energy reactions is that the primary interaction occurs more and more between *individual nucleons* of the two interacting nuclei rather than between the nuclei as a whole, as is to be expected on the basis both of the decreased deBroglie wavelength of the incident nucleons and the higher energy available to the system. Whereas in the lower energy range there is seldom enough energy available to eject more than one particle from the target nucleus, and even this generally at the expense of the capture of the incident particle, in the higher energy range there is sufficient energy for the ejection of several particles, either singly or as compound structures. A reaction in which several products of various size are produced is called a *spallation reaction*, and the several pieces are called *spallation products*.

H. High-energy Reactions. In the energy range above about 150 MeV, the spallation reactions just described merge more or less smoothly into new kinds of reactions in which new kinds of particles, in addition to protons and neutrons, emerge. First to appear are the π-mesons, which are believed to be the major source of interaction between nucleons, but at energies above 500 MeV or so a whole family of particles, the so-called *strange particles*, make their appearance also; and at energies above about 6000 MeV the *antinucleons* are produced. We shall discuss these various kinds of particles in Chap. 20 and some of the main features of nuclear reactions at extreme energies in Chap. 21. Although many of the results noted in the present chapter may be applied at higher energies, we shall restrict our attention mainly to the lower energy range, below about 50 MeV.

17-2. Common Modes of Formation of Excited Nuclei

We shall now briefly consider some simple examples of nuclear reactions by which given nuclei of interest may be studied. Let us assume that we are interested in the excited states of a certain nucleus (Z,A).

In order to obtain as many of the excited states of this nucleus as possible, we examine all possible ways of making the nucleus out of the raw materials that are available. The most common ways are as follows:

1. *By Coulomb excitation of the nucleus (Z,A).* If the nucleus of interest is itself available, it may be excited by direct bombardment with γ-rays, β-particles, or charged nuclei.
2. *By β-decay or electron capture.* If either of the nuclei $(Z \pm 1, A)$ is available and is unstable toward β-decay or electron capture, it sometimes happens that the product of the β-decay is not formed in its ground state, but in an excited state. It may then decay to its ground state by γ-ray emission.
3. *As a compound nucleus in a nuclear reaction.* In the preceding section we have seen that the compound nucleus formed by the union of two reacting particles is in an excited state. The most common types of reaction in which a given nucleus (Z,A) may be formed as a compound nucleus are those which involve a proton, neutron, deuteron, or α-particle as the bombarding particle (* = "excited"):

$$
\begin{array}{ll}
(a) & (Z - 1, A - 1) + \mathrm{p} \rightarrow (Z,A)^* \\
(b) & (Z, A - 1) + \mathrm{n} \rightarrow (Z,A)^* \\
(c) & (Z - 1, A - 2) + \mathrm{d} \rightarrow (Z,A)^* \\
(d) & (Z - 2, A - 4) + \alpha \rightarrow (Z,A)^*
\end{array}
\tag{1}
$$

4. *As an excited residual nucleus in a nuclear reaction.* It sometimes happens that the *residual nucleus* in a nuclear reaction is left in an excited state from which it subsequently decays. A number of reactions leading to the nucleus of interest as a residual nucleus can be written. A few of these are:

$$
\begin{array}{lll}
(a) & (Z,A) + \mathrm{x} \rightarrow (Z,A)^* + \mathrm{x}' & \text{inelastic scattering} \\
(b) & (Z - 1, A) + \mathrm{p} \rightarrow (Z,A)^* + \mathrm{n} & \\
(c) & (Z, A - 1) + \mathrm{d} \rightarrow (Z,A)^* + \mathrm{p} & \\
(d) & (Z - 2, A - 3) + \alpha \rightarrow (Z,A)^* + \mathrm{n} & \\
(e) & (Z,A) + \mathrm{p} \rightarrow (Z,A)^* + \mathrm{p} &
\end{array}
\tag{2}
$$

As an example of the above considerations, suppose one desires to study the excited states of ^{27}Al, the ordinary isotope of aluminum. Reference to a table of isotopes reveals that ^{24}Mg, ^{25}Mg, ^{26}Mg, and ^{23}Na are the nearest stable isotopes of smaller Z, and that ^{28}Si, ^{29}Si, and ^{30}Si are the nearest of larger Z. In addition, ^{27}Mg and ^{27}Si are β^-- and β^+-active, respectively. One might therefore expect to be able to study the excited states of ^{27}Al by means of the following reactions:

$$
\begin{aligned}
{}^{27}\mathrm{Mg} &\to ({}^{27}\mathrm{Al})^* + \mathrm{e}^- & {}^{25}\mathrm{Mg} + \mathrm{d} &\to ({}^{27}\mathrm{Al})^* \\
{}^{27}\mathrm{Si} &\to ({}^{27}\mathrm{Al})^* + \mathrm{e}^+ & {}^{25}\mathrm{Mg} + \alpha &\to ({}^{27}\mathrm{Al})^* + \mathrm{d} \\
{}^{27}\mathrm{Al} + \gamma &\to ({}^{27}\mathrm{Al})^* & {}^{24}\mathrm{Mg} + \alpha &\to ({}^{27}\mathrm{Al})^* + \mathrm{p} & (3) \\
{}^{26}\mathrm{Mg} + \mathrm{p} &\to ({}^{27}\mathrm{Al})^* & {}^{23}\mathrm{Na} + \alpha &\to ({}^{27}\mathrm{Al})^* \\
{}^{26}\mathrm{Mg} + \mathrm{d} &\to ({}^{27}\mathrm{Al})^* + \mathrm{n}
\end{aligned}
$$

Each of these reactions will provide information concerning the energy levels lying in a *certain range of excitation*. The lower limit to this range is, strictly speaking, determined by the *Q-value* of the reaction, which depends only upon the *masses* of the various particles involved. For practical purposes, however, one must also remember that it is necessary to have an *appreciable reaction probability* and that this may depend critically upon the energy of the bombarding particle, particularly if the particle must surmount a repulsive *Coulomb barrier* in order to penetrate sufficiently near the target nucleus to react with it. The upper limit to the range of excitation that can be studied with a given reaction is imposed by the maximum energy available in the bombarding particles.

Within the excitation energy range accessible with a given reaction, not every energy state is really accessible to study, for it may be impossible to form some of the states with the given reaction because of *selection rules*. Examples of limitations of this kind will be pointed out later.

17-3. Conservation Laws

We have so far considered only the qualitative features of nuclear reactions and of nuclear excited states. When one wishes to analyze a nuclear reaction quantitatively, or to identify the quantum properties of a given experimentally observed excited state of some nucleus, it is necessary to take into consideration some of the details of the *dynamics* of nuclear reactions. Because of our lack of knowledge of the fundamental interactions between nucleons, we are not able to describe the wave functions of a nucleus in any except a qualitative way, so that *calculations* of energy levels, transition probabilities, etc., are not yet possible. Certain symmetry properties of the wave functions can, however, be defined with great precision. The existence of such symmetry follows from the *invariance of the Hamiltonian operator* toward certain transformations, and this invariance is always associated with a conservation law for some physical quantity. Now, although most of the features of nuclear forces are unknown to us, we *do* know experimentally of many of the *invariance properties* of these forces, and we can therefore utilize the corresponding conservation laws in our analysis of the dynamics and the energy states of nuclei.

Some of the conservation laws that are believed to exist for nuclear reactions are already quite familiar to us, but others are not. Most of the laws can be shown to be intimately related to some invariant property of the Hamiltonian function of the nuclear system, but we cannot establish all such connections here because of limitations both of space and of quantum-mechanical background. We shall merely list the various conservation laws that appear to be valid in ordinary nuclear interactions.

CONSERVATION OF MOMENTUM. This is one of the most important and familiar of the dynamical conservation laws. It is of course fundamentally an *experimentally established* law, valid for dynamical systems which are isolated from outside influences. The law emerges, in the quantum-mechanical treatment of such a system, as a consequence of the invariance of the Hamiltonian function with respect to a *translational shift* of the origin of coordinates by which the position of the system as a whole is measured.

CONSERVATION OF MASS ENERGY. This conservation law, which has been experimentally verified with considerable precision, is related to the invariance of the Hamiltonian function with respect to the *time.* The conservation of momentum and of mass-energy together constitute a single law of the conservation of 4-vector momentum, as was discussed in Chap. 1, which in this form is related to the invariance of the Hamiltonian function of an isolated system with respect to a translational shift of all four space-time coordinates.

CONSERVATION OF ANGULAR MOMENTUM. Another quite fundamental dynamical conservation law is that of the conservation of *angular momentum.* This law is related to the invariance of the Hamiltonian function of an isolated system with respect to the *angular orientation* of the coordinate system used to describe it.

CONSERVATION OF STATISTICS (SPIN). This important conservation law asserts that the "spin character" of a closed system cannot change. That is, if a system consists of a number of particles of various spins, some integral and some half-integral, the *total number of particles of half-integral spin must remain either even or odd.* (Thus, Fermi particles can only appear or disappear in *multiples of two.*) This conservation law is in some respects a consequence of the conservation of angular momentum, since *orbital* angular momentum can have only integral quantum numbers and therefore cannot "carry away" all of the intrinsic angular momentum of a Fermi particle. We shall later see some of the important consequences of this law.

CONSERVATION OF CHARGE. This important and fundamental conservation law is related to the invariance of the Hamiltonian function with respect to the choice of a zero of electric potential. It appears also to be connected with the concept of isotopic spin (see below).

CONSERVATION OF PARITY. If we make the plausible assumption that the Hamiltonian function is invariant toward a change in sign of a coordinate, we can deduce that a nuclear system, like an atomic system, will have energy states which are eigenstates of the *parity operator*. Such states are said to have even $(+)$ or odd $(-)$ parity, according to whether the wave function is an even or odd function of the radial coordinates r_i of the particles of the system.

Each nuclear particle entering into a nuclear reaction may thus be considered to possess a characteristic parity, and in addition a certain parity is associated with the *orbital angular momentum* with which the nuclear particles move during the encounter. The total parity of the system is the product of these component parts, the parity associated with the orbital angular momentum of the system being even or odd according as the orbital angular-momentum quantum number is even or odd. The characteristic parity of a *nucleon*, called its *intrinsic parity, is assumed to be positive* (even).

Corresponding to the invariance property of the Hamiltonian function which is described by the parity of the wave function, there exists a *conservation law* for this quantity in an isolated system. We shall later see how this law can be fruitfully applied to the analysis of a nuclear reaction.

Although the conservation of parity appears to be valid in the kinds of nuclear reactions we are now discussing in which the reacting particles interact through the very strong *nuclear forces*, it is also known that in the so-called *weak interactions*, such as the one leading to β-decay, parity is not conserved. We shall discuss the nonconservation of parity in the weak interactions in Chap. 20.

CONSERVATION OF ISOTOPIC SPIN. In the study of the elementary interactions between the nucleons in a nucleus, clear evidence has been found that the intrinsically *nuclear* interactions between nucleons *are independent of the type of nucleon involved.* That is, it appears that the proton-proton interaction is identical with the neutron-neutron and neutron-proton interactions, except for the purely *electrical* interactions between charges and electric and magnetic moments, which of course differ for the two particles. Because of this "symmetry" of the nuclear forces, it has been suggested that, at least for nuclear considerations, the neutron and proton can be regarded merely as *different states of a single type of nuclear particle:* a nucleon. Thus a nucleon can exist in either of two "charge states."

The above situation brings to mind an analogous situation familiar in atomic physics, namely, the existence of precisely two *spin states of an electron*. This similarity—the existence of only *two* eigenvalues of the charge and of the z-component of spin—suggests a possible parallelism in the dependence of the *energy levels* of a system upon these physical quantities. It is thus natural to study the consequences of regarding the measurement of the charge of a nucleon as being actually a measurement of a "component" of a "spin vector" associated with the nucleon. This spin vector is called the *isotopic spin* T, and the component that is measured is said to be the ζ-component T_ζ, where the ζ-axis is one of the axes of isotopic-spin "space." The possible physical significance of the axes in this space is not yet completely clear, but this does not prevent successful and fruitful application of the isotopic-spin concept.

Thus a nucleon is said to have an isotopic spin $T = \frac{1}{2}$, and the two values, $+\frac{1}{2}$ and $-\frac{1}{2}$, of T_ζ correspond physically to a proton and a neutron, respectively. Starting from this rather obvious correlation, the analogy is extended next to include an *exclusion principle* in which isotopic-spin quantum numbers play a role exactly analogous to those of the ordinary spin, so that a given single-particle energy level can be occupied by precisely *four nucleons* with different isotopic-spin and ordinary-spin components. This clearly will lead to the observed results that:

1. Nuclei tend to contain equal numbers of protons and neutrons.
2. Two protons with opposite spins, and two neutrons with opposite spins, together seem to constitute a *complete subshell* (in the light elements), resulting in relatively high binding energy for nuclei with $A = 4n$.

Pursuing the analogy between isotopic spin and ordinary spin, the ordinary rules for the combination of *independent angular momenta* are adopted for the isotopic spin. This leads to the idea of *isotopic-spin multiplets* of nuclear energy levels analogous to the *ordinary*-spin multiplets familiar in atomic spectroscopy. For example, a *two-nucleon* system can exist in an isotopic-spin *singlet* ($T = 0$) or *triplet* ($T = 1$) state, for which $T_\zeta = 0$, or $T_\zeta = +1, 0, -1$, respectively. In the former case the two nucleons must have opposite values of T_ζ; that is, they must be a proton and neutron. In the latter case, $T_\zeta = +1$ corresponds to two protons, $T_\zeta = 0$ to a proton and neutron, and $T_\zeta = -1$ to two neutrons. Therefore we see that some of the *energy levels* of the proton-neutron system (deuteron) must be components of isotopic-spin *singlet* states, and others, of isotopic-spin *triplet* states. The energy levels of a two-proton or two-neutron system, on the other hand, must all be components of isotopic-spin *triplet* states.

For a general nuclear system of mass number A, the isotopic-spin quantum number T can have positive values up to $\frac{1}{2}A$ in integral steps, and the ζ component of the isotopic spin can have integrally spaced values from $+T$ to $-T$. The isobaric nuclei corresponding to various values of T_ζ would differ by *one unit of charge* from one to the next state, the state $T_\zeta = +\frac{1}{2}A$ being a nucleus consisting of *all* protons ($Z = A$, $N = 0$), and the state $T_\zeta = -\frac{1}{2}A$ consisting of *all* neutrons ($N = A$, $Z = 0$). In general, the atomic number of a nucleus is related to the isotopic-spin component T_ζ by the equation

$$Z = \frac{1}{2}A + T_\zeta \tag{1}$$

The invariance of the nuclear Hamiltonian function toward the charge character of the nucleons can be expressed analytically as an invariance toward *rotational shifts of the axes* in isotopic-spin space, and there should correspondingly exist a *conservation law for the isotopic spin of a nuclear system.*

In describing the properties of isotopic spin, we have so far neglected the actual difference between a proton and a neutron arising from the *electrical* interactions of the charge distributions on these particles. The electrical interaction may be treated as a small perturbation; it has the effect of *removing the isotopic-spin degeneracy* of a given level, in much the same way as the spin-orbit interaction removes the degeneracy of ordinary-spin multiplets. Thus the inclusion of the *Coulomb perturbation* causes the various component energy levels of a given isotopic-spin multiplet in a nuclear system of a given A to *differ* slightly in energy and has the further effect of causing the isotopic spin to be not quite a constant of the motion, exactly in analogy to the situation with ordinary spin when spin-orbit coupling is included. This also introduces the possibility of *intersystem transitions* between levels of different isotopic-spin multiplicities, a phenomenon often observed in connection with γ-ray deexcitation of an excited nucleus, and of *selection rules* which govern the change in the isotopic spin in such transitions.

The concept of isotopic spin has proved valuable in providing a framework for describing certain systematic features of nuclear energy levels, and it seems to be also of fundamental physical significance in a complete theory of nuclear structure. We shall have occasion to mention isotopic spin several times in our remaining discussion, both in connection with nuclear forces and energy levels and in connection with the so-called elementary particles.

CONSERVATION OF NUCLEONS. Finally, we mention a conservation law which seems experimentally to be valid, and is often assumed to be valid in theoretical treatments of elementary particles, but whose *exact* validity has not yet been conclusively demonstrated. This is the *law of conservation of nucleons*, which states that nucleons can be neither created nor destroyed, so that the number of nucleons minus the number of anti-nucleons in the universe remains constant. This conservation law is always observed to be satisfied for ordinary nuclear reactions, so that we need not question its ultimate validity here. We shall take this matter up again in connection with the elementary particles.

The above-described conservation laws provide a basis for analyzing many of the properties of nuclear reactions and of the transitions between nuclear energy levels.   Indeed, it is actually upon the analysis of nuclear reactions *using* these conservation laws that our present experimental

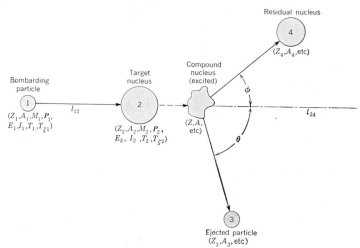

Fig. 17-1. Schematic diagram of a nuclear reaction.

knowledge of the properties of nuclear energy levels is ultimately based. Some of the relationships that connect the experimental observations of a nuclear reaction with the intrinsic properties of the nuclei will now be deduced, using the conservation laws.

Many nuclear reactions proceed as shown in Fig. 17-1.   A bombarding particle of atomic number Z_1, mass number A_1, nuclear angular momentum I_1, isotopic spin T_1, and rest mass M_1, moving with momentum P_1 and kinetic energy E_1, strikes, with relative orbital angular momentum l_{12}, a stationary target nucleus of atomic number Z_2, mass number A_2, etc.   These particles may combine to form an *excited compound nucleus* of atomic number Z, mass number A, etc., moving with momentum P and kinetic energy E.   (Note that the rest mass M includes the mass equivalent of the excitation energy of the compound nucleus.) From this reaction then emerge, with relative orbital angular momentum l_{34}, a particle of atomic number Z_3, mass number A_3, etc., moving with momentum P_3 at an angle θ with respect to P_1, and a residual nucleus of atomic number Z_4, mass number A_4, etc., moving with momentum $\boldsymbol{P}_4$ at an angle φ with respect to $\boldsymbol{P}_1$. The notation used here for kinetic energy (E) and total energy (W) is that commonly used in nuclear physics.

In the analysis of such a reaction the following quantities can be assumed to be known prior to the experiment or as a result of direct measurement in the experiment itself:

1. All of the parameters of the bombarding and target particles except their initial relative orbital angular momentum l_{12} and the spin orientations.
2. All of the parameters of the *ejected* particle except the relative orbital angular momentum l_{34} and the spin orientations.
3. Often, the ground-state rest mass of the compound nucleus and/or the residual nucleus is known from mass-spectroscopic measurements or from other nuclear reactions in which they play a role. Sometimes, however, the determination of one or the other of these quantities is an objective of the experiment itself.
4. Depending upon the particular circumstances, certain of the other parameters of the compound and residual nucleus may be known.

The object of the study of such a nuclear reaction as the one described is often to ascertain the nature of the excited states of the compound nucleus and/or of the residual nucleus. With the aid of the known information and the conservation laws, various of the unknown parameters of the two nuclei can be deduced. For example we may immediately write

(a) $Z = Z_1 + Z_2$ $Z_4 = Z_1 + Z_2 - Z_3$ conservation of charge

(b) $A = A_1 + A_2$ $A_4 = A_1 + A_2 - A_3$ conservation of nucleons

(c) $\boldsymbol{P} = \boldsymbol{P}_1$ $\begin{aligned} P_4 \cos \varphi &= P_1 - P_3 \cos \theta \\ P_4 \sin \varphi &= P_3 \sin \theta \end{aligned}$ conservation of momentum

(d) $E = E_1 + M_1 + M_2 - M$ $E_4 = E_1 + M_1 + M_2 - E_3 - M_3 - M_4$
conservation of mass-energy (energy units)   (2)

Suppose we wish to find the *excitation energy* of the compound nucleus, assuming its ground-state rest mass M_c to be known. Clearly this is just

$$E_{\mathrm{exc}} = M - M_c \qquad (3)$$

in energy units.

We can easily find M in terms of known parameters by using the relativistic relation $W^2 = P^2 + M^2$:

$$M = [(E_1 + M_1 + M_2)^2 - P_1^2]^{1/2}$$

so that $E_{\mathrm{exc}} = [(E_1 + M_1 + M_2)^2 - P_1^2]^{1/2} - M_c$ (4)

One usually wishes to know the excitation energy in terms of the kinetic energy E_1 of the bombarding particle and the rest masses M_1 and M_2 of the initial particles. If the bombarding particle is non-relativistic ($\beta_1 \ll 1$), the above expression may be transformed as follows:

$$\begin{aligned}
E_{\text{exc}} &= [M_1{}^2 + 2(E_1 + M_1)M_2 + M_2{}^2]^{\frac{1}{2}} - M_c \\
&= (M_1 + M_2)\left[1 + \frac{2E_1 M_2}{(M_1 + M_2)^2}\right]^{\frac{1}{2}} - M_c \\
&\approx (M_1 + M_2 - M_c) + \frac{E_1 M_2}{(M_1 + M_2)}
\end{aligned} \tag{5}$$

This shows that the excitation energy of the compound nucleus is made up of two parts, one contribution independent of the kinetic energy of the bombarding particle and the other proportional to this kinetic energy. The first of these quantities is just the Q-value of the reaction

$$(Z_1, A_1) + (Z_2, A_2) \rightarrow (Z_1 + Z_2,\ A_1 + A_2) + Q$$

and the second quantity is the kinetic energy available in the CM system.

The following exercises bring out further aspects of the dynamics of a nuclear reaction.

EXERCISES

17-1. Starting from Eq. (2), find a relativistically exact expression for the excitation energy of the residual nucleus, assuming its ground-state rest mass M_R to be known.

Ans.:
$$E_{\text{exc}} = [(W_1 + M_2 - W_3)^2 - P_1{}^2 + 2P_1 P_3 \cos\theta - P_3{}^2]^{\frac{1}{2}} - M_R \tag{6}$$

17-2. Assume that the kinetic energies E_1 and E_3 of the bombarding and ejected particles are measured and are small compared to the corresponding rest-mass energies. Find an approximate expression for the excitation energy of the residual nucleus for this case.

Ans.: $E_{\text{exc}} \approx (M_1 - M_2 - M_3 - M_R)$
$$+ \frac{E_1(M_2 - M_3) - E_3(M_1 + M_2) + 2(M_1 M_3 E_1 E_3)^{\frac{1}{2}} \cos\theta}{M_R} \tag{7}$$

17-3. In one method of measuring the lifetime of an excited residual nucleus against γ-ray deexcitation, the *speed* of the residual nucleus must be known. Show that, if the ejected nucleus comes out at 90°, the residual nucleus has a speed given by

$$\begin{aligned}
\beta = \frac{P_4}{W_4} &= \frac{(P_1{}^2 + P_3{}^2)^{\frac{1}{2}}}{W_1 + M_2 - W_3} \\
&\approx \frac{(P_1{}^2 + P_3{}^2)^{\frac{1}{2}}}{M_R}
\end{aligned} \tag{8}$$

17-4. If the residual nucleus is left in its ground state, it is possible to measure the Q-value of the reaction and thereby the *mass of the residual nucleus.* Find an approximate expression for the Q-value of the reaction in terms of the kinetic energies of the bombarding and ejected particles.

Ans.:

$$Q \approx \frac{E_1(M_3 - M_2) + E_3(M_1 + M_2) - 2(M_1 M_3 E_1 E_3)^{\frac{1}{2}} \cos \theta}{M_1 + M_2 - M_3} \qquad (9)$$

17-5. Since the neutron has no charge, its mass must be found indirectly. This can be done in a number of ways. From the following data calculate $M_n - M_H$ in three independent ways:

(1)	$^2H + {}^2H \rightarrow {}^3H + {}^1H + 4.031$ MeV
(2)	$^2H + {}^2H \rightarrow {}^3He + n + 3.267$ MeV
(3)	$^3H \rightarrow {}^3He + 0.0185$ MeV
(4)	$^{14}N + n \rightarrow {}^{14}C + {}^1H + 0.627$ MeV
(5)	$^{14}C \rightarrow {}^{14}N + 0.155$ MeV
(6)	$2M_H - M_{2H} = 1.4320$ MeV mass spectrometer
(7)	$^1H + n \rightarrow {}^2H + \gamma$ $E_\gamma = 2.225$ MeV

17-6. Find the threshold energy for the incident particle for the reactions (a) $^7\text{Li}(p,n)^7\text{Be}$, (b) $^{13}\text{C}(n,\alpha)^{10}\text{Be}$. *Ans.:* (a) 1.88 MeV, (b) 4.13 MeV.

17-7. What are the maximum and minimum CM energies of the α-particle from the reaction $^3\text{H}(^3\text{H},2n)\alpha$ in terms of the incident energy E_0 of the ^3H in the lab system? *Ans.:* $0 < E_\alpha^* < 3.74 + \frac{1}{6}E_0$.

17-8. In the reaction $^7\text{Li}(p,\gamma)^8\text{Be}$, a sharp resonance is observed at an incident-proton energy of 0.44 MeV. If one wishes to study this reaction by bombarding *stationary protons* with ^{7}Li nuclei, what acceleration voltage will be required to reach this resonance? *Ans.:* 1.03 MeV.

17-9. What reactions could be used to study the excited states of ^{31}P, using α-particles of energy up to 35 MeV, and what would be the highest excitation energy that could be reached with each? *Ans.:* $^{31}\text{P}(\alpha,\alpha')^{31}\text{P}$, 31.0 MeV; $^{31}\text{Si}(\alpha,{}^3\text{H})^{31}\text{P}$, 18.4 MeV; $^{29}\text{Si}(\alpha,{}^2\text{H})^{31}\text{P}$, 22.6 MeV; $^{28}\text{Si}(\alpha,p)^{31}\text{P}$, 28.7 MeV; $^{27}\text{Al} + \alpha \rightarrow {}^{31}\text{P}^*$, 40.2 MeV.

Analyses such as those outlined above, when combined with measurements of the reaction probability, provide information regarding the energy values and profiles of the various nuclear excited states. Other properties of the energy levels can be deduced by using the remaining conservation laws. Some of these further applications of the conservation laws are well illustrated by the following specific example:

If ^{7}Li is bombarded with protons, a sharp resonance absorption of the protons at a bombarding energy of 0.44 MeV is observed.   This is a *radiative absorption* of the proton, resulting in the ultimate production of two α-particles, and there is another process in which two α-particles are produced *without* a radiated γ-ray.   These two processes are represented by the reactions

$$^7\text{Li} + \text{p} \rightarrow (^8\text{Be})^* \rightarrow {}^8\text{Be} + \gamma$$
$$\phantom{^7\text{Li} + \text{p} \rightarrow (^8\text{Be})^* \rightarrow {}^8\text{Be}} \searrow {}^4\text{He} + {}^4\text{He} \qquad (10)$$

and $^7\text{Li} + \text{p} \rightarrow (^8\text{Be})^* \rightarrow {}^4\text{He} + {}^4\text{He}$

This situation is unusual in that, if the emission of heavy particles is energetically possible and is not impeded by a barrier, the lifetime for heavy-particle emission is generally so much shorter than that required for γ-ray emission that the γ-rays are not able to compete.   It is of interest to interpret this unusual situation in terms of the quantum properties of the various energy states of ^8Be, which can be done by considering the angular momentum, parity, and spin character (statistics) of the reaction.   The nuclear spin and parity of the ground states of the initial and final particles for the two reactions now under consideration are

^1H:spin $\frac{1}{2}$, parity $+$ ^7Li:spin $\frac{3}{2}$, parity $-$ ^4He:spin 0, parity $+$

We see from this that the spins of the proton and of the ^7Li can combine to give either 1 or 2, and these spins, combined with an orbital angular momentum l_{12}, will lead to both even and odd total angular momenta for any value of l_{12}.   Also, recalling that even values of l_{12} make an even contribution to the total parity and odd values an odd contribution, we find that the total parity of the ^8Be must be odd for even l_{12} and even for odd l_{12}.   It thus appears that one could *produce* states of ^8Be having any combination of total angular momentum and total parity by bombardment of lithium with protons.

When we consider the *decay* of the ^8Be into two α-particles, however, a new feature enters the problem.   This is the *indistinguishability* of the two α-particles which come from the decay of ^8Be.   Thus, being Bose particles, the α-particles must have symmetric wave functions. However, since the α-particles have spin 0, for which *only symmetric spin states exist*, the *space* wave function must *also be symmetric*.   The space symmetry of the wave function for two identical particles is described by the *parity*, however, and this is just that of their relative orbital angular momentum.   We thus arrive at the important conclusion that two α-particles can be formed *only from states of even parity and even angular momentum*.

The above result means that a reaction which leads *directly* to two α-particles must be one in which a state of even angular momentum and even parity is formed, whereas a reaction which leads first to a γ-ray and *then* to two α-particles must be one in which a state of odd or even angular momentum and *odd* parity is formed.   (This assumes that the γ-ray is of electric-dipole character, for which the total angular momentum may change either by one unit or zero, but for which the parity *must change*.)

Considerations such as these can be effectively applied in many cases, and the problem of assigning quantum designations to the various nuclear energy levels can thereby be considerably simplified.

17-4. Cross Section

One of the most important measurements that must be made in a nuclear reaction is of the *probability that the reaction will occur* under given experimental conditions.   For example the probability of occurrence of a given reaction may exhibit a very sharp peak in the vicinity of a certain energy.   This is called a *resonance reaction*.   Again, the reaction probability may be a rather smooth function of the energy.   The reaction probability is usually expressed in terms of an *effective area* presented by a nucleus toward the beam of bombarding particles, such that the number of incident particles that would strike such an area, calculated upon a purely geometrical basis, is the number observed to lead to the reaction in question.   This effective area is called the *cross section* for that reaction. The cross section may be defined in a number of ways.   For example, the cross section for any given event to occur is:

1. The probability that this event will occur when a single nucleus is exposed to a beam of particles of total flux equal to one particle per unit area.
2. The probability that the event will occur when a single particle is shot perpendicularly at a target consisting of one nucleus per unit area.
3. If the target is smaller than the extent of the beam, the ratio of the number of events which are observed to occur to the product of the integrated incident-beam flux in particles per unit area and the total number of "eligible" nuclei presented to the beam.
4. If the beam is smaller than the extent of the target, the ratio of the total number of events which are observed to occur to the product of the total integrated number of incident particles and the number of eligible nuclei per unit area of the target normal to the beam.

Definitions (1) and (2) are more easily visualized, but (3) and (4) are in more appropriate forms for experimental evaluation.   Form (3) is appropriate to situations where the area of the target is *less* than that of the beam, and (4) is appropriate to the reverse case where the area of the target is *greater* than that of the beam. Any of these definitions can be adapted either to a *"counting-rate"* experiment in which a uniform incident flux of particles produces reactions at a certain rate, or to a *"standard-run"* experiment in which a known *total number* of particles is shot at a target and the number of reactions produced by these particles is measured.

There are a great many kinds of cross sections which play an important role in nuclear physics; among them are the following:

TOTAL NUCLEAR-INTERACTION CROSS SECTION. This is the effective area possessed by a nucleus for removing the incident particles from a collimated beam by *all possible processes* involving an interaction between the nucleus and the incident particles. For a given nucleus it is a function of the type of incident particle and its kinetic energy. According to the definition used here, the *Coulomb scattering* of the incident particles by the nucleus is included, and since this cross section is infinite if arbitrarily small angular deflections are included, it is necessary to place a lower limit on the angular deflections that will be accepted as representing a nuclear interaction. The size of this limiting angle is usually chosen to fit the particular experimental situation, but is commonly taken to be a few degrees.

PARTIAL CROSS SECTION. The total cross section described above can be written as a sum of several *partial* cross sections which represent the contributions of the various distinct, independent processes which can remove particles from the incident beam. It is common practice to distinguish first between *scattering* processes and *nuclear reactions*, according to whether the outgoing particle (if any) is of the *same* or of *different* character than the incident particle. Thus we may write

$$\sigma_t = \sigma_{sc} + \sigma_n \tag{1}$$

where σ_{sc} is the contribution to the total cross section σ_t made by *scattering processes* and σ_n is that made by *nuclear reactions*.

One may wish to make a further subdivision of each of these partial cross sections into smaller components. Thus, one may wish to distinguish *inelastic* scattering processes, in which the outgoing particle has less energy than the incident one, and *elastic* scattering processes, where the outgoing and incident energies are the same. Thus one would write

$$\sigma_{sc} = \sigma_{el} + \sigma_{inel} \tag{2}$$

Still a further subdivision of each of these partial cross sections can be made; in this subdivision various independent processes contribute to the inelastic scattering cross section and to the elastic cross section.

Caution must be used when dealing with *elastic* scattering, however, since in this case two or more scattering processes which are *not independent of one another* may exist, and in such a case separate partial cross sections cannot be written for these processes because of the possibility of *interference* between them. Thus elastic scattering by the Coulomb field and by the non-Coulomb nuclear forces *will interfere* unless the spins become reoriented. This interference may be thought of as being analogous to the interference effects in optical diffractive scattering by obstacles which are separated by distances comparable to the wavelength of the incident light. On the other hand, all inelastic scattering processes are incoherent, and their cross sections are additive:

$$\sigma_{inel} = \sigma_{i1} + \sigma_{i2} + \cdots \cdot \tag{3}$$

In a similar way, the partial cross section for nuclear interaction may be subdivided into component parts associated with the various possible nuclear reactions which might occur:

$$\sigma_n = \sigma_{n1} + \sigma_{n2} + \sigma_{n3} + \cdots \cdot \tag{4}$$

Each partial cross section is a function of the energy of the incident particle. If it is a *sharply peaked function* of the energy, it is said to exhibit a *resonance*. It is of great importance to know the energy dependence of each of the partial cross sections entering into a given experiment; the determination of this energy dependence for a particular nuclear reaction is often one of the major objectives of the experiment.

DIFFERENTIAL CROSS SECTION. One quantity of great theoretical importance which can often be measured experimentally is the *angular distribution* of the particles emitted in a nuclear reaction. A knowledge of this angular distribution helps to define the character of the nuclear forces and also aids in identifying the quantum designations of nuclear excited states. The distribution in angle of the emitted particles in a nuclear reaction can be described in terms of a *cross section which is a function of the angular coordinates in the problem*. This cross section defines the number of particles which emerge from the reaction *per unit solid angle* at the angular coordinates (θ, φ). Cross sections which define a distribution of emitted particles with respect to some parameter such as solid angle are called *differential cross sections*. If α is a parameter with respect to which the particles are distributed, the corresponding differential cross section is written $d\sigma/d\alpha$. Clearly the partial cross section for a given process is given by

$$\sigma = \int \frac{d\sigma}{d\alpha}\, d\alpha \tag{5}$$

LABORATORY VS. CENTER-OF-MASS CROSS SECTION. For many purposes it is desirable to express the differential cross section for a given reaction in terms of angular coordinates measured in the CM system, in which the net momentum is zero. On the other hand, the direct experimental measurements are almost always made in the lab system in which the *target nucleus* is stationary. In order to transform from one of these coordinate systems to the other, it is necessary to take into account the relative motion of the two systems and the effect of this relative motion upon the various parameters of the experiment. Some of the factors which enter into such a transformation are considered in Exercises 17-11 to 17-13.

GEOMETRIC CROSS SECTION. An important cross section in high-energy nuclear interactions is the projected area of the nucleus given by the measured nuclear radius $\sigma_g = \pi R^2 = \pi R_0^2 A^{2/3}$, where $R_0 \approx 1.2 \times 10^{-15}$ m. The total interaction cross section of a nucleus toward very energetic particles is of the order of magnitude of this area. This cross section is called the *geometric cross section* of a nucleus. It will be discussed somewhat further in a later chapter.

Total or partial cross sections are commonly expressed either in square centimeters or in terms of a unit of area about equal to the geometrical area of a nucleus. This unit of area is the *barn*, equal to 10^{-24} cm², or 10^{-28} m². Differential cross sections are often expressed in *millibarns per steradian*.

EXERCISES

17-10. Using the notation described above, show that

$$\frac{d\sigma}{d\theta} = 2\pi \sin \theta \frac{d\sigma}{d\Omega} \tag{6}$$

if axial symmetry exists.

17-11. Find the speed of motion of the CM system of the nuclear reaction considered in Fig. 17-1. (Recall Exercise 1-48.)

Ans.: $$\beta_{CM} = \frac{P_1}{E_1 + M_1 + M_2} \approx \frac{(2M_1E_1)^{1/2}}{M_1 + M_2} \approx \frac{\beta_1 M_1}{M_1 + M_2} \tag{7}$$

17-12. If the ejected particle in Fig. 17-1 has a rest mass M_3 and a kinetic energy E_3 and is ejected at an angle θ with respect to the incident-particle direction, what kinetic energy and direction of motion does this particle have in the center-of-mass system?

Ans.:
$$E_3' = \frac{E_3 + M_3 - \beta_{CM}[(E_3 + M_3)^2 - M_3{}^2]^{1/2} \cos \theta}{(1 - \beta_{CM}^2)^{1/2}} - M_3 \quad (8)$$

$$\approx E_3 + \tfrac{1}{2} M_3 \beta_{CM}^2 - \beta_{CM}(2E_3 M_3)^{1/2} \cos \theta \quad (9)$$

$$\tan \theta' = \frac{(1 - \beta_{CM}^2)^{1/2} \tan \theta}{1 - \dfrac{\beta_{CM}(E_3 + M_3) \sec \theta}{[(E_3 + M_3)^2 - M_3{}^2]^{1/2}}}$$

$$\approx \frac{\tan \theta}{1 - M_3 \beta_{CM} \sec \theta / (2E_3 M_3)^{1/2}} \quad (10)$$

$$\approx \frac{\tan \theta}{1 - \beta_{CM}/\beta_3 \cos \theta}$$

From the above expressions such factors as a transformation of a differential solid angle from one system to the other can be evaluated, and the complete transformation of differential cross section can therefore be effected.

17-13. Show from the definition of the cross section for a given event that the following identity holds between the differential cross section in the CM and lab systems: $(d\sigma/d\Omega)^* \, d\Omega^* = (d\sigma/d\Omega) \, d\Omega$.

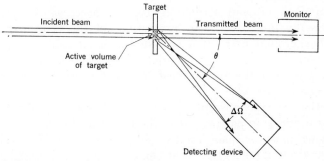

FIG. 17-2. Schematic arrangement of an experiment for measuring a cross section.

Before concluding this discussion of cross sections, it is well to describe briefly the nature of the experimental measurements which must be made in order to define a cross section accurately. Such a measurement can actually be made in a great variety of ways; we shall discuss only one of them. A common physical arrangement for measuring a differential cross section is shown schematically in Fig. 17-2, which depicts a so-called *thin-target* arrangement. A beam of monoenergetic particles impinges upon a thin target. Most of the particles traverse the target without significant modification in their direction of motion and enter a calibrated monitoring device which serves as a "dosage" meter. This beam monitor may be an ionization chamber, some sort of particle counter, or other device capable of registering the number of particles impinging upon it. A few particles undergo nuclear interactions with the target nuclei, and the ejected particles from the specific interaction being studied are measured by a detection device which is (ideally) sensitive only to the ejected particles from the interaction in question. This detection device accepts particles which emerge within a solid angle $\Delta\Omega$ at an angle θ with respect to the incident-beam direction.

The direct result of such an experiment, carried out for a range of incident-beam energies, is a curve which gives the *relative* numbers of ejected particles at an angle θ as a function of energy. The transformation of such a curve into an *absolute differential cross section* requires a number of measurements and corrections to be made. Among them are:

1. A measurement of the thickness of the target, to obtain the number of eligible nuclei per unit area normal to the beam.
2. Measurement of the absolute detection efficiency of the beam monitor and of the detection device for the transmitted or ejected particles.
3. Measurement of the solid angle $\Delta\Omega$ accepted by the detection system.
4. A correction for the lowering of the effective beam energy because of ionization energy loss suffered by the incident particles in passing through the target.
5. Measurement of, and correction for, any *background counting rate* not actually due to nuclear interactions of the type under study. Such background may often be reduced in intensity or importance by careful *shielding* of the detection device from all except particles originating from the target itself, or by the use of *anticoincidence counting arrangements*.
6. Correction for the absorption, or the reduction in energy, of the ejected particles in passing out of the target material.
7. Allowance must be made for the possible presence of reactions due to other kinds of nuclei than that being studied. It is seldom possible to provide a target which consists of a pure isotope of a single element; more usually a mixture of isotopes of an element must be used, and often a chemical compound containing the isotope of interest is used as a target material. In such a case, care must be used to distinguish the reaction of interest from other reactions leading to similar products of similar energy.

EXERCISE

17-14. A beam of thermal neutrons from a pile impinges upon a kilogram of ammonium nitrate: 10^{10} neutrons cm^{-2} strike the sample each second, and the sample is irradiated for one week. At the end of this time, the sample is removed and carbon is separated chemically. If the capture cross section of ^{14}N for thermal neutrons is 1.74 barns, what activity, in curies, should be observed for the carbon? Neglect the

attenuation of the neutron beam in passing through the sample. *Ans.:*
$R = 17$ μcurie.

17-5. Some Properties of Specific Types of Nuclear Reactions

With the preparation provided by the preceding discussion of the general features of nuclear reactions, we are now able to interpret some of the observed characteristics of certain specific types of nuclear reactions. The most common types studied under laboratory conditions and involving energies up to several MeV are those in which the incident and ejected particles are protons, deuterons, α-particles, neutrons, or γ-rays. If we limit our discussion to these possibilities, the possible reactions having a single ejected particle are as shown in Table 17-1. The reactions which appear on the diagonal of this table are *scattering reactions;* we shall consider only those which are true nuclear interactions in the sense that *a compound nucleus* is formed. This restriction places the scattering reactions upon the same basis as the other reactions,

TABLE 17-1. SOME COMMON KINDS OF NUCLEAR REACTION

Incident particle	Ejected particle				
	p	d	α	n	γ
p	(p,p)	(p,d)	(p,α)	(p,n)	(p,γ)
d	(d,p)	(d,d)	(d,α)	(d,n)	(d,γ)
α	(α,p)	(α,d)	(α,α)	(α,n)	(α,γ)
n	(n,p)	(n,d)	(n,α)	(n,n)	(n,γ)
γ	(γ,p)	(γ,d)	(γ,α)	(γ,n)	(γ,γ)

since a compound nucleus, once formed, may decay in any one of a number of possible ways, one of which is always by the ejection of a particle of the same type as the incident particle.

One of the most prominent factors which governs the relative ease with which the above nuclear reactions may be excited by particles of a given kinetic energy is the *Coulomb barrier*, which acts to oppose the approach *or escape* of a charged particle. The physical situation can be visualized most easily with the aid of Fig. 17-3, which shows the potential energy vs. distance (schematically) for a proton and a neutron in the vicinity of a nucleus. As indicated by this figure, the potential energy of a proton differs from that of a neutron because of the Coulomb energy of the proton in the field of the nucleus. The height of the Coulomb barrier is equal to the energy of the proton when it is at the "surface" of the nucleus. According to the discussion of the previous chapter, the

energy states available to the protons inside the nucleus must be occupied
up to the same level as are those available to neutrons, since otherwise
a lower energy state could be attained by a β-transformation of a proton
into a neutron, or vice versa.   The top of the occupied levels is located
about 8 MeV below $V = 0$.

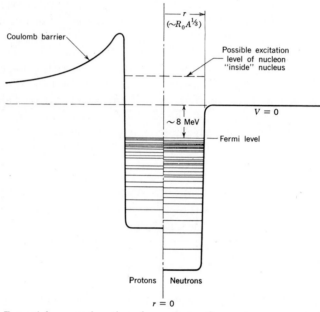

FIG. 17-3. Potential-energy function of a proton and a neutron near a nucleus (schematic).

If a proton approaches a nucleus, it must, classically, possess a kinetic
energy at least equal to the Coulomb-barrier height in order to penetrate
closely enough toward the nucleus to be captured by the nuclear forces.
A neutron, on the other hand, can literally "fall in" the potential well
represented by the nucleus.   Actually, of course, a charged particle
whose energy is somewhat below the barrier height may penetrate the
barrier, and, conversely, one whose energy is greater may be reflected.
When the transmission coefficient is evaluated numerically, one finds
that the transmission probability is approximately 0.1 or 0.2 for an
incident energy equal to the barrier height.

If the nucleus is excited to a sufficient extent that a single nucleon
possessing this excitation energy would be in a positive energy state, it
can be seen from Fig. 17-3 that a neutron of a given positive energy
could easily escape the nucleus, while a proton would have to surmount
the Coulomb barrier.   Thus neutrons of relatively low energy might be

emitted from an excited nucleus, but only rather energetic protons would be able to escape.

Similar considerations to the above hold for deuterons, α-particles, and other charged particles. For example, consider the situation that would hold if a deuteron were to try to escape from an excited nucleus. Since the binding energy of a deuteron is only 2 MeV, the level of excitation of the nucleus would have to be such that *two* nucleons could simultaneously occupy positive energy states. In this case, however, since the Coulomb barrier would act to retard a deuteron from escaping, the most likely result would be the escape of one or two neutrons, a neutron and a proton, or a single proton. The situation is somewhat similar with respect to the escape of an α-particle from an excited nucleus. On the one hand, the higher binding energy per nucleon (7 MeV) in an α-particle would tend to require less excitation of the nucleus to make α-particle emission energetically possible; on the other hand, the greater charge on the α-particle nearly doubles the Coulomb-barrier height, which tends to retard its escape.

EXERCISE

17-15. Using the equation $R = 1.2 \times 10^{-15} A^{1/3}$ m for the radius of a nucleus, compute the height of the Coulomb barrier for protons and for α-particles as a function of Z. Assume that $A = 2Z$ and express the result in MeV. Evaluate numerically the barrier heights of (a) carbon, (b) silver, and (c) lead for protons. *Ans.:* $B_p \approx \frac{1}{2}\beta_\alpha \approx 0.58 A^{2/3}$. (a) 3 MeV, (b) 13 MeV, (c) 20 MeV.

The above discussion, together with the fact that the formation and the decay of a compound nucleus are independent events, makes it possible to state some simple general rules concerning the various possible reactions represented in Table 17-1:

1. Except in light nuclei, the Coulomb-barrier height constitutes an approximate lower limit to the *threshold* energy of any reaction involving an incident charged particle, even if the reaction is nominally an exoergic one.
2. For a given state of excitation, a nucleus is usually most likely to emit a *neutron*, if this is energetically possible, rather than a charged particle. For this reason, (p,n), (d,n), (α,n), and (n,n) reactions are quite common, whereas reactions involving the ejection of a charged particle are considerably less so.
3. Nuclear reactions can be induced by very low-energy neutrons because of the absence of a Coulomb barrier. Charged particles are almost never emitted in such reactions for the reasons discussed above. There are a few exceptions to this rule, however, in which

α-particles or protons are emitted; for example, $^6\text{Li}(n,\alpha)^3\text{H}$, $^{10}\text{B}(n,\alpha)^7\text{Li}$, $^{10}\text{B}(n,p)^{10}\text{Be}$, $^{14}\text{N}(n,p)^{14}\text{C}$, and $^{35}\text{Cl}(n,p)^{35}\text{S}$ occur with thermal neutrons. By far the most common processes involving low-energy

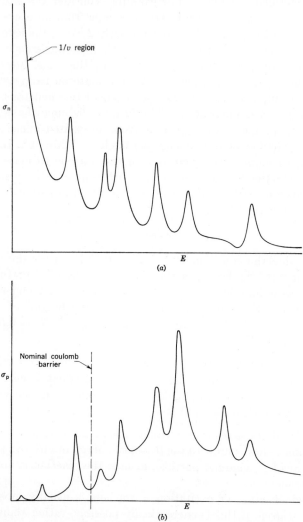

FIG. 17-4. Schematic illustration of cross-section curves for (a) neutrons and (b) protons.

neutrons are *elastic scattering* and *radiative capture* (n,n) and (n,γ). The exceptions cited almost all involve odd-odd nuclei, for which the binding energy of an additional neutron is greater than that of a proton or α-particle because of the odd-even effect discussed in the previous chapter.

In addition to the above general rules, there are a few special situations which deserve mention. One of these is that (d,p) reactions are much more commonly observed than would be expected upon the basis of the above considerations. The reason for this was deduced by Oppenheimer and Phillips (1935). When a deuteron approaches a nucleus, the repulsion between the nucleus and the proton causes the deuteron to become *polarized* with its proton end farther from the nucleus. The proton-

TABLE 17-2. COMMONLY OBSERVED NUCLEAR REACTIONS WITH $A > {\sim}25$‡

The order of the emergent particles is roughly that of the relative yields. Reactions whose yield is usually less than about 1 per cent of the leading one are omitted. Elastic scattering of charged particles is also omitted, since it cannot be separated from the nonnuclear Coulomb scattering. el = elastic, inel = inelastic, res = resonances.

Incident energy	Emergent particles of reactions in which incident particles are:			
	n	p	α	d
0–1 keV	n (el) γ (res)	(Coulomb barrier)	(Coulomb barrier)	(Coulomb barrier)
1–500 keV	n (el) γ (res)	n γ α (res)	n γ p (res)	p n
		(Coulomb barrier for heavier elements)		
0.5–10 MeV	n (el) n (inel) p α $(A < {\sim}80)$ γ $(A > {\sim}80)$ (res for lower energies for $A < {\sim}80$)	n p (inel) $\alpha(A < {\sim}80)$ γ $(A > {\sim}80)$ (res for lower energies for $A < {\sim}80$)	n p α (inel) $(A < {\sim}80)$ γ $(A > {\sim}80)$ (res for lower energies for $A < {\sim}80$)	p n np 2n d (inel) $(A < {\sim}80)$
10–50 MeV	2n n (inel) n (el) p np 2p α 3 or more particles	2n n p (inel) np 2p α 3 or more particles	2n n p np 2p α (inel) 3 or more particles	p 2n np 3n d (inel) ³H 3 or more particles

‡ Adapted by permission from Blatt and Weisskopf, "Theoretical Nuclear Physics," John Wiley & Sons, New York, 1952.

neutron bond distance for the deuteron is of such a size ($\sim 5 \times 10^{-15}$ m) that the neutron can be *inside the nucleus* before the proton has surmounted the Coulomb barrier. The weak bond (2 MeV) of the deuteron is easily broken, so that the proton can be ejected and the neutron retained.

Another situation of considerable interest and importance is the very large capture cross sections presented by some nuclei toward *slow neutrons*. In the lowest energy range, below a few hundred or thousand electron volts, this cross section varies *inversely with the speed of the neutron*, and it may attain sizes of thousands of times the geometric cross section of the nucleus. The physical basis for this v^{-1} law is to be found in two factors which were discussed in connection with particle scattering and with β-decay; one is that the effective impact parameter within which a neutron can be captured is proportional to its wavelength, so that the effective *area* is proportional to λ^2; the other is a limitation upon the probability of encounter because of the variation with v of the *density of momentum states* in which the neutron may approach the nucleus. These two effects combine to give a cross section proportional to $\lambda^2 v$, or v^{-1}.

To illustrate some of the properties of nuclear reactions outlined above, Fig. 17-4 shows schematically the variation of the total cross section with the incident energy for neutrons and protons. This figure shows the v^{-1} dependence for slow neutrons, the barrier effect for protons, and the existence of resonances, corresponding to various excited states of the respective nuclei for both cases.

Table 17-2 summarizes the types of nuclear reactions most commonly observed in intermediate ($25 < A < 80$) and heavy ($80 < A$) nuclei. Reactions involving the light nuclei cannot be similarly systematized because of the strong individuality of each light nucleus.

EXERCISES

17-16. The cross section for radiative capture of thermal neutrons by ^{113}Cd is 25,000 barns. What thickness of Cd metal would be required to reduce the intensity of a beam of thermal neutrons to 0.01 per cent of its initial value?

17-17. A certain thickness of boron reduces the intensity of a beam of thermal neutrons at $T = 27°$C to 90 per cent of its original intensity. What transmitted intensity would have been obtained if the neutrons had been thermalized at 327°C?   *Ans.:* 93 per cent.

17-18. It is desired to study the reaction

$$^{12}\text{C} + {}^{12}\text{C} \rightarrow (^{24}\text{Mg})^*$$

(a) What range of excitation of ^{24}Mg will be accessible to study if carbon nuclei of energy up to 30 MeV are available?   (b) If the incident carbon nuclei have 30-MeV energy, with what energy should α-particles be observed at 90° with respect to the incident beam?   *Ans.*: (a) $\sim$23.2 $< E_{exc} < 28.9$ MeV.   (b) 13.8 MeV.

17-6. Experimentally Observed Properties of Nuclear Excited States

As a final preparatory step to our discussion of nuclear models and nuclear forces, we shall now tabulate briefly the major observed features of the excited states of nuclei. Some of these features have a quite direct bearing upon the nuclear models that we shall later discuss.

SPACING OF ENERGY LEVELS. The average spacing of the energy levels of nuclei is given in Table 17-3 for three ranges of excitation: near the ground state, near the binding energy of a nucleon, and levels well above the binding energy of a nucleon. It will be seen from this table that the level density increases markedly with excitation energy and also with A.

WIDTHS OF ENERGY LEVELS. The lower excited states of nuclei, for which only γ-ray emission is possible, are generally quite narrow compared to the average level spacing. As the state of excitation increases, however, heavy-particle emission becomes possible and the level width

TABLE 17-3. AVERAGE LEVEL SPACING IN NUCLEI

	$E_{exc} \approx 0$	$E_{exc} \approx 8$ MeV	$E_{exc} \approx 15$ MeV
$A \approx 10$	$\sim 10^6$ eV	$\sim 10^4$–10^5 eV	$\sim 10^3$ eV
$A \approx 150$	$\sim 10^5$ eV	~ 10–100 eV	$\sim 10^{-2}$–1 eV

increases greatly. This increase in width, coupled with a considerable decrease in spacing at higher energies, results in a considerable overlapping of levels at the higher excitations.

REGULARITIES IN LEVEL SPACING. As more and more information concerning nuclear excited states has become available, it has become possible to seek evidence of regularities in the nuclear energy levels. Although the vast majority of the levels of a given nucleus do not appear to fit into any simple level scheme, some cases have been found (for example, ^{183}W) among heavy nuclei in which some of the excited states can be fitted by a series similar to that found for a rigid rotator, and a preliminary theory which seems to fit these levels quite well has been presented.[1] In addition to the agreement of the energy levels with this series, the spins and parities of the successive levels are also in good agreement with the rigid-rotator scheme. There is, however, some diffi-

[1] Bohr and Mottelson, *Phys. Rev.*, **89**, 316 (1953).

culty in accounting for the effective moment of inertia of the nucleus with that required by the level spacings. The moment of inertia of the nucleus, calculated by using various models for the nucleon motion, seems to differ from the empirical quantity by a factor of about 2. In spite of this discrepancy, however, this regularity can be considered to be the first important success in the search for series regularities in nuclear spectra (Fig. 17-5).

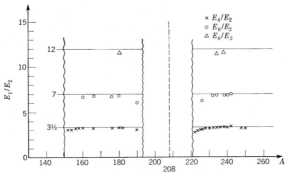

FIG. 17-5. Experimentally observed energy ratios for the rotational excited states of even-even nuclei. [After Mottelson, *Revs. Modern Phys.*, **29**, 186 (1957).]

ISOTOPIC-SPIN MULTIPLETS. The only other known regularity in nuclear energy levels is that having to do with *isotopic-spin multiplets*, which were described in connection with the conservation laws. This kind of regularity has been found only in the light elements, where widely spaced energy levels exist and where both positive and negative isotopic number $(N - Z)$ can be observed. The expected situation is illustrated schematically in Fig. 17-6, which shows the ground states and excited levels of some neighboring isobars in an even-A nucleus, neglecting effects due to the different Coulomb energies of the nuclei.

Figure 17-6 shows that the various levels of an even-A nucleus with $T_\zeta = 0$ (that is, $N = Z = \frac{1}{2}A$) will belong to isotopic-spin singlet, triplet, quintet, etc., systems, the remaining members of each multiplet being found in the various neighboring isobars. Thus, the isotopic singlet levels occur only in the nucleus $T_\zeta = 0$; isotopic-spin triplets occur in the three neighboring isobars $T_\zeta = +1, 0, -1$; quintets occur in the five neighboring isobars $T_\zeta = +2, +1, 0, -1, -2$; and so on. A similar situation should exist for odd-A nuclei, with isotopic-spin doublet, quartet, etc., systems.

The correspondence of isotopic-spin multiplet levels in neighboring isobars has been verified most satisfactorily in *mirror nuclei*. These are nuclei of odd A, and therefore of half-integral isotopic spin, for which the isobars $T_\zeta = +\frac{1}{2}$ and $T_\zeta = -\frac{1}{2}$ are available for study. An example

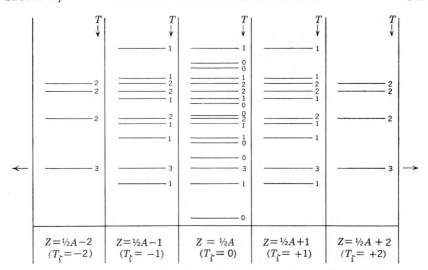

FIG. 17-6. Illustrating the concept of isotopic-spin multiplets in the excited states of isobaric nuclei of even A. In this example, the ground state of the nuclide $Z = \frac{1}{2}A$ is a member of an isotopic-spin singlet state of the nuclear system composed of A nucleons.

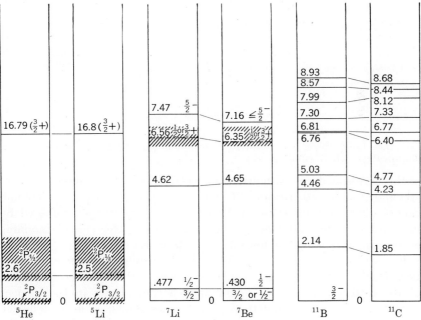

FIG. 17-7. Illustrating the correspondence of energy levels in the mirror nuclides ^{5}He, ^{5}Li; ^{7}Li, ^{7}Be; and ^{11}B, ^{11}C.

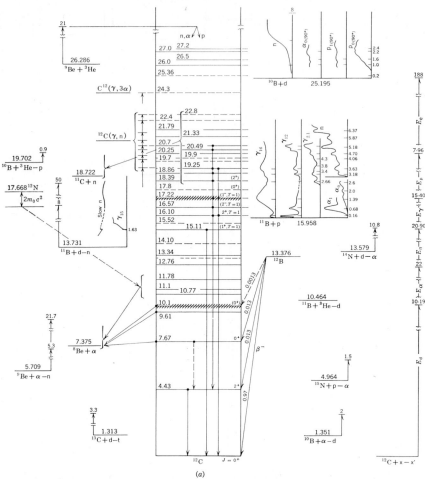

Fig. 17-8. Energy-level diagrams for ^{12}C and ^{14}N.   [Ajzenberg and Lauritsen, *Revs. Modern Phys.*, **27**, 77 (1955).]

of this correspondence is shown in Fig. 17-7, in which the lowest levels of ^{5}He and ^{5}Li, ^{7}Li and ^{7}Be, and ^{11}C and ^{11}B are compared.   The Coulomb energy differences for these nuclei have been essentially eliminated by bringing their ground states into coincidence.   The small remaining systematic differences are not unexpected theoretically.

Experimental tests of the validity of the isotopic-spin concept have been limited so far almost entirely to mirror nuclei, where every level of the nucleus $T_\zeta = +\frac{1}{2}$ should have a counterpart in the nucleus $T_\zeta = -\frac{1}{2}$.   A small amount of work has been done on even-A nuclei, but the results are still meager.   Much remains to be done on this aspect of nuclear energy levels.

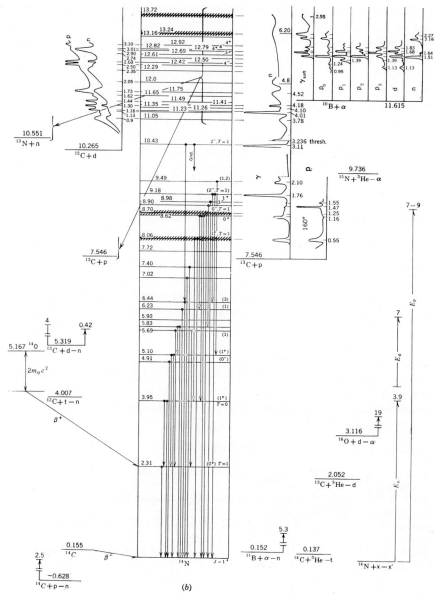

FIG. 17-8. (*Continued*)

The most precise studies of nuclear excited states have been carried out on the light nuclei, where the energy levels are well spaced and where the low Coulomb barriers permit relatively low-energy accelerators to be used effectively. It is in the light nuclei also where the most direct information concerning the fundamental nucleon-nucleon interaction laws is to be obtained. To illustrate the extent and precision of the quantum identification of the energy levels in the light nuclei, energy-level diagrams for ^{12}C and ^{14}N are shown in Fig. 17-8.

The brief discussion given here of the factors involved in the study of nuclear excited states serves to emphasize the similarities and the differences between atomic and nuclear spectroscopy. It is seen that the nuclear energy levels themselves, while exhibiting great quantitative differences with atomic levels, are still describable by similar qualitative quantum language. On the other hand, the physical phenomena available for the study of nuclear levels, and the precision attainable with existing techniques, are quite different in the two cases. Thus many phenomena, such as the Zeeman effect, which enabled rather direct quantum classification of atomic energy levels to be made, cannot be used in the analysis of nuclear levels. Furthermore, the nature of the nuclear forces appears to be such that a "small-perturbation" analysis of nuclear "fine structure" is impossible. From almost all points of view, nuclear spectroscopy is an almost hopelessly difficult art in comparison with optical spectroscopy. In spite of this, however, a great deal of interesting and useful knowledge has been attained through the study of nuclear spectroscopy, and it may be the means by which the fundamental character of nuclear forces will eventually be defined.

REFERENCES

Bethe, H. A., and P. Morrison: "Elementary Nuclear Theory," 2d ed., John Wiley & Sons, Inc., New York, 1956.

Blatt, J. M., and V. F. Weisskopf: "Theoretical Nuclear Physics," John Wiley & Sons, Inc., New York, 1952.

Devons, S.: "The Excited States of Nuclei," Cambridge University Press, New York, 1949.

Friedlander, G., and J. W. Kennedy: "Introduction to Radiochemistry," John Wiley & Sons, Inc., New York, 1949.

18

Nuclear Models

In the last two chapters we have examined some of the more prominent features of nuclear structure and excited states. In the course of this discussion we have gradually built up a mental picture of the nucleus as a droplet-like collection of nucleons, as a qualitative guide to our thinking. This process of trying to visualize an incompletely understood physical phenomenon in terms of familiar objects and ideas is a quite common one and is almost indispensable to progress in the theoretical treatment of physical phenomena. Indeed, the history of physics has been, in substance, a process of representing the physical world at each stage of our understanding by a *model* which incorporates certain fundamental qualitative ideas and which follows certain simple quantitative rules of behavior. By thus defining in a clear, simple, mathematically tractable form all of the features that are deemed essential to a phenomenon, the correspondence with experimental results can most critically be tested, and sometimes in this way new properties of the world have even been predicted and discovered. Any significant discrepancy between the model and reality necessarily requires a suitable change in the model, and in this way our view of nature gradually evolves toward higher and higher levels. The evolution of our ideas of atomic structure from Lorentz and J. J. Thomson, up through the Rutherford-Bohr-Sommerfeld theories to those of Schroedinger and Dirac, is an excellent example of this process.

In nuclear physics we seem at present still to be rather far from a simple complete nuclear theory; for this reason our concepts of nuclear structure are rather crude and the range of applicability of any given model of the nucleus is correspondingly rather restricted. Indeed, it sometimes has seemed that a separate model is needed to describe each nuclear property, and models which appear quite satisfactory within their designed range of applicability sometimes appear to contradict one another when used to describe certain other nuclear phenomena.   In this chapter we shall outline very briefly some of the nuclear models that have been applied with more or less success toward the description of various features of nuclear structure.

18-1.   The α-particle Model

One attractive possibility that suggests itself when one considers certain nuclei such as ^{12}C, ^{16}O, and ^{20}Ne, which consist of even numbers of protons and an equal number of neutrons, is that these nuclei might be treated as *groups of α-particles*.   The relatively high stability of these nuclei, and the large binding energy of ^{4}He, make it plausible to assume that pairs of protons and neutrons become associated together inside the nucleus, for short periods of time, as α-particles.   If one could show that these transient associations persist for periods of time considerably greater than would be expected upon the basis of pure chance, there would then be considerable justification for the α-particle model.

Superficially, it might appear that the observed binding energies of nuclei would support the α-particle model.   The binding energy per nucleon for an average nucleus is about 8 MeV, and that for an α-particle is 7 MeV.   Thus it seems plausible that a nucleus could be composed of a number of α-particles bound to one another with energies of about 4 MeV per α-particle, or 1 MeV per nucleon.   That is, about seven-eighths of the binding energy per nucleon would be accounted for if the nucleons were grouped into α-particles, and only relatively weak binding of the α-particles to one another would then be required.   A nucleus could thus be thought of as a sort of "polyatomic molecule" of α-particles.

When one attempts to apply the α-particle model quantitatively, serious difficulties appear.   For example, one might expect to be able to interpret some of the excited states of those nuclei which can be made up entirely of α-particles in terms of a rotation-vibration type of "molecular spectrum."   However, such a hypothesis has not been able to account for observed energy levels.   Another quite serious difficulty faced by the α-particle model comes from the experimental fact that the low-energy scattering of α-particles by α-particles apparently *cannot be satisfactorily interpreted in terms of an interaction potential between the particles*.   It seems necessary, rather, to assume that the nucleons of the two α-particles *become mixed together* and subsequently break up into two α-particles which are "different" from the original ones.   This indicates that the time during which the original α-particles retain their identities is actually quite small, and that the α-particle model is therefore not valid. In spite of these difficulties, the α-particle model is still sometimes used in connection with the excited states of light nuclei, but with results that are thus far not completely satisfactory.

18-2. The Liquid-drop Model; Fission

The fact that nuclear forces are of very short range, so that the nucleons of a nucleus can interact only with their nearest neighbors, suggests that a nuclear model which exploits this similarity with interatomic forces in chemistry might be fruitful. Thus, one might inquire whether a nucleus can be represented as a crystalline aggregate of nucleons. If one tries to do this, it turns out that the zero-point vibrations of the nucleons about their mean rest positions would be too violent for stability. That is, the individual nucleons must be able to move about within the nucleus much as does an atom of a liquid. One might therefore think of a nucleus as being like a *small drop of liquid*.

If one follows this analogy between nuclear structure and a liquid drop, one can visualize the excitation of a nucleus as a statistical "heating" of the collection of particles of which it is composed. The emission of nuclear particles from the excited nucleus can then be thought of as an *evaporation process*, the binding energies of the emitted particles being analogous to a *heat of vaporization*. These ideas are essentially those involved in Bohr's theory of the compound nucleus described in the previous chapter.

A further extension of the liquid-drop model of nuclear structure can be attempted, one in which certain low-lying states of excitation of a nucleus are interpreted as excitations of the various modes of oscillation of the droplet regarded as a continuous fluid held together by surface tension. Until recently, however, such an analysis of the collective motions of the nucleons did not appear to be valid. A particular case in which significant agreement with experiment has been obtained will be described below (Sec. 18-5).

One successful semiquantitative application of the liquid-drop model is to the interpretation of the phenomenon of *nuclear fission*. It was pointed out in connection with the binding-energy curve (Sec. 16-1) and the stability of nuclei (Sec. 16-4) that *energy would be released if a heavy nucleus could be split into two roughly equal parts*. Such a process actually occurs when certain nuclei are excited by external means, and even occurs spontaneously in certain other unexcited nuclei. This process is called *fission* and can be visualized qualitatively as follows, using the liquid-drop model:

A heavy nucleus is held together by the nuclear forces, and we picture

its stability against deformation to be a result of the tendency of a droplet to keep its surface area to a minimum.   There are forces, however,
which resist this tendency; namely, the *electric charges* carried by the
protons repel one another, and thus tend to magnify any distortion of the
nucleus that may occur.   Now with increasing A, the disruptive effects
of Coulomb repulsion increase and the stabilizing effects of surface ten-

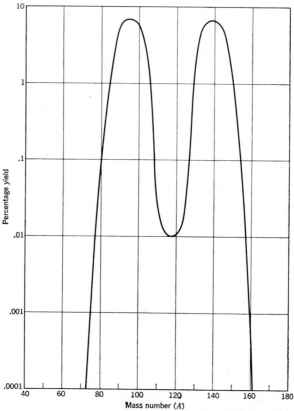

Fig. 18-1. Observed distribution of mass number for the fission fragments of ^{235}U.

sion decrease.   Thus, a sufficiently heavy nucleus should spontaneously
split into smaller fragments, and a nucleus somewhat less heavy than one
that is spontaneously fissionable should split if it is excited into the proper
kind of oscillation.   Such excitation can be brought about by the absorption of a slow neutron, in many cases.   The boundary line of spontaneous
fission is roughly at $A = 240$, according to one calculation.

The fission process exhibits certain interesting and important features
which merit some mention.   Perhaps the most important of these is the
sheer magnitude of the energy release accompanying the process.   This

energy release—about 200 MeV—is of such size that 1 kg of ordinary uranium is equivalent in its heating effect to 2.5 *million* kg of coal, and the cost of fuel is about 400:1 *in favor of uranium*. Another important property of fission is that the resulting fission fragments are *relatively rich in neutrons*, principally because of the different magnitudes of the Coulomb effect in the parent and in the fragments. Thus, the fragments must reduce their neutron number, either *by emitting neutrons directly*, which incidentally also serves to deenergize them from the rather high state of excitation in which they are formed, or by a series of β-decay processes. Sometimes a neutron is emitted only after a series of β-decays has occurred; these are called *delayed neutrons*. On the average, about two or three neutrons are actually emitted, the remaining neutron excess being adjusted by β-decay. It is this emission of neutrons which makes possible a *chain reaction* involving a fissionable material. Such chain reactions are utilized in devices called *nuclear reactors;* details of such practical applications of nuclear fission can be found in the many modern texts on nuclear engineering.

Another interesting feature of nuclear fission is that the fission fragments tend to be of unequal size. The reason for this has not yet been firmly established. Figure 18-1 shows the observed distribution of mass number for the fragments resulting from the fission of ^{235}U.

18-3. The Fermi Gas Model

A model which is closely related to the liquid-drop model is the *Fermi gas model*. According to this model the nucleons are regarded, in analogy with the electrons inside a metal, as noninteracting Fermi particles which are confined within a well of volume $\frac{4}{3}\pi R_0{}^3 A$ and of "depth" sufficient to give the uppermost occupied state an energy value of -8 MeV with respect to the exterior of the nucleus. This model is quite useful in applications where one wishes to take into account the energy or momentum distribution of the neutrons and protons inside the nucleus, as for example in considering the dynamics of very high-energy collisions in which only a single one of the nucleons participates. This Fermi motion, as it is called, can have a quite significant effect upon the threshold energy required for such a reaction, and must also be taken into account when the angular distributions in the CM system of the collision are being computed from measurements made in the lab system.

EXERCISES

18-1. Use the quantities W_a and ϵ_m discussed in Chap. 10 to deduce the effective potential-well depths V_p and V_n which give a binding energy of 8 MeV for the highest proton and neutron. Evaluate these numerically for the observed values of Z and N for Fe and Pb. *Ans.*: $V_p = 45.2(Z/A)^{2/3} + 8 = 33$ MeV (Pb); $V_n = 37$ MeV (Fe), 40 MeV (Pb).

18-2. Compute the approximate kinetic energy, momentum, and speed of the most energetic proton and neutron in an iron nucleus. *Ans.*: $\epsilon_p = 21$ MeV, $P_p = 200$ MeV/c, $\beta_p = 0.21$, $\epsilon_n = 29$ MeV, $P_n = 230$ MeV/c, $\beta_n = 0.24$.

18-3. (a) Compute the threshold bombarding energy for the nuclear reaction $p + p \rightarrow n + p + \pi^+$, given that the mass of a π^+ meson is 273 m_e and assuming that the target proton is stationary. (b) Compute the threshold bombarding energy, assuming the target proton to be moving toward the incident proton with a speed $\beta = 0.21$. *Ans.*: (a) 293 MeV. (b) 141 MeV.

The above exercises show that the effective well depth is about 40 MeV and that the kinetic energy of a proton or neutron may be as high as 20 or 30 MeV, respectively. While the model upon which these numbers are based is not quite correct, since in computing the energies the nucleus was assumed to have infinitely steep, *impenetrable walls* at the radius R_0, the results are sufficiently accurate for rough calculations.

One interesting feature of this model, which seems at first rather surprising, is that the neutrons should be distributed throughout a *larger volume* than are the protons, whereas one would expect the Coulomb repulsion of the protons to cause just the opposite effect. One reason for this is that the difference between the potential energy and the total energy—which determines the attenuation of the wave function outside the nucleus—is greater for the protons than for the neutrons because of the Coulomb barrier, so that the proton wave functions extend less far outside the nuclear boundary. Another reason is that the most energetic neutron has a greater kinetic energy than the most energetic proton has and thus, even on a classical basis, should penetrate farther beyond the nuclear "surface" than a proton should.

18-4. The Shell Model

In our consideration of the electronic structure of atoms, we found it possible to interpret the periodic nature of the chemical properties of the elements in terms of a *shell model* of the atom. The shell model not only provides a clear and simple explanation of this feature of atomic structure but also greatly simplifies the treatment of optical and X-ray spectra. It will be recalled that the fundamental situation which led to the shell model was that certain energy levels were available to a system of Fermi-Dirac particles, that each of these levels could be occupied only by two particles with opposite spins, and that the levels fall into various more or less distinct groups, each having a definite orbital angular momentum.   This approach was, of course, made possible by the fact that the central attractive field of the atomic nucleus is so much stronger than the mutual repulsions of the electrons that the independent-particle approximation could be used.

In the case of the nucleus, we have both a strong similarity and an important difference with the atomic case.   For on the one hand, the constituents of nuclei are neutrons and protons which are Fermi-Dirac particles, but on the other hand there is no single interaction which acts on all nucleons so strongly that their mutual interactions can be neglected. Thus, there appear to be good reasons why nuclei both should and should not exhibit shell structure.

It happens that a considerable body of experimental evidence strongly suggests that nuclei *do* possess a shell structure of some kind.   This evidence is somewhat analogous to the chemical evidence which led to the periodic system of classification of the elements and is based upon the fact that certain nuclear properties show a tendency toward periodic variation with Z and N.   The main properties which show such a variation are the nuclear stability (analogous to chemical inertness), the absolute abundances of the various nuclides, the distribution of the lowest-lying nuclear excited states, and the magnitudes of the nuclear electric-quadripole moments.   These properties indicate that the numbers 2, 8, 20, 50, 82, and 126 have special significance in nuclear structure.   A nucleus for which either Z or N is equal to one of these numbers is said to be a "magic-number nucleus," and if both Z and N are equal to one of these numbers (for example, $^{208}_{82}$Pb), the nucleus is said to be "doubly magic."   The numbers themselves are called magic numbers to signify that they have some significant, but not understood, connection with nuclear structure.   Briefly, the evidence for the significance of these numbers is as follows:

1. Even-Z elements occurring in nature usually possess at least two isotopes of about equal abundance.   That is, no single isotope of an even-Z element accounts for more than about half the total amount of that element.   The exceptions to this rule are

^{88}Sr ($N = 50$)	accounts for 82 per cent of all strontium
^{138}Ba ($N = 82$)	accounts for 72 per cent of all barium
^{140}Ce ($N = 82$)	accounts for 89 per cent of all cerium

2. The lightest isotopes of even-Z elements are usually rare, in the range $Z > 32$.   The most striking exceptions to this rule are that ^{92}Mo ($N = 50$) and ^{142}Nd ($N = 82$) account for 15 and 26 per cent, respectively, of all molybdenum and neodymium.

3. The largest number of stable isotones corresponds to $N = 82$, and the next largest group to $N = 50$. These two groups have seven and six members, respectively, whereas the usual number is about three or four.

4. Ca ($Z = 20$) has five stable isotopes, an unusually large number for this region of the periodic table.

5. Sn ($Z = 50$) has 10 stable isotopes, more than any other element.

6. ^{208}Pb ($Z = 82$, $N = 126$) and ^{209}Bi ($N = 126$) are the heaviest stable nuclei.

7. The absolute-abundance curve of the elements shows peaks at ^{90}Zr ($N = 50$), Sn ($Z = 50$), ^{138}Ba ($N = 82$), ^{208}Pb ($N = 126$, $Z = 82$).

8. Nuclei with one neutron more or less than one of these "closed-shell" values show exceptionally large level spacing in their lowest excited states.

9. The electric-quadripole moments of nuclei are a rough measure of the departure of the nuclear shape from sphericity. These show sharp minima at the "closed-shell" numbers, indicating that such nuclei are very nearly spherical.

The foregoing evidence leads one to try to devise some sort of single-particle model of a nucleus which will yield the magic numbers 2, 8, 20, 50, 82, and 126 as closed shells or subshells of states. Several models seem capable of this. In one of these, each nucleon is assumed to move inside a spherically symmetric "square-well" potential due to the average effects of all the remaining nucleons. That is, a nucleon is supposed to be subject to no forces if it is either outside or inside a certain sphere, but receives a sharp inward impulse whenever it attempts to cross the spherical boundary. This model, with the additional assumption of *strong spin-orbit splitting*, leads to the sequence of states shown in Table 18-1, which are grouped into shells and subshells to match the experimental magic numbers.

TABLE 18-1. AN ORDERING OF SUBSHELL STATES WHICH LEADS TO THE
OBSERVED MAGIC NUMBERS

Spectroscopic configuration	No. of states in shell	Total no.
$(1s_{1/2})^2$	2	2
$(2p_{3/2})^4$ $(2p_{1/2})^2$	6	8
$(3d_{5/2})^6$ $(3d_{3/2})^4$ $(2s_{1/2})^2$	12	20
$(4f_{7/2})^8$ $(4f_{5/2})^6$ $(3p_{3/2})^4$ $(3p_{1/2})^2$ $(5g_{9/2})^{10}$	30	50
$(5g_{7/2})^8$ $(4d_{5/2})^6$ $(4d_{3/2})^4$ $(3s_{1/2})^2$ $(6h_{11/2})^{12}$	32	82
$(6h_{9/2})^{10}$ $(5f_{7/2})^8$ $(5f_{5/2})^6$ $(4p_{3/2})^4$ $(4p_{1/2})^2$ $(7i_{13/2})^{14}$	44	126

A similar ordering of states can be obtained by using an isotropic harmonic-oscillator potential, so that the agreement with the experimental numbers should not be taken as establishing the validity of the

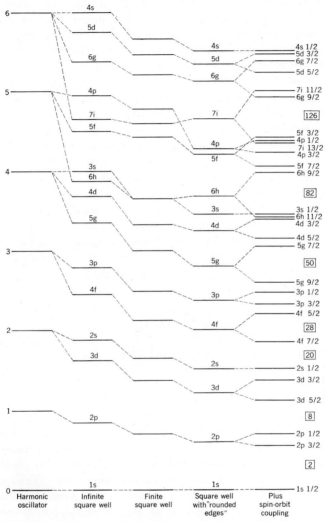

FIG. 18-2. Ordering of states according to the shell model, using various potentials (schematic). [After Feld, *Ann. Rev. Nuclear Sci.*, **2**, 239 (1953).]

spherical square-well potential (Fig. 18-2). Although the shell model agrees rather impressively with observation, there are as yet some shortcomings of the model. Certain of these inadequacies might be remedied by more detailed calculations using nonspherical wells for the nuclei of high spin, but other problems would even then remain. For example, the shell model in which the individual nucleons are supposed to ignore each other and respond only to an average potential is difficult to reconcile with Bohr's liquid-drop model, which is so successful in the theory of nuclear reactions. An essential feature of the latter model is, of course, that nucleons collide rather freely in a nuclear reaction and transfer their energy very readily to other nucleons. The complex potential model (Sec. 18-6) seems to throw some light on this question.

Perhaps the most convincing evidence of the essential correctness of the shell model is the detailed agreement between the predicted and observed *ground-state angular momenta of odd-A nuclei*. Such nuclei have an even number of nucleons of one kind and an odd number of the other. Now, there exists quite convincing evidence that pairs of nucleons of a given kind strongly tend to combine with one another to yield *zero angular momentum*. (Thus, all even-even nuclei have $I = 0$.) This suggests that the angular momentum of an odd-A nucleus should be just that of the odd nucleon, and indeed this is the case for virtually all odd-A nuclei whose spin is known. Further, the same argument applied to (presumably unstable) odd-odd nuclei suggests that their angular momentum ought to be some combination of the angular momenta of the two odd nucleons. Again this is observed to be the case, but, of course, this test is not nearly as definitive as the first one.

The shell model has also met with considerable success in providing a semiquantitative description of many of the properties of the low-lying *excited states* of most odd-A nuclides.

Another interesting phenomenon for which the shell model provides an illuminating semiquantitative explanation is that of *nuclear isomerism*. In the study of the excited states of nuclei it is found that certain nuclei can exist for quite extended periods of time in an excited state, from which they often finally decay by γ-ray emission. Such states may have decay lifetimes ranging up to many days. The nuclei which can exist in these so-called *metastable states* tend to be those having a nearly closed shell configuration of nucleons. The explanation of the long lifetimes of these metastable states lies in the fact that the ground state and the first excited state differ in angular momentum by several units of $\hbar$. Thus, many of the lower-order transitions, such as electric dipole, magnetic dipole, and electric quadripole, are forbidden, so that a transition of very high multipole order may be required. The probability of transition rapidly becomes smaller with increasing multipole order, so that a long lifetime for the state results.

Inspection of the above table of states reveals that near a closed-shell configuration the angular momentum of the nucleons is rather large. Since this table indicates roughly the order of states in the neighborhood of a given shell boundary, we may expect that the states of a high-angular-momentum subshell should also be adjacent to one another as the available states of excitation for a nucleus having a nearly closed shell structure. This then accounts for the high angular-momentum change which leads to metastable states.

Another nuclear property for which the shell model provides an explanation is the *parity of the ground state*, since by the arguments used in connection with angular momenta, the ground-state parity of even-even nuclei should be even, and of odd-A nuclei should correspond to the l-value of the odd nucleon. This rule is observed to lead to the correct parities for virtually all cases.

Finally, mention should be made of the application of the shell model to the prediction of the magnetic moments of nuclei. According to the shell model, one would interpret the magnetic moment of an odd-A, odd-Z nucleus as being due to the intrinsic moment of the odd proton, combined according to a Landé-type formula with the moment due to its orbital motion. For odd-A, odd-N nuclei, on the other hand, the orbital motion of the odd neutron contributes no moment. One finds in this way that the magnetic moment of each type of nucleus ought to have one of two values, according to whether $j = l + \frac{1}{2}$ or $j = l - \frac{1}{2}$. As a function of j, these two values are called the *Schmidt lines*, and it is observed that with one or two exceptions all known magnetic moments lie *between* these two limits and tend to cluster closer to the limit corresponding to the observed l-value in each case. The discrepancy between observation and the shell-model predictions is thought to be due to the reaction of the nucleus to the motion of the odd nucleon and to the fact that the magnetic moment of a nucleon inside a nucleus is not necessarily equal to its moment when free, because of the distortion of the nucleonic structure by the presence of the other nucleons.

In spite of its recognized shortcomings, the successes of the shell model are sufficiently impressive that it may properly be regarded as the most successful nuclear model now known. It shows great promise of even further improvement, and, as we shall see, some significant steps have already been taken in reconciling it with the apparently contradictory concepts of the statistical model and the compound nucleus.

EXERCISES

18-4. Find the lowest 13 energy levels of a particle inside an impenetrable sphere of radius a. Use the atomic-shell notation 1s, 2s, 2p, etc. to designate the n and l values. *Hints:* (1) To solve the radial equation, recall the results obtained in Sec. 14-6. (2) The functions $R_l(\rho)$ satisfy the recurrence relation

$$\frac{2l + 1}{\rho} R_l = R_{l-1} + R_{l+1} \tag{1}$$

(3) The eigenfunctions must satisfy appropriate boundary conditions at the origin and at the surface of the sphere. (4) $x^6 - 210x^4 + 4725x^2 - 10,395 = (x^2 - 2.47)(x^2 - 22.80)(x^2 - 184.73)$. (5) The graphs shown in Fig. 18-3 will be of help.

18-5. Examine the angular momenta of odd-A nuclides in Appendix G and show that the observed angular momenta can be obtained from the energy states of the preceding exercise by the introduction of a spin-orbit coupling which causes the state of higher j for each l-value to lie lower than the state of lower j. Show also that, by choosing the right magnitude for the spin-orbit effect, it is possible to produce relatively large "gaps" at the places required by the observed magic numbers.

18-6. Use a procedure similar to that by which the Landé formula (Sec. 8-8) was obtained to evaluate the expected maximum z-component of the magnetic moment of a proton or neutron in a state of orbital angular momentum l and total angular momentum j ($j = l \pm \frac{1}{2}$). Denote the intrinsic moments of the proton and neutron by μ_p and μ_n, measured in nuclear magnetons (that is, $\mu_p = 2.79 \mu_N$, $\mu_n = -1.91 \mu_N$).

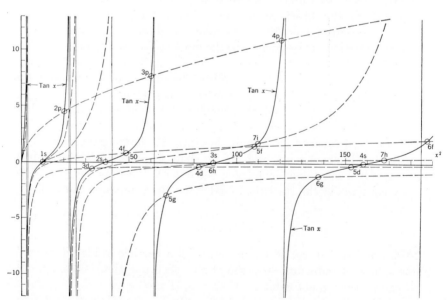

FIG. 18-3. Graphs for Exercise 18-4, showing the ordering of states for a particle in a spherical potential well with impenetrable walls.

Ans.. (a) $\mu_z = j - \frac{1}{2} + \mu_p$　　　　$j = l + \frac{1}{2}$　proton

(b)　　　$= (j + \frac{3}{2} - \mu_p)\dfrac{j}{j+1}$　　$j = l - \frac{1}{2}$　proton

(c)　　　$= \mu_n$　　　　　　$j = l + \frac{1}{2}$　neutron　　　(2)

　　　　$= -\mu_n\dfrac{j}{j+1}$　　　$j = l - \frac{1}{2}$　neutron

18-5. The Liquid-rotator Model

A most promising development in the search for valid nuclear models is the discovery[1] by A. Bohr and Mottelson (1951–1953) that some of the excited states of nuclei are analogous to those of a *rigid rotator*. The nuclei for which this is true are those having large electric-quadripole moments, i.e., those which deviate rather greatly from spherical symmetry. The essential feature of this model is that a nucleus which deviates from spherical symmetry can be considered to have a "shape"; and certain appropriate parameters which describe this shape and its directional orientation can be introduced. The complete motion of the nuclear particles can then be decomposed into a relatively slowly varying "orientation" wave function which describes *collective* motions of the nucleons in the nucleus and a wave function describing the *individual-particle states* occupied by the nucleons in the resultant average field of the remaining nucleons. This situation is quite analogous to the separation of the rotation-vibration states and the electronic excitation states in molecular spectroscopy. The principal differences between the nuclear and molecular cases are that the strong central potential of the molecular case justifies neglect of the electronic interactions and the relatively great mass of the atomic nuclei compared to the individual electrons permits the separation into rotation-vibration and electronic states as an exceedingly good

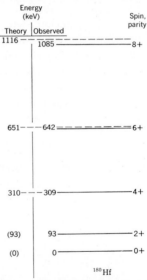

Fig. 18-4. Energy levels of ^{180}Hf, together with theoretical values predicted by the liquid-rotator model.

approximation. Neither of these simplifying factors is present in the nuclear case, for the nucleons all are equivalent to one another, both in strength of interaction and in mass.

The details of this theory are somewhat involved, since the collective

---

[1] Bohr and Mottelson, *Phys. Rev.*, **89**, 316 (1953).

motion of the nucleons must be analyzed in terms of quantum hydro-
dynamics. The changing orientation of the nonspherical nucleus results
from a continuous wavelike flow of nucleons about the "axis" of rotation,
so that the bulk of the nucleons do not partake of a true rigid-body type
of rotational motion. In spite of this, however, the character of the
motion is such that the energy levels follow a rigid-rotator type of spectrum, with a certain effective "moment of inertia" representing the
relation between angular momentum and the angular velocity with which
the shape of the nucleus rotates. A theoretical evaluation of this effective moment of inertia has not yet been completely successful, but even
so the main ideas of the theory appear to be quite valid and constitute
a significant advance in our understanding of nuclear dynamics. The
observed energy levels of ^{180}Hf are shown in Fig. 18-4, together with their
theoretical values.

18-6. The Complex-potential Model

Nuclear reactions which take place with neutrons of energy up to a
few MeV exhibit numerous closely spaced, strong resonances and often a
$1/v$ variation of cross section in the low-energy range. It might therefore
be concluded that the interaction of a neutron with a nucleus is so sensi-
tively dependent upon the detailed nature of the excited states of each
given compound nucleus that the interaction cross section should show
little or no systematic variation with mass number A. It is actually
observed, however, that the total cross section and the angular distribution of elastically scattered neutrons, *when averaged over the resonances,*
show quite marked systematic variations with A (Fig. 18-5). This suggests
that these properties depend upon some relatively gross property of the

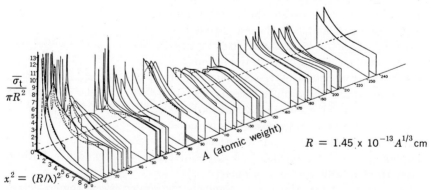

FIG. 18-5. The total cross section for elastic scattering of neutrons, as a function of E
and A. [After Feshbach, Porter, and Weisskopf, *Phys. Rev.*, **96**, 448 (1954).]

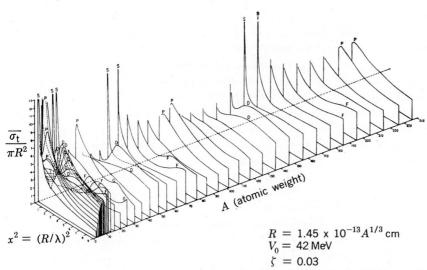

$$\frac{\overline{\sigma_t}}{\pi R^2}$$

$$x^2 = (R/\lambda)^2$$

A (atomic weight)

$R = 1.45 \times 10^{-13} A^{1/3}$ cm
$V_0 = 42$ MeV
$\zeta = 0.03$

FIG. 18-6. Neutron cross sections as calculated on the basis of the complex-potential model. [After Feshbach, Porter, and Weisskopf, *Phys. Rev.*, **96**, 448 (1954).]

nucleus, and not upon the particular arrangement of the nucleons within the nucleus.

A theory of nuclear structure[1] which seems capable of describing the average properties has been advanced, and it may provide a bridge between the contradictory aspects of the shell model and the liquid-drop model. This theory replaces the interior of a given nucleus, for the purpose of computing the (averaged) total cross section and angular distribution, by a spherical square-well potential, in which the potential is taken to be *complex*. The real part of this potential is taken to be about 40 MeV "deep," and the imaginary part to be a few per cent as large as the real part. It will be recognized, from the discussion of the one-dimensional free particle (Chap. 4) and particle scattering (Sec. 14-6), that the real part of this potential will lead to a refraction and reflection of an incoming plane-wave beam of neutrons and that the imaginary part of the potential will lead to an *attenuation* of the beam as it traverses the nucleus. This optical analogy has led to the name "cloudy crystal ball" for this model.

According to this model, the total cross section exhibits very broad "peaks" and "valleys" analogous to the resonance transmission of a particle through a rectangular potential well (Fig. 18-6). Since the size of the nucleus varies with A, the neutron energy corresponding to a given "resonance" condition will likewise vary with A. The role played by the imaginary component of the potential is that *the attenuation produced*

[1] Feshbach, Porter, and Weisskopf, *Phys. Rev.*, **96**, 448 (1954).

by this part is identified with the formation of a compound nucleus. The strength of this attenuation is such that a neutron would make several traversals of a nucleus before forming a compound nucleus. Thus, the "transparency" of the nucleus is rather great, as suggested by the shell model, but once a compound nucleus is formed, it can behave according to the liquid-drop model.

The particular potential assumed in this model is simple, and yet leads to remarkably good semiquantitative agreement with experiment. Further refinement promises to bring many of the general features of particle reactions within its scope.

18-7. Conclusion

The history of models of the nucleus is about as old as nuclear physics itself, and yet up until 1950, little quantitative success was attained in describing experimental facts by means of models of the nucleus. This situation appears now to be quite changed; great progress in our theoretical understanding of nuclear structure has been made since the introduction of the shell model, and even greater advances are doubtless in store for the future. The outlines of a comprehensive theory of the nucleus can perhaps even be perceived in the models that are now known.

REFERENCES

Bethe, H. A., and P. Morrison: "Elementary Nuclear Theory," 2d ed., John Wiley & Sons, Inc., New York, 1956.

Blatt, J. M., and V. F. Weisskopf: "Theoretical Nuclear Physics," John Wiley & Sons, Inc., New York, 1952.

Green, A. E. S.: "Nuclear Physics," International Series in Pure and Applied Physics, McGraw-Hill Book Company, Inc., New York, 1955.

Mayer, M. G., and J. H. D. Jensen: "Elementary Theory of Nuclear Shell Structure," John Wiley & Sons, Inc., New York, 1955.

19

Nuclear Forces

In our treatment of nuclear physics we have thus far avoided direct contact with what is perhaps the central problem of present-day physics, namely, that of the origin and nature of the elementary interaction between the nucleons of which nuclei are composed. We have, of course, made use of certain qualitative features of this interaction, such as its limited range of action and its approximate magnitude, but the precise quantitative form of the nuclear-force law has not been explicitly considered. In the present chapter we shall examine this problem in somewhat greater detail.

It can be stated at the outset that the exact form of the nuclear interaction is not yet known—which is, of course, one of the principal reasons why nuclear physics occupies such a prominent place in current experimental and theoretical research. Even so, it is desirable to examine the various approaches by which quantitative information concerning the nuclear interaction has been obtained and to see how this information has been fitted into a theoretical model of nuclear forces.

19-1. Bound States of the Two-nucleon System

The spectacular success of the Schroedinger theory in describing the energy states of a one-electron atom suggests that the search for a quantitative nuclear-force law might well begin with a study of the simplest nuclear system available: a nucleus composed of *two nucleons*. Three such nuclei might be considered: the two-neutron system, called a *dineutron;* the proton-neutron system ^{2}H, called a *deuteron;* and the two-proton system ^{2}He. Of these, only the deuteron exists in nature, the dineutron and ^{2}He being unstable.

Considering first the deuteron, we might set out experimentally to measure the various bound energy states of a proton and neutron. If this system were found to possess some simple series of energy levels, analogous to the Balmer series of hydrogen which started Bohr and others on the right track in atomic physics, it might be possible to deduce from this an appropriate potential function to represent the interaction between proton and neutron. Unfortunately this attack is not completely satisfactory because, experimentally, the deuteron exhibits but *one* bound level, its ground state, of -2.21 MeV energy.

Fortunately, on the other hand, the deuteron does possess other measurable properties which might serve as a guide in the search for the correct nuclear interaction. These properties are as follows:

1. The angular-momentum quantum number of the ground state (often called the nuclear spin) of the deuteron is 1.
2. The deuteron possesses a magnetic moment of $+0.85735\ \mu_N$ (nuclear magneton). This magnetic moment is presumably to be accounted for in terms of the intrinsic magnetic moments, $+2.79275\ \mu_N$ and $-1.91315\ \mu_N$, of the proton and neutron and the magnetic effect of their relative orbital angular momentum.
3. The deuteron also possesses a measurable *electric-quadripole moment*, of magnitude $Q = +0.0027 \times 10^{-24}$ cm². The size of this quantity is an indication of the departure from spherical symmetry of a charge distribution. The positive sign indicates that the charge distribution is prolate rather than oblate.

These several properties of the deuteron, together with the ground-state energy, serve to define certain properties of the ground-state wave function. For example, the unit ground-state angular momentum restricts the ground-state angular wave function to be some combination of $^3S_1$, $^3P_1$, $^3D_1$, and $^1P_1$ wave functions. Of these, the $^3S_1$ and $^3D_1$ have even parity and the $^3P_1$ and $^1P_1$, odd parity, so that the mixture should contain only $^3S_1$ and $^3D_1$ *or* $^3P_1$ and $^1P_1$, but not both. One can find what mixture of states describes the ground state by computing the magnetic moment given by each of the four possible states and then combining these values suitably to duplicate the observed moment of the deuteron. The procedure is similar to that used in the Zeeman effect in the calculation of $\boldsymbol{\mu} \cdot \boldsymbol{B}$. Thus, in the $^1P_1$ state the magnetic moment must be due entirely to the orbital motion of the proton and neutron, since the magnetic moments associated with the spins cannot enter if $S = 0$. In this case we therefore have

$$|\boldsymbol{\mu}| = \left| \frac{e}{4M_p} l \right| = \frac{e\hbar}{4M_p} = \tfrac{1}{2}\mu_N$$

(The factor-of-two difference with respect to the electronic orbital-moment expression comes from the fact that the neutron contributes half of the angular momentum but no current.) If the neutron and proton spins are parallel, on the other hand, we have

$$\mathbf{\mu} = \frac{e}{M}(2.79275\mathbf{\delta}_p - 1.913156_n + \tfrac{1}{4}\mathbf{l})$$

$$= \frac{e}{2M}(0.87960\mathbf{S} + \tfrac{1}{2}\mathbf{l}) \qquad \text{if } S = 1$$

$$= \mu_N(0.37960\mathbf{S} + \tfrac{1}{2}\mathbf{J})/\hbar$$

Since $\mathbf{\mu}$ is not parallel to J, we make the same approximation as was used in the Zeeman effect:

$$\langle \mu_z \rangle = \left\langle \frac{\mathbf{\mu} \cdot JJ_z}{|J|^2} \right\rangle$$

$$= \mu_N \left[\tfrac{1}{2} + 0.37960 \frac{J(J+1) + S(S+1) - L(L+1)}{2J(J+1)} \right] M_J$$

This formula gives, for the $^3S_1$, $^1P_1$, $^3P_1$, and $^3D_1$ states,

$^3S_1$: $\langle \mu_z \rangle = \mu_N(\tfrac{1}{2} + 0.37960) = 0.87960 \ \mu_N$

$^1P_1$: $\langle \mu_z \rangle = \mu_N(\tfrac{1}{2} + 0) = 0.50000 \ \mu_N$

$^3P_1$: $\langle \mu_z \rangle = \mu_N \left(\tfrac{1}{2} + 0.37960 \dfrac{1 \cdot 2 + 1 \cdot 2 - 1 \cdot 2}{2 \cdot 1 \cdot 2} \right) = 0.68980 \ \mu_N$

$^3D_1$: $\langle \mu_z \rangle = \mu_N \left(\tfrac{1}{2} + 0.37960 \dfrac{1 \cdot 2 + 1 \cdot 2 - 2 \cdot 3}{2 \cdot 1 \cdot 2} \right) = 0.31020 \ \mu_N$

Inspection of the above four magnetic moments shows that no combination of the $^3P_1$ and $^1P_1$ states of the form

$$\mu(D) = p\mu(^3P_1) + (1 - p)\mu(^1P_1)$$

will yield $\mu(D) = 0.85735 \ \mu_N$, since both $\mu(^3P_1)$ and $\mu(^1P_1)$ are less than $\mu(D)$, and p, the probability associated with the $^3P_1$ state, must lie between 0 and 1. On the other hand, we *can* find a probability between 0 and 1 which satisfies the equation $0.85735 \ \mu_N = p\mu(^3S_1) + (1 - p)\mu(^3D_1)$, namely,

$$p = \frac{0.85735 - 0.31020}{0.87960 - 0.31020} = \frac{0.54715}{0.56940} = 0.96$$

That is, the deuteron ground-state angular wave function should be a mixture of 0.96 $^3S_1$ and 0.04 $^3D_1$, and its parity should be even.

The above mixture of angular-momentum states also accounts satisfactorily for the electric-quadripole moment of the deuteron.

The result just obtained demonstrates that, if a potential function which represents the interaction between a proton and a neutron exists at all, it *cannot be a central-force potential*, since the ground-state wave function of any such potential would necessarily be a pure S state. That is, a valid potential function must show some dependence upon the spins or the orbital angular momentum, or both.

Further evidence of the spin-dependent character of the nucleon-nucleon force is furnished by the fact that only a parallel spin orientation yields a bound state for the proton-neutron system. If the nuclear force were not strongly spin-dependent, a bound state should then also exist for antiparallel spins. This incidentally provides a simple explanation for the fact that the dineutron and ^{2}He are not stable systems; for, being Fermi particles, two neutrons or two protons cannot exist with *parallel* spins in the lowest state of zero angular momentum. It is just this state, however, which enters most strongly into the ground state of the deuteron. The elimination of this state by the exclusion principle thus leaves no bound state available to the two-neutron and two-proton systems.

19-2.   Scattering Experiments

The failure of the bound two-nucleon system to yield a complete, quantitative nuclear-force law leads one to consider other means by which information concerning the nucleon-nucleon interaction can be studied. One such means is afforded by quantitative measurements of the *scattering* of protons by protons and of neutrons by protons. (The scattering of neutrons by neutrons cannot be carried out because of the unavailability of free neutrons as a target material for scattering experiments.) One might analyze the experimental data in the manner outlined in Chap. 14; from the observed variation of each phase shift with energy, one would hope to learn useful things about the nuclear-force law. We shall now discuss briefly what has so far been learned from such procedures.

In considering proton-proton scattering quantum-mechanically, allowance must of course be made for the *exchange antisymmetry requirement* which must be satisfied by the wave function, and for the fact that it is not possible experimentally to distinguish the incident and target protons after the collision. Thus, two protons in a singlet state (i.e., spins antiparallel) must have a space-symmetric wave function, which requires that their orbital angular-momentum quantum number be even; in the triplet spin state, their orbital angular momentum must be odd. Since at the lowest energies only S-wave scattering should be present, only the singlet spin state will participate in the low-energy scattering.

Although both p-p and n-p scattering have been measured with considerable precision for energies up to above 1000 MeV, the theoretical interpretation of the experimental results is by no means clear. The main results appear to be the following:

At low energies ($E <$ ~10 MeV), both p-p and n-p scattering are accurately isotropic in the CM system, as would be expected on the basis of the partial-wave theory of Chap. 14. Furthermore, the p-p and n-p interactions can both be interpreted in terms of the same numerical parameters in this energy range, which lends considerable support to the hypothesis of charge independence of nuclear forces.

In the energy range $10 < E < 400$ MeV, it is observed that the scattering is still very nearly isotropic, whereas one would expect that the higher partial waves should begin to appear at about 15 to 20 MeV and should dominate the scattering at energies above about 100 MeV. This surprising result, that the scattering appears to be nearly pure S-wave even at 400 MeV, is by no means clearly understood theoretically. However, it is certain that the observed data cannot be interpreted on the basis of the simple partial-wave analysis alone. For example, the expression for the total scattering cross section, assuming only S-wave contributions, was found in Chap. 14 to be

$$\sigma = \frac{4\pi}{k^2} \sin^2 \delta_0 \tag{1}$$

If we set $\delta_0 = \pi/2$, we find that the largest possible value for the total cross section is

$$\sigma_{\max} = \frac{4\pi}{k^2} = \frac{2\pi\hbar^2}{m_r E} \tag{2}$$

which now depends only upon well-known parameters. *It is actually found that the total cross section is several times larger than this value.* This can be taken as conclusive proof that other effects than pure S-wave scattering are present.

In fact, it has been concluded that the data cannot be completely accounted for in terms of *any* simple two-body potential function, even a non-spherically symmetric one. It appears that at least one of the following other possible types of interaction must be present in addition to whatever ordinary potential is present:

VELOCITY-DEPENDENT FORCES. These might be expected in any case because a simple potential function, even though it might suffice to interpret the lowest energy data, cannot be the sole interaction at all energies because *it is not Lorentz invariant.* In analogy with the electrical case, where the static (i.e., low-velocity) interaction of charges is described by a scalar potential alone, it may be that this apparent scalar is actually the time component of a 4-vector, the three space components of which involve the velocities of the nucleons.

STRONG SPIN-ORBIT OR SPIN-SPIN INTERACTIONS. We have already seen, in considering the bound states of two-nucleon systems, that nuclear forces must have some spin dependence, since otherwise ^{2}He and the dineutron would presumably be stable. These forces may, in fact, dominate the interaction.

EXCHANGE FORCES. Another possibility, which would also provide an explanation of the saturation properties of nuclear forces, is that there are interactions analogous to those which occur in *molecular binding*. As we saw in Chap. 9, the strength of a molecular bond is in part contributed by the so-called *exchange forces* which arise from the invariance of the Hamiltonian function with respect to an interchange of the two electrons of a covalent bond, and the consequent requirement that the wave function be purely symmetric or antisymmetric with respect to this interchange. A qualitatively similar situation might well be present in the nuclear case also.

OTHER TYPES OF INTERACTION. One should, of course, keep in mind the possibility that the nuclear interaction arises from some mechanism that lies completely outside our previous experience, and indeed may even require a new form of mechanics to describe it. If we recall the situation in atomic physics half a century ago, we see that many of the confusing inconsistencies and contradictions of that day were the result of the application of classical physics to situations which, as we now know, require quantum mechanics for their description. We are now so accustomed to quantum mechanics that it is difficult for us to imagine that any known phenomena might, in fact, lie outside *its* realm of applicability. Yet this might possibly be the case.

Of course one does not like to abandon a leaky ship for no ship at all, and it has so far not proved necessary to abandon the quantum theory. But one of the most important tasks is to find and apply simple, clear tests of the validity of quantum mechanics in the realm of the nucleus. If in the end it proves necessary to resort to fundamental changes in our outlook, the sooner we know it, the better.

19-3. The Meson Theory of Nuclear Forces

We shall now briefly examine some of the qualitative features of a quite different approach to the question of the origin of nuclear forces. In 1935, Yukawa advanced the hypothesis that the forces between nucleons might be the result of an *exchange of particles between them* in analogy with chemical forces, which arise from the sharing of electrons between two atoms. According to this hypothesis, the particles which are observed physically as neutron and proton are regarded as consisting of a basic nucleonic "core" surrounded by one, or more, particles called *mesons.* In the present-day theory it is further assumed that this "meson cloud" constitutes the *only* difference between neutron and proton and that these particles can be transformed into one another by the emission or absorption of mesons:

(a) $$\text{p} \leftrightarrow \text{n} + \text{M}^+$$
(b) $$\text{n} \leftrightarrow \text{p} + \text{M}^-$$
(1)

The forces between nucleons are supposed to result from an exchange of one or more mesons between them. Such an exchange can take place only by a violation of the conservation of energy if the nucleons are distant from one another, because mesons have a certain rest-mass energy which must be supplied in order to remove them from a nucleon. More will be said about this matter later.

Assuming the meson hypothesis to be essentially correct, what can be said about the properties of these particles?

1. Concerning the charges carried by mesons, we may infer, from the fact that p-p and n-n forces exist, that mesons must occur with *zero charge*. Charged mesons can give forces only between unlike nucleons by the mechanism (1). Uncharged mesons can give forces between both unlike and like nucleons by the mechanisms

(a) $$\text{p} \leftrightarrow \text{p}' + \text{M}^\circ$$
(b) $$\text{n} \leftrightarrow \text{n}' + \text{M}^\circ$$
(2)

2. To account for the possibility of β-decay transitions between protons and neutrons, it is assumed that a charged meson can undergo β-decay, perhaps in the manner

$$\text{M}^\pm \rightarrow \text{e}^\pm + \nu$$
(3)

3. It is possible to obtain a rough estimate of the *mass* of a meson on the basis of the observed range of nuclear forces and the uncertainty principle. As was mentioned previously, the conservation of energy must be violated in order for a meson to be exchanged between nucleons. The uncertainty principle limits the *time* duration over which such a violation may exist by the familiar relation

$$\Delta E \, \Delta t \approx \hbar$$
(4)

Thus, a large energy unbalance can exist—but only for a short time. If during this time the meson travels to the other nucleon and is absorbed by it, energy is again conserved for the system as a whole, and the meson has accomplished its purpose. If we assume that the quantity ΔE is roughly equal to the rest-mass energy of the meson and that the meson travels at about the speed of light for the time Δt, the distance it can travel cannot exceed about

$$R = c\,\Delta t \approx \frac{\hbar c}{Mc^2} = \frac{\hbar}{Mc} \tag{5}$$

or

$$M \approx \frac{\hbar}{Rc} \tag{6}$$

We may evaluate M in terms of an electron mass, as follows:

$$\frac{M}{m_e} \approx \frac{\hbar}{m_e cR} = \frac{\lambda_c}{2\pi R} \tag{7}$$

where λ_c is the Compton wavelength of the electron. If we set R equal to the known (approximate) range of nuclear forces, we find numerically

$$\frac{M}{m_e} \approx \frac{2426 \times 10^{-15}}{2\pi \times 1.4 \times 10^{-15}} = 275 \tag{8}$$

Thus mesons should have a mass a few hundred times that of an electron.

Yukawa's hypothesis was given great impetus by the experimental discovery in the cosmic radiation of particles having mass intermediate between that of an electron and a proton (Anderson and Neddermeyer, 1936). Subsequent measurement revealed that these particles, which were then called *mesotrons* (from the Greek *mesos*, intermediate), possess a mass of about $220m_e$ and undergo β-decay into an electron. It thus appeared that Yukawa's particle had been found.

In the ensuing several years—until 1947—theorists tried to fit together the Yukawa theory and the experimental particle. But certain difficulties appeared and caused great confusion and concern; among them the principal one was that, whereas the Yukawa particle by its very nature must exhibit quite strong interactions with nuclear matter, the experimental particle was found to have extremely little, *if any*, interaction other than that brought about by its electrical charge. In fact, the cosmic-ray meson is capable of penetrating literally kilometers of solid rock—where it should penetrate perhaps several centimeters—without suffering a catastrophic collision with a nucleus.

Hope of reconciliation was quite dim when, in 1947, Powell et al. discovered another particle, also in the cosmic radiation, which has proven indeed to be the Yukawa particle. This new particle, which is called a π-meson, has a mass embarrassingly close to that obtained above: $273m_e$. It also undergoes β-decay, but of a new type. The π-meson is observed to decay into the first type of meson, which is now called a μ-meson, and a neutrino:

$$\pi^{\pm} \rightarrow \mu^{\pm} + \nu \tag{9}$$

Subsequent to its initial discovery in cosmic radiation, the π-meson has been produced artificially in particle accelerators, and it is one of the most plentiful products of nuclear encounters at energies above a few hundred MeV.   There is now no doubt that the π-meson is the Yukawa particle.   Actually, the story of subatomic particles does not stop here, but that is the subject of the next chapter; we shall pause for the present to consider a few further aspects of the Yukawa theory.

One of the first questions one might ask is this: If the meson theory is indeed correct, might this not explain the difficulty that is encountered in interpreting the nucleon-nucleon scattering data in terms of a potential function?   Perhaps so, but recalling the case of molecular binding, the tremendously complex motions of the electrons in a covalent bond appear as a relatively simple potential energy acting between the two atoms.   Of course, the mass of a meson is not negligible in comparison with that of a nucleon, so that the approximations that were valid in molecular binding may not be valid for mesons, and the energies involved are sufficiently large compared with the meson mass energy that relativistic quantum mechanics must be used from the outset.

Actually, there is some reason to believe that low-energy interactions between nucleons might indeed be describable in terms of a potential, but, of course, at energies comparable with $m_\pi c^2$—that is, 140 MeV— one cannot expect this to be so.

In order to investigate the form such a potential might assume, let us compare the meson theory with a more familiar theory which it resembles even more closely than it does molecular binding, namely, the quantum theory of electromagnetic interactions, which will be discussed briefly in the next chapter.   According to the latter theory the electrical forces which act between charges have their origin in *quanta* which are emitted by each charge and absorbed by the other.   The momentum carried by these quanta is thus transferred from one charge to the other, which changes the momentum of each charge; that is, a *force* acts between them. The quanta involved in this process are not the ordinary quanta with which we are familiar, but are described by four components, which may however be combined to represent the familiar static (Coulomb) interaction and transversely polarized quanta.   Each charge is thus regarded as being the *source of a photon field*.

In a similar way each *nucleon* is regarded as being the source of a *meson field* which acts upon other nucleons.   And similarly also, the meson field possesses a static type of interaction which can be represented by a potential function.   A meson is thus a quantum of the nuclear field —a quantum with a nonzero rest mass.

In order to find a potential function to represent the static meson field, we recall that the Schroedinger equation (Chap. 2) may be obtained by replacing p_i by $(\hbar/i)(\partial/\partial x_i)$ and iH/c by $(\hbar/i)(\partial/\partial ict)$, a procedure which is Lorentz covariant. The Hamiltonian function for a free particle of rest mass m is

$$H = (p^2c^2 + m^2c^4)^{1/2} \tag{10}$$

If we square this to obtain

$$H^2 = p^2c^2 + m^2c^4 \tag{11}$$

and carry out the above substitutions, we obtain

$$\left(\nabla^2 - \frac{1}{c^2}\frac{\partial^2}{\partial t^2}\right)\psi = \frac{m^2c^2}{\hbar^2}\psi \tag{12}$$

This is called the *Klein-Gordon equation* for a free particle of spin 0.

If we set $m = 0$, we have just the familiar wave equation for the electromagnetic field, and we know that this field possesses both source-free wavelike solutions and static solutions in the presence of fixed sources. The latter are obtained by solving

$$\nabla^2\phi = 4\pi\delta(r) \tag{13}$$

The solution for an isolated point-charge is[1]

$$\phi = \frac{1}{r} \tag{14}$$

In the meson-field case, the corresponding equation is

$$\nabla^2\psi = \frac{m^2c^2}{\hbar^2}\psi + 4\pi\delta(r) \tag{15}$$

which is analogous to the equation for the electrostatic potential due to a charge distribution in which the charge density is proportional to the potential, plus a unit "point charge" at $r = 0$. If we assume that $\psi = \psi(r)$, we have

$$\frac{1}{r^2}\frac{d}{dr}\left(r^2\frac{d\psi}{dr}\right) = \frac{m^2c^2}{\hbar^2}\psi \qquad r \neq 0 \tag{16}$$

which with the substitution $\psi = \chi r^{-1}$ is readily integrated to yield

$$\psi(r) = \frac{A\exp[(-mc/\hbar)r]}{r} \tag{17}$$

The coefficient A may be evaluated by using Gauss's theorem, with the result $A = 1$. This is called the *Yukawa potential*. Clearly, this potential is sharply restricted in its action to distances of the order

---

[1] The wave function for a Bose particle corresponds to the potential function of the *field* which represents the particle.

$\hbar/mc$—the Compton wavelength of the meson—and it was on the basis of this property that Yukawa estimated the meson mass.

Unfortunately we cannot penetrate further into the meson theory without encountering rather formidable mathematical complications. The following brief remarks indicate some of the current ideas concerning mesons:

1. The original Yukawa theory referred to *charged mesons* which are describable by a *scalar field* in space time. Actually, mesons occur without charge also, and although they have zero spin as is required of a scalar particle, they are now known to have negative intrinsic parity relative to a nucleon. This means that their wave function changes sign when the signs of all three space axes are reversed. A quantity which has this property, but which is otherwise a scalar, is called a *pseudoscalar*. As was described in Chap. 1, a space-time pseudoscalar is, in fact, a completely antisymmetric fourth-rank tensor.

2. While the meson theory bears strong resemblance to the quantum theory of electrical forces, there is one quantitative difference which renders the theory almost hopelessly difficult to deal with in practice. In quantum electrodynamics the so-called fine-structure constant

$$\alpha = \frac{e^2}{4\pi\varepsilon_0\hbar c} = \frac{1}{137.037} \tag{18}$$

appears in the role of a dimensionless parameter in powers of which various electrodynamic effects are expressed. For example, the ground-state energy of a hydrogen atom is

$$E_1 = -\frac{me^4}{32\pi^2\varepsilon_0^2\hbar^2} = -\frac{1}{2}mc^2\left(\frac{e^2}{4\pi\varepsilon_0\hbar c}\right)^2 = \tfrac{1}{2}mc^2\,\alpha^2 \tag{19}$$

and the shift of this level by the fine-structure effects is

$$\Delta E_1 = \tfrac{5}{4}E_1\,\alpha^2 = -\tfrac{5}{8}mc^2\,\alpha^4 \tag{20}$$

Further corrections to the energy levels could be evaluated by using the perturbation theory; that these are not of practical significance is due to the quite small value of α^2, $\sim 10^{-4}$.

In the analogous case of the meson theory, the quantity which corresponds to the fine-structure constant is

$$A = \frac{g^2}{\hbar c} \tag{21}$$

where g is the strength of the mesic "charge" carried by a nucleon. The quantity A is not known precisely but is approximately equal to 15. Thus, any attempt to express nuclear properties in the form of a power series in A seems doomed to failure.

Another way of looking at this situation is that, in the electrical case, only one photon at a time is usually exchanged between two charges, two photons are exchanged only $\sim 10^{-4}$ as often, and so on. In the meson case, on the other hand, several mesons are "in the air" at once, which greatly complicates matters.

3. The existence of charged mesons inside nuclei, and the interconvertibility of protons and neutrons, provide a plausible explanation of the failure of the shell model to account quantitatively for the magnetic moments of nuclei. It must be assumed that the magnetic moment of a nucleon inside a nucleus is not the same as that of the free nucleon but is modified by the distortion of its meson cloud caused by the presence of other nucleons.

4. Experimental evidence that the proton possesses a finite structure is provided by the beautiful experiments of Hofstadter and his coworkers,[1] in which high-energy electrons are used to probe the nuclear structure by measuring the scattering of the electrons by the nuclear electric and magnetic fields. The experimental data are best interpreted in terms of a finite size for the proton charge distribution, and the best fit is obtained if an rms "radius" of 0.78×10^{-15} m is used. This may be taken to be the radius of the mesic charge cloud which surrounds the proton. Further evidence that a proton is a composite structure comes from measurements on the production and scattering of π-mesons from hydrogen.[2] These measurements have revealed the existence of several *excited states* of the meson cloud surrounding the proton. The best established of these is about 300 MeV above the ground state and is characterized by spin $I = \frac{3}{2}$ and isotopic spin $T = \frac{3}{2}$.

5. The status of the theory may be summed up as follows: On the one hand, no one seriously doubts that the meson theory is at least qualitatively correct, but on the other hand, not a single quantity has yet (1958) been calculated and measured with sufficient accuracy to constitute a convincing confirmation of its quantitative correctness.

REFERENCES

Bethe, H. A., and P. Morrison: "Elementary Nuclear Theory," 2d ed., John Wiley & Sons, Inc., New York, 1956.

Blatt, J. M., and V. F. Weisskopf: "Theoretical Nuclear Physics," John Wiley & Sons, Inc., New York, 1952.

Green, A. E. S.: "Nuclear Physics," International Series in Pure and Applied Physics, McGraw-Hill Book Company, Inc., New York, 1955.

[1] McAllister and Hofstadter, *Phys. Rev.*, **102**, 851 (1956).

[2] Bethe and de Hoffmann, "Mesons and Fields," vol. II, secs. 36c, 38b, c, Row, Peterson & Company, Evanston, Ill., 1955.

20

Particles

In the work of the last fifteen chapters, and especially of the last five, we have repeatedly come into contact with what is perhaps the most fundamental fact of the physical world: that most, if not all, of what goes on in the universe rests ultimately upon interactions between the various kinds of *particles* of which all matter and radiation are composed. Beginning first with the electron and proton, the retinue of known particles has grown steadily to include the photon, neutron, neutrino, positron, the μ- and π-mesons, and most recently a new class of particles called *strange particles*, whose role in the universe is not yet even qualitatively understood. Most of the known particles are not stable, and none are immutable; all can and often do change, either spontaneously or under suitable external stimulus, into other particles. Clearly, our understanding of the physical world cannot be considered complete until the properties of all particles and their relationships to each other have been brought within the compass of theory. One of the most active fields of research today is that of particle physics: on the experimental side, the discovery of particles and the measurement of their intrinsic properties and their interactions with other particles; on the theoretical side, the search for a workable theoretical framework within which each property is embodied. In the present chapter we shall enumerate the experimentally established particles and some of their known properties, and we shall then try to indicate some of the directions in which theoreticians have searched for a theory of particles.

Since each particle is related to all others through various kinds of interactions, or *couplings*, it will not be desirable to describe *all* of the known properties of a given particle at once; some properties are best reserved for discussion in connection with some other particle. Many of these relationships involve the decay processes of the unstable particles, and so we shall first discuss the stable particles and then the unstable particles in an approximate order of increasing mass.

20-1. Electrons and Positrons

The first and best known particle is, of course, the electron, discovered by J. J. Thomson in 1897. Our work of Chaps. 5 to 12 strongly emphasized the fact that the electron plays a direct role in almost all observed physical phenomena except those involving nuclei or gravitation, principally because of its relatively large value of e/m.

Measured Properties: The measured properties of the electron given in Table 20-1, and many of the properties of other particles that we shall discuss, are for the most part not *directly* measured but are "best values" of these quantities derived from certain primary measured quantities by least-squares adjustment.[1]

TABLE 20-1. PROPERTIES OF THE ELECTRON

Mass m_e...............	9.1083×10^{-31} kg
	0.510976 MeV$/c^2$
	0.548763×10^{-3} amu
Charge q_e...............	-1.60206×10^{-19} C
	-4.80286×10^{-10} esu
	-1 electron
Spin $\|m_s\|$...............	$\frac{1}{2}$
Magnetic moment $\|\mu_z\|$....	0.92837×10^{-23} Am^2
	1.001145358 μ_B (μ_B = Bohr magneton)
Mean life...............	Stable

Antiparticle: Intimately related to the electron is the positive electron or *positron*, discovered in the cosmic radiation by Anderson in 1932. The positron differs from the electron only in that it carries a positive charge, and would presumably be stable in the absence of electrons. At rest in the presence of ordinary matter, however, the positron is observed to undergo mutual annihilation with an electron, the products being *electromagnetic quanta*. Because of this annihilation property, the positron is said to be the *antiparticle* to the electron.

Before annihilation, a positron and an electron are often observed to form a neutral "atom" which is called *positronium*. The ground-state configuration of positronium can be either $^1S_0$ or $^3S_1$, and the annihilation process is different for these two states because of the operation of selection rules. The singlet state decays with a mean lifetime of 8×10^{-9} s

---

[1] Dumond et al., *Revs. Modern Phys.,* **27,** 363 (1955).

into two quanta of equal energy, and the triplet state decays with a mean lifetime of 7×10^{-6} s into *three* quanta whose total energy is $2m_e c^2$:

(a) $^1S_0 : e^+ + e^- \rightarrow 2h\nu$ $h\nu = 0.511$ MeV

(b) $^3S_1 : e^+ + e^- \rightarrow 3h\nu$ $\Sigma h\nu = 1.022$ MeV   (1)

EXERCISE

20-1. Evaluate the ground-state energy of positronium in electron volts. Should any of the transitions of this atomic system lie in the visible region of the spectrum? *Ans.:* $E_0 = -6.76$ eV. Balmer series limit $\approx 0.73 \, \mu$.

Positrons are copiously produced in high-energy collisions of photons with atoms by a process called *pair production*, in which a high-energy quantum interacts with the Coulomb field of an atomic nucleus to produce a positron and electron. The quantum disappears in this process, its energy being transformed into the mass energy and kinetic energy of the

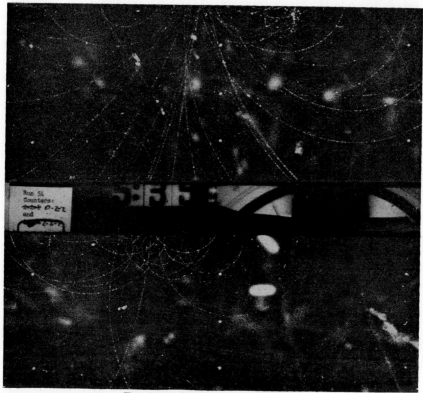

Fig. 20-1. Electrons and positrons.

positron and electron. (The atomic nucleus carries away only an insignificant amount of the energy, its main role being to supply momentum.)

In pair production, both the electron and the positron move in the same general direction as did the quantum that produced them. However, the two particles do not move along exactly the same line, nor do they move with the same speed; in *their* CM system (not that of the entire system) they may have (equal and opposite) momenta of the order of $m_e c$, oriented in any way with respect to their line of motion.

The above properties of pair production are well illustrated by the cloud-chamber photographs of electron pairs shown in Fig. 20-1.

EXERCISES

20-2. Find the order of magnitude of the angle between the paths of the two electrons of a pair produced by a quantum whose momentum is P ($P \gg m_e c$). *Ans.:* $\theta \approx m_e c / P$.

20-3. Find the order of magnitude of the ratio of the momenta of the two electrons of the above problem. *Ans.:* $p_{max}/p_{min} \approx 4$.

Interaction with Matter. An electron or positron passing through matter loses energy by ionization in the manner described in Chap. 14 and by occasional close encounters with atomic nuclei (bremsstrahlung). The latter process serves to reconvert the energy expended by a quantum in producing an electron pair back into electromagnetic form, but, of course, degraded into several quanta rather than just one. In fact, the processes of pair production and bremsstrahlung together give rise to a most important phenomenon called an electronic *cascade shower:* An incident quantum (say) produces an electron pair which in turn produces more quanta by bremsstrahlung; these quanta produce more pairs which produce more bremsstrahlung; and so on. The process continues to build up in a cumulative way, but soon the average energy per quantum becomes smaller than that required for pair production, whereupon the shower rapidly decays by ionization loss of the photoelectrons produced by absorption of the quanta. Cascade showers are of great importance in the interaction of cosmic-ray particles with the earth's atmosphere, in the operation and shielding of high-energy electron accelerators, and in several other practical applications. They will be discussed in somewhat greater detail in the next chapter.

One further feature of the energy-loss properties of positrons and electrons is of some practical importance. This is that, because atoms contain negative electrons, one cannot distinguish an electron which emerges from an atom from an incident electron, and the wave function used to describe this situation must be properly antisymmetric with

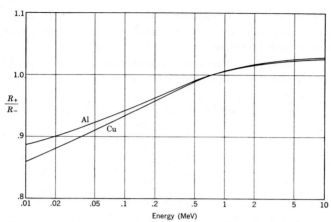

FIG. 20-2. Comparison of ranges of electrons and positrons in aluminum and copper. [From Energy Loss and Range of Electrons and Positrons, *Natl. Bur. Standards Circ.* 577, 1956. See also Rohrlich and Carlson, *Phys. Rev.*, **93**, 38 (1954).]

respect to the coordinates of all of the electrons. A positron, on the other hand, *can* be distinguished—by its charge—and no such anti-symmetry requirement need be made. Figure 20-2 shows the experimental difference in energy loss of electrons and positrons in aluminum.

20-2. Photons

The particle-like nature of electromagnetic radiation was first recognized by Einstein (1905), who extended Planck's quantum hypothesis to apply to radiation in order to explain certain puzzling features of the photoelectric effect. Most of the important properties of quanta have already been treated in our preceding work, so that it is hardly necessary to recount them here. A quantum has the properties given in Table 20-2.

TABLE 20-2. PROPERTIES OF THE PHOTON

Mass	0	Magnetic moment	0
Charge	0	Mean life	Stable
Spin	1		

Its interaction with other particles is through the charge and electric and magnetic moments of the particles.

20-3. Protons

Since early in the present century, the proton has been recognized as an essential component of matter. Although its discovery may formally be credited to J. J. Thomson, it is also in many respects the brain child of Ernest Rutherford. By far the greatest contribution to the mass of the universe is made by protons, because of the overwhelming cosmic abundance of hydrogen, and protons also comprise roughly half the mass of all other atomic nuclei.

Measured Properties. From our previous work, we are already familiar with many of the measured properties of protons given in Table 20-3.

TABLE 20-3. PROPERTIES OF THE PROTON

Mass.....................	1.67239×10^{-27} kg
	938.211 MeV
	1.007593 amu
	$1836.12 m_e$
Charge..................	1.60206×10^{-19} C
	+1 electron
Spin.....................	$\frac{1}{2}$
Magnetic moment.........	1.41044×10^{-26} Am^2
	2.79275 μ_N (nuclear magneton)
Isotopic spin.............	$\frac{1}{2}$
Isotopic-spin projection....	$+\frac{1}{2}$
Mean life................	Stable

Antiparticle. Like the electron, the proton has a counterpart of opposite charge with which it undergoes annihilation. This negative proton, or antiproton, was sought unsuccessfully for many years and was finally found,[1] in 1955, to be one of the products of high-energy nuclear reactions induced by 6200-MeV protons from the University of California Bevatron.

The annihilation of an antiproton with a proton or neutron differs from that of a positron and electron in that several charged and neutral π-mesons (to be described later) are usually produced, rather than two or three quanta. This result is by no means unexpected, however, since the coupling of nucleons with π-mesons is far stronger than that of charges with quanta.

Interaction with Matter. When a proton moves through matter, it loses energy by ionization as described in Chap. 14, and if it has sufficient energy, it also induces nuclear reactions. We have already discussed some aspects of the latter for relatively low proton energies, and we shall later discuss the kinds of nuclear reactions induced by protons of higher energy.

20-4. Neutrons

Closely related to the proton is the *neutron*, discovered in 1932 by Chadwick. As we have seen in previous chapters, neutrons and protons combine together in roughly equal numbers to form many stable and

[1] Chamberlain et al., *Phys. Rev.*, **100**, 947 (1955).

unstable nuclides, and under certain circumstances may be transformed one into the other.

Measured Properties. Some of the measured properties of the neutron are shown in Table 20-4.

TABLE 20-4. PROPERTIES OF THE NEUTRON

Mass.................... 939.505 MeV
                        1.008982 amu
                        $1838.65m_e$
Charge.................. 0
Spin.................... ½
Magnetic moment‡........ -1.91315 μ_N (nuclear magneton)
Isotopic spin........... ½
Isotopic-spin projection.... $-½$
Mean life............... 1.11×10^3 s
Decay modes............. $n \to p + e^- + \bar{\nu} + 0.783$ MeV

‡ Cohen, Corngold, and Ramsey, *Phys. Rev.*, **104**, 283 (1956).

Antiparticle. In the series of experiments which led to the detection of the antiproton, the *antineutron* was also found. Like the antiproton, the antineutron annihilates with either a proton or a neutron to yield several π-mesons.

Interaction with Matter. In addition to the extremely weak β-decay interaction which is responsible for the instability of the free neutron, protons and neutrons also interact quite strongly through the π-meson field and can be transformed into one another in nuclear reactions of the type

$$(a) \qquad p + p \to p + n + \pi^+$$
$$(b) \qquad p + n \to p + p + \pi^- \tag{1}$$

Further evidence of the interconvertibility of protons and neutrons is afforded by the prevalence in nuclear reactions of *charge-exchange scattering,* in which a proton collides with a stationary neutron and exchanges charge with it, leaving an almost stationary proton and a fast-moving neutron, or vice versa:

$$(a) \qquad p + n \to n' + p'$$
$$(b) \qquad n + p \to p' + n' \tag{2}$$

Because of its lack of charge, a neutron passing through matter does not lose energy by ionization, but its interaction with *nuclei* is, if anything, somewhat stronger than that of protons because of the absence of a Coulomb barrier. At low energies, in fact, neutrons react extremely strongly with certain nuclei, such as those which lack one neutron to complete a shell, and relatively weakly with others which already have a closed shell of neutrons. In passing through a substance which is composed entirely of nuclides inert toward neutrons, the neutrons still may undergo *elastic* nuclear collisions and thus be slowed down, since the lab-system kinetic energy of the neutron is diminished in each collision by an amount equal to that carried away by the struck nucleus. This mechanism is more effective as the mass of the struck nucleus is smaller, so that hydrogenous materials such as paraffin are most useful for preparing so-called "thermal" neutrons.

20-5. Neutrinos

As we saw in our discussion of β-decay, it is possible to avoid violations of the law of conservation of energy, linear and angular momentum, and spin by assuming the existence of a particle, called a *neutrino*, which has the properties shown in Table 20-5.

TABLE 20-5. PROPERTIES OF THE NEUTRINO

Mass................ $<5 \times 10^{-4}m_e$
                      (probably zero)
Charge.............. 0
Spin................ $\frac{1}{2}$
Magnetic moment.... $<10^{-8}$ μ_B (Bohr magneton)
                      (probably zero)
Mean life........... Stable

Antiparticle. Our experience with other particles leads us to expect that the neutrino too will have an antiparticle, the *antineutrino;* indeed, the theory of β-decay is made somewhat more symmetrical if we assume that an antineutrino is emitted in the decay of, say, a neutron, n → p + e⁻ + $\bar{\nu}$, since one can then transpose the antineutrino to the left side, making it a neutrino (an ingoing particle is equivalent to an outgoing antiparticle), or n + ν → p + e⁻. Since the neutrino has no charge, however, one may ask on what basis the antiparticle should differ from the particle. There are two possibilities:

1. The neutrino and antineutrino are *identical particles*, so that two neutrinos from the *same kind* of β-decay might annihilate one another.
2. The neutrino and antineutrino are *different particles*, as would be the case, for example, if the neutrino were to have a small magnetic moment.

It might appear rather academic to argue these two possibilities, but there appear to be ways to distinguish experimentally between them; in fact, tremendous new interest in the nature of the neutrino and antineutrino has been aroused by the startling discovery that parity is not conserved in the β-decay process.[1] One of the ways of distinguishing

[1] Lee and Yang, *Phys. Rev.*, **105**, 1671 (1957).

between the two possibilities makes use of a phenomenon, called *double
β-decay*, which is expected to occur. In this process a given nuclide
(Z,A) might emit *two* electrons, thus changing to the nuclide $(Z + 2, A)$.
If a neutrino is emitted for each electron, and if neutrinos cannot annihilate one another, two neutrinos must also be emitted; but if neutrino
and antineutrino are the same particle, the neutrinos could annihilate
"internally" without being emitted. The relative probabilities of these
two kinds of processes can be estimated sufficiently closely to distinguish
between them if the process can be detected at all; one finds from such
estimates, for example, that the mean lifetime of ^{48}Ca toward decay into
^{48}Ti should be about 10^{17} years if neutrinos are not emitted, and about
10^{24} years if neutrinos *are* emitted. Furthermore, if neutrinos are not
emitted, the two electrons should share the total available energy
between them, while if neutrinos are emitted, the electrons will have *less*
than this total energy. Experiments suggest that neutrinos *are* emitted,
and therefore that neutrino and antineutrino are not identical.

At this point it is appropriate to describe briefly the relatively recent
work dealing with the question of conservation of parity in processes
involving neutrinos, since this work strongly suggests a particularly
simple difference between neutrino and antineutrino. The original suggestion that parity might not be conserved in the so-called "weak"
interactions was made by Lee and Yang[1] who, along with many other
theorists, were greatly disturbed by the existence of two particles called
τ and θ, which we shall describe later, which seem to have identical
masses and lifetimes but are apparently of opposite intrinsic parity.
These authors pointed out that the conclusion that these two particles
must have opposite parity is based upon the assumption that the natural
laws are invariant toward the operation called *space inversion*, which
means the replacement of a right-handed system of space axes by a left-
handed system, or alternatively, a reflection of the space axes in a plane;
while it is well known that the so-called "strong" interactions (nuclear
forces) and the electromagnetic interactions *do* exhibit this kind of
invariance, the extrapolation to the weak interactions was so far untested
experimentally. Lee and Yang then went on to propose a number of
possible experimental tests of the hypothesis of space-inversion invariance.

Two different kinds of experiments almost immediately showed that
the suspicions raised by Lee and Yang were in fact well founded, and that
the β-decay interaction is *not* invariant under space inversion. In one
of these experiments,[2] the angular distribution of the electrons emitted
by the β-active isotope ^{60}Co was studied under conditions wherein the

[1] *Phys. Rev.*, **104**, 254 (1956).

[2] Wu et al., *Phys. Rev.*, **105**, 1413 (1957).

space orientation of the ^{60}Co spins was known. In this experiment it was observed that more β-particles are emitted with momentum components opposite the nuclear spin direction than along it, quite in contrast with the expected forward-backward equality which must hold if the β-decay inter- action is invariant under space inversion. In a second experiment,[1] the successive reactions $\pi^+ \rightarrow \mu^+ + \nu$ and $\mu^+ \rightarrow e^+ + \nu + \nu$ were studied, and again it was found that in both of these reactions a strong polariza- tion exists, contrary to the results expected on the basis of space-inversion invariance. (We shall discuss the results of this experiment in greater detail in connection with the properties of π- and μ-mesons.)

Closely following the experimental verification of nonconservation of parity, Lee and Yang reintroduced a theory of neutrinos and β-decay which had been examined many years earlier but rejected because it did not satisfy the assumed requirement of space-inversion symmetry.[2] According to this so-called *two-component theory* of the neutrino, the spin of a neutrino is always aligned in one sense *along the direction of its momentum* and the spin of an antineutrino is aligned in the *opposite* sense along the direction of *its* momentum. Thus in this theory the neutrino and antineutrino differ in the sense of their "helicity." A necessary consequence of such a theory is that the rest mass of a neutrino is exactly zero, since otherwise the correlation of spin direction with momentum direction could not be the same in all Lorentz frames; it also cannot possess an electric or magnetic moment. It now seems almost certain that the two-component theory is correct, and the present indica- tion is that the neutrino possesses "left-hand" helicity, its spin being oriented oppositely to its momentum vector.[3] For the most recent results, the current literature should be consulted.

Interaction with Matter. Because of its lack of charge and magnetic moment, a neutrino has essentially no interaction with matter except that which leads to inverse β-decay. This interaction is extremely weak. The cross section for inverse β-decay should be roughly the cross section for striking a nucleus times the probability of β-decay while the neutrino is inside the nucleus:

$$\sigma \approx \sigma_{\text{nucl}} \frac{T}{\tau} \tag{1}$$

where $\sigma_{\text{nucl}} \approx 10^{-28}$ m^2, $T \approx 10^{-21}$ s (traversal time of neutrino across nucleus), and $\tau \approx 10^{-1}$ s (mean life for β-decay). Thus

$$\sigma \approx 10^{-48} \text{ m}^2 \tag{2}$$

---

[1] Garwin, Lederman, and Weinrich, *Phys. Rev.*, **105**, 1415 (1957).

[2] Lee and Yang, *Phys. Rev.*, **105**, 1671 (1957).

[3] Feynman and Gell-Mann, *Phys. Rev.*, **109**, 193 (1958); Goldhaber, Grodzins, and Sunyar, *Phys. Rev.*, **109**, 1015 (1958).

EXERCISE

20-4. Compute the approximate mean free path of a neutrino through solid iron. *Ans.:* $\lambda \approx 1000$ light years.

The above exercise indicates that a neutrino, once set free, has virtually no chance of being reabsorbed even if it is emitted at the center of a star; the energy carried away by neutrinos is effectively lost from the universe.

20-6. μ-mesons (Muons)

In 1935, the Japanese theoretical physicist, Yukawa, proposed that the strong interaction between nucleons might be "transmitted" via particles of mass about $200m_e$. Less than two years later, charged particles of about this mass were found experimentally in the cosmic radiation.[1] These particles were at first called *mesotrons*, but with the passage of time this has become shortened to *mesons*. For several years these cosmic-ray mesons were thought to be the particles postulated by Yukawa, but it was later found that they are not. Another kind of particle of somewhat greater mass—the so-called π-meson—is apparently the true Yukawa particle. The meson found by Anderson and Neddermeyer is the most abundant component of the sea-level cosmic radiation and is called the μ-*meson*, or *muon*.

Measured Properties. Most of the properties of μ-mesons given in Table 20-6 were measured by using cosmic-ray mesons, but some have been obtained by using artificially prepared mesons.

TABLE 20-6. PROPERTIES OF THE MUON

Mass.................	$206.84m_e$	Magnetic moment‡.....	$1.0026e\hbar/2m_\mu$
	105.7 MeV	Mean life.............	2.22×10^{-6} s
Charge..............	± 1 electron	Decay mode..........	$\mu^{\pm} \rightarrow e^{\pm} + \nu + \bar{\nu}$ §
Spin.................	$\frac{1}{2}$		

‡ Coffin et al., *Phys. Rev.*, **109**, 973 (1958).

§ The conclusion that a neutrino and an antineutrino are emitted in the decay of a muon, rather than two neutrinos or two antineutrinos, is based upon the shape of the energy spectrum of the decay electrons.

Antiparticle. μ-mesons occur with both signs of charge, and these presumably correspond to particle and antiparticle. One can specify which is which if one assumes the validity of the so-called principle of *conservation of light Fermi particles*, which include electrons, neutrinos, and muons. (These particles are sometimes referred to as "leptons.") Thus the decay $\mu^{-} \rightarrow e^{-} + \nu + \bar{\nu}$ would require that, if e^{-} is defined as a

[1] Anderson and Neddermeyer, *Phys. Rev.*, **50**, 263 (1936).

particle, μ^- must likewise be a particle; e^+ and μ^+ are then *antiparticles*.

Interaction with Matter. A μ-meson interacts with matter through its electric charge and magnetic moment, and through the β-decay interaction. Therefore a muon loses energy through ionization of the matter through which it passes, as described in Chap. 14. In fact, ionization is virtually the *only* means by which muons lose energy, since their relatively large mass leads to quite small scattering by atomic *nuclei*, so that bremsstrahlung is quite unimportant, and their nuclear interaction is extremely weak. It is primarily this inertness of μ-mesons which leaves them as the major constituent ($\sim$80 per cent) of cosmic rays at sea level; electrons and quanta are rapidly removed by the cascade-shower process, and nucleons are removed by nuclear collisions. Indeed, so inert are μ-mesons that they have been detected in deep mines, under hundreds of meters of solid rock!

If a muon is brought to rest by traversing matter, its fate depends upon its sign of charge. A *positive* muon, being repelled by atomic nuclei and attracted by electrons, may borrow an outer electron from an atom and with it make a hydrogen-like "atom," the muon being the "nucleus." This phase of its life is short; for the muon decays into an electron and two neutrinos with a mean life expectancy of but 2 μs (Fig. 20-3a).

On the other hand, a *negative* muon is *attracted* by atomic nuclei and, in virtue of its mass being so much greater than that of an electron, it readily displaces an electron from an atom and forms what is called a *mesic atom*. The meson drops from one bound state to another, emitting quanta as it goes, until it attains the 1s state. This capture process takes only about 10^{-13} s, so that, once brought to rest in matter, nuclear capture is virtually certain. From this state the muon has two choices: it may decay into a negative electron and two neutrinos as it would do if free or it may induce a β-transition in the nucleus by the process

$$\mathrm{p} + \mu^- \rightarrow \mathrm{n} + \nu \tag{1}$$

The probability of occurrence of the so-called *forced decay* (1) is found experimentally to vary approximately as Z^4 for low and medium Z, and to compete equally with spontaneous decay at $Z \approx 11$.

The energies of the low-lying bound states of a mesic atom are virtually unaffected by the atomic electron cloud because the wave function of the meson is so strongly localized near the nucleus, but they are *greatly* modified by the penetration of the meson *inside the nucleus itself*. This property renders the μ-meson quite valuable as a "test particle" with which to probe the structure of the nuclear charge distribution.[1]

Another most interesting phenomenon which stems from the relatively

---

[1] Fitch and Rainwater, *Phys. Rev.*, **92**, 789 (1953); Cooper and Henley, *Phys. Rev.*, **92**, 801 (1953); Wheeler, *Phys. Rev.*, **92**, 812 (1953).

great mass and small nuclear interaction of the μ-meson is the mesic catalysis of nuclear reactions.[1] Consider a μ-mesic hydrogen atom composed of a proton and negative muon. This structure is quite small because of the large meson mass and, being neutral, is able to penetrate *inside* other atoms, just as a neutron can. When the mesic atom closely approaches another proton, the meson may be "shared" between the two protons to form a *mesic hydrogen molecule ion*. The interproton distance of such a molecule will be comparable with the size of the charge cloud in the mesic hydrogen atom, namely, about

$$A_0 = \frac{4\pi\varepsilon_0\hbar^2}{m_\mu e^2} = \frac{m_e}{m_\mu}\, a_0 = 2.56 \times 10^{-13} \text{ m} \tag{2}$$

Although this separation is rather large compared with the "range" of nuclear forces, there may still be an appreciable probability of nuclear reaction between the protons within the 2.2-μs lifetime of the muon. If such a nuclear reaction does occur, the muon should usually be expelled, and might repeat the whole process.

Nuclear reactions of the above general type have been observed in the liquid hydrogen of a bubble chamber (Fig. 20-3c), and are believed to result from the following sequence of events: A negative muon is brought to rest in the hydrogen and forms a mesic atom with a proton. This mesic atom wanders about until it encounters a deuterium atom (ordinary hydrogen contains about 1 per cent of deuterium), whereupon the meson leaves the proton and attaches itself to the deuteron. The mesic deuterium in turn wanders about until it forms an HD mesic molecule ion, and the proton and deuteron combine to form ^{3}He, ejecting the muon with several MeV energy.

EXERCISES

20-5. Calculate the maximum kinetic energy that a μ-meson decay electron can possess, if the muon decays at rest. *Ans.:* $E_{max} = 52.3$ MeV.

20-6. Find the approximate range in rock of a muon of 10^{12} eV energy. *Ans.:* $R \approx 10^3$ m.

20-7. A muon of kinetic energy $E \gg m_\mu c^2$ is slowed down in a medium of density ρ. Assuming the average "rate" of energy loss to be a constant R_0 MeV g^{-1} cm^2, find the probability that the muon will be slowed down to a kinetic energy $E = m_\mu c^2$ before it decays.

Ans.:
$$P \approx \left(\frac{W}{1.866 m_\mu c^2}\right)^{-(m_\mu c/R_0\rho\tau_\mu)}$$

[1] Alverez et al., *Phys. Rev.*, **105**, 1127 (1957).

20-8. Assuming that the probability of forced decay of a muon is a constant λ_f per unit time when the muon is inside the nucleus and zero when it is outside, and that $A \approx 2Z$, derive the Z^4 law for the probability of forced decay in light elements.

20-9. From the observed fact that the probability of forced decay equals that of spontaneous decay for $Z = 11$, calculate the interaction time $\tau_f = \lambda_f^{-1}$ and compare this roughly with the time required for the muon to traverse the nuclear diameter of sodium. *Ans.:* $\tau_f = 1.1 \times 10^{-8}$ s $\approx 10^{14}$ "traversals."

20-10. Compute the approximate momentum and kinetic energy of the neutron which results from the forced decay of a muon in hydrogen. *Ans.:* $p_n = 100$ MeV/c, $E_n = 5.3$ MeV.

20-11. Calculate the binding energy of a muon in mesic hydrogen and in deuterium, and thus show that it is energetically advantageous for the muon to transfer from a proton to a deuteron. *Ans.:* $E_{\scriptscriptstyle .1} = -2520$ eV, $E_D = -2670$ eV.

20-12. Assume that the mesic catalyzed reaction discussed above is

$$^1\text{H} + {}^2\text{H} + \mu^- \rightarrow {}^3\text{He} + \mu^-$$

where the three particles on the left are initially at rest. Find the kinetic energy with which the muon is ejected. *Ans.:* $E_\mu = 5.4$ MeV.

20-7. π-mesons (Pions)

The great penetrating power of the cosmic-ray μ-mesons discussed above was a source of considerable puzzlement to physicists, who wanted to identify it with Yukawa's particle. For to produce strong nuclear binding, Yukawa's meson must have quite strong nuclear interaction and therefore relatively small penetrating power. This difficulty was resolved by the discovery[1] in the cosmic radiation of another meson, called the π-meson, or *pion*. The π-meson is now regarded as being the true Yukawa particle—its mass is about $270m_e$, its spin integral, and its interaction with nuclei quite strong. Furthermore, π-mesons are found quite copiously in nuclear reactions above the threshold energy for their production.

The π-mesons of Lattes et al. are positive and negative *charged* mesons which when free are unstable, decaying into a μ-meson and a neutrino with a mean lifetime of 2.54×10^{-8} s. (This is in fact the source of the μ-mesons that one finds at sea level in the cosmic radiation; a high-energy nucleon incident at the top of the atmosphere collides with a nitrogen or oxygen nucleus and produces many π-mesons, some of which decay into

---

[1] Lattes, Occhialini, and Powell, *Nature*, **160**, 453 (1947).

μ-mesons.) In 1950 there was also discovered a *neutral* meson which is taken to be the uncharged meson required by the Yukawa theory to account for the nuclear forces between like nucleons. This meson is also unstable when free, but its decay is an electromagnetic one, yielding two quanta.[1]

Measured Properties. The intrinsic properties and interactions of pions have been studied quite intensively, mostly with pions produced in high-energy accelerators. Virtually all of the data given in Table 20-7 were obtained by using artificially produced pions.

TABLE 20-7. PROPERTIES OF THE PION

Mass:
  $\pi^\pm$.................... $273.23m_e$
                            139.6 MeV
  π^0.................... $264.4m_e$
                            135.1 MeV
Charge................ ±1, 0 electronic charge
Spin.................... 0
Magnetic moment... 0
Intrinsic parity.......... odd
Isotopic spin............ 1
Isotopic-spin projection:
  π^+.................... $+1$
  π^0....................   0
  π^-.................... -1
Mean life:
  $\pi^\pm$... .... 2.54×10^{-8} s
  π^0.................... $\sim10^{-15}$ s

Decay modes..........

$\pi^\pm \rightarrow \mu^\pm + \nu$	$\sim100\%$	Q = 33.9 MeV
$e^\pm + \nu$	$\sim10^{-2}\%$	Q = 139.1
$\pi^0 \rightarrow \gamma + \gamma$	98.8%	Q = 135.1
$e^+ + e^- + \gamma$	1.2%	Q = 134.1
$e^+ + e^- + e^+ + e^-$	0.0036%	Q = 133.1

Antiparticles. Although no mutual annihilation has been observed experimentally, there are strong theoretical reasons for believing that π^+- and π^--mesons are antiparticles to one another, and the two-quantum decay of the π^0 may be taken as experimental proof that this particle is identical with its antiparticle; that is, two π^0-mesons could mutually annihilate one another, leaving only quanta.

Interaction with Matter. In passing through matter, a charged π-meson loses energy by ionization and is strongly subject to nuclear collisions which result in scattering or absorption. The cross section for nuclear interaction is a sizable fraction of the nuclear geometrical cross section $(\pi R_0^2 \approx 5 \times 10^{-30}A^{2/3} \text{ m}^2 = 0.05A^{2/3} \text{ barn})$ so that nuclei may be considered to be essentially "black" toward pions. Because of their

---

[1] Steinberger, Panofsky, and Stellar, *Phys. Rev.*, **78**, 802 (1950).

extremely short lifetime, *neutral* pions move only a few atomic diameters before they decay into quanta, so that they are not perceptibly affected by the matter through which they pass; they are, of course, strongly subject to interaction with the nucleons of that nucleus in which they are produced, as are also the charged pions.

Although their interaction with nuclei is quite strong, the mean free path for nuclear collisions of charged pions in ordinary matter varies from about 80 g cm^{-2} in light elements ($Z < \sim 20$) to about 150 g cm^{-2} in heavy elements, so that, if their energy is not too great, there is some likelihood of their being brought to rest by ionization loss before they disappear either by nuclear collision or by spontaneous decay while in flight.   If they are brought to rest, their fate depends upon the sign of charge: Positive pions are not able to surmount the nuclear Coulomb barrier and therefore undergo spontaneous decay, while negative pions form π-mesic atoms and thence undergo catastrophic interaction with a proton in the nucleus to form a neutron.   Unlike the corresponding situation with μ-mesons, where a neutrino carries away most of the available energy, practically all of the rest-mass energy of a captured pion is transmitted to the nucleus, and this energy is usually expended in the disruption of the nucleus into several pieces having considerable kinetic energies.   Such a nuclear disruption is called a "star," because of its appearance in a nuclear emulsion or a cloud or bubble chamber, and each emergent charged particle having significant range is called a *prong* of the star (Fig. 20-3*b*).

A few special cases of the interaction of pions with matter are worthy of separate consideration:

1. Spin.   The inverse reactions

$$^{1}\text{H} + {}^{1}\text{H} \leftrightarrow {}^{2}\text{H} + \pi^{+} \tag{1}$$

have been used to determine the spin of the π^{+}-meson, using the measured cross sections of the two reactions together with the statistical-mechanical *principle of detailed balance*.[1] The latter principle is a fundamental result of statistical mechanics which requires that, under conditions of statistical equilibrium, inverse processes occur at equal rates.   The factors that would enter into the rates of Eq. (1) are their respective cross sections, the density of states for the outgoing particles, and the *statistical weights*.   The cross sections can be measured, the density of states calculated, and the relative statistical weights thus obtained.   In this way it is found that the statistical weight factor $2S + 1$ contributed by the spin S of the pion is

[1] Cheston, *Phys. Rev.*, **83**, 1118 (1951).

unity, within the errors of measurement, rather than three or more. Thus the π^+ spin is zero.

2. Parity. The absorption of π^--mesons by deuterium has been used to determine the *intrinsic parity* of the negative pion. It is observed that two decay modes compete on an approximately equal basis:

$$^2\mathrm{H} + \pi^- \; \nearrow \; \mathrm{n} + \mathrm{n}$$
$$\searrow \; \mathrm{n} + \mathrm{n} + \gamma \tag{2}$$

We shall be concerned with the first of these. Inasmuch as the pion is absorbed from the 1s state of the mesic atom and itself has spin zero, the total angular momentum of the system is that of the deuteron alone, namely, $I = 1$. But the exclusion principle requires that the resulting two neutrons be in an antisymmetric state: if their spins are symmetric ($S = 1$), their orbital angular momentum must be odd, and if antisymmetric ($S = 0$), even. Thus, out of the possible states $^3S_1$, $^3P_1$, $^3D_1$, and $^1P_1$, which are the only ones that can yield $I = 1$, all except $^3P_1$ are excluded by this requirement. Therefore, since the parity of the two neutrons is determined by their orbital angular momentum, their parity is *odd*. But the parity of the system before decay is that of the pion alone, since the parity of the deuteron is known to be even and the pion is in an S state. Thus, the intrinsic parity of the pion is found to be *odd*. On this basis the wave function for a pion is taken to be a *pseudoscalar*, since such a quantity changes sign when the three space axes are reversed, as must be the case for odd parity.

3. $\pi^- - \pi^\circ$ mass difference. The absorption of slow π^--mesons in hydrogen leads to the two processes

$$^1\mathrm{H} + \pi^- \; \nearrow \; \mathrm{n} + \gamma$$
$$\searrow \; \mathrm{n} + \pi^\circ$$
$$\phantom{\searrow \; \mathrm{n} + \pi^\circ} \; \llcorner\!\!\rightarrow \gamma + \gamma \tag{3}$$

The fact that the second reaction occurs (as is evidenced by the appearance of two nearly equally energetic quanta) demonstrates that the neutral pion is *less massive* than the charged pion by an amount greater than the neutron-proton mass difference. The energy spread of the emitted γ-rays may be used to evaluate the $\pi^- - \pi^\circ$ mass difference as follows: Because the π^--meson is absorbed by a stationary proton, the π° meson is emitted *isotropically* in the lab system and carries away most of the released energy. Similarly, in the decay of the π°, the two quanta are emitted isotropically *in the pion system*, and thus have a certain energy spread as seen in the lab system. The resultant energy distribution of the quanta is *uniform* between two limits, and by measuring this "line breadth" one can evaluate the desired mass difference, with the result

$$m_{\pi^-} - m_{\pi^\circ} = 4.5 \pm 0.3 \text{ MeV} \tag{4}$$

It is believed that this mass difference is entirely electromagnetic in nature, i.e., attributable to the energy in the electrostatic field of the charged pion.

EXERCISES

20-13. Calculate the momentum and kinetic energy of the muon in the CM system of a decaying charged pion. *Ans.*: $E_\mu = 4.1$ MeV, $p_\mu = 29.8$ MeV/c.

20-14. (*a*) Find the range in copper of a 200-MeV pion. (*b*) If such a pion decays in free flight, what is the range of the resulting muon if the latter is ejected in the direction of motion of the pion? *Ans.*: $R_\pi = 11.2$ cm (Cu). $R_\mu = 15.2$ cm (Cu).

20-15. Find the threshold bombarding energy for the production of a charged pion, assuming the target protons to be at rest, in the reactions (*a*) $p + p \rightarrow p + n + \pi^+$ and (*b*) $\gamma + p \rightarrow n + \pi^+$. *Ans.*: (*a*) 293 MeV, (*b*) 149 MeV.

20-16. Find the thresholds of the reactions of the preceding exercise if the target is a carbon nucleus whose protons have a Fermi energy of 20 MeV. *Ans.*: (*a*) 165 MeV, (*b*) 142 MeV.

20-17. A high-energy charged pion is produced in a nuclear collision of a cosmic-ray proton with a nitrogen atom at an altitude where the pressure is 8×10^4 dyn cm^{-2} and the temperature is 250°K. If the mean free path for nuclear absorption of the pion is 80 g cm^{-2} in air, what must be its energy if it is to have at least a 50 per cent chance of decaying to a muon before it undergoes nuclear absorption? Neglect the change in density of the air along the path of the pion. *Ans.*: $E < \sim 1.3 \times 10^{11}$ eV.

20-18. Show that the geometric mean energy of the two decay quanta of a moving π° which are emitted along and opposite the π° line of flight is $\frac{1}{2} m_{\pi^\circ}$.

20-19. Evaluate the $\pi^- - \pi^\circ$ mass difference Δm in terms of the breadth ΔE of the observed energy distribution of the π°-decay quanta from the absorption of negative pions in hydrogen. *Ans.:* $\Delta m \approx (\Delta E)^2/2m_\pi(1 + m_\pi/m_\text{p}) + 1.3$ MeV.

20-8. K-mesons (Thetons and Tauons)

The discovery of the pion in 1947 was triumphantly hailed by theoretical physicists as the capstone of the Yukawa theory and was regarded as possibly completing the list of the so-called *elementary particles* of which matter and energy are composed. But in that same year, Rochester and Butler[1] presented evidence in the form of two cloud-chamber photographs which suggested the existence of yet more unstable particles, both charged and neutral. Their observations have since been fully confirmed, and it is now known that there exists a whole family of particles, which have come to be called the *strange particles*, which are produced rather copiously in high-energy nuclear collisions but whose role in the scheme of things is, like that of the muon, obscure. For several years the strange particles were studied primarily as they occur in the cosmic radiation, but with the debut of ultra-high-energy particle accelerators, more and more of the properties of these particles have been measured under the well-controlled laboratory conditions available with accelerators.

Two general classes of strange particles are known: K-*mesons*, whose masses are about $1000m_e$, and *hyperons*, whose masses are greater than the mass of a proton. As we shall later see, these two classes are intimately related; we treat them separately here only as a matter of convenience. Each class of strange particle is observed to encompass several different particle types, some of which may merely be different phenomenological manifestations of a given actual particle. The different types of K-mesons are as follows:

1. The first K-meson to be discovered, and one of the most accurately measured so far, is the *τ-meson, or tauon*. This meson is characterized by its decay into *three charged pions:*

$$\tau^\pm \rightarrow \pi^\pm + \pi^+ + \pi^- + 75.0 \text{ MeV} \tag{1}$$

2. The next best established K-meson is the *θ°-meson, or theton*, a neutral particle whose characteristic decay is into *two charged pions:*

$$\theta^\circ \rightarrow \pi^+ + \pi^- + 214.6 \text{ MeV} \tag{2}$$

[1] *Nature*, **160**, 855 (1957).

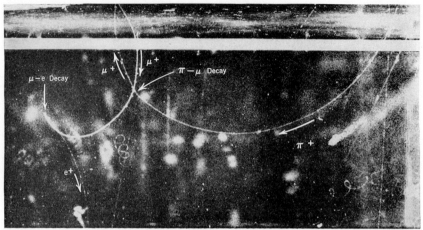

FIG. 20-3. Examples of the production, interaction, and decay of mesons and hyperons. (a) A π^+-μ^+-e^+ decay. A positive pion decays into a muon which strikes the wall of the cloud chamber and is scattered through almost 180°. The muon finally stops in the gas of the cloud chamber and decays into a positron.

3. A charged counterpart of the θ°-meson is also well substantiated. This particle undergoes decay into two pions, one charged and one neutral:

$$\theta^+ \rightarrow \pi^+ + \pi^\circ + 219.1 \text{ MeV} \tag{3}$$

4. In addition to the above types of K-mesons which may or may not actually be distinct from one another, several other decay schemes which have been observed experimentally are almost certainly but alternate decay modes of these particles. Some of them are:

$$
\begin{array}{lll}
(a) & \mathrm{K}^+_{\mu_3} \rightarrow \mu^+ + (\pi^\circ + \nu) & \\
(b) & \mathrm{K}^+_{\mu_2} \rightarrow \mu^+ + (\nu) & \\
(c) & \mathrm{K}^+_{e_3} \rightarrow e^+ + (\pi^\circ + \nu) & (4) \\
(d) & \mathrm{K}^+_{\pi_3} \rightarrow \pi^+ + (\pi^\mp + \pi^\circ) &
\end{array}
$$

$$
\begin{array}{lll}
(a) & \theta^\circ_{\mathrm{anom}} \rightarrow \pi^\pm + \mu^\mp + (\nu) & \\
(b) & \theta^\circ_{\mathrm{anom}} \rightarrow \pi^\pm + e^\mp + (\nu) & (5)
\end{array}
$$

Parentheses are used to indicate that the identities of the neutral particles are not firmly established.

Measured Properties. In spite of the manifold *decay schemes* which are observed for K-mesons, measurements of the properties of the parent particles themselves indicate that, if more than one charged K-meson actually exists, these are very nearly degenerate in mass and lifetime. Table 20-8 assumes only one charged K-particle; further measurement may require more. Two distinct neutral K-mesons are known, of nearly the same mass but widely different lifetime.

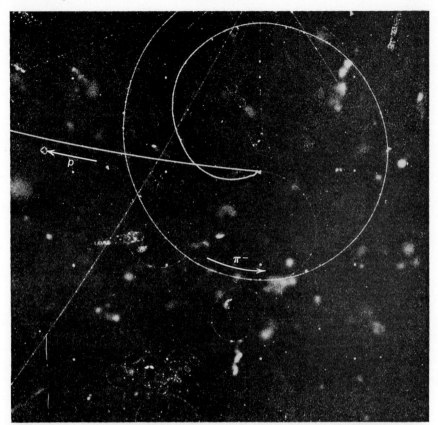

FIG. 20-3. (*b*) π^--capture. A slow negative pion comes to rest in the gas of the cloud chamber and is captured by a carbon or argon nucleus, from which a 65-MeV proton emerges. The remaining mass energy of the pion presumably was carried away by one or more neutrons.

TABLE 20-8. PROPERTIES OF K-MESONS

Mass:		Spin............ 0
K^+...........	$966.6 \pm 0.4m_e$	Mean life:
	494.0 MeV	K^+........... $\sim 1 \times 10^{-8}$ s
K°...........	$974.4 \pm 1m_e$	K°........... $\sim 1 \times 10^{-10}$ s, $\sim 10^{-7}$ s
	497.9 MeV	Decay modes.... See text
Charge:		
K^+...........	1 electron	
K°...........	0	

Antiparticles. In the case of the *charged* K-mesons, the relatively rarely observed K⁻-meson is regarded as the antiparticle of the K⁺-meson. In addition, a *neutral* K-meson ($\overline{K^\circ}$) is supposed to exist as antiparticle to the commonly observed K°-meson. We shall later discuss this matter

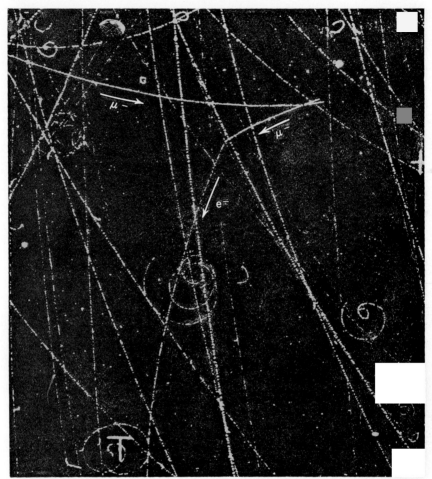

FIG. 20-3. (c) A μ^--catalyzed nuclear reaction. A negative muon stops in liquid hydrogen and is captured by a proton. Near this point the mesic catalysis takes place, with the expulsion and subsequent decay of the muon.

further, from the standpoint of a theory of the production and decay of the strange particles.

Interaction with Matter. K$^+$- and K$^-$-mesons of course lose energy by ionization as they pass through matter, but in addition each has a certain interaction with atomic nuclei. A K$^+$-meson undergoes relatively little nuclear scattering or absorption, and if brought to rest in matter undergoes spontaneous decay as it would in the absence of matter. In contrast with this, a K$^-$-meson undergoes quite strong nuclear scattering and absorption; if it comes to rest in matter, it forms a K-mesic atom

FIG. 20-3. (d) A double μ^--catalyzed nuclear reaction. Same as (c), except that the muon catalyzes two nuclear reactions (at 1 and 2) before it decays at 3.

and is absorbed, producing a rather large star. This star may exhibit some rather remarkable properties, as will be described later.

Neutral θ-mesons do not appear to have very strong interaction with nuclei and, of course, do not lose energy by ionization. Their relatively short mean life makes it most likely that they will decay in flight, even

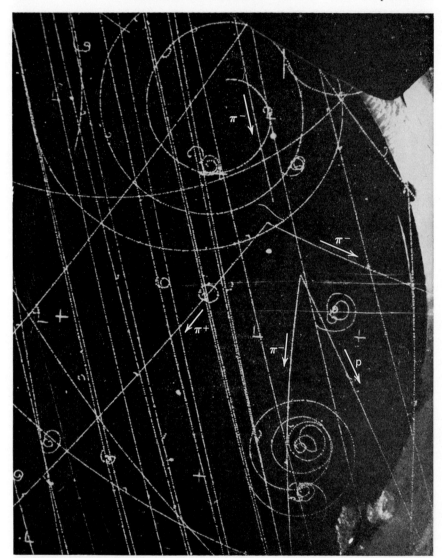

FIG. 20-3. (e) Associated production of Λ° and θ°.   A negative pion, traveling through liquid hydrogen, strikes a proton and produces a Λ°-hyperon and a θ°-meson, both of which decay inside the bubble chamber.

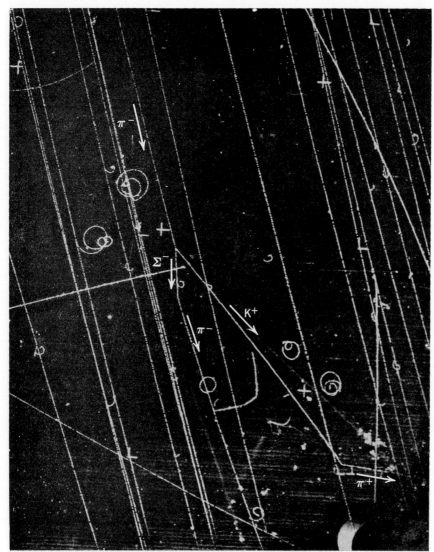

FIG. 20-3. (f) Associated production of $\Sigma^- + K^+$. A negative pion, traveling through liquid hydrogen, strikes a proton and produces a Σ^--hyperon and a K^+-meson, both of which subsequently decay.

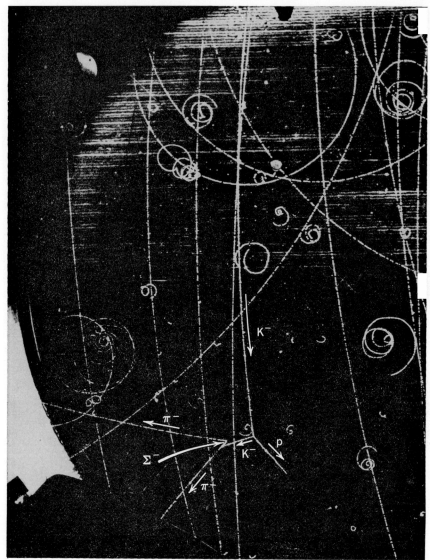

FIG. 20-3. (g) K⁻-scatter and absorption. A slow K⁻-meson, traveling through liquid hydrogen, is scattered by a proton just before coming to rest. The stationary K⁻ is absorbed by a proton to yield a Σ⁻-hyperon and a π⁺-meson, which move in opposite directions from the point of absorption. The Σ⁻-hyperon decays after traveling a short distance. [(a) and (b) are cloud-chamber photographs taken at California Institute of Technology; (c) to (g) are photographs of the 10-in. liquid-hydrogen bubble chamber of the Radiation Laboratory of the University of California, furnished through the courtesy of Prof. Hugh Bradner.]

when moving through dense matter. Anti-θ°-mesons ($\overline{\theta^\circ}$) should have relatively strong nuclear interaction, similar to the K^-; the reason for this also will be discussed later.

20-9. Hyperons

Closely related to the K-mesons are a second class of strange particles whose masses are greater than the mass of a nucleon and which yield a nucleon as a decay product. These particles are called *hyperons*, and they comprise the following principal types:

1. The first and best known hyperon is a neutral unstable particle called a Λ°-particle, which decays with a mean life of about 2.8×10^{-10} s into a proton and a negative pion with an energy release of about 37 MeV:

$$\Lambda^\circ \rightarrow p + \pi^- + 37 \text{ MeV} \tag{1}$$

The symbol Λ° was chosen for this particle because of the characteristic "inverted V" appearance of its decay in a cloud chamber (Fig. 20-3e).

2. A second type of hyperon is the charged unstable particle called a Σ-hyperon. Σ-hyperons occur both positively and negatively charged, and there are strong indications of the existence of a neutral Σ-hyperon as well. The charged Σ-hyperons undergo decay according to the schemes (Fig. 20-3f).

$$
\begin{array}{ll}
(a) & \Sigma^+ \nearrow p + \pi^\circ + 116.1 \text{ MeV} \\
 & \searrow n + \pi^+ + 110.3 \text{ MeV} \\
(b) & \Sigma^- \rightarrow n + \pi^- + 118.5 \text{ MeV}
\end{array}
\tag{2}
$$

The mean life of the Σ^+ is shorter than that of the Σ^- because of the additional decay channel available to it.

The Σ°-hyperon, if it exists, is supposed to decay into a Λ° hyperon:

$$\Sigma^\circ \rightarrow \Lambda^\circ + \gamma + (\sim 80 \text{ MeV}) \tag{3}$$

3. A third well-established hyperon is the Ξ^--hyperon, which undergoes the decay

$$\Xi^- \rightarrow \Lambda^\circ + \pi^- + 66.5 \text{ MeV} \tag{4}$$

The existence of a neutral counterpart of the Ξ^--particle was predicted theoretically well in advance of its experimental discovery. Its decay is difficult to observe because no charged particles are involved:

$$\Xi^\circ \rightarrow \Lambda^\circ + \pi^\circ + (\sim 70 \text{ MeV}) \tag{5}$$

Measured Properties. At the time of the present writing, only the masses and the mean lives of the hyperons have been measured. These particles may, of course, possess other properties such as magnetic moments, electric-quadripole moments, etc. (Table 20-9).

TABLE 20-9. PROPERTIES OF THE HYPERONS

Mass:
Λ°........... 2181.4m_e
　　　　　1115.2 MeV
Σ^+........... 2327.7m_e
　　　　　1189.4 MeV
Σ^-........... 2343.2m_e
　　　　　1196.5 MeV
Ξ^-........... 2584m_e
　　　　　1320.5 MeV

Charge:
Λ°........... 0
Σ^+........... +1 electron
Σ^-........... −1 electron
Ξ^-........... −1 electron

Spin:
Λ°........... ½
Σ^+........... Half-integral
Σ^-........... Half-integral
Ξ^-........... Half-integral

Mean life:
Λ°........... 2.8×10^{-10} s
Σ^+........... $\sim 5 \times 10^{-11}$ s
Σ^-........... $\sim 10^{-10}$ s
Ξ^-........... $\sim 10^{-10}$ s

Decay modes:

Λ°
$\nearrow$ p + π^-　　>50%
$\searrow$ n + π°　　<50%
p + π°　　$\sim$50%

Σ^+
$\nearrow$
$\searrow$ n + π^+　　$\sim$50%

$\Sigma^- \to$ n + π^-
$\Xi^- \to \Lambda^\circ + \pi^-$

Antiparticles. Antiparticles corresponding to each of the above particles presumably exist, but since these all involve antinucleons, they are not produced in sufficiently great numbers to be experimentally accessible.

Interaction with Matter. In addition to the usual ionization energy loss characteristic of charged particles, the hyperons have been found to possess significant nuclear interaction. Foremost among these nuclear interactions is that of the Λ°, which has been observed to enter into bound states with various nuclei to form relatively stable structures called *hyperfragments.* Hyperfragments have mostly been observed in nuclear emulsions as products of the nuclear absorption at rest of K⁻- or Σ^--particles, but they are also sometimes found among the products of a nuclear collision.

If a K⁻-meson is absorbed by a nucleus, it is believed to rapidly undergo one of the reactions

(a)　　　　　K⁻ + p → Λ° + kinetic energy
(b)　　　　　　→ Λ° + π°
(c)　　　　　　→ Λ° + π^+ + π^-
(d)　　　　　　→ Σ^- + π^+　　　　　　(6)
(e)　　　　　　→ Σ^+ + π^-
(f)　　　　　　→ Σ° + π°

(a)	$K^- + n \rightarrow \Lambda^\circ + \pi^-$
(b)	$\rightarrow \Lambda^\circ + \pi^- + \pi^\circ$
(c)	$\rightarrow \Sigma^- + \text{kinetic energy}$
(d)	$\rightarrow \Sigma^- + \pi^\circ$
(e)	$\rightarrow \Sigma^\circ + \pi^-$

(7)

Similarly, a Σ^- absorbed by a nucleus is believed to rapidly undergo the reaction

$$\Sigma^- + p \rightarrow \Lambda^\circ + n \tag{8}$$

In all of the above reactions considerable energy is released, and the nucleus in which the absorption occurs therefore breaks up into several pieces, one of which may be a hyperfragment. If a hyperfragment is formed, the Λ° in it must eventually decay into a proton and pion, each of which may either interact with the residual nucleus or escape with its full energy. By measuring the energies of the proton and pion in cases in which both escape, it has been determined that the binding energy of a Λ° in a hyperfragment is quite small—perhaps 2 or 3 MeV—indicating that the Λ° acts somewhat differently than a neutron with respect to the nuclear forces.

20-10. Theoretical Interpretation of Particle Properties

The foregoing brief description of the known particles has been limited to a bare enumeration of some of their more prominent properties, particularly those that can be regarded as being established directly by experiment. Unfortunately, space does not permit even a partial description of the excellent experimental work that has led to our present knowledge in this field. It must suffice to say that nearly all this work has involved precise measurements of such quantities as the range, magnetic curvature, specific ionization, and multiple Coulomb scattering of charged particles in cloud chambers, photographic emulsions, scintillating media, and bubble chambers. Such measurements, combined with the conservation laws, have led to the elucidation of one decay process after another, and they are still of great use even in experiments carried out with high-energy particle accelerators.

We now turn from these experimentally established relationships between the various particles to a consideration of some of the more promising advances that have been made toward a theoretical understanding of these relationships. Of course, inasmuch as the theory of particles is not yet complete, we cannot expect all of the current ideas to be correct, but it seems quite likely that at least some of them will be incorporated into the "ultimate" theory.

A. Classification of Particle Types and Interactions. We begin by observing that the known particles fall into four general classes:[1]

1. Relatively massive particles of half-integral spin, called *baryons*, which contain a nucleon: p, n, Λ°, Σ^+, Σ^-, Σ°, Ξ^-, and their antiparticles, the antibaryons.
2. Particles of intermediate mass and integral spin, called *mesons*, which interact strongly with the baryons: π^+, π°, π^-, K^+, K°, K^-.
3. The lightest particles of half-integral spin, called *leptons*, whose interaction with baryons and mesons is considerably weaker than that *between* baryons and mesons: e^+, e^-, ν, μ^+, μ^-. (Note that in this new definition the muon is not considered to be a meson.)
4. The photon: γ.

Similarly, the interactions between the various particles appear to fall into four classes:

1. The strongest interactions, which occur exclusively among the baryons, antibaryons, and mesons. These are the so-called *nuclear forces* which are responsible for almost all that happens in nuclear *collisions*.
2. The *electromagnetic* interaction, much weaker than the nuclear interaction. This connects together all charged particles, and particles having electric or magnetic moments, through the intermediary of electromagnetic quanta (photons).
3. An interaction, much weaker still than the electromagnetic interaction, which leads to the β-decay of the muon, neutron, and K-meson and to the μ-decay of the pion and K-meson, i.e., to those decays in which neutrinos take part.
4. An interaction, of about the same strength as the β-decay interactions, which leads to the decay of the strange particles.

The relative strengths of the above interactions may be estimated roughly on the basis of the lifetimes of excited systems whose decay is governed by them. Thus, one finds that the strong interactions, the electromagnetic interactions, and the weak (β-decay) interactions are characterized by reaction times of about 10^{-23}, 10^{-18}, and 10^{-8} s, respectively, unless a selection rule, centrifugal barrier, or other limiting factor decreases the transition probability. On the basis of this great range of interaction strength, one expects that the main features of the structure of particles should be defined by the strong interactions, and that the electromagnetic and weak interactions should have only a small perturbing effect upon this structure. On this basis, then, the zero-order descrip-

---

[1] Our treatment follows closely that presented by Gell-Mann, *Proc. Pisa Conference*, 1955. See also Gell-Mann and Rosenfeld, Hyperons and Heavy Mesons: Systematics and Decay, *Ann. Rev. Nuclear Sci.*, **7**: 407 (1957).

tion of particles would ignore all electromagnetic interactions, β-decays, and (presumably) strange-particle decays; that is, all particles would be *stable* except those that can react by a strong interaction. (E.g., $\pi^- + p \rightarrow n + KE$ would still occur, but π^- would be stable in the absence of other baryons or mesons.)

In the next approximation the electromagnetic interactions would be included. The photon (which does not exist in the zero-order approximation) now appears and induces the electromagnetic decay of the π°, γ-emission deexcitation of nuclear excited states, and other more subtle effects such as a mass difference between p and n, between $\pi^\pm$ and π°, and between $\Sigma^\pm$ and Σ°. In this approximation $\pi^\pm$, $\mu^\pm$, and the strange particles are still stable. Finally, the introduction of the weak interactions leads to the remaining instabilities, and also causes a small additional shift in mass of some of the particles.

B. Phenomenological Model of the Baryons and Mesons (Theory of Strangeness). Because of the great range of interaction strength also, one does not expect either electromagnetic or a fortiori the weak interactions to be effective in nuclear collisions, and in fact one does not observe electrons or photons (or neutrinos) as the direct products of nuclear collisions. By the same token, one expects that a baryon or meson which is unstable toward decay into lighter baryons and/or mesons should undergo this decay *quite rapidly* through the strong interaction. Here, contrary to expectation, we find that, although K-mesons do in fact decay into pions, and hyperons into pions and nucleons, these decays occur fantastically slowly compared with the expected rate for the strong interactions. This poses a serious question of the mechanism by which the decay of these particles is slowed from an expected mean life of $\sim 10^{-23}$ s to the observed values of $\sim 10^{-9}$ s.

One possibility is that the particles might possess quite large angular momenta—in the neighborhood of five or six units—which, together with the nuclear attractive potential, would yield an unstable system having a *potential barrier* to retard its decay. Experimental tests of this hypothesis have failed to reveal any of the expected consequences of such higher angular momenta, however.

Various other suggestions have been advanced to reconcile the rapid production and slow decay of the strange particles. Among these, one stands out at present both for its simplicity and for its ability to relate all known baryons and mesons to one another. This is the so-called *strangeness theory* advanced by Gell-Mann[1] and also by Nishijima. One of the basic concepts of this theory is that the strange particles cannot be produced singly in a nuclear collision but must be produced in association

[1] *Proc. Pisa Conference*, 1955.

with other strange particles, *each strange particle carrying away some quantized quantity which the other needs in order to decay.* Thus in nuclear collisions the total amount of this quantity is conserved, but its conservation (like that of isotopic spin) is not rigorous, and indeed must be violated, in the decay of the particle.

More specifically, the strangeness theory is based upon a generalization of the concept of isotopic spin (the assumption of charge independence of nuclear forces). According to the ordinary idea of isotopic spin, each nuclear particle (nucleon or pion) possesses a certain total isotopic spin, and each possible projection of this isotopic spin along a certain axis (the "ζ axis") appears to us as a different charge state of the corresponding particle. Analytically, if A is the mass number of the particle, T its isotopic-spin quantum number, T_ζ the ζ component of T, and Q the charge in units of the electron charge, the relation between these quantities is

$$Q = T_\zeta + \tfrac{1}{2}A \qquad -T \leq T_\zeta \leq +T \text{ (integer steps)} \qquad (1)$$

Inserting the values $T = \tfrac{1}{2}$, $A = 1$ for nucleons and $T = 1$, $A = 0$ for pions, one obtains

$$
\begin{array}{lll}
(a) \text{ proton:} & Q_p = +\tfrac{1}{2} + \tfrac{1}{2} = +1 & T_\zeta = +\tfrac{1}{2} \\
(b) \text{ neutron:} & Q_n = -\tfrac{1}{2} + \tfrac{1}{2} = 0 & T_\zeta = -\tfrac{1}{2}
\end{array} \qquad (2)
$$

$$
\begin{array}{lll}
(a) \text{ positive pion:} & Q_{\pi^+} = +1 + 0 = +1 & T_\zeta = +1 \\
(b) \text{ neutral pion:} & Q_{\pi^0} = 0 + 0 = 0 & T_\zeta = 0 \\
(c) \text{ negative pion:} & Q_{\pi^-} = -1 + 0 = -1 & T_\zeta = -1
\end{array} \qquad (3)
$$

The ordinary theory of isotopic spin (charge independence) goes on to assert that in all nuclear collisions the total isotopic spin (obtained by combining the isotopic spins of the component particles according to the rules for combination of quantized angular momenta) is conserved, as are its various components, and that the cross section for a given collision reaction is the same for all ζ-projections of its isotopic-spin vector, for given isotopic spin. As an example, the reactions

$$
\begin{array}{ll}
(a) & \pi^+ + p \rightarrow \pi^+ + \pi^0 + p \\
(b) & \pi^- + n \rightarrow \pi^- + \pi^0 + n
\end{array} \qquad (4)
$$

must have the same cross section, since they both belong to the same total isotopic-spin state $T = \tfrac{3}{2}$.

In the strangeness theory the relation (1) is replaced by

$$Q = T_\zeta + \tfrac{1}{2}A + \tfrac{1}{2}S \qquad (5)$$

where S is an integral quantum number corresponding to that property possessed by the strange particles in virtue of which they cannot decay rapidly.   It is then further postulated that T, Q, and A (and therefore S) are conserved in *all* nuclear collisions, including those which involve the production of strange particles.   The theory then asserts that *all baryons and mesons must possess definite values for T, A, and S.*   Because $S = 0$ corresponds to the ordinary nucleons and pions, and $S \neq 0$ to the strange particles, S is called the *strangeness*.   For given A, S, and T, the various projections T_ζ give rise to an isotopic-spin multiplet, each component of which corresponds to a different charge state.   The various baryons and mesons are thus analogous to the fine-structure multiplets of optical spectroscopy.

Let us now see what kinds of particle types can be formed by various choices of T, A, and S, bearing in mind that *no experimental evidence has yet indicated the existence of multiple charges for elementary particles:* We are already familiar with the particles having no strangeness ($S = 0$); if $A = 1$ and $T = \frac{1}{2}$, we obtain $Q = +1$ or 0 (nucleons), and if $A = 0$ and $T = 1$, we obtain $Q = +1, 0, -1$ (pions).   One might also have a neutral meson with $A = 0$, $T = 0$; such a particle has not yet been found experimentally, but might be quite difficult to detect.

Turning next to particles with $S = +1$, we see that multiple charges can be avoided only by taking $A = 1$ and $T = 0$ or $A = 0$ and $T = \frac{1}{2}$. The former would correspond to a baryon singlet with $Q = +1$, and the latter to a meson doublet with $Q = +1, 0$.   The former is not known, but the latter could correspond to the K$^+$ and K$^\circ$ mesons.

If $S = -1$, there are three possibilities which do not involve multiple charges:

1. $A = 0$, $T = \frac{1}{2}$ yields $Q = 0, -1$, which may be identified with anti-particles of K$^+$ and K$^\circ$: the K$^-$ and $\overline{\text{K}^\circ}$.
2. $A = 1$, $T = 0$ yields a neutral singlet baryon, which may be identified with the Λ°, since no charged baryon of mass close to that of Λ° has been observed.
3. $A = 1$, $T = 1$ yields a baryon triplet, which evidently corresponds to the Σ^+ and Σ^- if a Σ° is added.   The Σ°, being considerably more massive than the Λ° and of the same strangeness, would presumably decay rapidly into Λ° in matter, or by electromagnetic coupling into $\Lambda^\circ + \gamma$ if free.

Finally, if $S = -2$, one can obtain a doublet baryon with $A = 1$ and $T = \frac{1}{2}$ ($Q = 0, -1$) or singlet or triplet *double* baryons with $A = 2$, $T = 0$ and $A = 2$, $T = 1$, respectively.   Strange particles having $A = 2$ have not been observed, but the doublet $S = -2$, $A = 1$, $T = \frac{1}{2}$ is taken to represent the Ξ hyperon.

The above results may be conveniently plotted diagrammatically as shown in Fig. 20-4.

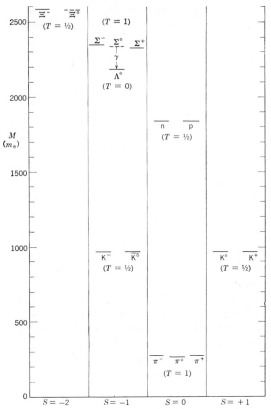

FIG. 20-4. Particle states according to the strangeness theory.

The assumption that T, A, Q, and S must be conserved in all fast reactions (collisions) leads to some important consequences which have been strikingly confirmed experimentally. Some of them are:

1. *Strange particles can be produced only in certain combinations or associations.* In a nuclear collision in which the ingoing particles are ordinary ($S = 0$) particles, one should find strange particles among the products only in the combinations (Fig. 20-3e and f)

$$
\begin{aligned}
&(a) &&\Lambda^\circ + K^{\circ,+} \\
&(b) &&\Sigma^{+,\circ,-} + K^{\circ,+} \\
&(c) &&K^{-,\circ} + K^{\circ,+} \\
&(d) &&\Xi^- + 2K^{\circ,+}
\end{aligned}
\tag{6}
$$

The agreement of these predictions with experiment is excellent: Numerous examples of the associated production $\Lambda^\circ + \theta^\circ$, $\Sigma^\pm + \theta^\circ$, $\Lambda^\circ + K^+$ have been observed in cloud chambers and bubble chambers, and a few examples of $K^- + K^{+,\circ}$, $K^\circ + K^\circ$, and $\Xi^- + 2K^\circ$ have also been found; on the other hand, no cases of $\Lambda^\circ + \Lambda^\circ$, $\Lambda^\circ + K^-$, $\Sigma^\pm + K^-$ or any other combination not given by (6) have been seen.

In a collision of a strange particle with an ordinary particle, only reactions which conserve strangeness are permitted. These include simple scattering of the incident particle, certain charge-exchange scatterings such as $K^\circ \leftrightarrow K^+$, $\Sigma^+ \leftrightarrow \Sigma^-$, etc., and various reactions such as (Fig. 20-3g)

$$
\begin{array}{llr}
(a) & K^- + p \rightarrow \Sigma^- + \pi^+ & \\
(b) & \Sigma^+ + n \rightarrow \Lambda^\circ + p & (7) \\
(c) & \Xi^- + p \rightarrow \Lambda^\circ + \Lambda^\circ &
\end{array}
$$

It should be noted that the strangeness theory predicts quite unsymmetrical properties for K^+ and K^- interacting with ordinary matter: A K^+ can undergo only simple scattering or charge-exchange scattering, while a K^- can undergo both these and reactions in which Λ° or $\Sigma^{+,-,\circ}$ are formed. This asymmetry is fully borne out by observation and constitutes a strong confirmation indeed of the essential correctness of the strangeness theory.

2. If a negatively charged strange particle is absorbed at rest by a nucleus, only those reactions which conserve strangeness can occur rapidly. These include the reactions 20-9(6) and (7) and in addition

$$
\begin{array}{llr}
(a) & \Sigma^- + p \rightarrow \Lambda^\circ + n & \\
(b) & \Xi^- + p \rightarrow \Lambda^\circ + \Lambda^\circ & (8)
\end{array}
$$

Several cases which can be interpreted most easily in terms of one of these reactions have been observed.

3. In addition to restricting the production of strange particles to certain combinations, the conservation of strangeness also prohibits the free decay of a strange particle. Here, one introduces the assumption that $\Delta S = 0$ holds rigorously only for the strong and the electromagnetic interactions but that there exists a weak interaction which permits transitions with $\Delta S = \pm 1$ to take place slowly. In this way the decays of all strange particles are placed upon the same basis, which then accounts in a most satisfactory way for the approximate equality of most known strange-particle lifetimes.

Perhaps the most striking example of the operation of the selection rule $\Delta S = \pm 1$ is seen in the decay of the Ξ^--hyperon, which proceeds in two steps, first to a Λ° and then to a proton.

From the above discussion we see that the strangeness theory is able to account in a relatively straightforward way for many of the most prominent and, at first sight, most puzzling properties of the baryons and mesons. While the theory may not yet be in its best form, there seems to be considerable promise that it is essentially correct.

C. Nature of the Weak Interactions, Universal β-Interaction. Except for our brief consideration of β-decay in Chap. 15, we have given our attention almost exclusively to the electromagnetic and the nuclear interactions, which together account for nearly all of what goes on in the visible universe. But the weak interactions, primarily through the phenomenon of β-radioactivity, also play an important role in shaping the character of the universe as we see it. Therefore it is important to try to measure the weak interactions quantitatively and thus determine their basic physical nature. We cannot penetrate this subject very deeply here, but it will be interesting to recount a few of the salient features of the weak interactions.

In our previous treatment of β-decay we learned that the great majority of the observed β-decay transitions follow the simple Fermi theory, but that some involve the so-called Gamow-Teller selection rules. These observed properties are of considerable help in establishing the qualitative character of the nuclear β-decay interaction, and measurements of the $fT_{1/2}$ values, especially for transitions involving mirror nuclides, establish the *strength* of the interaction.

In the present chapter we have seen that many of the so-called *elementary particles* also undergo β-decay. It is of great interest to compare the type and the strength of the β-decay interaction as it appears in these decays with that determined from nuclear data. From studies of selection rules and angular distributions of the particles in nuclear β-decay, it appears that a coupling of the form *V-A* (page 542), of "strength" $G = 1.41 \times 10^{-49}$ erg cm³ (page 543), is responsible for nuclear β-decay, free muon decay, and forced muon decay. In fact, one calculation has indicated that the strengths of the nuclear β-decay, the free muon decay, and the forced muon decay are all equal within the *experimental* error, which in the case of the first two quantities is only one or two per cent![1] This fact, together with the observation that β-decay competes effectively with decay into pions for some of the strange particles, make it appear that *the weak interactions may be basically of a single universal type.*

The various manifestations of the β-decay and strangeness coupling may be systematized schematically in terms of an interaction between certain pairs of particles as is shown pictorially in Fig. 20-5, in which pairs of particles associated with the vertices of a square or tetrahedron are connected by weak

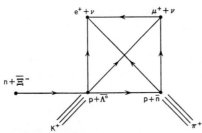

FIG. 20-5. Schematic diagram of the weak interactions.

[1] Feynman and Gell-Mann, *Phys. Rev.*, **109**, 193 (1958).

interactions, and those at two of the vertices are also strongly coupled with the mesons.

For example, we may write for the free decay of the muon

$$\mu^+ + \nu \rightarrow e^+ + \nu \tag{9}$$

or, exchanging the ν on the left for a $\bar{\nu}$ on the right (an ingoing particle is equivalent to an outgoing antiparticle),

$$\mu^+ \rightarrow e^+ + \nu + \bar{\nu} \tag{10}$$

Similarly, for the forced decay of the muon,

$$p + \bar{n} \rightarrow \mu^+ + \nu \tag{11}$$

or, exchanging n on the right for ñ on the left and μ^- on the left for μ^+ on the right,

$$p + \mu^- \rightarrow n + \nu \tag{12}$$

If the analytic forms as well as the strengths of the various couplings are assumed, one can calculate the branching ratios of the various decay modes available to a given particle. Thus one should be able to evaluate the fraction of π^+ mesons which decay into $\mu^+ + \nu$ as against $e^+ + \nu$, or the fraction of Λ° hyperons decaying into $p + \pi^-$, $p + \mu^- + \nu$, and $p + e^- + \nu$. For a time, the electronic decay of the pion

$$\pi^+ \rightarrow e^+ + \nu \tag{13}$$

was predicted theoretically to occur about 0.01 per cent as often as the familiar decay into a muon, but was not found experimentally. A carefully designed experiment[1] has revealed this decay mode, however, and the observed branching ratio appears to agree satisfactorily with the theoretical value. The branching ratios for hyperon decays have not yet (1958) been calculated or measured precisely, and the character of the coupling for the decay of the strange particles is not yet clear.

Another decay, $\mu^+ \rightarrow e^+ + \gamma$, which might be expected to occur extremely rapidly compared with the usual two-neutrino muon decay, apparently does not occur at all. Its absence is of course quite reasonable in the scheme shown in Fig. 20-5, where the interactions occur only between certain *pairs* of particles, one particle and one antiparticle being needed at each vertex.

D. Nonconservation of Parity in Weak Interactions. Speculations such as the above are most interesting and suggestive, but they are, of course, likely to remain inconclusive in the absence of adequate experimental information against which to test specific coupling schemes. In this connection we recall again the experimental measurements on *con-*

[1] Fazzini et al., *Phys. Rev. Let.*, **1**, 247 (1958).

servation of parity in β-decay, which undoubtedly penetrate deeply into the fundamental nature of the weak interactions.

As was described previously, the measurements were carried out as a result of suggestions made by the theorists Lee and Yang, who pointed out that there was as yet no adequate experimental proof that parity is conserved in decay processes involving the weak interactions and that the puzzling coincidence that the τ^+- and θ^+-mesons have the same mass and lifetime would be removed if these could be in fact *the same particle*— which on the basis of their observed decay schemes can only be possible if a particle need not conserve parity in its decay.

Now it had heretofore been universally assumed that every interaction of nature is symmetrical with respect to "right and left" or "forward and backward," so that the wave function of any isolated system must possess either even or odd symmetry with respect to the reflection of any or all space-time coordinate axes through the origin. On this basis every elementary particle must possess a definite parity, and every decay process must be symmetrical with respect to the "forward" and "backward" directions as defined by the spin vector of the decaying system.

But almost immediately after Lee and Yang put forward their suggestion, it was shown in two quite different experiments that systems which undergo β-decay may *show a marked forward-backward asymmetry in the angular distribution of the decay products:*

1. In the experiment of Wu et al., a sample of the β-active nuclide ^{60}Co was subjected to a strong magnetic field at liquid-helium temperature, in order to align the nuclei. The relative numbers of electrons emitted along and contrary to the field direction were compared and were found to be quite different.

2. In the experiment of Garwin et al., the decay properties of muons obtained from the decay in flight of monoenergetic charged pions were studied. Those muons which are thrown forward in the CM frame of the pion possess a greater range than do the pions (see Exercise 20-14) and can thus be separated from undecayed pions by inserting just the right thickness of stopping material into the pion beam: the pions are stopped and the muons emerge with a small residual range. The muons are in turn stopped in a second absorber, where they are subjected to a weak magnetic field (transversely oriented with respect to their original direction of motion) in which they are allowed to precess. A delayed coincidence circuit counts, as a function of the time after stopping, the number of muons which send their decay electrons into a given direction perpendicular to the field. If the muon itself has no preferential spin orientation as it emerges from the pion decay, and if a muon shows no forward-back-

ward decay asymmetry, the delayed coincidences should show a pure exponential time dependence with the usual 2.22-μs mean lifetime. But the observations showed instead an exponential decay with a *sinusoidally varying component*, which can only mean that the original muons showed a preferential spin orientation *and* that a muon emits its decay electron preferentially forward or backward. Thus, both the pion decay and the muon decay fail to conserve parity.

These experiments have given rise to quite extensive new theoretical and experimental work concerning the nature of β-decay and the weak interactions. Much of the theoretical work has been concerned with the question of just what symmetries *are* valid for the weak interactions. A possible conclusion, so far consistent with the experimental data, is that the weak interactions are invariant toward the operations of time reversal alone (symbolized T) or toward a combination of space reflection P and reversal of sign of charge C. Thus, whereas it had previously been assumed that a possible state of matter, when "reflected in a mirror," would still be a possible state of matter, the new results indicate that this is not true, but that such a state *would* be a possible state of *antimatter*.

At the time of writing, new experimental and theoretical results are emerging so rapidly that it is not possible to see clearly the character of the new structure that is being uncovered. One thing *is* clear, however— we still have much to learn about the ways of nature!

20-11. The Dirac Theory of the Electron

We now go on to consider briefly a theory of the only particle which is so far considered to be well understood: the electron. This theory, which is called the *Dirac theory*, constitutes a relativistically invariant description of spin-$\frac{1}{2}$ particles, and it is probably also correct for the muon as well.

There are many possible ways in which the quantum mechanics of a charged particle in an electromagnetic field can be extended to relativistic energies. The celebrated Dirac equation is the result of a particular approach to this problem. In devising a Lorentz covariant form of quantum mechanics, it would be desirable to build upon postulates that are themselves Lorentz covariant, and this can in fact be done. Unfortunately, space does not permit us to exploit this approach, and we shall instead proceed in a different way which does, however, lead to Lorentz-covariant equations.

As a suitable Hamiltonian function to represent the particle in an electromagnetic field, it is customary to use

$$H = c[(\boldsymbol{p} - e\boldsymbol{A})^2 + m^2c^2]^{1/2} + e\phi \tag{1}$$

where $\boldsymbol{A}$ is the magnetic vector potential and ϕ is the scalar potential.

EXERCISES

20-20. Show that the Lagrangian function

$$L = -mc^2 \left(1 - \frac{v^2}{c^2}\right)^{\frac{1}{2}} + e\boldsymbol{v} \cdot \boldsymbol{A} - e\phi \tag{2}$$

yields correct equations of motion for a charged particle in an electro-magnetic field.

20-21. Define three momenta conjugate to x, y, and z by the equations $p_i = \partial L/\partial \dot{x}_i$ and form the Hamiltonian function

$$H = \sum_{i=1}^{3} p_i \dot{q}_i - L$$

Ans.: Eq. (1).

The expression (1) may be put into more symmetrical form by squaring:

$$(H - e\phi)^2 - (\boldsymbol{p} - e\boldsymbol{A})^2 c^2 = m^2 c^4 \tag{3}$$

One might now expect that, if the usual substitutions

$$\boldsymbol{p} \to \frac{\hbar}{i} \nabla \qquad \text{and} \qquad H \to -\frac{\hbar}{i} \frac{\partial}{\partial t} \tag{4}$$

are used in (3), the resulting equation

$$\left[\left(-\frac{\hbar}{i} \frac{\partial}{\partial t} - e\phi\right)^2 - \left(\frac{\hbar}{i} \nabla - e\boldsymbol{A}\right)^2 c^2\right] \psi = m^2 c^4 \psi \tag{5}$$

might be a suitable relativistic generalization of the Schroedinger equation. This equation is called the *Klein-Gordon* equation, and it is now believed to be a valid relativistic equation for charged particles having no intrinsic angular momentum. It was first discovered by Schroedinger, but was almost immediately rejected because it leads to the seemingly impossible or meaningless result that *a particle can exist in states of negative as well as positive total energy.* This can readily be seen by finding the possible energy eigenvalues for a free particle which is moving with given momentum $\boldsymbol{p}$. The wave function for such a particle should evidently be the solution of

$$\nabla^2 \psi - \frac{1}{c^2} \frac{\partial^2 \psi}{\partial t^2} = \frac{m^2 c^2}{\hbar^2} \psi \tag{6}$$

The usual method of separation of variables leads to solutions of the form

$$\psi = A \exp\left[(i/\hbar)(\boldsymbol{p} \cdot \boldsymbol{r} - Wt)\right] \tag{7}$$

where the separation constants p_x, p_y, p_z, and W must satisfy

$$W^2 - p^2 c^2 = m^2 c^4 \tag{8}$$

Thus there are two possible energy eigenvalues for given momentum

p, one positive and one negative:

$$W = \pm (p^2c^2 + m^2c^4)^{1/2} \tag{9}$$

EXERCISE

20-22. Verify that Eq. (7) is a solution of Eq. (6) corresponding to momentum p, and that Eq. (8) must hold.

Dirac sought to avoid the negative energy states by finding a linear, Hermitian operator to represent Eq. (1) directly. His great step lay in his method of accomplishing this. He assumed that the radical appearing in Eq. (1) must be *symmetrical and linear* in the momenta:

$$[(p - eA)^2 + m^2c^2]^{1/2} = \boldsymbol{\alpha} \cdot (p - eA) + \beta mc \tag{10}$$

Now clearly such an expression cannot be valid if the quantities $\boldsymbol{\alpha}$ and β are regarded as an ordinary vector and scalar, since squaring both sides of Eq. (10) and equating coefficients of similar terms leads to the apparently contradictory requirements

(a)
(b)

$$\alpha_x{}^2 = \alpha_y{}^2 = \alpha_z{}^2 = \beta^2 = 1$$
$$\alpha_x\alpha_y + \alpha_y\alpha_x = \alpha_x\alpha_z + \alpha_z\alpha_x = \alpha_x\beta + \beta\alpha_x \tag{11}$$
$$= \alpha_y\alpha_z + \alpha_z\alpha_y = \alpha_y\beta + \beta\alpha_y$$
$$= \alpha_z\beta + \beta\alpha_z$$
$$\equiv 0$$

However, it is possible to satisfy Eq. (10) if α_x, α_y, α_z, and β are not numbers but *operators which anticommute with one another*. It is customary to take these as 4×4 matrix operators, specifically,

$$\hat{\alpha}_x = \begin{pmatrix} 0 & 0 & 0 & 1 \\ 0 & 0 & 1 & 0 \\ 0 & 1 & 0 & 0 \\ 1 & 0 & 0 & 0 \end{pmatrix} = \begin{pmatrix} \hat{0} & \hat{\sigma}'_x \\ \hat{\sigma}'_x & \hat{0} \end{pmatrix}$$

$$\hat{\alpha}_y = \begin{pmatrix} 0 & 0 & 0 & -i \\ 0 & 0 & i & 0 \\ 0 & -i & 0 & 0 \\ i & 0 & 0 & 0 \end{pmatrix} = \begin{pmatrix} \hat{0} & \hat{\sigma}'_y \\ \hat{\sigma}'_y & \hat{0} \end{pmatrix}$$

$$\hat{\alpha}_z = \begin{pmatrix} 0 & 0 & 1 & 0 \\ 0 & 0 & 0 & -1 \\ 1 & 0 & 0 & 0 \\ 0 & -1 & 0 & 0 \end{pmatrix} = \begin{pmatrix} \hat{0} & \hat{\sigma}'_z \\ \hat{\sigma}'_z & \hat{0} \end{pmatrix} \tag{12}$$

$$\hat{\beta} = \begin{pmatrix} 1 & 0 & 0 & 0 \\ 0 & 1 & 0 & 0 \\ 0 & 0 & -1 & 0 \\ 0 & 0 & 0 & -1 \end{pmatrix} = \begin{pmatrix} \hat{1} & \hat{0} \\ \hat{0} & -\hat{1} \end{pmatrix}$$

where $\acute{\sigma}_x'$, $\acute{\sigma}_y'$, and $\acute{\sigma}_z'$ are the Pauli spin operators and $\hat{1}$ and $\hat{0}$ are the unit and null 2×2 matrices.

EXERCISES

20-23. Show that squaring Eq. (10) and equating coefficients of like terms leads to Eq. (11).

20-24. Show that the matrices (12) satisfy Eq. (11), where in the latter, 1 is taken to mean the unit 4×4 matrix.

In the Dirac theory the wave functions are taken to be four-component column matrices[1] whose components are ordinary (complex) variables:

$$\psi = \begin{pmatrix} \psi_1 \\ \psi_2 \\ \psi_3 \\ \psi_4 \end{pmatrix} \tag{13}$$

The expectation value of a dynamical quantity F is obtained in terms of ψ and the Hermitian adjoint wave function

$$\psi\dagger = (\psi_1^* \psi_2^* \psi_3^* \psi_4^*) \tag{14}$$

according to the formula

$$\langle F \rangle = \int \psi\dagger \hat{F} \psi \, dV \tag{15}$$

where we assume ψ to be normalized so that

$$\int \psi\dagger \psi \, dV = \int (\psi_1^*\psi_1 + \psi_2^*\psi_2 + \psi_3^*\psi_3 + \psi_4^*\psi_4) \, dV = 1 \tag{16}$$

With the above definitions and the substitution (4), we obtain the *Dirac equation* for a charged particle in an electromagnetic field:

$$\hat{H}\psi = \left[c\hat{\mathbf{\alpha}} \cdot \left(\frac{\hbar}{i} \nabla - e\mathbf{A} \right) + \hat{\beta}mc^2 + e\phi \right] \psi = -\frac{\hbar}{i} \frac{\partial \psi}{\partial t} \tag{17}$$

Equation (18) shows this formula written in component form.

(a)　　　　$\left(\dfrac{\hbar}{i} \dfrac{\partial}{\partial t} + e\phi + mc^2 \right) \psi_1$

(b)　　　　　　　　　　　　　　　　$\left(\dfrac{\hbar}{i} \dfrac{\partial}{\partial t} + e\phi + mc^2 \right) \psi_2$

(c)　$c \left(\dfrac{\hbar}{i} \dfrac{\partial}{\partial z} - eA_z \right) \psi_1 + c \left[\left(\dfrac{\hbar}{i} \dfrac{\partial}{\partial x} - eA_x \right) - i \left(\dfrac{\hbar}{i} \dfrac{\partial}{\partial y} - eA_y \right) \right] \psi_2$

(d)　$c \left[\left(\dfrac{\hbar}{i} \dfrac{\partial}{\partial x} - eA_x \right) + i \left(\dfrac{\hbar}{i} \dfrac{\partial}{\partial y} - eA_y \right) \right] \psi_1$　　　　$- c \left(\dfrac{\hbar}{i} \dfrac{\partial}{\partial z} - eA_z \right) \psi_2$

---

[1] It is also possible to cast the Dirac theory into a form in which the wave functions have the form of *two*-component matrices, but we shall not consider this form here because it is not the one in most common use. [See Feynman and Gell-Mann, *Phys. Rev.*, **109**, 193 (1958).]

EXERCISE

20-25. Verify that Eq. (18) is the expanded form of the Dirac equation (17).

The above simultaneous, linear, first-order partial differential equations for the four wave-function components have been solved exactly for several important problems, including that of the hydrogen atom. However, our treatment must be limited to but a few of the simplest aspects of the Dirac equation. Let us first investigate the energy and momentum eigenstates of a *free particle* by finding for what values of W we can find wave functions of the form

$$\boldsymbol{\psi} = \exp\left[(i/\hbar)(\boldsymbol{p} \cdot \boldsymbol{r} - Wt)\right] \begin{pmatrix} C_1 \\ C_2 \\ C_3 \\ C_4 \end{pmatrix} \tag{19}$$

with at least one of the four amplitude constants not equal to zero. Substitution into Eq. (18) and cancellation of the exponential factor results in

$$
\begin{aligned}
(-W + mc^2)C_1 \quad &+ \quad & cp_zC_3 + \;\; c(p_x - ip_y)C_4 &= 0 \\
(-W + mc^2)C_2 + &c(p_x + ip_y)C_3 - & cp_zC_4 &= 0 \\
cp_zC_1 + c(p_x - ip_y)C_2 + &(-W - mc^2)C_3 & &= 0 \\
c(p_x + ip_y)C_1 - &cp_zC_2 & + (-W - mc^2)C_4 &= 0
\end{aligned}
\tag{20}
$$

Now in order that at least one amplitude constant be finite, it is necessary that the determinant of the coefficients be zero. This determinant reduces to $(W^2 - m^2c^4 - p^2c^2)^2 = 0$, or

$$W = \pm(p^2c^2 + m^2c^4)^{1/2} \tag{21}$$

From this we conclude that, in the Dirac theory also, *the permissible energy values for a free particle range from* $+mc^2$ *to* $+\infty$ *and from* $-mc^2$ *to*

$$
\begin{aligned}
&+ c\left(\frac{\hbar}{i}\frac{\partial}{\partial z} - eA_z\right)\psi_3 + c\left[\left(\frac{\hbar}{i}\frac{\partial}{\partial x} - eA_x\right) - i\left(\frac{\hbar}{i}\frac{\partial}{\partial y} - eA_y\right)\right]\psi_4 = 0 \\
+ c\left[\left(\frac{\hbar}{i}\frac{\partial}{\partial x} - eA_x\right) + i\left(\frac{\hbar}{i}\frac{\partial}{\partial y} - eA_y\right)\right]\psi_3 \qquad\qquad\; - c\left(\frac{\hbar}{i}\frac{\partial}{\partial z} - eA_z\right)\psi_4 &= 0 \\
+ \left(\frac{\hbar}{i}\frac{\partial}{\partial t} + e\phi - mc^2\right)\psi_3 \qquad\qquad\qquad\qquad\qquad &= 0 \\
+ \left(\frac{\hbar}{i}\frac{\partial}{\partial t} + e\phi - mc^2\right)\psi_4 &= 0
\end{aligned}
\tag{18}
$$

$-\infty$. The first of these results is of course just what we expect for a free particle—that its total energy can have any value greater than its rest energy. But the second result is quite puzzling, since it implies the existence of states of *negative total energy*.

Furthermore, there is nothing to prevent—and indeed the theory requires—the occurrence of *transitions between states of positive and negative energy.* These rather remarkable features of the Dirac theory were not understood until particles were found in the cosmic radiation (Anderson, 1932) which are similar to ordinary electrons except that they appear to carry a *positive charge.* These positive electrons, or *positrons,* are identified with the negative energy states of the Dirac theory in the following way: To avoid the unpleasant prospect of having all the electrons in the universe fall into the infinitely "deep" sea of negative energy states, it is customarily postulated (1) that all of the negative energy states are already occupied by electrons,[1] (2) that the exclusion principle prevents any more electrons from occupying negative energy states, and (3) that this infinite density of particles is not directly observable. Positrons are then regarded as being "bubbles" or "holes" left by the removal of an electron from a negative energy state into one of positive energy. Such a "hole" would appear to move in the opposite direction to that of the applied force on a negative charge because all of the electrons would tend to move in the direction of the force, squeezing the "hole," like a bubble in a liquid, in the opposite direction.

According to this so-called *hole theory* of positrons, then, the elevation of an electron from one of the fully occupied negative energy states into an unoccupied positive energy state appears experimentally as the *creation of a positive and negative electron pair*—since one sees both the *electron,* which was initially invisible, and also the *hole* it leaves in the negative energy states. The reverse of the creation of an electron-positron pair can also occur: an electron which falls from a positive energy state into a hole in the negative energy states appears experimentally as the *mutual annihilation* of a positive and a negative electron.

EXERCISES

20-26. Verify that the vanishing of the determinant of the coefficients of the C's in Eq. (20) leads to Eq. (21).

20-27. Use the procedure of Sec. 2-4C to show that the operator which represents the x-component of velocity is $c\hat{\alpha}_x$.

20-28. Use the result of the preceding exercise, plus the fact that $\hat{\alpha}_x^2 = \hat{1}$, to show that the only possible results of a *precise* measurement of an electron's x-velocity are $\pm c$.

20-29. Show that there are four linearly independent wave functions of the form (19) which describe a particle moving with momentum p in the z direction.

[1] Later we shall encounter another possible means of avoiding this difficulty.

Ans.:

$$\psi = A \exp\left[(i/\hbar)(pz \mp Wt)\right]\left\{\begin{pmatrix} pc \\ 0 \\ \pm W - mc^2 \\ 0 \end{pmatrix} \text{ or } \begin{pmatrix} 0 \\ -pc \\ 0 \\ \pm W - mc^2 \end{pmatrix}\right\} \quad (22)$$

where $W = +\sqrt{p^2c^2 + m^2c^4}$. In these expressions the upper sign corresponds to positive total energy and the lower to negative total energy.

We now examine another feature of the Dirac theory which is in spectacular agreement with observation: the existence of an *intrinsic angular momentum, or spin,* for an electron. In the nonrelativistic treatment of the one-electron atom—or indeed of any one-particle problem having a central-field potential—the orbital angular-momentum operators $\hat{L}_x$, $\hat{L}_y$, and $\hat{L}_z$, and the square of the total orbital momentum $\hat{L}^2$, all commute with the Hamiltonian operator and are therefore *constants of the motion of the system.* We shall now show that in the Dirac theory this is not the case, but that another set of quantities replaces these as constants of the motion. These new quantities are just the three components and the square of what we have previously called the total angular momentum $j = l + s$.

Let us first evaluate the commutator operator of $\hat{L}_z$ and $\hat{H}$. (We assume only a central electrostatic field is present.)

$$\hat{L}_z = x\hat{p}_y - y\hat{p}_x = \frac{\hbar}{i}\left(x\frac{\partial}{\partial y} - y\frac{\partial}{\partial x}\right)$$

and

$$\hat{H} = c\frac{\hbar}{i}\left(\hat{\alpha}_x\frac{\partial}{\partial x} + \hat{\alpha}_y\frac{\partial}{\partial y} + \hat{\alpha}_z\frac{\partial}{\partial z} + i\hat{\beta}\frac{mc}{\hbar}\right) + e\phi(x,y,z)$$

Now $\hat{L}_z$ clearly commutes with $\hat{\alpha}_z(\partial/\partial z)$, $i\hat{\beta}(mc/\hbar)$, and $e\phi$, if ϕ represents a central field. Thus we have

$$\hat{L}_z\hat{H} = -c\hbar^2\left(x\frac{\partial}{\partial y} - y\frac{\partial}{\partial x}\right)\left(\hat{\alpha}_x\frac{\partial}{\partial x} + \hat{\alpha}_y\frac{\partial}{\partial y}\right)$$

$$= -c\hbar^2\left(\hat{\alpha}_x x\frac{\partial^2}{\partial x\,\partial y} + \hat{\alpha}_y x\frac{\partial^2}{\partial y^2} - \hat{\alpha}_x y\frac{\partial^2}{\partial x^2} - \hat{\alpha}_y y\frac{\partial^2}{\partial x\,\partial y}\right)$$

and

$$\hat{H}\hat{L}_z = -c\hbar^2\left(\hat{\alpha}_x\frac{\partial}{\partial x} + \hat{\alpha}_y\frac{\partial}{\partial y}\right)\left(x\frac{\partial}{\partial y} - y\frac{\partial}{\partial x}\right)$$

$$= -c\hbar^2\left(\hat{\alpha}_x\frac{\partial}{\partial y} + \hat{\alpha}_x x\frac{\partial^2}{\partial x\,\partial y} + \hat{\alpha}_y x\frac{\partial^2}{\partial y^2} - \hat{\alpha}_x y\frac{\partial^2}{\partial x^2} - \hat{\alpha}_y\frac{\partial}{\partial x}\right.$$
$$\left. - \hat{\alpha}_y y\frac{\partial^2}{\partial x\,\partial y}\right)$$

so that

$$\hat{L}_z\hat{H} - \hat{H}\hat{L}_z = -c\hbar^2\left(\hat{\alpha}_y\frac{\partial}{\partial x} - \hat{\alpha}_x\frac{\partial}{\partial y}\right) \neq 0 \quad (23)$$

which means that L_z, and by similar arguments L_x, L_y, and $|\mathbf{L}|^2$, *are not constants of the motion.* But if we introduce the matrix operators

$$\acute{\sigma}_x = \begin{pmatrix} \hat{0} & \hat{1} \\ \hat{1} & \hat{0} \end{pmatrix} \hat{\alpha}_x = \begin{pmatrix} \acute{\sigma}_x' & \hat{0} \\ \hat{0} & \acute{\sigma}_x' \end{pmatrix} \qquad \acute{\sigma}_y = \begin{pmatrix} \hat{0} & \hat{1} \\ \hat{1} & \hat{0} \end{pmatrix} \hat{\alpha}_y = \begin{pmatrix} \acute{\sigma}_y & \hat{0} \\ \hat{0} & \acute{\sigma}_y \end{pmatrix}$$

and $\quad \acute{\sigma}_z = \begin{pmatrix} \hat{0} & \hat{1} \\ \hat{1} & \hat{0} \end{pmatrix} \hat{\alpha}_z = \begin{pmatrix} \acute{\sigma}_z' & \hat{0} \\ \hat{0} & \acute{\sigma}_z' \end{pmatrix}$ $\qquad\qquad$ (24)

and if we evaluate the commutator operator of, say, $\acute{\sigma}_z$ and $\hat{H}$, we find

$$\acute{\sigma}_z \hat{H} = \frac{c\hbar}{i} \acute{\sigma}_z \left(\hat{\alpha}_x \frac{\partial}{\partial x} + \hat{\alpha}_y \frac{\partial}{\partial y} + \hat{\alpha}_z \frac{\partial}{\partial z} + i\hat{\beta}\frac{mc}{\hbar} \right) + \acute{\sigma}_z V$$

$$= c\hbar \left(\hat{\alpha}_y \frac{\partial}{\partial x} - \hat{\alpha}_x \frac{\partial}{\partial y} \right)$$

and $\quad \hat{H}\acute{\sigma}_z = \dfrac{c\hbar}{i} \left(\hat{\alpha}_x \dfrac{\partial}{\partial x} + \hat{\alpha}_y \dfrac{\partial}{\partial y} + \hat{\alpha}_z \dfrac{\partial}{\partial z} + i\hat{\beta}\dfrac{mc}{\hbar} \right) \acute{\sigma}_z + V\acute{\sigma}_z$

$$= -c\hbar \left(\hat{\alpha}_y \frac{\partial}{\partial x} - \hat{\alpha}_x \frac{\partial}{\partial y} \right)$$

so that $\qquad\qquad \acute{\sigma}_z \hat{H} - \hat{H}\acute{\sigma}_z = 2c\hbar \left(\hat{\alpha}_y \dfrac{\partial}{\partial x} - \hat{\alpha}_x \dfrac{\partial}{\partial y} \right)$ $\qquad\qquad$ (25)

Comparing this result with Eq. (23), we therefore conclude that

$$\left(\hat{L}_z + \frac{\hbar}{2}\acute{\sigma}_z \right) \hat{H} - \hat{H}\left(\hat{L}_z + \frac{\hbar}{2}\acute{\sigma}_z \right) = 0 \qquad\qquad (26)$$

and therefore that *the dynamical quantity whose operator is*

$$\hat{J}_z = \hat{L}_z + \frac{\hbar}{2}\acute{\sigma}_z \qquad\qquad (27)$$

is a constant of the motion. By similar procedures applied to the x and y components and to $|\mathbf{J}|^2$, one finds that the Dirac theory *automatically endows the particle it describes—in this case the electron—with the property of spin angular momentum of just the amount that is required by experimental observation.*

Furthermore, if one solves the Dirac equation for a particle in a magnetic field, he finds that the possible energy states include a term which corresponds to the familiar orientation energy of a particle having a spin of ½ *and a magnetic moment equal to one Bohr magneton.*

In summary, the Dirac equation, using only the charge and mass as given data, provides a complete, quantitative description *of all other intrinsic properties of electrons,* including the existence of an *antiparticle* of opposite charge, the positron. Understandably, the Dirac theory is considered one of the greatest successes of theoretical physics.

EXERCISES

20-30. Verify that Eq. (23) is the correct commutator operator of $\hat{L}_z$ and $\hat{H}$. (Recall that the unit 4×4 matrix commutes with all 4×4 matrices.)

20-31. Show that $\hat{\sigma}_x{}^2 = \hat{\sigma}_y{}^2 = \hat{\sigma}_z{}^2 = \hat{1}$ (the unit 4×4 matrix).

20-32. Show that $\frac{1}{2}\hbar\hat{\sigma}_x$, $\frac{1}{2}\hbar\hat{\sigma}_y$, and $\frac{1}{2}\hbar\hat{\sigma}_z$ satisfy the commutation relations for angular momentum (see Sec. 5-7). *Ans.:* $(\hbar^2/4)(\hat{\sigma}_x\hat{\sigma}_y - \hat{\sigma}_y\hat{\sigma}_x)$ $= i\hbar(\frac{1}{2}\hbar\hat{\sigma}_z)$, etc.

20-33. Show that $\hat{\sigma}_z$ commutes with $\hat{\alpha}_z$ and $\hat{\beta}$ and that

$$\hat{\sigma}_z\hat{\alpha}_x - \hat{\alpha}_x\hat{\sigma}_z = 2i\hat{\alpha}_y \qquad \hat{\sigma}_z\hat{\alpha}_y - \hat{\alpha}_y\hat{\sigma}_z = -2i\hat{\alpha}_x \qquad (28)$$

20-34. Show that the free-particle wave functions (22) are also eigenfunctions of $\hat{\sigma}_z$, corresponding to spin parallel or antiparallel to p, respectively. An electronic momentum eigenstate is thus quadruply degenerate, the four substates corresponding to positive or negative total energy and spin orientation along or opposite its direction of motion.

20-35. Show that, for $p \ll mc$ and $W > 0$, the eigenfunctions (22) reduce to

$$\psi = C' \exp\left[(i/\hbar)(pz - Wt)\right] \left\{ \begin{pmatrix} 1 \\ 0 \\ 0 \\ 0 \end{pmatrix} \text{ or } \begin{pmatrix} 0 \\ 1 \\ 0 \\ 0 \end{pmatrix} \right\}$$

which then correspond to the spin eigenfunctions of the Pauli theory:

$$\psi = C' \exp\left[(i/\hbar)(pz - Wt)\right] \left\{ \begin{pmatrix} 1 \\ 0 \end{pmatrix} \text{ or } \begin{pmatrix} 0 \\ 1 \end{pmatrix} \right\}$$

20-12. Quantum Electrodynamics

The Dirac theory presented in the preceding section describes the behavior of electrons under the stimulus of a given externally applied electromagnetic field. However, it is well known that electrons are themselves often the principal contributors to this field. In particular, an electron which undergoes acceleration in an electromagnetic field radiates photons by virtue of this acceleration, and the radiation of excited atomic systems is in large measure brought about by the dynamics of the electronic charge cloud. Thus if we could find, as a complement to the Dirac equation, an equation which describes *the behavior of the electromagnetic field under the stimulus of given electronic motions*, these two sets of equations should provide a quite broad description of the behavior of both electrons and radiation.

The equations we seek are already well known; they are called *Maxwell's equations*. Maxwell's equations are most simply expressed in terms of the 4-vector potential A_μ, as was outlined in Chap. 1:

$$\nabla^2 A_\mu - \frac{1}{c^2}\frac{\partial^2 A_\mu}{\partial t^2} = -\mu_0 J_\mu \tag{1}$$

In this expression the 4-current J_μ—which is, of course, the source of the field—must be evaluated in terms of the electronic motions. Classically, we recall that the current density i due to a charge distribution ρ moving at velocity v is

$$i = \rho v \tag{2}$$

However, we also know that, in the quantum theory, the charge of an electron is "smeared out" into a probability density distribution $\psi^*\psi$ and into a "velocity density" distribution $\psi^*\hat{v}\psi$. Thus the proper expressions for the charge and current densities should be

(a) $\qquad\qquad\qquad\qquad \rho = e\psi^*\psi$
(b) $\qquad\qquad\qquad\qquad i = e\psi^*\hat{v}\psi$ $\qquad\qquad\qquad\qquad$ (3)

In terms of the four-component wave functions of the Dirac theory, the corresponding expressions are

$$\frac{c}{i}J_4 = \rho = e\psi^\dagger\psi = e(\psi_1^*\psi_1 + \psi_2^*\psi_2 + \psi_3^*\psi_3 + \psi_4^*\psi_4) \tag{4}$$

and $\qquad\qquad J_i = ec\psi^\dagger\hat{\alpha}_i\psi \qquad i = 1, 2, 3 \qquad\qquad\qquad$ (5)

where we have used the relation $\hat{v}_i = c\hat{\alpha}_i$ (see Exercise 20-27). Inserting these into Eq. (1), we obtain the equations of quantum electrodynamics:

(a) $\qquad\left[c\hat{\alpha}\cdot\left(\frac{\hbar}{i}\nabla - eA\right) + \hat{\beta}mc^2 + e\phi\right]\psi = -\frac{\hbar}{i}\frac{\partial\psi}{\partial t}$

(b) $\qquad\qquad\qquad \nabla^2 A - \frac{1}{c^2}\frac{\partial^2 A}{\partial t^2} = -\mu_0 e\psi^\dagger\hat{\alpha}\psi \qquad\qquad$ (6)

(c) $\qquad\qquad\qquad \nabla^2\phi - \frac{1}{c^2}\frac{\partial^2\phi}{\partial t^2} = -\frac{e}{\varepsilon_0}\psi^\dagger\psi$

EXERCISES

20-36. Write out J_x in terms of the four components of ψ. *Ans.:*
$J_x = ec(\psi_1^*\psi_4 + \psi_2^*\psi_3 + \psi_3^*\psi_2 + \psi_4^*\psi_1)$.

20-37. Find the charge and current densities represented by the wave function

$$\psi = f(x,y,z)\begin{pmatrix} A \\ 0 \\ B \\ 0 \end{pmatrix}$$

Ans.: $\rho = e(A^*A + B^*B)f^*f$, $J_x = J_y = 0$, $J_z = ec(A^*B + B^*A)f^*f$.

Because of the mathematical complexity of Eq. (6), it is not possible to write down exact solutions even for simple cases. Instead, it has so far proved necessary to solve these equations by techniques involving *successive approximation*. Even this involves quite formidable mathematics, however, and it will be more enlightening if we limit ourselves to an examination of a possible *physical interpretation* of these approximations.

The quantitative description of the interaction of charges with the radiation field and with one another is based upon Eq. (6) through the familiar *perturbation theory*, according to which the probability of a given transition is proportional to the absolute square of a *matrix element* which connects the initial and final states: The matrix element for a first-order transition is written

$$(f|V|i) = \frac{i}{h} \int \psi_f^\dagger \hat{V} \psi_i \, d^N q \tag{7}$$

Now the usual interpretation of an equation of this kind is that the perturbation potential $\hat{V}$ causes the system to change "gradually" from the initial state ψ_i to the final state ψ_f. However, Feynman[1] has propounded an equally acceptable alternative point of view according to which the potential $\hat{V}$ is regarded as acting *discontinuously* on the system. According to this viewpoint, the integrand of Eq. (7) is interpreted as the probability amplitude for the occurrence of a certain sequence of events, each term in the integrand being the amplitude for one of the events. Specifically, ψ_i is the probability amplitude that the system *arrives* at the "point" $q_1 \cdots q_N$ in the state i, $\hat{V}$ is the amplitude that the potential "acts" upon the system, and $\psi_f^\dagger$ is the amplitude that the system *leaves* the "point" $q_1 \cdots q_N$ in the state f. The total effect of the potential is obtained by integrating over all possible "points" $q_1 \cdots q_N$ at which the potential might have acted.

One of the main advantages of this new interpretation is that it enables one to write down by inspection the expressions for the *higher-order* perturbation effects. By the older viewpoint, these involve the continuous transition of the system from the initial state to the final state through one or more intermediate states. The new interpretation would be that the system arrives at some point where it is acted upon by the potential, leaving it in a new state; it then arrives at a second point where it is acted on a second time by the potential and left in a third state, and so on, the "order" of the transition being equal to the number of times the potential acts in shifting the system from the initial to the final state.

[1] *Revs. Modern Phys.*, **20**, 367 (1948) (nonrelativistic); *Phys. Rev.*, **76**, 749, 769 (1949) (relativistic).

This simple pictorial representation of the response of a system to a perturbing potential is especially well suited to the treatment of the electromagnetic interaction of charged particles. In this instance the operator $\hat{V}$, which represents the interaction between the material system and the electromagnetic field, is interpreted in terms of the *emission and absorption of photons* by the material system, and the "order" of the perturbation is correspondingly equal to the number of photons involved in the transition. (However, if a static field is also involved in the problem, this is usually best treated as a fixed potential in the usual way, rather than as a cloud of photons.)

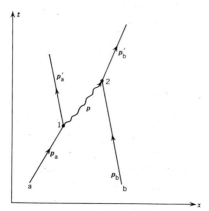

Fig. 20-6. Feynman diagram for electron-electron scattering.

We shall now describe qualitatively some of the problems to which quantum electrodynamics provides a solution. We shall not, of course, attempt actually to carry out any of the rather complicated integrals which appear in these problems, as this would require delving into the mathematical foundations of quantum electrodynamics considerably more deeply than space permits.

As a first example, consider the scattering of one electron by another. A possible sequence of events leading to such scattering is shown in Fig. 20-6, which is called the *Feynman diagram* for the process. (We ignore for simplicity the important fact that the two electrons might exchange roles in the process.) The interpretation of the various parts of the integrand of Eq. (7) for such a scattering process is as follows:

1. ψ_i is the probability amplitude that electrons a and b *arrive* at space-time points 1 and 2 with momenta p_a and p_b;
2. The operator $\hat{V}$ is the product of the probability amplitude (operators) for three independent events: (*a*) particle a *emits* a photon of momentum p, (*b*) a photon of momentum p goes from 1 to 2, and (*c*) particle b *absorbs* a photon of momentum p.
3. $\psi_f^\dagger$ is the probability amplitude that electrons a and b depart from points 1 and 2 with momenta p_a' and p_b'.

The total amplitude for the mutual scattering of two electrons between the given states is then the *integral* of the ordered product of the amplitudes for the events of the above sequence, taken over all points of emission and absorption of the photon in space time.

The probability amplitude for any observable process is found to equal zero unless energy and momentum are conserved. In the intermediate unobserved events one can also consider momentum and energy formally to be conserved, but the relation $W^2 = p^2c^2 + m^2c^4$ is not then valid; one calls such a process a *virtual process*. The various component amplitudes also possess certain important *symmetry properties* with respect to the space-time axes:

1. The emission of a photon is equivalent to the absorption of a photon of opposite momentum.
2. The propagation of a photon between two space-time points is equivalent to propagation in the reverse direction with reversed momentum.
3. The arrival of an electron is equivalent to the departure of a positron, and vice versa.

These symmetries appear to be a fundamental property of nature and are found to connect together certain physical phenomena that appear experimentally quite different from one another, as we shall later see.

We have so far used the term "photon" rather loosely; actually there are *four different kinds* of photons that can be exchanged between the electrons, which correspond to the four possible directions of polarization x, y, z, and t. By a suitable transformation of the representation, these can be replaced by an *instantaneously acting* Coulomb interaction and *two* kinds of photons which are of the familiar kind, polarized transverse to the direction of motion and propagated at the speed of light. It is thus found that the inverse-square law of static interaction and the delayed dynamic action between charges are both accounted for by the *single process* of the transmission of four-component "photons" between the charges.

The results given above provide a basis for treating other physical processes and for analyzing the higher-order terms in the perturbation expansion, because, as it turns out, the association of a certain kind of term with a particular event can be made in a consistent way independently of the over-all physical process in which the event occurs. Let us first consider some other physical processes as they would be described by the above theory. Figure 20-7a shows how a *Compton scattering* might be visualized. A possible sequence of events for this process is

1. An electron of momentum p_1 and a photon of momentum p_2 arrive at the point 1.
2. The electron absorbs a photon of momentum p_2.
3. The electron propagates with momentum $p = p_1 + p_2$ from point 1 to point 2.
4. The electron emits a photon of momentum p_2'.
5. The electron of momentum p_1' and a photon of momentum p_2' leave point 2.

The Compton-scattering process could also occur in a different sequence, depicted in Fig. 20-7b.   This sequence differs from the above one only in that the electron first emits the final quantum of momentum p_2' and propagates with momentum $p = p_1' - p_2'$ from 1 to 2.

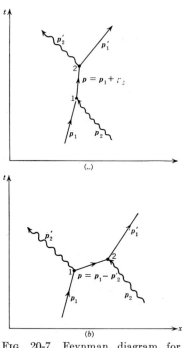

FIG. 20-7.   Feynman   diagram   for Compton scattering.

The total amplitude for Compton scattering would be obtained by multiplying together the amplitudes for the events depicted in Fig. 20-7a, adding this to the corresponding product for the sequence of Fig. 20-7b, and integrating the result over all values of the coordinates of points 1 and 2.   This leads to the *Klein-Nishina formula* 12-3(17).

Now consider the process called *bremsstrahlung* in which an electron emits a photon as it passes near a nucleus.   The essential features of this process are shown in the diagram of Figs. 20-8a and b, in which the nucleus is represented by its Coulomb potential alone.   In these sequences the electron arrives at 1, emits a photon, propagates from 1 to 2, and is scattered by the nuclear field V; or, is scattered by the nucleus, propagates from 1 to 2, and emits a photon at 2.   The amplitudes for these two possible sequences are again added and integrated over the locations of points 1 and 2.

We shall now describe two other processes which appear quite different from the above ones experimentally but which are in fact intimately related through the space-time symmetry of the elementary events which comprise the process.   First consider the process called *pair production*. In this process an energetic quantum creates an electron and a positron in the Coulomb field of a nucleus, and the quantum disappears.   The two possible diagrams for pair production are shown in Fig. 20-9a and b. In a, the quantum creates a pair at 1, the electron propagates to 2 and there is scattered by the nucleus; in b, the quantum creates a pair at 1, the *positron* propagates from 1 to 2 and there is scattered by the nucleus. Now, the space-time symmetry of the emission, absorption, and propagation of photons and electrons makes this process *exactly equivalent to bremsstrahlung*.   Specifically, the absorption of the photon in pair production is equivalent to the emission of the photon in bremsstrahlung;

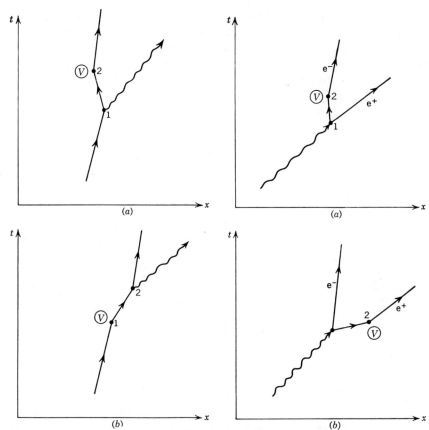

FIG. 20-8. Feynman diagram for brems-strahlung.

FIG. 20-9. Feynman diagram for pair pro-duction.

and *the departure of a positron is equivalent to the arrival of an electron.* In fact, one may consistently regard a positron as being *an electron traveling backward in time,* so that it is not necessary to adhere to the unsymmetrical and perhaps philosophically unsatisfying "hole" theory of positrons. According to this new idea, then, an electron can be scattered so violently that its direction of travel along the time axis becomes reversed, whereupon it appears experimentally as a *positron* traveling forward in time. Consideration of the process of two-quantum pair annihilation, as depicted in Fig. 20-10a and b, shows that this is equivalent, in the above sense, to the process of *Compton scattering!*

The life history of a positron might be as shown in Fig. 20-11, in which position is plotted as a function of time (for simplicity, only one space coordinate is shown). For times less than t_2, one sees an electron and a photon. At t_2, the photon disappears and is replaced by an electron and a positron, so that one now sees *two* electrons and one positron. At time t_1 the positron collides with and annihilates the first electron, yielding two photons.   At times later than t_1, one again sees a single electron and *two* quanta.

According to the new interpretation of positrons as electrons moving backward in time, one would regard this process as follows: An electron undergoes at time t_1 a Compton scattering in which it is scattered *back-*

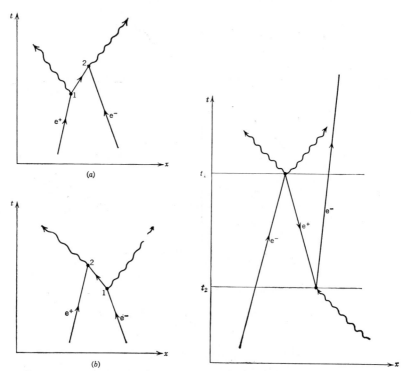

FIG. 20-10. Feynman diagram for pair annihilation.   FIG. 20-11. Life history of a positron.

ward in time; it then proceeds (backward) to time t_2, when it undergoes a violent bremsstrahlung collision in which it emits a photon which moves backward in time and is itself again scattered forward in time. In this way one interprets both of the electrons, and the positron also, as being a single electron which can be in several places at the same time![1]

Let us now consider a few further qualitative features of quantum electrodynamics from the point of view of photon exchange.   We have so far considered the interaction of two electrons and the interaction

---

[1] Indeed, it might even be that the reason all electrons are identical is that they are *all the same electron* which is at a tremendous number of places at the same time!

between an incident photon and an electron. But in addition to these processes the theory requires the existence of other interactions such as between an electron and the photons that *it itself emits*.
These interactions are called *self-interactions*, and they lead to the conclusion that the measured mass, charge, and magnetic moment of an electron are not the "true" values which should appear in the Dirac equation but *differ slightly from these values*. The correction of mass and charge is called *mass and charge renormalization*, and the magnetic-moment correction is called the *anomalous magnetic moment*.

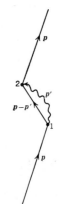

FIG. 20-12. Self-interaction of an electron.

As examples of processes which lead to the necessity of renormalizing mass and charge, we may cite those shown in Figs. 20-12 and 20-13. In Fig. 20-12 we have a process in which a single electron, initially having momentum p, emits a photon and reabsorbs it; thus the electron is left *in the same state as it was initially*. From the results of the first-order perturbation theory given in Chap. 2, we recall that perturbation integrals of the type (7) in which initial and final state are the same lead to an *energy shift* from the unperturbed value. This energy shift appears in the present context as a *mass correction*.

Figure 20-13 depicts the scattering of an electron by an external potential V such as the Coulomb field of a nucleus. Instead of acting directly upon the electron, however, the external potential may create an electron pair (in a virtual state) which annihilate, leaving a (virtual) photon which is absorbed by the electron. This leads to a slight reduction in the effect of the external potential, i.e., to a *smaller effective charge* for the electron. This shielding effect of so-called "empty" space, brought about by the virtual creation of electron pairs, is aptly called *vacuum polarization*.

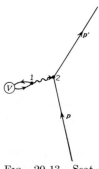

FIG. 20-13. Scattering of an electron by a nucleus.

If the necessary corrections to the observed mass and charge of an electron are calculated by the straightforward application of the perturbation theory, one obtains *infinite answers*. These divergences arise in such cases from the fact that there is no upper limit to the momenta of photons or electron pairs that may be emitted in a virtual process. But finite results have been obtained by "cutting off" the perturbation integrals (i.e., by limiting the momentum that a virtual photon or pair can possess) in some way. A procedure which is relativistically invariant and which is adequate to remove the divergences in all orders of the perturbation expansion has been found.   Quantum electrodynamics may be considered exact only if these "cutoff" procedures are incorporated into the theory.   Even with such a cutoff, however, it is not possible to find the "true" mass or charge of an electron, but only to evaluate the *difference* between the observed value for a free electron and for an electron acted upon by a potential; one can also evaluate the effective magnetic moment for a free or bound electron.   The latter quantity can be calculated as a power series in the fine-structure constant α; to terms in α^2 it is[1]

$$\mu = \frac{e\hbar}{2m} \left(1 + \frac{\alpha}{2\pi} - 0.328 \frac{\alpha^2}{\pi^2} \right) = 1.0011596\mu_0 \qquad (8)$$

This is in excellent agreement with the observed value[2]

$$\mu = (1.001156 \pm 0.000012)\mu_0 \qquad (9)$$

The first and as yet the most critical test of quantum electrodynamics is that of the so-called Lamb shift[3] in hydrogen and deuterium: the atomic states $2\ ^2S_{1/2}$ and $2\ ^2P_{1/2}$ of a one-electron atom are supposed to be degenerate in the Schroedinger theory and in the Dirac theory, but are observed actually to differ slightly in energy, the frequency of the quantum associated with the transition being

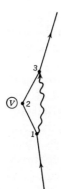

(a) For H: $\nu_2 = 1057.77 \pm 0.10 \times 10^6 \text{ s}^{-1}$

(b) For D: $\nu_2 = 1059.00 \pm 0.10 \times 10^6 \text{ s}^{-1}$ (10)

This energy difference is now known to arise from the second- and higher-order self-interaction effects discussed above, notably those corresponding to the Feynman diagrams of Figs. 20-13 and 20-14.   The total effect is attributed to several different terms, as shown in Table 20-10.

F I G . 2 0 - 1 4 . Second-order contributions to the Lamb shift.   The agreement between this calculated value and the observed value (10a) is truly impressive, and is regarded as a most convincing confirmation of the correctness of quantum electrodynamics.   In fact, because of this remarkable agreement, it is now believed that quantum electrodynamics provides an exact description of all physical phenomena which do not directly involve nuclear forces, the weak interactions, or gravitation: Nearly all of the data that appear in handbooks of physics and chemistry could, in principle, be calculated *from first principles* if suffi-

[1] The term $\alpha/2\pi$ was first obtained by Schwinger, *Phys. Rev.*, **73**, 416 (1948); *Phys. Rev.*, **76**, 790 (1949); and the next term by Sommerfield, *Phys. Rev.*, **107**, 328 (1957).

[2] Koenig, Pradell, and Kusch, *Phys. Rev.*, **88**, 191 (1952).

[3] Lamb and Retherford, *Phys. Rev.*, **72**, 241 (1947).

ciently powerful mathematical techniques were known! With the rapid advances that are being made in particle physics, perhaps it is not too much to expect that in a few more decades *all* physical phenomena will be equally well understood.

TABLE 20-10. CONTRIBUTIONS TO THE LAMB SHIFT IN HYDROGEN

Type of correction	Frequency shift, 10^6 s^{-1}		
	$2\ ^2S_{\frac{1}{2}}$	$2\ ^2P_{\frac{1}{2}}$	$2\ ^2S_{\frac{1}{2}} - 2\ ^2P_{\frac{1}{2}}$
Second order:			
General..............................	1015.52	4.07	1011.45
Magnetic moment.....................	50.86	−16.96	67.82
Vacuum polarization..................	−27.13	0	−27.13
Relativistic..........................	7.14	0	7.14
Fourth order:			
General..............................	0.24	0	0.24
Magnetic moment.....................	−0.71	0.23	−0.94
Vacuum polarization..................	−0.24	0	−0.24
Fine-structure correction.................	0.38	−0.02	0.40
Mass correction.........................	....	....	−1.57
Nucleon-electron direct interaction.........	0.03	0	0.03
Total..............................	....	....	1057.20 (±0.2)

REFERENCES

Allen, J. S.: "The Neutrino," Princeton University Press, Princeton, N.J., 1958.
Bethe, H. A., and P. Morrison: "Elementary Nuclear Theory," 2d ed., John Wiley & Sons, Inc., New York, 1956.
Heitler, W.: "The Quantum Theory of Radiation," 3d ed., Oxford University Press, New York, 1954.
Jauch, J. M., and F. Rorlich: "The Theory of Positrons and Electrons," Addison-Wesley Publishing Company, Reading, Mass., 1956.
Marshak, R. E.: "Meson Physics," International Series in Pure and Applied Physics, McGraw-Hill Book Company, Inc., New York, 1952.
Rojanski, V.: "Introductory Quantum Mechanics," Prentice-Hall, Inc., Englewood Cliffs, N.J., 1938.
Thorndike, A. M.: "Mesons: A Summary of Experimental Facts," International Series in Pure and Applied Physics, McGraw-Hill Book Company, Inc., New York, 1952.

Modern Physics in Nature

Looking back over the course of physics during the twentieth century, one is impressed by the increasing subtlety of the phenomena that have engaged the attention of the physicist and the apparently greater and greater remoteness of these phenomena from the everyday working of the universe. We seem to have arrived at a level of inquiry where, for all we can tell, the universe might get along just as well (or perhaps even better!) in the absence of a certain physical effect as it does in its presence. Of course the physicist must have as a basic faith the conviction that every phenomenon is somehow "needed" in the architecture of the universe, even though its special role may not be clear. With such thoughts in mind it is of great interest to ask, concerning each effect or phenomenon, whether there may be some place in the universe where this effect has played a perceptible, or perhaps a dominant, part in determining the local condition of matter. And indeed one finds that, far from being confined to earth-bound laboratories, the phenomena of modern physics are at work all around us in the universe or have left a characteristic mark upon many familiar parts of the world as we see it. In this final chapter we shall try to view the world from this standpoint and gain an appreciation for the large-scale importance of some of the phenomena of modern physics.

21-1. Cosmic Rays

It is most appropriate that we should begin our discussion with cosmic rays, both because of their truly cosmic character and because of the important role they have played as a research tool in the study of particles. Cosmic rays were first detected by Elster and Geitel (1899) and by C. T. R. Wilson (1900). In measuring the rate of discharge of a carefully insulated electroscope, it was found that pure, dry air possesses a small conductivity which presumably results from the presence in the laboratory or in the air of *ionizing radiations*. These radiations were for some years assumed to be caused by small amounts of radioactive substances in the apparatus, the laboratory, or the air itself, but in 1911 Hess showed, by sending electroscopes aloft in balloons, that this ionization effect is *more pronounced* at high altitudes rather than less so, contrary to what was then expected. To explain his rather remarkable results Hess advanced the hypothesis that the ionizing radiations responsible for the observed effects *are incident upon the earth's atmosphere from outside*. Many basic experimental studies of cosmic rays were carried out in the years that followed, and gradually many of the detailed properties of the cosmic rays have become known. We shall not describe these historical developments, but shall instead start with the well-established properties of cosmic rays and interpret them from the standpoint of the phenomena of modern physics.

A. The Primary Radiation. The cosmic rays incident upon the earth's atmosphere from outer space are called the *primary cosmic rays*, or simply *primaries*. These consist almost entirely of *energetic atomic nuclei* which have been completely stripped of electrons in their passage through the interstellar or interplanetary medium. The composition of the primaries with respect to atomic number corresponds roughly with the relative cosmic abundance of the elements (Table 21-1). Thus most

TABLE 21-1. COMPOSITION OF THE PRIMARY COSMIC RAYS

Z	Flux‡	Particles, %	Nucleons, %
1	360	77	37
2	100	21	41
3–5	0.6	0.1	0.5
6–10	7.6	1.6	13
>10	2.2	0.6	8

‡ Particles s^{-1} m^{-2} sr^{-1} having energies ≥ 5 GeV.

of the primaries are protons, most of the remainder are α-particles, and the rest consist of nuclei of the heavier elements up to at least iron. Beyond iron the abundances become so low that the presence of the elements in the cosmic rays has not yet been established with certainty, but on the whole, the elements with $Z \geq 10$ seem relatively somewhat more abundant in the cosmic rays than in the universe at large.

The *energy spectrum* of the primaries depends somewhat upon the latitude at which it is measured, because the earth's magnetic field acts to exclude low-energy particles, especially at the equator. Above an energy of a few GeV, however, the earth's field has no perceptible effect; in this range the energy spectrum has approximately the form

$$\frac{dN}{dE} = \frac{A}{(E + 5.3)^{2.75}} \tag{1}$$

where E is the *energy per nucleon*, measured in GeV. As this equation indicates, the number of particles per GeV falls off as the energy increases; effects which have been observed indicate the presence of particles whose energies are at least 10^9 GeV or 10^{18} eV—nearly a tenth of a joule!

Two principal questions might be asked concerning the cosmic rays: where do they come from and what happens to them after they strike the earth's atmosphere? Let us consider first the fate of a nucleus as it moves through matter. Such a nucleus will of course lose energy by ionizing the air; in traveling vertically through the entire depth of the atmosphere, a proton will lose about 2 GeV, an oxygen nucleus about 130 GeV, and an iron nucleus about 1600 GeV. But these figures are rather misleading because no nuclear particle has much chance of penetrating the entire atmosphere—it will instead be removed or greatly modified by *nuclear collisions* with the atmospheric nitrogen and oxygen. The mean free path for nuclear collision depends both upon the composition of the air and upon the size of the incident particle; it is about equal to the *geometric* mean free path calculated on the basis of nuclear radii given by the formula $R = 1.2A^{1/3} \times 10^{-15}$ m. Observed collision mean free paths range from 80 g cm^{-2} of air for protons down to 23 g cm^{-2} for iron nuclei, so that the probability of penetration of the full atmospheric depth is quite small for protons and completely negligible for heavy nuclei.

EXERCISES

21-1. The interstellar matter within the galaxy consists mostly of gaseous hydrogen at a density of about 1 atom cm^{-3}, and the galactic diameter is about 10^5 light years. Compute the amount of matter traversed by a cosmic-ray particle which passes diametrically through the galaxy. *Ans.:* $\sim$0.1 g cm^{-2}.

21-2. In a certain laboratory it is estimated that the rate of incidence of primary protons having energy ≥ 10 GeV is $\sim 10^{-1}$ s^{-1} for a certain apparatus. What is the incidence rate of protons having energy $\geq 10^3$ GeV? *Ans.:* $\sim$5 per day.

21-3. Compute the geometric mean free path of a proton and an iron nucleus in air. *Ans.:* $\lambda_p = 77$ g cm^{-2}, $\lambda_{Fe} = 22$ g cm^{-2}.

21-4. What fraction of the primary protons vertically incident upon the earth's atmosphere should penetrate to sea level without suffering a nuclear collision?   *Ans.:* $\sim$4 $\times$ 10^{-6}.

21-5. Assuming the atmosphere to be isothermal at a temperature of 290°K, by what factor should the rate of incidence of primary protons be

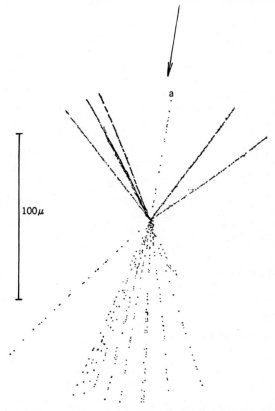

FIG. 21-1. (*a*) A primary cosmic-ray proton undergoes a collision with a stationary nucleus and produces a large number of mesons.

increased over its value at sea level at altitudes of 2000 m (6600 ft) and 4000 m (13,200 ft)?   *Ans.:* 13.5; 110.

21-6. Find the average particle energy corresponding to the distribution (1). [Assume (1) to be valid for all energies $\geq$0.]   *Ans.:* E_{av} = 7 GeV.

21-7. Evaluate the coefficient A of distribution (1), using the data of Table 21-1.   *Ans.:* A = 5.2 s^{-1} m^{-2} sr^{-1} GeV$^{1.75}$.

21-8. Assuming that the data of Table 21-1 represent an average for the entire earth's surface, compute the total number of particles striking the atmosphere per second.   *Ans.:* $\sim$8 $\times$ 10^{17} s^{-1}.

21-9. What is the approximate incident power due to cosmic rays, per square kilometer?   Over the entire earth?   *Ans.:* $P \approx 2 \times 10^9$ W.

21-10. What total electric current is carried to the earth by the cosmic rays?   *Ans.:* $I \approx 0.2$ A.

21-11. At what rate would a 0.2-A net current increase the static potential of the earth?   *Ans.:* $dV/dt = 280$ V s^{-1}.

B. Nuclear Collisions of Primary Protons.   Let us now consider the effects of the nuclear collisions of the primary-cosmic-ray protons.   We are here dealing with average particle energies of about 10 GeV—rather greater than would ordinarily be encountered in the laboratory—and at such energies a proton has so small a wavelength that its interaction with a nucleus is essentially geometric in character.   That is, one can treat the proton as a small sphere which traverses some well-defined path through the nucleus, and only those nucleons that lie near this path will be directly affected by its passage.   Hence for a nitrogen or oxygen nucleus, only about $15^{1/3} \approx 2.5$ nucleon impacts are likely to occur, but these may of course be quite violent in view of the high energies involved.   Empirical observations of proton-nucleus collisions at these energies are in general agreement with the following simplified picture:

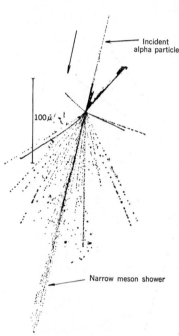

Incident
alpha particle

100 μ

Narrow meson shower

Fig. 21-1. (*b*) An incident α-particle strikes a nucleus and produces many mesons, which emerge in a narrow cone and a broader cone.   The residual target nucleus is left in an excited state, from which it decays by evaporating a number of slow (heavily ionizing) particles. [(*a*) and (*b*) by courtesy of Prof. M. F. Kaplon, University of Rochester.]

1. A proton, colliding with a nucleus, may interact violently with one of its nucleons; in this collision several pions will be produced, distributed roughly equally among positive, neutral, and negative charge. Strange particles may also be produced, and occasionally antinucleons as well.

2. In the CM system of this proton-nucleon encounter, the pions emerge mostly forward and backward, but they may have lateral components of momentum of the order of $m_\pi c$, namely, $\sim$100 to 200 MeV/c.

3. Because of the CM motion, the two nucleons and the pions all possess

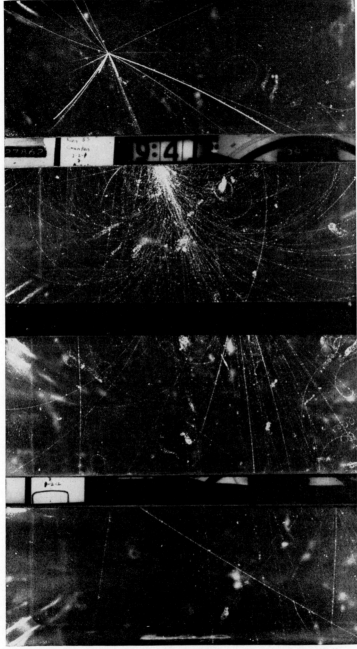

FIG. 21-1. (c) Cloud-chamber photographs of a high-energy proton colliding with an argon nucleus. Note the narrow cone of charged mesons which proceeds along the direction of the incoming proton, and the numerous evaporation fragments from the

a general forward motion which gives them rather high energy in the lab system.

4. Each of these secondary particles is capable of initiating another such collision inside the same nucleus, if the primary collision occurs sufficiently near the side of the nucleus where the primary proton enters. Thus several nucleons and many mesons may emerge from the nucleus as a result of this cascade-like multiplication process.

5. Only a few of the nucleons of a nucleus participate directly in the collision as described above. However, the sudden removal of one or a few nucleons from the nucleus leaves the residual nucleus in such a high state of excitation that several nuclear fragments, called *spallation fragments*, may evaporate from it. These fragments tend to be emitted more or less isotropically in the lab system because the residual nucleus is not given much forward velocity by the transient proton. Several neutrons will also be evaporated from the residual nucleus and perhaps from some of the larger fragments as well.

The above features of a proton-nucleus collision lead to a characteristic appearance for these events, as typified by Fig. 21-1. A great many fast, singly charged particles emerge within a narrow cone whose axis is that of the incoming proton, and as many more fast, singly charged particles emerge within a broader cone approximately coaxial with the first. In addition, several slow, singly or multiply charged spallation fragments emerge in all directions. The narrow cone is populated by nucleons and mesons which were emitted in the forward hemisphere, and the broader cone by those emitted in the backward hemisphere, of the proton-nucleon CM system. The heavier tracks are due to evaporation fragments. In addition to the charged particles which leave visible tracks, there are of course many fast and slow neutrons and neutral pions also among the secondaries.

EXERCISES

21-12. Calculate the wavelength of a 10-GeV proton and compare this with the internucleon spacing of a nucleus. *Ans.:* $\lambda = 0.021 \times 10^{-15}$ m $\approx 0.016\ R_0$.

---

argon nucleus. The evaporation fragments are mostly protons and α-particles, and together they account for practically the entire charge of the argon, indicating that this nucleus was completely disrupted by the impact.

Note also the appearance of an electronic shower below the first lead plate (3.8 cm = 6.7 radiation units thickness). This shower was evidently generated by the energetic γ-rays from the decay of *neutral* pions which were produced along with the charged ones in the original interaction. The charged pions undergo numerous secondary nuclear interactions in the three lead plates. The total energy of the incoming particle was on the order of 10^{11} eV. (Photograph from the 48-in.-magnet cloud chamber at the California Institute of Technology.)

21-13. A proton of momentum $P = \gamma\beta M_p c$ strikes a proton in a nucleus. Find γ_{CM} and β_{CM} for the motion of the CM system through the lab. *Ans.:* $\beta_{CM} = \gamma\beta/(\gamma + 1)$, $\gamma_{CM} = [(\gamma + 1)/2]^{1/2}$.

21-14. Assume that pions are emitted symmetrically in the forward and backward hemispheres of the CM system (although not necessarily

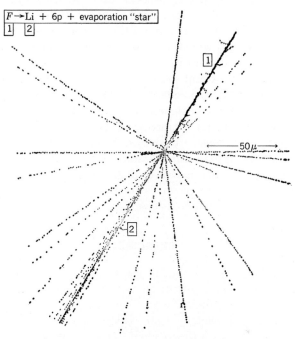

Fig. 21.2. A primary cosmic-ray fluorine nucleus strikes an emulsion nucleus peripherally and breaks up into a lithium nucleus and six protons, all of which emerge within a narrow cone. The target nucleus is left in a highly excited state, and it, too, breaks up into a great many fragments. (Courtesy of Prof. M. F. Kaplon, University of Rochester.)

isotropically) with speed $\beta^* c$. Find the half angle of the lab-system cone which contains (statistically) half the pions. *Ans.:* $\tan \theta_{1/2} = \beta^*/\gamma_{CM}\beta_{CM} \approx 2^{1/2}\beta^*/\gamma^{1/2}$.

21-15. It is often useful to use the measured value of $\theta_{1/2}$ to find the energy of the incoming proton. Find an appropriate relation between these quantities, assuming $\gamma \gg 1$ and $\theta_{1/2} \ll 1$. *Ans.:* $E \approx 2\beta^* M_p/\theta_{1/2}^2$.

C. Nuclear Collisions of Heavy Nuclei. Next we consider the collision of a primary *heavy nucleus* with a nitrogen or oxygen nucleus of the atmosphere. If the impact is squarely head-on—which is rather unlikely —several pairs of nucleons should undergo the catastrophic meson-producing collisions described above, and several times as many fast nucleons and mesons should result. Such a collision would be character-ized by a heavily ionizing ($Z^2 \times$ minimum) incoming track, which sud-denly breaks up into numerous fast, minimum ionizing tracks, with few if any slow evaporation tracks.

Much more likely than a head-on collision is a grazing or peripheral encounter, in which only small parts of each nucleus come into direct contact. In this case only a few nucleon pairs would undergo the meson-producing type of impact, and large portions of both the incident nucleus and the target nucleus would be only indirectly affected. From the residual target nucleus would then emerge several spallation fragments and evaporation protons and neutrons, ejected approximately isotrop-ically; and the residual incident nucleus should behave similarly except that, because of its tremendous forward momentum, the frag-ments would emerge in an extremely narrow cone in the lab system. The nucleon-meson "cones" described previously and the spallation fragments of the disrupted incident heavy nucleus thus differ in that the latter are confined to much narrower angles and may also include many neutrons and one or more particles having $Z > 1$. Figure 21-2 illus-trates these features of such collisions.

EXERCISE

21-16. A Ca nucleus of momentum $P = \gamma\beta M$ collides peripherally with a silver nucleus in a photographic emulsion and is left in an excited state. A spallation ^{8}Be nucleus is then ejected from the residual incident nucleus with a kinetic energy of 50 MeV, at right angles to its line of flight. At what angle with the incident line of flight does the ^{8}Be nucleus move? *Ans.:* $\theta_{\text{Be}} \approx 0.11/\gamma$, where $\gamma \gg 1$.

D. General Features of the Secondary Radiation. The result of the nuclear collisions of the primary cosmic rays is that many *secondary particles* are produced, most of them sufficiently energetic that they can themselves undergo further meson-producing collisions. Most of these secondaries play an essential role in phenomena which characterize the cosmic radiation at lower altitudes. The principal features of the second-ary radiation are the following:

1. Those secondary nucleons and charged pions which have sufficient energy continue to multiply in successive generations of nuclear collisions until the energy per particle drops below that required for multiple meson production ($\sim 10^3$ MeV). This process is called a *nucleonic cascade;* it leads to the eventual production of many pions and nucleons for each incident primary particle, to the production of strange particles, and to the formation of antinucleons.

2. The secondary *protons* lose energy by ionization, and most of those whose energy is less than about 1 GeV are brought to rest.

3. Most of the secondary *neutrons* are relatively low-energy "evaporation" particles.   They undergo elastic and inelastic collisions with air nuclei and are thereby further slowed down; ultimately, almost every neutron is absorbed by a nucleus of atmospheric ^{14}N to produce ^{15}N or ^{14}C.

4. Many of the *charged pions* decay in flight into muons.   The latter, having virtually no nuclear interaction, are slowed down only by ionization.   Many of the muons are produced high in the atmosphere in the early generations of the nucleonic cascades, because the low air density at high altitudes favors pion *decay* over nuclear collisions. These muons can therefore possess very high energy; those that do are very penetrating indeed, arriving at sea level with very little fractional loss of energy and even penetrating hundreds of meters into the earth's crust.   At sea level, about 80 per cent of the particles are muons.

5. In every generation of a nucleonic cascade virtually all of the *neutral* pions that are produced *decay in flight into two γ-rays* because of the short lifetime of the π°.   These γ-rays in turn generate *electronic cascade showers* in the atmosphere.

The composition of the cosmic radiation at various altitudes and latitudes is indicated in Tables 21-2 and 21-3.   We shall now consider some of these secondary effects in somewhat greater detail.

TABLE 21-2. COMPOSITION OF THE COSMIC RADIATION AT VARIOUS ALTITUDES
AT 50° GEOMAGNETIC LATITUDE
(Vertical Intensity in Particles $s^{-1} m^{-2} sr^{-1}$)

Altitude, km	Penetrating component‡	Electronic component	Total	Latitude effect§
0	100	30	130	0.10
2	120	120	240	0.15
4.5	200	500	700	0.25
10	500	2500	3000	0.45
16	800	4200	5000	0.75
30	1300	200	1500	0.85
> ~100¶	1000	~0	1000	0.90

‡ Nucleons, pions, and muons.
§ Latitude effect = $(I_{50^\circ} - I_{0^\circ})/I_{50^\circ}$.
¶ Cosmic-ray intensity measurements made in artificial satellites indicate the existence of belts of very-high-intensity, low-energy cosmic rays above about 600 km. These are most likely secondary particles produced inside the earth's atmosphere and trapped in the earth's magnetic field.

TABLE 21-3. CHARACTERISTICS OF COSMIC RAYS AT THE TOP OF THE
ATMOSPHERE AS A FUNCTION OF GEOMAGNETIC LATITUDE‡

Characteristic	Geomagnetic latitude		
	3°	39°	52°
Incident energy per cm² per s.........	1×10^9 eV	1.7×10^9 eV	3.2×10^9 eV
Lower energy cutoff, due to earth's magnetic field.......................	15×10^9 eV	8×10^9 eV	2×10^9 eV
Average energy per incident particle...	30×10^9 eV	16×10^9 eV	8.8×10^9 eV
Number of incident particles per cm² per s..........................	0.032	0.108	0.27
Total number of ions formed per s in a 1-cm² column.....................	3×10^7	5.4×10^7	7.4×10^7

‡ Data taken from Smithsonian Physical Tables.

E. The Structure of Electronic Showers. One of the most characteristic features of cosmic rays in the atmosphere, indeed the dominant feature at altitudes of 15 to 20 km, is that of the growth and decay of *electronic showers.* As was described in the preceding chapter, these showers can be initiated by either a photon or electron of sufficiently high energy, and are built up by the complementary mechanisms of *pair production* and *bremsstrahlung.* Most of the electronic showers in the atmosphere are initiated by the γ-rays from π^o decay, but some, particularly those at lower elevations, are initiated by high-energy knock-on electrons ejected from atoms by energetic muons, or by the electrons which result from muon decay in flight.

The growth and decay of a shower is a tremendously complex statistical process in which several physical mechanisms participate; the net result of the process is that the energy of the initiating particle is expended in ionization, and ultimately as heat. If the initiating particle has an energy of but a few MeV, the shower can contain but few particles and is most difficult to describe statistically; but if the initiating particle has several GeV of energy, the shower will be sufficiently large that a statistical description becomes quite useful, even though subject to rather large fluctuations. The physical processes which play a more or less important role at some stage of development of a shower are:

1. *Bremsstrahlung,* in which an electron or positron radiates a γ-ray in the field of a nucleus. This is by far the dominant process by which high-energy (>1 GeV) electrons lose energy.
2. *Pair production,* in which a photon "materializes" into an electron-positron pair in the field of a nucleus. This is the most important mechanism for the removal of *photons* of energy above about 1 GeV.

3. *Electron-electron collisions,* in which an incident electron undergoes an elastic (Rutherford) collision with an atomic electron and is effectively replaced by *two* electrons of lower energy. This process is important in the energy range below a few hundred MeV in air or a few tens of MeV in lead.
4. *Compton collisions,* in which photons are scattered elastically by the atomic electrons, transferring part of their energy to the electrons.

The quantitative treatment of a shower is quite complex even after many simplifying approximations are introduced, and we must therefore limit our discussion to the features which can be described in simplest terms. Two fundamental quantities are utilized in all treatments of showers: the *radiation length* and the *critical energy.* The radiation length is a certain length, characteristic of the material medium in which the shower occurs, which defines the *longitudinal scale* of the shower: If x is the distance along the shower axis from the point of initiation and if x_0 is the radiation length for the medium, the populations of photons and electrons at x, and their energy distributions, depend upon the medium only through the dimensionless parameter

$$T = \frac{x}{x_0} \qquad (2)$$

Both bremsstrahlung and pair production occur with a probability approximately proportional to Z^2 per nucleus, so that x_0 varies approximately as Z^{-2}. More precisely, the radiation length in grams per square centimeter for an elemental substance is given by the formula

$$\frac{1}{x_0} = \frac{e^2}{\pi \varepsilon_0 \hbar c} \frac{N_0}{A} Z(Z+1) r_0^2 \ln \frac{183}{Z^{1/3}} \qquad (3)$$

where r_0 is the classical electron radius, N_0 is Avogadro's number, and A is the atomic weight of the substance. Table 21-4 gives numerical values for the radiation length in both centimeters and grams per square centimeter for various materials. The physical significance of x_0 is that the energy of an electron or photon will on the average be reduced by a factor e in traveling through a distance x_0.

The *critical energy,* here called E_c, is a particle energy of the order of 50 MeV, above which only bremsstrahlung and pair production are significant mechanisms of energy loss, and below which electron collisions (ionization) and perhaps Compton collisions must be included. Values of this quantity for various materials are also given in Table 21-4.

The most important features of an electronic shower can be deduced qualitatively on the basis of a quite elementary model of the shower

TABLE 21-4. RADIATION LENGTHS x_0 AND CRITICAL ENERGIES E_c FOR VARIOUS MEDIA‡

Substance	Z	A	x_0		E_c, MeV
			g cm^{-2}	cm	
Carbon..............	6	12	44.6	~20	102
Nitrogen..............	7	14	39.4	....	88.7
Oxygen..............	8	16	35.3	....	77.7
Aluminum............	13	27	24.5	9.1	48.8
Argon...............	18	40	19.8	....	35.2
Iron.................	26	56	14.1	1.8	24.3
Copper..............	29	63.6	13.1	1.5	21.8
Lead................	82	207	6.5	0.57	7.8
Air..................	7.4	15	37.7	31,000 (S.L.)	84.2
Water..............	7.2	14.3	37.1	37.1	83.8

‡ Reproduced by permission from "High Energy Particles" by Bruno Rossi, p. 295. Copyright, 1952, by Prentice-Hall, Inc., Englewood Cliffs, N.J.

process, as follows:[1] Assume that a photon travels a distance $x_0 \ln 2$ and then produces an electron pair which equally share the original photon energy. Similarly, assume that each electron travels a distance $x_0 \ln 2$ and then suffers a radiative collision in which half its energy is communicated to a photon. Thus, starting with an initial photon of energy E, we have at $T_0 = x_0 \ln 2$ two electrons of energy $\frac{1}{2}E$; at $2T_0$, we have two electrons and two photons, each of energy $\frac{1}{4}E$; and so on. At a thickness NT_0 we have 2^N particles, about one-third photons and two-thirds electrons, whose average energy is $E2^{-N}$. This process can continue only so long as the energy per particle is above the critical energy E_c; subsequently, the photons do not readily make pairs and the electrons do not lose energy by bremsstrahlung alone but more and more by ionization, and the shower therefore rapidly disappears. The distance $N_m T_0$ at which this "crossover" occurs is that for which

$$E2^{-N_m} = E_c \qquad \text{or} \qquad N_m \ln 2 = \ln \frac{E}{E_c} \qquad (4)$$

At this distance, the number of photons plus electrons is equal to E/E_c, and at greater distances the number decays rapidly because of the ionization losses.

We thus see that (1) the growth of a shower should be exponential in its early stages, (2) the maximum number of particles should be attained

[1] See W. Heitler, "Quantum Theory of Radiation," 3d ed., Sec. 38, Oxford University Press, New York, 1954.

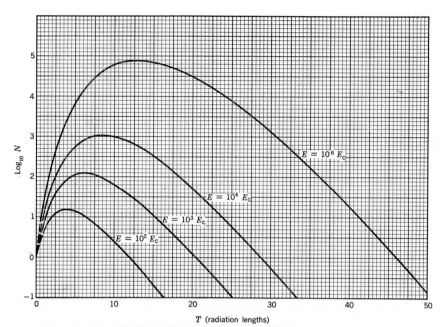

FIG. 21-3. The total number N of electrons in a shower initiated by an electron of energy E, as a function of the depth T in radiation lengths. E_c is the critical energy of the material. [After Rossi and Greisen, *Revs. Modern Phys.*, **13**, 240 (1941).]

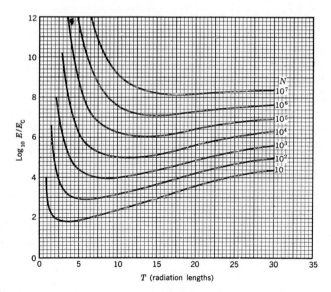

FIG. 21-4. Energy E of an initial electron needed to produce a shower having a given total number N of electrons at given depth T. (Reprinted by permission from "High Energy Particles," by Bruno Rossi, p. 259. Copyright 1952 by Prentice-Hall, Inc., Englewood Cliffs, N.J.)

within a few radiation lengths, (3) the maximum number of particles shculd be proportional to the energy of the incident particle, and (4) the post-maximum attenuation should be rapid. Of course in the actual shower process a spectrum of energies is obtained, both for the electrons of a pair and for the bremsstrahlung photons, so that the various sharp features exhibited by the crude model used above become smoothed out. The results of a rather complete treatment are indicated in Figs. 21-3 and 21-4. The qualitative features derived above can be seen in these more precise results also.

EXERCISES

21-17. Use the simplified model of a shower to deduce that the number of photons of a given energy ought to be about half the number of electrons of that energy.

21-18. A certain cosmic-ray-shower detector registers a pulse which signifies that 3×10^6 electrons passed through it. What was the minimum probable energy of the initiating particle?
Ans.: $E \approx 3.5 \times 10^{15}$ eV.

21-19. A vertically incident primary proton of energy 10^{16} eV produces a π° of energy 10^{15} eV after traversing 100 g cm^{-2} of the atmosphere. How many electrons should be observed from this event at sea level?
Ans.: $\sim 3 \times 10^5$.

In the preceding treatment of the longitudinal development of a shower, the *instantaneous* lateral and longitudinal structure has been ignored. Actually, a lateral spreading must occur because, both in bremsstrahlung and in pair production, the secondary particles are produced with finite components of momentum transverse to the path of their parent particle. Since the transverse momentum is of the order $m_e c$, the angle between the paths of a secondary and the parent particle is of the order $m_e c/p$; therefore, particles of low momentum will tend to move at larger angles with the shower axis, and a fortiori, so will their progeny. Thus the highest-energy particles should be concentrated near the line of motion of the initiating particle, and at increasing distances from this axis or *core*, the particles should have less and less energy. This effect is clearly shown in Figs. 21-5 and 21-6. Theoretical treatments of lateral spread can be found in textbooks on cosmic rays and high-energy particles, and in the literature.

Along with the lateral spreading of a shower, one also expects a spread in the longitudinal structure, brought about both by the difference in speed and by the difference in path length due to scattering for various particles in the shower. Since the slower particles also tend to have

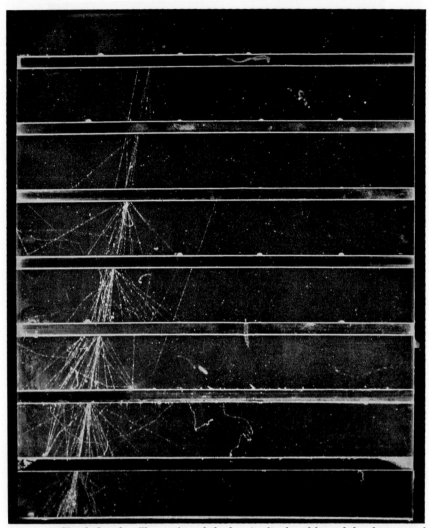

FIG. 21-5. Cloud-chamber illustration of the longitudinal and lateral development of an electronic shower in lead. The shower is initiated by an electron incident upon the top plate. Note that the electrons which emerge from a plate at large angles with the shower axis do not penetrate the following plate; these are low-energy electrons. (Courtesy of Prof. W. B. Fretter, University of California.)

greater scattering, this effect is greatest for the low-energy particles in the "wings" of the shower and less for the particles in the core. Thus one expects to find both a finite "thickness" for a shower, and also a "radius of curvature." Both of these have been measured;[1] the thickness is 2 or 3 m, and the radius of curvature about 2 or 3 km.

[1] Bassi, Clark, and Rossi, *Phys. Rev.*, **92**, 441 (1953).

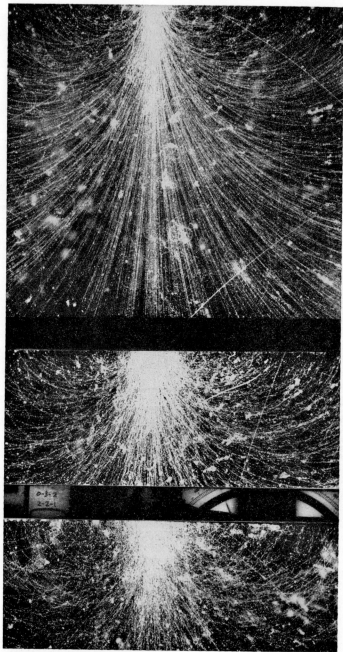

Fig. 21-6. Cloud-chamber photograph of the energy distribution of the electrons in a shower. The electrons are "spread out" by a 7500-gauss magnetic field according to their momenta. (Photograph from the 48-in.-magnet cloud chamber at the California Institute of Technology.)

Finally, we consider the *distribution* of shower particles throughout the atmosphere. At sea level, one finds experimentally that about 20 per cent of the ionization due to cosmic rays is caused by electronic showers; but the contribution of the electronic component (called the *soft component* because of its rapid absorption by a few centimeters of lead) increases rapidly with increasing altitude, until at heights of 15 to 20 km it contributes the greatest part of the ionization. The reason for this is not hard to find: the atmosphere is about 10 nuclear interaction lengths, or about 26 radiation lengths, deep, so that most showers are initiated at altitudes corresponding to one or two interaction lengths of the top of the atmosphere, reach their maximum within a few shower lengths, and then decay. Almost all energetic showers which reach sea level are actually the weak residues of showers which reached their maximum at great altitude, whereas most showers are near their maximum size at altitudes of 15 to 20 km.

EXERCISE

21-20. Use the graphs of Fig. 21-3 to predict the rate of increase of the intensity of the soft component with altitude near sea level. *Ans.:* $I(h) \approx I_0 e^{h/600}$, where h is in meters.

F. Radioactivity Induced by Cosmic Rays. We shall now consider briefly a small but scientifically useful effect of cosmic rays: the production of β-radioactive ^{14}C and ^{3}H in the earth's atmosphere. Most of the neutrons produced in the atmosphere by the cosmic radiation are eventually absorbed by nuclei of ^{14}N, mostly by the reaction

$$^{14}N + n \rightarrow {}^{14}C + {}^{1}H \tag{5}$$

Furthermore, about 5 per cent of the neutrons having energy greater than 4 MeV undergo the endoergic reaction

$$^{14}N + n \rightarrow {}^{12}C + {}^{3}H \tag{6}$$

and about 10 per cent of the charged spallation fragments from excited residual air nuclei are tritons (^{3}H). The total rate of formation of ^{14}C and ^{3}H atoms has been estimated[1] as 2.23 cm^{-2} s^{-1} and 0.2 cm^{-2} s^{-1}, respectively, and, if this formation has been proceeding at a steady rate for a duration long compared with the mean lives of these nuclei, an equal number of nuclei must be disintegrating each second. Where these disintegrations will occur depends upon what happens to the radioactive atoms after they are produced.

The ^{14}C and ^{3}H are formed high in the atmosphere, at about 10 to

[1] Libby, "Radiocarbon Dating," University of Chicago Press, Chicago, 1952; Fireman and Rowland, *Phys. Rev.,* **97,** 780 (1955).

15 km altitude; in the presence of atmospheric oxygen, both should rapidly become oxidized to $^{14}CO_2$ and ^3HOH, should then mix with the natural CO_2 and H_2O in the atmosphere, and thereafter should participate in the same physical and chemical processes as do the ordinary carbon dioxide and water. The particular processes that have so far proved of scientific value are (1) the assimilation of the $^{14}CO_2$ by the plant life along with ordinary CO_2 and its subsequent transfer to animals and (2) the precipitation of ^3HOH in rain or snow. Now most of the carbon that is in active circulation is contained as carbonates in sea water, or as organic carbon on land and in the sea, and by far most of the water is in the oceans or in the land as ground water and ice caps. Thus we see that, although these radioactive atoms are *formed* high in the atmosphere, they will almost surely *decay* on the surface of the earth or in the ocean.

The radiocarbon formed by the cosmic rays becomes, then, part of the reservoir of carbon which participates in the life cycles of all living things and thus endows all living tissue with a degree of radioactivity in proportion to the amount of carbon it contains. This radioactivity amounts to about 15.5 disintegrations per minute per gram of carbon. Now when a plant or animal dies, a part of the carbon it contains may remain "out of circulation" for hundreds or thousands of years—or in extreme cases for millions of years, as in petroleum or coal deposits. Since this carbon does not participate in processes by which it can mix with freshly formed radiocarbon, its *activity decays with the characteristic 5568-year half-life of* ^{14}C. It then follows that a measurement of the specific activity of a sample of dead organic matter can (ideally) be used to determine the elapsed time since the sample lost contact with the reservoir of cycling carbon. As a practical matter it is rather difficult to distinguish the relatively weak activity even of "live" carbon from the ever-present background of cosmic radiation and mineral radioactivity,[1] but by exercising great care in reducing the effects of these unwanted radiations, it has proved possible to "date" organic objects as much as 25,000 years old.

The activities of many objects of known age have been measured, and these generally agree quite well with the expected activities. This provides strong support for the assumptions underlying the radiocarbon dating method, the most critical of which are the following:

1. The formation of radiocarbon has proceeded at an essentially constant rate during at least the past 25,000 years.
2. The radiocarbon formed at high altitude becomes well mixed with the

---

[1] These problems are discussed by Anderson and Hayes, *Ann. Rev. Nuclear Sci.*, **6**, 303 (1956).

world-wide reservoir of carbon which participates in the life cycles of living things.

3. After the death of an organism, that part of its carbon which is not actually dissolved or otherwise carried away is isolated from this reservoir and subsequently diminishes in radioactivity at a rate characteristic of the 5568-year half-life of ^{14}C.

The successful application of the ^{14}C dating method by Libby and his coworkers immediately established it as a powerful tool in the study of archaeology. Similar procedures have been used to study the activity due to tritium as it occurs in water and ice. The 12.46-year half-life of ^{3}H adapts radiotritium dating to several important problems connected with rainfall and meteorology, such as the relation between ground water present at a given locality and the local rainfall, the speed of vertical mixing of air in the stratosphere, the age of agricultural products, and the source of water flowing from springs.

The uses of ^{14}C and ^{3}H outlined above are the most important ones that have so far been found for cosmic-ray-induced radioactivity, and it is of course possible that other important applications for these or other radioisotopes formed by cosmic rays will be discovered in the future. Some work has already been done on the β-active radioisotope ^{7}Be of 53-day half-life;[1] measurement of the distribution of this isotope in the atmosphere should be of great use in the study of atmospheric mixing.

G. Information Relevant to Cosmic-ray Origin. Let us now return to the first of the two main questions that were propounded above concerning the cosmic rays; namely, where do they come from? The following experimental information seems relevant to this problem:

COMPOSITION. As has already been pointed out, the cosmic radiation arriving at the top of the earth's atmosphere consists of atomic nuclei that are largely, if not completely, stripped of electrons. There is no significant flux of either electrons or γ-rays among the primary cosmic rays. The various atomic nuclei are present in abundances roughly comparable to those found in the visible universe, with perhaps a tendency for the heavier nuclei to be relatively somewhat more abundant in the cosmic rays.

The presence of Li, Be, and B among the primary cosmic rays, coupled with the scarcity of these elements in the universe as a whole, is evidence that the "true" primary particles have been somewhat modified by nuclear collisions with the interstellar or interplanetary medium, various light elements including Li, Be, and B having been produced as *spallation fragments* in these collisions.

---

[1] Arnold and Al-Salih, *Science*, **121**, 451 (1955).

ENERGY SPECTRUM.   The energy spectrum 21-1(1) is observed to be approximately correct over an energy range from 10^9 to 10^{18} eV.   At the lowest energies ($E < 10^9$ eV), the effects of the earth's magnetic field and upper atmosphere have prevented accurate measurement of the energy distribution, and for energies $>10^{18}$ eV the intensity is so low that statistical errors become a limiting factor in determining the spectrum.

TOTAL ENERGY FLUX.   The total energy flux at the top of the atmosphere due to cosmic radiation, 4×10^{-6} W m^{-2}, *is about equal to the radiant energy flux received from all stars excluding the sun*: 3×10^{-6} W m^{-2}.

CONSTANCY IN TIME.   Many experiments have been performed to test whether the cosmic radiation varies systematically with time, say with the period of a solar day or a sidereal day, with the rotational period of the sun, or with the sunspot cycle.   Clearly, the existence of a sizable periodic variation would have great significance in connection with the origin of the radiation.   While no such periodic effect has as yet been definitely established, more sensitive measurements of greater statistical accuracy are in progress.   The absence of a systematic variation in the *primary* radiation does not preclude periodic variations of cosmic-ray intensity within the atmosphere; the largest such variation, a daily one, is well accounted for in terms of the day-to-night variation of barometric pressure and atmospheric thickness.

In contrast with *periodic* variations which are certainly very small, if indeed they exist at all, many instances of cosmic-ray intensity fluctuations *correlated in time with solar activity* have been observed.   There appear to be two types of such effects present:

1. A *decrease* in intensity of a few per cent, mainly in the low energy range ($< \sim 10^9$ eV); these decreases are apparently associated with the same general kind of solar activity as is responsible for the familiar magnetic storms which produce auroras and which interfere with radio transmission on the earth.   This activity is usually evidenced by the presence of a large sunspot group on the visible solar surface. It is often observed to have a recurrent effect with the 27-day period of the solar rotation, as though a "beam" of radiation or magnetic field were attached to the center of activity and swept through space by the solar rotation; a magnetic storm and/or decrease in cosmic-ray intensity results if the earth is enveloped by such a "beam."   In addition to these sporadic, relatively short period decreases, evidence is mounting which indicates that the low-energy cosmic rays which reach the earth near its geomagnetic poles undergo a slow modulation in step with the sunspot cycle.   The phase of this modulation is such that the low-energy cosmic rays are more abundant near sunspot minimum.[1]

---

[1] Neher and Anderson, *Phys. Rev.*, **109**, 608 (1958).

2. An *increase* in intensity of from a few per cent to severalfold, again mostly in the lower energy range ($<\ \sim 10^{10}$ eV); these increases are associated with the relatively rare *solar flares* which sometimes burst forth from active sunspot areas.   The intensity increase follows rather closely the optical intensity change—a rather abrupt rise to maximum followed by a quasi-exponential decay with half-life of an hour or so— except that *the cosmic-ray changes are delayed in time by different amounts at different points on the earth's surface*, as though streams of energetic particles were transmitted via different *curved paths*.

H. Possible Cosmic-ray Origins.   What can now be said concerning the origin of cosmic rays?   Many possibilities present themselves; some have merited quite serious consideration, but none appears as yet to be completely satisfactory.   Let us consider some of the possibilities:

THE SUN.   The increase in cosmic-ray intensity associated with solar flares demonstrates quite clearly that *some* cosmic rays are of solar origin.   Are they all?   There appear to be two main objections to such a hypothesis:

1. *The absence of a prominent diurnal variation of intensity.*   This objection would be weakened considerably, however, if a magnetic field of sufficient strength to "trap" cosmic-ray particles were present throughout the solar system.   The trapped particles might then become randomized in direction through "collisions" with the planetary magnetic fields, and thus strike the earth more or less equally from all directions.   This field would thus act as a storage volume for the cosmic rays and would have the effect of smoothing out both in *direction* and in *time* the particles emitted by the sun during its sporadic periods of flare activity.   The time delays between the arrival of radiation and particles, and the variability of these delays over the earth, speak strongly for some kind of magnetic curvature of the particle trajectories, but there is as yet no quantitative experimental information concerning the strength and distribution of magnetic fields within the solar system.

2. *The absence of a systematic variation with the sunspot cycle:* Even if there were a sufficiently strong magnetic field to "smooth out" the cosmic rays as described above, there should be a slow loss of cosmic rays by collisions with the planets, the sun, and interplanetary matter, and by diffusion away from the solar system.   Since the only known emission of cosmic rays by the sun occurs in connection with sunspot activity, however, the "average" rate of replenishment of these lost cosmic rays ought to follow the 11-year cycle of solar activity.   If this cyclic replenishment is not to affect the total cosmic-ray intensity, therefore, it is necessary that the "time constant" of the storage vol-

ume against leakage and collision loss be long compared with 11 years, say, a thousand years or so. Such a requirement of course places an even more strict condition upon the strength and distribution of the interplanetary fields, the collision loss rate, and the rate of emission of cosmic rays by the sun than does the condition of isotropy alone.

Despite the difficulties just discussed, it has not yet been shown to be impossible that all cosmic rays received by the earth originate in the sun, and the hypothesis of solar origin must still be considered.

COSMIC-RAY STARS. If the sun is not the source of all cosmic rays, the possibility suggests itself that there might be stars which are much more efficient generators of cosmic rays than is the sun, and that these so-called *"cosmic-ray stars"* are the source of most of the cosmic rays we observe. Clearly, there must be stars more active than the sun, and these might well emit more cosmic rays; but the following objections can be raised against the hypothesis that these produce the cosmic rays which we observe:

1. Assume first that practically all stars emit cosmic rays and that the total emission adds up to the observed cosmic-ray intensity. Then one is faced with the obvious impossibility that almost every star must emit cosmic-ray energy *as efficiently as it emits light.*

2. It has also been suggested that some *special class* of stars, such as the so-called *flare stars*, which exhibit sudden, sporadic increases in brightness; the *magnetic variables*, discovered by H. W. Babcock, which are surrounded by alternating magnetic fields of several thousand gauss amplitude; or *supernovae*, which occur about once per millennium and which briefly outshine all the rest of the galaxy combined, may be responsible for the bulk of the cosmic rays. However, one still must face the fact that such objects would have to be incredibly strong cosmic-ray sources in order to yield a cosmic-ray energy density equal to that of starlight; and the smaller the class of supposed cosmic-ray stars, the greater must be their strength.

3. If the cosmic rays originate in the galaxy and propagate rectilinearly, one should observe an *intensity variation with sidereal time.*

Just as for the case of solar origin, of course, a *galactic magnetic field* capable of trapping the particles might serve the purpose of rendering their motions isotropic and their intensity relatively constant in time. But even more important, a slower rate of loss of particles by leakage out of the galaxy would correspondingly reduce the source strength needed to maintain the observed energy density. Now there is in fact strong experimental evidence—the polarization of starlight by aligned interstellar dust grains—that a general galactic magnetic field does exist, probably of sufficient strength to trap particles even of the highest known energy.   Furthermore, the existence of measurable amounts of Li, Be, and B among the cosmic-ray primaries, presumably as spallation fragments, strongly suggests that the cosmic rays striking the earth's atmosphere have previously traversed 1 or 2 g cm^{-2} of interstellar matter. Now the total amount of such matter lying on a 10^5 light-year diameter of the galaxy is only about 0.1 g cm^{-2}, so that the average primary particle has evidently been circulating within the galaxy for a million years or so, but certainly not longer than 10 million years.   This ten- or hundredfold increase in lifetime within the galaxy reduces the necessary source strength by an equal factor, but does not really make much more plausible the idea that the cosmic rays are produced by stars alone.

INTERSTELLAR ACCELERATION.   One of the earliest suggestions of the origin of cosmic rays was that they might have been accelerated by interstellar electric fields.   It might seem plausible that large electrostatic potential differences could exist between one body and another, especially over distances of galactic magnitude.   However, enough of the interstellar matter is in an ionized state to render it electrically conducting; because of this conductivity, the time required to neutralize any static charge that might exist is only a few thousand years.

On the other hand, there is no such limitation upon the presence of a *magnetic field* or of *transient electromagnetic fields*, and there appear to be possible mechanisms by which charged particles could be accelerated by such fields.   Some of the simplest of these are:

1. *The betatron mechanism.*   If a charged particle is situated within a changing magnetic field, it is in general acted upon by a transient electric field induced by the changing magnetic field, and is thus accelerated.   This is the principle used in electron accelerators called *betatrons.*   For the case of cosmic rays, one can easily show that a charged particle, initially moving with momentum p at right angles to a uniform magnetic field B_0 whose strength is slowly changing, satisfies the condition

$$p = p_0 \left(\frac{B}{B_0}\right)^{1/2} \qquad (7)$$

so that, if B increases, p also increases.   However, the radius r of its circular path decreases:

$$r = r_0 \left(\frac{B_0}{B}\right)^{1/2} \qquad (8)$$

Thus such a particle might very well gain energy, but in doing so it would be restricted to remain within a smaller region of the field. It is difficult to imagine how particles could acquire energies of 10^{18} eV by a single such process, but it seems quite possible that betatron-type acceleration is of importance for the cosmic rays emitted by the sun.

2. *Other induction processes.* The betatron mechanism is by no means the only possible electromagnetic acceleration process. In fact, there seem to be so many possibilities that one is faced with the absurd situation of having too many explanations for the phenomenon —most of which are probably wrong. Most of the proposed mechanisms involve the interaction between a conducting medium—the atmosphere or corona of a star, or the interstellar medium—and a magnetic field. As was mentioned above, ordinary interstellar space, while being by laboratory standards a fantastically high vacuum, is by astronomical standards not so, and is in fact an almost perfectly conducting medium for electromagnetic phenomena on a cosmic scale. Now one important property of a perfect conductor is that magnetic field "lines" become "frozen in" and cannot move transversely through it; if any transverse motion is to occur, the conductor and the field must move together. (This is the principle of operation of an a-c induction motor.) What actual motion takes place when a magnetic field is imbedded in a stellar atmosphere or corona, or in an interstellar gas cloud, depends upon the relative *energy density* $\frac{1}{2}\boldsymbol{B} \cdot \boldsymbol{H}$ of the magnetic field and kinetic-energy density $\frac{1}{2}\rho V^2$ of the matter in the gas cloud: If $\boldsymbol{B} \cdot \boldsymbol{H} \gg \rho V^2$, the magnetic field will change almost as if no matter were present, and the matter will be dragged along as a small inertial load; but if $\rho V^2 \gg \boldsymbol{B} \cdot \boldsymbol{H}$, the matter moves almost as if no field were present, and the lines of force are dragged along with it. Various electromagnetic acceleration schemes that have been proposed make use of these effects in stellar atmospheres or coronas, where collisions of gas clouds carrying magnetic fields occur, but none of these schemes has so far proved entirely satisfactory.

3. *The Fermi mechanism.* Fermi proposed[1] that the acceleration of cosmic rays might result from "collisions" between the cosmic ray particles and the magnetic fields carried by turbulent interstellar gas clouds. Charged particles moving randomly about would encounter the magnetic fields carried by drifting gas clouds and would be "reflected" by them with increased energy if the cloud were initially drifting toward the particle and with decreased energy if away. Since more collisions per unit time will occur with the cloud moving toward than away from the particle, there should be a *net gain in energy* by the particle, on the average. (These collisions effectively tend to establish an *equipartition of energy* between the particles and the gas clouds.) This so-called *Fermi mechanism* merits our special consideration because it leads naturally to a power-law

---

[1] *Phys. Rev.*, **75**, 1169 (1949); *Astrophys. J.*, **119**, 1 (1954).

energy spectrum; whether its several shortcomings can be satisfactorily remedied is a subject of considerable interest.

Consider for simplicity a head-on collision of a particle of mass M, momentum P, and energy W with an infinitely massive object moving with speed V, and assume that in this collision the particle rebounds with the same relative speed as it had when it struck. Let $\beta_c = V/c$ and $\beta = P/W$. With what energy will the particle rebound? In the CM system—that of the massive object—the momentum and energy of the particle are

(a)
$$W' = \gamma_c(W + \beta_c P)$$
(b)
$$P' = \gamma_c(P + \beta_c W)$$
(9)

In the collision W' is unchanged and P' is reversed. Thus in the original system we have, after the collision,

$$W'' = \gamma_c(W' + \beta_c P') = \gamma_c^2[W(1 + \beta_c^2) + 2\beta_c P]$$
$$= W + 2\gamma_c^2\beta_c(\beta_c + \beta)W \tag{10}$$

Now, if such a particle makes collisions at random with massive objects which may be moving with speed V in either the same or the opposite direction as the particle, what will be the average probable change in energy of the particle per collision? This will be the change in energy for a head-on collision times the probability of such a collision, plus that for an "overtaking" collision times the corresponding probability. But, ignoring collisions, the particle would encounter more objects moving opposite itself than in its own direction; the probabilities are in fact proportional to the relative velocities $\beta + \beta_c$ and $\beta - \beta_c$, respectively:

(a)   Probability of head-on encounter $= \dfrac{\beta + \beta_c}{2\beta}$

(b)   Probability of overtaking encounter $= \dfrac{\beta - \beta_c}{2\beta}$

(11)

Thus we have

$$(\Delta W)_{\text{av}} = 2\gamma_c^2 W\left[\beta_c(\beta + \beta_c)\frac{(\beta + \beta_c)}{2\beta} - \beta_c(\beta - \beta_c)\frac{(\beta - \beta_c)}{2\beta}\right]$$
$$= +4\gamma_c^2\beta_c^2 W \approx 4\beta_c^2 W \tag{12}$$

Thus on the average the particle should *gain energy*, and after N collisions should have an energy

$$W = W_0 \exp{(4\beta_c^2 N)} \tag{13}$$

where W_0 is the initial energy. According to this equation, then, the particle energy increases exponentially with distance traversed, as long as it continues to undergo the same general type of reflective collisions with the gas clouds. However, there are several physical effects that have so far been neglected, and their action will modify the acceleration process.

The first such effect is the loss of energy by ionization as the particle passes through the interstellar medium. Since the ionization loss rate is *greater* for low speeds ($I \approx \beta^{-2}$) and the rate of energy gain by Fermi collisions is *less* for low speeds, there will be a critical energy for each type of particle below which ionization losses will predominate and thus slow the particle down. For the observed density of interstellar material, collision rate, and gas-cloud speed β_c, this critical kinetic energy, called the *injection energy*, is about 200 MeV for protons, 1 GeV for α-particles, and about 300 GeV for iron. The mechanism by which particles having such energies are generated is not known, but it is possible that they could result from some of the stellar mechanisms discussed above. The quite large injection energies that are required for heavy nuclei are difficult to reconcile with the observed prevalence of these nuclei among the primary cosmic rays.

Another effect that we have neglected is the removal of particles by nuclear collisions with the interstellar medium. This effect, together with the acceleration mechanism, leads to a power-law energy spectrum: Suppose a particle travels a mean distance d between acceleration collisions and a mean distance D between nuclear collisions. The average number of acceleration collisions a particle should undergo before being removed by a nuclear collision should then be $N_0 = D/d$, and with this number of collisions it will have acquired an energy

$$W_1 = W_0 \exp (4\beta_c^2 N_0) \tag{14}$$

However, some particles will suffer a nuclear collision after having traversed a longer or shorter path than the average, and will in fact be distributed exponentially in path length so that the number which survive a distance greater than Δ will be

$$n(\Delta) = n_0 \exp (- \Delta/d) \tag{15}$$

where n_0 is the total number of particles of all energies. Now those particles which survive for a greater path length than Δ will have an energy greater than

$$W = W_0 \exp (4\beta_c^2 \, \Delta/d)$$
$$= W_0 \left(\frac{W_1}{W_0}\right)^{\Delta/D} \tag{16}$$

Therefore, the number of particles having energy greater than W

will be

$$n(W) = n_0 \left(\frac{W}{W_0}\right)^{-[\ln(W_1/W_0)]^{-1}}$$
$$= A W^{-\alpha} \tag{17}$$

Experimentally, the exponent α of the integral energy distribution spectrum is 1.7.   Thus we find

$$\frac{1}{\ln(W_1/W_0)} = 1.7 \quad \text{or} \quad \ln\left(\frac{W_1}{W_0}\right) = 0.6 = 4\beta_c^2 \frac{D}{d} \tag{18}$$

If we then adopt $V = \beta_c = 30$ km s^{-1} as the mean random drift speed of a gas cloud and $D = 80$ g cm$^{-2} \times 10^{24}$ cm^3 g^{-1} for the absorption mean free path for protons, we find

$$d = \frac{4}{0.6} \times 10^{-8} \times 8 \times 10^{25} = 5 \times 10^{18} \text{ cm}$$
$$= 5 \text{ light years} \tag{19}$$

as the mean distance between acceleration encounters—a quite reasonable value.

A third effect that has been ignored is the fact that the nature of the Fermi encounters might vary with the energy of the particle. Thus if the energy becomes so great that the particle penetrates completely through a gas cloud, it will not gain energy as described by Eq. (13).   In fact, the interstellar magnetic fields ($B \approx 10^{-5}$ gauss) are probably unable to keep protons of energy greater than about 10^{20} eV within the 1000-light-year thickness of the galaxy itself, so that there should be fewer such particles present in the cosmic radiation and their motion should not be isotropic.   Experimental study of these high-energy particles would no doubt throw considerable light upon the question of the origin and acceleration mechanism of the cosmic rays.

The various acceleration mechanisms mentioned above are only a small fraction of those that have been proposed; most of the mechanisms fall into the same general classes as those we have discussed but differ from one another in certain details.   No theory has yet proved equal to all tests, and indeed the cosmic rays may arise from a series of mechanisms, or even by several different processes.   This question is understandably one of the most interesting ones in physics, and is the subject of considerable theoretical and experimental effort.

EXERCISES

21-21. Derive Eqs. (7) and (8).
21-22. Verify Eqs. (10) and (12).
21-23. Verify Eq. (17).

21-2. Stellar Energy

As a second example of phenomena in which modern physics plays a dominant role, let us consider the source of the energy which stars radiate so profusely into space. The radiation flux from a star is indeed prodigious; our own sun—an average star—radiates 6.25 kW from *each square centimeter*, or 3.8×10^{26} W from its entire surface, and we have ample evidence that it has been doing so for several billion years! Where does this energy come from?

Early speculations on this question attempted to account for the sun's heat as residual heat from an even hotter past, as chemical heat from burning a fuel, or as gravitational energy released by a contraction in size. But all such mechanisms fall far short of accounting for the tremendous *time span* over which the sun has been known to shine much as it shines today.

Among the first to recognize the true nature of the stellar energy-generation process was Eddington,[1] who, working from Aston's measurements of the atomic masses of ^{1}H and ^{4}He, suggested that the main source of solar energy might be the fusion of four atoms of ^{1}H to form one atom of ^{4}He. The energy release in this process, 7 MeV per nucleon, or 6.7×10^{11} joules per gram atom of ^{1}H consumed, is millions of times greater than that of any ordinary chemical reaction, and is quite sufficient to permit the sun to shine for many times its known 5-billion-year age.

Of course, four ^{1}H atoms cannot collide to form a ^{4}He atom in a single step; two β-decays must intervene to transform two of the protons into neutrons. Furthermore, even at the 30-million-degree temperature that exists at the center of the sun, the mean particle kinetic energy is only 4 or 5 keV—hardly in the range usually thought of as leading to nuclear reactions. Actually the process takes place in a series of ordinary two-body reactions, the almost vanishingly small reaction probability at each step being compensated by the overwhelming rate at which collisions occur.

The first such possible series was suggested by Bethe;[2] it is called the *carbon cycle* because the first reaction involves an atom of ^{12}C which

---

[1] *Brit. Assoc. Advance Sci., Rept.,* Cardiff (1920).
[2] *Phys. Rev.,* **55,** 434 (1939). See also Weizsacker, *Physik. Z.,* **39,** 633 (1938).

is successively transmuted by the addition of protons and by β^+-decays, finally to yield an α-particle *and a* ^{12}C *atom:*

$$
\begin{aligned}
(a) \qquad\qquad & ^{12}\text{C} + \text{p} \to {}^{13}\text{N} + \gamma \\
(b) \qquad\qquad & ^{13}\text{N} \to {}^{13}\text{C} + \beta^+ + \nu \\
(c) \qquad\qquad & ^{13}\text{C} + \text{p} \to {}^{14}\text{N} + \gamma \\
(d) \qquad\qquad & ^{14}\text{N} + \text{p} \to {}^{15}\text{O} + \gamma \\
(e) \qquad\qquad & ^{15}\text{O} \to {}^{15}\text{N} + \beta^+ + \nu \\
(f) \qquad\qquad & ^{15}\text{N} + \text{p} \to {}^{12}\text{C} + {}^4\text{He}
\end{aligned} \qquad (1)
$$

Using an obvious notation, the above reactions may be abbreviated

$$^{12}\text{C}(\text{p},\gamma)\,^{13}\text{N}(\beta^+\nu)\,^{13}\text{C}(\text{p},\gamma)\,^{14}\text{N}(\text{p},\gamma)\,^{15}\text{O}(\beta^+\nu)\,^{15}\text{N}(\text{p},\alpha)\,^{12}\text{C}$$

More recently, another series of reactions, in which only hydrogen is needed as the original reactant, has been devised. This series is called the proton-proton cycle. In the abbreviated notation used previously, it is

$$^1\text{H}(\text{p},\beta^+\nu)\,^2\text{H}(\text{p},\gamma)\,^3\text{He}(^3\text{He},2\text{p})\,^4\text{He}$$
$$\qquad\qquad \hookrightarrow (\alpha,\gamma)\,^7\text{Be}(\text{e}^-,\nu)\,^7\text{Li}(\text{p},\gamma)\,^8\text{Be}(\alpha)\,^4\text{He} \qquad (2)$$
$$\qquad\qquad\qquad \hookrightarrow (\text{p},\gamma)\,^8\text{B}(\beta^+\nu)\,^8\text{Be}(\alpha)\,^4\text{He}$$

The first reaction of this series amounts to a β^+-decay of ^{2}He formed by the collision of the two protons: Even though the lifetime of ^{2}He against decay back into two protons is extremely short, the stupendous number of collisions that occur inside a star allows a modest rate of formation of deuterium. The second reaction is well known and occurs rapidly, even at the low bombardment energies prevailing in stellar interiors. Since ^{3}He does not react with either ^{1}H or ^{2}H, its concentration builds up until further reactions remove it as fast as it is formed. The reaction ^{3}He(^{3}He,2p)^{4}He was first suggested by C. C. Lauritsen and has been detected in the laboratory. The participation of the reactions starting with ^{3}He(α,γ)^{7}Be in stellar energy generation processes has been treated by Fowler[1] who points out that they are probably the dominant ones for the sun.

Other helium-forming series than the above are known, but they do not contribute appreciably to the energy generation in ordinary stars. They are, however, of importance in the synthesis of elements in stellar interiors, as will be described below.

Almost all stars are believed to utilize one or the other of the above hydrogen-burning mechanisms, but it is not yet possible to state with certainty to what extent each contributes to the total energy output of a given star. It now seems quite certain that the sun is supported mostly

[1] Fowler, *Astrophys. J.*, **127**, 551 (1958).

by the proton-proton cycle, but only order-of-magnitude estimates of the
relative reaction rates at stellar temperatures have so far been possible.
This rapidly developing subject can best be followed by referring to the
current literature.

21-3.   The Formation of the Elements

As a final topic in our treatment of modern physics in nature, it is
fitting that we examine what is perhaps the most fascinating problem

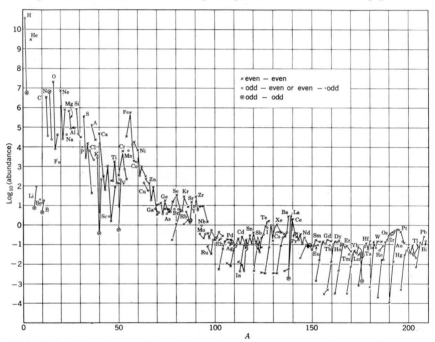

FIG. 21-7a. Experimentally observed abundances of stable nuclides relative to
Si = 10⁶, with each nuclide represented individually.   [After Suess and Urey, *Revs.
Modern Phys.*, **28**, 53, 1956).]

that faces the physicist: that of clearing away the mystery that surrounds
the origin of the nuclear species that exist on the earth, in the sun, and
in the universe as a whole.   As we shall see, nuclear physics has played a
determining role in establishing the mixture of nuclides we find about us,
and is even today at work modifying this mixture.

Until relatively recently most theories of the formation of the elements
have placed great emphasis upon the fact that several lines of evidence
seem to point to a cataclysmic "beginning" of the universe as we know
it, about 5×10^9 years ago.   Foremost among these indications is that
of the "red shift" of the galaxies, in which the light now reaching us from

a distant galaxy appears systematically reddened by an amount proportional to its apparent distance, and independent of its direction from us. If this reddening is interpreted as a Doppler shift, one concludes that the nebulae are in a state of homogeneous, isotropic expansion; and if this

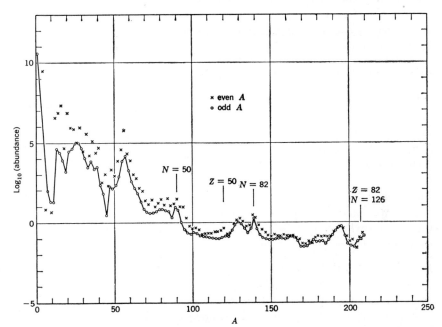

FIG. 21-7b. Abundances of even-A and odd-A isobaric nuclides. [After Suess and Urey, *Revs. Modern Phys.*, **28**, 53 (1956).]

expansion is extrapolated backward in time one finds that some 5 to 10×10^9 years ago the galaxies would have been much more densely packed than they are now. The further fact that the earth and the meteorites are of substantially this same age is taken as a strong indication that the elements were formed in an explosive, prestellar state of the universe a few billion years ago. Several attempts have been made to interpret the present relative abundances of the nuclides in terms of the short-term action of known nuclear processes within a hot, dense, expanding protouniverse, and a subsequent modification by the radioactive decay of the many unstable species formed in this relatively sudden process.

Since about 1950, however, there has been amassed an impressive body of evidence which strongly indicates that, quite independently of a possible cataclysmic beginning, nuclear species are *at present* undergoing significant modification inside and at the surfaces of certain types of stars. The basic experimental facts that must be explained quantitatively are the relative cosmic abundances of the nuclides, shown in Fig. 21-7. The gross features exhibited by the measurements are:

1. A steep decline in abundance with increasing A up to $A \approx 80$
2. Several local "peaks," the most prominent of which is in the neighborhood of iron
3. A fairly constant abundance for $A > 80$
4. A systematic overabundance of even-A relative to odd-A nuclides
5. A definite cyclic tendency with period $\Delta A = 4$ in the region $A < 40$
6. A marked deficiency of Li, Be, and B

Detailed examination of the abundances of nuclides in the neighborhood of given A reveals a distinct but not perfect correlation between nuclear stability—say, the binding energy per particle—and the abundance. This, coupled with the great local peak near iron, the most stable nucleus of all, suggests that the observed distribution might be the result of a *thermodynamic equilibrium* distribution, modified by the subsequent decay of the radioactive species initially present. A number of attempts have been made to interpret the observed abundances in this way; such theories are called *equilibrium theories*.[1] The main conclusions to which they lead are as follows:

1. Abundances up to $A \approx 60$ can be accounted for as an equilibrium distribution corresponding to a temperature $T \approx 10^{10}$ degrees.
2. The nuclides with $A > 60$ are far more abundant than permitted by the equilibrium theory using the above temperature.

Various modifications and corrections to the theory have been made to allow for (1) the statistical weights introduced by the nuclear spins; (2) the effects of low-lying excited states which would also contribute to statistical weights; and (3) electrostatic effects which, under conditions of high temperature and density, tend to drive the equilibrium toward fewer, heavier particles—i.e., heavy, neutron-rich nuclides. Even with these additions, however, the theory still grossly underestimates the abundance of the heavier nuclides.

In another class of theories, which may be called *nonequilibrium theories*, the present abundances are supposed to have resulted from a dynamic process of element building, perhaps starting with neutrons, which was "frozen" in a partially complete state by the rapid initial expansion of the universe. One such theory, due to Gamow[2] and his coworkers, postulates that the universe initially consisted of a single ball of neutrons and radiation of such high energy density that rapid expansion

---

[1] For a review of various theories up to 1950, see ter Haar, *Revs. Modern Phys.*, **22**, 119 (1950); and Alpher and Herman, *Revs. Modern Phys.*, **22**, 153 (1950).

[2] *Phys. Rev.*, **70**, 572 (1946). See also Alpher and Herman, *loc. cit.*

immediately set in.   As this great mass expanded, some of the neutrons decayed, while others underwent neutron-capture reactions with the protons so produced and in this way built up the heavy elements within the first few minutes of the beginning of the expansion.

If such a process were actually operative, one should find an inverse relation between the abundance of a nuclide and its cross section for neutron capture.   And this is indeed found to be true: Measured neutron-capture cross sections show a general rise with increasing A up to $A \approx 100$ and become approximately constant for $A > 100$.   Also, those nuclides which have a magic number of neutrons, and which therefore have abnormally small cross sections for capturing an additional neutron, *are observed to be abnormally abundant in nature*.   In spite of this semiquantitative agreement with observation, the Gamow theory faces several grave difficulties, the most important of which has to do with the mechanism by which nuclides having mass number 5 or 8 are bridged in forming heavy nuclei by neutron capture.   To see the nature of these difficulties, let us consider the build-up of the first few nuclides starting with neutrons:

Because of their radioactivity, some of the neutrons would β-decay to form ^{1}H.   Some of this would capture a neutron to form ^{2}H, and a second neutron to form ^{3}H.   This would β-decay to form ^{3}He, which in turn would capture a neutron to form ^{4}He.   However, ^{4}He is completely inert toward neutrons and protons alike, so that ^{5}He and ^{5}Li cannot be formed by neutron or proton capture.   Now the brief time scale of the entire capture phase does not permit reactions involving charged particles to contribute appreciably; furthermore, even if elements such as ^{6}Li or ^{7}Li were formed, ^{8}Li would be produced by neutron capture and would almost immediately decay to ^{8}Be, which in turn would decay almost immediately into two α-particles.   Only by capturing *two* neutrons in quick succession could ^{7}Li avoid this fate.

These two stumbling blocks in the chain of neutron reactions have so far proved insuperable to the Gamow theory and have caused considerable resistance to its acceptance.   Fortunately, another possibility has appeared, one which avoids these difficulties and appears to fit in well with certain astronomical observations which are otherwise most difficult to interpret.   The possibility, which was first studied intensively by Hoyle,[1] is that the heavy nuclides may be formed in significant amounts in the interiors and/or in the surface layers of certain kinds of stars. This is of course a logical extension of the helium-building cycles which are believed to furnish the radiant energy of most stars.

Although the details of this theory of *stellar nucleogenesis* are rather involved and not yet completely worked out, the main ideas of the

---

[1] *Astrophys. J., Suppl.*, **1**, 121 (1954); also Burbridge et al., *Revs. Modern Phys.*, **29**, 547 (1957).

theory are relatively simple. They are that a star is formed by condensation of the dust and gas of interstellar space, and when its central
temperature becomes high enough by the release of gravitational energy,
a cycle of nuclear reactions which effectively builds heavy elements out of
lighter ones sets in. The reaction products are believed to remain at the
center of the star rather than be mixed throughout its volume, and thus
the internal composition of the star gradually becomes richer in the
reaction products. It is assumed that an instability eventually sets in,
perhaps brought about by the exhaustion of the fuel and a consequent
gravitational collapse, which might initiate a catastrophic nuclear
explosion leading to a nova or supernova, or perhaps by a centrifugal
effect accompanying the contraction of a rapidly rotating star. At
any rate, it is assumed that by some mechanism the material of a sizable
fraction of the stars becomes ejected again into space where it becomes
well mixed and available for further condensation into new stars. The
galaxy is thus pictured in this theory as being populated by stars of
various ages and indeed of various "generations."

Ideally, this process of element building might proceed as far as iron,
the most stable of the elements. To account for the building up of the
elements above iron, it is necessary to resort again to the old idea of
neutron-capture reactions. According to the theory, neutrons supplied
by (α,n) reactions which are known to occur with several lighter nuclear
species are assumed to be available in the necessary (relatively small)
numbers to form nuclides of successively greater mass number. The
resulting abundances of the heavier nuclides should therefore be roughly
in inverse ratio to the neutron-capture cross sections, as is found to be
true.

In addition to gradual synthesis at the centers of stars, two other
processes are believed to play a significant role in element building:
(1) a small but appreciable formation in the *atmospheres* of certain stars,
perhaps through nuclear reactions of charged particles accelerated by
electromagnetic (betatron) processes in regions of rapidly changing
magnetic field and (2) sudden, catastrophic formation during the initial
few seconds of a *supernova eruption*. Let us now consider in somewhat
greater detail some of the more important features of the stellar-synthesis
theory.

Although it is by no means crucial to the theory, it is customary to
assume, for lack of more specific information, that the matter in the
universe originally was pure gaseous hydrogen of such temperature and
density that condensation into galaxies and stars could occur; one then
follows the life history of stars of various initial masses as time proceeds.

The stars which condense out of this pure ^1H medium are called *first-
generation stars*, and their energy must at first come entirely from the

proton-proton cycle. This cycle sets in at a central temperature of some 5 million degrees and continues until a sizable fraction of the hydrogen is used up; the star is left at this stage with a core composed almost entirely of ^{4}He, but with no source of energy. Theoretical computations indicate that, contrary to what might be expected, the core of the star gradually contracts and *becomes hotter*, while the outer envelope expands and *becomes cooler;* the star becomes a *red giant*. At a temperature of about 10^8 degrees, a new process is believed to set in: two ^{4}He nuclei combine to form the very short-lived nuclide ^{8}Be, and it may sometimes happen that, during the 10^{-14} lifetime of the latter, a third ^{4}He nucleus strikes with sufficient energy to penetrate the Coulomb barrier and form a nucleus of ^{12}C:

$$(a) \qquad \qquad ^4\text{He} + {}^4\text{He} \rightarrow {}^8\text{Be}$$
$$(b) \qquad \qquad ^4\text{He} + {}^8\text{Be} \rightarrow {}^{12}\text{C} \tag{1}$$

This step is the most critical one in the element-building process, for it circumvents the two barriers at mass numbers 5 and 8 that were discussed previously. This reaction can of course occur with significant probability only under relatively steady conditions where a long time is available for element building, because of the extremely low probability of reaction; it probably cannot contribute significantly in a short-time (explosive) process except under conditions of very high particle density. It also accounts for the relative scarcity of Li, Be, and B among the light elements.

With this formation of ^{12}C in an environment of ^{4}He, one also expects subsequent formation of the higher α-particle nuclei ^{16}O, ^{20}Ne, etc., up to a point where the Coulomb barrier prohibits further reaction. When the ^{4}He is used up in these reactions, a further slow collapse of the core should follow, accompanied by even higher temperatures; ultimately, the central temperature might reach a value above which heavy-particle reactions can occur, and the elements up to iron might then be formed.

It is conceivable that the processes so far described might all occur simultaneously inside a single star: the outer envelope consists almost entirely of ^{1}H, and at some "combustion interface" inside the star the proton-proton cycle is perhaps at work, converting ^{1}H to ^{4}He. Inside this boundary the matter is essentially pure ^{4}He, and at a deeper, hotter level the ^{4}He might be "burning" to form ^{12}C and some of the heavier α-particle nuclei. At a deeper level still, the temperature might reach the multibillion-degree heat at which heavy-particle reactions set in; and so on, the very center of the star being composed of the most stable elements—iron and its neighbors.

To explain the presence of elements other than hydrogen in the outer layers of stars, it seems necessary to assume that the material of a first-generation star is somehow returned to interstellar space where it participates in the condensation of new stars or is absorbed by already existing stars, since it is not believed that mixing can otherwise occur to the necessary extent in the ordinary course of stellar evolution. The observed occurrence of the *supernovae* seems to answer the purpose; in any case this mechanism, whatever its ultimate cause, does seem to eject into space substantially the entire mass of a star, including a large amount of matter heavier than ^{12}C; whether all supernovae can be equated to the death of first-generation stars is not yet established. Supernovae are believed to occur once or twice per 1000 years per galaxy, which seems to be a sufficient rate to account for the observed distribution of heavy elements in the stars.

Granting that, by some mechanism, hydrogen and some of the heavier elements find themselves distributed *throughout* a star, new kinds of nuclear reactions can occur. These are the carbon cycle, discussed in connection with stellar energy sources, which takes place at temperatures above 10 or 15 million degrees and another, analogous cycle, building on ^{20}Ne, which occurs at higher temperatures. Although these cycles do not consume the ^{12}C, ^{16}O, and ^{20}Ne upon which they depend, they do involve the establishment of a certain *equilibrium concentration* of the stable isotopes ^{13}C, ^{14}N, ^{15}N, ^{17}O, ^{21}Ne, and ^{23}Na, and these stable isotopes will remain even after the ^1H fuel is exhausted and replaced by ^4He. In the subsequent slow contraction of the stellar core, still other reactions become possible when the temperature becomes sufficiently high. Among these is an important type which leads to the liberation of *neutrons*. Because of the weak binding of the odd neutron in ^{13}C, ^{17}O, and ^{21}Ne, exoergic (α,n) reactions occur:

$$(a) \qquad\qquad {}^{13}C + {}^4He \rightarrow {}^{16}O + n$$
$$(b) \qquad\qquad {}^{17}O + {}^4He \rightarrow {}^{20}Ne + n \qquad\qquad (2)$$
$$(c) \qquad\qquad {}^{21}Ne + {}^4He \rightarrow {}^{24}Mg + n$$

The great importance of these reactions is that the liberated neutrons are preferentially absorbed by the heavier elements and serve to generate many of the nuclides above iron. A second-generation star is thus characterized in its later life by the appearance in its core of virtually all nuclear species, perhaps even including some radionuclides. The theory assumes that these second-generation stars, too, somehow become unstable and return a large portion of their matter to the interstellar medium. It may be that this process is what is actually seen in supernova explosions.[1]

---

[1] There is quite ample observational evidence of loss of mass from stars—several dozen ordinary novae occur in the galaxy each year, and each of them ejects a few

The galaxy is believed to be at present in a stage of its evolution wherein many second-generation stars have returned their matter to the interstellar medium, so that any stars now being formed start with all stable elements distributed throughout their volumes. Furthermore, many stars situated near a supernova outburst may well have become contaminated in their surface layers by accretion of the debris from the supernova.

The foregoing course of events is supposed to describe the evolutionary history of a great majority of the stars in the galaxy, but it is not yet possible to point to a given star and state in detail what its particular history has been. Knowledge in this field is increasing so rapidly, however, that it may soon prove possible to do just that. Further discussion of this rapidly developing field is unwarranted at this time in view of the tentative nature of the theory.

We shall close with a few experimental facts that apparently lend general support to the theory of stellar nucleogenesis. Most important, perhaps, is the fact that many of the most conspicuous stars in the heavens are very young stars in the sense that their present rate of energy generation cannot have been going on for more than a mere million years or so; in fact, there is fast-growing evidence that these stars were formed within the relatively recent past and that other stars are even now being born out of cosmic gas and dust clouds such as the Nebula of Orion.

Another important observational fact is that many stars show anomalously large abundances of certain elements in their spectra. Thus some stars show hydrogen and little else; others show principally helium. In still others, carbon or nitrogen are many times more prominent than usual, or certain metals may show abnormal abundances. Some of these cases are believed to result from nuclear reactions in the outermost envelope of the star, but some, at least, probably represent an enhanced abundance throughout the star. Perhaps the most exciting observation of abnormal abundances in stellar atmospheres is the discovery by P. W. Merrill[1] that some stars contain significant quantities of the radioactive element Tc (technetium), whose longest-lived isotope has a half-life of but 2×10^5 years. Clearly, this element—and therefore, no doubt, many others—are currently being made in stellar atmospheres, most likely by neutron absorption.

Finally, we mention the remarkable coincidence, pointed out by

---

per cent of a stellar mass into space. *Two* types of supernovae are known, and in each of them a large fraction of a stellar mass is ejected. It is also known that many kinds of stars eject matter more or less continuously from their outer layers and that, over long periods of time, large amounts of mass may be lost.

[1] *Science,* **115**, 484 (1952).

Burbidge et al.,[1] between the decay in brightness of supernovae and the half-life of the radionuclide ^{254}Cf.   It is observed that the post-maximum brightness of the so-called type I supernovae diminishes exponentially with a half-life of 55 ± 1 nights.   This happens to correspond exactly with the half-life of ^{254}Cf, which has been observed as a residue in H-bomb tests.[2]   In addition to this agreement in half-life, the particular nuclide ^{254}Cf possesses certain distinctive properties which, taken together, strongly suggest that the supernova eruption is a gigantic nuclear explosion in which a copious supply of neutrons is available to build up heavy elements by neutron capture, that a large amount of ^{254}Cf would be formed in this processs, and that the decay of this nuclide could, energy-wise at least, dominate the situation after the main outburst.   Briefly, these properties are:

1. If one starts with iron and successively adds neutrons at such a fast rate that β-decay cannot completely readjust the charge, one builds a sequence of neutron-rich nuclides in which $N \approx 0.7A$.   The mechanism of this process is that, when enough neutrons are suddenly added to a nucleus, a point is reached where more neutrons cannot be bound; β-decay then causes enough readjustment of the nuclear structure that another one or two neutrons can be added, and so on. The build-up is therefore effectively limited by *the rate at which the charge can be adjusted by β-decay*.   Since the β-decay lifetimes are about 0.1 s, the time scale for building up to the transuranic nuclides from iron is about 7 s.   However, as A increases, the β-decay lifetimes become longer and the build-up rate correspondingly slower; and in crossing the nuclear shell boundaries at $Z = 82$ and $N = 152$, the much greater β-decay lifetimes encountered may quite reasonably limit the build-up process to the neighborhood of ^{254}Cf.

2. ^{254}Cf is unique among its neighbors in undergoing *fission* with essentially no α- or β-decay.   The tremendous energy release in this process—about 200 MeV—permits a great contribution to the total energy of the post-maximum decline of the supernova to be made by relatively small amounts of this nuclide.   Other radionuclides having similar half-lives, 54-day ^{7}Be and 55-day ^{89}Sr, both suffer from the difficulty that, to account for the observed energy release, great quantities would have to be present in each supernova and, as a cumulative effect from many such explosions, would be about 100 times as abundant as actually observed in the galaxy as a whole.

The details of the build-up of ^{254}Cf and its neighbors from iron in a supernova eruption are not fully worked out, but the broad features of the

[1] *Phys. Rev.*, **103**, 1145 (1956).

[2] Fields et al., *Phys. Rev.*, **102**, 180 (1956).

process as they are described by Burbidge et al. seem quite plausible. These workers also point out that the ordinary slow, quasi-equilibrial element-building processes at work in the later-generation stars are not capable of forming the parent nuclides of the naturally occurring radio-active series, but that these and even higher members (for example, ^{254}Cf) can readily be formed *in quantity* in the abrupt explosive outburst of a supernova and that this outburst also serves to distribute these nuclides —along with all others—throughout interstellar space.   The theory of stellar nucleogenesis therefore seems quite capable of accounting for the presence of all observed nuclides, and considerable progress has even been made in quantitatively accounting for the relative abundances of the nuclear species.

It seems most fitting that we should conclude our discussion of the role of modern physics in the operation of the universe with this brief treatment of the problem of the origin of the elements, because surely no other example that can be cited touches more closely our immediate surroundings on this planet as well as the farthest reaches of the visible universe, both in space and in time.   The processes of modern physics are indeed at work in the world!

REFERENCES

Richtmyer, F. K., E. H. Kennard, and T. Lauritsen: "Introduction to Modern Physics," 5th ed., International Series in Pure and Applied Physics, McGraw-Hill Book Company, Inc., New York, 1955.

Rossi, B.: "High Energy Particles," Prentice-Hall, Inc., Englewood Cliffs, N.J., 1952.

APPENDIX A

A TABLE OF PHYSICAL CONSTANTS

The constants given below are evaluated both in MKSA units (bold-face) and in certain other systems of units which are often used. For a more complete list, including errors, see DuMond et al., *Revs. Modern Phys.*, **27**, 363 (1955).

1. Speed of light
$$c = \mathbf{2.997930 \times 10^8\ m\ s^{-1}}$$
$$= 2.997930 \times 10^{10}\ cm\ s^{-1}$$

2. Avogadro's number (physical scale)
$$N = \mathbf{6.02486 \times 10^{26}\ (kg\ mole)^{-1}}$$
$$= 6.02486 \times 10^{23}\ (g\ mole)^{-1}$$

3. Loschmidt's number (physical scale)
$$L_0 = \mathbf{2.68719 \times 10^{25}\ m^{-3}}$$
$$= 2.68719 \times 10^{19}\ cm^{-3}$$

4. Electronic charge
$$e = \mathbf{1.60206 \times 10^{-19}\ C}$$
$$= 4.80286 \times 10^{-10}\ statcoulomb$$

5. Electron rest mass[1]
$$m = \mathbf{9.1083 \times 10^{-31}\ kg} = 0.510976\ MeV = 5.48763 \times 10^{-4}\ amu$$
$$= 9.1083 \times 10^{-28}\ g$$

6. Proton rest mass
$$M_p = \mathbf{1.67239 \times 10^{-27}\ kg} = 938.211\ MeV = 1.007593\ amu$$
$$= 1.67239 \times 10^{-24}\ g$$

7. Neutron rest mass
$$M_n = \mathbf{1.67470 \times 10^{-27}\ kg} = 939.505\ MeV = 1.008983\ amu$$
$$= 1.67470 \times 10^{-24}\ g$$

8. Planck's constant
$$h = \mathbf{6.62517 \times 10^{-34}\ J\ s}\ (or\ \mathbf{m^2\ kg\ s^{-1}}) = 4.1354 \times 10^{-15}\ eV\ s$$
$$= 6.62517 \times 10^{-27}\ erg\ s$$
$$\hbar = h/2\pi = \mathbf{1.05443 \times 10^{-34}\ J\ s} = 6.5817 \times 10^{-16}\ eV\ s$$
$$= 1.05443 \times 10^{-27}\ erg\ s$$

---

[1] In discussions involving particles, m_e is often used to designate the electron rest mass.

9. Faraday constant (physical scale)

$F = \mathbf{9.65219 \times 10^7 \ C \ (kg \ mole)^{-1}}$

$= 9652.19 \ \text{emu} \ (\text{g mole})^{-1}$

10. Charge-to-mass ratio of electron

$e/m = \mathbf{1.75890 \times 10^{11} \ C \ kg^{-1}}$

$= 1.75890 \times 10^7 \ \text{emu} \ \text{g}^{-1}$

11. Ratio h/e

$h/e = \mathbf{4.1354 \times 10^{-15} \ V \ s}$ (or $\mathbf{m^2 \ kg \ s^{-1} \ C^{-1}}$)

$= 1.37942 \times 10^{-17} \ \text{erg s esu}^{-1}$

12. Fine-structure constant

$\alpha = e^2/4\pi\varepsilon_0\hbar c = 7.29729 \times 10^{-3}$

$= e^2/\hbar c$ (e in esu)

$1/\alpha = 137.0373$

13. Ratio proton mass to electron mass

$M_\text{p}/m = 1836.12$

14. Reduced mass of electron in H

$$\frac{mm_{\text{H}^+}}{m_\text{H}} = \mathbf{9.1034 \times 10^{-31} \ kg}$$

$= 9.1034 \times 10^{-28} \ \text{g}$

15. Schroedinger constant for fixed nucleus

$2m/\hbar^2 = \mathbf{1.63836 \times 10^{38} \ J^{-1} \ m^{-2}}$ (or $\mathbf{m^{-4} \ kg^{-1} \ s^2}$)

$= 1.63836 \times 10^{27} \ \text{erg}^{-1} \ \text{cm}^{-2}$

16. First Bohr radius

$a_0 = 4\pi\varepsilon_0\hbar^2/me^2 = \mathbf{5.29172 \times 10^{-11} \ m}$

$= \hbar^2/me^2 = 5.29172 \times 10^{-9} \ \text{cm}$ (e in esu)

17. Classical electron radius

$r_0 = e^2/4\pi\varepsilon_0 mc^2 = \mathbf{2.81785 \times 10^{-15} \ m}$

$= e^2/mc^2 = 2.81785 \times 10^{-13} \ \text{cm}$ (e in esu)

18. Compton wavelength of electron

$\lambda_\text{ce} = h/mc = \mathbf{2.42626 \times 10^{-12} \ m}$

$= 2.42626 \times 10^{-10} \ \text{cm}$

19. Compton wavelength of proton

$\lambda_\text{cp} = h/M_\text{p}c = \mathbf{13.2141 \times 10^{-16} \ m}$

$= 13.2141 \times 10^{-14} \ \text{cm}$

20. Boltzmann constant

$k = \mathbf{1.38044 \times 10^{-23} \ J \ °K^{-1}}$ (or $\mathbf{m^2 \ kg \ s^{-2} \ °K^{-1}}$)

$= 8.6164 \times 10^{-5} \ \text{eV} \ °\text{K}^{-1}$

$= 1.38044 \times 10^{-16} \ \text{erg} \ °\text{K}^{-1}$

$1/k = 11{,}605.4 \ °\text{K eV}^{-1}$

21. Gas constant

$R = \mathbf{8.31662 \times 10^3 \ J \ (kg \ mole)^{-1} \ °K^{-1}}$

$= 8.31662 \times 10^7 \ \text{erg} \ (\text{g mole})^{-1} \ °\text{K}^{-1}$

22. Wien's displacement law constant
$$\lambda_{max} T = \mathbf{0.289782 \times 10^{-2} \ m \ °K}$$
$$= 0.289782 \text{ cm } °K$$

23. Stefan-Boltzmann constant
$$\sigma = (\pi^2/60)(k^4/\hbar^3 c^2) = \mathbf{0.56687 \times 10^{-7} \ W \ m^{-2} \ °K^{-4}} \text{ (or kg s}^{-3} \text{ °K}^{-4})$$
$$= 0.56687 \times 10^{-4} \text{ erg s}^{-1} \text{ cm}^{-2} \text{ °K}^{-4}$$

24. Bohr magneton
$$\mu_B = \hbar e/2m = \mathbf{0.92731 \times 10^{-23} \ A \ m^2} \text{ (or m}^2 \text{ s}^{-1} \text{ C})$$
$$= \hbar e/2mc = 0.92731 \times 10^{-20} \text{ erg G}^{-1} \text{ (}e \text{ in esu)}$$

25. Magnetic moment of electron
$$\mu_e = \mathbf{0.92837 \times 10^{-23} \ A \ m^2} \text{ (or m}^2 \text{ s}^{-1} \text{ C})$$
$$= 1.001145358 \ \mu_B$$

26. Nuclear magneton
$$\mu_N = \hbar e/2M_p = \mathbf{0.505038 \times 10^{-26} \ A \ m^2} \text{ (or m}^2 \text{ s}^{-1} \text{ C})$$
$$= \hbar e/2M_p c = 0.505038 \times 10^{-23} \text{ erg G}^{-1}$$

27. Proton magnetic moment
$$\mu_p = 2.79275 \ \mu_N$$
$$= \mathbf{1.41044 \times 10^{-26} \ A \ m^2} \text{ (or m}^2 \text{ s}^{-1} \text{ C})$$
$$= 1.41044 \times 10^{-23} \text{ erg G}^{-1}$$

28. Mass-energy conversion factors

1 kg = 5.61000×10^{29} MeV	1 M_n = 939.505 MeV
1 m_e = 0.510976 MeV	1 amu = 931.141 MeV
1 M_p = 938.211 MeV	

29. Quantum conversion factors
$$1 \text{ eV} = 1.60206 \times 10^{-19} \text{ J}$$
Wavelength of 1-eV photon = 1.239767×10^{-6} m = 12,397.67 Å

30. The Rydberg

$R = \mathbf{10{,}973{,}730.9 \ m^{-1}}$	$R_H = \mathbf{10{,}967{,}757.6 \ m^{-1}}$
$= 109{,}737.309 \text{ cm}^{-1}$	$R_{4He} = \mathbf{10{,}972{,}226.7 \ m^{-1}}$

31. Capacitivity of vacuum
$$\varepsilon_0 = \mathbf{8.85434 \times 10^{-12} \ F \ m^{-1}}$$

32. Permeability of vacuum
$$\mu_0 = \mathbf{4\pi \times 10^{-7} \ H \ m^{-1}}$$

APPENDIX B

BLACK-BODY DISTRIBUTION FUNCTION

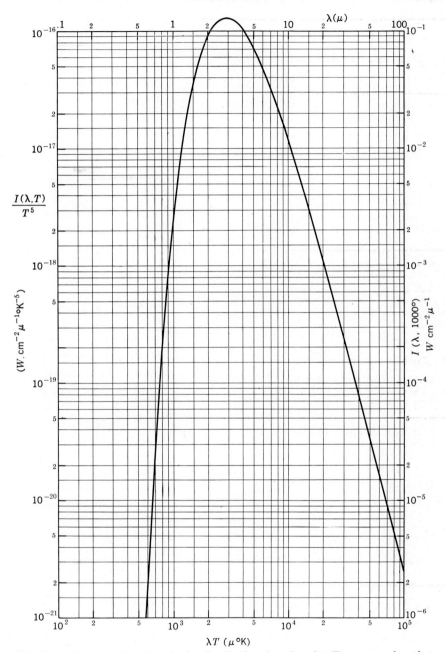

$\lambda(\mu)$

$\dfrac{I(\lambda,T)}{T^5}$

$(W.\,cm^{-2}\,\mu^{-1}{}^{\circ}K^{-5})$

$I\,(\lambda,\,1000^{\circ})$ $W\,cm^{-2}\,\mu^{-1}$

$\lambda T\;(\mu\,{}^{\circ}K)$

Fig. B-1. Graph of the black-body distribution function $I(\lambda,T)$ vs. wavelength λ at a temperature of $1000^{\circ}K$ (upper and right-hand scales). Also shown is T^{-5} $I(\lambda,T)$ plotted vs. λT (bottom and left-hand scales).

APPENDIX C

SOME ELECTRONIC PROPERTIES OF ATOMS

This table gives the densities and the electron configuration, spectroscopic terms, and ionization potentials of the ground states of the neutral and singly ionized atoms. In each case the electronic configuration of the singly ionized atom is obtained by removing one electron from the last-named shell in the configuration of the neutral atom, except as indicated. (Example: Sc ion ground state is 3d 4s $^3D_1$.) g = gas, l = liquid, IP = ionization potential.

Z	Element	Density g cm^{-3} at 0°C, 1 atm	Ground state			Ion ground state	
			Elect config	Term	IP, eV	Term	IP, eV
1	H	8.988×10^{-5} (g)	1s	$^2S_{1/2}$	13.595		
2	He	1.785×10^{-4} (g)	1s^2	$^1S_0$	24.580	$^2S_{1/2}$	54.503
3	Li	0.534	[He] 2s	$^2S_{1/2}$	5.390	$^1S_0$	75.619
4	Be	1.85	2s^2	$^1S_0$	9.320	$^2S_{1/2}$	18.206
5	B	2.535	2s^2 2p	$^2P_{1/2}$	8.296	$^1S_0$	25.149
6	C	3.52 (dia.) 2.25 (graph.)	2s^2 2p^2	$^3P_0$	11.264	$^2P_{1/2}$	24.376
7	N	1.257×10^{-3}(g)	2s^2 2p^3	$^4S_{3/2}$	14.54	$^3P_0$	29.605
8	O	1.428×10^{-3}(g)	2s^2 2p^4	$^3P_2$	13.614	$^4S_{3/2}$	35.146
9	F	1.695×10^{-3}(g)	2s^2 2p^5	$^2P_{3/2}$	17.418	$^3P_2$	34.98
10	Ne	0.9005×10^{-3}(g)	2s^2 2p^6	$^1S_0$	21.559	$^2P_{3/2}$	41.07
11	Na	0.9712	[Ne]3s	$^2S_{1/2}$	5.138	$^1S_0$	47.29
12	Mg	1.741	3s^2	$^1S_0$	7.644	$^2S_{1/2}$	15.03
13	Al	2.70	3s^2 3p	$^2P_{1/2}$	5.984	$^1S_0$	18.823
14	Si	2.42	3s^2 3p^2	$^3P_0$	8.149	$^2P_{1/2}$	16.34
15	P	wh 1.83, red 2.20 met 2.34, blk 2.69	3s^2 3p^3	$^4S_{3/2}$	10.55	$^3P_0$	19.65
16	S	2.0–2.1	3s^2 3p^4	$^3P_2$	10.357	$^4S_{3/2}$	23.4
17	Cl	3.16×10^{-3}(g)	3s^2 3p^5	$^2P_{3/2}$	13.01	$^3P_2$	23.80
18	A	1.782×10^{-3}(g)	3s^2 3p^6	$^1S_0$	15.755	$^2P_{3/2}$	27.62
19	K	0.870	[A] 4s	$^2S_{1/2}$	4.339	$^1S_0$	31.81
20	Ca	1.54	4s^2	$^1S_0$	6.111	$^2S_{1/2}$	11.87
21	Sc	2.5	3d 4s^2	$^2D_{3/2}$	6.56	$^3D_1$	12.80
22	Ti	4.5	3d^2 4s^2	$^3F_2$	6.83	$^2F_{3/2}$	13.57
23	V	5.6	3d^3 4s^2	$^4F_{3/2}$	6.74	(3d^4) $^5D_0$	14.65

Z	Element	Density g cm^{-3} at 0°C, 1 atm	Ground state			Ion ground state	
			Elect config	Term	IP, eV	Term	IP, eV
24	Cr	6.93	3d^5 4s	^{7}S$_3$	6.764	^{6}S$_{5/2}$	16.49
25	Mn	7.3	3d^5 4s^2	^{6}S$_{5/2}$	7.432	^{7}S$_3$	15.64
26	Fe	7.86	3d^6 4s^2	^{5}D$_4$	7.90	^{6}D$_{9/2}$	16.18
27	Co	8.71	3d^7 4s^2	^{4}F$_{9/2}$	7.86	(3d^8) ^{3}F$_4$	17.05
28	Ni	8.8	3d^8 4s^2	^{3}F$_4$	7.633	(3d^9) ^{2}D$_{5/2}$	18.15
29	Cu	8.90	3d^{10} 4s	^{2}S$_{1/2}$	7.724	^{1}S$_0$	20.29
30	Zn	7.0	3d^{10} 4s^2	^{1}S$_0$	9.391	^{2}S$_{1/2}$	17.96
31	Ga	5.93	3d^{10} 4s^2 4p	^{2}P$_{1/2}$	6.00	^{1}S$_0$	20.51
32	Ge	5.46	3d^{10} 4s^2 4p^2	^{3}P$_0$	7.88	^{2}P$_{1/2}$	15.93
33	As	5.73	3d^{10} 4s^2 4p^3	^{4}S$_{1/2}$	9.81	^{3}P$_0$	20.2
34	Se	4.82	3d^{10} 4s^2 4p^4	^{3}P$_2$	9.75	^{4}S$_{3/2}$	21.5
35	Br	3.12(l)	3d^{10} 4s^2 4p^5	^{2}P$_{3/2}$	11.84	^{3}P$_2$	21.6
36	Kr	3.736 × 10^{-3}(g)	3d^{10} 4s^2 4p^6	^{1}S$_0$	13.996	^{2}P$_{3/2}$	24.56
37	Rb	1.532	[Kr] 5s	^{2}S$_{1/2}$	4.176	^{1}S$_0$	27.5
38	Sr	2.60	5s^2	^{1}S$_0$	5.692	^{2}S$_{1/2}$	11.027
39	Y	3.8	4d 5s^2	^{2}D$_{3/2}$	6.377	^{1}S$_0$	12.233
40	Zr	6.44	4d^2 5s^2	^{3}F$_2$	6.835	^{4}F$_{3/2}$	12.916
41	Nb	8.4	4d^4 5s	^{6}D$_{1/2}$	6.881	^{5}D$_0$	13.895
42	Mo	9.0	4d^5 5s	^{7}S$_3$	7.131	^{6}S$_{5/2}$	15.72
43	Tc	9.0	4d^5 5s^2	^{6}S$_{5/2}$	7.23	^{7}S$_3$	14.87
44	Ru	12.6	4d^7 5s	^{5}F$_5$	7.365	^{4}F$_{9/2}$	16.597
45	Rh	12.44	4d^8 5s	^{4}F$_{9/2}$	7.461	^{3}F$_4$	15.92
46	Pd	12.16	4d^{10}	^{1}S$_0$	8.33	^{2}D$_{5/2}$	19.42
47	Ag	10.5	4d^{10} 5s	^{2}S$_{1/2}$	7.574	^{1}S$_0$	21.48
48	Cd	8.67	4d^{10} 5s^2	^{1}S$_0$	8.991	^{2}S$_{1/2}$	16.904
49	In	7.28	4d^{10} 5s^2 5p	^{2}P$_{1/2}$	5.785	^{1}S$_0$	18.828
50	Sn	7.29	4d^{10} 5s^2 5p^2	^{3}P$_0$	7.332	^{2}P$_{1/2}$	14.63
51	Sb	6.65	4d^{10} 5s^2 5p^3	^{4}S$_{3/2}$	8.639	^{3}P$_0$	19
52	Te	6.25	4d^{10} 5s^2 5p^4	^{3}P$_2$	9.01	^{4}S$_{3/2}$	21.5
53	I	4.94	4d^{10} 5s^2 5p^5	^{2}P$_{3/2}$	10.44	^{3}P$_2$	19.0
54	Xe	5.85 × 10^{-3}(g)	4d^{10} 5s^2 5p^6	^{1}S$_0$	12.127	^{2}P$_{3/2}$	21.21
55	Cs	1.873	[Xe] 6s	^{2}S$_{1/2}$	3.893	^{1}S$_0$	25.1
56	Ba	3.78	6s^2	^{1}S$_0$	5.210	^{2}S$_{1/2}$	10.001
57	La	6.15	5d 6s^2	^{2}D$_{3/2}$	5.61	(5d^2) ^{3}F$_2$	11.43
58	Ce	6.79	4f 5d 6s^2	^{3}H$_5$	......	(4f^2 6s) ^{4}H$_{7/2}$	
59	Pr	6.48	4f^3 6s^2	^{4}I$_{9/2}$	......	^{5}I$_4$	
60	Nd	7.00	4f^4 6s^2	^{5}I$_4$	6.3	(4f^5 6s) ^{6}I$_{7/2}$	
61	Pm	....	4f^5 6s^2	^{6}H$_{5/2}$			
62	Sm	7.7	4f^6 6s^2	^{7}S$_3$	5.6	^{8}F$_{1/2}$	11.2
63	Eu	....	4f^7 6s^2	^{8}S$_{7/2}$	5.67	^{9}S$_4$	11.24
64	Gd	....	4f^7 5d 6s^2	^{9}D$_2$	6.16	^{10}D$_{5/2}$	12 +
65	Tb	....	4f^8 5d 6s^2				
66	Dy						
67	Ho						
68	Er	4.77					
69	Tm	....	4f^{13} 6s^2	^{2}F$_{7/2}$	......	^{3}F$_4$	
70	Yb	....	4f^{14} 6s^2	^{1}S$_0$	6.22	^{2}S$_{1/2}$	12.10
71	Lu	....	4f^{14} 5d 6s^2	^{2}D$_{3/2}$	6.15	^{1}S$_0$	14.7
72	Hf	13.3	4f^{14} 5d^2 6s^2	^{3}F$_2$	5.5	^{2}D$_{3/2}$	14.9
73	Ta	16.6	4f^{14} 5d^3 6s^2	^{4}F$_{3/2}$	7.7	^{5}F$_1$	
74	W	19.3	4f^{14} 5d^4 6s^2	^{5}D$_0$	7.98	^{6}D$_{1/2}$	
75	Re	....	5f^{14} 5d^5 6s^2	^{6}S$_{5/2}$	7.87	^{7}S$_3$	
76	Os	22.5	4f^{14} 5d^6 6s^2	^{5}D$_4$	8.7		

Z	Element	Density g cm^{-3} at 0°C, 1 atm	Ground state			Ion ground state	
			Elect config	Term	IP, eV	Term	IP, eV
77	Ir	22.42	$4f^{14}\,5d^7\,6s^2$	$^4F_{9/2}$	9.2		
78	Pt	21.37	$4f^{14}\,5d^9\,6s$	$^3D_3$	9.0	$^2D_{5/2}$	18.56
79	Au	19.3	$4f^{14}\,5d^{10}\,6s$	$^2S_{1/2}$	9.22	$^1S_0$	20.5
80	Hg	13.596(l)	$4f^{14}\,5d^{10}\,6s^2$	$^1S_0$	10.434	$^2S_{1/2}$	18.751
81	Tl	11.86	$4f^{14}\,5d^{10}\,6s^2\,6p$	$^2P_{1/2}$	6.106	$^1S_0$	20.42
82	Pb	11.342	$4f^{14}\,5s^{10}\,6s^2\,6p^2$	$^3P_0$	7.415	$^2P_{1/2}$	15.028
83	Bi	9.75	$4f^{14}\,5d^{10}\,6s^2\,6p^3$	$^4S_{3/2}$	7.287	$^3P_0$	19.3
84	Po	. . . .	$4f^{14}\,5d^{10}\,6s^2\,6p^4$	$^3P_2$	8.43		
85	At	. . . .	$4f^{14}\,5d^{10}\,6s^2\,6p^5$				
86	Em	. . . .	$4f^{14}\,5d^{10}\,6s^2\,6p^6$	$^1S_0$	10.745		
87	Fr	. . . .	[Em] 7s	$^2S_{1/2}$			
88	Ra	5(?)	$7s^2$	$^1S_0$			
89	Ac	. . . .	$6d\,7s^2$	$^2D_{3/2}$	. . . . . .	$^1S_0$	
90	Th	11.00	$6d^2\,7s^2$	$^3F_2$	. . . . . .	$^4F_{3/2}$	
91	Pa	. . . .	$6d^3\,7s^2$				
92	U	18.7	$5f^3\,6d\,7s^2$	$^5L_6$	4	$^4I_{9/2}$	
93	Np						
94	Pu						
95	Am						
96	Cm						
97	Bk						
98	Cf						
99	E						
100	Fm						
101	Mv						

APPENDIX D

CONSTANTS OF SOME DIATOMIC MOLECULES

This table gives the following information concerning a number of rather commonly encountered diatomic molecules. The numbers are the column numbers of the table.

1. Chemical formula and state of ionization.
2, 3. Mass numbers of the two atoms.
4. Reduced mass m_r (physical scale).
5. The quantity $B = \hbar/4\pi c m_r r_0^2$, the wave number of the basic rotational term. (See Secs. 9-4 and 9-5.)
6. The quantity α, which describes the change in the quantity B as a function of the vibrational excitation level and is defined by the equation $B_v = B - \alpha(v + \frac{1}{2})$, where B_v is the value of the rotational wave number B in the vibrational state v.
7. r_0, the equilibrium separation of the two atoms.
8. The vibrational wave number $\tilde{\nu}_v : \tilde{\nu}_v = \omega/2\pi c$.
9. The so-called *anharmonic constant* x, which describes the change in vibrational-level spacing with $v : E_v = \tilde{\nu}_v(v + \frac{1}{2}) - x(v + \frac{1}{2})$.
10. Dissociation energy D_0.

Molecule	m_1	m_2	m_r, amu	B, cm^{-1}	α, cm^{-1}	r_0, 10^{-8} cm	$\tilde{\nu}_v$, cm^{-1}	x, cm^{-1}	D_0, eV
(1)	(2)	(3)	(4)	(5)	(6)	(7)	(8)	(9)	(10)
C$_2$	12	12	6.00194	1.6326	0.01683	1.3117	1641.35	11.67	(3.6)
CaH	40	1	0.983332	4.2778	0.0963	2.0020	1299	19.5	≤1.70
CaO	40	16	10.4265	0.445	0.00335	1.822	732.1	4.81	5.9
CH	12	1	0.930024	14.457	0.534	1.1198	2861.6	64.3	3.47
CH$^+$	12	1	0.930021	14.1767	0.4898	1.13083	....	....	3.6
Cl$_2$	35	35	17.48942	0.2438	0.0017	1.988	564.9	4.0	2.475
Cl$_2^+$	35	35	17.48928	0.2697	0.0018	1.891	645.3	2.90	(4.4)
CN	12	14	6.46427	1.8996	0.01735	1.1718	2068.705	13.144	....
CO	12	16	6.85841	1.9314	0.01748	1.1282	2170.21	13.461	11.108
CO$^+$	12	16	6.85823	1.9772	0.01896	1.1151	2214.24	15.164	(9.9)

Molecule	m_1	m_2	m_r, amu	B, cm^{-1}	α, cm^{-1}	r_0, 10^{-8} cm	$\tilde{\nu}_v$, cm^{-1}	x, cm^{-1}	D_0, eV
(1)	(2)	(3)	(4)	(5)	(6)	(7)	(8)	(9)	(10)
CsH	133	1	1.00054	2.709	0.057	2.494	890.7	12.6	(1.9)
CuF	63	19	14.5979	0.3803	0.0046	1.743	622.65	3.95	(3.0)
CuH	63	1	0.992242	7.938	0.249	1.463	1940.4	37.0	2.89
H$_2$	1	1	0.504066	60.809	2.993	0.74166	4395.24	117.995	4.4763
H$_2^+$	1	1	0.503928	29.8	1.4	1.06	2297	62	2.6481
HCl	1	35	0.979889	10.5909	0.3019	1.27460	2989.74	52.05	4.430
He$_2$	4	4	2.00193	7.664	0.131	1.0483	1811.2	39.2	(2.6)
HF	1	9	0.957347	20.939	0.7705	0.9171	4138.52	90.069	6.40
HgH	200	1	1.00309	5.549	0.312	1.7404	1387.09	83.01	0.376
HI	1	127	1.000187	6.551	0.183	1.6041	2309.53	39.73	3.0564
I$_2$	127	127	63.4665	0.03736	0.000117	2.6660	214.57	0.6127	1.5417
LiH	7	1	0.881506	7.5131	0.2132	1.59535	1405.609	23.200	2.5
MgH	24	1	0.967480	5.8181	0.1668	1.7306	1495.7	31.5	2.49
MgO	24	16	9.5989	0.5743	0.0050	1.749	785.1	5.18	(3.7)
N$_2$	14	14	7.00377	2.010	0.0187	1.094	2359.61	14.456	9.756
N$_2^+$	14	14	7.00363	1.932	0.020	1.1162	2207.19	16.136	8.724
NaH	23	1	0.96579	4.9012	0.1353	1.8873	1172.2	19.72	(2.2)
O$_2$	16	16	8.00000	1.445666	0.015791	1.207398	1580.361	12.0730	5.080
OH	16	1	0.94838	18.871	0.714	0.9706	3735.21	82.81	4.35
OH$^+$	16	1	0.94837	16.793	0.732	1.0289	(2955)	. . . .	4.4
PbO	208	16	14.8534	0.3073	0.0019	1.922	72.18	3.70	(4.3)
PbS	208	32	27.7213	0.10605	0.000873	2.394	428.14	1.201	(4.7)
S$_2$	32	32	15.99126	0.2956	0.0016	1.889	725.68	2.852	4.4
SiO	28	16	10.18013	0.7263	0.00494	1.5101	1242.03	6.047	(7.4)
SO	32	16	10.66472	0.7089	0.005622	1.4933	1123.73	6.116	5.146
TiO	48	16	11.9979	0.5355	0.0031	1.620	1008.4	4.61	(6.9)
ZrO	90	16	13.5836	0.6187	0.0070	(1.416)	936.6	3.45	(7.8)

APPENDIX E

SOME PROPERTIES OF THE ATMOSPHERE (LATITUDE 45°)‡

Height h (m or km)	Height h (ft or mi)	Apparent gravity g, cm s⁻²	Temp, °K	Pressure p, dyn cm⁻²	Density g cm⁻³	Density Particles cm⁻³	Mean speed, km s⁻¹	Mean free path L, cm (mean dia 3×10⁻⁸ cm)	Mean coll freq, s⁻¹	N	O	He	H	Mean mol wt
0	0	980.7	288.0	1.014×10^6	1.223×10^{-3}	2.568×10^{19}	0.459	9.74×10^{-6}	4.71×10^9	78	21	$\sim10^{-4}$	$\sim3 \times 10^{-6}$	28.9
1524	5000	980.2	278.1	8.43×10^5	1.055×10^{-3}	2.21×10^{19}	0.451	1.13×10^{-5}	3.99×10^9					
3048	10000	979.7	268.2	6.97×10^5	9.05×10^{-4}	1.90×10^{19}	0.443	1.32×10^{-5}	3.36×10^9					
6096	20000	978.8	248.4	4.67×10^5	6.54×10^{-4}	1.37×10^{19}	0.426	1.82×10^{-5}	2.34×10^9					
9144	30000	977.8	226.6	3.02×10^5	4.60×10^{-4}	9.65×10^{18}	0.409	2.59×10^{-5}	1.58×10^9					
10769	35000	977.4	218.0	2.36×10^5	3.77×10^{-4}	7.91×10^{18}	0.400	3.17×10^{-5}	1.26×10^9					
13716	45000	976.5	218.0	1.49×10^5	2.38×10^{-4}	5.00×10^{18}	0.400	5.01×10^{-5}	7.97×10^8					
16764	55000	975.5	218.0	9.29×10^4	1.48×10^{-4}	3.11×10^{18}	0.400	8.05×10^{-5}	4.96×10^8					
22860	75000	973.7	218.0	3.61×10^4	5.76×10^{-5}	1.21×10^{18}	0.400	2.07×10^{-4}	1.93×10^8					
27432	90000	972.3	218.0	1.79×10^4	2.85×10^{-5}	5.99×10^{17}	0.400	4.18×10^{-4}	9.57×10^7					
32000	105000	970.9	218.0	8.90×10^3	1.42×10^{-5}	2.98×10^{17}	0.400	8.40×10^{-4}	4.76×10^7					
39.6	24.6	968.5	276.0	3.15×10^3	3.94×10^{-6}	8.32×10^{16}	0.450	3.01×10^{-3}	1.50×10^7					
50.0	31.1	965.4	355.0	1.05×10^3	1.02×10^{-6}	2.17×10^{16}	0.512	1.16×10^{-2}	4.43×10^6	82	18			28.7
60.0	37.3	962.4	355.0	4.14×10^2	4.02×10^{-7}	8.50×10^{15}	0.512	2.94×10^{-2}	7.96×10^5					
68.6	42.6	959.8	300.2	1.74×10^2	2.00×10^{-7}	4.23×10^{15}	0.471	5.91×10^{-2}	2.80×10^5					
78.0	48.5	957.0	240.0	5.47×10^1	7.86×10^{-8}	1.66×10^{15}	0.421	1.51×10^{-1}	1.41×10^5					
83.0	51.6	955.5	240.0	2.75×10^1	3.96×10^{-8}	8.37×10^{14}	0.421	2.99×10^{-1}	3.86×10^4	82	18			28.7
93.0	57.8	952.5	276.4	7.94×10^0	9.51×10^{-9}	2.10×10^{14}	0.461	1.19×10^0	1.66×10^4					
100.6	62.5	950.3	304.2	3.53×10^0	3.71×10^{-9}	8.46×10^{13}	0.492	2.96×10^0	2.96×10^3					
120.0	74.6	944.6	375.0	6.7×10^{-1}	5.22×10^{-10}	1.30×10^{13}	0.571	1.93×10^1	3.32×10^2	70	30			24.4§
152.4	94.7	935.2	505.5	8.67×10^{-2}	5.02×10^{-11}	1.25×10^{12}	0.663	2.00×10^2	1.88×10^2					
213.4	132.6	917.9	751.0	5.99×10^{-3}	2.33×10^{-12}	5.83×10^{10}	0.807	4.29×10^3	3.93×10^1					
259.1	161.0	905.2	935.2	1.40×10^{-3}	4.38×10^{-13}	1.09×10^{10}	0.901	2.29×10^4	1.25×10^1					
300.0	186.4	894.1	1100	4.84×10^{-4}	1.29×10^{-13}	3.21×10^9	0.978	2.79×10^4	1.24×10^{-1}	70	30	1.4×10^{-4}	6×10^{-6}	24.4
400	248.5	867.7	1500	9.70×10^{-5}	1.12×10^{-14}	4.72×10^8	1.48	1.20×10^6	4.6×10^{-2}	78	22	2.6×10^{-4}	1.3×10^{-5}	14.4
500	310.7	842.7	1900	4.06×10^{-5}	3.09×10^{-15}	1.56×10^8	1.67	3.63×10^6	2.41×10^{-3}	80	20	6.5×10^{-4}	9.3×10^{-5}	14.3
700	435.4	795.4	2500	1.16×10^{-5}	8.00×10^{-16}	3.39×10^7	1.92	1.66×10^7	2.97×10^{-5}	83	17	2.03×10^{-3}	1.88×10^{-4}	14.3
1000	621.4	732.0	2500	2.96×10^{-6}	1.65×10^{-17}	7.03×10^6	1.93	8.01×10^7	1.22×10^{-7}	85	15	4.8×10^{-2}	1.1×10^{-2}	14.3
2000	1234	567.7	2500	2.96×10^{-8}	2.01×10^{-18}	8.62×10^4	1.94	6.52×10^9		92	8			14.1
4000	2485	370.0	2500	9.75×10^{-11}	4.29×10^{-21}	2.84×10^2	2.41	1.98×10^{10}		89.3	3.4	3.9	3.4	9.1

‡ Adapted from Grimminger, *Rand Corp. Report* (1948). See also Smithsonian Tables.
§ The atmospheric gases become progressively more dissociated with increasing altitude above ~100 km.

APPENDIX F

PLOTS OF RANGE VERSUS ENERGY AND MOMENTUM

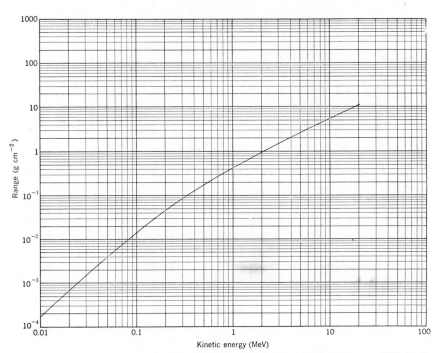

FIG. F-1. Range vs. energy for electrons in aluminum. [After Katz and Penfold, *Revs. Modern Phys.*, **24**, 28 (1952).]

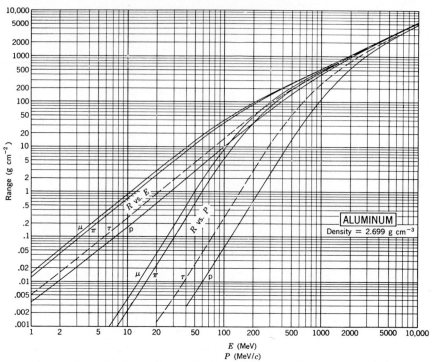

Fig. F-2. Range vs. energy and range vs. momentum for various particles in aluminum.

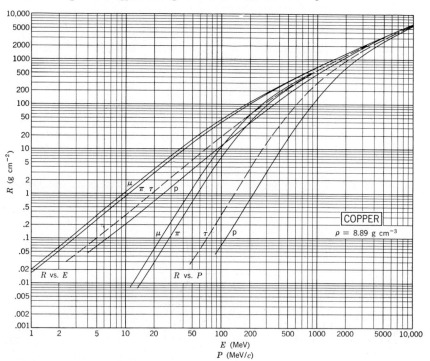

Fig. F-3. Range vs. energy and range vs. momentum for various particles in copper.

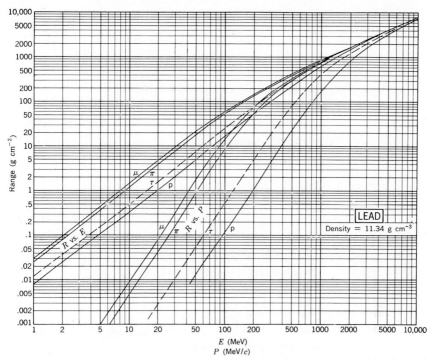

Fig. F-4. Range vs. energy and range vs. momentum for various particles in lead

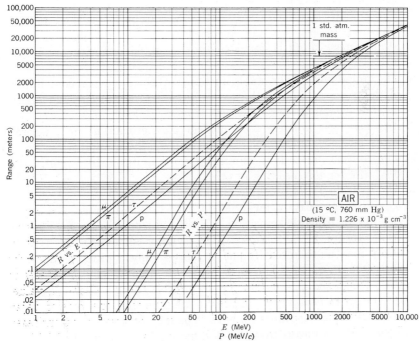

Fig. F-5. Range vs. energy and range vs. momentum for various particles in air.

735

APPENDIX G

TABLE OF NUCLEAR SPECIES

This table gives the following information concerning the known nuclear species. The numbers are those of the table columns.

1. The atomic number Z.
2. The chemical symbol. Below this, the *name* of the element is given.
3. The mass number A. The first entry in this column for each element is the *atomic weight* for the isotopic mixture found in nature.
4. The neutron number N.
5. The ground-state spin quantum number I.
6. The ground-state parity π.
7. The ground-state shell configuration, according to the shell model.
8. The mass excess $M - A$, expressed in MeV. (The mass excess in atomic mass units is equal to the tabulated quantity divided by 931.141.)
9. The abundances of the stable nuclides: (*a*) the *percentage* contribution of the given nuclide to the naturally occurring mixture and (*b*) the *absolute abundance* per 10^6 atoms of elemental silicon.
10. The magnetic moment μ in nuclear magnetons. $\mu = gI$, where g is the nuclear g-factor and I is the spin quantum number.
11. The nuclear quadripole moment Q_m, in units of 10^{-24} cm^2. (See Sec. 13-4 for the meaning of Q_m.)
12. The half-life $T_{1/2}$ of unstable nuclides.
13. The type of decay of unstable nuclides ($\alpha = \alpha$-particle, $\beta = \beta$-particle, IT = isomeric transition, K = K-electron capture, L = L-electron capture, fiss = fission). The symbol indicating the type of decay may be followed by one or more numbers which indicate the approximate energy release in MeV. A number in parentheses in this column signifies the maximum possible energy release (Q-value) for the given decay type, in MeV. If more than one type of decay is observed, the Q-values for each type may be given in parentheses. Thus $(2.75+)$ means that the Q-value for β^+-decay is 2.75 MeV.

14. The binding energy of the most weakly bound neutron, in MeV. If this amount of energy is supplied, a neutron may be evaporated, leaving the residual nucleus in its ground state.
15. The total cross section for "thermal" neutrons.

Numbers in Parentheses. In columns 5, 6, and 7, numbers in parentheses generally indicate either uncertain or theoretically indicated values, in columns 8, 10, and 11, they indicate the uncertainty in the last digit of the stated value.

Boldface Type. Boldface type is used for the over-all properties (atomic weight, cosmic abundance, and thermal-neutron cross section) of each *element* and for *all stable nuclides.*

Isomeric States. The isomeric states of *unstable* nuclides are indicated by multiple entries in columns 12 and 13; of *stable* nuclides, by single or multiple entries in columns 12 and 13.

Example. The element *germanium* ($Z = 32$) has an atomic weight of 72.60, a cosmic abundance of 50.5 atoms per 10^6 atoms of Si, and a total thermal-neutron-absorption cross section of 2.3×10^{-24} cm² (2.3 barns). Thirteen separate nuclides are listed; of them, five ($A = 70, 72, 73, 74,$ and 76) are *stable.* The stable isotope $A = 73$ has 41 neutrons, a ground-state spin quantum number $I = \frac{9}{2}$, even parity, the corresponding shell state of the odd neutron being $g_{9/2}$. It has a mass *defect* of 49.9 ± 0.1 MeV, or 53.6 mmu, so that its mass is

$$73.0000 - 0.0536 = 72.9464 \text{ amu}$$

It accounts for 7.8 per cent of all germanium, or 3.84 atoms per 10^6 atoms of Si. Its magnetic moment is $\mu = -0.8767 \pm 0.0001$ nuclear magneton, so that its g-factor is $g = -0.8767 \div \frac{9}{2} = -0.1946$; and its electric-quadripole moment is $Q = -0.2 \pm 1.1 \times 10^{-24}$ cm². The neutron-absorption cross section of this isotope is 14 barns. The isotope also can exist in an isomeric state whose half-life is 0.53 s, from which it decays by an isomeric transition γ-ray whose quantum energy is 54 keV.

REFERENCES

"American Institute of Physics Handbook," McGraw-Hill Book Company, Inc., New York, 1957.

General Electric Company, Knolls Atomic Power Laboratory: "Chart of the Nuclides," 5th ed., April, 1956.

Strominger, Hollinger, and Seaborg: Table of Isotopes, *Revs. Modern Phys.*, **30**, 585 (1958).

Suess and Urey: Abundances of the Elements, *Revs. Modern Phys.*, **28**, 53 (1956).

(1) Z	(2) Chem symbol	(3) A	(4) N	(5) I	(6) π	(7) Ground-state config	(8) M-A, MeV	(9) Abundance Per cent	(9) Abundance Cosmic
0	n	1	1	1/2	+	$s_{1/2}$	8.367(2)		
1 Hydrogen	H	1.0080							4.00 × 10¹⁰
		1	0	1/2	+	$s_{1/2}$	7.584(2)	99.985	4.00 × 10¹⁰
	D	2	1	1	+	$(1/2, 1/2)_1$	13.725(3)	0.015	5.7 × 10⁶
	T	3	2	1/2			15.835(5)		
2 Helium	He	4.003							3.08 × 10⁹
		3	1	1/2	(+)	$s_{1/2}$	15.817(5)	0.00013	—
		4	2	0	+		3.607(3)	~100	3.08 × 10⁹
		5	3				12.93(3)		
		6	4				19.40(3)		
3 Lithium	Li	6.940							100
		5	2				13.0(3)		
		6	3	1	+	$(3/2, 3/2)_1$	15.862(5)	7.4	7.4
		7	4	3/2	(—)	$p_{3/2}$	16.977(6)	92.6	92.6
		8	5				23.310(5)		
		9	6						
4 Beryllium	Be	9.013							20
		7	3				17.840(5)		
		8	4				7.309(5)		
		9	5	3/2	(—)	$p_{3/2}$	14.010(6)	100	20
		10	6				15.566(7)		
5 Boron	B	10.82							24
		8	3				24.9(4)		
		9	4				15.081(6)		
		10	5	3	+	$(3/2, 3/2)_3$	15.010(6)	18.8	4.5
		11	6	3/2	(—)	$p_{3/2}$	11.914(6)	81.2	19.5
		12	7				16.917(6)		
6 Carbon	C	12.011							3.54 × 10⁶
		10	4				18.8(1)		
		11	5				13.895(7)		
		12	6	0	(+)	—	3.541(5)	98.89	3.50 × 10⁶
		13	7	1/2	—	$p_{1/2}$	6.963(5)	1.11	3.92 × 10⁴
		14	8	0			7.157(3)		
		15	9				13.19(5)		
7 Nitrogen	N	14.008							6.60 × 10⁶
		12	5				21.2(1)		
		13	6				9.185(5)		
		14	7	1	+	$(1/2, 1/2)_1$	7.002(3)	99.63	6.58 × 10⁶
		15	8	1/2	—	$p_{1/2}$	4.528(5)	0.37	2.41 × 10⁴
		16	9				10.40(1)		
		17	10				13.0(2)		
8 Oxygen	O	16.000							2.14 × 10⁷
		14	6				12.17(2)		
		15	7				7.233(6)		
		16	8	0	+		0(std)	93.759	2.13 × 10⁷
		17	9	5/2	+	$d_{5/2}$	4.222(5)	0.037	8.00 × 10³
		18	10	0	+		4.521(8)	0.204	4.36 × 10⁴
		19	11				8.93(1)		
9 Fluorine	F	19.00							1600
		17	8				6.989(4)		
		18	9				6.19(1)		
		19	10	1/2	+		4.142(7)	100	1600
		20	11				5.90(1)		
		21	12						

(10) μ (nucl mag)	(11) Quad mom, 10^{-24} cm^2	(12) Half-life ($T_{1/2}$)	(13) Decay — Type and energy	(Q)	(14) Binding energy of last neutron E_n, MeV	(15) σ_{tn}, 10^{-24} cm^2	
$-1.91315(7)$		13 m	β^- 0.78	(0.78)	—		
$+2.7926(1)$					2.25	0.33 0.33	H 1
$+0.85735(1)$	$+0.00274(2)$				2.225	0.00057	2
$+2.9788$		12.26 y	β^- 0.018	(0.018)			3
$-2.127414(3)$					6.255	0.007 5400	He 3
	(0)				20.58	0	4
		2×10^{-21} s	n $+\ \alpha$				5
		0.82 s	β^- 3.50	(3.50)	1.90		6
		$\sim 10^{-21}$ s	p $+\ \alpha$			71	Li 5
$+0.82189(4)$	$(+0.02)$				5.3	950(nα)	6
$+3.2559(1)$					7.24	0.033	7
		0.84 s	β^- 13; 2α 3	(16.0)	2.034		8
		0.17 s	β^-; n $+ 2\alpha$	(14)			9
		53 d	K	(0.86)		0.010	Be 7
		$<4 \times 10^{-15}$ s	2α 0.09		18.88	$\sim 10^4$(np)	8
$-1.177(1)$	(0.02)				1.66	0.010	9
		2.7×10^6 y	β^- 0.56	(0.56)	6.816		10
		0.5 s	β^+ 14; 2α 3	(18)		755	B 8
		$>3 \times 10^{-19}$ s	p $+ 2\alpha$		18.4		9
$+1.800(1)$	$+0.074(5)$				8.5	4020(np)	10
$+2.6886(3)$	$+0.036(2)$				11.46	<0.05	11
		0.025 s	β^- 13.4, 9.0	(13.4)	3.364		12
		19 s	β^+ 1.9	(3.6)		0.0032	C 10
		20.5 m	β^+ 0.96	(1.98)	13.38		11
					18.77	0.0032	12
$+0.7023(2)$					4.95	0.0009	13
		5568 y	β^- 0.158	(0.158)	8.169	$<10^{-6}$	14
		2.3 s	β^- 4.3, 9.8	(9.8)	2.2		15
		0.012 s	β^+ 16.7; 3α $\sim$4	(17.7)		1.88	N 12
$+0.40365(3)$	$+0.02$	10.0 m	β^+ 1.20	(2.22)	20.45		13
$-0.28299(3)$					10.55	1.8(np)	14
					10.83	0.00002	15
		7.4 s	β^- 4.10, . . .	(10.4)	2.6		16
		4.14 s	β^- 3.7; (n 1.0)	(8.8)	5.6		17
		72 s	β^+ 1.83	(5.15)		<0.0002	O 14
		2.1 m	β^+ 1.7	(2.7)	13.2		15
$-1.8935(2)$	-0.027				15.6	<0.00002	16
					4.14	0.5(nα)	17
					8.06	0.00022	18
		29 s	β^- 3.2, 4.4	(4.79)	3.96		19
		66 s	β^+ 1.75	(2.77)		0.009	F 17
		1.87 h	β^+ 0.65	(1.67)	9.13		18
$+2.6273$					10.3	0.009	19
		11 s	β^- 5.42	(7.05)	6.60		20
		5 s	β^-	(5.7)			21

(1) Z	(2) Chem symbol	(3) A	(4) N	(5) I	(6) π	(7) Ground-state config	(8) M-A, MeV	(9) Abundance Per cent	(9) Abundance Cosmic
10 *Neon*	**Ne**	**20.183**							**8.6 × 10⁶**
		18	8				10.4(2)		
		19	9				7.40(1)		
		20	**10**	**(0)**	**(+)**		**−1.15(1)**	**90.8**	**7.74 × 10⁶**
		21	**11**	**3/2**	**(+)**	$(d_{5/2})^3_{3/2}$	**0.46(1)**	**0.26**	**2.58 × 10⁴**
		22	**12**	**(0)**	**(+)**		**−1.53(1)**	**8.9**	**8.36 × 10⁵**
		23	13				1.64(1)		
		24	14						
11 *Sodium*	**Na**	**22.991**							**4.38 × 10⁴**
		20	9				14.2(2)		
		21	10				3.99(4)		
		22	11	3			1.31(2)		
		23	**12**	**3/2**	**+**	$(d_{5/2})^3_{3/2}$	**−2.74(1)**	**100**	**4.38 × 10⁴**
		24	13	4			−1.34(2)		
		25	14				−2.1(2)		
12 *Magnesium*	**Mg**	**24.32**							**9.12 × 10⁵**
		23	11				1.35(1)		
		24	**12**	**(0)**	**(+)**		**−6.85(2)**	**78.8**	**7.21 × 10⁵**
		25	**13**	**5/2**	**(+)**	$d_{5/2}$	**−5.82(2)**	**10.1**	**9.17 × 10⁴**
		26	**14**	**(0)**	**(+)**		**−8.57(2)**	**11.1**	**1.00 × 10⁵**
		27	15				−6.64(2)		
		28	16				−6.78(3)		
13 *Aluminum*	**Al**	**26.98**							**9.48 × 10⁴**
		24	11				7.2(3)		
		25	12				−1.57(6)		
		26	13				−4.54(2)		
		27	**14**	**5/2**	**(+)**	$d_{5/2}$	**−9.23(2)**	**100**	**9.48 × 10⁴**
		28	15				−8.59(2)		
		29	16				−9.4(1)		
14 *Silicon*	**Si**	**28.09**							**1.00 × 10⁶**
		26	12						
		27	13				−4.41(2)		
		28	**14**	**(0)**	**(+)**		**−13.25(2)**	**92.17**	**9.22 × 10⁵**
		29	**15**	**1/2**	**(+)**	$s_{1/2}$	**−13.35(2)**	**4.71**	**4.70 × 10⁴**
		30	**16**	**(0)**	**(+)**		**−15.60(2)**	**3.12**	**3.12 × 10⁴**
		31	17				−13.83(2)		
		32	18				−14.77(6)		
15 *Phosphorus*	**P**	**30.975**							**1.00 × 10⁴**
		28	13				+0.5(3)		
		29	14				−8.30(2)		
		30	15				−11.28(5)		
		31	**16**	**1/2**	**(+)**	$s_{1/2}$	**−15.31(2)**	**100**	**1.00 × 10⁴**
		32	17	1			−14.87(3)		
		33	18				−16.62(3)		
		34	19				−14.8(2)		
16 *Sulfur*	**S**	**32.066**							**3.75 × 10⁵**
		31	15				−9.87(8)		
		32	**16**	**0**	**+**		**−16.58(3)**	**95.0**	**3.56 × 10⁵**
		33	**17**	**3/2**	**+**	$d_{3/2}$	**−16.86(3)**	**0.75**	**2.77 × 10³**
		34	**18**	**0**	**+**		**−19.89(5)**	**4.2**	**1.57 × 10⁴**
		35	19	3/2			−18.54(4)		
		36	**20**	**(0)**	**(+)**		**−20.1(1)**	**0.017**	**51**
		37	21				−16.7(3)		
17 *Chlorine*	**Cl**	**35.457**							**8850**
		32	15				−3.5(4)		
		33	16				−11.4(2)		
		34	17				−14.37(6)		

(10)	(11)	(12)	(13)		(14)	(15)	
μ (nucl mag)	Quad mom, 10^{-24} cm²	Half-life ($T_{1/2}$)	Decay		Binding energy of last neutron E_n, MeV	σ_{tn}, 10^{-24} cm²	
			Type and energy	(Q)			
		1.6 s	β^+ 3.2	(4.2)		<1	**Ne** 18
		18.5 s	β^+ 2.2	(3.2)			19
					16.86		**20**
−0.6614					6.75		21
					10.36		**22**
		40 s	β^- 4.2, 3.8, . . .	(4.2)	5.19		23
		3.4 m	β^- 1.95, . . .	(2.42)			24
		0.3 s	β^+; $\alpha > 2$	(15)		0.53	**Na** 20
		23 s	β^+ 2.50	(3.52)			21
+1.746		2.6 y	β^+ 0.54, . . . ; K	(2.84)			22
+2.2161					12.25	0.53	**23**
+1.69		15.0 h	β^- 1.39, . . .	(5.51)	6.95		24
		60 s	β^- 4.0, 3.4, . . . , 2.6	(4.0)	9.2		25
		12 s	β^+ 3.0	(4.0)		0.063	**Mg** 23
					16.6	0.03	**24**
−0.8552					7.33	0.27	25
					11.11	0.03	**26**
		9.5 m	β^- 1.75, 1.57	(2.59)	6.44		27
		21.3 h	β^- 0.45	(1.83)			28
		2.1 s	β^+ 8.5, . . .	(14)		0.23	**Al** 24
		7.3 s	β^+ 3.24	(4.26)	17.0		25
		{ 6.5 s / ~10^6 y }	β^+ 3.21 (4.23); β^+ 1.2 (4.0)		11.5		26
+3.6408(4)	+0.149(2)				12.99	0.23	**27**
		2.30 m	β^- 2.87	(4.65)	7.72		28
		6.6 m	β^- 2.5, 1.4	(3.8)			29
						0.13	**Si** 26
		1.7 s	β^+				27
		4.4 s	β^+ 3.8	(4.8)	16.8	0.1	**28**
−0.55492(4)					8.47	0.3	29
					10.60	0.11	**30**
		2.62 h	β^- 1.48, . . .	(1.48)	6.59		31
		~300 y	β^- 0.1	(0.1)			32
		0.28 s	β^+ 11, 8, . . .	(14)		0.20	**P** 28
		4.5 s	β^+ 3.94	(4.96)			29
		2.5 m	β^+ 3.3	(4.3)	11.2		30
+1.1316(2)					12.1	0.20	**31**
−0.2523		14.5 d	β^- 1.71	(1.71)	7.93		32
		25 d	β^- 0.25	(0.25)	10.09		33
		12.4 s	β^- 5.1, 3.2	(7.2)			34
		2.6 s	β^+ 4.4	(5.4)		0.49	**S** 31
					1.47		**32**
+0.6429(2)	−0.06(1)				8.64	0.002(np)	**33**
					10.9	0.26	34
	+0.045	87 d	β^- 0.167	(0.167)	7.0		35
					9.2	0.14	**36**
		5.0 m	β^- 4.7, 1.6	(4.7)			37
						33	**Cl** 32
		0.31 s	β^+ 10, 8, . . .	(13)			33
		2.8 s	β^+ 4.2	(5.2)			34
		{ 32.4 m / 1.5 s }	β^+ 2.5, 1.4, . . . ; IT 0.14 (5.5); β^+ 4.5		10.8		

(1)	(2)	(3)	(4)	(5)	(6)	(7)	(8)	(9)	
								Abundance	
Z	Chem symbol	A	N	I	π	Ground-state config	$M\text{-}A$, MeV	Per cent	Cosmic
17	Cl	35.457							
		35	18	3/2	+	$d_{3/2}$	−18.71(3)	75.53	6670
		36	19	2			−18.91(4)		
		37	20	3/2	(+)	$d_{3/2}$	−20.91(5)	24.47	2180
		38	21				−18.66(6)		
		39	22				−18.79(8)		
		40	23						
18	A	39.944							1.50 × 10⁵
	Argon	35	17				−13.30(6)		
		36	18	(0)	(+)		−19.63(4)	0.337	1.26 × 10⁵
		37	19				−20.10(5)		
		38	20	(0)	(+)		−23.48(5)	0.063	2.4 × 10⁴
		39	21				−21.75(7)		
		40	22	(0)	(+)		−23.23(5)	99.60	−
		41	23				−20.92(6)		
		42	24						
19	K	39.100							3160
	Potassium	38	19				−17.60(6)		
		39	20	3/2	(+)	$d_{3/2}$	−22.31(7)	93.2	2940
		40	21	4	(−)	(3/2, 7/2)₄	−21.74(6)	0.0119	0.38
		41	22	3/2	+	$d_{3/2}$	−23.50(6)	6.8	219
		42	23	2			−22.51(7)		
		43	24				−23.90(8)		
		44	25				−22.4(2)		
		45	26						
20	Ca	40.08							4.90 × 10⁴
	Calcium	39	19				−15.5(4)		
		40	20	(0)	(+)		−23.07(6)	96.9	4.75 × 10⁴
		41	21				−23.07(6)		
		42	22	(0)	(+)		−26.18(6)	0.64	314
		43	23	7/2	(−)	$f_{7/2}$	−25.75(8)	0.14	64
		44	24	(0)	(+)		−28.55(8)	2.1	1040
		45	25				−27.61(8)		
		46	26	(0)	(+)			0.0032	1.6
		47	27				−28.4(1)		
		48	28	(0)	(+)		−30.1(2)	0.18	87.7
		49	29				−26.9(2)		
21	Sc	44.96							2.8
	Scandium	40	19				−9.1(5)		
		41	20				−17.11(7)		
		42	21						
		43	22				−23.53(8)		
		44	23				−24.90(8)		
		45	24	7/2	(−)	$f_{7/2}$	−27.87(8)	100	2.8
		46	25				−28.41(8)		
		47	26				−30.5(1)		
		48	27				−30.3(1)		
		49	28				−32.1(1)		
		50	29						
22	Ti	47.90							2440
	Titanium	43	21						
		44	22						
		45	23				−25.82(8)		
		46	24	(0)	(+)		−30.77(8)	8.0	194
		47	25	5/2	(−)	$(f_{7/2})^5_{5/2}$	−31.19(8)	7.4	189
		48	26	(0)	(+)		−34.34(8)	73.8	1790
		49	27	7/2	(−)	$f_{7/2}$	−34.09(8)	5.5	134
		50	28	(0)	(+)		−36.71(8)	5.3	130
		51	29				−34.8(1)		

(10) μ (nucl mag)	(11) Quad mom, 10^{-24} cm²	(12) Half-life ($T_{1/2}$)	(13) Decay Type and energy	(13) (Q)	(14) Binding energy of last neutron E_n, MeV	(15) σ_{in}, 10^{-24} cm²	
							Cl
+0.8219(2)	−0.07894(2)				12.8	{ 44 , 0.30(np)	**35**
+1.284	−0.017	3.1 × 10⁵y	β⁻ 0.71; K	(0.71⁻)		∼90	**36**
+0.6841(2)	−0.06213(2)				9.9	0.56	**37**
		{ 1 s , 37.3 m	IT 0.66, β⁻ 4.8, 2.7, 1.1	(4.8) }	6.3		**38**
		55 m	β⁻ 3.0, 1.7	(3.3)	8.4		**39**
		1.4 m	β⁻ ∼7				**40**
						0.6	**A 35**
		1.8 s	β⁺ 4.95	(5.97)			**35**
					14.7	6	**36**
		35 d	K, L	(0.82)	8.82		**37**
					11.8	0.8	**38**
		260 y	β⁻ 0.57	(0.57)	6.9		**39**
					10.0	0.53	**40**
		1.82 h	β⁻ 1.20, 2.49	(2.49)	6.1	>0.06	**41**
		>3.5 y	β⁻				**42**
						2.0	**K**
		{ 0.95 s , 7.7 m	β⁺ 5.1 , β⁺ 2.7	(6.1) , (5.8) }			**38**
+0.39087(1)	+0.1						
					13.2	1.9	**39**
−1.2982(4)		1.3 × 10⁹y	β⁻ 1.33; K	(1.33⁻)	7.80	70	**40**
+0.21453(3)					10.2	1.1	**41**
−1.137		12.5 h	β⁻ 3.5, 2.0, . . .	(3.5)	7.34		**42**
		22 h	β⁻ 0.83, . . .	(1.84)			**43**
		22 m	β⁻ 4.9, 1.5				**44**
		34 m	β⁻				**45**
						0.43	**Ca 39**
		1.0 s	β⁻ 5.7	(6.7)			**39**
					15.4	0.2	**40**
		1.1 × 10⁵y	K	(0.44)	8.37		**41**
					11.4	40	**42**
−1.3152(2)					7.93		**43**
						0.6	**44**
		160 d	β⁻ 0.25	(0.25)			**45**
						0.3	**46**
		4.7 d	β⁻ 0.7, 2.0	(2.0)			**47**
						1.1	**48**
		8.7 m	β⁻ 2.0, 1.0	(5.1)	5.0		**49**
						23	**Sc 40**
		0.2 s	β⁺ 0.9	(14)			**40**
		0.87 s	β⁺ 5	(6)			**41**
		0.66 s	β⁺ ∼5	(∼6)			**42**
		3.9 h	β⁺ 1.19, 0.82, 0.39	(2.21)			**43**
		{ 2.4 d , 4.0 h	IT 0.271 , β⁺ 1.47; K	(3.65) }			**44**
+4.7563(1)						23	**45**
		{ 20 s , 85 d	IT 0.14 , β⁻ 0.36, . . .	(2.36) }	8.8	0.25	**46**
		3.4 d	β⁻ 0.44, 0.60	(0.60)	10.5		**47**
		44 h	β⁻ 0.64	(4.0)	7.98		**48**
		57 m	β⁻ 2.0	(2.0)	10.3		**49**
		1.7 m	β⁻ ∼3.5	(∼6.3)			**50**
						6.0	**Ti 43**
		0.6 s	β⁺				**43**
		>20 y	K				**44**
		3.08 h	β⁺ 1.02; K	(2.04)			**45**
					13.3	0.6	**46**
−0.7871(1)					8.7	1.6	**47**
					11.4	7.8	**48**
−1.1022					8.1	1.8	**49**
					10.8	0.14	**50**
		5.80 m	β⁻ 2.1, 1.5	(2.4)	6.3		**51**

(1)	(2)	(3)	(4)	(5)	(6)	(7)	(8)	(9)	
Z	Chem symbol	A	N	I	π	Ground-state config	$M\text{-}A$, MeV	Abundance	
								Per cent	Cosmic
23 **V** *Vanadium*		**50.95**							220
		45	22						
		46	23				$-23.4(4)$		
		47	24				$-28.28(8)$		
		48	25				$-30.31(8)$		
		49	26	7/2			$-33.48(8)$		
		50	27	6	(+)	$(7/2, 7/2)_6$	$-34.3(2)$	0.25	0.55
		51	**28**	**7/2**	**(−)**	$\mathbf{f_{7/2}}$	$\mathbf{-37.2(1)}$	**99.75**	**220**
		52	29				$-36.1(1)$		
		53	30						
		54	31						
24 **Cr** *Chromium*		**52.01**							7800
		46	22						
		47	23						
		48	24						
		49	25				$-30.92(8)$		
		50	**26**	(0)	(+)		$\mathbf{-35.7(2)}$	**4.4**	**344**
		51	27				$-36.5(1)$		
		52	**28**	(0)	(+)		$\mathbf{-40.0(1)}$	**83.7**	**6510**
		53	**29**	3/2	(−)	$\mathbf{p_{3/2}}$	$\mathbf{-39.6(1)}$	**9.5**	**744**
		54	**30**	(0)	(+)		$\mathbf{-41.0(1)}$	**2.4**	**204**
		55	31				$-38.7(2)$		
25 **Mn** *Manganese*		**54.94**							6850
		50	25				$-27.9(5)$		
		51	26				$-33.2(1)$		
		52	27	6?			$-35.3(1)$		
		53	28	7/2			$-39.0(1)$		
		54	29				$-39.8(2)$		
		55	**30**	5/2	(−)	$\mathbf{(f_{7/2})^5_{5/2}}$	$\mathbf{-41.5(2)}$	**100**	**6850**
		56	31				$-40.4(2)$		
		57	32						
26 **Fe** *Iron*		**55.85**							6.00×10^5
		52	26				$-33.3(2)$		
		53	27				$-35.2(2)$		
		54	**28**	(0)	(+)		$\mathbf{-40.4(1)}$	**5.9**	3.54×10^4
		55	29				$-41.3(2)$		
		56	**30**	(0)	(+)		$\mathbf{-44.1(2)}$	**91.6**	5.49×10^5
		57	**31**	(1/2)	(−)	$\mathbf{p_{1/2}}$	$\mathbf{-43.4(2)}$	**2.20**	1.35×10^4
		58	**32**	(0)	(+)		$\mathbf{-45.2(2)}$	**0.33**	**1980**
		59	33				$-43.2(3)$		
		60	34						
		61	35						
27 **Co** *Cobalt*		**58.94**							1800
		54	27				$-31.5(7)$		
		55	28				$-37.8(2)$		
		56	29	4			$-39.5(2)$		
		57	30	7/2			$-42.9(3)$		
		58	31				$-42.9(2)$		
		59	**32**	7/2	(−)	$\mathbf{f_{7/2}}$	$\mathbf{-44.8(3)}$	**100**	**1800**
		60	33				$-43.9(3)$		
		61	34				$-45.4(3)$		
		62	35				$-43.8(4)$		
28 **Ni** *Nickel*		**58.71**							2.74×10^4
		56	28						
		57	29				$-39.6(3)$		
		58	**30**	(0)	(+)		$\mathbf{-43.0(3)}$	**68.0**	1.86×10^4
		59	31				$-43.7(3)$		
		60	**32**	(0)	(+)		$\mathbf{-46.7(3)}$	**26.2**	**7170**

(10) μ (nucl mag)	(11) Quad mom, 10^{-24} cm²	(12) Half-life ($T_{1/2}$)	(13) Decay — Type and energy	(Q)	(14) Binding energy of last neutron E_n, MeV	(15) σ_{tn}, 10^{-24} cm²	
						4.9	**V**
		~1 s	β^+				45
		0.4 s	β^+ > 6				46
		31 m	β^+ 1.89	(2.89)			47
		16.2 d	β^+ 0.69; K	(4.02)			48
		~1 y	K	(0.62)	11.5		49
+3.3412(3)		4 × 10^{14} y	K		9.1	~100	50
+5.1478(5)	+0.3(2)				11.0	4.5	51
		3.77 m	β^- 2.6	(4.0)	7.3		52
		2.0 m	β^- 2.50				53
		55 s	β^- 3.3				54
						3.1	**Cr**
		1.1 s	β^+				46
							47
		23 h	K				48
		42 m	β^+ 1.54	(2.56)			49
					13.4	16	50
		27 d	K	(0.75)	9.1		51
					11.8	0.8	52
−0.4735(6)					7.93	18	53
					9.72	0.37	54
		3.6 m	β^- 2.8	(2.8)	5.9		55
						13.3	**Mn**
		0.28 s	β^+ > 6				50
		45 m	β^+ 2.2	(3.2)			51
		{ 21 m	β^+ 2.7; IT 0.39		10.5		52
		{ 5.7 d	K; β^+ 0.6	(4.7)			
±5.05		~140 y	K	(0.60)	12.0		53
		{ 2 m	IT?		8.9		54
		{ 300 d	K	(1.38)			
+3.4681(4)	+0.5				10.1	13.3	55
		2.58 h	β^- 2.8, 1.0, 0.7	(3.7)	7.3		56
		1.7 m	β^- 2.6	(2.7)			57
						2.5	**Fe**
		8 h	β^+ 0.80; K	(2.21)			52
		9 m	β^+ 2.6	(3.6)	10.5		53
					13.8	2.2	54
		2.9 y	K	(0.22)	9.3		55
					11.1	2.6	56
+0.05					7.6	2.4	57
					10.1	0.9	58
		45 d	β^- 0.46, 0.27, 1.56	(1.56)	6.4		59
		~3 × 10^5 y	β^-				60
		5.5 m	β^-				61
						37	**Co**
		0.18 s	β^+ > 7				54
±3.85		18 h	β^+ 1.50, 1.0, . . . ; K	(3.45)			55
±4.65		77 d	K; β^+ 1.50, . . .	(4.62)			56
		267 d	K	(0.5)			57
		{ 9 h	IT 0.025		9.0		58
+4.648(2)	+0.5(2)	{ 71 d	K; β^+ 0.48, . . .	(2.31)	10.2	37	59
		{ 10.5 m	IT 0.059; β^- 1.5		7.49	{ ~100 }	60
		{ 5.2 y	β^- 0.31, . . .	(2.81)		{ 6 }	
		1.65 h	β^- 1.22	(1.29)	9.96		61
		{ 1.6 m	β^-				62
		{ 14 m	β^- 2.8				
						4.6	**Ni**
		6.4 d	K				56
		36 h	K; β^+ 0.84	(3.24)			57
					11.7	4.3	58
		8 × 10^4 y	K	(1.07)	9.00		59
					11.4	2.6	60

(1)	(2)	(3)	(4)	(5)	(6)	(7)	(8)	(9)	
								Abundance	
Z	Chem symbol	A	N	I	π	Ground-state config	$M\text{-}A$, MeV	Per cent	Cosmic
28	Ni	58.71							
		61	33	(3/2)	(−)	$p_{3/2}$	−46.8(3)	1.1	342
		62	34	(0)	(+)		−48.8(3)	3.7	1000
		63	35				−47.0(2)		
		64	36	(0)	(+)		−48.3(2)	1.0	318
		65	37				−46.0(2)		
		66	38						
29	Cu	63.54							212
Copper		58	29				−33(1)		
		59	30						
		60	31	2			−40.4(3)		
		61	32	3/2			−44.6(3)		
		62	33				−44.9(3)		
		63	34	3/2	−	$p_{3/2}$	−47.1(2)	69.0	146
		64	35	1			−46.6(2)		
		65	36	3/2	−	$p_{3/2}$	−48.05(2)	31.0	66
		66	37				−46.8(2)		
		67	38				−47.4(2)		
		68	39						
30	Zn	65.38							486
Zinc		60	30						
		61	31						
		62	32				−43.2(3)		
		63	33				−43.7(2)		
		64	34	(0)	(+)		−47.2(2)	48.9	238
		65	35				−46.7(2)		
		66	36	(0)	(+)		−49.4(2)	27.8	134
		67	37	5/2	−	$f_{5/2}$	−48.0(2)	4.1	20.0
		68	38	(0)	(+)		−49.8(2)	18.6	90.9
		69	39				−47.9(2)		
		70	40	(0)	(+)		−48.9(2)	0.63	3.35
		71	41				−46.7(3)		
		72	42						
31	Ga	69.72							11.4
Gallium		64	33				−39.9(6)		
		65	34				−43.6(2)		
		66	35				−44.2(2)		
		67	36	3/2			−46.9(2)		
		68	37				−46.9(2)		
		69	38	3/2	−	$p_{3/2}$	−48.8(2)	60.1	6.86
		70	39				−48.3(2)		
		71	40	3/2	−	$p_{3/2}$	−49.0(2)	39.9	4.54
		72	41	3			−47.6(3)		
		73	42				−48.5(2)		
32	Ge	72.60							50.5
Germanium		66	34						
		67	35				−42.5(4)		
		68	36						
		69	37				−46.6(2)		
		70	38	(0)	(+)		−49.9(2)	20.5	10.4
		71	39				−48.8(2)		
		72	40	(0)	(+)		−51.6(3)	27.4	13.8
		73	41	9/2	+	$g_{9/2}$	−49.9(1)	7.8	3.84
		74	42	(0)	(+)		−51.6(1)	36.5	18.65
		75	43				−49.7(1)		
		76	44	(0)	(+)		−50.9(2)	7.8	3.87
		77	45				−48.4(1)		
		78	46						

(10) μ (nucl mag)	(11) Quad mom, 10^{-24} cm²	(12) Half-life ($T_{1/2}$)	(13) Decay — Type and energy	(Q)	(14) Binding energy of last neutron E_n, MeV	(15) σ_{tn}, 10^{-24} cm²	
							Ni
					8.53	2	61
						15	62
		80 y	β^- 0.063	(0.063)	6.7		63
					9.7	2	64
		2.56 h	β^- 2.10, 0.6, 1.0	(2.10)	6.0		65
		56 h	β^- 0.3				66
						3.7	**Cu**
		{ 9.5 m	β^+ <0.7	}			58
		{ 3 s	β^+ ~8	}			
		81 s	β^+ 3.7				59
		24 m	β^+ 2.0, 3.0, 3.9	(6.3)			60
		3.3 h	β^+ 1.22, . . . ; K	(2.24)	10.6		61
		9.9 m	β^+ 2.9	(3.9)			62
+2.2262(4)	−0.157				10.7	4.4	63
± 0.40		12.8 h	K; β^- 0.57; β^+ 0.66	$\begin{bmatrix}(0.57-)\\(1.68+)\end{bmatrix}$	7.91		64
+2.3845(4)	−0.145				9.8	2.2	65
							6
		5.1 m	β^- 2.63, 1.59	(2.63)	7.1	140	6
		61 h	β^- 0.40, 0.48, 0.58	(0.58)	9.1		67
		32 s	β^- 3.0				68
						1.10	**Zn**
		2.1 m					60
		1.5 m					61
		9 h	K; β^+ 0.66	(1.7)			62
		38 m	β^+ 2.36, 1.40, . . . ; K	(3.38)	9.0		63
					11.8	0.5	64
		245 d	K; β^+ 0.33	(1.35)	7.9		65
					11.1		66
+0.8738(1)	+0.18				7.0		67
					10.1	1.1	68
		{ 14 h	IT 0.44	}			69
		{ 52 m	β^- 0.90	(0.90) }	6.3		
					9.2	0.09	70
		{ 3 h	β^- 1.5	(3.0) }			71
		{ 2.2 m	β^- 2.4	(2.9) }			
		49 h	β^- 0.3, 1.6				72
						2.9	**Ga**
		2.5 m	β^+ ~5	(7)			64
		{ 15 m	IT 0.052; β^+ 2.5	}			65
		{ 8 m	β^+ 2.2	(3.2) }			
		9.4 h	β^+ 4.15, . . . ; K	(5.17)			66
+1.84	+0.21	78 h	K	(1.00)	11.2		67
		68 m	β^+ 1.88, . . . , 0.78; K	(2.90)	8.3		68
+2.16(1)	+0.232(2)				10.1	1.9	69
		21 m	β^- 1.65, . . .	(1.65)			70
+2.561	+0.146(2)				9.1	4.6	71
± 0.12		14.1 h	β^- 0.64, . . . , 3.17	(4.00)			72
		5 h	β^- 1.4	(1.5)			73
						2.3	**Ge**
		2.5 h	K; β^+				66
		19 m	β^+ 3.4				67
		250 d	K				68
		40 h	K; β^+ 1.21, . . . , 0.6, . . .				69
					3.4		70
		12 d	K	(0.23)		1.0	71
							72
−0.8767(1)	−0.2(11)	(0.53 s)	IT 0.054	)	14		73
						0.7	74
		{ 49 s	IT 0.14	}			75
		{ 82 m	β^- 1.18, 0.92, . . .	(1.18) }	6.5		
						0.32	76
		{ 52 s	β^- 2.9, . . . ; IT 0.16	}			77
		{ 12 h	β^- 2.20	(2.7) }			
		86 m	β^- 0.9				78

(1) Z	(2) Chem symbol	(3) A	(4) N	(5) I	(6) π	(7) Ground-state config	(8) M-A, MeV	(9) Abundance Per cent	(9) Abundance Cosmic
33	As	74.91							4.0
Arsenic		68	35						
		69	36						
		70	37						
		71	38				−46.8(2)		
		72	39				−47.2(3)		
		73	40				−49.5(1)		
		74	41				−49.0(1)		
		75	42	3/2	−	p₃/₂	−50.8(1)	100	4.0
		76	43	2			−49.8(1)		
		77	44				−51.1(1)		
		78	45				−49.8(2)		
		79	46				−50.2(2)		
		80	47						
34	Se	78.96							67.6
Selenium		70	36						
		71	37						
		72	38						
		73	39				−46.7(1)		
		74	40	0	(+)		−50.4(1)	0.93	0.649
		75	41	5/2			−50.0(1)		
		76	42	(0)	(+)		−52.8(1)	9.1	6.16
		77	43	1/2	−	p₁/₂	−51.8(1)	7.5	5.07
		78	44	0	(+)		−53.9(1)	23.6	16.0
		79	45	7/2			−52.5(2)		
		80	46	(0)	(+)		−54.0(2)	49.9	33.8
		81	47				−52.5(2)		
		82	48	(0)	(+)		−53.4(2)	9.0	5.98
		83	49						
		84	50						
35	Br	79.916							13.4
Bromine		74	39						
		75	40				−47.2(1)		
		76	41				−48.2(1)		
		77	42				−50.5(1)		
		78	43				−50.4(2)		
		79	44	3/2	−	p₃/₂	−52.7(2)	50.6	6.88
		80	45				−52.1(2)		
		81	46	3/2	−	p₃/₂	−53.9(2)	49.4	6.62
		82	47	5			−53.4(2)		
		83	48				−54.6(2)		
		84	49				−52.9(2)		
		85	50				−53.4(2)		
		86	51						
		87	52				−46.7(6)		
		88	53						
		89	54						
36	Kr	83.80							51.3
Krypton		76	40						
		77	41				−47.6(1)		
		78	42	(0)	(+)		−51.3(2)	0.35	0.175
		79	43				−51.1(2)		
		80	44	(0)	(+)		−54.1(2)	2.27	1.14
		81	45				−53.8(3)		
		82	46	(0)	(+)		−56.4(2)	11.6	5.90
		83	47	9/2	(+)	g₉/₂	55.6(2)	11.5	5.89

(10) μ (nucl mag)	(11) Quad mom, 10^{-24} cm²	(12) Half-life ($T_{1/2}$)	(13) Decay — Type and energy	(Q)	(14) Binding energy of last neutron E_n, MeV	(15) σ_{tn}, 10^{-24} cm²	
						4.3	**As**
		~7 m	β^+				68
		15 m	β^+ 2.9				69
		50 m	β^+ 1.4, 2.5	(6.6)			70
		62 h	β^+ 0.81, . . .				71
		26 h	K; β^+ 2.50, 3.34, . . .	(4.36)			72
		76 d	K;	(0.37)			73
		17 d	K; β^- 1.36, 0.72; β^+ 0.93, 1.53	$\begin{bmatrix}(1.36-)\\(2.55+)\end{bmatrix}$	10.2		74
+1.4347(3) −0.906	**+0.3(2)** +1.1	(0.018 s)	IT 0.28	)	10.2	4.3	**75**
		26.7 h	β^- 2.96, 2.41, . . .	(2.96)	7.3		76
		39 h	β^- 0.69, . . .	(0.69)			77
		90 m	β^- 4.1, . . .	(4.1)			78
		9 m	β^- 2.3	(2.4)			79
		~36 s	β^-				80
						13	**Se**
		44 m	β^+				70
							71
		9.7 d	K				72
		⎰ 7.1 h ⎱ 44 m	β^+ 1.29, 1.65 β^+ 1.7	(2.74) }			73
						40	**74**
		127 d	K				75
						85	**76**
+0.53326(5)		(17 s	IT 0.16	)	7.4	41	**77**
					10.5	0.4	**78**
−1.015	+0.8	⎰ 3.9 m ⎱ 7 × 10⁴ y	IT 0.096; β^- 0.16	(0.16) }	7.0		79
					9.3	0.5	**80**
		⎰ 57 m ⎱ 18 m	IT 0.103; β^- 1.38	(1.38) }	6.8		81
					9.8	0.05	**82**
		⎰ 69 s ⎱ 25 m	β^- 3.4 β^- 1.5	}			83
		~2 m	β^-				84
						6.6	**Br**
		36 m	β^+; K				74
		1.6 h	K; β^+ 1.70, . . .				75
		17 h	β^+ 3.57; K	(4.59)			76
		57 h	K; β^+ 0.34	(1.36)	10.7		77
		⎰ 6.4 m ⎱ <6 m	IT β^+ 2.4	(3.4) }	8.4		78
+2.1058(4)	**+0.26(8)**				10.6	11.4	**79**
		⎰ 4.6 h ⎱ 18 m	IT 0.05 β^- 2.0, 1.4; K; β^+ 0.86	$\begin{bmatrix}(2.0-)\\(1.9+)\end{bmatrix}$ }	7.3		80
+2.2696(5)	**+0.21(7)**				10.1	2.6	**81**
±1.6	±0.7	35.9 h	β^- 0.46	(3.1)			82
		2.3 h	β^- 0.94, . . .	(0.98)			83
		32 m	β^- 4.68, . . .				84
		3.0 m	β^- 2.5	(2.8)			85
							86
		56 s	β^- 2.6, 8.0; n 0.3;	(8.0)			87
		16 s	β^-				88
		4.5 s	β^-; n 0.5				89
						28	**Kr**
		10 h	K				76
		1.2 h	β^+ 1.86, 1.67; K				77
						>2	**78**
		⎰ 55 s ⎱ 34 h	IT 0.13 K; L; β^+ 0.60, 0.34	(1.62) }			79
					11.3	90	**80**
		⎰ 13 s ⎱ 2 × 10⁵ y	IT 0.19 K	}			81
						40	**82**
−0.9671	**+0.15**	(1.86 h	IT 0.032	)		200	**83**

(1) Z	(2) Chem symbol	(3) A	(4) N	(5) I	(6) π	(7) Ground-state config	(8) M-A, MeV	(9) Abundance Per cent	Cosmic
36	**Kr**	84	48	(0)	(+)		−57.7(2)	57.0	29.3
		85	49	9/2			−56.2(2)		
		86	50	(0)	(+)		−57.5(2)	17.3	8.94
		87	51				−54.7(2)		
		88	52				−53.6(2)		
		89	53				−51.4(4)		
		90	54						
		91	55						
		92	56						
		93	57						
		94	58						
37 *Rubidium*	**Rb**	**85.48**							6.5
		81	44	3/2			−51.6(3)		
		82	45				−52.6(5)		
		83	46	5/2					
		84	47	2			−55.0(2)		
		85	48	5/2	−	$f_{5/2}$	−56.9(2)	72.2	4.73
		86	49				−57.2(2)		
		87	50	3/2	−	$p_{3/2}$	−58.8(2)	27.8	1.77
		88	51				−56.5(2)		
		89	52				−55.4(4)		
		90	53				−54.1(4)		
		91	54						
		92	55						
38 *Strontium*	**Sr**	**87.63**							18.9
		81	43						
		82	44						
		83	45						
		84	46	(0)	(+)		−56.0(3)	0.55	0.106
		85	47				−55.9(5)		
		86	48	(0)	(+)		−59.0(2)	9.8	1.86
		87	49	9/2	+	$g_{9/2}$	−59.1(2)	7.0	1.33
		88	50	(0)	(+)		−61.7(2)	82.7	15.6
		89	51				−59.9(2)		
		90	52				−59.8(2)		
		91	53				−57.1(2)		
		92	54				−56.2(2)		
		93	55						
39 *Yttrium*	**Y**	**88.92**							8.9
		82	43						
		83	44						
		84	45						
		85	46						
		86	47				−54.8(2)		
		87	48				−57.4(3)		
		88	49				−57.9(2)		
		89	50	1/2	−	$p_{1/2}$	−61.4(2)	100	8.9
		90	51				−60.3(2)		
		91	52				−59.8(2)		
		92	53				−58.1(2)		
		93	54				−57.1(3)		
		94	55				−54.4(5)		
		95	56						

(10) μ (nucl mag)	(11) Quad mom, 10^{-24} cm²	(12) Half-life ($T_{1/2}$)	(13) Decay — Type and energy	(13) (Q)	(14) Binding energy of last neutron E_n, MeV	(15) σ_{tn}, 10^{-24} cm²	
						0.16	**Kr 84**
−1.001	+0.25	{ 4.4 h / 10.4 y	β⁻ 0.83; IT 0.31 / β⁻ 0.67, . . .	}	5.95	<15	85
						0.06	**86**
		78 m	β⁻ 3.8, 1.3, . . .	(4.2)	5.53	<600	87
		2.8 h	β⁻ 0.52, 2.7, . . .	(2.9)	6.8		88
		3.2 m	β⁻ 4.2,2				89
		33 s	β⁻ 3.2				90
		10 s	β⁻ 3.6				91
		3 s	β⁻				92
		2 s	β⁻				93
		1 s	β⁻				94
						0.7	**Rb 81**
+2.05		4.7 h	K; β⁺ 1.0	(2.2)			81
		{ 6.3 h / 75 s	K; β⁺ 0.77, . . . / β⁺ 3.2	}			82
+1.42		83 d	K				83
+1.32		{ 21 m / 33 d	IT 0.23, 0.46; K / K; β⁺ 1.7, . . . ; β⁻ .4	[(0.4−)] / [(2.7+)]			84
+1.3532(4)	**+0.3**					**0.8**	**85**
−1.69		{ 1.0 m / 18.6 d	IT 0.56 / β⁻ 1.77, 0.7	(1.77)			86
+2.7501(5)	+0.13	4.3 × 10¹⁰ y	β⁻ 0.27	(0.27)	10.0	0.14	87
		18 m	β⁻ 5.2, 3.3, 2	(5.2)	6.0	<200	88
		15 m	β⁻ 3.9, . . .	(3.9)	7.4		89
		2.7 m	β⁻ 5.7				90
		{ 1.7 m / 14 m	β⁻ 4.6 / β⁻ 3.0	}			91
		∼80 s	β⁻				92
						1.3	**Sr 81**
		29 m	β⁺				81
		26 d	K; β⁺ 3.2				82
		33 h	β⁺ 1.2; K				83
						1	**84**
		{ 70 m / 65 d	IT 0.007; K / K	}	7.5		85
					9.5	>1.3	86
−1.089(2)		(2.8 h	IT 0.39	)	8.4		87
					11.1	0.005	88
		{ ∼10 d / 54 d	IT / β⁻ 1.48, . . .	(1.48)	6.6	<130	89
		28 y	β⁻ 0.54	(0.54)	7.6	1	90
		9.7 h	β⁻ 0.61, . . .	(2.67)	5.7		91
		2.7 h	β⁻ ∼0.55	(1.9)			92
		7 m	β⁻				93
						1.3	**Y 82**
		70 m	β⁺ 2				82
		3.5 h					83
		3.7 h	β⁺ 2.0; K				84
		5 h					85
		15 h	β⁺ 1.80, 1.2	(6.01)			86
		{ 14 h / 80 h	IT 0.38; / K; β⁺ 0.7	(2.1)	10.5		87
−0.14		105 d	K; β⁺ 0.83	(3.7)	9.4		88
		(16 s	IT 0.91	)	11.7	1.3	**89**
		64.0 h	β⁻ 2.27;	(2.27)	6.7	6	90
		{ 50 m / 58 d	IT 0.55; / β⁻ 1.54	(1.54)	7.8		91
		3.5 h	β⁻ 3.60, 2.7, 1.3	(3.60)	6.6		92
		10 h	β⁻ 3.1	(3.1?)	6.8		93
		17 m	β⁻ 5.4	(5.4?)			94
		10 m	β⁻				95

(1) Z	(2) Chem symbol	(3) A	(4) N	(5) I	(6) π	(7) Ground-state config	(8) M-A, MeV	(9) Abundance Per cent	(9) Abundance Cosmic
40	Zr	91.22							54.5
	Zirconium	86	46						
		87	47				−53.9(3)		
		88	48						
		89	49				−58.5(2)		
		90	50	(0)	(+)		−62.5(2)	51.5	28.0
		91	51	5/2	+	$d_{5/2}$	−61.3(2)	11.2	6.12
		92	52	(0)	(+)		−61.6(2)	17.1	9.32
		93	53				−60.2(2)		
		94	54	(0)	(+)		−59.8(4)	17.4	9.48
		95	55				−57.8(4)		
		96	56	(0)	(+)		−57.2(5)	2.8	1.53
		97	57				−54.5(5)		
41	Nb	92.91							1.00
	Niobium (also called *Columbium*, Cb)	89	48						
		90	49				−58.1(2)		
		91	50				−59.9(3)		
		92	51				−60.0(3)		
		93	52	9/2	+	$g_{9/2}$	−60.3(2)	100	1.00
		94	53				−59.1(3)		
		95	54				−58.9(4)		
		96	55				−57.5(4)		
		97	56				−57.2(5)		
		98	57						
		99	58						
42	Mo	95.95							2.42
	Molybdenum	90	48				−55.3(3)		
		91	49				−56.3(3)		
		92	50	(0)	(+)		−60.3(3)	15.7	0.364
		93	51				−59.8(3)		
		94	52	(0)	(+)		−61.1(3)	9.3	0.226
		95	53	5/2	+	$d_{5/2}$	−59.9(4)	15.7	0.382
		96	54	(0)	(+)		−60.6(4)	16.5	0.401
		97	55	5/2	+	$d_{5/2}$	−59.2(5)	9.5	0.232
		98	56	(0)	(+)		−59.1(5)	23.8	0.581
		99	57				−55.9(9)		
		100	58	(0)	(+)		−57.5(5)	9.5	0.234
		101	59						
		102	60						
43	Tc	92	49				−53.9(8)		0
	Technetium	93	50				−56.7(3)		
		94	51				−56.8(4)		
		95	52				−58.2(4)		
		96	53				−57.5(6)		
		97	54						
		98	55						
		99	56	9/2			−57.3(9)		
		100	57						
		101	58						
		102	59						

(10) μ (nucl mag)	(11) Quad mom, 10⁻²⁴ cm²	(12) Half-life (T₁/₂)	(13) Decay: Type and energy	(Q)	(14) Binding energy of last neutron E_n, MeV	(15) σ_{tn}, 10⁻²⁴ cm²	
						0.18	**Zr**
		~17 h	K				86
		1.6 h	β^+ 2.10; K	(3.50)			87
		85 d	K				88
		{ 4.4 m	IT 0.59; β^+ 0.9, 2.4				89
		{ 79 h	K; β^+ 0.90	(2.83)			89
−1.2980		(0.8 s	IT 2.30	)	12.0	0.1	90
					7.2	1	91
					8.7	0.2	92
		9 × 10⁵ y	β^- 0.063, . . .	(0.063)	6.7	<5	93
						0.1	94
		65 d	β^- 0.36, 0.39, 0.88	(1.12)	6.4		95
						0.1	96
		17 h	β^- 1.91, . . .	(2.66)	3.7		97
						1	**Nb**
		{ ~2 h	β^+				89
		{ 1.9 h	β^+ 2.9	(3.9)			89
		{ 24 s	IT 0.12				90
		{ 14.6 h	β^+ 1.50, . .	(3.80)			90
		{ 62 d	IT 0.105				91
		{ long(?)	K	(>1.1)	10.0		91
		{ 13 h	K		8.5		92
+6.1659	**−0.3**	{ 10 d	K	(1.9)			92
		(3.7 y	IT 0.029	)	8.7	>1	93
		{ 6.6 m	IT 0.041; β^- 1.3		7.2	~15	94
		{ 2 × 10⁴ y	β^- 0.5	(2.1)			94
		{ 84 h	IT 0.23		9.2		95
		{ 35 d	β^- 0.16	(0.92)			95
		23 h	β^- 0.7, 0.4	(3.1)	6.9		96
		{ 1 m	IT 0.75		8.1		97
		{ 72 m	β^- 1.27	(1.93)			97
		30 m	β^-				98
		2.5 m	β^- 3.2	(3.2)			99
						2.5	**Mo**
		5.7 h	K; β^+ 1.2				90
		{ 66 s	IT 0.65; β^+ 2.45, 2.78, 3.99				91
		{ 15.6 m	β^+ 3.44	(4.46)			91
					13.3		92
		{ 6.9 h	IT 0.26		7.9		93
		{ >2 y	K	(0.49)			93
					9.9		94
−0.9140(2)					8.0	14	95
					9.1	1	96
−0.9332(1)					6.9	2	97
					8.3	0.13	98
		67 h	β^- 1.23, 0.45, . . .	(1.37)			99
						0.2	100
		15 m	β^- 1.2, 2.2	(2.4)			101
		11 m	β^- 1				102
							Tc
		4.3 m	β^+ 4.1				92
		{ 44 m	IT 0.39; K				93
		{ 2.7 h	K; β^+ 0.80, 0.6	(3.1)			93
		53 m	β^+ 2.41, . . . ; K	(4.30)	8.7		94
		{ 60 d	K; β^+ 0.6; IT 0.039		9.5		95
		{ 20 h	K	(1.6)			95
		{ 52 m	IT 0.034; K; β^+				96
		{ 4.3 d	K	(3.0)			96
		{ 91 d	IT 0.099				97
		{ ~10⁵ y	K				97
		~10⁴ y	β^- 0.3	(1.7)			98
		{ 6.0 h	IT 0.002, 0.142			20	99
+5.657		{ 2.1 × 10⁵ y	β^- 0.29	(0.29)			99
		16 s	β^- 2.8		7.1		100
		15 m	β^- 1.2	(1.5)			101
		5 s	β^- 4				102

(1) Z	(2) Chem symbol	(3) A	(4) N	(5) I	(6) π	(7) Ground-state config	(8) M-A, MeV	(9) Abundance Per cent	(9) Abundance Cosmic
44	Ru	101.1							1.49
Ruthenium		94	50						
		95	51				−56.1(5)		
		96	52	(0)	(+)		−57.8(5)	5.6	0.0846
		97	53						
		98	54	(0)	(+)		−58.5(7)	1.9	0.0331
		99	55	5/2	+	$d_{5/2}$	−57.5(9)	12.7	0.191
		100	56	(0)	(+)			12.7	0.189
		101	57	5/2	+	$d_{5/2}$		17.0	0.253
		102	58	(0)	(+)		−59.2(5)	31.5	0.467
		103	59				−57.1(5)		
		104	60	(0)	(+)		−57.9(7)	18.6	0.272
		105	61				−55.1(5)		
		106	62				−55.7(5)		
		107	63						
		108	64						
45	Rh	102.91							0.214
Rhodium		97	52						
		98	53						
		99	54						
		100	55						
		101	56						
		102	57				−57.0(5)		
		103	58	1/2	−	$p_{1/2}$	−57.8(5)	100	0.214
		104	59				−56.2(5)		
		105	60				−57.1(5)		
		106	61				−55.7(5)		
		107	62				−55.6(4)		
		108	63						
46	Pd	106.4							0.675
Palladium		98	52						
		99	53						
		100	54						
		101	55						
		102	56	(0)	(+)		−58.2(5)	1.0	0.0054
		103	57				−57.3(5)		
		104	58	(0)	(+)		−58.8(5)	11.0	0.0628
		105	59	5/2	+	$d_{5/2}$	−57.7(5)	22.2	0.1536
		106	60	(0)	(+)		−55.2(5)	27.3	0.1839
		107	61				−56.8(4)		
		108	62	(0)	(+)		−57.9(4)	26.7	0.180
		109	63				−55.4(4)		
		110	64	(0)	(+)		−56.2(5)	11.8	0.0911
		111	65				−53.2(5)		
		112	66				−53.3(5)		
		113	67						
47	Ag	107.880							0.26
Silver		102	55						
		103	56						
		104	57	2			−55.1(5)		
		105	58	1/2			−55.4(8)		
		106	59				−56.3(5)		
		107	60	1/2	−	$p_{1/2}$	−56.9(4)	51.4	0.134
		108	61				−55.8(4)		
		109	62	1/2	−	$p_{1/2}$	−56.5(4)	48.6	0.126

(10) μ (nucl mag)	(11) Quad mom, 10^{-24} cm²	(12) Half-life ($T_{1/2}$)	(13) Decay: Type and energy	(Q)	(14) Binding energy of last neutron E_n, MeV	(15) σ_{tn}, 10^{-24} cm²	
						2.5	**Ru** 94
		1 h	K				95
		98 m	K; β^+ 1.2	(2.2)		0.01	96
		2.9 d	K				97
							98
−0.6					7.1		99
					9.5		100
−0.7							101
						1.2	102
		40 d	β^- 0.20, 0.13, 0.69, . . .	(0.73)	6.5		103
						0.7	104
		4.5 h	β^- 1.15	(2.02)			105
		1.0 y	β^- 0.04, . . .	(0.04)			106
		4.5 m	β^- 4.3				107
		~4 m	β^-				108
						150	**Rh** 97
		35 m	β^+				97
		9 m	β^+ 3.3				98
		{ 15 d	β^+				99
		{ 4.5 h	β^+ 0.74				
		21 h	β^+ 2.62, . . . ; K	(3.64)	8.0		100
		{ ~5 y	IT?				101
		{ 4.5 d	K				
		220 d	K; β^- 1.15; β^+ 1.24, . . .	$\left[\begin{array}{c}(1.15-)\\(2.26+)\end{array}\right]$			102
−0.0879		(54 m	IT 0.040		9.4	150	103
		{ 4.4 m	IT 0.077; β^-		6.79		104
		{ 42 s	β^- 2.5, . . .	(2.5)			
		{ 30 s	IT 0.130				105
		{ 36 h	β^- 0.25, 0.56	(0.56)	9.1		
		{ 2 h	β^- ~1				106
		{ 30 s	β^- 3.53, . . .	(3.53)			
		22 m	β^- 1.2, ~2				107
		18 s	β^- ~4				108
						8	**Pd** 98
		17 m	β^+				98
		24 m					99
		4.0 d	K				100
		8 h	K; β^+ 0.5, 2.3(?)	(1.5?)			101
						4.8	102
		17 d	K	(0.57)			103
							104
−0.57(5)		(23 s	IT 0.2	)	7.1		105
							106
							107
		7 × 10⁶ y	β^- 0.04	(0.04)			108
					9.4	(0.17)	108
		{ 4.8 m	IT 0.17				109
		{ 13.6 h	β^- 1.0	(1.1)			109
						>0.4	110
		{ 5.5 h	IT; β^-				111
		{ 22 m	β^- 2.14				111
		21 h	β^- 0.28	(0.30)			112
		1.5 m	β^-				113
						60	**Ag** 102
		16 m	β^+				102
		1.1 h	β^+ 1.3; K				103
		27 m	β^+ 2.70				104
		40 d	K				105
		{ 24 m	β^+ 1.96; K	(2.98)			106
		{ 8.3 d	K				
−0.11304(1)		(44 s	IT 0.093	)		30	107
		2.3 m	β^- 1.77, . . . ; K; β^+ 0.8	$\left[\begin{array}{c}(1.77-)\\(1.8+)\end{array}\right]$	7.3		108
−0.12996(1)		(40 s	IT 0.088	)	9.1	84	109

(1)	(2)	(3)	(4)	(5)	(6)	(7)	(8)	(9)	
								Abundance	
Z	Chem symbol	A	N	I	π	Ground-state config	M-A, MeV	Per cent	Cosmic
47	Ag	107.880							
		110	63				−54.6(4)		
		111	64	1/2			−55.3(5)		
		112	65				−53.6(5)		
		113	66				−53.7(5)		
		114	67						
		115	68				−51.0(6)		
48	Cd	112.41							0.89
Cadmium		104	56				−53.1(5)		
		105	57				−52.4(8)		
		106	58	(0)	(+)		−56.3(5)	1.22	0.0109
		107	59				−55.5(4)		
		108	60	(0)	(+)		−55.5(4)	0.88	0.0079
		109	61				−56.3(4)		
		110	62	(0)	(+)		−57.5(4)	12.4	0.111
		111	63	1/2	+	$s_{1/2}$	−56.4(5)	12.8	0.114
		112	64	(0)	(+)		−57.5(5)	24.0	0.212
		113	65	1/2	+	$s_{1/2}$	−55.6(5)	12.3	0.110
		114	66	(0)	(+)		−56.3(5)	28.8	0.256
		115	67				−54.0(5)		
		116	68	(0)	(+)		−54.1(5)	7.6	0.068
		117	69				−51.4(6)		
		118	70						
49	In	114.82							0.11
Indium		107	58				−52.1(5)		
		108	59						
		109	60				−54.6(4)		
		110	61				−53.5(4)		
		111	62				−55.5(8)		
		112	63				−55.0(5)		
		113	64	9/2	+	$g_{9/2}$	−55.8(5)	4.2	0.0046
		114	65				−54.4(6)		
		115	66	9/2	+	$g_{9/2}$	−55.4(5)	95.8	0.105
		116	67				−53.5(5)		
		117	68				−54.2(5)		
		118	69						
		119	70						
50	Sn	118.70							1.33
Tin		108	58						
		109	59						
		110	60						
		111	61				−53.0(8)		
		112	62	(0)	(+)		−55.0(5)	1.02	0.0134
		113	63						
		114	64	(0)	(+)		−55.4(6)	0.69	0.0090
		115	65	1/2	+	$s_{1/2}$	−55.9(5)	0.38	0.00465
		116	66	(0)	(+)		−56.8(5)	14.3	0.189
		117	67	(1/2)	(+)	$s_{1/2}$	−55.7(5)	7.6	0.102
		118	68	(0)	(+)		−56.5(5)	24.1	0.316
		119	69	1/2	+	$s_{1/2}$	−54.9(5)	8.5	0.115
		120	70	(0)	(+)		−55.8(5)	32.5	0.433

(10) μ (nucl mag)	(11) Quad mom, 10^{-24} cm²	(12) Half-life ($T_{1/2}$)	(13) Decay — Type and energy	(Q)	(14) Binding energy of last neutron E_n, MeV	(15) $\sigma_{t\,n}$, 10^{-24} cm²	
							Ag
−0.145		⎰ 270 d	β⁻ 0.53, 0.1, . . . , IT 0.12				110
		⎱ 24 s	β⁻ 2.22, 2.88, . . .	(2.88)			
		⎰ 75 s	IT				111
		⎱ 7.5 d	β⁻ 1.04, 0.7, . . .	(1.04)			
		3.12 h	β⁻ 3.5, 4.1, 2.7, . . .	(4.1)			112
		5.3 h	β⁻ 2.0	(2.0)	8.6		113
		2 m	β⁻				114
		21 m	β⁻ 3	(3)			115
						3300	**Cd**
		59 m	K				104
		55 m	K; β⁺ 1.69, . . .				105
						1	106
		6.7 h	K; β⁺ 0.32	(1.43)			107
					10.2		108
		1.3 y	K; L	(0.15)			109
						>0.2	110
−0.5949(1)		(49 m	IT 0.150	)			111
						>0.03	112
−0.6224(1)		(5 y	β⁻ 0.58	)	6.4	2.7 × 10⁴	113
					9.05	1.2	114
		⎰ 43 d	β⁻ 1.63, . . .	(1.63)	5.6		115
		⎱ 54 h	β⁻ 1.11, . . .	(1.45)			
						1.4	116
		⎰ 3.0 h	IT				117
		⎱ 50 m	β⁻ 1.6, 3.0				
		∼30 m	β⁻ 4				118
						190	**In**
		30 m	β⁺ 2				107
		50 m	β⁺ 2.3				108
		4.3 h	K; β⁺ 0.7				109
		⎰ 5.0 h	K; IT 0.12				110
		⎱ 66 m	β⁺ 2.25	(3.93)			
		2.8 d	K				111
		⎰ 21 m	IT				112
		⎱ 2.5 s	IT 0.15				
		14 m	β⁻ 0.66; K; β⁺ 1.52				
+5.486(3)	+1.144	(1.73 h	IT 0.39	)	9.2	63	113
		⎰ 49 d	IT 0.190; K	[(1.98−)]			114
		⎱ 72 s	β⁻ 1.98, . . . ; K; β⁺ ∼1	[(2.3+)]			
+5.5095(1)	+1.161	⎰ 4.5 h	IT 0.33 β⁻ 0.83				115
		⎱ 6 × 10¹⁴ y	β⁻ 0.6		9.1	197	
		⎰ 54 m	β⁻ 1.00, . . .	(3.36)	6.6		116
		⎱ 13 s	β⁻ 3.3	(3.3)			
		⎰ 1.9 h	β⁻ 1.77, 1.61; IT 0.31				117
		⎱ 1.1 h	β⁻ 0.74	(1.46)			
		⎰ 4.5 m	β⁻ 1.5				118
		⎱ <1 m	β⁻ 4				
		18 m	β⁻ 2.7	(2.7)			119
						0.6	**Sn**
		4 h	K				108
		18 m	K, β⁺				109
		4 h	K				110
		35 m	K, β⁺ 1.51	(2.53)			111
						1.3	112
		112 d	K, L				113
							114
−0.9178(1)					7.7		115
					8.9	>0.006	116
−0.9998(1)		(14 d	IT 0.159	)			117
					9.3	>0.01	118
−1.0460(1)		(275 d	IT 0.065	)	6.6		119
						0.1	120

(1)	(2)	(3)	(4)	(5)	(6)	(7)	(8)	(9)	
								Abundance	
Z	Chem symbol	A	N	I	π	Ground-state config	$M\text{-}A$, MeV	Per cent	Cosmic
50	Sn	118.70							
		121	71				−53.6(5)		
		122	**72**	**(0)**	**(+)**		**−53.9(5)**	**4.8**	**0.063**
		123	73				−51.6(5)		
		124	**74**	**(0)**	**(+)**		**−51.6(5)**	**6.1**	**0.079**
		125	75				−49.0(5)		
		126	76						
		127	77						
51	Sb	121.76							0.246
Antimony		116	65				−52.1(5)		
		117	66						
		118	67				−52.4(6)		
		119	68						
		120	69				−53.1(5)		
		121	**70**	**5/2**	**+**	$d_{5/2}$	**−54.0(5)**	**57**	**0.141**
		122	71				−52.4(5)		
		123	**72**	**7/2**	**+**	$g_{7/2}$	**−53.0(5)**	**43**	**0.105**
		124	73				−51.0(5)		
		125	74				−51.3(5)		
		126	75						
		127	76						
		128	77						
		129	78						
		130	79						
		131	80						
		132	81						
52	Te	127.61							4.07
Tellurium		117	65						
		118	66						
		119	67						
		120	**68**	**(0)**	**(+)**		**−53.5(5)**	**0.091**	**0.00420**
		121	69						
		122	**70**	**(0)**	**(+)**		**−54.4(5)**	**2.5**	**0.115**
		123	**71**	**1/2**	**+**	$s_{1/2}$	**−52.8(9)**	**0.88**	**0.0416**
		124	**72**	**(0)**	**(+)**		**−53.9(5)**	**4.6**	**0.221**
		125	**73**	**1/2**	**+**	$s_{1/2}$	**−52.1(5)**	**7.0**	**0.328**
		126	**74**	**(0)**	**(+)**		**−52.5(5)**	**18.7**	**0.874**
		127	75				−50.6(5)		
		128	**76**	**(0)**	**(+)**		**−50.2(5)**	**31.8**	**1.48**
		129	77				−48.8(5)		
		130	**78**	**(0)**	**(+)**		**−48.6(5)**	**34.4**	**1.60**
		131	79				−46.5(5)		
		132	80				−46.3(5)		
		133	81				−43.9(7)		
		134	82						
53	I	126.91							0.80
Iodine		119	66						
		120	67						
		121	68						

(10) μ (nucl mag)	(11) Quad mom, 10^{-24} cm²	(12) Half-life ($T_{1/2}$)	(13) Decay — Type and energy	(Q)	(14) Binding energy of last neutron E_n, MeV	(15) σ_{tn}, 10^{-24} cm²	
							Sn
		>1 y / 27 h	β^- 0.42 / β^- 0.38	(0.38)	6.2		**121**
						0.2	**122**
		130 d / 40 m	β^- 1.42 / β^- 1.26	(1.41)			**123**
					8.5	0.2	**124**
		9.5 m / 10 d	β^- 2.1, … / β^- 2.4	(2.4) / (2.4)	5.75		**125**
		50 m	β^-				**126**
		1.5 h	β^-				**127**
						5.5	**Sb**
		15 m / 60 m	β^+ 2.4 / β^+ 1.4	(4.7) / (4.7)			**116**
		2.8 h	K				**117**
		3.5 m / 5.1 h	β^+ 3.1; IT 0.11 / K; β^+ 0.7	(4.1)			**118**
		38 h	K	(0.59)			**119**
		5.8 d / 17 m	K / β^+ 1.70	(2.72)			**120**
+3.360	**−0.7**				9.2	7.0	**121**
		3.5 m / 2.8 d	IT 0.075 / β^- 1.4, 1.98, 0.73; K; β^+ 0.5	(1.98−) / (1.5+)	6.8		**122**
+2.547	**−0.8**				9.3	3.46	**123**
		21 m / 1.3 m	IT 0.018; β^- / β^- 3; IT 0.012		6.6		**124**
		60 d	β^- 2.31, …	(2.91)	8.6		**125**
		2.7 y	β^- 0.30, 0.12, …	(0.76)			**126**
		9 h / 28 d	β^- 1 / β^- 1.9				**127**
		93 h / ~1 h	β^- 0.86, 1.57, 1.11 / β^-				**128**
		4.6 h	β^- 0.92, …				**129**
		10 m / 40 m	β^- 2.9				**130**
		22 m	β^- 1.1				**131**
		2 m	β^-				**132**
						4.6	**Te**
		2.5 h	β^+ 2.5				**117**
		6.0 d	K; β^+ 3.1				**118**
		16 h / 4.5 d	K / K				**119**
						<140	**120**
		150 d / 17 d	IT 0.082 / K				**121**
						3	**122**
−0.73188(4)		(104 d	IT 0.088	)		400	**123**
						7	**124**
−0.88235(4)		(58 d	IT 0.11	)	6.5	1.5	**125**
					7.2	0.9	**126**
		110 d / 9.3 h	IT 0.008 / β^- 0.68	(0.68)			**127**
						0.16	**128**
		33 d / 72 m	IT 0.106 / β^- 1.46, 1.01, …	(1.49)			**129**
						0.2	**130**
		30 h / 25 m	β^- 0.4; IT 0.18 / β^- 2.1	(2.3)			**131**
		77 h	β^- 0.22, …				**132**
		63 m / 2 m	IT 0.4 / β^- 1.4, 2.4	(3.0)			**133**
		44 m	β^-				**134**
						6.3	**I**
		18 m	β^+				**119**
		>1.3 h	β^+ 4.0				**120**
		1.4 h	β^+ 1.13, …				**121**

(1)	(2)	(3)	(4)	(5)	(6)	(7)	(8)	(9)	
Z	Chem symbol	A	N	I	π	Ground-state config	$M\text{-}A$, MeV	Abundance	
								Per cent	Cosmic
53	I	126.91							
		122	69				−50.3(5)		
		123	70	5/2					
		124	71	2			−50.7(5)		
		125	72	5/2			−51.9(5)		
		126	73				−50.4(5)		
		127	**74**	**5/2**	**+**	$d_{5/2}$	**−51.4(5)**	**100**	**0.80**
		128	75	1			−49.7(5)		
		129	76	7/2			−50.5(5)		
		130	77				−48.6(5)		
		131	78	7/2			−48.7(5)		
		132	79				−46.7(5)		
		133	80				−46.9(7)		
		134	81				−45.4(7)		
		135	82						
		136	83				−40.1(5)		
		137	84						
54	Xe *Xenon*	131.30							4.0
		121	67						
		122	68						
		123	69						
		124	**70**	**(0)**	**(+)**		**−50.9(5)**	**0.094**	**0.00380**
		125	71						
		126	**72**	**(0)**	**(+)**		**−51.7(5)**	**0.092**	**0.00352**
		127	73				−50.4(9)		
		128	**74**	**(0)**	**(+)**		**−51.7(5)**	**1.92**	**0.0764**
		129	**75**	**1/2**	**+**	$s_{1/2}$	**−50.7(5)**	**26.4**	**1.050**
		130	**76**	**(0)**	**(+)**		**−51.6(5)**	**4.1**	**0.162**
		131	**77**	**3/2**	**+**	$d_{3/2}$	**−49.7(5)**	**21.2**	**0.850**
		132	**78**	**(0)**	**(+)**		**−50.3(5)**	**26.9**	**1.078**
		133	79				−48.7(7)		
		134	**80**	**(0)**	**(+)**		**−48.8(5)**	**10.4**	**0.420**
		135	81						
		136	**82**	**(0)**	**(+)**		**−46.5(5)**	**8.9**	**0.358**
		137	83				−41.7(4)		
		138	84						
		139	85						
		140	86						
		141	87						
55	Cs *Cesium*	132.91							0.456
		125	70						
		126	71				−46.9(7)		
		127	72	1/2			−48.3(9)		
		128	73				−47.6(7)		
		129	74	1/2			−49.6(8)		
		130	75				−48.6(5)		
		131	76	5/2			−49.4(5)		
		132	77	2			−48.5(7)		
		133	**78**	**7/2**	**+**	$g_{7/2}$	**−49.2(7)**	**100**	**0.456**
		134	79	$\left\{\begin{smallmatrix}8\\4\end{smallmatrix}\right\}$			−47.5(8)		
		135	80	7/2					
		136	81						
		137	82	7/2			−45.7(10)		
		138	83				−42.9(10)		
		139	84						
		140	85						

(10) μ (nucl mag)	(11) Quad mom, 10^{-24} cm²	(12) Half-life ($T_{1/2}$)	(13) Decay — Type and energy	(13) (Q)	(14) Binding energy of last neutron E_n, MeV	(15) σ_{tn}, 10^{-24} cm²	
							I
		3.5 m	β⁺ 3.12	(4.14)			122
		13 h	K				123
		4.5 d	K; β⁺ 2.20, . . .	(3.22)			124
± 2.6	−0.66	60 d	K; L	(0.15)			125
		13.3 d	K; β⁻ 0.87, . . . ; β⁺ 1.11, . . .	[(1.25−)]			126
+2.8090(4)	**−0.6(2)**			[(2.13+)]	**9.1**	**6.3**	**127**
+2.603		25.0 m	β⁻ 2.12, 1.67, . . . ; K	(2.12−)	6.6		128
	−0.49	1.7 × 10⁷ y	β⁻ 0.15	(0.19)		30	129
		12.6 h	β⁻ 1.02, 0.60, . . .	(2.95)			130
	−0.35	8.05 d	β⁻ 0.61, . . .	(0.97)		∼600	131
		2.3 h	β⁻ 0.9, . . .	(3.57)			132
		21 h	β⁻ 1.3, 0.4	(1.8)			133
		52 m	β⁻ 1.5, 2.5	(3.4)			134
		6.7 h	β⁻ 1.0, 0.5, 1.4				135
		86 s	β⁻ 6.4, 5.0, 3.6	(6.4)			136
		22 s	β⁻; n 0.6				137
						35	**Xe**
							121
		40 m	β⁺				122
		19 h	K; β⁺ 3.12				123
		1.8 h	K; β⁺ 1.7				**124**
		{ 55 s	IT 0.075	}			125
		{ 18 h	K	}			
							126
		{ 75 s	IT 0.175	}			127
		{ 36.4 d	K	}			
						<5	**128**
−0.77255(2)		(8 d	IT 0.196	)		45	**129**
						<5	**130**
+0.68680(2)	**−0.12**	(12 d	IT 0.164	)		120	**131**
						<5	**132**
		{ 2.3 d	IT 0.233	}			133
		{ 5.27 d	β⁻ 0.35;	(0.43) }			
						<5	**134**
		{ 15 m	IT 0.52	}		3.2 × 10⁶	135
		{ 9.2 h	β⁻ 0.91, 0.54	(1.16) }			**136**
						0.15	
		3.8 m	β⁻ 3.5				137
		17 m	β⁻ 2.4				138
		41 s	β⁻				139
		16 s	β⁻				140
		3 s	β⁻				141
						31	**Cs**
		45 m	K; β⁺ 2.05				125
		1.6 m	β⁺ 3.8, . . . ; K	(4.8)			126
± 1.41		6.2 h	K; β⁺ 1.06, 0.7, . . .				127
		3.8 m	β⁺ 3.0, 2.5, 1.5	(4.0)			128
± 1.47		31 h	K				129
		30 m	β⁺ 1.97; K; β⁻ 0.44	[(0.44−)] [(2.99+)]			130
+3.48		9.7 d	K; L	(0.35)			131
+2.20		6.2 d	K				132
+2.5771	**−0.003**				**9.0**	**31**	**133**
+2.973		{ 3.1 h	IT 0.13; β⁻ 0.55	}	6.7		134
		{ 2.3 y	β⁻ 0.65	(2.05) }			
+2.7134		2.0 × 10⁶ y	β⁻ 0.21	(0.21)		∼15	135
		13 d	β⁻ 0.34, 0.66				136
+2.8219		30 y	β⁻ 0.52, 1.18	(1.18)	7.1	<2	137
		32 m	β⁻ 3.40, . . .		4.9		138
		9.5 m	β⁻ ∼4				139
		66 s	β⁻				140

(1) Z	(2) Chem symbol	(3) A	(4) N	(5) I	(6) π	(7) Ground-state config	(8) M-A, MeV	(9) Abundance Per cent	(9) Abundance Cosmic
56	Ba	137.36							3.66
Barium		126	70						
		127	71						
		128	72						
		129	73				−47.0(8)		
		130	74	(0)	(+)		−49.0(5)	0.101	0.00370
		131	75						
		132	76	(0)	(+)			0.097	0.00356
		133	77						
		134	78	(0)	(+)		−49.6(8)	2.42	0.0886
		135	79	3/2	+	d$_{3/2}$		6.6	0.241
		136	80	(0)	(+)			7.8	0.286
		137	81	3/2	+	d$_{3/2}$	−46.9(10)	11.3	0.414
		138	82	(0)	(+)		−47.8(10)	71.7	2.622
		139	83				−44.6(10)		
		140	84				−42.5(10)		
		141	85						
		142	86						
		143	87						
57	La	138.92							2.00
Lanthanum		131	74						
		132	75						
		133	76						
		134	77				−45.9(8)		
		135	78						
		136	79						
		137	80						
		138	81	5		(7/2, 3/2)	−46.5(10)	0.089	0.0018
		139	82	7/2	(+)	g$_{7/2}$	−47.0(10)	99.911	2.00
		140	83				−43.8(10)		
		141	84				−42.5(10)		
		142	85						
		143	86						
58	Ce	140.13							2.26
Cerium		133	75						
		134	76						
		135	77						
		136	78	(0)	(+)			0.19	0.0044
		137	79						
		138	80	(0)	(+)		−47.5(10)	0.26	0.00566
		139	81				−46.9(10)		
		140	82	(0)	(+)		−47.6(10)	88.47	2.00
		141	83	7/2			−45.0(10)		
		142	84	(0)	(+)		−43.7(10)	11.08	0.250
		143	85				−40.5(10)		
		144	86				−38.7(10)		
		145	87						
		146	88				−33.2(10)		
59	Pr	140.92							0.40
Praseodymium		135	76						
		136	77						
		137	78						
		138	79				−44.0(10)		
		139	80						
		140	81				−44.3(10)		
		141	82	5/2	+	d$_{5/2}$	−45.5(10)	100	0.40
		142	83				−43.0(10)		
		143	84				−41.9(10)		
		144	85				−39.0(10)		
		145	86						
		146	87				−34.3(10)		

(10)	(11)	(12)	(13)		(14)	(15)	
μ (nucl mag)	Quad mom, 10^{-24} cm²	Half-life ($T_{1/2}$)	Decay		Binding energy of last neutron E_n, MeV	σ_{tn}, 10^{-24} cm²	
			Type and energy	(Q)			
		97 m	K; β^+ 3.8			1.2	**Ba** 126
		~12 m					127
		2.4 d	K; β^+ 3.0, . . .				128
		1.9 h	β^+ 1.6				129
		11.6 d	K			6	**130**
							131
						3	**132**
		{ 39 h	IT 0.276	}			133
		{ 8 y	K				
+0.835(3)		(29 h)	IT 0.268	)		<4	**134**
						5	**135**
+0.935(3)		(2.60 m)	IT 0.662	)		<1	**136**
					6.8	4	**137**
					8.6	0.55	**138**
		85 m	β^- 2.22, 0.8, 2.38	(2.38)	5.2	4	139
		12.8 d	β^- 1.02, 0.48				140
		18 m	β^- 2.8				141
		6 m	β^-				142
		<0.5 m	β^-				143
		58 m	β^+ 1.6			8.9	**La** 131
		4.5 h	β^+ 3.5				132
		4 h	K; β^+ 1.2				133
		6.5 m	K; β^+ 2.7	(3.7)			134
		19 h	K				135
		9 m	K; β^+ 2.1				136
		>10⁵ y	K				137
+3.685	±0.9	2 × 10¹¹ y	K; β^- 1.0				138
+2.776(3)	**+0.3(1)**				8.8	8.9	**139**
		40.2 h	β^- 1.34, 0.8, . . .	(3.75)	5.1	3	140
		3.8 h	β^- 2.43, 0.9	(2.43)	6.9		141
		77 m	β^- >2.5				142
		~19 m	β^-				143
		6.3 h	K; β^+ 1.3			0.7	**Ce** 133
		72 h	K				134
		22 h	K; β^+ 0.8				135
		{ 35 h	IT 0.26	}		~22	**136**
		{ 9 h	K				137
						1	**138**
		140 d	K	(0.3)			139
					9.0	0.6	**140**
±0.89		32 d	β^- 0.43, 0.57	(0.57)	5.5		141
					7.1	1	**142**
		33 h	β^- 1.09, 0.3, . . .	(1.44)	5.1	6	143
		285 d	β^- 0.30, . . .	(0.30)			144
		3.0 m	β^- 2.0				145
		14 m	β^- 0.7				146
		22 m	β^+ 2.5			11	**Pr** 135
		70 m	β^+ 2.0				136
							137
		2.0 h	β^+ 1.4				138
		4.5 h	β^+ 1.0				139
		3.4 m	β^+ 2.3; K	(3.3)	7.8		140
+3.8(4)	**−0.054**				9.4	11	**141**
		19.1 h	β^- 2.16, 0.6	(2.16)		20	142
		13.8 d	β^- 0.92	(0.92)			143
		17 m	β^- 2.98, 0.8, 2.3	(2.98)			144
		5.9 h	β^- 1.7				145
		24 m	β^- 3.7, 2.3	(4.2)			146

(1) Z	(2) Chem symbol	(3) A	(4) N	(5) I	(6) π	(7) Ground-state config	(8) M-A, MeV	(9) Abundance Per cent	(9) Abundance Cosmic
60	**Nd**	**144.27**							**1.44**
Neodymium		138	78						
		139	79						
		140	80				−44.2(10)		
		141	81				−43.8(10)		
		142	**82**	**(0)**	**(+)**		**−45.2(10)**	**27.1**	**0.39**
		143	**83**	**7/2**	**(−)**	**f$_{7/2}$**	**−42.8(10)**	**12.2**	**0.175**
		144	84	(0)	(+)		−42.0(10)	23.9	0.344
		145	**85**	**7/2**	**(−)**	**f$_{7/2}$**		**8.3**	**0.119**
		146	**86**	**(0)**	**(+)**		**−38.5(10)**	**17.2**	**0.248**
		147	87	5/2			−35.8(10)		
		148	**88**	**(0)**	**(+)**		**−33.5(10)**	**5.7**	**0.0824**
		149	89				−31.0(10)		
		150	**90**	**(0)**	**(+)**		**−29.9(10)**	**5.6**	**0.0806**
		151	91						
61	**Pm**								
Promethium		141	80						
		142	81						
		143	82						
		144	83						
		145	84						
		146	85				−38.2(10)		
		147	86				−36.7(10)		
		148	87				−33.3(10)		
		149	88				−32.7(10)		
		150	89				−28.8(10)		
		151	90						
62	**Sm**	**150.35**							**0.664**
Samarium		143	81						
		144	**82**	**(0)**	**(+)**		**−41.0(16)**	**3.1**	**0.0108**
		145	83						
		146	84				−38.9(10)		
		147	85	7/2	(−)	f$_{7/2}$	−37.0(10)	15.0	0.100
		148	**86**	**(0)**	**(+)**		**−36.0(10)**	**11.2**	**0.0748**
		149	**87**	**7/2**	**(−)**	**(f$_{7/2}$)**	**−34.0(10)**	**13.8**	**0.0920**
		150	**88**	**(0)**	**(+)**		**−34.1(10)**	**7.4**	**0.0492**
		151	89						
		152	**90**	**(0)**	**(+)**		**−30.4(10)**	**26.8**	**0.176**
		153	91						
		154	**92**	**(0)**	**(+)**		**−27.5(10)**	**22.7**	**0.150**
		155	93				−24.7(11)		
		156	94				−23.2(10)		
63	**Eu**	**152.0**							**0.187**
Europium		144	81						
		145	82						
		146	83						
		147	84						
		148	85						
		149	86						
		150	87				−31.5(10)		
		151	**88**	**5/2**	**(+)**	**d$_{5/2}$**		**47.8**	**0.0892**
		152	89	3					
		153	**90**	**5/2**	**+**	**d$_{5/2}$**		**52.2**	**0.0976**
		154	91	3			−25.2(10)		
		155	92				−26.9(10)		
		156	93				−24.1(10)		
		157	94				−25.1(10)		
		158	95						
		159	96						

(10) μ (nucl mag)	(11) Quad mom, 10^{-24} cm²	(12) Half-life ($T_{1/2}$)	(13) Decay — Type and energy	(Q)	(14) Binding energy of last neutron E_n, MeV	(15) σ_{tn}, 10^{-24} cm²	
						48	**Nd** 138
		22 m	β^+ 2.4				139
		5.5 h	K; β^+ 3.1	(4.1)			140
		3.3 d	K	(0.1)			141
		2.4 h	K; β^+ 0.7	(1.7)		17	**142**
−1.0(2)					5.0	320	**143**
		1.5×10^{15} y	α 1.8			5	144
−0.6(1)						44	**145**
						2	**146**
±0.6		11.6 d	β^- 0.81, 0.37	(0.90)			147
						4	**148**
		1.8 h	β^- 1.5	(1.6)			149
						3	**150**
		15 m	β^- 1.9	(2.0)			151
							Pm 141
		20 m	β^+ ~2.6				142
		~300 d	K				143
		~300 d	K				144
		25 y	K; L				145
		~2 y	β^- 0.7				146
		2.6 y	β^- 0.23	(0.23)		~60	147
		{ 5.3 d { 42 d	β^- 2.5 β^- 0.6, 2.4	}			148
		50 h	β^- 1.05	(1.34)			149
		2.7 h	β^- 2.0, 3.0				150
		27 h	β^- 1.1				151
							Sm 143
		8 m	β^+ 2.3	(3.3)		(10⁴)	**144**
						~0.03	145
		1.0 y	K				146
−0.7(1)		~5×10^7 y 1.3×10^{11} y	α 2.5 α 2.18				147
							148
−0.6(1)						6.6×10^4	**149**
							150
		80 y	β^- 0.076	(0.096)		1.2×10^4	151
						140	**152**
		47 h	β^- 0.71, 0.64, 0.81, . . .	(0.81)			153
						5	**154**
		23 m	β^- 1.8	(2.2)	5.6		155
		~10 h	β^- 0.9				156
						4300	**Eu** 144
		18 m	β^+ 2.4				144
		5 d	K				145
		38 h	K				146
		24 d	K; α 2.9				147
		58 d	K				148
		120 d	K				149
		14 h	β^- 1.1				150
+3.6	+1.2					8600	**151**
±2.0		{ 9.3 h { 13 y	β^- 1.88 K; β^- 0.70, . . .	(1.88) }		5000	152
+1.6	+2.5					400	**153**
±2.1		16 y	β^- 1.5			1400	154
		1.7 y	β^- 0.15, 0.25	(0.25)		1.3×10^4	155
		15 d	β^- 0.4, 2.4	(2.4)			156
		15 h	β^- 1.0				157
		60 m	β^- 2.6				158
		20 m	β^-				159

(1) Z	(2) Chem symbol	(3) A	(4) N	(5) I	(6) π	(7) Ground-state config	(8) M-A, MeV	(9) Abundance	
								Per cent	Cosmic
64	Gd	157.26							0.684
	Gadolinium	148	84				−34.1(16)		
		149	85						
		150	86				−32.5(10)		
		151	87						
		152	88	(0)	(+)			0.20	0.00137
		153	89						
		154	90	(0)	(+)		−28.2(10)	2.15	0.0147
		155	91	(3/2)			−27.1(11)	14.7	0.101
		156	92	(0)	(+)		−26.5(10)	20.5	0.141
		157	93	(3/2)			−26.8(10)	15.7	0.107
		158	94	(0)	(+)		−24.7(10)	24.9	0.169
		159	95						
		160	96	(0)	(+)		−20.7(10)	21.9	0.149
		161	97						
65	Tb	158.93							0.0956
	Terbium	149	84						
		150	85						
		151	86						
		152	87						
		153	88						
		154	89						
		155	90						
		156	91						
		157	92						
		158	93						
		159	94	3/2	(+)	$d_{3/2}$		100	0.0956
		160	95				−20.7(20)		
		161	96						
		162	97						
66	Dy	162.51							0.556
	Dysprosium	156	90	(0)	(+)			0.052	0.00029
		157	91						
		158	92	(0)	(+)			0.090	0.00050
		159	93						
		160	94	(0)	(+)		−22.5(20)	2.29	0.0127
		161	95	(5/2)				18.9	0.105
		162	96	(0)	(+)		−21.5(20)	25.5	0.142
		163	97	(5/2)				25.0	0.139
		164	98	(0)	(+)		−18.2(20)	28.2	0.157
		165	99				−16.4(20)		
		166	100						
67	Ho	164.94							0.118
	Holmium	160	93						
		161	94						
		162	95						
		163	96						
		164	97				−16.0(10)		
		165	98	7/2	+	$g_{7/2}$	−17.6(20)	100	0.118
		166	99						
		167	100						
		168	101						
		169	102						
68	Er	167.27							0.316
	Erbium	160	92						
		161	93						
		162	94	(0)	(+)			0.136	0.000316
		163	95						
		164	96	(0)	(+)		−17.0(10)	1.56	0.00474

(10) μ (nucl mag)	(11) Quad mom, 10^{-24} cm²	(12) Half-life ($T_{1/2}$)	(13) Type and energy	(Q)	(14) Binding energy of last neutron E_n, MeV	(15) σ_{tn}, 10^{-24} cm²	
		>35 y	α 3.2			3.8×10^4	**Gd** 148
		9 d	K; α 3.0				149
		>10^5 y	α 2.7				150
		~150 d	K				151
						<180	**152**
		236 d	K				153
							154
±0.3	+1.1					~7×10^4	155
							156
±0.4	+1.0					1.8×10^5	157
						4	158
		18 h	β^- 0.95, 0.60, ...	(0.95)			159
						0.8	160
		3.7 m	β^- 1.6	(2.0)			161
		4.1 h	α 3.95			45	**Tb** 149
		19 h	α 3.4				150
							151
							152
		5.1 d	K				153
		7 h	K; β^-				154
		17.2 h	K; β^- 2.75, 1.66				
							155
		5 h	β^- 0.14				156
		5 d	K; β^- 0.6, 0.2				
							157
							158
						45	159
		72 d	β^- 0.56, 0.85, ...	(1.81)		~600	160
		7 d	β^- 0.55, ...	(0.6)			161
		14 m					162
						1100	**Dy 156**
		8.2 h	K				157
							158
		134 d	K; L				159
							160
−0.37	+1.1						161
							162
+0.51	+1.3						**163**
						2700	164
		1.3 m	IT 0.108; β^- 0.1			4700	165
		2.32 h	β^- 1.25, ...	(1.25)			
		82 h	β^- 0.3	(0.3)			166
		5.0 h	K			64	**Ho** 160
		2.5 h	K				161
		22 m	K; β^+ 1.3				162
							163
		37 m	K; β^- 0.99, 0.90	(0.99)			164
±3.3						64	**165**
		>30 y	β^- 0.2, ...	(2.4)			166
		27.2 h	β^- 1.85	(1.85)			
		3.0 h	β^- 1.0, 0.28				167
							168
		1.6 h	β^-				169
						170	**Er** 160
		29 h	K				161
		3 h	K; β^+ 1.2				**162**
		75 m	K				163
							164

(1)	(2)	(3)	(4)	(5)	(6)	(7)	(8)	(9)	
								Abundance	
Z	Chem symbol	A	N	I	π	Ground-state config	$M\text{-}A$, MeV	Per cent	Cosmic
68	**Er**	**167.27**							
		165	97						
		166	**98**	**(0)**	**(+)**			**33.4**	**0.104**
		167	**99**	**7/2**	**(−)**	$(f_{7/2})$		**22.9**	**0.770**
		168	**100**	**(0)**	**(+)**		**−15.0(10)**	**27.1**	**0.0850**
		169	101						
		170	**102**	**(0)**	**(+)**		**−9.9(20)**	**14.9**	**0.0228**
		171	103						
69	**Tm**	**168.94**							**0.0318**
	Thulium	165	96						
		166	97						
		167	98						
		168	99						
		169	**100**	**1/2**	**(+)**	$s_{1/2}$		**100**	**0.0318**
		170	101						
		171	102						
		172	103						
		173	104						
		174	105						
70	**Yb**	**173.04**							**0.220**
	Ytterbium	166	96						
		167	97						
		168	**98**	**(0)**	**(+)**			**0.14**	**0.00030**
		169	99						
		170	**100**	**(0)**	**(+)**			**3.03**	**0.00666**
		171	**101**	**1/2**	**(−)**	$p_{1/2}$		**14.3**	**0.0316**
		172	**102**	**(0)**	**(+)**		**−15.0(30)**	**21.8**	**0.0480**
		173	**103**	**5/2**	**(−)**	$f_{5/2}$		**16.2**	**0.0356**
		174	**104**	**(0)**	**(+)**		**−17.9(30)**	**31.8**	**0.0678**
		175	105						
		176	**106**	**(0)**	**(+)**			**12.7**	**0.0278**
		177	107						
71	**Lu**	**174.99**							**0.050**
	Lutecium	170	99						
		171	100						
		172	101						
		173	102						
		174	103						
		175	**104**	**7/2**	**(+)**	$g_{7/2}$		**97.40**	**0.0488**
		176	105			(11/2, 13/2)	−2.4(20)	2.60	0.0013
		177	106						
		178	107						
		179	108						
72	**Hf**	**178.50**							**0.438**
	Hafnium	170	98						
		171	99						
		172	100						
		173	101						
		174	**102**	**(0)**	**(+)**			**0.18**	**0.00078**
		175	103						
		176	**104**	**(0)**	**(+)**		**−3.4(20)**	**5.2**	**0.0226**
		177	**105**	**7/2**				**18.5**	**0.0806**
		178	**106**	**(0)**	**(+)**		**+0.2(20)**	**27.1**	**0.119**
		179	**107**	**9/2**				**13.8**	**0.0604**
		180	**108**	**(0)**	**(+)**		**+1.9(20)**	**35.2**	**0.155**
		181	109				+3.7(20)		

(10) μ (nucl mag)	(11) Quad mom, 10^{-24} cm²	(12) Half-life ($T_{1/2}$)	(13) Decay — Type and energy	(Q)	(14) Binding energy of last neutron E_n, MeV	(15) σ_{tn}, 10^{-24} cm²	
		10 h	K				**Er** 165
							166
±0.5	±10						167
					2		**168**
		9.4 d	β⁻ 0.33	(0.33)			169
					9		**170**
		7.5 h	β⁻ 1.06, 0.67, 1.48	(1.48)			171
					125		**Tm** 165
		25 h	K				165
		7.7 h	K; β⁺ 2.1				166
		9.6 d	K				167
		87 d	K; β⁻ 0.5				168
					125		**169**
		129 d	β⁻ 0.97, 0.89	(0.97)			170
		1.9 y	β⁻ 0.10	(0.10)	2000		171
		19 m	β⁻				172
							173
		~2 d	β⁻				174
					37		**Yb** 166
		54 d	K				166
		18 m	K				167
					1.1 × 10⁴		**168**
		32 d	K				169
							170
+0.45							171
							172
−0.65	+3.9(4)						173
					~60		**174**
		4.2 d	β⁻ 0.47, 0.07, 0.36	(4.7)			175
					7		**176**
		2.0 h	β⁻ 1.30, . . .	(1.30)			177
					111		**Lu** 170
		1.7 d	K				170
		8.5 d	K				171
		1.6 y	K				171
		4.0 h	β⁺ 1.2				172
		6.7 d	K				172
		1.4 y	K				173
		165 d	K; β⁻ 0.6				174
+2.6	+5.9				18		**175**
		3.7 h	β⁻ 1.2	(1.2)			176
+3.8	+7(1)	7.5 × 10¹⁰ y	β⁺ 0.42; K	(1.02)	3600		176
		6.8 d	β⁻ 0.50, 0.18, 0.39, . . .	(0.50)			177
		22 m	β⁻				178
		~5 h	β⁻				179
					105		**Hf** 170
		1.8 h	β⁺ 2.4				170
		16 h	K				171
		~5 y	K				172
		24 h	K				173
					1000		**174**
		70 d	K				175
					<30		**176**
+0.61	+3				370		**177**
					80		**178**
−0.47	+3	(19 s	IT 0.16	)	65		**179**
					10		**180**
		(5.5 h	IT	)			181
		46 d	β⁻ 0.41	(1.02)			

(1)	(2)	(3)	(4)	(5)	(6)	(7)	(8)	(9)	
								Abundance	
Z	Chem symbol	A	N	I	π	Ground-state config	$M\text{-}A$, MeV	Per cent	Cosmic
73 Ta *Tantalum*		180.95							0.065
		176	103						
		177	104						
		178	105						
		179	106						
		180	**107**				**2.0(20)**	**0.012**	**0.000007**
		181	**108**	7/2	+	$g_{7/2}$	**2.7(20)**	**99.988**	**0.065**
		182	109				5.0(20)		
		183	110						
		184	111						
		185	112						
		186	113						
74 W *Tungsten*		183.86							0.49
		176	102						
		177	103						
		178	104						
		179	105						
		180	**106**	(0)	(+)		**1.3(20)**	**0.14**	**0.0006**
		181	107						
		182	**108**	(0)	(+)		**3.3(20)**	**26.2**	**0.13**
		183	**109**	1/2	(−)	$p_{1/2}$	**4.9(20)**	**14.3**	**0.070**
		184	**110**	(0)	(+)		**5.9(20)**	**30.7**	**0.15**
		185	111						
		186	**112**	(0)	(+)		**9.7(20)**	**28.7**	**0.14**
		187	113				11.6(22)		
		188	114						
75 Re *Rhenium*		186.22							0.135
		180	105						
		181	106						
		182	107						
		183	108						
		184	109						
		185	**110**	5/2	+	$d_{5/2}$		**37.1**	**0.0500**
		186	111				10.0(23)		
		187	112	5/2	+	$d_{5/2}$	10.3(22)	62.9	0.0850
		188	113				15.2(20)		
		189	114						
		190	115						
		191	116						
76 Os *Osmium*		190.2							1.00
		182	106						
		183	107						
		184	**108**	(0)	(+)			**0.018**	**0.00018**
		185	109						
		186	**110**	(0)	(+)		**8.9(23)**	**1.59**	**0.0159**
		187	**111**	(1/2)	(−)	$(p_{1/2})$	10.3(22)	**1.64**	**0.0164**
		188	**112**	(0)	(+)		**13.1(20)**	**13.3**	**0.133**
		189	**113**	3/2	(−)	$p_{3/2}$	**16.9(20)**	**16.1**	**0.161**
		190	**114**	(0)	(+)		**16.2(20)**	**26.4**	**0.264**
		191	115				20.0(21)		
		192	**116**	(0)	(+)		**21.0(30)**	**41.0**	**0.410**
		193	117				24.6(20)		
		194	118						

(10) μ (nucl mag)	(11) Quad mom, 10^{-24} cm²	(12) Half-life $(T_{1/2})$	(13) Decay — Type and energy	(Q)	(14) Binding energy of last neutron E_n, MeV	(15) σ_{tn}, 10^{-24} cm²	
						22	**Ta** 176
		8.0 h	K				177
		2.2 d	K				178
		⎰ 9.3 m	K; β⁺ 1.1	⎱			178
		⎱ 2.1 h	K; β⁺ ∼1	⎰			179
		∼600 d	K				**180**
		(8.1 h	K; β⁻ 0.70, . . .	)			180
+1.9	+5.9	(0.33 s	IT	)	7.6	22	181
		⎰ 16 m	IT 0.18	⎱	6.0	∼2 × 10⁴	182
		⎱ 112 d	β⁻ 0.51	(1.7) ⎰			182
		5.2 d	β⁻ 0.62, . . .	(1.07)			183
		8.7 h	β⁻ 1.26, 0.15				184
		49 m	β⁻ 1.7				185
		10 m	β⁻ 2.2				186
						18	**W** 176
		1.3 h	K; β⁺ ∼2				176
		2.2 h	K				177
		21 d	K				178
		30 m	K				179
		(5 ms	IT	)		<20	**180**
		140 d	K; L	(0.9)			181
						20	**182**
		(5.5 s	IT	)	6.2	11	**183**
						2.0	**184**
		⎰ 1.7 m	IT	⎱			185
		⎱ 74 d	β⁻ 0.43	(0.43) ⎰			**186**
					7.1	36	
		24 h	β⁻ 0.62, 1.31	(1.31)		∼80	187
		65 d	β⁻				188
						86	**Re** 180
		2.4 m	β⁻ 1.1				181
		⎰ 13 h	K	⎱			182
		⎱ 64 h	K				182
		150 d	K				183
		⎰ 50 d	K	⎱			184
		⎱ 2.2 d	K				**185**
+3.1714(6)	(2.8)					105	
		91 h	β⁻ 1.07, 0.93, . . . ; K	(1.07)			186
+3.2039(6)	+2.6	∼5 × 10¹⁰ y	β⁻ <0.008		7.3	75	187
		⎰ 20 m	IT 0.10	⎱		<3	188
		⎱ 17 h	β⁻ 2.12				189
		∼200 d	β⁻ 0.2				190
		3 m	β⁻ 1.7				191
		10 m	β⁻ 1.8				
						15	**Os** 182
		24 h	K				183
		⎰ 10 h	K	⎱			183
		⎱ 15 h	K				**184**
		95 d	K; L			∼20	185
							186
+0.12		(35 h	IT	)	6.6		**187**
							188
+0.651	+0.6						**189**
		⎰ 9 m	IT 0.62	⎱		8	190
		⎱ 6 h	IT				
		⎰ 14 h	IT 0.074	⎱			191
		⎱ 16 d	β⁻ 0.14	(0.31)			192
						1.6	193
		31 h	β⁻ 1.10, . . .	(1.10)		200	194
		∼2 y	β⁻				

(1)	(2)	(3)	(4)	(5)	(6)	(7)	(8)	(9)	
								Abundance	
Z	Chem symbol	A	N	I	π	Ground-state config	$M\text{-}A$, MeV	Per cent	Cosmic
77 *Iridium*	**Ir**	**192.2**							**0.821**
		187	110						
		188	111						
		189	112						
		190	113						
		191	**114**	**3/2**	**+**	$d_{3/2}$	**19.9(21)**	**38.5**	**0.316**
		192	115				23.0(20)		
		193	**116**	**3/2**	**+**	$d_{3/2}$	**23.5(20)**	**61.5**	**0.505**
		194	117				24.6(15)		
		195	118				26.7(15)		
		196	119						
		197	120						
		198	121				31.4(20)		
78 *Platinum*	**Pt**	**195.09**							**1.625**
		187	109						
		188	110						
		189	111						
		190	112					0.012	0.0001
		191	113						
		192	**114**	**(0)**	**(+)**		**21.5(20)**	**0.78**	**0.0127**
		193	115				23.6(15)		
		194	**116**	**(0)**	**(+)**		**22.3(15)**	**32.8**	**0.533**
		195	**117**	**1/2**	**—**	$p_{1/2}$	**24.6(15)**	**33.7**	**0.548**
		196	**118**	**(0)**	**(+)**		**25.0(15)**	**25.4**	**0.413**
		197	119				27.3(30)		
		198	**120**	**(0)**	**(+)**		**27.0(20)**	**7.2**	**0.117**
		199	121						
79 *Gold*	**Au**	**197.0**							**0.145**
		191	112						
		192	113	1					
		193	114	3/2					
		194	115	1					
		195	116						
		196	117				26.2(30)		
		197	**118**	**3/2**	**+**	$d_{3/2}$	**26.5(30)**	**100**	**0.145**
		198	119	2			28.4(30)		
		199	120	3/2			28.9(30)		
		200	121				32.0(30)		
		201	122				33.1(30)		
		202	123						
		203	124						
80 *Mercury*	**Hg**	**200.61**							**0.284**
		191	111						
		192	112						
		193	113						
		194	114						
		195	115						
		196	**116**	**(0)**	**(+)**		**25.5(30)**	**0.15**	**0.00045**
		197	117	1/2					
		198	**118**	**(0)**	**(+)**		**27.0(30)**	**10.0**	**0.0285**
		199	**119**	**1/2**	**—**	$p_{1/2}$	**28.4(30)**	**16.9**	**0.0481**
		200	**120**	**(0)**	**(+)**		**29.7(30)**	**23.1**	**0.0656**

(10) μ (nucl mag)	(11) Quad mom, 10^{-24} cm²	(12) Half-life ($T_{1/2}$)	(13) Decay — Type and energy	(Q)	(14) Binding energy of last neutron E_n, MeV	(15) σ_{tn}, 10^{-24} cm²	
						460	**Ir**
		12 h	K; β^+ 2.2				**187**
		41 h	K; β^+ 2				**188**
		11 d	K				**189**
		⎰ 3 h	K; β^+ 2.0				**190**
		⎱ 11 d	K				
+0.17(3)	+1.2(7)	(5 s	IT 0.042	)		1000	**191**
		⎰ 1.4 m	IT 0.057; β^-				**192**
+0.18(3)	+1.0(5)	⎱ 74 d	β^- 0.67, . . . ; K	(1.58)	7.8	120	**193**
		19 h	β^- 2.24, 1.191, . . .	(2.24)			**194**
		2.3 h	β^- 1.2, 2.1, . . .	(2.1)	6.5		**195**
		9.7 d	β^- 0.08				**196**
		7 m	β^- 1.6, . . .				**197**
		50 s	β^- 3.6	(4.4)			**198**
						10	**Pt**
		3 h					**187**
		10 d	K				**188**
		11 h	K				**189**
		$\sim 10^{12}$ y(?)	α 3.3(?)			$\sim$90	**190**
		3.0 d	K				**191**
						8	**192**
		⎰ 3.4 d	IT 0.135				**193**
		⎱ "long"	L				
+0.60592(8)		($\sim$6 d	IT 0.130	)	9.5	1.2	**194**
					6.1	27	**195**
					8.2	1.2	**196**
		⎰ 1.4 h	IT				**197**
		⎱ 19 h	β^- 0.67, 0.48, 0.47	(0.75)		4	**198**
		30 m	$\beta^- \sim$1.2	($\geq$1.7)			**199**
						98	**Au**
		3 h	K				**191**
		4.8 d	K; β^+ 1.9				**192**
+0.1		⎰ 4 s	IT 0.032				**193**
		⎱ 17 h	K				
+0.07		39 h	K; β^+ 1.55, 1.22	(2.57)			**194**
		⎰ 30 s	IT 0.057, 0.318				**195**
		⎱ 180 d	K	(0.27)			
		⎱ 14 h	K or IT				**196**
		⎱ 5.6 d	K; β^- 0.27	(0.70−)			
+0.13(1)	+0.56	(7.4 s	IT 0.130, 0.409	)	7.9	98	**197**
+0.50		2.70 d	β^- 0.96, 0.28, 1.37	(1.37)	6.4	2.6×10^4	**198**
+0.24		3.15 d	β^- 0.30, 0.25, 0.46	(0.46)		$\sim$30	**199**
		48 m	β^- 2.2	(2.2)			**200**
		26 m	β^- 1.5	(1.5)			**201**
		$\sim$25 s	β^-				**202**
		55 s	β^- 1.9	(1.9)			**203**
						350	**Hg**
		57 m	K				**191**
		6 h	K; β^+ 1.2				**192**
		⎰ 12 h	IT 0.101; K				**193**
		⎱ 5 h	K				
		⎰ 0.4 s	IT?				**194**
		⎱ $\sim$130 d	K				
		⎰ 40 h	K IT 0.128				**195**
		⎱ 9.5 h	K				
						2500	**196**
+0.52		⎰ 25 h	IT 0.165; K				**197**
		⎱ 65 h	K				
							198
+0.50413(3)		(43 m	IT 0.368			2000	**199**
						<50	**200**

(1) Z	(2) Chem symbol	(3) A	(4) N	(5) I	(6) π	(7) Ground-state config	(8) M-A, MeV	(9) Abundance Per cent	(9) Abundance Cosmic
80	Hg	200.61							
		201	121	3/2	(—)	$p_{3/2}$	31.7(30)	13.2	0.0375
							Based on ^{16}O std $\uparrow$		
							Based on ^{208}Pb std $\downarrow$		
		202	122	(0)	(+)		32.9(6)	29.8	0.0844
		203	123				34.0(4)		
		204	124	(0)	(+)		34.7(2)	6.8	0.0194
		205	125				37.6(2)		
81 Thallium	Tl	204.39							0.108
		195	114						
		196	115						
		197	116						
		198	117						
		199	118						
		200	119						
		201	120						
		202	121				33.9(5)		
		203	122	1/2	+	$s_{1/2}$	33.5(4)	29.5	0.0319
		204	123				35.1(1)		
		205	124	1/2	(+)	$s_{1/2}$	35.8(1)	70.5	0.0761
RaE″		206	125				37.67(2)		
AcC″		207	126				39.24(4)		
ThC″		208	127				43.77(2)		
		209	128				47.19(6)		
RaC″		210	129				51.79(3)		
82 Lead	Pb	207.21							0.47
		197	115						
		198	116						
		199	117						
		200	118						
		201	119						
		202	120				34.0(5)		
		203	121				34.8(5)		
		204	122	(0)	(+)		34.3(1)	1.3	0.0063
		205	123				35.89(5)		
		206	124	(0)	(+)		36.15(1)	26	0.122
		207	125	1/2	—	$p_{1/2}$	37.79(1)	21	0.0995
		208	126	(0)	(+)		38.774(std)	52	0.243
		209	127				43.27(5)		
RaD		210	128				46.40(2)		
AcB		211	129				51.0(1)		
ThB		212	130				54.17(3)		
		213	131				58.7(4)		
RaB		214	132				62.00(8)		
83 Bismuth	Bi	209.00							0.144
		198	115						
		199	116						
		200	117						
		201	118						
		202	119						
		203	120						
		204	121						
		205	122						
		206	123				39.7(2)		
		207	124				40.19(4)		
		208	125				41.70(7)		
		209	126	9/2	(—)	$h_{9/2}$	42.64(5)	100	0.144

(10) μ (nucl mag)	(11) Quad mom, 10^{-24} cm²	(12) Half-life $(T_{1/2})$	(13) Decay Type and energy	(Q)	(14) Binding energy of last neutron E_{n_1} MeV	(15) $\sigma_{t\,n_1}$ 10^{-24} cm²	
—0.5990(1)	+0.6				6.3	<50	**Hg** **201**
						3	**202**
		48 d	β^- 0.21	(0.49)			203
						0.4	**204**
		5.2 m	β^- 1.6, 1.4	(1.6)			205
						3.3	**Tl** **195**
							196
		1.2 h	K				
		~4 h					196
		⎰ 0.54 s	IT?				197
		⎱ 2.8 h	K				
		⎰ 1.9 h	K; IT 0.26				198
		⎱ 5 h	K				
		7.4 h	K				199
		27 h	K; β^+				200
		3.0 d	K				201
		12 d	K; L				202
+1.6117(1)					8.8	11	**203**
± 0.07		4.1 y	β^- 0.76; K	$\left[\begin{array}{c}(0.76-)\\(0.34+)\end{array}\right]$	6.5		204
+1.6275(1)					7.6	0.11	**205**
		4.20 m	β^- 1.51	(1.51)	6.2		206
		4.78 m	β^- 1.45	(1.45)	6.8		207
		3.1 m	β^- 1.79, 1.28, . . .	(4.99)	3.8		208
		2.2 m	β^- 1.8, 2.3	(3.9)	4.8		209
		1.32 m	β^- 1.9	(5.4)	3.9		210
						0.17	**Pb** **197**
		42 m	K; IT(?)				197
		2.3 h	K				198
		⎰ 12 m	IT 0.42				199
		⎱ 1.5 h	K				
		21 h	K				200
		⎰ 1.0 m	IT 0.66				201
		⎱ 9 h	K				
		⎰ 3.5 h	IT 0.79, 0.13; K				202
		⎱ ~10⁵ y	L				
		⎰ 6 s	IT 0.86				203
		⎱ 52 h	K	(1.4)			
		(68 m	IT 0.91	)		0.9	**204**
		>10⁶ y	L		6.4	0.03	205
					8.1	0.03	**206**
+0.58750(7)		(0.80 s	IT 1.06	)	6.7	0.73	**207**
					7.4	0.00045	**208**
		3.3 h	β^- 0.62	(0.62)	3.9		209
		20 y	β^- 0.020	(0.067)	5.2		210
		36.1 m	β^- 1.4, 0.5	(1.4)	3.8		211
		10.64 h	β^- 0.34, 0.58	(0.58)	5.2		212
							213
		26.8 m	β^- 0.7, . . .	(1.0)			214
						0.033	**Bi** **198**
		7 m	K; α 5.83				198
		~25 m	K; α 5.47				199
		35 m	K				200
		⎰ 1.0 h	K; α 5.15				201
		⎱ ~2 h	K				
		1.6 h	K				202
		12 h	K; α 4.85				203
		12 h	K				204
		14 d	K				205
		6.4 d	K	(3.6)			206
		8.0 y	K; L	(2.4)			207
							208
+4.082	—0.4				7.4	0.033	**209**

(1) Z	(2) Chem symbol	(3) A	(4) N	(5) I	(6) π	(7) Ground-state config	(8) $M\text{-}A$, MeV	(9) Abundance Per cent	(9) Abundance Cosmic
83	**Bi**	**209.00**							
RaE		210	127	1			46.34(2)		
AcC		211	128				49.60(5)		
ThC		212	129				53.58(3)		
		213	130				56.77(6)		
RaC		214	131				61.01(4)		
		215	132				64.2(3)		
84	**Po**								
Polonium		201	117						
		202	118						
		203	119						
		204	120						
		205	121						
		206	122						
		207	123						
		208	124				43.1(1)		
		209	125	1/2			44.47(9)		
RaF		210	126				45.17(2)		
AcC'		211	127				48.98(2)		
ThC'		212	128				51.33(1)		
		213	129				55.38(6)		
RaC'		214	130				57.84(3)		
AcA		215	131				62.1(1)		
ThA		216	132				64.68(3)		
		217	133				69.0(3)		
RaA		218	134				71.72(8)		
85	**At**								
Astatine		203	118						
		204	119						
		205	120						
		206	121						
		207	122						
		208	123						
		209	124						
		210	125				49.0(2)		
		211	126				49.77(5)		
		212	127				53.0(6)		
		213	128				55.6(1)		
		214	129				58.90(6)		
		215	130				61.36(6)		
		216	131				65.14(5)		
		217	132				67.52(8)		
		218	133				71.4(2)		
		219	134				74.2(3)		
86	**Em**								
Emanation		206	120						
		207	121						
		208	122						
		209	123						
		210	124						
		211	125						
		212	126				53.1(1)		
		213	127						
		214	128				58.1(3)		
		215	129				61.4(1)		
		216	130				63.11(4)		
		217	131				66.88(8)		
		218	132				68.7(4)		
Actinon: *An*		219	133				72.6(1)		
Thoron: *Tn*		220	134				74.69(4)		
		221	135				78.7(3)		
Radon: *Rn*		222	136				80.92(8)		

(10) μ (nucl mag)	(11) Quad mom, 10^{-24} cm²	(12) Half-life ($T_{1/2}$)	(13) Decay — Type and energy	(Q)	(14) Binding energy of last neutron E_n, MeV	(15) σ_{tn}, 10^{-24} cm²	
							Bi
		⎰ 5.0 d	β⁻ 1.17; α	(1.17) ⎱	4.7		210
		⎱ 2.6 × 10⁶ y	α 4.94; β⁻	(1.14) ⎰			210
		2.15 m	α 6.62, 6.27; β⁻	(0.63−)	5.1		211
		60.5 m	β⁻ 2.25; α 6.05, 6.09, . . .	(2.25−)	4.4		212
		47 m	β⁻ 1.39, 0.96, α 5.9	(1.39)	5.1		213
		19.7 m	β⁻ 1.6, 3.17; α 5.5	(3.17)	4.2		214
		8 m	β⁻				215
							Po
		18 m	K; α 5.67				201
		56 m	K; α 5.59				202
		47 m	K				203
		3.8 h	K; α 5.37				204
		1.5 h	K; α 5.2				205
		9 d	K; α 5.22, . . .				206
		5.7 h	K; α 5.10				207
		2.9 y	K; α 5.11				208
		~100 y	α 4.88; K		6.6		209
		138.40 d	α 5.30, . . .		7.7		210
		⎰ 25 s	α 7.1, 8.7, . . .		4.6		211
		⎱ 0.52 s	α 7.43, . . .	⎰			211
		0.30 μs	α 8.78		6.0		212
		4 μs	α 8.34		4.3		213
		160 μs	α 7.68		5.9		214
		1.8 ms	α 7.36; β⁻		4.1		215
		0.16 s	α 6.77		5.8		216
		<24 m	α 6.5				217
		3.05 m	α 6.00; β⁻				218
							At
		7 m	α 6.10				203
		~25 m	K				204
		25 m	K; α 5.90				205
		2.6 h	K				206
		1.8 h	K; α 5.75				207
		⎰ 1.7 h	K; α 5.65	⎰			208
		⎱ 6 h	K				208
		5.5 h	K; α 5.64				209
		8.3 h	K; α 5.35				210
		7.5 h	K; L; α 5.86				211
		0.22 s	α				212
		<2 s	α 9.2				213
		<5 s	α 8.78				214
		~100 μs	α 8.00		5.9		215
		~300 μs	α 7.79		4.6		216
		0.018 s	α 7.02		5.9		217
		1.3 s	α 6.63		4.6		218
		0.9 m	α 6.27; β⁻				219
							Em
		7 m	α 6.22; K				206
		11 m	K; α 6.12				207
		23 m	K; α 6.14				208
		30 m	K; α 6.04				209
		2.7 h	α 6.04; K				210
		16 h	K; α 5.78, 5.85, 5.61				211
		23 m	α 6.26				212
							213
							214
		<1 m	α 8.6				215
		<9 m	α 8.01		6.7		216
		~1 ms	α 7.74		4.6		217
		19 ms	α 7.13, 6.53		6.6		218
		3.92 s	α 6.82, 6.56, . . .		4.4		219
		52 s	α 6.28, 5.75		6.3		220
		25 m	β⁻; α				221
		3.825 d	α 5.48			0.7	222

(1)	(2)	(3)	(4)	(5)	(6)	(7)	(8)	(9)	
								Abundance	
Z	Chem symbol	A	N	I	π	Ground-state config	$M\text{-}A,$ MeV	Per cent	Cosmic
87 Fr *Francium*		217	130				67.7(2)		
		218	131				70.5(1)		
		219	132				72.41(8)		
		220	133				75.56(6)		
		221	134				77.55(8)		
		222	135				81.1(3)		
AcK		223	136				83.4(2)		
88 Ra *Radium*		219	131				73.1(2)		
		220	132				74.29(5)		
		221	133				77.32(9)		
		222	134				78.99(5)		
AcX		223	135				82.2(1)		
ThX		224	136				84.08(4)		
		225	137				87.4(1)		
Ra		226	138				89.39(9)		
		227	139				93.4(2)		
MsTh₁		228	140				95.50(9)		
		229	141				99.1(2)		
		230	142				101.6(3)		
89 Ac *Actinium*		221	132				79.1(2)		
		222	133				81.2(1)		
		223	134				82.78(9)		
		224	135				85.45(9)		
		225	136				87.06(9)		
		226	137				90.2(4)		
Ac		227	138	3/2			92.1(1)		
MsTh₂		228	139				95.46(6)		
		229	140				97.3(3)		
		230	141				100.7(5)		
90 Th *Thorium*		223	133				84.4(2)		
		224	134				85.16(6)		
		225	135				87.6(1)		
		226	136				89.06(6)		
RdAc		227	137				92.0(1)		
RdTh		228	138				93.22(5)		
		229	139				96.1(1)		
Io		230	140				97.8(1)		
UY		231	141				101.1(1)		
Th		232	142	(0)	(+)		103.17(6)	100	∼0.06
		233	143				106.5(1)		
UX₁		234	144				108.7(1)		
		235	145				112.6(3)		
91 Pa *Protoactinium*		225	134				90.0(3)		
		226	135				91.8(2)		
		227	136				92.97(9)		
		228	137				95.26(9)		
		229	138				96.5(1)		
		230	139				99.3(4)		
Pa		231	140	3/2			100.8(1)		
		232	141				103.53(6)		
		233	142				105.2(1)		
UX₂ *UZ*		234	143				108.5(4)		
		235	144				110.8(3)		
		236	145				114.4(3)		
		237	146				116.5(5)		
92 U *Uranium*		227	135				94.9(3)		
		228	136				95.56(8)		
		229	137				97.8(1)		
		230	138				98.66(6)		

(10) μ (nucl mag)	(11) Quad mom, 10^{-24} cm²	(12) Half-life ($T_{1/2}$)	(13) Decay — Type and energy	(Q)	(14) Binding energy of last neutron E_n, MeV	(15) σ_{tn}, 10^{-24} cm²	
							Fr
		<2 s	α 8.3				217
		<5 s	α 7.85				218
		0.02 s	α 7.30		6.5		219
		28 s	α 6.69		5.2		220
		4.8 m	α 6.30, 6.07		6.3		221
		15 m	β⁻; α				222
		22 m	β⁻ 1.0, 1.3; α 5.3				223
						20	**Ra**
		<1 m	α 8.0				219
		<9 m	α 7.43		7.2		220
		30 s	α 6.71		5.3		221
		38 s	α 6.55, 6.23		6.7		222
		11.6 d	α 5.70, 5.60, 5.42, . . .		5.2	125	223
		3.64 d	α 5.68, 5.44;		6.4	12	224
		14.8 d	β⁻ 0.32	(0.36)	5.1		225
		1620 y	α 4.78, 4.59		6.3	20	226
		41 m	β⁻ 1.30	(1.30)	4.6		227
		6.7 y	β⁻ <0.02	(<0.05)	6.1	~36	228
		<5 m	β⁻				229
		1 h	β⁻ 1.2				230
						520	**Ac**
		<2 s	α 7.6				221
		5.5 s	α 6.96				222
		2.2 m	α 6.64; K		6.8		223
		2.9 h	K; α 6.17		5.7		224
		10.0 d	α 5.80		6.6		225
		29 h	β⁻ 1.2	(1.2)	5.4		226
+1.1	−1.7	22 y	β⁻ 0.046; α 4.94		6.6	~520	227
		6.13 h	β⁻ 1.11, 0.45, −2.18	(2.18)	4.8		228
		66 m	β⁻				229
		<1 m	β⁻ 2.2				230
						7.5	**Th**
		<1 m	α 7.5				223
		<9 m	α 7.13		7.6		224
		8 m	α 6.57; K		5.9		225
		31 m	α 6.34, 6.23, 6.10		7.0		226
		18.2 d	α 5.97, 5.65–6.03		5.4	1500 (fiss)	227
		1.90 y	α 5.42, 5.34		7.0	120	228
		7300 y	α 4.85, 4.94, 5.02		5.5	45 (fiss)	229
		8 × 10⁴ y	α 4.68, 4.61, . . .		6.7	35	230
		25.6 h	β⁻ 0.09, 0.30, 0.22	(0.32)	5.2		231
		1.39 × 10¹⁰ y	α 3.99, 3.93; spon fiss		6.2	7.5	232
		23.3 m	β⁻ 1.23, . . .	(1.23)	5.2	1400	233
		24.10 d	β⁻ 0.19, 0.10	(0.19)	6.0	1.8	234
		<5 m	β⁻				235
						260	**Pa**
		2.0 s	α				225
		1.8 m	α 6.81				226
		38 m	α 6.46; K		7.1		227
		22 h	K; α 6.09, 5.85		6.1		228
		1.5 d	K; α 5.69		7.0		229
		17 d	K; β⁻ 0.40; β⁺ 0.2, 0.4; α			1500 (fiss)	230
		3.4 × 10⁴ y	α 5.00, 4.63–5.05			200	231
		1.31 d	β⁻ 0.28, 0.4–1.24	(1.24)	5.3	⎡0.1 (fiss)⎤ ⎣700⎦	232
		27.4 d	β⁻ 0.26, 0.14, 0.57	(0.57)	6.9	⎡700 (fiss)⎤ ⎣66⎦	233
		⎰ 1.18 m	β⁻ 2.31, . . . ; IT	⎱			234
		⎱ 6.66 h	β⁻ 0.5, . . .		5.4		234
		24 m	β⁻ 1.4	(1.4)	5.9		235
							236
		10 m	β⁻				237
						3.5 + 4.2 (fiss)	**U**
		1.3 m	α 6.8				227
		9.3 m	α 6.67; K		7.73		228
		58 m	K; α 6.42		6.1		229
		21 d	α 5.89, 5.82, 5.66		7.6	~25 (fiss)	230

(1)	(2)	(3)	(4)	(5)	(6)	(7)	(8)	(9)	
								Abundance	
Z	Chem symbol	A	N	I	π	Ground-state config	$M\text{-}A$, MeV	Per cent	Cosmic
92 **U**									
		231	139				101.1(2)		
		232	140				102.24(5)		
		233	141	5/2			104.7(1)		
	UII	234	142	(0)	(+)		106.2(1)	0.0055	
	AcU	235	143	7/2	(+)		109.4(2)	0.72	
		236	144				111.4(1)		
		237	145				114.3(2)		
	UI	238	146	(0)	(+)		116.6(1)	99.27	∼0.02
		239	147				120.3(2)		
		240	148				122.7(1)		
93 **Np** *Neptunium*		231	138				103.0(1)		
		232	139				104.9(2)		
		233	140				105.7(1)		
		234	141				108.3(4)		
		235	142				109.6(1)		
		236	143				112.21(8)		
		237	144	5/2			113.8(2)		
		238	145	2			116.7(1)		
		239	146				119.0(2)		
		240	147				122.4(1)		
94 **Pu** *Plutonium*		232	138				105.9(1)		
		233	139				107.9(2)		
		234	140				108.6(1)		
		235	141				110.7(2)		
		236	142				111.70(8)		
		237	143				114.0(2)		
		238	144				115.4(1)		
		239	145	1/2			118.2(2)		
		240	146				120.2(1)		
		241	147	5/2			123.0(2)		
		242	148				125.2(1)		
		243	149				128.6(2)		
		244	150				131.0(3)		
		245	151				134.2(5)		
		246	152				137.0(6)		
95 **Am** *Americium*		237	142				115.4(2)		
		238	143				117.9(5)		
		239	144				119.0(2)		
		240	145				121.6(2)		
		241	146	5/2			123.0(2)		
		242	147				125.9(1)		
		243	148	5/2			128.0(2)		
		244	149				131.2(2)		
		245	150				133.0(2)		
		246	151				136.7(3)		
96 **Cm** *Curium*		238	142				118.8(1)		
		239	143				120.8(2)		
		240	144				121.7(1)		
		241	145				123.9(2)		
		242	146				125.2(1)		
		243	147				128.0(2)		

μ (nucl mag)	Quad mom, 10^{-24} cm²	Half-life ($T_{1/2}$)	Type and energy	(Q)	Binding energy of last neutron E_n, MeV	σ_{tn}, 10^{-24} cm²	
							U
		4.3 d	K; α 5.45		5.8	~400 (fiss) ~300	231
		74 y	α 5.32, 5.26, 5.13; spon fiss		7.2	80 (fiss)	232
± 0.55	± 3.4	1.62 × 10⁵ y	α 4.82, 4.78, 4.73		6.0	[60 / 520 (fiss)]	233
		2.5 × 10⁵ y	α 4.78; spon fiss		6.7	80	234
± 0.34	± 4.0	7.1 × 10⁸ y	α 4.40, 4.58; spon fiss		5.4	[108 / 590 (fiss)]	235
		2.39 × 10⁷ y	α 4.50, spon fiss		6.3	8	236
		6.75 d	β⁻ 0.24	(0.51)	5.5		237
		4.51 × 10⁹ y	α 4.18, . . . spon fiss		6.0	2.8 [22 / ~12 (fiss)]	238
		23.5 m	β⁻ 1.21	(1.28)	4.9		239
		14 h	β⁻ 0.36	(0.36)	5.8		240
						170 + 0.019 (fiss)	**Np**
		50 m	α 6.28				231
		~13 m	K				232
		35 m	K; α 5.53				233
		4.4 d	K; L; β⁺ 0.8			~900 (fiss)	234
		1.1 y	L; K; α 5.06				235
		{ 2.2 h { ≥5000 y	β⁻ 0.52; K β⁻	(0.52)	5.6	2800 (fiss)	236
	± 6	2.2 × 10⁶ y	α 4.79; 4.52–4.87		6.9	[170 / 0.019 (fiss)]	237
		2.10 d	β⁻ 1.26, 0.27, . . .	(1.30)	5.2	1600 (fiss)	238
		2.33 d	β⁻ 0.33–0.72	(0.72)	6.4	80	239
		{ 7.3 m { 60 m	β⁻ 2.16 β⁻ 0.90	(2.16) (2.06)	4.8		240
							Pu
		36 m	K; α 6.58				232
							233
		9 h	K; α 6.19				234
		26 m	L; K; α 5.85				235
		2.7 y	α 5.75; . . . ; spon fiss		6.2		236
		40 d	K		7.3		237
		90 y	α 5.49, 5.45, . . . ; spon fiss			[450 / 18 (fiss)]	238
± 0.02	± 0.4	2.43 × 10⁴ y	α 5.15, 5.14, 5.10; spon fiss		5.7	[300 / 730 (fiss)]	239
		6600 y	α 5.16, 5.12, . . . ; spon fiss		6.3	~510	240
± 1.4		13 y	β⁻ 0.02; α 4.89, 4.85	(0.020)	5.6	[~380 / ~1100 (fiss)]	241
		3.8 × 10⁵ y	α 4.90, 4.85; spon fiss		6.1	23	242
		5.0 h	β⁻ 0.57, 0.48	(0.57)	5.2	~100	243
		8 × 10⁷ y	α; spon fiss			~1.4	244
		11 h	β⁻			~260	245
		11 d	β⁻ 0.15, . . .				246
							Am
		~1 h	K; α 6.01				237
		1.9 h	K				238
		12 h	K; α 5.78				239
		47 h	K				240
+1.4	+4.9	470 y	α 5.48, . . . ; spon fiss			[750 / 3.2 (fiss)]	241
		{ 16.0 h { ~100 y	β⁻ 0.62; K β⁻ 0.59, . . .; K; α	(0.62)	5.4	[~2500 (fiss) / ~4500 / ~3500 (fiss)]	242
+1.4	+4.9	~8000 y	α 5.27, 5.22, 5.17–5.34		6.4	82	243
		26 m	β⁻ 1.5; K	(0.15−)			244
		2.0 h	β⁻ 0.90				245
		25 m	β⁻ 1.22	(2.4)			246
							Cm
		2.3 h	K; α 6.50				238
		~3 h	K				239
		27 d	α 6.25; spon fiss				240
		35 d	K; α 5.95				241
		163 d	α 6.11, 6.07, . . . ; spon fiss			~20	242
		35 y	α 5.78, 5.73–5.99		5.7	~300	243

(1)	(2)	(3)	(4)	(5)	(6)	(7)	(8)	(9)	
Z	Chem symbol	A	N	I	π	Ground-state config	$M\text{-}A$, MeV	Abundance	
								Per cent	Cosmic
96	**Cm**	244	148				129.7(1)		
		245	149				132.3(3)		
		246	150				134.2(1)		
		247	151				137.3(3)		
		248	152				139.8(5)		
		249	153				143.0(7)		
97	**Bk** *Berkelium*	243	146				129.5(2)		
		244	147				131.8(2)		
		245	148				133.1(2)		
		246	149				135.7(3)		
		247	150				137.5(4)		
		248	151				140.3(4)		
		249	152				142.1(2)		
		250	153				145.9(2)		
98	**Cf** *Californium*	244	146				132.6(1)		
		245	147				134.6(3)		
		246	148				135.7(1)		
		247	149				138.2(3)		
		248	150				139.7(1)		
		249	151				142.0(3)		
		250	152				144.0(2)		
		251	153				147.1(4)		
		252	154				149.6(5)		
		253	155				152.6(7)		
		254	156				155.3(5)		
99	**E** *Einsteinium*	246	147				139.8(6)		
		247	148				140.5(2)		
		248	149				142.6(4)		
		249	150				143.6(4)		
		250	151				145.9(5)		
		251	152				147.7(5)		
		252	153				150.6(2)		
		253	154				152.4(2)		
		254	155				156.0(2)		
		255	156				158.1(7)		
		256	157				162.2(7)		
100	**Fm** *Fermium*	250	150				146.8(5)		
		251	151				149.0(6)		
		252	152				150.5(5)		
		253	153				152.9(4)		
		254	154				154.9(2)		
		255	155				158.0(4)		
		256	156				160.2(6)		
101	**Mv** *Mendeleevium*	256	155				161.9(7)		

(10) μ (nucl mag)	(11) Quad mom, 10^{-24} cm²	(12) Half-life $(T_{1/2})$	(13) Decay — Type and energy (Q)	(14) Binding energy of last neutron E_n, MeV	(15) σ_{tn}, 10^{-24} cm²	
						Cm
		18 y	α 5.80, 5.76; spon fiss	6.5	$\sim$15 $\left[\begin{array}{l}\sim200 \\ 1800 \text{ (fiss)}\end{array}\right]$	244
		1.1×10^4 y	α 5.34			245
		4000 y	α 5.36; spon fiss		$\sim$15	246
		$\gg$1 y			180	247
		4×10^5 y	α 5.05			248
		"short"	β^-			249
						Bk
		4.5 h	K; α 6.55, 6.72, 6.20			243
		4.4 h	K; α 6.66			244
		50.0 d	K; α 6.17, 6.33, 5.90			245
		1.8 d	K			246
		7000 y	α 5.50, 5.67, 5.30			247
		$\sim$18 h	β^- 0.67; K			248
		290 d	β^- 0.09; α 5.40, 5.08; spon fiss		$\sim$500	249
		3.1 h	β^- 0.9, 1.9			250
						Cf
		25 m	α 7.17			244
		44 m	K; α 7.11			245
		36 h	α 6.75, 6.71, . . . ; spon fiss			246
		2.5 h	K			247
		225 d	α 6.26; spon fiss			248
		500 y	α 5.81, 6.00, . . . ; spon fiss		$\left[\begin{array}{l}\sim270 \\ \sim600 \text{ (fiss)}\end{array}\right]$	249
		10 y	α 6.02, 5.98; spon fiss		$\sim$1500	250
		$\sim$700 y	α		$\sim$3000	251
		2.2 y	α 6.11, 6.07; spon fiss		$\sim$30	252
		18 d	β^-			253
		54 d	spon fiss		$\leq$2	254
						E
		"short"	K			246
		7.3 m	α 7.35; K(?)			247
						248
		2 h	α 6.76			249
						250
		1.5 d	K; α 6.48			251
		$\sim$150 d	α 6.64			252
		20 d	α 6.64, 6.60, 6.55, . . . ; spon fiss		$\sim$200	253
		37 h	β^- 1.1; K		$\leq$15	254
		$\sim$1 y	α 6.44; spon fiss			
		$\sim$30 d	β^-		$\sim$40	255
		"short"	β^-			256
						Fm
		30 m	α 7.43			250
						251
		30 h	α 7.1			252
		3 d	K			253
		3.4 h	α 7.22, 7.18, . . . ; spon fiss			254
		20 h	α 7.08		<100	255
		3.1 h	spon fiss			256
						Mv
		1 h	K			256

Index

Page references in **boldface** type indicate main discussions of topics